5th International Workshop on Multi-Rate Processes and Hysteresis 2010

(MURPHYS 2010)

Journal of Physics: Conference Series Volume 268

Pecs, Hungary
31 May – 3 June 2010

Editors:

Amalia Ivanyi
Peter Ivanyi
Dmitrii Rachinskii
Vladimir A Sobolev

ISBN: 978-1-61782-370-1
ISSN: 1742-6588

Printed from e-media with permission by:

Curran Associates, Inc.
57 Morehouse Lane
Red Hook, NY 12571

Some format issues inherent in the e-media version may also appear in this print version.

Printed by Curran Associates, Inc. (2011)

For permission requests, please contact the Institute of Physics
at the address below.

Institute of Physics
Dirac House, Temple Back
Bristol BS1 6BE UK

Phone: 44 1 17 929 7481
Fax: 44 1 17 920 0979

techtracking@iop.org

Additional copies of this publication are available from:

Curran Associates, Inc.
57 Morehouse Lane
Red Hook, NY 12571 USA
Phone: 845-758-0400
Fax: 845-758-2634
Email: curran@proceedings.com
Web: www.proceedings.com

TABLE OF CONTENTS

PREFACE

The International Workshop on Multi-Rate Processes and Hysteresis conference series focuses on singular perturbation problems and hysteresis as common strongly nonlinear phenomena occurring in mathematical, physical, economical, engineering and information systems. The term 'strongly nonlinear' means, in particular, that linearization will not encapsulate the observed phenomena. Singular perturbation problems and hysteresis can be manifested at different stages of the same or similar processes. Furthermore, a number of fundamental hysteresis models can be considered as a limit of time relaxation processes, or admit an approximation by a differential equation, which is singular with respect to a particular parameter. However, interaction between researchers in the areas of systems with time relaxation and systems with hysteresis (and between the 'multi-rate' and 'hysteresis' research communities) has so far been limited, and there is little cross-fertilization of ideas. It is the aim of the conference series to fill this gap.

The 5th International Workshop on Multi-Rate Processes and Hysteresis (MURPHYS 2010) was hosted by the Pollack Mihály Faculty of Engineering, University of Pécs, Hungary, from 31 May to 3 June 2010, on the occasion of Pécs being the Cultural Capital of Europe in 2010. The workshop was organized in cooperation with University College Cork, Ireland, which hosted all of the previous Workshops:

- International Workshop on Multi-rate Processes and Hysteresis (University College, Cork, Ireland, 31 March–5 April 2008). Proceedings are published in *Journal of Physics: Conference Series* volume 138. See http://euclid.ucc.ie/appliedmath/murphys2008/murphys2008.htm;
- International Workshop on Multi-rate Processes and Hysteresis (University College, Cork, Ireland, 3–8 April 2006). Proceedings are published in *Journal of Physics: Conference Series* volume 55. Further information is available at http://Euclid.ucc.ie/murphys2006.htm;
- International Workshop on Hysteresis and Multi-scale Asymptotic (University College, Cork, Ireland, 17–21 March 2004). Proceedings are published in *Journal of Physics: Conference Series* volume 22. Further details are available at http://Euclid.ucc.ie/hamsa2004.htm;
- International Workshop on Relaxation Oscillations and Hysteresis (University College, Cork, Ireland, 1–6 April 2002). The related collection of invited lectures was published as a volume Singular Perturbations and Hysteresis, SIAM, Philadelphia, 2005.
- International Workshop on Geometrical Methods of Nonlinear Analysis and Semiconductor Laser Dynamics (University College Cork, Ireland, 5–6 April 2001). A collection of invited papers has been published as a special issue of Proceedings of the Russian Academy of Natural Sciences:

Nonlinear dynamics of laser and reacting system, available at http://euclid.ucc.ie/appliedmath/gmna2001/ProcGMNA2001p1.pdf.

Among the aims of this and previous workshops were:

- to bring together the leading experts in singular perturbation and hysteresis phenomena in applied problems;
- to discuss important problems in the areas of reacting systems, semiconductor lasers, shock phenomena, economic modelling, fluid mechanics, electrical engineering and modelling biological systems with emphasises on hysteresis and singular perturbations;
- to learn and share modern techniques in areas of common interest.

The International Workshop on Multi-rate Processes and Hysteresis (Pollack Mihály Faculty of Engineering, University of Pécs, Hungary, 31 May–3 June 2010) brought together about 50 scientists who are actively researching the areas of dynamical systems with hysteresis and singular perturbations with applications to physical, engineering and economic systems. The countries represented at the Workshop included the Czech Republic, Germany, Hungary, Ireland, Israel, Italy, Poland, Romania, Russia, the United Kingdom and USA.

Workshop photo 31 May 2010

Sponsorship of the Workshop by the Pollack Mihály Faculty of Engineering, University of Pécs (Hungary), University College Cork (Ireland), University of Pécs (Hungary), The University of Texas at Dallas (USA), and the Cultural Capital of Europe 2010, Pécs (Hungary), is gratefully acknowledged.

The Editors and Organizers of the Workshop are sincerely grateful to Dr Géza Várady, Ms Andrea Zseni and Mr Ádám Schiffer of the Pollack Mihály Faculty of Engineering, University of Pécs, and Dr Alexander Pimenov of University College Cork for managing the organization of the conference and for the assistance in formatting of all the manuscripts.

More information about the workshop can be found at
http://murphys5.pmmk.pte.hu/

Amalia Ivanyi, Péter Iványi, Dmitrii Rachinskii and Vladimir A Sobolev
Editors

MURPHYS 2010, PMMK PTE, 31 May – 3 June 2010

Sponsored by

POLLACK MIHÁLY FACULTY OF ENGINEERING, UNIVERSITY OF PÉCS

ucc

Coláiste na hOllscoile Corcaigh, Éire
University College Cork, Ireland

PÉCSI TUDOMÁNYEGYETEM UNIVERSITY OF PÉCS

UT DALLAS
The University of Texas at Dallas

EURÓPA KULTURÁLIS FŐVÁROSA
2010

Cultural Capital of Europe 2010, Pécs, Hungary

International Steering Committee

Z I Balanov	*Israel*
M Brokate	*Germany*
R Cross	*UK*
K Dahmen	*USA*
M Dimian	*Romania*
G Friedman	*USA*
A Ivanyi (Co-Chairman)	*Hungary*
P Iványi (Co-Chairman)	*Hungary*
L Kalachev	*USA*
P Krejčí	*Czech Republic*
R O'Malley (Co-Chairman)	*USA*
A Pokrovskii (Co-Chairman)	*Ireland*
N Popovic	*UK*
D Rachinskii (Co-Chairman)	*Ireland*
S S Sazhin	*UK*
V Sobolev (Co-Chairman)	*Russia*
S Szabó	*Hungary*
C Visone	*Italy*

International Program Committee

G Almási	*Hungary*
Z Balanov	*Israel*
M Brokate	*Germany*
R Cross	*UK*
K Dahmen	*USA*
M Dimian	*Romania*
G Friedman	*USA*
A Ivanyi (Co-Chairman)	*Hungary*
P Iványi (Co-Chairman)	*Hungary*
S Jenei	*Hungary*
G Kádár	*Hungary*

L Kalachev	*USA*
R Kersner	*Hungary*
G Kovács	*Hungary*
P Krejčí	*Czech Republic*
P M Kuczmann	*Hungary*
P P O'Kane	*Ireland*
R O'Malley (Co-Chairman)	*USA*
A Pokrovskii (Co-Chairman)	*Ireland*
N Popovic	*UK*
D Rachinskii (Co-Chairman)	*Ireland*
B V H Topping	*UK*
V C Visone	*Italy*

Journal of Physics: Conference Series **268** (2011) 011002 doi:10.1088/1742-6596/268/1/011002

In Memory of Alexei Pokrovskii: Systems with Hysteresis & Murphys, 10 Years Later

D Rachinskii[1], E Shchepakina[2] and V Sobolev[2]

[1]Department of Applied Mathematics, University College Cork, Ireland
[2]Department of Technical Cybernetics, Samara State Aerospace University, Samara, Russia

E-mail: d.rachinskii@ucc.ie

02.06.1948 – 01.09.2010

This volume is dedicated to Professor Alexei Pokrovskii, a founder and the main driver of MURPHYS conference, who died unexpectedly on the 1st of September 2010. He was an outstanding scientist, an excellent mathematician and a faithful friend.

Alexei was an eclectic mathematician who never stopped surprising his colleagues with the depth of his knowledge and the breadth of his scientific interests. The main body of his work belongs to the areas of nonlinear dynamical systems (including systems with discontinuous and nonsmooth nonlinearities and, of course, hystereis), control theory, nonlinear functional analysis and applied mathematical modelling. But the remarkable diversity of his research was much broader and included, at different stages of his work, contributions to game theory, stochastic systems, complexity, general functional analysis. As one illustration, Andrei Kolmogorov highly valued Alexei's results on measures of unpredictability of information flow.

The scientific legacy of Alexei will receive a dedicated account elsewhere; the list of his publications is presented as a separate section of this volume. However, if one main theme, invariably recurrent in his research, is to be identified, that would be hysteresis.

From 1969, when Alexei was yet a 4th year student (but already an author of 4 publications), the Functional Analysis Seminar of Mark Krasnosel'skii at University of Voronezh started to discuss the phenomenon of hysteresis, first presented to the seminar by a physicist Boris Darinskii. The first paper on a hysterant operator (renamed later to hysteron) appeared next year [1]. That was the beginning. Soon after, in the early 70s, Mark Krasnosel'skii with part of his group, including Alexei, moved to Moscow to the Institute for Control Problems of the Russian Academy of Sciences where the new environment stimulated their interest in problems of control as well as reinforcing the research of hysteretic systems and providing a new perspective on them. This environment included applied mathematicians, physicists and engineers. The language and paradigm of systems theory fused with the ideology of nonlinear analysis to create the fundamentals of the theory of hysteresis operators.

The seminal monograph of M. A. Krasnosel'skii and A. V. Pokrovskii "Systems with Hysteresis" [2] appeared in 1983, more than a decade after the first paper [1]. It became the bible of the mathematical hysteresis community paving the way for the research that followed by laying the foundations of the mathematical theory of hysteresis operators, forming the modern concepts of mathematical modelling of hysteretic systems and opening the door to systematic application of mathematical tools to analysis of dynamical systems with hysteretic components. The density of new ideas and concepts in the book, which developed a mathematical theory on the basis of analysis of many phenomenological models of hysteresis from diverse disciplines, is remarkable. Literally every chapter of the book and sometimes even a section or remark pioneered a branch of the mathematical theory of systems with hysteresis such as regularity of hysteresis operators, identification theorems, vibrostable differential equations, links with the theory of differential inclusions and sweeping processes, geometric interpretation of state dynamics of complex hysteresis models (such as Preisach and Prandtl-Ishlinskii operators), composition and inversion of hysteresis operators and compensators.

The publication of an (extended) English translation of the book in 1989 [3] was followed by an explosion of interest in systems with hysteresis in the mathematical and applied mathematics communities. It resulted in a flow of publications of several monographs [4–8] and hundreds of papers in the 1990s continuing into the first decade of this century. Several groups of researchers across Europe, the USA and Japan contributed to the vibrant development of the mathematical theory. Of critical importance was the convergence of the mathematical community with researchers in hysteretic systems from different fields (electromagnetism, phase transitions, mechanics, engineering, economics and others) on the basis of common mathematical language and common understanding of hysteresis phenomena. This convergence enriched the theory by new problems, methods and links to other disciplines (thermodynamics, statistical physics, multi-rate systems, stochastic systems, optimal control, etc.), leading to the versatile science of hysteresis, as we know it now, which has many faces depending on application [9].

The interaction of hysteresis researchers with different backgrounds greatly benefited from regular mathematical and interdisciplinary meetings such as the series of conferences in Trento (and later Berlin) and Hysteresis and Micromagnetic Modelling symposium series. It was at a "Models of Hysteresis" meeting, organised by Augusto Visintin in Trento in 1991, where Alexei first met colleagues from the western hysteresis community, with many of whom he collaborated closely, and became friends, afterwards.

During the period of Alexei's work in Australia in 1993-1997, his research focused on the dynamics of discretisations of chaotic systems. The starting point was the observation that the dynamics of the logistic map is qualitatively different from those of its discretised version (computer realisation) no matter how accurate the discretisation may be. Understanding of this effect was achieved when it became clear that the discretisation acts as a randomising factor. As a result, Alexei and his colleagues developed phenomenological models based on the theory of a class of random maps which were capable of an accurate description of the effect of discretisation on the original system dynamics. In order to ensure that a computer model accurately mimics the dynamics of a map (i.e., robustness to discretisation) new mathematical tools such as split-hyperbolicity and bi-shadowing have been developed on the basis of the degree theory [10].

Brisbane, 1995: Ferry to University—absorbed in another mathematical problem.

However far the focus of this research seems to be from the hysteresis theme, presumably, Alexei was returning in his thoughts to systems with hysteresis, the subject which was always close to his heart. In particular, he tested the applicability of the split-hyperbolicity technique to the analysis of complex dynamics of hysteretic systems [11]. Later, already in Ireland, this technique was extended and used for rigorous computer-based proofs of chaos in hysteretic systems (as well as in applications to laser and population dynamics). The importance of hysteresis research for Alexei might have been part of the reason for moving from Australia to Europe, closer to the centres of active 'hysteretic life'.

By the beginning of the 21st century the mathematical theory of systems with hysteresis had reached a certain stage of maturity. Alexei's vision of its future would be that the evolution vector should now point to transcending new disciplinary boundaries, making the theory available to, and useful for, new applications and open to new problems (in the same way, as the theory has been stimulated by, and proven already successful in, applications to magnetism [8, 12, 13], plasticity, smart materials, mechanical engineering and control design). Alexei's research over the last decade was inspired by challenges of modelling hysteresis in economics, hydrology, biology (population dynamics) and multi-rate systems. The research themes of the last cohort of his PhD students included bifurcations and chaos in systems with Preisach hysteresis operator (Oleg Rasskazov), dynamics of macroeconomic models with hysteresis (Hugh McNamara), modelling hysteresis in hydrology (Denis Flynn), canard solutions and chaos in nonsmooth singularly perturbed systems (Andrew Zhezherun) and modelling seasonal dynamics of wild bird populations (Suzanne O'Reagan).

Two articles in this volume of MURPHYS proceedings contain accounts of Alexei's impact on modelling hysteresis in economics and hydrology. The MURPHYS conference itself originated from the idea of exploring the links between the methods of the theory of multi-rate systems and the theory of systems with hysteresis, Van der Pol relaxation oscillations being the classical basic example of such a link. The history of the conference started in 1999 at the Industrial Mathematics Congress in Edinburgh where Alexei met Vladimir Sobolev, a friend since their student years at the Mathematical Faculty in the University of Voronezh, Elena Shchepakina and Robert E. O'Malley. Alexei became very enthusiastic about unifying the efforts of the singular perturbation and hysteresis communities for solving new interesting mathematical and applied problems. An interdisciplinary workshop would have been a crucial means for starting a dialogue and maintaining regular contacts. After a pilot seminar on Geometrical Methods of Nonlinear Analysis and Semiconductor Laser Dynamics organised in Cork in 2001 [14], Robert E. O'Malley agreed to chair the Scientific Committee and Michael P. Mortell was convinced to become the Chair of the Organising Committee of a larger scale conference in 2002. Since then, the conference has grown to a series of bi-annual meetings and gradually acquired the acronym MURPHYS, which by 'pure coincidence' is also the name of an Irish stout brewed in Cork. From the very start, the conference has been truly multi-disciplinary providing a forum for researchers in mathematics, applied mathematics, engineering, control, physics, hydrology, combustion processes, economics, financial mathematics and, later, biology, epidemiology and, even, history.

One characteristic quality of Alexei, the scientist, was his extraordinary imagination, the ability to see a problem from a different original perspective, combined with fresh interest in the world around us in all its variety and diversity, whether aspects of life, physics or society. Always eager to learn, explore and understand phenomena by building a mathematical model, he was extremely rich in ideas and absolutely generous in sharing them with his students and colleagues. He was proud and sincerely happy whenever they solved a problem or made progress, especially younger colleagues and students. Sharing your discovery with Alexei was pure joy as his response was fantastically encouraging and stimulating (no matter whether the credit should have really gone to him!). His priority meant little, if anything, to him. The way of doing research he enjoyed and cultivated was collective, which is apparent from his publications. He had a talent to identify and consolidate interests of colleagues and involve them in research projects. Naturally, wherever Alexei worked, he was an attractor of excellent research and scientific life, a generator of novel ideas and a leading mind, most esteemed by colleagues and most beloved by students. MURPHYS seminar has grown to a mature conference series, in many ways, due to his attracting effect, enthusiasm and energy. Similarly, his talk at any conference and seminar attracted a big audience, being not only interesting and entertaining but always full of stimulating nontrivial ideas. A feedback from him was known to be extremely valuable and could have provided an unexpected angle on, or insight in, the problem, suddenly opening a door to a new application or extension of the research. He was truly interested in, and enthusiastic about, research done by others, and we have witnessed on multiple occasions how his remarks to a seminar presentation were received with great interest by the speaker.

A special atmosphere of MURPHYS seminars had a very clear imprint of Alexei's warm-hearted personality. Undoubtedly, a most remarkable of Alexei's qualities was his friendliness and sympathy to people. Unassuming himself, he was willing to support everybody around him and help with any, even a little, problem. Eugene Asarin who has known Alexei for almost 30 years since being a student in Mark Krasnosel'skii's laboratory, probably, pin-pointed a unique recipe of Alexei's warm-heartedness, so familiar to anybody who knew him: "Alexei, himself a strongest and sharpest researcher, admired everybody (which is not so common in the competitive research community). He looked at everybody he met with huge sympathetic interest and invariably discovered a special talent or a special merit worthwhile of admiration in them—this could have been a skill in proving theorems, researching literature, playing chess, speaking to authorities or telling funny stories". And if Alexei met a mathematician, he would have spent with them many hours trying to understand new problems. This sincere sympathy to people combined with Alexei's erudition, intelligence and open-mindedness, his ability to get the gist of the problem, and good humour made him a centre of attention of any company, though he never sought to be it. He was proud of, loyal and very thankful to, colleagues in his Department and broader scientific community he worked in for their support and help.

Caring for young colleagues and students, supporting, encouraging and inspiring them, was a very clear priority for Alexei. He involved students in his research orbit by turning science into a compelling quest in Wonderland full of miracles, intriguing problems and unexpected facts, where it is so natural to make discoveries. His charisma was irresistible. No wonder that he was a legend among undergraduate and postgraduate students.

Whichever research area Alexei turned to, he initiated fascinating original work bringing his students and collaborators to new *terra incognita* with him. During the last few years Alexei put a lot of effort into cultivating biologically oriented mathematical research in his Department. With his usual contagious enthusiasm and driving energy he set up a regular meeting working group of applied mathematicians and biologists, led several joint research projects [15–19], built collaboration with mathematical biology researchers in Ireland and the USA, co-supervised two PhD theses in environmental science and epidemiology, increased the presence of biological modelling theme in BSc and MSc applied mathematics courses and launched a mathematical module for Systems Biology students. In one of the latest works he formulated principles of modelling hysteretic response of human population to epidemics [20].

Given Alexei's method of work and mathematical universality, the huge number of collaborators he had worldwide comes as no surprise. Maintaining collaboration with colleagues from Russia was of special meaning to him during the Australian and Irish periods of his life. Many Russian mathematicians visited Alexei at Universities of Brisbane, Geelong and Cork and enjoyed very welcoming warm hospitality of Pokrovskii's family and Alexei's wife Natasha, in the first place—among them, Victor Kozyakin, Alexander Krasnosel'skii, Nikolai Kuznetsov, Anatoly Neishtadt, Vladimir Sobolev, Elena Shchepakina, Alexander Vladimirov. Alexei travelled to Moscow on multiple occasions for collaboration and presenting lectures at seminars. He was working for a few weeks in Moscow in winter 2009 and summer 2010. He took a train from Moscow to Voronezh to see his sister Olga in August 2010.

Alexei's life was inseparable from mathematics and science. He never stopped thinking and could have told you even at a party or in a pub '...look what I have found...' or '...I think I finally understood...' And only when spending time with his family, his wife and children whom he infinitely adored, he would have devoted himself to them completely and absolutely.

Alexei's daughter Olya is completing the degree of Medical Doctor continuing Alexei's family tradition. Alexei's son Alexey followed in his father's footsteps. He is pursuing PhD in London School of Economics after receiving the degree in Mathematics from Cambridge.

Acknowledgments

The authors thank Eugene Asarin, Victor Kozyakin, Alexander Krasnosel'skii, Petr P. Zabreiko, Martin Brokate and Pavel Krejčı for contributing to this article in various ways.

References

[1] Krasnosel'ski M A, Darinskii B M, Emelin I V, Zabreiko P P, Lifsic E A, and Pokrovski A V 1970 Hysterant operator *Soviet Math. Dokl.* **11** 29–33

[2] Krasnosel'skii M A and Pokrovskii A V 1983 *Systems with Hysteresis* (Moscow: Nauka), in Russian

[3] Krasnosel'skii M A and Pokrovskii A V 1989 *Systems with Hysteresis* (Springer)

[4] Mayergoyz I D 1991 *Mathematical Models of Hysteresis* (New York: Springer)

[5] Visintin A 1994 *Differential Models of Hysteresis* (New York: Springer)

[6] Brokate M and Sprekels J 1996 *Hysteresis and Phase Transitions* (Springer)

[7] Krejčí P 1996 *Hysteresis, Convexity and Dissipation in Hyperbolic Equations* (Tokyo: Gakkōtosho)

[8] Mayergoyz I D 2003 *Mathematical Models of Hysteresis and their Applications* (Academic Press)

[9] I Mayergoyz and G Bertotti, eds 2006 *The Science of Hysteresis*, three volume set (Elsevier, Academic Press)

[10] Diamond P, Kloeden P E, Kozyakin V S and Pokrovskii A V *Semi-Hyperbolic Lipschitz Mappings*, monograph (in preparation)

[11] Diamond P, Kloeden, P E, Kozyakin V S, Krasnosel'skii M A and Pokrovskii A V 1995 Robustness of trajectories of dynamical systems with respect to a class of hysteresis perturbations *Doklady Akademii Nauk* **343** 1–3

[12] Bertotti G 1998 *Hysteresis in Magnetism* (Elsevier, Academic Press)

[13] Della Torre E 1999 *Magnetic Hysteresis* (IEEE Press)

[14] http://euclid.ucc.ie/appliedmath/gmna2001/gmna2001.htm

[15] Gavin C, Prentice M, Sobolev V and Pokrovskii A V 2006 Dynamics of a Lotka-Volterra type model with applications to marine phage population dynamics *J. Phys.: Conf. Ser.* **55** 80–93

[16] O'Reagan S, Kelly T C, Korobeinikov A, O'Callaghan M J A and Pokrovskii A V 2008 Qualitative and numerical investigations of the impact of a novel pathogen on a seabird colony *J. Phys.: Conf. Ser.* **138** 012018

[17] O'Reagan S, Kelly T C, Korobeinikov A, O'Callaghan M J A and Pokrovskii A V 2009 The impact of an emerging avian influenza virus (H5N1) in a seabird colony *Global View of Fight Against Influenza*, ed P M Mitrasinovic (New York: Nova Science Publishers)

[18] O'Reagan S, Kelly T C, Korobeinikov A, O'Callaghan M J A and Pokrovskii A V 2010 Lyapunov functions for SIR and SIRS epidemic models *Applied Mathematics Letters* **23** 446–448

[19] O'Regan S M, Flynn D, Kelly T C, O'Callaghan M J A, Pokrovskii A V and Rachinskii D A hybrid model of the population dynamics of woodpigeon (*Columba palumbus*) *Ecological Modelling* submitted

[20] Pimenov A, Kelly T C, Korobeinikov A, O'Callaghan and M J A and Pokrovskii A V 2010 Systems with hysteresis in mathematical biology via a canonical example *Mathematical Modeling, Clustering Algorithms and Applications*, ed C L Wilson (New York: Nova Science Publishers)

Journal of Physics: Conference Series **268** (2011) 011003 doi:10.1088/1742-6596/268/1/011003

Alexei Pokrovskii: a biographical sketch

F Holland

Department of Mathematics, University College Cork, Ireland

E-mail: f.holland@ucc.ie

Alexei Vadimovich Pokrovskii, an unassuming eclectic mathematician, who occupied the Chair of Applied Mathematics at University College, Cork, for the last nine years, died unexpectedly at his home in Carrigaline on September 1, 2010, aged 62. His passing will be mourned by his family, colleagues at UCC and his many friends around the world.

Alexei was born and reared in Voronezh, a city in Central Russia about 500 kilometres south of Moscow. Alexei's family came from a medical background: his grandfather Alexei Ivanovich Pokrovskii was a professor and Chair of Ophthalmology at Voronezh Medical Academy; his father Vadim Alexeevich Pokrovskii was a professor and Chair of Hygiene in Voronezh Medical Academy. Alexei's uncle Alexei Alexeevich Pokrovskii was an academician, a vice-president of the Academy of Medical Science of the USSR and the director of the Institute of Nutrition in Moscow. Alexei's mother Angelina was a teacher of English.

Alexei attended Voronezh State University, founded in 1918, where he received his BSc and MSc degrees. His early academic career was notable for the fact that he started publishing original mathematics while still a teenager. In 1969 he moved with Mark Krasnosel'skii—a world-renowned expert in Nonlinear Functional Analysis, and one of Russia's foremost mathematicians of the last century—to Moscow. By the time he completed his PhD there in 1974, under the direction of Mark Krasnosel'skii, with whom, amongst many others, he had a long and productive professional career, his published work numbered 16 articles. Together, they developed a mathematical theory of Hysteresis, and, in 1983, published *Systems with Hysteresis*, a highly influential book that was the first in the field. It was later translated into English, and had a huge bearing on the subsequent development of the subject.

In 1983 Alexei was awarded the prestigious Andronov prize by the USSR Academy of Sciences. He obtained his Dr. Sci (Dr. habil) at the Institute for Control Problems, Russian Academy of Sciences in 1989.

Until 1992, when he moved with his family to Australia, Alexei remained in Moscow, where he became head of the centre responsible for developing mathematical methods in Control, at the Institute for Information Transmission Problems, Russian Academy of Sciences. His research work continued apace at the Institute, and his collaboration with others was wide-ranging and prolific.

For the period of his stay in Australia during 1993-1997, Alexei divided his time between being an Adjunct Professor at Deakin University, Geelong, and Director for European Operations at the Centre for Applied Dynamical Systems, Mathematical Analysis and Probability, at the University of Queensland, Brisbane. Even though he didn't have security of tenure at either place, nevertheless, the time spent in Australia was a productive one for him, and a happy time for his family, who retain fond memories of their stay. Indeed, his daughter has just returned there to commence her medical career.

Alexei first came to Cork in the Spring of 1997 to be interviewed for the Chair of Applied Mathematics at University College, Cork (UCC). As part of the interview process he delivered a lecture about his research interests, during the course of which he paid tribute to the work of his mentor and teacher, Mark Krasnosel'skii, who had just passed away. Later that day he gave an impromptu lecture at a meeting of the student Mathematical Society, regaling the assembled students with a selection of witty problems. One such problem was the following: How should Vladimir in Leningrad mail a gold ring to his beloved Irene in Moscow knowing full well that any letter or parcel, thought to contain something valuable, would be opened as it passed through the Russian postal system, and its contents stolen? The students were captivated by him. On the same occasion, he offered a bottle of wine to the first student who could solve essentially the following problem: Is it possible to place 26 non-overlapping unit squares in a square of side 5.1 units? This anecdote encapsulates his capacity to draw students to him, and to open their minds to the wonders of mathematics.

While his bid for the Chair wasn't successful on that occasion, his exceptional ability as a research scientist of the first rank was recognised—by that time his research output ran to over 100 papers—and he was offered and accepted a research position in the Institute of Non-linear Science in UCC, which he held for three or four years. This post, though not a permanent one, provided a modicum of security for himself and his family, which they used to good effect to settle in Cork. During that period, his expertise in several branches of mathematics, and his capacity to interact productively with a range of experts working in fields outside pure and applied mathematics, such as statistics, computer simulation, economics and physics, became known within the College. Accordingly, when the Chair of Applied Mathematics became vacant again, and Alexei was an applicant, it came as no surprise that his star-quality was finally acknowledged by the College, and in 2001 he was appointed to this Chair, which he richly merited.

Following his appointment, while continuing to work with former colleagues elsewhere in different parts of the world, he developed fruitful collaborations with people in his own department, and in the UCC Departments of Computer Science, Civil, Electrical and Food Engineering, Mathematics, Microbiology, Statistics, Physics, and Zoology, which led to a raft of joint publications. In doing so, he displayed a commanding knowledge of several branches of science which enabled him to appreciate the relevance of pure mathematical advances to the world around us, and the ability to apply them in a host of different areas. He had a child-like curiosity and wonderment for the scientific world, a deep knowledge of several disparate areas which, combined with a penetrating mind, enabled him to make significant progress in whatever problem that took his interest. But he also took a keen interest in other people's work, and whenever somebody shared a surprising new fact with him, his countenance would alter, his eyes would sparkle with delight, and one would get the "thumbs up", signifying his pleasure. Such a response was very encouraging to the person sharing the information, especially to a young researcher, still unsure of his or her own ability. He was immensely generous with his time and talents, and warm-hearted in attributing to others ideas that were very often his alone, qualities which endeared him to his students. In truth, he was a polymath of the first rank.

From the beginning of his time in UCC, Alexei took a keen interest in the UCC Mathematics Enrichment Programme for secondary school students, and in the fortunes of Ireland's International Mathematical Olympiad team. He often devoted his Saturday mornings to tell pre-Leaving certificate pupils about the amazing Pigeon-hole Principle. This simple principle is illustrative of the kind of research problem that was close to Alexei's heart, namely, existence problem. When solving existence problems, he often used topological degree theory, of which he made substantial applications.

Journal of Physics: Conference Series **268** (2011) 011003 doi:10.1088/1742-6596/268/1/011003

During his headship of the UCC Department of Applied Mathematics, he organised several conferences and workshops on Hysteresis, Multi-rate Processes and Dynamical Systems, the proceedings of which were co-edited by him, and published in book form. These stimulating meetings attracted specialists in diverse areas from many places across the world, and established his department as a centre of research excellence in applied mathematics.

Alexei was widely read and knowledgeable about a range of subjects, and could be hugely entertaining: for instance, he could discuss the influence the novel *The Gadfly*, by the Cork woman Ethel Voynich, had in Russia, and compare and contrast the art work of Kandinsky with that of Klee. When he could be persuaded to drag himself away from doing mathematics—such as at Christmas time when he couldn't get into his office!—he was the perfect host. He took special delight in telling his friends about the ruses and ploys people got up to make life more bearable under the communist system.

Alexei was a mathematical genius, with a special way with people, both young and old. He was loved by everyone who knew him; he will be sadly missed.

He is survived by his wife Natasha, his daughter Olya, his son Alesha and his sister Olga.

In memory of Alexei Pokrovskii: his hysteresis effect on economics

R Cross

Department of Economics, University of Strathclyde, Sir William Duncan Building, 130 Rottenrow, Glasgow G4 0GE, Scotland, UK

E-mail: `rod.cross@strath.ac.uk`

Abstract. This memorial note assesses the impact of Alexei Pokrovskii, a pioneer in the mathematical analysis of systems with hysteresis, on economic analysis. Apart from the seminal contributions made, Alexei was a source of inspiration and encouragement to others seeking to understand the imprints of hysteresis on economic behaviour.

I first came across Alexei's name as the co-author of *Systemy s Gisteresisom* (Krasnosel'skii and Pokrovskii, 1983), having obtained a photocopy of this seminal work by way of the inter-library loan service. At the time, the mid-1980s, I was looking for a mathematical analysis of systems with hysteresis that was more interesting than the unit root special case of a linear difference equation that I had been using since the late 1970s in an attempt to understand economic fluctuations, particularly those in unemployment. Apart from the problems in applying unit root analysis to a bounded variable such as the rate of unemployment, the elephantine memory properties were implausible. My knowledge of the Russian language, however, was even more lacking than my knowledge of the mathematics.

It took a chance conversation with Marek Niezgodka in Warsaw in 1990, when I was working for the Polish National Bank, to set me on the path leading to Alexei. Marek had translated *Systemy s Gisteresisom* into *Systems with Hysteresis* (Krasnosel'skii and Pokrovskii, 1989). This book paved the way for an economist such as myself to see hysteresis as a general systems property that could maybe be applied to economic systems. I wrote papers that used the brilliant Isaac Mayergoyz (1991) staircase partition simplification of the Krasnosel'skii - Pokrovskii analysis to show how economic systems could be haunted by a selective, erasable memory of demand shocks (see, for example, Cross, 1993). I sent the papers to Alexei who responded with his characteristic enthusiasm. He subsequently became not only my mentor on matters relating to hysteresis but also a good friend.

In this memoir I will attempt to sketch how Alexei came to make a valuable contribution to understanding how economic systems respond to shocks such as the financial bubble that imploded on the world stage in 2007. To fill in the background to this interdisciplinary aspect of Alexei's work, it is useful to begin with the coining of the term 'hysteresis'. I will then discuss the flaws in conventional economic analysis that hysteresis explanations attempt to redress. Alexei's contributions are then considered in this context.

1. The Coining of the Term Hysteresis

The intellectual curiosity across a wide range of disciplines that was the hallmark of Alexei was evident in the coining of the term 'hysteresis'. Many of the key players were from Scotland and Alexei took a keen interest in this background. The point of departure was James Clerk Maxwell's formalisation of the concept of a field of force. Incredibly, Maxwell had become redundant when the University of Aberdeen merged its two colleges and decided that one of the two physics professors had to go! Various experimenters observed violations of Maxwell's equations in magnetic fields involving ferric metals, the fields not returning to their previous states once a temporary magnetising force was removed (see Glazebrook, 1935 for this background).

James Alfred Ewing, originally from Dundee, was one of the researchers who observed violations of homeostasis and reversibility in such fields. He coined the term 'hysteresis' to describe such effects in a paper published in 1881: "...these curves exhibit, in a striking manner, a persistence of previous state... to this action... the author now gives the term Hysteresis." (J A Ewing, 1881, p. 22). His assessor for the Royal Society of London paper was the University of Glasgow professor, Sir William Thomson (later Lord Kelvin). Thomson objected to the use of the mysterious term 'hysteresis', suggesting that the phrase 'effects of retentiveness' be used instead. Ewing stuck to his guns, arguing that 'hysteresis' was a generic phenomenon that would be observed in a wider range of contexts than electromagnetism (see A W Ewing, 1939).

Ewing's prescience became evident in Alexei's work on formalising hysteresis as a general systems property (Krasnosel'skii and Pokrovskii, 1983, 1989), and in applying the properties to a wide set of problems such as soil wetting and drying processes (Applebe et al, 2009) and trade cycles (McNamara and Pokrovskii, 2008).

2. Mainstream Economic Analysis

Mainstream economic analysis, which has encountered particularly searching criticism since the world financial crisis and subsequent recessions began in 2007, does not incorporate hysteresis effects. The analytical methods used in neoclassical economics involve conservation and reversibility properties. At the micro level this means that supply-side shocks, such as the burning of cotton fields during the US Civil War, have no lasting effects on the structure of demand, consumer tastes reverting to the status quo ante once the cotton fields are re-planted, despite the plausible shifts in demand arising from learning-by-consuming non-cotton materials during the disturbance. And demand shifts, such as those arising form increased demand for air travel in the 1960s, have no lasting effects on the supply conditions, despite the learning-by-producing efficiency gains in aircraft production. At the macro level this means that recessions have no lasting deleterious effects on potential output and employment, despite the debilitating effects of recessions on bankrupted firms and unemployed workers. And booms in economic activity also have no lasting beneficial effects on potential output, despite the new capital investments made and learning-by-working of employees. Financial markets are also taken to be efficient, asset prices accurately reflecting all the relevant information, and not being haunted by waves of optimism or pessimism about the unknowable future.

Dissidents have argued that this neoclassical system does not explain how actual economic systems behave, and have attempted to inject realism into the complacency of the neoclassical framework. John Maynard Keynes was a notable dissident, arguing that pessimistic 'animal spirits', leading consumers and firms to cut back on consumption and investment, can allow economies to become trapped in low output, high unemployment equilibria. While it is not too difficult to formalise Keynes' precepts in hysteretic terms, Keynes himself did not use the terminology. Keynes and Ewing, the coiners of the term 'hysteresis', were contemporaries at King's College, Cambridge. Whilst there is evidence of conversations between the two great men on international affairs, there is no indication that hysteresis was discussed (A W Ewing,

1939). Alexei, as was his wont, was fascinated by this tale of what might have been. Ewing, after establishing Engineering as an academic discipline at the University of Cambridge, had headed the enemy communications intercept group at the Admiralty during the 1914-18 war, and gone on to become the Principal of the University of Edinburgh. Keynes also had developed a wide portfolio as a man of affairs, so there would have been a variety of other topics of common interest.

3. Allowing for Hysteresis in Economic Systems

Alexei's initial role in applying hysteresis to economic systems came in the invaluable encouragement he provided to those who were engaged in this task. I first met Alexei in person when he arranged for me to give a talk at a SIAM conference at the University of Edinburgh. Giving a paper at a mathematics conference could have been a daunting experience for a non-mathematician such as myself, but Alexei made the occasion more of an enjoyable next step rather than an ordeal with his interventions during the discussion of the paper. The ordeal, in fact, proved to be one of attempting to explain to a large group of mathematicians the relative merits of the different beers on offer in a real ale bar in Leith. There were at least a dozen brews on offer!

Alexei would pose fundamental questions, such as whether it was plausible to suppose that rate independence characterised the responses of economic agents to input perturbations. And he would guide me through the development of the hysteresis literature, patiently explaining how the Preisach model had been developed, for example.

Around this time I came to supervise the Ph.D. work of Laura Piscitelli at the University of Strathclyde, jointly with Michael Grinfeld of the Department of Mathematics. This thesis, "Hysteresis in Economics" (Piscitelli, 1997) reflected the brilliance of Laura, the valuable supervision of Michael, and that of Harbir Lamba, then a postdoctoral fellow at Strathclyde. A key innovation in the thesis was to devise a means of testing for the existence of hysteresis in time series regressions designed to explain unemployment. Harbir gave important guidance in implementing the programme designed to detect the selective memory of demand and supply side shocks affecting unemployment. Alexei also played an important role in this work, suggesting how sensitivity tests of the Preisach weighting function might help get round the problem of the absence of experimental data regarding agent switching points. At Alexei's instigation, an extension of this work was later published as Darby, Cross and Piscitelli (2006).

There followed from Alexei an invitation to give a seminar at University College Cork. At the time Alexei had a temporary position there. After the seminar I got talking to Michael Mortell who, at the time, was President of UCC. Mike made some typically searching comments on the paper I had presented, but then we started talking about Alexei's importance in the academic world. Alexei must have subsequently heard a summary of this conversation from Mike. I must have said something along the following lines: "First there was Isaac Newton, then James Clerk Maxwell, next Albert Einstein, now there is Alexei Pokrovskii!". The implication was that UCC should offer Alexei a permanent position forthwith! Alexei took great delight in this story. This was not because he had the heightened sense of self-importance that plagues some academics. He did not suffer from such an affliction, being down-to-earth in the manner of all the truly great academics I have known. Instead, Alexei's delight in this story reflected his keen sense of humour and appreciation of the absurd.

There followed regular visits to Cork for the MURPHYS conferences organised by Alexei and his co-researchers. Alexei established a research team of truly international significance that provided a welcome to interdisciplinary researchers. The trips to the bars of Cork, and to the fishmongers for raw fish to ingest with the beer, Russian-style, played an important role in establishing an esprit de corps for the Cork group and the leading world experts who gravitated to this centre of excellence.

But these meetings were also a serious engine of discovery. To take but one example in which I was directly involved, prolonged discussion revealed that the Preisach model needed to be revised for application to economic systems by allowing agent switching points to vary over time. The outcome, for which Alexei and Sasha Krasnosel'skii generously offered me co-authorship, was Cross, Krasnosel'skii and Pokrovskii (2001). This was an example of Alexei's genius in being able to modify the mathematical analysis to deal with the circumstances of the application.

4. Alexei and Economics

An international forum that Alexei introduced me to was the bi-annual Hysteresis and Micromagnetics conference. He arranged an invitation for me to give a paper at the Washington DC conference in 2001. Problems with Alexei's visa meant that he arrived late, but just in time to hear my presentation. This applied the Preisach model to the Economic and Monetary Union project initiated in Europe in 1999 (Cross, 2000). One of the implications was that interest rates would be too low in booming economies such as Ireland, with inevitable dire consequences once the ensuing property bubble burst. As with many implications of economic models, this was treated with suspicion– and took until 2007 to be borne out! Again, this was Alexei opening doors for others, allowing me to meet other greats of the hysteresis profession such as Isaac Mayergoyz.

The Micromagnetics and Hysteresis conference in Naples in 2007 honoured Alexei with a presidential address. It was at about this time that Alexei had started to expand the economics component of his research portfolio. Maybe it was the growing signs of major turbulence in the world economy that stimulated his heightened interest. Alexei chose to unveil some new economic models in his presidential address (Cross, Pokrovskii and Raschinskii, 2008, and Cross, McNamara and Pokrovskii, 2008). The models were based on a modification of Darcy's Law, with output flows responding to differences between potential and actual demand. Hysteresis in output flows arose from differences between behaviour in booms - the 'wetting' phase - and in recessions - the 'drying' phase. Alexei showed that these models followed the 'hydraulic' tradition in economic modelling, surprising many in the audience with photographs of working models that had been built by the economists Irving Fisher and Bill Phillips to illustrate the determination of relative prices, and of aggregate output. The difference was that Alexei's equations involved non-reversible paths for the flows within economic systems rather than the reversible flows involved in the Fisher and Phillips' models.

Alexei's supervision of Hugh McNamara's Ph.D thesis at Cork involved extending these models. By this time recessions had come to afflict many economies with a vengeance after the implosion of the financial bubble in 2007. Ireland, for example, experienced the largest cumulative drop in output, in percentage terms, suffered by an advanced economy since the Great Depression of the early 1930s. Mainstream economic models assume that such recessions have only temporary effects, with output subsequently recovering to its previous growth path. The models that Alexei designed do not have this felicitous property but instead have recessions bringing lasting curses by displacing potential output levels downwards. The evidence from the last three years suggests that Alexei was right. One implication of this analysis is that fiscal retrenchment in recessions can be self-defeating. In a recent paper, 'Memory of Recessions', (Cross, McNamara and Pokrovskii, 2010), we chronicled the evidence on how the recessions that usually follow financial crises displace potential output paths downwards, and showed how hysteretic models, in which the rate of change of output responds to demand gradients, can mimic the key features of this evidence. Alexei's work may well play an important role in the reformulation, with Keynesian properties, of economic models of macroeconomic fluctuations.

The last time I saw Alexei in person was when he attended an Institute of Advanced Studies workshop in Glasgow in the summer of 2009. This "Limits to Rationality in Financial Markets" workshop was co-organised with Michael Grinfeld and Harbir Lamba. Alexei was brimming with ideas as to how hysteresis models could throw light on how the financial crisis had emerged, and on its aftermath. Along with Gary Friedman, we had begun to work on the application of return-point memory models to interbank interest rates and their effects on bank lending.

Alexei had a most distinctive manner of speaking. Instead of using full stops, he would punctuate his sentences with a variety of smiles, ranging from the rueful to the exuberant. When we talked on the telephone I could always picture the type of smile on his face. I will miss those conversations.

In good biblical tradition, Alexei saved some of the best until the last. The day before he died he sent me the last draft of a paper entitled 'Periodic Sequences of Arbitrage: A Tale of Four Currencies' (2010) which Victor Kozyakin, Brian O'Callaghan, Alexei's son Alexey Pokrovskiy, myself and Alexei had been working on over the summer. Mainstream theories in economics and finance assume that all profitable arbitrage opportunities are exploited so that the 'law of one price' always holds. The evidence suggests significant and persistent deviations of actual goods and asset prices from this law. This paper uses combinatorial analysis to show that arbitrage sequences can be periodic in nature, rather than involving a smooth convergence towards a law of 'one price'. The implications of these results for equilibrium theorising in economics and finance are profound. The mathematical techniques for analysing asynchronous systems used in this paper could well prove to have a wide set of applications to other problems in economics and finance.

The peer review jury is still out, but one eminent US economist has described the results as 'novel' and 'beautiful'. These adjectives provide a fitting description of the contributions Alexei has made to the difficult task of understanding how economic systems behave. Others will continue Alexei's work. We will be sadder for his absence, but much wiser for the legacy he has bequeathed. Alexei's ship may have passed beyond the horizon but, in a very real sense, the ship still sails on.

References

[1] Applebe B, Flynn D, McNamara H, O'Kane P, Pimenov A, Pokrovskii A V, Rachinskii D and Zhezherun A 2009 Rate-Independent Hysteresis in Terrestial Hydrology *IEEE Central Systems Magazine* **29** (1) 44–69

[2] Cross R 1993 On the Foundations of Hysteresis in Economic Systems *Economics and Philosophy* **9** (1) 53–74

[3] Cross R 2000 Hysteresis and EMU *Metroeconomica* **51** (4) 367–379

[4] Cross R, Krasnosel'skii A M and Povrovskii A V 2001 A Time Dependent Preisach Model *Physica B* **403** 231–236

[5] Cross R, McNamara H and Pokrovskii A V 2008 Modelling Macroeconomic Flows Related to Large Ensembles of Elementary Exchange Operations *Physica B* **403** 451–455

[6] Cross R, McNamara H, Pokrovskii A V and Rachinskii D 2008 A New Paradigm for Modelling Hysteresis in Macroeconomic Flows *Physica B* **403** 231–236

[7] Cross R, McNamara H and Pokrovskii A V 2010 Memory of Recessions, forthcoming *Journal of Post Keynesian Economics*

[8] Cross R, Kozyakin V, O'Callaghan B, Pokrovskii A V and Pokrovskiy A 2010 Periodic Sequences of Arbitrage: A Tale of Four Currencies *University of Strathclyde Discussion Paper.*

[9] Darby J, Cross R and Piscitelli L 2006 Hysteresis and Unemployment: a Preliminary Investigation *The Science of Hysteresis* vol I, ed I Mayergoyz and G Bertotti (Elsevier, Academic Press) pp 667–699

[10] Ewing J A 1881 On the Production of Transient Electric Currents in Iron and Steel Conductors by Twisting them when Magnetised or by Magnetising them when Twisted *Proceedings of the Royal Society of London* **33** 21–23

[11] Ewing A W 1939 *The Man of Room 40: the Life of Sir Alfred Ewing* (London: Hutchinson)

[12] Glazebrook R T 1935 James Alfred Ewing 1855-1935 *Obituary Notices of the Royal Soc. of London I* 475–492

[13] Krasnosel'skii M A and Pokrovskii A V 1983 *Systemy s Gysteresisom* (Moscow: Nauka)

[14] Krasnosel'skii M A and Pokrovskii A V 1989 *Systems with Hysteresis* (Berlin: Springer-Verlag)

[15] Mayergoyz I D 1991 *Mathematical Models of Hysteresis* (Berlin: Springer-Verlag)

[16] McNamara H and Povrovskii A V 2008 Hysteresis and the Trade Cycle *Physica B* **372** 202–206

[17] Piscitelli L 1997 Hysteresis in Economics *University of Strathclyde Ph D. thesis*

Journal of Physics: Conference Series **268** (2011) 011005 doi:10.1088/1742-6596/268/1/011005

Alexei Pokrovskii – a tribute with some open problems

J Philip O'Kane
Emeritus Professor of Civil Engineering

1. The beginning

I first met Alexei at a seminar he gave in UCC towards the end of the last century. At that time Alexei did not have a permanent academic position and the seminar was a prolegomenon to his eventual appointment as Professor of Applied Mathematics. As I try to recall that seminar it appears to me now as belonging to a different age. The communist world had imploded at its Moscow centre in 1990, and in the ensuing chaos, had dispatched many of its brilliant physicists and mathematicians to the research institutes and universities of the world. It was the great fortune of UCC that Alexei came to Cork, very far from his native Russia, or Australia, which had first given him sanctuary.

I believe Alexei settled in Cork because of the climate: physical, intellectual and social.

Having experienced extreme continental climates throughout his life, he never tolerated critical comments by colleagues and friends on the benign Irish climate. The rebuke, when it came, was always made with a smile that said "You do not know how fortunate you are!".

The mathematicians and physicists of the University created the intellectual climate that allowed Alexei "to be himself": a passionate mathematician, devoted to his students, respected by his peers, and always willing to explain his beloved theory of hysteresis, which he had begun as a graduate student with his professor, Mark Alexandrovich Krasnosel'skii[1]. We listened and learned, and our appreciation of this magnificent academic grew and grew.

Navigating the social climate of Cork, an ancient port city[2] open to the world, is not easy for new arrivals. Alexei's wife Natasha found interests in the University; their children Olya and Alesha found places in Cork secondary schools to their liking; and Alexei and his family stayed and made their home in Carrigaline, high above the Abhann Buí - the yellow or golden river - that flows into Cork's great harbour.

2. The invitation

I cannot recall the time or place after his appointment as professor when Alexei said to me

"Philip, I hear there is hysteresis in soil-water!"

to which I replied

"Alexei, would you like access to the largest soil-water database[3] in the world, assembled by

[1] Марк Александрович Красносельский (1920-1997) a Soviet, Russian and Ukrainian mathematician renowned for his work on nonlinear functional analysis and its applications, and Head of Laboratory (1974-1990) at the Institute of Control Sciences of the USSR Academy of Science in Moscow. http://en.wikipedia.org/wiki/Mark_Krasnosel'skii

[2] D. Dickson (2005) Old World Colony - Cork and South Munster 1630-1830. Cork University Press.

[3] R. Haverkamp, C. Zammit, F. Bouraoui, K. Rajkai, J.L. Arrúe and N. Heckmann (1998) GRIZZLY, Grenoble catalogue of soils: Survey of soil field data and description of particle-size, soil water retention and hydraulic conductivity functions. Laboratoire d'Etude des Transferts en Hydrologie et Environnement, Grenoble, France. http://ltheln21.hmg.inpg.fr/LTHE/index.php

Randel Haverkamp and colleagues at Grenoble?"
and so began our joint exploration of hysteretic flows in unsaturated porous media, especially in soils, at scales ranging from the laboratory bench to whole catchments, as they respond to rain, wind and sun. We exchanged our books, both co-authored with the professors under whom we had each studied hysteresis[4] and hydrology[5] respectively.

The collaboration produced many joint papers with members of Alexei's circle, and at UCC, one Masters thesis: "Preisach Operator-Differential Equations with a Degenerate Density Function" by Brian Applebe (2007), and two PhD theses: "Modelling the flow of water through multiphase porous media with the Preisach model" by Dennis Flynn (2008), and "Mathematical modelling of discontinuous hysteretic flows through porous media" by Andrew Zhezherun (2010).

The most fitting tribute I can make in this brief article is to list some open problems in Alexei's memory which I believe would have excited his interest and won his approval. They are high-lighted in bold type in the following short discourse.

3. Hysteresis in hydrology and soil physics

All terrestrial life ultimately depends on soil and water[6]. The layers of soil that cover much of the Earth consist of living and dead granular material continuously reworked chemically and mechanically by geological, climatic and hydrological forces in the longer term, and by bacteria, funghi, plant roots, worms, and other animals in the short term. The smaller living material consists of aggregates of bacteria and fungi that decompose the dead parts of plants and animals, making inorganic nutrients available for recycling in a "water-air-soil particle" solution. These nutrients are taken up again through the roots of plants that grow in the soil harvesting the energy of the sun. The soil ecosystem is an open system with a complex set of feedback loops that govern the physics, chemistry and biology of the soil on multiple time-scales.

Close to the surface, soil layers may contain both air and water flowing through a matrix of particles, a mixture of sand, silt and clay, ultimately derived from underlying rock. In contrast to sand and silt, the clay particles have a very large chemically reactive surface. They swell with increasing water content, crack when dry, and retain the memory of such fissures over many years. Soils, and their vertical profiles in the ground, are classified differently in Soil Science and in Geotechnical Engineering, for different purposes. However, the sand, silt, clay and organic fractions are always important discriminants.

When air is present, soil water is bound to the organic and inorganic particles by surface forces. The unsaturated soil exhibits internal air-water interfaces of very large area. These interfaces are dynamic, may coalesce or fragment suddenly, are extremely convoluted in shape, and of variable curvature implying small to very large tensile forces exerted by the soil water on the soil matrix.

The flow of water is driven internally by unbalanced surface forces and by gravity. It is also driven externally by infiltration of rain, by evaporation from the surface of the ground, and by transpiration of water upwards through the root hairs and plant xylem cells. The plant controls transpiration to win carbon dioxide from the air by satisfying the evaporative demand of the atmosphere for water. There is a double diffusion through the open stomata of the leaves that are continuously connected to the xylem cells and root hairs in the soil: carbon dioxide diffuses in, soil water out, in the vapour phase. The size of the openings is controlled by guard cells in a feedback loop that responds to the availability of water in the root zone. During drought the

[4] M.A. Krasnosel'skii, A.V. Pokrovskii (1989) and M. Niezgodka (Translator) Systems with Hysteresis (translation). Springer, Berlin. 410 pages.

[5] J.C.I. Dooge and J.P. O'Kane (2002) Deterministic Methods in Systems Hydrology: IHE Delft Lecture Note Series (UNESCO-IHE Delft Lecture Note Series). A.A. Balkema, Lisse (2003). Taylor & Francis Group (2002). 309 pages.

[6] D. Hillel (1991) Out of the Earth - Civilization and the Life of the Soil. University of California Press, Berkeley, Los Angeles. 321 pages.

plant conserves its own water by shutting down transpiration.

The contact angles between the solid particles and the air-water interfaces (outside the root hairs) are believed (a) to flip suddenly in time, or jump in space, or both, whenever the flow of water changes direction, and (b) to do so much faster than any change in the volumetric rate of flow of water in the soil.

At any moment in time the setting of these spatially distributed contact angles constitutes a record of past changes in the direction of flow. This complex rate-independent memory, operating over time in a non-local manner, controls the flow of water. This is believed to be the physcial mechanism underlying the phenomenon of hysteresis that has been observed for 70 years in the flow of water through unsaturated porous media such as soil.

Rate-independence remains to be demonstrated in the laboratory - an open problem.

Hysteresis is significant in the irrigation of agricultural soils that have a high content of sand particles.

In the discipline of Soil Physics the matrix of solid particles, through which both air and water flow, is usually assumed to be perfectly rigid in spite of the very large surface forces that act on it. When the particles are formed from clay minerals, the rigid approximation breaks down completely. The plastic clay particles swell with increasing water content and exhibit a second kind of hysteresis even when no air is present.

The mathematical formulation of double hysteresis in *deformable and swelling* unsaturated porous media is an open problem.

4. Mathematical treatment of unsaturated flow

The mathematical treatment of unsaturated flow through a rigid porous matrix is usually stated as a problem in non-linear potential theory. The flow rate of water is assumed to be proportional to the gradient in a suitable total potential which is the sum of several partial potentials, such as the gravitational, matric, osmotic, and other partial potentials. The proportionality factor is known as the unsaturated hydraulic conductivity. When this dynamic equation is combined with the equation for the conservation of water, a relationship between matric potential and water content is required to provide closure, the so-called soil-moisture characteristic. The well-known convective-diffusion, concentration-dependent equation of Richards is the result in one dimension, and has been extensively studied for unidirectional flows. The generalisation to three dimensions shows that the flow, as a specific flux through a unit area, is a tensor quantity.

The matric potential relates the mechanical work required to extract water from a porous body, to its water content during drying, and conversely during wetting. The relationship between potential energy measured as mechanical work and water content is multi-valued and hysteretic. An outer envelope is defined by a complete wetting, or a complete drying, with an infinite number of possible scanning curves in between, as wetting and drying are reversed in an arbitrary manner. The unsaturated hydraulic conductivity K also exhibits hysteresis. It is more marked when K is related to moisture content, and less so when related to matric potential.

5. The GRIZZLY database and the Preisach model

Dennis Flynn successfully fitted the Preisach model to all the soil-moisture characteristics and their associated scanning curves in the GRIZZLE database.

The datbase does not report the methods of measurement used in each case. In no case was there an attempt to measure the initial non-local memory of soil samples. In addition, maesurement devices which are themselves porous also exhibit hysteresis. It is very important to control the non-local memory of these devices, if accurate measurements are to be made.

The insights provided by the theory of the Preisach model demand an examination of existing methods of measurement of water content and flow through hysteretic unsaturated porous media, and probably the invention of new apparatus and measurement protocols - an open problem.

6. The FEST model

The acronym FEST stands for "fully-vegetated slab of uniform soil transpiring to the atmosphere". I invented the FEST model to provide Alexi with an ordinary differential equation, rather than the full Richards equation, which could be combined with the Preisach operators identified by him and Dennis Flynn, and which could be analysed with full rigour. We were both surprised by the result, in our different ways.

The combined Preisach-differential operator in FEST was different from any that had already been analysed in the hysteresis literature, much to Alexei's delight. The examination of solutions for existence, uniqueness, continuous dependence on the given data, and high-accuracy numerical methods, led to several papers and the theses referred to already.

The next step is to proceed to a combined Richards-Preisach model in one or more space dimensions - an open problem.

Of particular interest is the case of two dimensions, vertical and horizontal, in which the phenomenon of fingering occurs. Preferential flow downwards on reproducible paths through layers of homogeneous porous media have been observed in two-dimensional laboratory experiments. It is believed that the hysteretic properties of the media steer the flow exhibiting strong non-local memory. Once an initial instability has marked a particular preferential path as a finger, future flows do likewise. We may ask:

Would the presence of the third dimension lead to further instability, breaking the pattern of reproducible fingers in the downward flow? - an open problem, which also has implications for measurement.

A second surprise was the discovery in the FEST model of two stable equilibria under particular circumstances, with high and low moisture contents respectively, separated by one unstable equilibrium. I called this the bifurcation to desert, since the vegetation dies at the low moisture content.

The open question is whether the bifurcation to desert is an artefact of the missing spatial variation in the FEST model - open problem.

In conclusion,
with hope and without fear,
let you the reader and I the author,
say of Alexei

Ar dheis Dé go raibh a anam.

Journal of Physics: Conference Series **268** (2011) 011006 doi:10.1088/1742-6596/268/1/011006

List of publications of Professor Alexei Pokrovskii

Books

1. M.A. Krasnosel'skii and A.V. Pokrovskii, "*Systems with Hysteresis*", Nauka, Moscow 1983 (271 p.), in Russian.
2. Extended English translation of "*Systems with Hysteresis*": Springer-Verlag, Berlin, New York, 1989 (410 p.).
3. N.A. Bobylev, V.V. Boltyanskii, S.Yu. Vsechsvyatskii, V.V. Kalashnikov, V.B. Kolmanovskii, V.S. Kozyakin, A.A. Kravchenko, A.M. Krasnosel'skii and A.V.Pokrovskii, "*Mathematical Systems Theory*", Nauka, Moscow 1986 (150 p.), in Russian.
4. P. Diamond, P.E. Kloeden, V.S. Kozyakin and A.V. Pokrovskii, "*Semi-Hyperbolic Lipschitz Mappings*", (in preparation).

Book Chapters

1. P. Diamond, P.E. Kloeden, A. Pokrovskii, 'Interval stochastic matrices and simulation of chaotic dynamics', In: *Chaotic Numerics*, American Mathematical Society, Providence, 1994, pp. 203-217.
2. A. Krasnosel'skii, A. Pokrovskii, 'Large sub-harmonics of pendulum like equations', In: *The first 60 years of Nonlinear Analysis of Jean Mawhin*, M. Delgado, J. Lopez-Gomez, R. Ortega, A. Suarez (eds), World Scientific Publishing, 2004, pp. 103-116.
3. A. Pokrovskii, O. Rasskazov, R. Studdert, 'Split-hyperbolicity, hysteresis and Lang-Kobayashi equations', In: *Singular Perturbations and Hysteresis*, SIAM, 2005, pp. 299-342.
4. A. Pokrovskii A., V. Sobolev, 'Naive view on singular perturbation and hysteresis', In: *Singular Perturbations and Hysteresis*, SIAM, 2005, pp. 1-60.
5. M. Brokate, A. Pokrovskii, D. Rachinskii, O. Rasskazov, 'Differential equations with hysteresis via a canonical example', In: *The Science of Hysteresis*, G. Bertotti, I. Mayergoyz (eds), Elsevier, Academic Press, 2006, vol. I, pp. 125-291.
6. D. Flynn, H. McNamara, J.P. O'Kane, A. Pokrovskii, 'Application of the Preisach model to soil-moisture hysteresis', In: *The Science of Hysteresis*, G. Bertotti, I. Mayergoyz (eds), Elsevier, Academic Press, 2006, vol. III, pp. 689-744.
7. S. O'Reagan, C. Kelly, A. Korobeinikov, M.J.A. O'Callaghan and A.V. Pokrovskii, 'The impact of an emerging avian influenza virus (H5N1) in a seabird colony', In: *Global View of Fight Against Influenza*, P. M. Mitrasinovic, Ed, Nova Science Publishers, New York, 2009.
8. A. Pimenov, T. Kelly, A. Korobeinikov, M.J.A. O'Callaghan, A.V. Pokrovskii, 'Systems with hysteresis in mathematical biology via a canonical example', In: *Clustering Algorithms and Mathematical Modeling*, C. L. Wilson, Ed, Nova Science Publishers, New York, 2010 (in print).

Research Papers

1967

1. A theorem of A. F. Timan, *Functional Analysis and its Applications*, **1** (1967), No 3, pp. 250 - 251.

1968

2. On the problem of bifurcation points, In: "*Problems of Mathematical Analysis of Complex Systems* ", **2** (1968), pp. 41 - 56, Voronezh State University Press, Voronezh, in Russian. Co-authored with P.P. Zabrejko and M.A. Krasnosel'skii.
3. On positive almost periodic functions, In: "*Problems of Mathematical Analysis of Complex Systems*", **3** (1968), Voronezh State University Press, Voronezh, in Russian.

1969

4. Inertial games, *Soviet Mathematics*, **10** (1969), No 5, pp. 1060 - 1064. Co-authored with I.V.Emelin and M.A. Krasnosel'skii.

1970

5. Hysterant operator, *Soviet Mathematics*, **11** (1970), No 1, pp. 29 - 33. Co-authored with B.M. Darinskii, I.V.Emelin, M.A. Krasnosel'skii, E.A. Livshitz and P.P. Zabrejko.
6. Vibrostability of solutions of differential equations, *Soviet Mathematics*, **11** (1970), No 6, pp. 1510 - 1512. Co-authored with M.A. Krasnosel'skii.

1971

7. Systems of hysterons, *Soviet Mathematics*, **12** (1971), No 5, pp. 1388 - 1391. Co-authored with M.A. Krasnosel'skii.

1972

8. Vibrationally stable hysterons, *Soviet Mathematics*, **13** (1972), No 5, pp. 1305 - 1309. Co-authored with M.A. Krasnosel'skii and V.S. Kozjakin.
9. Vibrationally stable differential equations with a continuous right side, *Transactions of the Moscow Mathematical Society*, **27** (1972), pp. 95 - 116. Co-authored with M.A. Krasnosel'skii.

1973

10. Nonlocal continuability of solutions of vibrationally stable differential equations, *Soviet Mathematics*, **14** (1973), No 1, pp. 267 - 271.
11. On the theory of hysteresis nonlinearities, *Soviet Mathematics*, **14** (1973), No 3, pp. 896 - 900.

1974

12. Periodic oscillations in systems with relay nonlinearities, *Soviet Mathematics*, **15** (1974), No 3, pp. 873 - 877. Co-authored with M.A. Krasnosel'skii.

1976

13. Regular solutions of equations with discontinuous nonlinearities, *Soviet Mathematics*, **17** (1976), No 1, pp. 128 - 132. Co-authored with M.A. Krasnosel'skii.
14. Simulation of transformers with hysteresis by means of continuum of relay nonlinearities, *Soviet Mathematics*, **17** (1976), No 2, pp. 447 - 451. Co-authored with M.A. Krasnosel'skii.

1977

15. The discontinuous superposition operators, *Uspehi Matem. Nauk*, **32** (1977), No 1, pp. 169 - 170, in Russian. Co-authored with M.A. Krasnosel'skii.
16. The principle of absence of bounded solutions in the problem of absolute stability, *Soviet Mathematics*, **18** (1977), No 2, pp. 388 - 392. Co-authored with M.A. Krasnosel'skii.
17. A new method for absolute stability investigation, *News of USSR Academy., Mechanics of solid body*, **3** (1977), p. 154, in Russian. Co-authored with M.A. Krasnosel'skii.
18. Natural solutions of stochastic differential equations, *News of USSR Academy. Mechanics of solid body*, **4** (1977), p. 193, in Russian. Co-authored with M.A. Krasnosel'skii.

1978

19. Natural solutions of stochastic differential equations, *Soviet Mathematics*, **19** (1978), No 3, pp. 578 - 582. Co-authored with M.A. Krasnosel'skii.
20. On absolute stability of systems with discrete time, *Automation and Remote Control*, **39** (1978), No 2, pp. 189 - 198. Co-authored with M.A. Krasnosel'skii.

1979

21. Principle of the absence of bounded solutions in the absolute stability of equations of neutral type, *Serdica*, **4** (1979), pp. 61 - 69, in Russian. Co-authored with R.R. Akhmerov and N.G. Kazakova.
22. Equations with discontinuous nonlinearities, *Soviet Mathematics Doklady*, **20** (1979), No 5, pp. 1117 - 1120. Co-authored with M.A. Krasnosel'skii.
23. The method of block diagrams in control theory, In "Theory and technic of control", Moscow, Institut of Control Problems, 1979, pp. 5-14. In Russian. Co-authored with M.A. Krasnosel'skii and L.I. Rozonoer.
24. Continuity points of monotone operators, *Special issue dedicated to W. Orlich on the occasion of his 75-th birthday, Comment. Math. Special Issue*, **2**, pp. 205 - 216. "*Orlicz volume*", Warsaw, 1979. Co-authored with M.A. Krasnosel'skii.
25. The limit norm of a linear operator and its applications, *Soviet Mathematics Doklady*, Akad. Nauk SSSR, **249** (1979), No 3, pp. 548 - 551; **20**, No 6, pp. 1314 - 1317.

1980

26. Bifurcation values of parameters in variational problems, *Soviet Mathematics Doklady*, **22** (1980), No 3, pp. 682 - 686. Co-authored with M.A. Krasnosel'skii and E.M. Muchamadiev.
27. Positivity conditions for pulse-frequency responses, *Automation and Remote Control*, **41** (1980), No 2, pp. 35 - 44. Co-authored with M.A. Krasnosel'skii and V.B. Prival'skii.
28. On new applications of frequency criteria, *News of USSR Academy. Mechanics of solid body*, **3** (1980), p. 170, in Russian. Co-authored with M.A. Krasnosel'skii.

1981

29. Vector hysteresis nonlinearities of the von Mizes-Tresca type, *Soviet Physics Doklady*, **26** (1981), No 3, pp. 581 - 583. Co-authored with A.F. Klepcyn, M.A. Krasnosel'skii, E.A. Livshitz and A.A. Vladimirov.
30. Principle of absence of bounded solutions in a problem of absolute stability, In "*Stability of Motion. Analytical Mechanic. Control of Motion*", Nauka, (Moscow), 1981, pp. 156 - 169, in Russian. Co-authored with M.A. Krasnosel'skii.

31. On general systems and systems with initial conditions, *Avtomatica i Telemekhanika*, **8** (1981), in Russian.
32. Infinitely repeating games with inertia, In "*Dynamic Game Systems*", VNIISI (Moscow), 1981, pp 15-27, in Russian. Co-authored with M.A. Krasnosel'skii and A.A. Vladimirov.
33. Dissipative property of a class of multicriterion systems, *Automation and Remote Control*, **42** (1981), No 4, Part 1, pp. 420 - 424.
34. A class of discontinuous systems, *Automation and Remote Control*, **42** (1981), No 8, pp. 1009 - 1012.

1982

35. Dynamics of control systems with several nonlinearities, *Differential Equations and Applications*, I, II, Tech. Univ. Russe (Bulgar.), 1982, pp. 366 - 376. Co-authored with V.S. Kozyakin, A.M. Krasnosel'skii and M.A. Krasnosel'skii.
36. Forced vibrations of an oscillator with hysteresis taken into account, *Soviet Mathematics Doklady*, **25** (1982), No 2, pp. 424 - 427. Co-authored with T.S. Gil'man.

1983

37. Effect of small perturbations of input signals, *Soviet Mathematics Doklady*, **27** (1983), No 2, pp. 423 - 426. Co-authored with M.A. Krasnosel'skii.
38. The absolute stability of systems with several nonlinear components, *Soviet Mathematics Doklady*, **28** (1983), No 1, pp. 267 - 269. Co-authored with M.A. Krasnosel'skii.
39. The principle of absence of bounded modes in the problem of absolute stability of distributed systems, *Automation and Remote Control*, **44** (1983), No 3, pp. 283 - 289. Co-authored with M.A. Krasnosel'skii and A.V. Zarubin.
40. Sliding modes and stochastic differential equations, In: "*Dynamics of Nonhomogeneous Systems*", VNIISI (Moscow), 1983, in Russian.
41. On hysteresis nonlinearities, *Russian Mathematical Surveys*, **38**, No 6, 1983, pp. 156-157, in Russian. Co-authored with M.A. Krasnosel'skii.

1984

42. On applications of guiding potentials method to systems with hysteresis, In: "*Dynamics of Nonhomogeneous Systems*", VNIISI (Moscow), 1984, in Russian. Co-authored with A.M. Krasnosel'skii and M.A. Krasnosel'skii.
43. Regular solutions of equations with strong nonlinearities, *Soviet Mathematics Doklady*, **29** (1984), No 1, pp. 107 - 110.

1985

44. Limit norm of a linear element, *Automation and Remote Control*, **46** (1985), No 11, pp. 1397 - 1404.
45. Absolute dissipation of systems with hysteresis, In "*Applied Methods of Functional Analysis*", Voronezh State University Press, Voronezh, 1985, in Russian.
46. Structure of outputs of discontinuous systems with quasi-periodic and desynchronized inputs, In "*Applied methods of functional analysis*", Voronezh State University Press, Voronezh, 1985, in Russian.

1986

47. Application of Kolmogorov's complexity to analysis of control systems, *Automation and Remote Control*, **47** (1986), No 1, pp. 21 - 28. Co-authored with E.A. Asarin.
48. Existence and calculation of stable modes in switching systems, *Automation and Remote Control*, **47** (1986), No 4, pp. 451 - 458.

1987

49. Averaging and limit hysteresis nonlinearities, *Ukrainian Math. Journal*, **39** (1987), No 1, pp. 30 - 38. Co-authored with M.A. Krasnosel'skii and A.A. Vladimirov.

1988

50. Functional analysis and dynamics of nonlinear control systems, *Selected Papers from 10th Triennial World Congress of the International Federation of Automatic Control, Munich, West Germany, 27 - 31 July, 1987*, Oxford, UK: Pergamon, 1988, **8**, pp. 11 - 15. Co-authored with N.A. Bobylev, A.M. Krasnosel'skii, M.A. Krasnosel'skii and A.A. Vladimirov.
51. The averaging principle for systems with hysteresis, In: "*Asymptotic Methods in Mathematical Physics*", "Naukova Dumka", Kiev, 1988, pp. 126 - 133, in Russian. Co-authored with M.A. Krasnosel'skii.

1989

52. Measures of unpredictability of binary sequences, *Soviet Physics Doklady*, **34** (1989), No 7, pp. 594 - 596.
53. Global criteria for the existence of self-oscillatory regimes, *Soviet Physics Doklady*, **34** (1989), No 10, pp. 867 - 868.
54. New ideas in nonlinear analysis, *Nonlinear Boundary Problems*, Kiev, **1**, (1989), pp. 22-26, in Russian. Co-authored with N.A. Bobylev, M.A. Krasnosel'skii and A.M. Krasnosel'skii.
55. Self-induced oscillations in systems with hysteresis, *Soviet Physics Doklady*, **34** , (1989), No 12, pp. 1066 - 1068. Co-authored with I.V. Fomenko and M.A. Krasnosel'skii,

1990

56. Identification of linear system on the basis of two tests, *Soviet Physics Doklady*, **35** (1990), No 2, pp. 109 - 111. Co-authored with N.I. Grachev N.A. Kuznetsov and M.A. Krasnosel'skii.
57. Stable periodic regimes in control systems with monotone nonlinearities, *Automation and Remote Control*, **51** (1990), No 2, Part 1, pp. 158 - 164. Co-authored with M.E. Semenov.

1991

58. New theorems on forced periodic oscillations and bounded solutions, *Soviet Physics Doklady*, **36** (1991), No 11, pp. 743 - 745. Co-authored with M.A. Krasnosel'skii and J. Mawhin.

1992

59. Operators of a problem on forced oscillation in systems with hysteresis, *Soviet Mathematics Doklady*, **44** (1992), No 1, pp. 228 - 232. Co-authored with M.A. Krasnosel'skii.
60. The role of controllability-type properties in the study of the stability of desynchronized dynamical systems, *Soviet Physics Doklady*, **37** (1992), No 5, pp. 213 - 215. Co-authored with V.S. Kozyakin.

61. Mathematical models of hysteresis, In: "Numerical methods and optimization", Tallin, **3** (1992), pp. 37 - 45. Co-authored with Krasnosel'skii M.A.

62. On the dynamics of control systems described by equations of parabolic type with hysteresis nonlinearities, *Automation and Remote Control*, **53** (1992), No 11, Part 1, pp. 1705 - 1711. Co-authored with M.A. Krasnosel'skii, V.V. Chernorutskii and G. Tronel.

63. Nonexistence of weakly improving monotone shifts in semi-ordered sets with nontrivial topological structure, *Automation and Remote Control*, **53** (1992), No 12, Part 2, pp. 1951 - 1959. Co-authored with D.I. Rachinskii.

64. Asymptotical classification of aliasing structures. *Journal of Appl. Math. and Stoch. Anal.*, **5** (1992), No 3, pp. 193 - 204. Co-authored with R.N. Izmailov.

1993

65. Chaotic phenomena in desynchronized systems and stability analysis. *Computer Mathematics and Applications*, **25** (1993), No 1, pp. 81 - 87. Co-authored with E.A. Asarin, P. Diamond, I.V. Fomenko.

66. On the existence of a fixed points of the operator acting in the space of continuous functions, *Nonlin. Anal. TMA*, **20** (1993), pp. 1257 - 1259. Co-authored with D. Rachinskii.

67. Methods of desynchronized systems theory in investigation of dynamics of information networks, In: *"Perspective Means of Telecommunication and Integrate Informatics Systems. Project: Methods of Synthesis of Integrate Information Transmission Systems"*, **1**, Russian Academy of Sciences, Institute for Information Transmission Problems, Moscow, 1993, pp. 1 - 30, in Russian. Co-authored with E. Asarin, A. Vladimirov, V. Kozyakin, A. Mandelbaum.

68. Shuttle algorithm in analysis of systems with hysteresis nonlinearities, In: *"Models of Hysteresis", Pitman Research Notes in Mathematics Series*, **286** , Longman Sci. Tech., Harlow, 1993, pp. 124 - 142.

69. Variable states of continual relay systems, *Russian Acad. Sci. Dokl. Math*,**47** (1993), No 3, pp. 513 - 517. Co-authored with M.A. Krasnosel'skij, I. Meyergoyz and D.I. Rachinskii.

70. Stable fixed points of monotone operators, *Russian Acad. Sci. Dokl. Math*,**47** (1993), No 3, pp. 427 - 430. Co-authored with M.A. Krasnosel'skij and A.V. Lusnikov.

71. Differential equations with hysteresis nonlinearities of vector relay systems type, *Russian Acad. Sci. Dokl. Math*, **48** (1993), No 1, pp. 105 - 109. Co-authored with M.A. Krasnosel'skij, I.D. Meyergoyz and D.I. Rachinskii.

72. On a class of games of finite automata, *Automation and Remote Control*, **54** (1993), No 8, Part 2, pp. 1304 - 1309. Co-authored with R.N. Izmajlov and V.V. Chernorutsckii.

73. Numerical modelling of toroidal dynamical systems with invariant Lebesgue measure, *Journal of Complexity*, **7** (1993), pp. 415 - 422. Co-authored with P. Diamond and P.E. Kloeden.

1994

74. Operators of hysteresis nonlinearities generated by continual relay systems, *Automation and Remote Control*, **55** (1994), No 7, Part 1, pp. 964 - 973. Co-authored with M.A. Krasnosel'skij, I.D. Meyergoyz and D.I. Rachinskii.

75. Chaos, entropy and a generalized extension principle, *Fuzzy Sets and Systems*, **61** (1994), No 3, pp. 277 - 283. Co-authored with P. Diamond

76. Weakly chain recurrent points and spatial discretizations of dynamical systems, *Random & Computational Dynamics*, **2** (1994), No 1, pp. 97 - 110. Co-authored with P. Diamond and P.E. Kloeden.

77. An invariant measure arising in computer simulation of a chaotic dynamical system, *Journal of Nonlinear Science*, **4** (1994), pp. 59 - 68. Co-authored with P. Diamond and P.E. Kloeden.

78. Generalized guiding functions in a problem on high frequency forced oscillations, *Nonlin. Anal. TMA*, **22** (1994), No 11, pp. 1357 - 1372. Co-authored with A.M. Krasnosel'skii, M.A. Krasnosel'skii and J. Mawhin.

1995

79. Analysis of an algorithm for computing invariant measures. *Nonlin. Anal. TMA*, **24** (1995), No 3, pp. 323 - 336. Co-authored with P. Diamond and P.E. Kloeden.

80. Degree theory on finite lattices, *Numerical Functional Analysis and Optimization*, **1-2** (1995), pp. 43 - 52. Co-authored with P. Diamond and P.E. Kloeden.

81. Elliptic equations with discontinuous nonlinearities, *Doklady Akademii Nauk*, **51** (1995), No 3, pp. 415 - 418. Co-authored with M.A. Krasnosel'skii.

82. Expansivity of semi-hyperbolic Lipschitz mappings, *Bulletin of the Australian Mathematical Society.*, **51** (1995), No 2, pp. 301 - 308. Co-authored with P. Diamond, P.E. Kloeden and V.S. Kozyakin.

83. Semi-hyperbolic mappings, *Journal of Nonlinear Science*, **5** (1995), No 5, pp. 419 - 431. Co-authored with P. Diamond, P.E. Kloeden and V.S. Kozyakin.

84. Robustness of trajectories of dynamical systems with respect to a class of hysteresis perturbations, *Doklady Akademii Nauk*, **343** (1995), No 1, pp. 1 - 3. Co-authored with P. Diamond, P.E. Kloeden, V.S. Kozyakin and M.A. Krasnosel'skii.

85. Cycles of spatial discretizations of shadowing dynamical systems, *Mathematische Nachrichten*, **171** (1995), pp. 415 - 422. Co-authored with P. Diamond and P.E. Kloeden.

86. Collapsing effects in numerical simulation of a class of chaotic dynamical systems and random mappings with a single attracting centre, *Physica D*, **86** (1995), pp. 559-571. Co-authored with P. Diamond, P.E. Kloeden and A.A. Vladimirov.

87. On the fragmentary complexity of symbolic sequences, *Theoretical Computer Science*, **148** (1995), pp. 1-17. Co-authored with P. Diamond, P.E. Kloeden and V.S. Kozyakin.

88. Periodic trajectories of nonsmooth perturbations of systems with chaotic behaviour, *Automation and Remote Control*, **56** (1995), No 5, Part 1, pp. 637 - 643. Co-authored with P. Diamond, P.E. Kloeden, V.S. Kozyakin and M.A. Krasnosel'skii.

89. Robustness of observed behaviour of semi-hyperbolic dynamical systems, *Avtomatica i Telemekhanika* **11** (1995), No 11, pp. 148 - 159. (Special issue dedicated to M. Krasnosel'skii's 75th anniversary), in Russian. English translation in *Automation and Remote Control.* Co-authored with P. Diamond, P.E. Kloeden and V.S. Kozyakin.

90. On elliptic equations with discontinuous nonlinearities, *Doklady RAN*,**342** (1995), No 6, pp. 731 - 734, in Russian. Co-authored with M.A. Krasnosel'skii.

91. Computer robustness of semi-hyperbolic mappings, *Random & Computational Dynamics*, **3**, (1995), pp. 47 - 70. Co-authored with P. Diamond, P.E. Kloeden and V.S. Kozyakin.

92. Interval stochastic matrices, a combinatorial lemma and the computation of invariant measures of dynamical systems, *Journal on Dynamics and Differential Equations*, **7** (1995), No. 2, pp. 341 - 364. Co-authored with P. Diamond and P.E. Kloeden.

93. The minimal center of attraction of measurable systems and their discretizations, *J. of Mathematical Analysis and Applications*, **200** (1995), No. 1, pp. 207 - 223. Co-authored with P. Diamond and P.E. Kloeden.

1996

94. Robustness of dynamical systems to a class of nonsmooth perturbations, *Nonlin. Anal. TMA***26** (1996), pp. 351 - 361. Co-authored with P. Diamond, P.E. Kloeden, V.S. Kozyakin and M.A. Krasnosel'skii.
95. Visualization of Polynomials, *Computers & Graphics,* **20** (1996), No 1, pp. 95 - 105. Co-authored with R. Izmailov and A. Vladimirov.
96. Superior information is insufficient to win in games with finite automata, *SIAM Journal on Control and Optimization,* **34** (1996), No 2, pp. 542 - 553. Co-authored with R.N. Izmajlov and V.V. Chernorutsckii.
97. Bi-shadowing and delay equations, *Dynamics and Stability of Systems,* **11** (1996), No 2, pp. 121 - 129. Co-authored with A. Al-Nayef, P. Diamond, P.E. Kloeden and V.S. Kozyakin.
98. Basin of attraction of cycles of space discretizations of dynamical systems with SRB invariant measure, *Journal of Statistical Physic,* **84** (1996), No 3/4, pp. 713 - 733. Co-authored with P. Diamond, P.E. Kloeden and A.D. Klemm.
99. Statistical properties of discretizations of a class of chaotic dynamical systems, *Computers and Mathematics with Applications,* **31** (1996), No 11, pp. 83 - 95. Co-authored with P. Diamond, P.E. Kloeden and M. Suzuki.
100. Statistical laws for computational collapse of discretized chaotic mappings, *International Journal of Bifurcation and Chaos in applied Sciences and Engineering,* **6** (1996) No 12A, pp. 2389 - 2399. Co-authored with P. Diamond.

1997

101. Expansivity of nonsmooth functional differential equations, *J. Math. Anal. Applns,* **208**, pp. 453 - 461. Co-authored with A. Al-Nayef and P.E. Kloeden.
102. Absolute retracts and a general fixed point theorem for fuzzy sets, *Fuzzy Sets and Systems,* **86** (1997), pp. 377 - 380. Co-authored with P. Diamond and P.E. Kloeden.
103. The structure and Lyapunov stability of invariant sets of a fuzzy relation, *"Fuzzy Logic and Soft Computing",* B. Bouchon-Meunier, R.R. Yager and L.A. Zadeh (eds.), World Scientific, Singapore 1995. Co-authored with P. Diamond.
104. Semi-hyperbolic mappings, condensing operators and neutral type delay equations, *Journal of Differential Equations,* **137** (1997), pp. 320 - 329. Co-authored with A. Al-Nayef and P.E. Kloeden.
105. A model for roundoff and collapse of chaotic dynamical systems. *Mathematics & Computers in Simulation,* **44** (1997), pp. 163 - 185. Co-authored with P. Diamond, P.E. Kloeden and V.S. Kozyakin.
106. Some problems in analysis of discretizations of continuous dynamical systems, *Nonlinear Analysis, Theory, Methods and Applications* (Proc. 2nd World Congress of Nonlinear Analysts), **30** (1997), No 2, pp. 767 - 778. Co-authored with I.G. Vladimirov, N.A. Kuznetsov and V.S. Kozyakin.
107. Bifurcation at infinity for equations in spaces of vector-valued functions, *Journal of the Australian Mathematical Society, Series A,* **63** (1997), pp. 263 - 280.
108. Quasi-controllability and estimation of the amplitudes of transient regimes in discrete systems. *Transactions of RANS, series MMMIC,* **1** (1997), No 3, pp. 128 - 150. Co-authored with V.S. Kozyakin.
109. Topological shadowing and split-hyperbolicity, *Functional Differential Equations* (special issue dedicated to M.A. Krasnosel'skii), **4** (1997), No 3-4, pp. 335 - 360.

110. Semi-hyperbolic mappings, condensing operators and neutral type delay equations, *Journal of Differential Equations,* **137** (1997), pp. 320 - 329. Co-authored with A. Al-Nayef and P. Kloeden.

1998

111. Monotonic dynamical systems under spatial discretization, *Proceedings of AMS,* **126** (1998), No 7, pp. 2169 - 2174. Co-authored with P. Diamond, P.E. Kloeden and V.S. Kozyakin.
112. Asymptotically stable oscillations in systems with hysteresis nonlinearities, *Journal of Differential Equations,* **150** (1998), pp. 98 - 123. Co-authored with M. Brokate.
113. Boundedness of truncated rotations on uniform planar lattices, *Mathematische Nachrichten,* **91** (1998), pp. 59 - 81. Co-authored with P. Diamond, P.E. Kloeden and V.S. Kozyakin.

1999

114. A spectral property for the symbolic dynamical system, *Irish Math. Soc. Bull.*, **42** (1999), pp. 11 - 14. Co-authored with K. Abodayeh.
115. Stability and shadowing in convex discrete time systems, *Proc. of the Centre for Mathematics and its Applications, Australian National University,* **37** (1999), pp. 63 - 75. Co-authored with P. Diamond.
116. Continuous and inverse shadowing, *Funct. Differ. Equ.*, **6** (1999), No 1-2, pp. 137 - 153. Co-authored with P. Kloeden and J. Ombach.
117. Robustness of an analog dynamic memory system to a class of information transmission channels perturbations, *Functional Differential Equations,* **6** (1999), No 3-4, pp. 411 - 438. Co-authored with F. Holland, J. McInerney, M. M. Suzuki and T. Suzuki.
118. Random mappings with a single absorbing centre and combinatorics of discretizations of the logistic mapping, *Journal of Applied Mathematics and Stochastic Analysis,* **12** (1999), No 3, pp. 205 - 221. Co-authored with T. Klemm.
119. Statistical properties of some spatially discretized dynamical systems, *Zeitschrift fur angewandte Mathematik und Physik ZAMP,* **50** (1999), pp. 638 - 660. Co-authored with P. Kloeden and J. Mustard.

2000

120. A mathematical analysis of the effect of input variability on the output mean for a class of thechnological processes, *Proc. Russian Acad of Natural Sci.*, **4** (2000), No 1-2, pp. 214 - 235. Co-authored with K. Abodayeh and K.A. Cronin.
121. Topological chaos, *Nonlinear Analysis. Theory, Methods & Applications,* **42** (2000), No 6, pp. 1011 -1016. Co-authored with K. Abodayeh.
122. Asymptotically stable almost periodic oscillations in systems with hysteresis nonlinearities, *Zeitschrift fuer Analysis und ihre Anwendungen (ZAA),* **19** (2000), No 2, pp. 469 - 487. Co-authored with M. Brokate, J. Collins and F. Stagnitti.
123. Chaotic dynamics in nonsmooth perturbations of bi-shadowing system, *Arab. J. Math. Sci.*, **6** (2000), No 1, pp. 41 - 74. Co-authored with P. Diamond, P.E. Kloeden and M.A. Krasnosel'skii.
124. Weak resonances for Hopf bifurcations in control systems with nonpolynomial nonlinearities, *Doklady Mathematics,* **62** (2000), No 3, pp. 432-436. Co-authored with A.M. Krasnosel'skii and J. McInerney.
125. Mixed moments of random mappings and chaotic dynamical systems, *Proc. of Royal Soc.*, **456** (2000), pp. 2465-2487. Co-authored with A. Kent and J. McInerney.

126. Properties of nonequiprobable random transformations, *Ann. Appl. Prob.*, **10** (2000), No 8, pp. 1793 - 6578. Co-authored with C. O'Cinneide.

2001

127. Time-dependend Preisach nonlinearities and their properties, *Doklady Mathematics*, **64** (2001), No 3, pp. 435 - 439. Co-authored with A.M. Krasnosel'skii and R. Cross.
128. Rigorous analysis of complicated behaviour in a trancated Lang-Kobayashi model, *Nonlinear Dynamics of Laser and Reacting Systems*, **5** (2001), No 1-2, pp. 205 - 258. Co-authored with O. Rasskazov, G. Huyet and J. McInerney.
129. Topological degree in locating homoclinic structures for discrete dynamical systems, *Nonlinear Dynamics of Laser and Reacting Systems*, **5** (2001), No 1-2, pp. 152-185. Co-authored with S.J. Szybka and J. McInerney.
130. On subharmonics bifurcation in equations with homogeneous nonlinearities, *Discrete and Continuous Dynamical Systems - Series A*, **7** (2001), No 7, pp. 100 - 114. Co-authored with A.M. Krasnosel'skii.
131. The asymptotic behavior of elementary symmetric functions on a probability distribution, *J. Appl. Math. Stochastic Anal.*, **14** (2001), pp. 237 - 248. Co-authored with V.S. Kozyakin.
132. Synchronized double frequency oscillations in a class of weakly resonant systems. *Discrete and Continuous Dynamical Systems - Series A*, **7** (2001), No 4, pp. 747 - 762. Co-authored with A.M. Krasnosel'skii.
133. Recurrent oscillations in systems with hysteresis nonlinearities. *Phys. Rev. B*, **306** (2001), No 1-4, pp. 191 - 194. Co-authored with K. Abodayeh and J. McInerney.

2002

134. Combinatorial problem arising in chaotic control, *Math. Probl. Eng.*, **4** (2002), No 5, pp. 295 - 309. Co-authored with V.A. Bondarenko and J. Mustard.

2003

135. On large-amplitude sub-harmonics in semilinear Duffing oscillator, *Doklady Mathematics*, **391** (2003), No 4, pp. 448 - 452. Co-authored with A.M. Krasnosel'skii.

2004

136. On the use of the topological degree theory in broken orbits analysis, *Proc. Amer. Math. Soc.*, **132** (2004), pp. 567 - 577. Co-authored with O. A. Rasskazov.
137. Synchronized double-frequency oscillations in a class of weakly resonant systems, *Nonlinear Anal., Theory Methods & Appl.*, **57** (2004), No 7-8, pp. 929 - 949. Co-authored with A.M. Krasnosel'skii and J. McInerney.
138. Guaranteed convergence rate estimates for a class of iterative procedures: their comparison, *Automat. Remote Control*, **65** (2004), No 10, pp. 1635 - 1640. Co-authored with A.M. Krasnosel'skii and D. Rachinskii.

2005

139. On chaotic wave patterns in periodically forced steady-state KdVB and extended KdVB equation, *Proc. R. Soc. A*, **461** (2005), pp. 2857 - 2885. Co-authored with E.A. Cox, M.P. Mortell and O. Rasskazov.

2006

140. Asymptotic stability of continuum sets of periodic solutions to systems with hysteresis, *J. Math. Anal. Appl.*, **319** (2006), No 1, pp. 94 - 109. Co-authored with M. Brokate and D. Rachinskii.
141. Dissipativity of nonresonant systems with Preisach friction via a canonical example, *Physica B: Condensed Matter*, **372** (2006), pp. 13 - 16. Co-authored with A.M. Krasnosel'skii.
142. A symmetric solution to semi-linear Duffing equation with Preisach nonlinearity. *Physica B: Condensed Matter*, **372** (2006), pp. 30 - 32. Co-authored with O. Rasskazov.
143. Dissipativity of a nonresonant pendulum with ferromagnetic friction, *Automation and Remote Control*, **67** (2006), No 2, pp. 221 - 232. Co-authored with A.M. Krasnosel'skii.
144. Hysteresis in the trade cycle, *Physica B: Condensed Matter*, **372** (2006), pp. 202 - 206. Co-authored with H. McNamara.

2007

145. Topologically stable subharmonics of large periods and amplitudes, *Differential equations*, **43** (2007), No 2, pp. 218 - 245. Co-authored with A.M. Krasnosel'skii.
146. Structure of index sequences for mappings with an asymptotic derivative, *Discrete and Continuous Dynamical Systems - Series A*, **17** (2007), No 3, pp. 653 - 670. Co-authored with O. Rasskazov.
147. Homoclinic trajectories and chaotic behaviour in a piecewise linear oscillator, *Discrete and Continuous Dynamical Systems - Series B*, **8** (2007), No 4, pp. 943 - 970. Co-authored with O. Rasskazov and D. Visetti.

2008

148. Modelling macroeconomic flows related to large ensembles of elementary exchange operations, *Physica B: Condensed Matter*, **403** (2008), pp. 451 - 455. Co-authored with R. Cross and H. McNamara.
149. Modeling discontinuous flow through porous media using ODEs with Preisach operator, *Physica B: Condensed Matter*, **403** (2008), pp. 440 - 442. Co-authored with D. Flynn, A. Zhezherun and J.P. O'Kane.
150. ODEs involving the derivative of a Preisach operator with a degenerated density, *Physica B: Condensed Matter*, **403** (2008), pp. 443 - 447. Co-authored with B. Appelbe.
151. Non-local branches of cycles, bi-stability, and mixed mode oscillations, *Chaos*, **18** (2008), No 1, 015109. Co-authored with E. Bouse, A.M. Krasnosel'skii and D. Rachinskii.
152. Topological degree in analysis of chaotic behavior in singularly perturbed systems, *Chaos*, **18** (2008), No 2, pp. 23 - 130. Co-authored with A. Zhezherun.
153. A new paradigm for modelling hysteresis in macroeconomic flows, *Physica B: Condensed Matter*, **403** (2008), pp. 231 - 236. Co-authored with R. Cross, H. McNamara and D. Rachinskii.

2009

154. Rate-independent hysteresis in terrestrial hydrology, *IEEE Control Systems Magazine*, **29** (2009), No 1, pp. 44 - 69. Co-authored with B. Appelbe, D. Flynn, H. McNamara, J.P. O'Kane, A. Pimenov, D. Rachinskii and A. Zhezherun.

2010

155. Lyapunov functions for SIR and SIRS epidemic models, *Applied Mathematics Letters*, **23** (2010), pp. 446 - 448. Co-authored with S. O'Reagan, T.C. Kelly, A. Korobeinikov and M.J.A. O'Callaghan.

To appear

156. Topological degree in analysis of canard-type trajectories in 3-D systems, *Applicable Analysis: Int. J.*, in press. Co-authored with D. Rachinskii, V. Sobolev and A. Zhezherun.
157. Energy dissipation in hydrological systems due to hysteresis, *Environmental Modeling & Assessment*, in press. Co-authored with M. Brokate, S. McCarthy, A. Pimenov and D. Rachinskii.
158. Mathematical models of hydrological systems with Preisach hysteresis, *Physica D*, submitted. Co-authored with P. Krejci, J.P. O'Kane and D. Rachinskii.
159. A hybrid model of the population dynamics of woodpigeon (Columba palumbus), *Ecological Modeling,* submitted. Co-authored with S.M. O'Regan, D. Flynn, T.C. Kelly, M.J.A. O'Callaghan and D. Rachinskii.
160. Effect of positive feedback on Devil's staircase input-output relationship, *Journal of Nonlinear Science,* submitted. Co-authored with D. Rachinskii.
161. Hysteresis in the fundamentals of macroeconomics, Working Papers 2010-08, University of Strathclyde Business School, Department of Economics. Co-authored with R. Cross, H. McNamara and L. Kalachev.
162. Periodic sequences of arbitrage: a tale of four currencies, Working Papers 2010-19, University of Strathclyde Business School, Department of Economics. Co-authored with R. Cross, V.S. Kozyakin, B. O'Callaghan and A.A. Pokrovskiy.
163. Memory of recessions, Working Papers 2010-09, University of Strathclyde Business School, Department of Economics. Co-authored with R. Cross and H. McNamara.

Selected Conference Proceedings

1. Regular fixed points of monotonic operators, In: *Proc. of 7-th ICNO*, DDR Acad. Sci., Berlin.
2. Operators of hysteresis nonlinearities, In: "*Operators Theory in Function Spaces*", *Proc. of Novosibirsk School 1975*, Nauka, Moscow, 1977, pp. 167 - 187, in Russian. Co-authored with M.A. Krasnosel'skii.
3. A new method for absolute stability investigation, *Proc. of Chetaev Conference on Stability of Motion, Analytical Mechanic, Control of Motion*, Irkutsk 1977, p. 57, in Russian. Co-authored with M.A. Krasnosel'skii.
4. The principle of absence of bounded modes in the problem of absolute stability of control systems, *Proc. of 7-th USSR Conference on Control Problems*, **1**, Moscow - Minsk, 1977, pp. 6 - 7, in Russian. Co-authored with M.A. Krasnosel'skii.
5. Identification of some nonlinear transformers, In: "*Identification and System Parameter Estimation*", *Proc. of 4-th IFAC Symposium*, Tbilisi 1976, part 3, Nord-Holland, Amsterdam, 1978, pp. 1797 - 1814. Co-authored with M.A. Krasnosel'skii.
6. Regular solutions of elliptic equations with discontinuous nonlinearities, *Proc. of the conference dedicated to the 75-th birthday of I.G. Petrovskii*, Moscow, MGU, 1978, pp. 346 - 347, in Russian. Co-authored with M.A. Krasnosel'skii.
7. Periodic oscillations in systems with complex nonlinearities, *Proc. of 9-th International Conf. on Nonlinear Oscillations*, **1**, Kiev 1981, Naukova Dumka, Kiev, 1984, pp. 189 - 193. Co-authored with M.A. Krasnosel'skii.

8. Method of block diagrams in mathematical modeling of systems incorporating complex nonlinearities, *Proc. of IFAC Congress*, Japan 1981, VI-137 - VI-142. Co-authored with M.A. Krasnosel'skii.
9. Vibrational robustness of some nonlinearities of hysteresis type, In: "*Dynamics of Nonhomogeneous Systems*", VNIISI, Moscow, 1982, 11 p., in Russian.
10. Absolute stability of multi-loop control systems, *Proc. 5-th Conf. on Multi-Loop Systems*, Acad. Sci. Gruz. SSR, Tbilisi 1983, in Russian. Co-authored with M.A. Krasnosel'skii.
11. Frequency estimations of outputs of control systems, *Proc. of 10-th National Control Conf.*, Acad. Sci. Kaz. SSR, Alma-Ata 1986, in Russian.
12. On new types of mathematical models of complex systems, In: "*Modeling and Adaptive Control*", *Proc. of IIASA Conf.*, Sopron, Hungary 1986, Springer, Berlin, 1986. Co-authored with E.A. Asarin, R.N. Izmailov and M.A. Krasnosel'skii.
13. Averaging principle for equations with hysteresis nonlinearities, *Proc. of Conference on Nonlinear Problems of Differential Equations and System Physics*, Ternopol 1981, **1**, 1989, pp. 221 - 223. Co-authored with M.A. Krasnosel'skii.
14. On modeling of controllable systems by integral equations with discontinuous nonlinearities, *Proc. of 3-rd Ukrainian Conference on Integral Equations and Applied Modeling*, Kiev 1989, pp. 79 - 80. Co-authored with R. Izmailov and M.A. Krasnosel'skii.
15. Stable periodic oscillations in systems with monotonous hysteresis nonlinearities, *Proc. of 7-th Czechoslovak Conference on Differential Equations and their Applications (Equadiff 7)*, Prague 1989, Teubner-text zur Mathematik, Leipzig, **118**, 1990, pp. 92 - 95. Co-authored with M.A. Krasnoselsky and M.E. Semenov.
16. New results in the theory of periodic oscillations in nonlinear control systems, *Proc. of 10-th Triennial World Congress of the International Federation of Automatic Control*, Tallin 1990, Oxford, Pergamon, 1991, **3**, pp. 311 - 316. Co-authored with N.A. Bobylev, V.V. Chernorutskii, I.V. Fomenko, A.M. Krasnosel'skii, M.A. Krasnosel'skii, M.E. Semenov and E.G. Vainikko.
17. Self-induced oscillations in systems with hysteresis, *Proc. of International Conference on Differential Equations and Applications*, Russe 1989, 1991, pp. 2 - 11. Co-authored with I.V. Fomenko and M.A. Krasnosel'skii.
18. Effect of controllability-type properties on reliability and stability of desynchronized systems, *Preprints of IFAC Workshop Automatic Control for Quality and Productivity*, Istanbul, Turkey, 1992, 8 p. Co-authored with V.S. Kozyakin and N.A. Kuznetsov.
19. Transients in quasi-controllable systems. Overshooting, stability and instability, *Preprints 12th World IFAC Congress*, Sydney, Australia, 1993, **8**, pp. 465 - 468. Co-authored with V.S. Kozyakin and N.A. Kuznetsov.
20. Lyapunov stability of the invariant hull of fuzzy relation. *Actes/Proc. of 5-th International Conference on Information Processing and Management of Uncertainty in Knowledge-Based Systems.* Paris, 1994, **1**, pp. 193 - 198. Co-authored with P. Diamond.
21. Collapsing effects in numerical simulation of chaotic dynamical systems, *Proc. of 94-th Korea Automatic Control Conference. International Session*, Taejon, Korea, 1994, pp. 753 - 757. Co-authored with P. Diamond, P.E. Kloeden and M. Suzuki.
22. Shadowing and approximation in dynamical systems, *Proc. of the Centre for Mathematics and its Applications the Australian National University*, **33**, 1994. G. Martin and H.B. Thompson (eds), pp. 47 - 60. Co-authored with P. Diamond and P.E. Kloeden.
23. A phenomenological model for discretizations of chaotic dynamical systems, *Proc. of conference dedicated to 75-th anniversary of Mark Krasnosel'skii*, Voronezh, 1995, pp. 31

- 32, Voronezh State University Press, Voronezh, 1995, in Russian. Co-authored with P. Diamond, P.E. Kloeden and V.S. Kozyakin.

24. Superior information is insufficient to win in games with finite automata, *Proc. of 34-th IEEE Conf on Decision and Control*, New Orleans, 1995, pp. 3044 - 3048. Co-authored with R.N. Izmajlov and V.V. Chernorutsckii.
25. Combinatorial problems in spatial discretization of chaotic dynamical systems, *Proc. of 21-st Australian Conference on Combinatorial Mathematics and Combinatorial Computing (21ACCMCC)*. Co-authored with P. Diamond and P. Kloeden
26. Multivalued spatial discretization of dynamical systems, *Proc. of the Centre for Mathematics and its Applications the Australian National University*, **33**, 1994, G. Martin and H.B. Thompson (eds), pp. 61 - 70. Co-authored with P. Diamond and P.E. Kloeden.
27. Computational roundoff and collapse in chaotic systems, *Proc. of IPMU'96*, Granada, Spain, 1996. Co-authored with P. Diamond.
28. Some problems in analysis of discretizations of continuous dynamical systems, *Proc. of 2-nd World Congress in Nonlinear Analysis*, Athens, 1996, 10 p. Co-authored with V.S. Kozyakin, N.A. Kuznetsov and I.G. Vladimirov.
29. Quasi-controllability and estimates of amplitudes of transient regimes in discrete systems, *Proc. of CESA'98 IMACS-IEEE Multiconference*, Hammamet, Tunisia, 1998, P. Borne, M. Ksouri and A. El Kamel (eds), CD-ROM ISBN 2-9512309-0-7, 1, pp. 266 - 271. Co-authored with V.S. Kozyakin and N.A. Kuznetsov.
30. Dynamics of a Lotka-Volterra type model with applications to marine phage population dynamics, Proc. of MURPHYS 2006, *Journal of Physics: Conference Series*, **55** (2006), pp. 80 - 93. Co-authored with C. Gavin, M. Prentice and V. Sobolev.
31. Differentiability of evolution operators for dynamical systems with hysteresis, Proc. of MURPHYS 2006, *Journal of Physics: Conference Series*, **55** (2006), pp. 171 - 190. Co-authored with T. Power, D. Rachinskii and A. Zhezherun.
32. Canard doublet in a Lotka-Volterra type model, Proc. of MURPHYS 2008, *Journal of Physics: Conference Series*, **138** (2008), 012019. Co-authored with E. Shchepakina and V. Sobolev.
33. Qualitative and numerical investigations of the impact of a novel pathogen on a seabird colony, Proc. of MURPHYS 2008, *Journal of Physics: Conference Series*, **138** (2008), 012018. Co-authored with S. O'Reagan, T.C. Kelly, A. Korobeinikova and M.J.A. O'Callaghan.
34. Stability results for a soil model with singular hysteretic hydrology, Proc. of MURPHYS 2010, *Journal of Physics: Conference Series* (2010). Co-authored with: P. Krejci, J.P. O'Kane and D. Rachinskii.

Educational Papers

1. Functional analysis pioneer turns 75, *SIAM News*, **28** (1995), No 6, p. 3. Co-authored with P.E. Kloeden and V.S. Kozyakin.
2. Hysteresis, *Encyclopedia of Mathematics*, Supplement Volume I, M. Hazewinkel (ed), Kluwer Acad. Publ., 1997, p. 310. Co-authored with A.M. Krasnosel'skii and M.A. Krasnosel'skii.
3. Hysteron, *Encyclopedia of Mathematics*, Supplement Volume I, M. Hazewinkel (ed), Kluwer Acad. Publ., 1997, pp. 310 - 311. Co-authored with A.M. Krasnosel'skii and M.A. Krasnosel'skii.

4. Nonideal Relay, *Encyclopedia of Mathematics*, Supplement Volume I, M. Hazewinkel (ed), Kluwer Acad. Publ., 1997, p. 384. Co-authored with A.M. Krasnosel'skii and M.A. Krasnosel'skii.
5. Obituary: Mark Alexandrovich Krasnosel'skii, *J. Appl. Math. Stochastic Anal.*, **10** (1997), No 2, pp. 119 - 126, Co-authored with N.A. Bobylev, V.S. Kozyakin, B.N. Sadovskii *et al.*
6. In Memoriam Mark A.Krasnosel'skii, *Mitt. Ges. Angew. Math. Mech.*, **20** (1997), No 1, pp. 13 - 14. Co-authored with N.A. Bobylev, P.E. Kloeden, V.S. Kozyakin, A.M. Krasnosel'skii, B.N. Sadovskii.

Editing

1. *Singular Perturbations and Hysteresis,* Proceedings of International Workshop on Relaxation Oscillations and Hysteresis 2002, M.P. Mortell, R.E. O'Malley, A. Pokrovskii, V. Sobolev (eds), SIAM, 2005.
2. *Proceedings of International Workshop on Hysteresis & Multi-scale Asymptotics 2004,* M.P. Mortell, R.E. O'Malley, A.V. Pokrovskii, V.A. Sobolev (eds), *Journal of Physics: Conference Series* vol. 22, 2005.
3. *Proceedings of International Workshop on Multi-Rate Processes and Hysteresis 2006,* M.P. Mortell, R.E. O'Malley, A.V. Pokrovskii, V.A. Sobolev (eds), *Journal of Physics: Conference Series* vol. 55, 2006.
4. *Proceedings of International Workshop on Multi-Rate Processes and Hysteresis 2008,* M.P. Mortell, R.E. O'Malley, A. Pokrovskii, D. Rachinskii, V.A. Sobolev (eds), *Journal of Physics: Conference Series* vol. 138, 2008.

5th International Workshop on Multi-Rate Processes and Hysteresis (MURPHYS 2010) IOP Publishing
Journal of Physics: Conference Series **268** (2011) 012001 doi:10.1088/1742-6596/268/1/012001

Distribution of return point memory states for systems with stochastic inputs

A Amann[1], M Brokate[2], D Rachinskii[3] and G Temnov[4]
[1]Tyndall National Institute, University College Cork, Ireland
[2]Zentrum Mathematik - M6, Technische Universität München, Germany
[3]Department of Applied Mathematics, University College Cork, Ireland
[4]Department of Mathematics, University College Cork, Ireland

E-mail: d.rachinskii@ucc.ie

Abstract. We consider the long term effect of stochastic inputs on the state of an open loop system which exhibits the so-called return point memory. An example of such a system is the Preisach model; more generally, systems with the Preisach type input-state relationship, such as in spin-interaction models, are considered. We focus on the characterisation of the expected memory configuration after the system has been effected by the input for sufficiently long period of time. In the case where the input is given by a discrete time random walk process, or the Wiener process, simple closed form expressions for the probability density of the vector of the main input extrema recorded by the memory state, and scaling laws for the dimension of this vector, are derived. If the input is given by a general continuous Markov process, we show that the distribution of previous memory elements can be obtained from a Markov chain scheme which is derived from the solution of an associated one-dimensional escape type problem. Formulas for transition probabilities defining this Markov chain scheme are presented. Moreover, explicit formulas for the conditional probability densities of previous main extrema are obtained for the Ornstein-Uhlenbeck input process. The analytical results are confirmed by numerical experiments.

1. Introduction

1.1. Preisach memory models driven by random input

The response of systems characterised by internal state variables or memory to stochastic inputs, noise and random fluctuations is an area that has received much recent attention in both theoretical and applied studies. Nontrivial effects such as stochastic resonance [1–3] and coherence resonance [4] can be observed in the simplest noise driven bistable or oscillatory systems. In particular, two-state systems such as a non-ideal relay (also known as a non-ideal switch, Schmitt trigger, rectangular loop operator and binary memory element in different applications) or a particle in a double well potential exhibit stochastic resonance in response to stochastic perturbations. This effect underpins the methods of dithering widely used in engineering design, which exploits the constructive role of noise for improving the quality of the output in image processing and signal processing [5].

Various models of systems with memory can be constructed using the formalism of the phenomenological *Preisach model* as a weighted superposition of many two-state non-ideal relays that are individually and independently driven by the same input [6]. For example, the response of the Preisach model to stochastic inputs has been studied in [7, 8] in order to model the *viscosity* or *after-effect* in ferromagnetic materials, which is detrimental for magnetic recording technologies. A similar approach to modelling *creep* in semiconductor materials by means of stochastically driven Preisach memory systems has been adopted in [9]. While in these examples the source of the stochastic component of the input is the internal noise caused by thermal fluctuations, in other areas of engineering and physics the input is stochastic due to its random nature or because it includes an external noise component superimposed on a deterministic input signal. Ubiquitous applications of the Preisach model in magnetizm, superconductivity, piezoelectricity, elastoplasticity, shape memory alloys, economics and finance, modelling capillary hysteresis in flows through porous media such as soils etc. (see, for example, the surveys in [10, 11]) motivated further analysis of the Preisach operator input-output relationship for stochastic inputs of various types including uncorrelated inputs and general diffusion processes [12–17]. This research was focused mainly on finding characteristics of the output time series such as the spectral density and the autocorrelation function. The method was based on the decomposition of the Preisach model into the sum (integral) of two-state non-ideal relays, *i.e.*, independent elementary hysteretic transducers, each producing a binary stochastic output in response to the common input. The decomposition reduced the problem to analysis of the cross-correlation function of the output of a pair of relays, which was then averaged over the set of relays composing the Preisach model. Key computations in this analysis were based on the relation between randomly induced switching of relays and the well studied escape (exit) problems for stochastic processes and, furthermore, partly involved some ingenious techniques such as, for example, the theory of diffusion processes on graphs [18, 19]. This approach resulted, in particular, in analytic and numerical estimates for the spectral density of the output time series, asymptotic formulas for the long-time tales in its autocorrelation, and analysis of $1/f$ noise in the frequency domain.

1.2. Problem statement

Modelling of real world systems is often complicated by the existence of internal degrees of freedom which depend on the history of the system and give rise to hysteretic behaviour. In addition it is often the case that the system of interest has been subject to a random input in the past, but neither this input nor the resulting system evolution are known at a given point in time. Examples of systems with internal memory and random input include magnetization dynamics of magnetic rock in geology, folding dynamics of proteins in biology and modelling of trends, fads and opinions in sociology. Under these circumstances the question arises, *whether it is possible to characterise the internal state or memory structure of the system at a given instant T assuming that this state results from the effect of a random input on the system between some moment in the past and the moment T.* It is the purpose of this paper to investigate this question.

In the following, we will focus on systems with *rate-independent* input-output relationship and *return point memory*, which are characteristic for many real-world systems. These two properties are important idealisations which effectively reduce the relevant memory structure of the system to a specific sequence of shock values (known as *main extrema*) of the input experienced in the past and can generically be modelled by the Preisach operators and their generalisations. Rate-independence of an input-output relationship means that this relationship is invariant with respect to the action of the group of affine transformations of the time scale, implying that shock values of the input have permanent effect on the output in the future and thus are memorised by the system. The magnetization of ferromagnetic material by a fluctuation of external magnetic field which creates a permanent magnet is one example of such memory. In the mathematical theory of hysteretic systems, hysteresis nonlinearities are defined (omitting a few technicalities) as deterministic rate-independent operators with non-local memory [20, 21]. This general definition entails a set of non-trivial properties of hysteresis operators which are sufficient for developing formal concepts with various applications [22–26]. The rate-independence property distinguishes hysteretic memory from other memory types such as in delay differential equations or convolution operators. It is this property that allows one to describe constitutive hysteretic relationships in terms of 2-D input-output diagrams such as the familiar picture showing the nested structure of magnetization curves and hysteresis loops, which present the relationship between the magnetization and the applied magnetic field in ferromagnetic materials.

The *return point memory* (also known as the wiping out memory) property of a rate-independent system ensures that hysteresis loops corresponding to periodic inputs are closed on the input-output diagram (hence there is no effects like ratcheting), which is a reasonable approximation for many real world hysteretic systems. We will use an alternative definition due to [25] whereby the memory state $\omega(T)$ at a given instant T is a vector (with changing dimension), or an infinite sequence, of certain extremum values of the input achieved prior to this instant; the evolution of the state in response to the evolution of the input is defined by a set of simple updating rules dictating which extremum values are included in the vector $\omega(t)$. In particular, for piecewise monotone inputs (or discrete time inputs), each new extremum is included in $\omega(t)$ as the last component at the instant it is achieved; while, whenever the input reaches the value equal to the last component of the vector $\omega(t)$, the last two components are removed from $\omega(t)$. The memory state $\omega(T)$ is a concise record of all those events which happened prior to the moment T and can influence dynamics of the system after this moment in the sense that the state $\omega(T)$ and the input after the moment T completely define the state $\omega(t)$ and the output of the system for $t > T$.

For stochastic inputs which drive the system between moments t_0 and T the memory state $\omega(T)$ is a multidimensional (or infinite-dimensional) random variable. *The goal of this paper is to find, or to characterise, the distribution of the memory state $\omega(T)$ and its parameters at a given moment T for several classes of inputs such as a random walk, Wiener process, Ornstein-Uhlenbeck process and a general stationary diffusion process. An important assumption we make is that the time interval between the moments t_0 and T is large meaning that the initial state $\omega(t_0)$ has negligible effect on the distribution of $\omega(T)$.* More technically, we consider the probability that an input trajectory is confined to a prescribed interval for all times between t_0 and T and assume that the time interval is long enough to make this probability negligibly small.

Thus, our focus is on a *multidimensional memory configuration*, *i.e.*, a detailed description of the system's state at a *single* time moment. This compares to the problem described in the previous subsection focusing on the description of the output of the Preisach model, *i.e.*, a *scalar-valued output time series*. The output value at a given instant can be viewed as an average value of the components of the vector state at this instant. Hence, the two problems are complementary. The most detailed description of system's dynamics would provide information about the multi-dimensional stochastic process with the values in the space of memory states, *i.e.*, description of the stochastic evolution of the state in response to the evolution of a stochastic input. However, the problem in this generality appears to be difficult to approach.

In applications, the state of the Preisach model described in terms of the binary states of all its relay components often models the physical state of the system. The memory vector ω, at a given moment, contains a compact description of the binary function representing the physical state. More precisely, ω can be mapped straightforwardly to this binary function by a procedure proposed in [22] whereby the components of ω encode the staircase boundary separating the domain which represents the relays that are 'on' and the domain which represents the relays that are 'off' on the Preisach plane.

Summarising and extending the discussion of this subsection, let us assume that we control or observe the input of the Preisach model only after a certain instant T. As the output and the state at any instant $t > T$ are defined both by the dynamics of the input between the instants T and t and by the memory state $\omega(T)$, which has recorded the shock values (the main extrema) of the input obtained at times prior to T, information about the state $\omega(T)$ is important for prediction and control of the dynamics of the system after the moment T. In particular, the return point memory property of the Preisach model ensures that the shock values stored as components of the vector $\omega(T)$ will be felt by the systems at the moments $t > T$ when the input reaches again those recorded values. At these moments, the composition of the memory vector $\omega(t)$ changes and, as a result, those moments are characterised by a switch to another input-output branch on the input-output diagram of the model, such as the switch to another magnetization curve on the diagrams depicting the dependence of the magnetization of ferromagnetic materials on the external magnetic field. The jump to another input-output branch is accompanied by a jump of the time-derivative of the output, which is smooth at other times for smooth input time series. Therefore information about the state $\omega(T)$ translates directly to the list of input values for which these jumps (possibly, representing a strong impact on the system) will happen after the moment T, allowing one to predict them.

If the input was not controlled or measured prior to the instant T, then the state $\omega(T)$ is not known or controllable, neither it is typically measurable. However, we make the assumption that the memory state $\omega(T)$ is the result of a random input with known parameters, which has effected the system prior to the instant T, and characterise the distribution and typical parameters of the memory state $\omega(T)$ under this assumption. Because the Preisach nonlinearity has the return point memory property, an equivalent question to ask is what is the distribution of the main extrema of the random process which creates the memory state $\omega(T)$. We approach this question assuming that the random input was effecting the system for a long period of time prior to the moment T.

1.3. Method and results

The two main ingredients of analysis developed in the work, which we refer to in Subsection 1.1, are (a) some form of the time reversion applied to the input; and, (b) reduction of the problem to a one-dimensional escape (exit) problem, or a series of such problems, for a diffusion process. The same two ingredients underpin the approach we adopt here. However, the escape problem component is adapted appropriately as we work with the memory states in the compact form ω of a vector of the past shock values of the input rather than the binary functions representing the 'physical state' of the Preisach model. In particular, we do not resort to the decomposition of the model into the superposition of binary relays.

In order to describe the distribution of the random vector (sequence) $\omega = \omega(T)$, we introduce a Markov chain scheme where the running index n of the component of ω plays the role of the fictitious time. Using the above two ingredients of the analysis, we derive explicit expressions for the transition probabilities of this Markov chain in terms of the SDE defining the input. For the random walk and continuous time Wiener process inputs, along this line of argument, we obtain explicitly the distribution function for the state $\omega(T)$. We then show how to use this result to derive distributions of several random variables naturally associated with $\omega(T)$. In particular, we show that the number q_d of the components of $\omega(T)$ which have values in a given interval I is an exponentially (Poisson) distributed random quantity, and discuss scaling of $q_d = q_d(I)$ with the ends of the interval I, thus obtaining a chrcterisation of the dimension of a typical memory state $\omega(T)$. The random quantity q_d can be viewed as a measure of the amount of information stored in $\omega(T)$ (complexity of $\omega(T)$), or a measure of the effect of the past history of the input on dynamics after the instant T. It is also directly related to the description of the system's state in terms of the input-output diagram. In this description, one distinguishes the input-output curves of different order, such as the primary magnetization curve, secondary magnetization curves and k-th order magnetization curves of any higher order k in the magnetic hysteresis applications. A particular value of q_d places the system to the input-output curve of order $k = q_d$ (exactly, or with controllable accuracy) at the moment T.

Considering general diffusion input processes and using the expressions for the transition probability densities of the subsidiary Markov chain, we derive formulas for the conditional probability density of the n-th component ω^n of the memory vector $\omega = \omega(T)$ conditioned on the values of the two previous components ω^{n-1} and ω^{n-2}. These formulas can be used to obtain the joint probability distribution of those components ω^n whose values are confined to a given interval I. The formulas are specified for the Ornstein - Uhlenbeck processes and compared to their counterparts for the Wiener process with and without drift.

We confirm analytical results by numerical experiments.

Our results are equally applicable to all rate-independent systems with return-point memory as such systems are characterised by the same input-state relationship mapping the input to the memory state $\omega(t)$ (see, [25]). The state-output relationship is model specific. For example, the state-output relationship of the Preisach model ensures that for any periodic input all the hysteresis loops on the input-output diagram are congruent; according to Mayergoyz's theorem this property identifies the Preisach model among all the return-point memory systems, ensuring the decomposition into the sum of relay operators [23]. Another important class of systems with return point memory are spin interaction models stemming from the classical Ising model of the array of coupled spins. The Ising model based approach is quite general and is widely used in many disciplines to study such effects as avalanches, Barkhausen noise, clustering, emergent hysteresis behaviour, and order-disorder phase transitions which are universal to many systems; examples include ferromagnetic materials and earthquake fault systems [27], sand piles [28], capillary effects and percolation in partially saturated porous media [29], phase transitions in solids [30], random networks [31,32], cellular automata [33], and multi-agent models in economics and finance [34]. The roots of this universality, and the universal scaling laws characterising such systems, have been recently revealed in [35,36] (see also the review [37] and references therein). Sethna et al. showed that any system of spins with positive interactions (such as, for example, those in ferromagnetic materials) has the return point memory property [38]. The return point memory state ω recording the shock values of the input can be mapped to the 'physical state' of the model describing the binary state of each spin.

It is worthwhile to note that the so-called *rainflow counting method* which is widely used in damage and fatigue estimates for engineering applications is based on tracking the evolution of a return point memory state $\omega(t)$ driven by a stochastic input [39]. However, it applies to input processes with smooth trajectories, while the focus of this paper is on nonsmooth processes.

This paper, as well as the literature cited in Subsection 1.1, refers to open loop systems. Closed loop stochastic models of simple hydrological systems involving the Preisach operator have been studied in [40] by a combination of analytic and numerical methods. The input of the Preisach operator in these models is yet smooth. Closed loop models involving systems of many switches with jumping thresholds driven by nonsmooth stochastic inputs have been recently proposed and investigated numerically in [41–45] in the context of modelling economics and finance.

The rest of the paper is organised as follows. In the next section, we present the definition of the return point memory states ω and define their evolution in response to the evolution of the input. In Section 3, we discuss systems driven by the random walk input. The purpose of this section is to illustrate the main steps of our approach on the simplest example. Section 4 presents the results for systems driven by continuous time diffusion processes.

2. Memory structure

Consider a continuous scalar-valued input $x(t)$ of the Preisach nonlinearity on the interval $[t_0, T]$. At every time moment t, the memory of the system consists of two arrays of numbers m_k^t and M_k^t $(k = 1, 2, \ldots)$. Here M_1^t, m_1^t are the running global maximum value and global minimum value of x over the interval $[t_0, t]$ defined via

$$M_1^t = \max_{t_0 \le \tau \le t} x(\tau), \quad m_1^t = \min_{t_0 \le \tau \le t} x(\tau). \tag{1}$$

The other entries of the memory at the moment t are defined recursively. In the case where the global maximum is obtained before the global minimum, *i.e.*

$$\tau^t(M_1^t) \le \tau^t(m_1^t), \tag{2}$$

where

$$\tau^t(c) := \max\{\tau \in [t_0, t] : x(\tau) = c\}, \tag{3}$$

we define

$$m_k^t = \min_{\tau^t(M_k^t) \le \tau \le t} x(\tau), \quad M_{k+1}^t = \max_{\tau^t(m_k^t) \le \tau \le t} x(\tau), \quad k = 1, 2, \ldots \tag{4}$$

In other words, M_2^t is the maximum value of x on the time interval $[\tau^t(m_1^t), t]$ from the moment $\tau^t(m_1^t)$ when the input achieves the value m_1^t for the last time to the moment t; m_2^t is the minimum value of x on the time interval $[\tau^t(M_2^t), t]$ from the moment $\tau^t(M_1^t)$ when the value M_2^t is achieved for the last time to the moment t, and so on. We will refer to M_k^t and m_k^t as the *main extrema* of x on the interval $[t_0, t]$ or as the *memory elements* of the system, and will call the set of these elements the *memory array* $\omega = \omega(t)$.

Similarly, if the global minimum is obtained before the global maximum, *i.e.*

$$\tau^t(m_1^t) < \tau^t(M_1^t) \tag{5}$$

we define

$$M_k^t = \max_{\tau^t(m_k^t) \le \tau \le t} x(\tau), \quad m_{k+1}^t = \min_{\tau^t(M_k^t) \le \tau \le t} x(\tau), \quad k = 1, 2, \ldots \tag{6}$$

Directly from the definition (and under assumption (2)) we observe the following inequality chains

$$\tau^t(M_1^t) \le \tau^t(m_1^t) \le \tau^t(M_2^t) \le \tau^t(m_2^t) \le \cdots \le t \tag{7}$$
$$M_1^t \ge M_2^t \ge \cdots \ge x(t) \tag{8}$$
$$m_1^t \le m_2^t \le \cdots \le x(t) \tag{9}$$
$$M_1^t - m_1^t \ge M_2^t - m_2^t \ge \cdots \ge 0. \tag{10}$$

If equality between any two elements in equations (7)–(9) holds, then these two elements as well as all the elements to the right of them are equal to the final value of the process (or the final element of the chain).

We will also consider discrete time input sequences $x = x(t)$ with integer $t = 0, 1, 2, \ldots$ To define the main extrema for them, one considers the piecewise linear extension of $x(t)$ (which is linear between any neighbouring integer times) to a segment of the real line, and applies formulas (2)–(6) to this extension.

For any given moment t, the number of different memory elements can either be finite, for example if x is a discrete time input, or infinite, which is typical for realisations x of a continuous Markov process. The number of different memory elements in the memory array, if finite, is generally variable in time. With increasing time the memory array is updated according to definition (2)–(6) by adding and deleting the appropriate memory elements. Assume for example that we want to obtain the memory array at time $t+\Delta$ starting from a known set of main extrema

M_k^t and m_k^t at time t. Let k_0 be the largest index such that $x(t')$ with $t' \in [t, t+\Delta]$ does not leave the interval of the associated main extrema, *i.e.*

$$k_0 = \max_k \left\{ k : x([t, t+\Delta]) \subset \left[m_k^t, M_k^t \right] \right\}. \tag{11}$$

Then the memory elements for indices less or equal k_0 are unchanged

$$M_k^{t+\Delta} = M_k^t, \quad m_k^{t+\Delta} = m_k^t \quad \text{for } k = 1, \ldots, k_0 \tag{12}$$

and only the memory elements with indices larger than k_0 need to be updated according to (4), (6). These definitions naturally extend the rules for deleting and adding memory elements in response to piecewise monotone inputs to any continuous inputs, see, for example, a detailed discussion in [22]. In particular these definitions also apply if the input is not piecewise monotone, which is typically the case for stochastic inputs.

In this paper we are concerned with stochastic inputs and our aim is to characterise the distribution of the memory array, which becomes a multi-dimensional random quantity (sequence) in this case. In the following section we will first consider a simple classical random walk input process before we turn to more general Markov processes in Section 4.

3. Memory distribution for random walk

3.1. Time reversion and reduction to escape problem

Consider a classical random walk W_t on the integer grid as an input $x(t)$ of the Preisach memory operator, hence the time t is discrete. The point W_t is assumed to move one step left or right with equal probability 1/2, *i.e.*,

$$P[W_{t+1} = W_t + 1] = P[W_{t+1} = W_t - 1] = 1/2.$$

For the purposes of this section, we use the value $x(t)$ as a reference point and characterise the memory at the moment t by the array of values $m_k^t - x(t), M_k^t - x(t)$. For a finite time interval there exist only a finite number of different memory elements

$$m_1^t < m_2^t < \cdots < m_{L_t}^t < x(t) < M_{N_t}^t < \cdots < M_2^t < M_1^t \tag{13}$$

with $|L_t - N_t| \le 1$, while $m_k^t = M_n^t = x(t)$ for all $k > L_t$, $n > N_t$. Moreover, we reorder the entries of the finite memory array (13) to obtain the *memory string*

$$\mathfrak{M}_t = \{\mathfrak{m}_{-L_t}^t, \ldots, \mathfrak{m}_{-1}^t, \mathfrak{m}_1^t, \ldots, \mathfrak{m}_{N_t}^t\} \tag{14}$$

with

$$\mathfrak{m}_{-L_t}^t < \cdots < \mathfrak{m}_{-2}^t < \mathfrak{m}_{-1}^t < 0 < \mathfrak{m}_1^t < \mathfrak{m}_2^t < \cdots < \mathfrak{m}_{N_t}^t$$

where

$$\mathfrak{m}_{-k}^t = m_{L_t-k+1}^t - x(t), \qquad \mathfrak{m}_n^t = M_{N_t-n+1}^t - x(t)$$

for $k = 1, \ldots, L_t$, $n = 1, \ldots, N_t$, $|L_t - N_t| \le 1$. Thus, $\mathfrak{M}_t$ is a reordered memory array of the input $y(\tau) = x(\tau) - x(t)$ on the interval $t_0 \le \tau \le t$; $\mathfrak{m}_k^t$ are local maximum values of y for $k > 0$ and local minimum values for $k < 0$. If τ_k^t is the last moment when y reaches the value $\mathfrak{m}_k^t$ before the moment t, *i.e.*, $\tau_k^t = \max\{\tau \in [t_0, t] : y(\tau) = \mathfrak{m}_k^t\}$, then either $t > \tau_1^t > \tau_{-1}^t > \tau_2^t > \tau_{-2}^t > \cdots$ or $t > \tau_{-1}^t > \tau_1^t > \tau_{-2}^t > \tau_2^t > \cdots$, hence the ordering of the moments τ_k^t is defined by the sign of the difference $\tau_1^t - \tau_{-1}^t$. As the ordering is a part of the memory, we will distinguish between the memory strings with the same set of values (14) for $\sigma^t = 1$ and $\sigma^t = -1$ where

$$\sigma^t = \text{sign}(\tau_1^t - \tau_{-1}^t).$$

Note that with this notation the number N_t of positive entries of the memory string and the number L_t of negative entries are related by either $N_t = L_t$ or $N_t = L_t + \sigma^t$.

We will be interested in the distribution of the memory string entries $\mathfrak{m}_k^T$ which lie in a certain interval $-K \leq \mathfrak{m}_k^T \leq K$. If we continue the input after the moment T, then the other memory string entries do not change until $x(t) - x(T)$ reaches one of the values $\pm K$ at some moment[1] $\tau > T$. Thus, for a vector of integers j_k, σ satisfying

$$-K < j_{-L} < \cdots < j_{-2} < j_{-1} < 0 < j_1 < j_2 < \cdots < j_N < K; \quad |\sigma| = 1$$

with either $L = N$ or $L = N + \sigma$, we are interested in the probability $P_{j_{-L},\ldots,j_N,\sigma}$ of the event

$$\mathfrak{m}_{-L}^T = j_{-L}, \ldots, \mathfrak{m}_{-1}^T = j_{-1}, \mathfrak{m}_1^T = j_1, \ldots, \mathfrak{m}_N^T = j_N; \quad \text{sign}(\tau_1^T - \tau_{-1}^T) = \sigma \tag{15}$$

which implies $L \leq L_T, N \leq N_T$.

To obtain the desired distribution, we reverse the time and consider a sequence of escape problems for the reversed process.

Consider the reversed process $\hat{W}_t = W_{T-t} - W_T$ for $0 \leq t \leq T - t_0$. This is also a random walk with the probability 1/2 to go one step left or right starting from zero. Consider the hitting times

$$\hat{\tau}_k = \min\{t \geq 0 : \hat{W}_t = k\} \tag{16}$$

of $\hat{W}_t$ for integer $k \neq 0$ with $|k| \leq K$. We assume that $T \gg t_0$, and hence identify $\hat{W}_t$ with a random walk on the infinite interval $t \geq 0$. More precisely, we neglect the probability that $|W_t - W_T|$ never reaches the value K over the time interval $t_0 \leq t \leq T$. In this approximation, all the hitting times (16) are well-defined and finite, as they are almost surely for the random walk on the infinite time interval $t \geq 0$.

We now observe that under this assumption the event (15) is equivalent to the event

$$\hat{\tau}_{j_1} < \hat{\tau}_{-1} \leq \hat{\tau}_{j_{-1}} < \hat{\tau}_{j_1+1} \leq \hat{\tau}_{j_2} < \hat{\tau}_{j_{-1}-1} \leq \hat{\tau}_{j_{-2}} < \hat{\tau}_{j_2+1} \leq \cdots \leq \hat{\tau}_{j_N} < \hat{\tau}_{j_{-L}-1} < \hat{\tau}_{j_N+1} \tag{17}$$

for $N = L + 1$, and to the event

$$\hat{\tau}_{j_1} < \hat{\tau}_{-1} \leq \hat{\tau}_{j_{-1}} < \hat{\tau}_{j_1+1} \leq \hat{\tau}_{j_2} < \hat{\tau}_{j_{-1}-1} \leq \hat{\tau}_{j_{-2}} < \hat{\tau}_{j_2+1} \leq \cdots \leq \hat{\tau}_{j_{-L}} < \hat{\tau}_{j_N+1} < \hat{\tau}_{j_{-L}-1} \tag{18}$$

for $N = L$ if $\sigma = 1$. Similarly, if $\sigma = -1$, then (15) is equivalent to

$$\hat{\tau}_{j_{-1}} < \hat{\tau}_1 \leq \hat{\tau}_{j_1} < \hat{\tau}_{j_{-1}-1} \leq \hat{\tau}_{j_{-2}} < \hat{\tau}_{j_1+1} \leq \hat{\tau}_{j_2} < \hat{\tau}_{j_{-2}-1} \leq \cdots \leq \hat{\tau}_{j_N} < \hat{\tau}_{j_{-L}-1} < \hat{\tau}_{j_N+1} \tag{19}$$

for $N = L$, and to

$$\hat{\tau}_{j_{-1}} < \hat{\tau}_1 \leq \hat{\tau}_{j_1} < \hat{\tau}_{j_{-1}-1} \leq \hat{\tau}_{j_{-2}} < \hat{\tau}_{j_1+1} \leq \hat{\tau}_{j_2} < \hat{\tau}_{j_{-2}-1} \leq \cdots \leq \hat{\tau}_{j_{-L}} < \hat{\tau}_{j_N+1} < \hat{\tau}_{j_{-L}-1} \tag{20}$$

for $N = L - 1$. For example, (17) ensures that a realisation of the process $\hat{W}_t$ hits the positive level j_1 before it hits the level -1; then it hits the level $j_{-1} \leq -1$ before it hits the level $j_1 + 1$, and so on. Hence the values of such a realisation belong to the interval $[0, j_1] = [\hat{W}_0, \hat{W}_{\hat{\tau}_{j_1}}]$ for $0 \leq t \leq \hat{\tau}_{j_1}$; then to the interval $[j_1, j_{-1}] = [\hat{W}_{\hat{\tau}_{j_1}}, \hat{W}_{\hat{\tau}_{j_{-1}}}]$ for $\hat{\tau}_{j_1} \leq t \leq \hat{\tau}_{j_{-1}}$, etc.

As the random walk $\hat{W}_t$ is a time-homogeneous Markov chain, the probability $P_{j_{-L},\ldots,j_N,1}$ of the event (17) equals the following product of the probabilities

$$P_{j_{-L},\ldots,j_N,1} = P[\hat{\tau}_{j_1}^0 < \hat{\tau}_{-1}^0] P[\hat{\tau}_{j_{-1}}^{j_1} < \hat{\tau}_{j_1+1}^{j_1}] P[\hat{\tau}_{j_2}^{j_{-1}} < \hat{\tau}_{j_{-1}-1}^{j_{-1}}] \cdots P[\hat{\tau}_{j_{-L}-1}^{j_N} < \hat{\tau}_{j_N+1}^{j_N}],$$

[1] In many applications of the Preisach model, the entries $\mathfrak{m}_k^T$ with $|\mathfrak{m}_k^T| > K$ do not affect the dynamics as long as they are constant, *i.e.*, until the moment τ in this case. Moreover, large in absolute value entries $\mathfrak{m}_k^t$ typically have either no or little effect on the future.

where $\hat{\tau}^i_j = \min\{t \geq 0 : \hat{W}^i_t = j\}$ is the hitting time at the level j of a random walk $\hat{W}^i_t$ starting at the level $\hat{W}^i_0 = i$. Moreover, for the random walk,

$$P[\hat{\tau}^i_j < \hat{\tau}^i_k] = \frac{k-i}{k-j}, \quad P[\hat{\tau}^i_j > \hat{\tau}^i_k] = \frac{i-j}{k-j} \quad \text{for} \quad j < i < k$$

(see, for example, [46]), hence the probability of the event (15) for $\sigma = 1$ is defined by

$$\frac{1}{P_{j_{-L},\dots,j_N,1}} = (1+j_1)(1+j_1-j_{-1})(1+j_2-j_{-1})(1+j_2-j_{-2})\cdots(1+j_N-j_{-L})(2+j_N-j_{-L}). \quad (21)$$

Similarly, for $\sigma = -1$,

$$\frac{1}{P_{j_{-L},\dots,j_N,-1}} = (1-j_{-1})(1+j_1-j_{-1})(1+j_1-j_{-2})\cdots(1+j_N-j_{-L})(2+j_N-j_{-L}). \quad (22)$$

For example, the probability that $\mathfrak{m}^T_1 = j, \sigma = 1$ equals $P_{j,1} = (j+1)^{-1}(j+2)^{-1}$ for $j = 1, \dots, K-1$, and the event $\mathfrak{m}^T_1 = -j, \sigma = -1$ has the same probability $P_{-j,-1} = (j+1)^{-1}(j+2)^{-1}$. These probabilities sum up to

$$2\sum_{j=1}^{K-1} \frac{1}{(j+1)(j+2)} = 1 - \frac{2}{K+1}.$$

The complimentary event is that either $\hat{\tau}_K < \hat{\tau}_{-1}$ or $\hat{\tau}_{-K} < \hat{\tau}_1$ with $P[\hat{\tau}_K < \hat{\tau}_{-1}] = P[\hat{\tau}_{-K} < \hat{\tau}_1] = 1/(K+1)$.

A simple numerical check of equation (21) is shown in figure 1, where for $\sigma = 1$ the probabilities $P_{j,k,1}$ as a function of j are plotted for a range of values of k. We observe the expected excellent agreement between the analytical formula (21) and the numerical results.

3.2. Examples

Relations (21), (22) can be used to derive distributions and mean values of parameters of the memory string $\mathfrak{M}_T$. As an illustration, we consider a few examples.

Example 1. Suppose an initial memory state $\mathfrak{M}_T = \mathfrak{M}$ has been created by the random walk input $x(t) = W_t$, $t_0 < t < T$ with $t_0 \ll T$, and we control the input after the moment T. Suppose we increase the input. We ask at which input value the system switches to a new memory branch, that is when the memory string element $\mathfrak{m}_1 = \mathfrak{m}^T_1$ will be deleted. This happens when the increment $\Delta = x(t) - x(T) > 0$ of the input reaches the value $\mathfrak{m}_1$. Thus, we are interested in the probabilities

$$P[\hat{\tau}_1 < \hat{\tau}_{-1}, \mathfrak{m}_1 = m], \quad P[\hat{\tau}_1 > \hat{\tau}_{-1}, \mathfrak{m}_1 = m]$$

for positive integers m. These probabilities are different as the distribution of $\mathfrak{m}_1$ depends on whether $\hat{\tau}_1$ is less or greater than $\hat{\tau}_{-1}$.

According to the previous subsection, the first of these probabilities is $P[\hat{\tau}_1 < \hat{\tau}_{-1}, \mathfrak{m}_1 = m] = P_{m,1} = (m+1)^{-1}(m+2)^{-1}$. Therefore, the corresponding joint probability that $\hat{\tau}_1 < \hat{\tau}_{-1}$ and that the switch happens for $\Delta \geq m$ is

$$P[\hat{\tau}_1 < \hat{\tau}_{-1}, \mathfrak{m}_1 \geq m] = \sum_{k=m}^{\infty} \frac{1}{(k+1)(k+2)} = \frac{1}{m+1}.$$

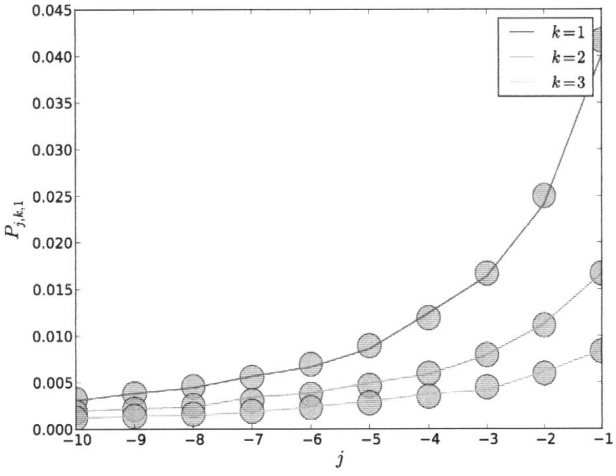

Figure 1. Comparison of analytical results from equation (21) (orange circles) with the averaged outcomes of 10^5 independent simulations of the random walk (solid lines). We plot the joint probability $P_{j,k,1}$ versus the last main minimum j for various values of the last main maximum k.

For $m = 1$, this probability equals $P[\hat{\tau}_1 < \hat{\tau}_{-1}] = 1/2$.

In the case $\hat{\tau}_1 > \hat{\tau}_{-1}$, $P[\hat{\tau}_1 > \hat{\tau}_{-1}, \mathfrak{m}_{-1} = -k, \mathfrak{m}_1 = m] = P_{-k,m,-1} = (k+1)^{-1}(k+m+1)^{-1}(k+m+2)^{-1}$. Summing over k, we obtain

$$P[\hat{\tau}_1 > \hat{\tau}_{-1}, \mathfrak{m}_1 = m] = \sum_{k=1}^{\infty} \frac{1}{(k+1)(k+m+1)(k+m+2)},$$

and for the corresponding cumulative distribution

$$P[\hat{\tau}_1 > \hat{\tau}_{-1}, \mathfrak{m}_1 \geq m] = \sum_{j=m}^{\infty}\sum_{k=1}^{\infty} \frac{1}{(k+1)(k+j+1)(k+j+2)} = \frac{1}{m}\sum_{i=2}^{m+1}\frac{1}{i}.$$

For $m = 1$, this cumulative probability equals $P[\hat{\tau}_1 > \hat{\tau}_{-1}] = 1/2$.

Example 2. The memory string $\mathfrak{M}_t$ changes when the input increment $\Delta = x(t) - x(T)$ reaches the value $\mathfrak{m}_1 = \mathfrak{m}_1^T$ in Example 1 and the system switches to a new memory branch. For instance, if $\hat{\tau}_1 < \hat{\tau}_{-1}$, then the entry $\mathfrak{m}_{-1}^t$ of $\mathfrak{M}_t$ changes from the value $x(T) - x(t)$ to the value $\mathfrak{m}_{-1}^T + x(T) - x(t)$ when Δ reaches the value $\mathfrak{m}_1$. Suppose the jump amplitude $\mathfrak{m}_{-1} = \mathfrak{m}_{-1}^T$ of this entry is used as a measure of the effect of the jump to a new memory branch on the system. Hence, the probability

$$P[\hat{\tau}_1 < \hat{\tau}_{-1}, \mathfrak{m}_1 = m, \mathfrak{m}_{-1} = -k]$$

is of interest, as in our interpretation this is the probability that a jump of strength k happens when Δ reaches the value m. According to the previous subsection, this probability equals $P_{-k,m,1} = (m+1)^{-1}(m+k+1)^{-1}(m+k+2)^{-1}$. Therefore, the probability that $\hat{\tau}_1 < \hat{\tau}_{-1}$ and the strength of the jump is greater or equal than k when Δ reaches the value m is

$$P[\hat{\tau}_1 < \hat{\tau}_{-1}, \mathfrak{m}_1 = m, \mathfrak{m}_{-1} \leq -k] = \sum_{j=k}^{\infty} \frac{1}{(m+1)(m+j+1)(m+j+2)} = \frac{1}{(m+1)(m+k+1)}.$$

For $k = 1$, this is the probability $P_{m,1}$ considered in Example 1.

Further random quantities associated with the random memory string $\mathfrak{M}_T$ such as the number of entries $\mathfrak{m}_k$ in a given interval are considered in the next section.

4. Continuous time stochastic inputs

In this section we consider the situation where the input of the Preisach operator is given by a one-dimensional continuous Markov process $x(t)$, $t_0 \leq t \leq T$, defined by the Ito stochastic differential equation (SDE)

$$dx = A(x)\, dt + \sqrt{B(x)}\,(x)\, dW(t). \tag{23}$$

Here $A(x)$ and $B(x) \geq 0$ are continuous (not explicitly time dependent) bounded functions of x and $W(t)$ is the standard Wiener process [47]. In addition we assume that there exists a stationary probability distribution whose density $p_s(x)$ fulfills the stationary Fokker-Planck equation,

$$A(x)\, p_s(x) - \frac{1}{2}\partial_x \Big(B(x)\, p_s(x) \Big) = 0; \tag{24}$$

here and henceforth the notation ∂_x is used for the ordinary and partial derivatives. We stipulate that the distribution of the initial point $x(t_0)$ follows $p_s(x)$. As a direct consequence, $x(t)$ follows the stationary distribution at each given instant, *i.e.*,

$$p(x,t) = p_s(x) \quad \text{for } t \in [t_0, T] \tag{25}$$

Following Section 2 we assume that the stochastic process $x(t)$ is used as input of the Preisach nonlinearity. To avoid notational clumsiness, let us from now on assume (unless otherwise stated) that condition (2) holds, *i.e.*, the global maximum is assumed to have occurred before the global minimum. Often we will be interested in the memory array at the final time T only, and we introduce the quantities $M_k, m_k, \tau_k^M, \tau_k^m$ to simplify the notation using

$$\mathbf{M}_k = M_k^T, \quad \tau_k^M = \tau^T(M_k^T), \quad \mathbf{m}_k = m_k^T, \quad \tau_k^m = \tau^T(m_k^T), \quad k = 1, 2, \ldots \tag{26}$$

In figure 2, a realisation of the Ornstein-Uhlenbeck process is plotted, allowing to illustrate the definitions given above. Since the Markov process defined by (23) is everywhere continuous, but in general not differentiable at points with $B(x) > 0$, it follows that infinitely many different main extrema are possible (generally, the memory array is almost surely infinite). It is our aim to characterise this memory structure.

Here and in the following we will use bold letters for random variables, when the potential for confusion with their corresponding values exists.

In particular we are interested in quantifying the probability distribution of a main extremum (memory element) which was attained *before* a pair of main extrema with known values. For example, we might ask about the probability distribution of a main minimum $\mathbf{m}_{k_0-1}$ under the condition that the immediately following main maximum and minimum have the values $\mathbf{M}_{k_0}$ and $\mathbf{m}_{k_0}$, respectively. This type of question is motivated by the situation, often encountered in practice, that a process is observed for a limited period of time and thus extrema are known up to a certain instant in the past. The problem is then to deduce from those known extrema the probability distribution of previous extreme events for which no recorded data exists. Another related question is about the expected sequence of main extrema in the past. For example if the value of the stochastic process $x(T)$ at time T is known, but not its history, one might be interested to know whether the last main maximum bigger than a certain threshold value $M^0 > x(T)$ has occurred before or after the last main minimum which was below a different threshold value $m^0 < x(T)$.

In order to derive these probability distributions we will first introduce a Fokker-Planck like equation for transition probabilities in reverse time, and then reduce the problem of finding the distribution of previous extrema to a well defined escape problem for the reversed process.

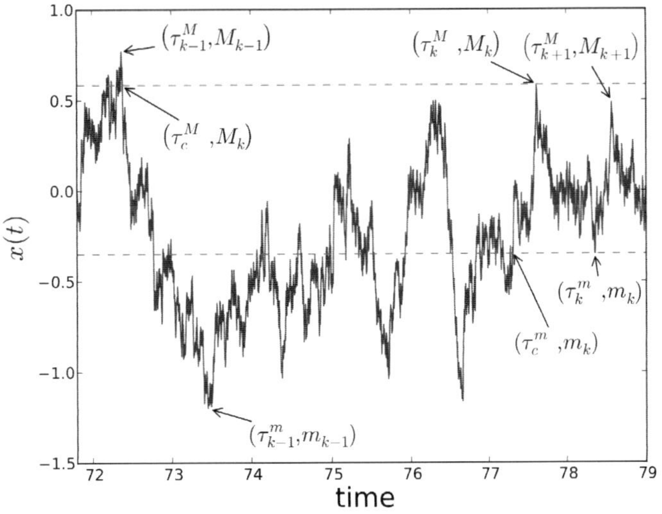

Figure 2. Illustration of the positions of consecutive maxima and minima for a realisation of the Ornstein-Uhlenbeck process with $K = 1$ and $D = 1$. Here M_{k-1} occurring at time τ_{k-1}^M is the global maximum in the shown time interval, and the following main extrema m_{k-1}, M_k, m_k and m_{k+1} are then defined recursively according to (4) and (26). τ_c^M and τ_c^m are defined via $\tau_c^M = \max\{t \in [\tau_{k-1}^M, \tau_{k-1}^m] : x(t) = M_k\}$, $\tau_c^m = \max\{t \in [\tau_{k-1}^m, \tau_k^M] : x(t) = m_k\}$

4.1. Formulation as escape problem

Let us define a reverse transition probability density for $t' < t$ via

$$p_r\left(y, t'|x, t\right) = \frac{p\left(x, t; y, t'\right)}{p_s\left(x\right)} = \frac{p\left(x, t|y, t'\right) p\left(y, t'\right)}{p_s\left(x\right)} = \frac{p\left(x, t|y, t'\right) p_s\left(y\right)}{p_s\left(x\right)}, \tag{27}$$

where equation (25) has been used.

It can be shown that this function fulfills the Fokker-Planck equation in backward time

$$-\partial_y\Big(A(y)p_r(y, t'|x, t)\Big) + \frac{1}{2}\partial_y^2\Big(B\left(y\right) p_r\left(y, t'|x, t\right)\Big) = -\frac{\partial}{\partial t'} p_r\left(y, t'|x, t\right). \tag{28}$$

Note that equation (28) is different from the *backward FPE* for example in the sense of [47]. Instead, it can be regarded as a forward FPE for the reverse transition probability p_r.

Using equation (28) for p_r we can now formulate the problem of finding the distribution of a previous main extremum by using the well developed escape formalism.

Let us consider the case as illustrated in figure 2, and let us assume that some recording of this particular process has started shortly before the instant τ_c^m and thus has established the M_k and m_k as the first known main extrema (realisations of $\mathbf{M}_k$ and $\mathbf{m}_k$). The question is then, what can be assumed about the probability distribution of $\mathbf{m}_{k-1}$. Let us write the probability that $\mathbf{m}_{k-1}$ is below a value $\bar{m}$ under the condition that the next known main extrema are given by M_k and m_k formally as $P(\mathbf{m}_{k-1} < \bar{m}|M_k, m_k)$. This probability can be related to an escape problem in reverse time. Consider the history of the process shown in figure 2 up to τ_c^m. It is clear that we have $\tau_c^M < \tau_{k-1}^m < \tau_c^m$, where $\tau_c^M = \tau^{\tau_c^m}(M_k)$ is the largest time less than τ_c^m at which M_k is obtained. Therefore $\mathbf{m}_{k-1} < \bar{m}$ implies that $\tau^{\tau_c^m}(\bar{m}) > \tau_c^M$. In other words the last time at which the process has entered the interval $[\bar{m}, M_k]$ before reaching m_k at time τ_c^m it has entered this interval from the lower end. On the other hand if $m_{k-1} > \bar{m}$, then the process did last enter the interval $[\bar{m}, M_k]$ at time τ_c^M from the upper end.

In the reverse time formulation as in (28), the problem of last entering an interval is equivalent to the more familiar *first exit* problem from a given interval in forward time [47]. Therefore the probability that m_{k-1} is less than a certain value $\bar{m}$ can be expressed as the probability

$$P(\mathbf{m}_{k-1} < \bar{m} | \mathbf{M}_k = M_k, \mathbf{m}_k = m_k) = \pi(m_k; \bar{m}, M_k) \tag{29}$$

where $\pi(m_k; \bar{m}, M_k)$ is the probability that the reverse time process given by (28) exits the interval $[\bar{m}, M_k]$ through the lower end at $\bar{m}$ when starting at position m_k.

This type of escape problem is well studied in the literature and we here follow the notation of [47] where it is shown that $\pi(m_k; \bar{m}, M_k)$ fulfills the differential equation

$$A(x)\,\partial_x \pi_{\bar{m}}(x) + \frac{1}{2} B(x)\,\partial_x^2 \pi_{\bar{m}}(x) = 0 \tag{30}$$

with the boundary conditions $\pi_{\bar{m}}(\bar{m}) = 1$, $\pi_{\bar{m}}(M_k) = 0$. The explicit solution is given by

$$\pi(m_k; \bar{m}, M_k) = P(\mathbf{m}_{k-1} < \bar{m} | \mathbf{M}_k = M_k, \mathbf{m}_k = m_k) = \frac{I(M_k) - I(m_k)}{I(M_k) - I(\bar{m})} \tag{31}$$

with

$$I(y) \;=\; \int_0^y \frac{dx}{B(x) p_s(x)} = \mathcal{N} \int_0^y \exp\left(-\int_0^x \frac{2A(x')}{B(x')} dx'\right) dx. \tag{32}$$

Here the normalization constant $\mathcal{N}$ is determined by the normalization condition $\int p_s(x)dx = 1$. Note that $I(y)$ is continuous and monotonic with $I(0) = 0$, but is in general not defined for all values $y \in \mathbb{R}$, since $I(y)$ may diverge to plus or minus infinity at some finite values $y = y_I^M > 0$ or $y = y_I^m < 0$, respectively. If this happens, the domain $\mathcal{O}$ of $I(y)$ is an open interval limited by y_I^M from above or y_I^m from below. Points outside $\mathcal{O}$ cannot be reached by the Markov process $x(t)$ starting at zero, which means that if, for example, we consider two points $a < 0$ and $b > 0$ such that $a \le y_I^m$ while $I(b)$ is finite, than a process starting at zero will have zero probability of leaving the interval (a, b) at the lower end a. It follows that the function $I(y)$ is invertible and its inverse $I^{-1}(z)$ is well defined for any $z \in \mathbb{R}$.

The probability in equation (31) can also be interpreted as the probability that a (forward time) Wiener process which starts at $I(m_k)$ will first leave the interval $(I(\bar{m}), I(M_k))$ at the lower end boundary at $\bar{m}$, *i.e.*

$$\pi(m_k; \bar{m}, M_k) = \pi^W(I(m_k); I(\bar{m}), I(M_k)) \tag{33}$$

with $\pi^W(x; a, b) = (b - x)/(b - a)$. Therefore the function $I(x)$ relates the characterisation of the main extrema of an arbitrary process with an escape problem of the Wiener process. This connection will be made more explicit in Subsection 4.4 where the memory structure of the Wiener process is studied in greater detail.

To simplify the notation let us also introduce for $x \in \mathcal{O}$

$$\psi(x) = \partial_x I(x) = \frac{1}{B(x) p_s(x)} = \mathcal{N} \exp\left(-\int_0^x \frac{2A(x')}{B(x')} dx'\right). \tag{34}$$

Formula (31) for the cumulative conditional probability $P(\mathbf{m}_{k-1} < \bar{m} | \mathbf{M}_k = M_k, \mathbf{m}_k = m_k)$ for m_{k-1} being less than $\bar{m}$ implies that the conditional probability density $p(m_{k-1}|M_k, m_k)$ for the main minimum $\mathbf{m}_{k-1}$ under the condition that the following main extrema are given by M_k and m_k equals

$$p(m_{k-1}|M_k, m_k) = \left.\frac{\partial}{\partial \bar{m}}\right|_{\bar{m}=m_{k-1}} P(\mathbf{m}_{k-1} < \bar{m} | \mathbf{M}_k = M_k, \mathbf{m}_k = m_k) = \frac{I(M_k) - I(m_k)}{[I(M_k) - I(m_{k-1})]^2} \psi(m_{k-1}). \tag{35}$$

We note that there is no explicit k dependence in equation (35), other than via the realisations of the main extrema M_k, m_k and m_{k-1}, which means that formula (35) can be used to calculate the probability density for a main minimum under the condition that the following two main extrema are known independent of k.

Similarly, the conditional density distributions of the previous maximum $\mathbf{M}_{k-1}$ is given (assuming condition (2)) explicitly by

$$p(M_{k-1}|m_{k-1}, M_k) = \frac{I(M_k) - I(m_{k-1})}{\left[I(M_{k-1}) - I(m_{k-1})\right]^2}\psi(M_{k-1}). \tag{36}$$

The generalisation to the case $\tau_k^M > \tau_k^m$ is straightforward.

A combination of equations (35) and (36) can be given in the form of

$$\begin{aligned} p(M_{k-1}, m_{k-1}|M_k, m_k) &= p(M_{k-1}|m_{k-1}, M_k)p(m_{k-1}|M_k, m_k) \\ &= \frac{I(M_k) - I(m_k)}{I(M_k) - I(m_{k-1})}\frac{\psi(m_{k-1})\psi(M_{k-1})}{\left[I(M_{k-1}) - I(m_{k-1})\right]^2} \end{aligned} \tag{37}$$

In this form we can regard $p(M_{k-1}, m_{k-1}|M_k, m_k)$ as the transition probability density from one pair of main extrema (m_k, M_k) to a preceding pair (m_{k-1}, M_{k-1}). This transition probability can be thought to originate from a time discrete two-dimensional master equation on the (m, M) space which proceeds in negative k direction, *i.e.*, $-k$ is the fictitious time. Equation (37) therefore represents the Markov transition chain for this process. This idea is illustrated in figure 3.

The generalization of (37) is straightforward and allows us to write closed form expression for the joint probability of all previous main extrema

$$\begin{aligned} p(M_{k-l}, m_{k-l}, \quad \ldots \quad &, M_{k-1}, m_{k-1}|M_k, m_k) \\ &= p(M_{k-l}, m_{k-l}|M_{k-l+1}, m_{k-l+1})\cdots p(M_{k-1}, m_{k-1}|M_k, m_k) \\ &= \frac{I(M_k) - I(m_k)}{I(M_{k-l}) - I(m_{k-l})}\prod_{i=1}^{l}\frac{\psi(m_{k-i})\psi(M_{k-i})}{\left(I(M_{k-i+1}) - I(m_{k-i})\right)\left(I(M_{k-i}) - I(m_{k-i})\right)}. \end{aligned} \tag{38}$$

which gives formally a complete characterisation of the memory structure before a pair of known main extrema M_k and m_k.

Another related characterisation of past main extrema concerns the sequence in which they have historically occurred. Consider the situation where the input $x(T) = x_0$ of the Preisach nonlinearity is only known at the final time T without any information on the past history. We can then ask the question, whether the last main maximum which is bigger than a certain value $M^0 > x(T)$ has occurred before of after the last main minimum with value less than $m^0 < x(T)$. In other words we want to determine the probability $P\left(\tau^T(M^0) < \tau^T(m^0)|x(T) = x_0\right)$. An escape problem argument as before then leads to the answer that this probability equals the probability of leaving (in reverse time formulation) the interval $[m^0, M^0]$ at the lower end when starting at x_0. Explicitly we find (see equation (31))

$$P\left(\tau^T(M^0) < \tau^T(m^0)|x(T) = x_0\right) = \frac{I(M^0) - I(x_0)}{I(M^0) - I(m^0)}. \tag{39}$$

This equation allows us to predict in which sequence the main extrema are likely to have occurred in the past. The similarity of the expression in (39) with equation (31) allows us to formally write

$$P\left(\tau^T(M^0) < \tau^T(m^0)|x(T) = x_0\right) = P(\mathbf{m}_{k-1} < m^0|\mathbf{M}_k = M^0, \mathbf{m}_k = x_0). \tag{40}$$

This concludes our theoretical investigation on the characterisation of the previous main extrema for a continuous Markov process of the form (23). Next, we discuss a few examples.

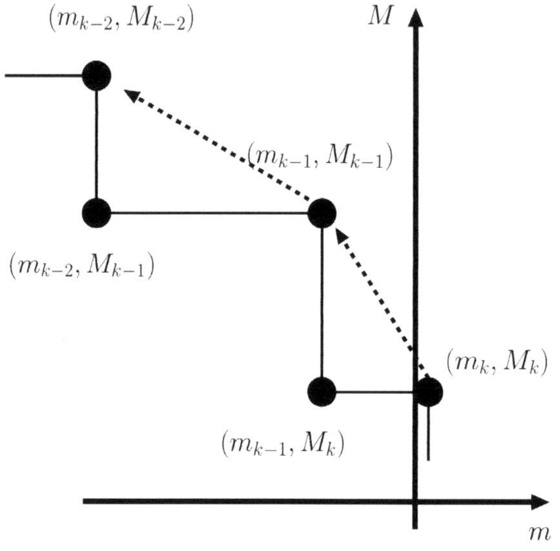

Figure 3. Illustration of the Markov chain between pairs of extrema $(m_k, M_k) \to (m_{k-1}, M_{k-1}) \to (m_{k-2}, M_{k-2}) \to \cdots$ The transition probability density between different pairs of extrema in the (m, M) plane is given by (37). This plane can be interpreted as the Preisach plane, see [22], in which case the staircase line connecting the elements of the chain is the border line between the domain representing the relays which are 'on' and the domain representing the relays which are 'off'.

4.2. Example I: The Ornstein-Uhlenbeck process

Let us consider the case where the input of the Preisach nonlinearity is given by the Ornstein-Uhlenbeck process defined by (23) with $A(x) = -\kappa x$, $B(x) = D$. Then (34) implies

$$\psi(x) = \frac{1}{Dp_s(x)} = \sqrt{\frac{\pi}{\kappa D}} \exp\left(\frac{\kappa x^2}{D}\right) \tag{41}$$

which, according to (31), yields for the probability that m_{k-1} is below $\bar{m}$ for known subsequent main extrema M_k and m_k:

$$P(\mathbf{m}_{k-1} < \bar{m} | \mathbf{M}_k = M_k, \mathbf{m}_k = m_k) = \frac{\int_{m_k}^{M_k} \exp\left(\frac{\kappa}{D} y^2\right) dy}{\int_{\bar{m}}^{M_k} \exp\left(\frac{\kappa}{D} y^2\right) dy}. \tag{42}$$

The probability density $p(m_{k-1}|M_k, m_k)$ is then again obtained as in equation (35). The probability concerning the sequence of main extrema can then be obtained via equation (38).

As an interesting limit case we now expand the integrand in (42) around m_k via

$$\exp\left(\frac{\kappa}{D} y^2\right) \approx \exp\left(\frac{\kappa}{D} m_k^2\right) \left[1 + (y - m_k) \frac{2\kappa}{D} m_k\right] + \cdots \tag{43}$$

which leads to the simple approximation

$$P(\mathbf{m}_{k-1} < \bar{m} | \mathbf{M}_k = M_k, \mathbf{m}_k = m_k) \approx \frac{M_k - m_k}{M_k - \bar{m}} \left[1 + \frac{\kappa m_k}{D}(m_k - \bar{m})\right]. \tag{44}$$

This approximation is valid if $\sqrt{\frac{\kappa}{D}}(M_k - \bar{m}) \ll 1$ and $\sqrt{\frac{\kappa}{D}} m_k$ stays bounded as $\sqrt{\frac{\kappa}{D}}(M_k - \bar{m})$ decreases.

4.3. Example II: Wiener process with constant drift

If the input for the Preisach nonlinearity is given by a Wiener process with drift, *i.e.*, $A(x) = a$, $B(x) = 1$, then the problem arises that the stationary probability distribution $p_s(x)$ does not exist. The existence of $p_s(x)$ was however a perquisite in the derivation of the Fokker-Planck equation in backward time (28). Nevertheless it turns out that the Wiener process with drift represents an important small scale limit for arbitrary continuous Markov processes. The reason is that for x sufficiently close to m_k we can approximate

$$\psi(x) = \psi(m_k) \exp\left(-\int_{m_k}^{x} \frac{2A(x')}{B(x')} dx'\right) \approx \psi(m_k) \exp\left(-\frac{2A(m_k)}{B(m_k)}(x - m_k)\right). \qquad (45)$$

This expression is equivalent to the formula for $\psi(x)$ formally obtained from (34) for the Wiener process with drift $a = A(m_k)/B(m_k)$. Note that the Wiener process itself does not possess a stationary distribution, which was required to derive equation (34). Nevertheless we observe that the Wiener process with drift appears naturally in the small scale limit and is therefore a useful abstraction for the dynamics at small scale.

The quantities of interest are not $\psi(x)$ explicitly, but rather the probabilities constructed via $\psi(x)$. For the Wiener process we formally find from (31)

$$\begin{aligned} P(\mathbf{m}_{k-1} < \bar{m} | \mathbf{M}_k = M_k, \mathbf{m}_k = m_k) &= \frac{\int_{m_k}^{M_k} \exp(-2ax)\, dx}{\int_{\bar{m}}^{M_k} \exp(-2ax)\, dx} = \frac{\exp(-2aM_k) - \exp(-2am_k)}{\exp(-2aM_k) - \exp(-2a\bar{m})} \\ &\approx \frac{M_k - m_k}{M_k - \bar{m}} \left[1 - a(m_k - \bar{m})\right]. \qquad (46) \end{aligned}$$

This last expression is valid up to the first order in $a(M_k - \bar{m})$. We observe that this approximation agrees as expected with the corresponding approximation (44) for the Ornstein-Uhlenbeck process for $a = -\kappa m_k/D$.

In a region around a point x_0 where the ratio $2A(x)/B(x)$ of a given process is about constant, we can characterise the infinitely many main extrema sufficiently close to x_0 by formulas derived from a Wiener process with drift $a = 2A(x_0)/B(x_0)$. At an even smaller scale the drift term can also be neglected and we obtain a characterisation of the main extrema on the basis of the classical Wiener process without drift, *i.e.*, $a = 0$, which we discuss in the next subsection.

4.4. Example III: Wiener Process without drift

In the limit of $a = 0$ the function I defined in (32) becomes formally simply the identity function and we obtain $\psi(x) = 1$. Therefore all previous formulas involving I simplify considerably. For example, formulas (35), (36) for the Wiener process without drift take the form

$$p(m_{k-1}|M_k, m_k) = \frac{M_k - m_k}{(M_k - m_{k-1})^2}, \qquad p(M_{k-1}|M_k, m_k) = \frac{M_k - m_{k-1}}{(M_{k-1} - m_{k-1})^2} \qquad (47)$$

and the transition probability (37) in the previously discussed Markov chain is given by

$$p(M_{k-1}, m_{k-1}|M_k, m_k) = \frac{M_k - m_k}{(M_k - m_{k-1})(M_{k-1} - m_{k-1})^2}. \qquad (48)$$

This provides the universal characterisation of the main extrema at small scale. Interestingly from equation (48) we can deduce the closed form expression for the joint probability (38) of all previous main extrema

$$\begin{aligned} & p(M_{k-l}, m_{k-l}, \ldots, M_{k-1}, m_{k-1}|M_k, m_k) \\ &\quad = \frac{M_k - m_k}{(M_k - m_{k-1})(M_{k-1} - m_{k-1})(M_{k-1} - m_{k-2}) \cdots (M_{k-l} - m_{k-l})^2}. \qquad (49) \end{aligned}$$

Choosing for example $M_k = 1$ and $m_k = 0$ the structure of this formula is directly comparable with the corresponding formula (22) for the discrete random walk process for $\sigma = -1$.

Again, as in the previous section, we remark that there is no stationary probability distribution for the Wiener process and in contrast to our requirement for the derivation of (28) the initial condition of the process cannot be neglected even in the long term limit. Nevertheless, the formulas formally obtained for the Wiener process accurately characterise the memory state of an arbitrary process on a scale at which I is approximately linear.

We can also view the situation from the following perspective. Since the function I is invertible, the main extrema M_k and m_k for a general Markov process $x = x(t)$ are completely characterised if the distribution of the mapped extrema defined as

$$M_k^I = I(M_k), \quad m_k^I = I(m_k) \tag{50}$$

is characterised. Formulas (50) define the main extrema of the process $I(x)$. From (31) it follows that

$$P(\mathbf{m}_{k-1}^I < \bar{m}^I | \mathbf{M}_k^I = M_k^I, \mathbf{m}_k^I = m_k^I) = \frac{M_k^I - m_k^I}{M_k^I - \bar{m}^I}$$

and therefore the conditional probability density for the process $I(x)$ is given by

$$p^I(m_{k-1}^I | M_k^I, m_k^I) = \frac{M_k^I - m_k^I}{\left(M_k^I - m_{k-1}^I\right)^2}$$

which agrees with the corresponding formula (47) for the Wiener process. On the other hand, using

$$p^I(m_{k-1}^I | M_k^I, m_k^I) \partial_y I(y = m_{k-1}) = p(m_{k-1} | M_k, m_k)$$

we directly recover (35). Thus the problem of characterising the distribution of the main extrema of an arbitrary continuous Markov process can be mapped to the problem of characterising the memory structure of the Wiener process without drift by using the invertible function I.

Due to the discussed importance of the Wiener process without drift for the characterisation of main extrema in arbitrary continuous Markov processes let us now consider the differences between main extrema of the Wiener process in more detail. As an alternative equivalent description of the memory array, let us define the strictly monotonically decreasing sequence $\{\mathbf{d}_k\}_{k\geq 1}$ for the differences between the main extrema via

$$\mathbf{d}_k = \begin{cases} \mathbf{M}_{(k+1)/2} - \mathbf{m}_{(k+1)/2} & \text{for } k \text{ odd} \\ \mathbf{M}_{k/2+1} - \mathbf{m}_{k/2} & \text{for } k \text{ even} \end{cases}$$

(cf. (10)) where (2) is assumed. According to formulas (47), in this notation, the conditional probability density for preceding elements in the sequence $\{\mathbf{d}_k\}$ is given via

$$p\left(d_{k-1} | d_k\right) = \frac{d_k}{\left(d_{k-1}\right)^2} \quad \text{for} \quad \mathbf{d}_{k-1} > \mathbf{d}_k; \qquad p\left(d_{k-1} | d_k\right) = 0 \quad \text{for} \quad \mathbf{d}_{k-1} \leq \mathbf{d}_k. \tag{51}$$

We are now interested in the number $\mathbf{q}_d$ of elements of the sequence $\{\mathbf{d}_k\}$ whose value lie in the interval between some maximal value K and some minimal value ϵ. For every $K > \epsilon > 0$, we define the integer valued random variable $\mathbf{q}_d = \mathbf{q}_d(K, \epsilon)$ by

$$\mathbf{q}_m = \min\left\{k : \mathbf{d}_k < \epsilon\right\}, \quad \mathbf{q}_M = \max\left\{k : \mathbf{d}_k < K\right\}, \quad \mathbf{q}_d = \mathbf{q}_m - \mathbf{q}_M.$$

Using the escape problem argument as above we obtain the probability density distribution of the least element $\mathbf{d}_k = \mathbf{d}_{q_m-1}$ satisfying $\mathbf{d}_{q_m-1} \geq \epsilon$ as

$$p\left(d_{q_m-1}\right) = \frac{\epsilon}{\left(d_{q_m-1}\right)^2} \quad \text{for} \quad \mathbf{d}_{q_m-1} > \epsilon\,; \qquad p\left(d_{q_m-1}\right) = 0 \quad \text{for} \quad \mathbf{d}_{q_m-1} \leq \epsilon. \tag{52}$$

A numerical confirmation of formula (52) is shown in figure 4.

We can now interpret d_k as an inverse Markov process with backward transition probabilities given by (51). Combining formulas (51) for k and $k-1$, we obtain $p\left(d_{k-2}|d_k\right) = 0$ for $d_{k-2} \leq d_k$ and

$$\begin{aligned} p\left(d_{k-2}|d_k\right) &= \int p\left(d_{k-2}|d_{k-1}\right) p\left(d_{k-1}|d_k\right)\, \mathrm{d}d_{k-1} \\ &= \int_{d_k}^{d_{k-2}} \frac{d_{k-1}}{\left(d_{k-2}\right)^2} \frac{d_k}{\left(d_{k-1}\right)^2}\, \mathrm{d}d_{k-1} = \frac{d_k}{\left(d_{k-2}\right)^2} \ln \frac{d_{k-2}}{d_k} \quad \text{for} \quad d_{k-2} > d_k. \end{aligned}$$

Similarly, by induction in i, for each $i < k$

$$p\left(d_i|d_k\right) = \int p\left(d_i|d_{i+1}\right) p\left(d_{i+1}|d_k\right)\, \mathrm{d}d_{i+1} = \frac{d_k}{2\left(d_i\right)^2} \left(\ln \frac{d_i}{d_k}\right)^2 \quad \text{for} \quad d_i > d_k$$

with $p\left(d_i|d_k\right) = 0$ for $d_i \leq d_k$.

Relations (52) imply

$$P\left(\mathbf{q}_d = 0\right) = P\left(\mathbf{d}_{q_m-1} > K\right) = \epsilon/K,$$

that is with the probability ϵ/K the memory array does not contain any elements $\mathbf{d}_k$ with values in the range between ϵ and K. Similarly in the case of $\mathbf{q}_d = 1$ there is precisely one $\mathbf{d}_k$, namely $\mathbf{d}_{q_m-1}$, between ϵ and K. With the use of (51), the probability of this event can be expressed as

$$\begin{aligned} P\left(\mathbf{q}_d = 1\right) &= \int_K^\infty \mathrm{d}d_{q_m-2} \int_\epsilon^K p\left(d_{q_m-2}|d_{q_m-1}\right) p\left(d_{q_m-1}|d_{q_m} = \epsilon\right) \mathrm{d}d_{q_m-1} \\ &= \int_K^\infty \mathrm{d}d_{q_m-2} \int_\epsilon^K \frac{d_{q_m-1}}{\left(d_{q_m-2}\right)^2} \frac{\epsilon}{\left(d_{q_m-1}\right)^2}\, \mathrm{d}d_{q_m-1} = \frac{\epsilon}{K} \ln \frac{K}{\epsilon}. \end{aligned}$$

Continuing by induction in k, we obtain the following probability distribution of $\mathbf{q}_d$

$$P\left(\mathbf{q}_d = k\right) = \frac{\epsilon}{k!K} \left(\ln \frac{K}{\epsilon}\right)^k \tag{53}$$

which is plotted for $\epsilon/K = 0.1$ in figure 5. That is, for the Wiener process, the number $\mathbf{q}_d$ of the memory elements $\mathbf{d}_k$ between two thresholds ϵ and K has the Poisson distribution with the mean value (intensity)

$$\langle\, \mathbf{q}_d \,\rangle = \ln(K/\epsilon).$$

Acknowledgments

This publication has emanated from research conducted with the financial support of Federal Programme 'Scientists of Innovative Russia', grant 2009-1.5-507-007, Science Foundation Ireland under Grant Number 09/SIRG/I1615, Science Foundation Ireland research grant 07/MI/008 associated with Edgeworth Centre for Financial Mathematics, and Russian Foundation for basic Research, grant 10-01-93112.

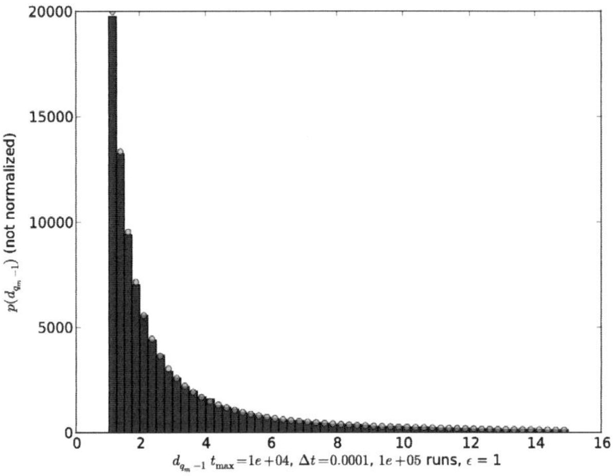

Figure 4. Confirmation of (52) via numerical integration of a Wiener process with time step $\Delta t = 0.0001$ from 0 to $t_{\mathrm{max}} = 10000$. We plot a histogram of the last $\mathbf{d}_m$ which is bigger than $\epsilon = 1$ using a total of 10^5 runs. The circles show the analytical expectation according to (52).

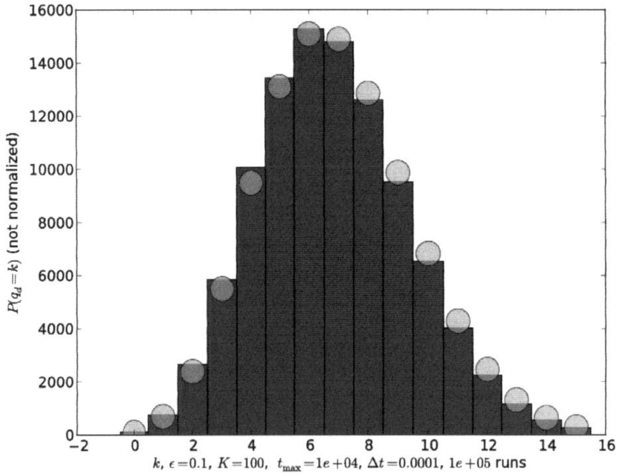

Figure 5. Probability distribution of the number of min-max differences $\mathbf{d}_k$ occurring in the interval $[\epsilon = 0.1, K = 100]$. For the numerical data we integrated a Wiener process with time step $\Delta t = 0.0001$ from 0 to $t_{\mathrm{max}} = 10000$ and recorded the $\mathbf{d}_m$. The circles are the expected values according to (53) with $\epsilon/K = 0.1$.

References

[1] Benzi R, Sutera A and Vulpiani A 1981 *J. Phys. A: Math. Gen.* **14** L453–L457
[2] Benzi R, Parisi G, Sutera A and Vulpiani A 1982 *Tellus* **34** 10-16
[3] Sagues F, Sancho J and García-Ojalvo J 2007 *Rev. Mod. Phys.* **79** 829–882
[4] Lindner B, García-Ojalvo J, Neiman A and Schimansky-Geier L 2004 *Phys. Rep.* **392** 321
[5] Mitaim S and Kosko B 1998 *Proc. IEEE: Special Issue on Intelligent Signal Processing* **86** 2152–2183
[6] Preisach F 1935 *Z. Phys.* **94** 277–302
[7] Mayergoyz I D and Korman C E 1991 *J. Appl. Phys.* **69** 2128–2134
[8] Rugkwamsook P, Korman C, Bertotti G and Pasquale M 1999 *J. Appl. Phys.* **85** 4361–4363
[9] Mayergoyz I D, Adly A, Huang M W and Kraft C 2000 *IEEE Trans. Magn.* **36** 3208
[10] 2006 *The Science of Hysteresis*, three volume set, ed I Mayergoyz and G Bertotti (Elsevier, Academic Press)
[11] 2009 *Hysteresis*, ed R Iyer and X Tan (*IEEE Control Systems Magazine* **1**)
[12] Mayergoyz I D 2006 *The Science of Hysteresis* vol I, ed I Mayergoyz and G Bertotti (Elsevier, Academic Press) pp 529–604
[13] Korman E C 2006 *The Science of Hysteresis* vol I, ed I Mayergoyz and G Bertotti (Elsevier, Academic Press) pp 639–665
[14] Radons G 2008 *Phys. Rev. Lett.* **100** 240602
[15] Radons G 2008 *Phys. Rev. E* **77** 061133
[16] Adedoyin A, Dimian M and Andrei P 2009 *IEEE Trans. Magn.* **45** 3934–3937
[17] Dimian M, Coca E and Popa V 2009 *J. Appl. Phys.* **105** 07D515
[18] Mayergoyz I and Dimian M 2003 *J. Appl. Phys.* **93** 6826–6828
[19] Dimian M and Mayergoyz I D 2004 *Phys. Rev. E* **70** 046124
[20] Visintin A 1994 *Differential Models of Hysteresis* (New York: Springer)
[21] 1997 *Kluwer Encyclopedia of Mathematics* supplement I, ed M Hazewinkel (Kluwer Acad. Publ) pp 310–384
[22] Krasnosel'skii M A and Pokrovskii A V 1989 *Systems with Hysteresis* (Springer)
[23] Mayergoyz I D 1991 *Mathematical Models of Hysteresis* (New York: Springer)
[24] Mayergoyz I D 2003 *Mathematical Models of Hysteresis and Their Applications: Electro-magnetism* (Academic Press)
[25] Brokate M and Sprekels J 1996 *Hysteresis and Phase Transitions* (Springer)
[26] Krejčí P 1996 *Hysteresis, Convexity and Dissipation in Hyperbolic Equations* (Tokyo: Gakkōtosho)
[27] Dahmen K and Ben-Zion Y 2009 *Enciclopedia of Complexity and Systems Science* vol 5, ed C Marchetti and R A Meyers (Springer) pp 5021–5037
[28] Mehta A and Barker G C 2001 *Europhys. Lett.* **56** 626-632
[29] Guyer R A and McCall K R 1996 *Phys. Rev. B.* **54** 18–21
[30] Dahmen K A, Ben-Zion Y and Uhl J T 2009 *Phys. Rev. Lett.* **102** 175501
[31] Gleeson J P 2009 *Phys. Rev. E* **80** 036107
[32] Watts D J 2002 *Proc. of the National Acad. Sci. of USA* **99** 5766–5771
[33] Goicoechea J and Ortin J 1994 *Phys. Rev. Lett.* **72** 2203
[34] R. Cross, M. Grinberg and H. Lamba 2009 *IEEE Control Systems Magazine* **29** 30–43
[35] Sethna J P, Dahmen K and Myers C R 2001 *Nature* **410** 242–250
[36] Dahmen K 2005 *Nature Physics* **1** 13–14
[37] Sethna J P, Dahmen K A and Perkovic O 2005 *The Science of Hysteresis* vol 2, ed G Bertotti and I Mayergoyz (Elsevier) pp 107–168
[38] Sethna J P, Dahmen K, Kartha S, Krumhansl J A, Robetrs B W and Shore J D 1993 *Phys. Rev Lett.* **70** 3347
[39] Rychlik I and Gupta S 2007 *Int. J. of Fatigue* **29** 406–420
[40] Brokate M, McCarthy S, Pimenov A, Pokrovskii A and Rachinskii D 2010 *Environ. Model. Assess.* in press
[41] Lamba H and Seaman T 2008 *Int. J. of Theoretical and Applied Finance* **11** 717–737
[42] Cross R, Grinfeld M, Lamba H and Seaman T 2007 *The European Phys. Journal B* **57** 213–218
[43] Cross R, Grinfeld M, Lamba H and Seaman T 2005 *Physica A* **354** 463–478
[44] Lamba H and Seaman T 2009 arXiv:0709.4242, *Physica A* in press
[45] Lamba H 2009 arXiv:0908.0949v1
[46] Borodin A N and Salminen P 2002 *Handbook of Brownian Motion - Facts and Formulae* (Birkhäuser)
[47] Gardiner C 2009 *Stochastic Methods* (Springer)

Journal of Physics: Conference Series **268** (2011) 012002 doi:10.1088/1742-6596/268/1/012002

On the equivariant Hopf bifurcation in hysteretic networks of van der Pol oscillators

Z Balanov[1,2], W Krawcewicz[2] and D Rachinskii[3]

[1] Department of Mathematics and Computer Sciences, Netanya Academic College, 42365, Netanya, Israel

[2] Department of Mathematical Sciences, University of Texas at Dallas, Richardson,Texas, 75080 US

[3] Department of Applied Mathematics, University College Cork, Ireland

E-mail: d.rachinskii@ucc.ie

Abstract. The standard approach to study symmetric Hopf bifurcation phenomenon is based on the usage of the equivariant singularity theory developed by M. Golubitsky et al. In this paper, we present the equivariant degree theory based method which is complementary to the equivariant singularity approach. Our method allows systematic study of symmetric Hopf bifurcation problems in non-smooth/non-generic equivariant settings. The exposition is focused on a network of eight identical van der Pol oscillators with hysteresis memory, which are coupled in a cube-like configuration leading to S_4-equivariance. The hysteresis memory is the source of non-smoothness and of the presence of an infinite dimensional phase space without local linear structure. Symmetric properties and multiplicity of bifurcating branches of periodic solutions are discussed showing a direct link between the physical properties and the equivariant topology underlying this problem.

1. Introduction

1.1. Subject

Given a family of dynamical systems parameterized by a real parameter α, for example,

$$\dot{x} = f(\alpha, x) \qquad (\alpha \in \mathbb{R},\ x \in \mathbb{R}^n), \tag{1}$$

where $f : \mathbb{R} \oplus \mathbb{R}^n \to \mathbb{R}^n$ is a continuous map, and a curve of trivial stationary solutions $(\alpha, 0)$, the Hopf bifurcation is a phenomenon occurring when α crosses some critical value α_o (for which the linearization admits a purely imaginary eigenvalue) and resulting in appearance of small amplitude periodic solutions near $(\alpha_o, 0)$. Many problems in population dynamics, neural networks, fluid dynamics, solid mechanics, elasticity, chemistry, electrical engineering, etc., lead to studying the Hopf bifurcation in networks of identical oscillators coupled symmetrically. This feature is expressed as the so-called *equivariance* of system (1), i.e. $\mathbb{R}^n$ is a representation of a (compact Lie) group Γ and $f(\alpha, \gamma x) = \gamma f(\alpha, x)$ for all $x \in \mathbb{R}^n$, $\gamma \in \Gamma$ and $\alpha \in \mathbb{R}$ (cf. [1–9] and references therein). Typical local problems include: (i) *occurrence of the Hopf bifurcation*, (ii) *classification of symmetry properties of bifurcating branches of periodic solutions*, (iii) *estimation of a minimal number of these branches (if Γ is finite)*, (iv) *finding their asymptotics*, and (v) *providing the stability analysis*. In many cases, these problems have been effectively studied under certain smoothness conditions.

For the last decades, the interest in the bifurcation phenomena in *non-smooth* systems with different sources of non-smoothness is growing rapidly (we refer to [10–12] that are in the context relevant to our discussion). In the present paper, we are concerned with the equivariant Hopf bifurcation in systems where the non-smoothness comes from the hysteresis effect in several system components. The system is described by a set of differential equations which are coupled to hysteresis operator relationships between some of the variables. We thus use the mathematical setting provided by the theory of systems with hysteresis memory operators (see [13–17], for a recent survey of applications in several disciplines we refer to [18, 19]). We would like to stress that the hysteresis memory operators are *intrinsically non-smooth.*

Specifically, we discuss the equivariant Hopf bifurcation in networks of van der Pol type oscillators coupled symmetrically, where each oscillator has a hysteretic memory component described by the Preisach operator (see, for example, [13, 14]). To simplify our exposition, we are focused on a concrete electrical network of eight oscillators admitting the octahedral symmetry group (see figures 1, 2). The memory of each oscillator is related to the assumption that the constitutive relationship between the magnetic field and magnetization in the inductance element of each oscillator is hysteretic. In this setting, the above mentioned problems (i), (ii) and (iii) on the equivariant Hopf bifurcation are completely settled.

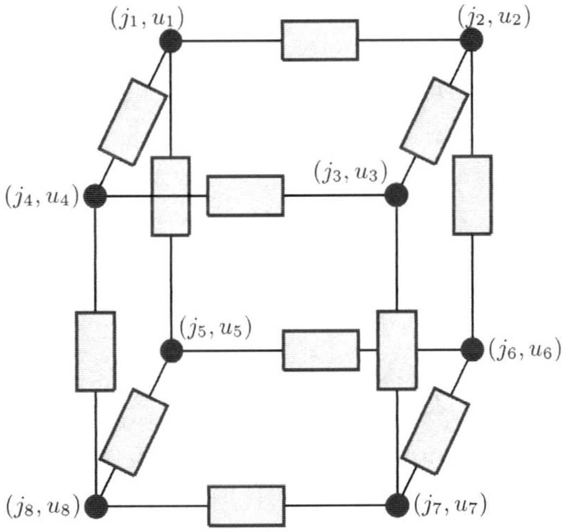

Figure 1. Symbolic graph representation of eight van der Pol oscillators arranged into a cube-like configuration. Yellow box symbolizes an electronic component used to connect the oscillators. The symmetry group of the cube, which is the octahedral group S_4, is the symmetry group of this network of oscillators.

1.2. Methodology.

The standard way to study the equivariant Hopf bifurcation is rooted in the equivariant singularity theory: assuming the system to satisfy several reasonable conditions (for example, f in (1) is smooth and equivariantly generic around the bifurcation point), one can combine the equivariant normal form classification with Center Manifold Theorem/Lyapunov-Schmidt Reduction. For a detailed exposition of this concept and related techniques, we refer to [6, 7] (see also [9, 20, 21]). Several examples with symmetry groups important from the viewpoint of the present paper, have been considered in [22, 23].

(i) Phase space approach. The phase space of a system with the Preisach hysteresis operator includes an *infinite dimensional* component in the space of states (memory configurations) of the Preisach model. This component is a metric space *without a local linear structure*, which

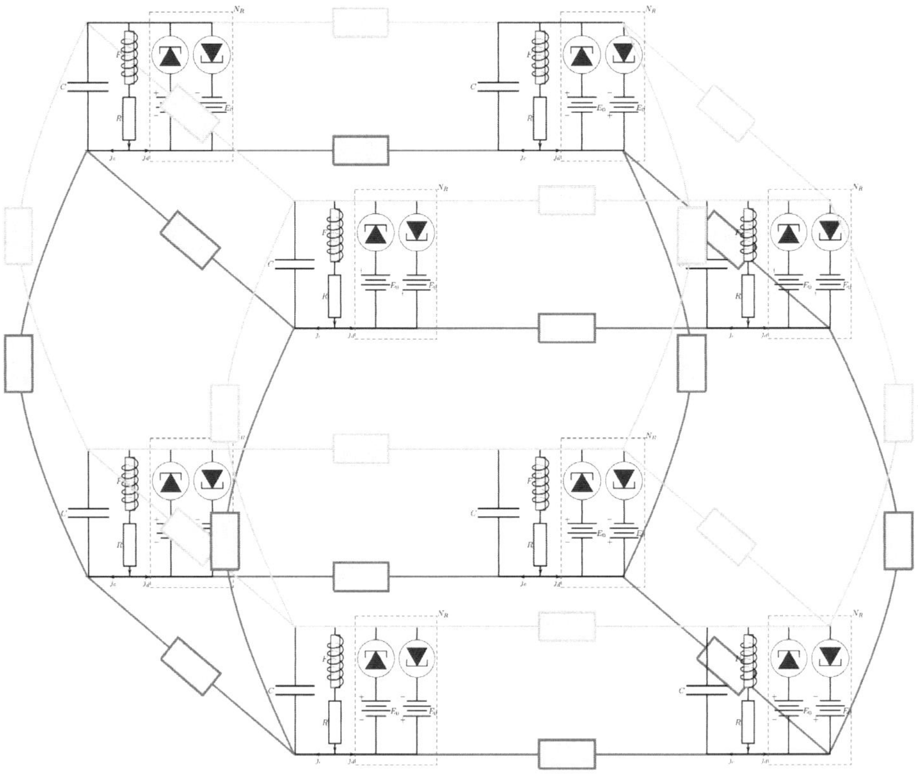

Figure 2. A network of identical van der Pol oscillators with octahedral symmetries corresponding to the graph shown in figure 1. The connecting electronic components are marked as rectangles.

is a common situation for dynamical systems with complex hysteresis operators. Therefore, the existing theory of such dynamical systems misses important tools of the smooth theory of ODEs, like invariant manifolds, dimension reduction (including Center Manifold Theorem), and normal forms. Actually, systems with hysteresis operators exhibit dynamical and bifurcation scenarios different from those observed in smooth systems of ODEs (see, for example, [11, 24]). By the same reason (lack of local linear structure in the state space and non-smoothness of the hysteresis input-state-output operators), the methods based on the linearization of the evolution operator, such as linear stability analysis, cannot be generally applied to systems with hysteresis.

Several alternative approaches to study stability of solutions (still based on the phase space analysis) developed in the last decades involve: contraction mapping principle [25], energy consideration [17, 26, 27] and, for scalar non-autonomous differential equations with the Preisach operator, monotonicity argument combined with upper and lower solutions method [28, 29]. A major simplification can be achieved when the evolution operator has an invariant *finite-dimensional* set in the phase space. For example, the Hopf bifurcation from *infinity* gives rise to the zero-dimensional invariant set as only one (saturated) state of the hysteresis operator is possible at a moment when the input achieves a sufficiently large value [30]. Another example is related to solutions with piecewise monotone components which do not have extremum points with equal values and do not have degenerate extremum points [31]. However, this simplification breaks down on *local periodic problems* (being the subject of the present paper). Even for a scalar non-autonomous differential equation coupled with the Preisach operator relationship, the analysis of dynamics in the phase space, such as the one used in [32] to obtain stability criteria for periodic solutions by invoking the contraction mapping principle and to study simple saddle-node bifurcation of such solutions, is tricky.

(ii) Functional-analytic approach. By the above reasons, in the existence problems, the functional-analytic setting whereby an equivalent fixed point problem is formulated in an appropriate function space, seems to be an appealing alternative to the phase space analysis. Under this approach, a periodic solution is sought in a space of periodic functions, rather than among trajectories of the dynamical systems or their initial values. However, the operator of the fixed point equation, say F, inherits non-smooth features from the dynamical system: it is differentiable on stationary functions, but has points of non-differentiability in arbitrary vicinity of those. In particular, any *non-constant* periodic function with repeating maximum/minimum values at local maxima/minima *inside* the period admits a cone of directions for which F does not have the Gateux derivative. Moreover, the structure of nonlinear terms in "Taylor decomposition" of F around stationary solutions is rather complicated in nature since these terms involve hysteretic operators. All this makes us believe that, to obtain the existence results related to local Hopf bifurcation in systems with hysteresis, one should analyze *topological* properties of F.

It should be pointed out that F is completely continuous that suggests to use homotopy invariants of degree type to study periodic problems (see overview in [24] for results using this general approach). In particular, the existing results on Hopf bifurcation in systems with the Preisach operator, including the model with *one* van der Pol type oscillator, have been obtained using Leray-Schauder degree (see [33,34]). However, the Leray-Schauder degree based methods meet serious difficulties when applied to study networks of oscillators. In addition, these methods seem to be ineffective to study *symmetric* properties of bifurcating periodic solutions to systems related to *symmetric networks* of oscillators (with/without hysteresis).

(iii) Twisted equivariant degree based method. During the last twenty years the *equivariant degree theory* emerged in non-linear analysis (for the detailed exposition of this theory, including historical remarks, we refer to recent monographs [8,35] and surveys [36,37]; for the prototypal invariants, see [38–40]). In short, the equivariant degree is a topological tool allowing "counting" orbits of solutions to symmetric equations in the same way as the usual Brouwer degree does, but according to their symmetry properties. In particular, the equivariant degree theory has all the attributes allowing its application in non-smooth/non-generic equivariant settings related to equivariant dynamical systems having, in general, infinite dimensional phase spaces with lack of linear structure. In fact, the equivariant degree has different faces reflecting a diversity of symmetric equations related to applications. To study symmetric properties of *periodic* solutions to dynamical systems, Z. Balanov and W. Krawcewicz introduced in [41] the so-called *twisted equivariant degree* (see also [8,37]). This is the *main technical tool* used to prove the results of the present paper.

At first glance, the twisted degree looks like a sophisticated topological tool. However, similarly to the usual Brouwer degree, the twisted degree admits an *axiomatic definition* (see [8], Chapter 4; see also [36,37,41]), and therefore, can be effectively used outside its theoretical context and foundations. Actually, for its usage one needs no more than basic representation theory, usual Brouwer degree theory, and the so-called Recurrence Formula (see [8], Section 4.6 or [37], Theorem 4.5, Property (T6)) which is a kind of Borsuk-Ulam mod p Theorem for equivariant maps with one free parameter. Since once done computations can be compiled into a database (that can be conveniently used by means of eTools), the twisted equivariant degree theory becomes as simple as any other symbolic computational package being used for analysis of dynamical systems.

1.3. Overview

After the Introduction the paper is organized as follows. In Section 2, we present the known Hopf bifurcation results related to *one* van der Pol oscillator (both in cases with and without hysteretic relationship between the magnetic induction and magnetic field being taken into account). In Section 3, we give an informal introduction to the Preisach model. Finally, in Section 4, the basic example of the network of eight van der Pol circuits with Preisach memory coupled symmetrically in cubic configuration is considered (cf. figure 2). In particular, the mathematical model is deduced (see (20)), the corresponding equivariant bifurcation result is presented (see Theorem 1 and Remark 3), and a sketch of the proof of Theorem 1 is outlined. For the equivariant topology (resp. G-representations) background, we refer to [42, 43] (resp. [44]).

2. Hopf bifurcation in one van der Pol oscillator with hysteretic element

2.1. Van der Pol oscillator without hysteresis

The van der Pol oscillator is one of the simplest non-conservative oscillators with non-linear damping used in electrical engineering. The prototypal electrical circuit consists of an LCR contour and a negative feedback loop, which can be implemented on the basis of a triode, as in the classical van der Pol circuit, or, for instance, by using tunnel diodes as shown on figure 3. Here, the negative feedback element consists of two tunnel diodes connected in parallel, each with a DC power source connected in series to bias the diode. It has a current-voltage characteristic curve of the form

$$j_d = -\sigma_1 u + \sigma_2 u^3 \tag{2}$$

with $\sigma_1 > 0$ and $\sigma_2 \geq 0$.

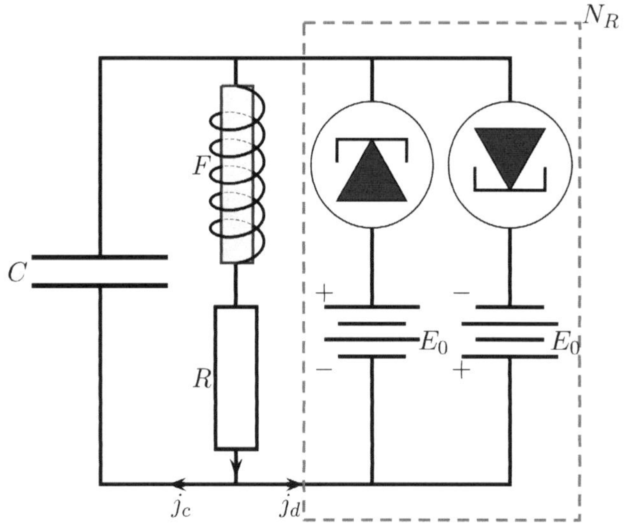

Figure 3. A van der Pol circuit. F denotes an inductor with ferromagnetic core; R an ohmic resistor; C a capacitor; N_R a resistor with negative resistance for small voltages; E_0 a DC power source. j_d, j_c denote the current flowing in the different branches of the circuit.

If one ignores hysteresis effects, then the EMF in the inductor can be expressed as

$$E_{ind} = L\frac{dj}{dt} \tag{3}$$

where $j = j_c + j_d$ is the current flowing through the resistance and the inductance elements. Applying simple circuit analysis rules results in the system

$$j' = -\frac{R}{L}j + \frac{1}{L}u, \qquad u' = -\frac{1}{C}j + \frac{\sigma_1}{C}u - \frac{\sigma_2}{C}u^3, \tag{4}$$

which can be rewritten as the classical van der Pol equation. Here C, R and L stand for the capacitance, resistance and inductance of the inductor, resp., while u represents the voltage drop across the resistor and inductor, i.e., $u = Rj + E_{ind}$; prime denotes the differentiation with respect to time.

In a standard way, using $\alpha := \sigma_1$ as a bifurcation parameter, one can easily study the Hopf bifurcation in (4) (cf., for example, [45]). Namely, the eigenvalues of the Jacobian

$$A = \begin{pmatrix} -\frac{R}{L} & \frac{1}{L} \\ -\frac{1}{C} & \frac{\alpha}{C} \end{pmatrix} \tag{5}$$

of the right hand side are

$$\xi = \frac{L\alpha - RC \pm \sqrt{(RC - \alpha L)^2 - 4CL(1 - R\alpha)}}{2CL}. \tag{6}$$

Clearly, they cross the imaginary axis at the points $\xi = \pm i\frac{\sqrt{1-\frac{R^2C}{L}}}{\sqrt{CL}}$ for

$$\alpha = \frac{RC}{L} \tag{7}$$

provided that

$$\frac{R^2C}{L} < 1. \tag{8}$$

Put $\xi(\alpha) := \mathfrak{u}(\alpha) + i\mathfrak{v}(\alpha)$. Then,

$$\left.\frac{d\mathfrak{u}}{d\alpha}\right|_{\alpha=\frac{RC}{L}} = \frac{1}{2C} > 0. \tag{9}$$

Thus, condition (8) ensures that the Hopf bifurcation occurs at the critical parameter value (7).

The amplitude of the small cycle of equation (4) near the bifurcation point scales as $\sqrt{|\alpha|}$ for $\sigma_2 \neq 0$. Stability of cycles born via the Hopf bifurcation is determined by the sign of the cubic term (cf., for example, [45]), namely: for $\sigma_2 > 0$ (resp. $\sigma_2 < 0$), the Hopf bifurcation is supercritical, i.e., the small cycles are stable (resp. subcritical, i.e., the small cycles are unstable).

2.2. Van der Pol oscillator with hysteresis

If the inductance element contains a ferromagnetic core, the ferromagnetic material can introduce a hysteresis relation between the magnetic induction B and magnetic field H. A typical picture of hysteresis loops is presented in figure 4. Here the instantaneous value of B depends not only on the value of H at the same moment, but also on some previous values of H. Hence the constitutive relationship between B and H is an operator relationship. While this hysteresis effect can often be neglected, there are specific applications, where it is essential to include it in the model. Examples of such applications include power transformers and dynamics of power electronics circuits in the presence of the ferroresonance phenomenon. Furthermore, hysteresis effect can be the main effect responsible for the energy dissipation. The models, using the Preisach operator to describe the relation between B and H have been shown to provide

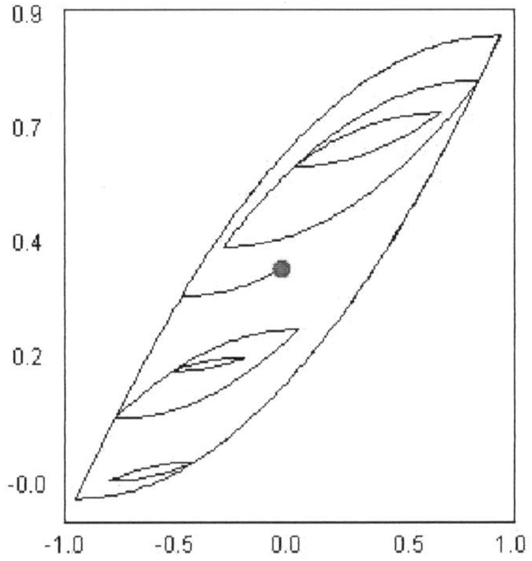

Figure 4. Hysteresis loops in relationship between the external magnetic field H (horizontal axis) and the magnetic induction B of a ferromagnetic material (vertical axis).

a better agreement between the numerical results and experimental data in [10] (see Section 3 for an informal discussion of the Preisach model and, for example, [13, 14] for a rigorous mathematical treatment of the model).

To be more specific, if the relationship between the magnetization M and the magnetic field H is modeled by the Preisach input-output operator, then

$$B = \nu H + M = \nu H + \Theta[\eta_0]H. \tag{10}$$

Here $\nu > 0$, the Preisach input-output operator $\Theta[\eta_0]$ maps the variable magnetic field $H(t)$ to the variable magnetization $M(t) = (\Theta[\eta_0]H)(t)$ of the ferromagnetic material, and the infinite-dimensional parameter η_0 (called the initial state of the Preisach model) describes the physical state of the ferromagnetic material at the initial moment. As the field H (resp. EMF induced in the inductor) is proportional to the current j (resp. rate of change of the magnetic induction B), the EMF is now given by

$$E_{ind} = L\frac{d}{dt}(j + \Theta[\eta_0]j), \tag{11}$$

where the Preisach density function (see Section 3) is properly rescaled when we pass from (10) to (11). System (4) changes accordingly to the system

$$(j + \Theta[\eta_0]j)' = -\frac{R}{L}j + \frac{1}{L}u, \qquad u' = -\frac{1}{C}j + \frac{\sigma_1}{C}u - \frac{\sigma_2}{C}u^3 \tag{12}$$

involving the Preisach operator. System (12) has the zero equilibrium $j = u = 0$. The Hopf bifurcation for system (12) and similar higher order systems with the Preisach operator was studied in [33, 34] (the impact of symmetries was not addressed). The *Leray-Schauder degree theory methods* used in [33, 34] allowed the authors to localize the Hopf bifurcation point and to show the existence of a global branch of cycles stretching from the equilibrium to infinity. In particular, similarly to the case of ODEs, it was shown that the Hopf bifurcation point is determined by the Jacobian of the right-hand side of the system. This is explained by the asymptotically small Preisach nonlinearity (in a proper sense) compared to the linear term near the equilibrium, which implies that the hysteresis term does not affect the critical value of the parameter. For example, if we assume $R = 0$, then system (4) without hysteresis term and system (12) with the Preisach hysteresis term have the Hopf bifurcation at the same critical value $\sigma_1 = 0$ of the parameter σ_1.

However, the hysteresis effect has a crucial impact on the asymptotics and stability of cycles born via the Hopf bifurcation. In [33, 34], a numerical evidence is given that the Hopf bifurcation in system (12) is supercritical *regardless* the sign of σ_2 due to the effect of the Preisach nonlinearity which dominates the cubic term near the equilibrium point. In particular, system (12) exhibits the supercritical Hopf bifurcation for $\sigma_2 = 0$, i.e., when the negative feedback is linear and the Preisach term is the only nonlinearity and the only source of dissipation in the system. The authors are not aware of the literature where the stability of small cycles of system (12) was rigourously proved, however they believe that the stability analysis method proposed in [32] for problems with forced oscillations can be used for this purpose after a proper modification.

The asymptotics of small cycles born via the Hopf bifurcation is also different for systems (4) and (12). The amplitude of the cycle of system (12) near the bifurcation point is asymptotically proportional to σ_1 rather than $\sqrt{|\sigma_1|}$ (for $\sigma_2 \neq 0$), and the coefficient of proportionality is defined by parameters of the Preisach nonlinearity irrespective of the value of σ_2.

Remark 1 (i) Loosely speaking, the asymptotics and stability of small cycles of system (12) are the same as those of the van der Pol type ODE obtained from (4) by replacing the cubic nonlinearity with a non-smooth quadratic term such as $u'|u|$.

(ii) Observe that several aspects of dynamics of oscillators modeled by differential equations (coupled with the Preisach operator relationship) different from (12), have been studied, for example, in [24,26]. Maxwell's equations coupled with the Preisach constitutive relation between B and H in spatially distributed models were considered in [16, 46, 47].

3. Preisach model

Preisach-type models are at the heart of the hysteresis modeling paradigm. We use this model as the basis of a quantitative description of ferromagnetic hysteresis. For the reader's convenience, in this section, we present an informal description of the Preisach model allowing us to formulate the main result (see Theorem 1).

3.1. Non-ideal relay

An elementary building block of the Preisach model is the simplest hysteretic transducer called non-ideal relay. This transducer is characterized by two threshold values α and β, where $\alpha < \beta$. Its output $\eta(t)$ can take one of two values 0 or 1 at any moment, meaning that the relay is either switched off or switched on. The output varies in response to changes of the input $x(t)$.

The dynamics of the relay are described by figure 5. The variable output of the relay, traditionally denoted by

$$\eta(t) = (R_{\alpha,\beta}[t_0, \eta_0]x)(t), \qquad t \geq t_0, \tag{13}$$

depends on the input $x(t)$, $t \geq t_0$, and on the initial state η_0 of the relay. The input $x(t)$ is an arbitrary continuous scalar function, and η_0 is either 0 or 1. The output $\eta(t)$ has at most a finite number of jumps between the values 0 and 1 on any finite interval $t_0 \leq t \leq t_1$. The output behaves lazily, preferring to be unchanged as long as the phase pair $(x(t), \eta(t))$ belongs to the union of the bold lines in figure 5. The equalities $\eta(t) = 1$ for $x(t) \geq \beta$ and $\eta(t) = 0$ for $x(t) \leq \alpha$ always hold for $t \geq t_0$. In the context of modeling magnetic materials in a one-dimensional magnetic field, a non-ideal relay is related to dynamics of the magnetic moment of an individual domain in the domain structure created by the field in the ferromagnetic material. In this interpretation, the equality $\eta(t) = 1$ (resp. $\eta(t) = 0$) means that the magnetic moment points in the direction of the applied field (resp. in the opposite direction).

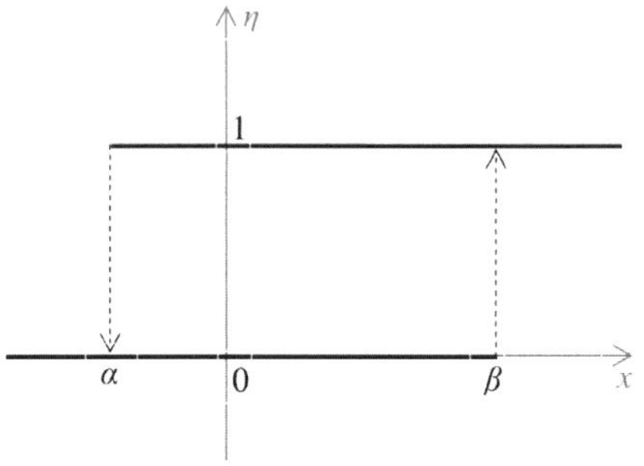

Figure 5. Non-ideal relay.

3.2. Preisach operator

The main assumption made in the Preisach model is that the system can be thought of as the sum (or superposition, or parallel connection) of a continuum of weighted non-ideal relays $R_{\alpha,\beta}$ with common input $x(t)$, where the weight of each relay is $\mu(\alpha, \beta)$.

The relays contributing to the system can be represented by points on the two-dimensional half-plane $\Pi = \{(\alpha, \beta) : \beta > \alpha\}$, which is also known as the Preisach plane (see figure 6). The colored area $S = S(t)$ in figure 6 is the set of threshold values (α, β) for which the output of the corresponding relay $R_{\alpha,\beta}$ is 1 at a given moment t. The output of the remaining relays at the same moment is 0. Denote by $\eta(t) = \eta(t; \alpha, \beta)$ the characteristic function of the set $S(t)$. The output of the Preisach model is then represented by the formula

$$\begin{aligned} \theta(t) &= \int\limits_{\alpha<\beta} \mu(\alpha, \beta)(R_{\alpha,\beta}[t_0, \eta_0(\alpha, \beta)]x)(t)\, d\alpha d\beta \\ &= \int\limits_{\alpha<\beta} \mu(\alpha, \beta)\eta(t, \alpha, \beta)\, d\alpha d\beta = \int\limits_{S(t)} \mu(\alpha, \beta)\, d\alpha d\beta, \end{aligned} \tag{14}$$

where the *Preisach density function* $\mu(\alpha, \beta)$ is an integrable non-negative function defined on Π. As we see from (14), the output $\theta(t)$ depends on both the input and the initial states $\eta_0(\alpha, \beta)$ of all relays. Therefore, we use the notation

$$\theta(t) = (\Theta[\eta_0]x)(t) \quad (\eta_0 = \eta_0(\alpha, \beta), \quad t \geq t_0) \tag{15}$$

for the input-output operator of the Preisach model. The set $S(t)$, its characteristic function $\eta(t; \alpha, \beta)$, and its boundary $L(t)$ (see figure 6) are called a state of the Preisach nonlinearity.

In magnetism, θ is interpreted as the magnetization of the ferromagnetic material; all the magnetic moments of domains modeled by the relays contribute to θ.

Remark 2 (i) The evolution of the input defines the evolution of the state $S(t)$ and thus the evolution of the output (14) by a few simple rules. We refer to `http://euclid.ucc.ie/hysteresis/node17.htm` where an online applet explaining and implementing these evolution rules is available.

(ii) The state $S(t)$ stores the sequence of certain maximum and minimum values of the input over the time interval between the moments t_0 and t (the so-called sequence of main extrema or shock values of the input, see [14]). This sequence defines the curve followed by the input-output pair on the diagram in figure 4 on some time interval after the moment t (i.e., from t until the moment when the input reaches one of its past extremum values stored in $S(t)$). There are infinitely many input-output curves passing through each point of this diagram. The evolution rules of S define switches from one to another input-output curve.

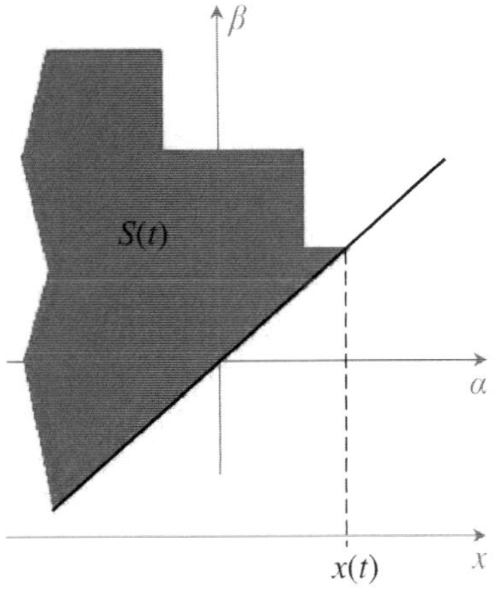

Figure 6. The Preisach plane. The colored area $S = S(t)$ is the set of threshold values (α, β) for which the corresponding relays $R_{\alpha,\beta}$ are switched on at a given moment t. The remaining relays are off.

4. Main result: model, formulation and discussion

4.1. Model

We are interested in classifying symmetry properties and estimating minimal number of branches of periodic solutions occurring as a result of the Hopf bifurcation in a cube-like network of identical van der Pol oscillators admitting the S_4-symmetry group (see figure 1). We give a complete equivariant topological classification of small amplitude periodic solutions bifurcating from the trivial one at Hopf bifurcation points.

An electrical circuit realization of the network we are interested in is shown in figure 2. We make the following assumption.

Assumption A. (a) An individual oscillator at each vertex of the cube has a ferromagnetic core in the inductance element and is described by the same equations (12) as the oscillator shown in figure 3; in particular, a Preisach density function μ is given and, for any initial state of the Preisach model η_0, the Preisach input-output operator $\Theta[\eta_0]$ taking the magnetic field to the magnetization of the ferromagnetic material, is defined by (14)—(15).

(b) The couplings between the oscillators are done by means of resistors only. For simplicity, all the wires connecting the oscillators are assumed to have the same resistance r.

In order to derive the equations describing the network shown in figure 2, we essentially use Kirhhoff's current and voltage laws (in short, KCL and KVL, respectively).

Denote by J_{ik} the current in a red wire from the oscillator i to the oscillator k, where $i < k$. Let $\tilde{J}_{ik}$ be the current in a green wire from the oscillator i to the oscillator k, where $i < k$. According to KCL, the total current which flows into an oscillator from all the connecting red and green wires must be zero. For the oscillator No. 1 this gives

$$J_{12} + J_{14} + J_{15} + \tilde{J}_{12} + \tilde{J}_{14} + \tilde{J}_{15} = 0. \tag{16}$$

A similar equation holds for each of the eight oscillators (in these equations a current is taken with the positive sign if it flows from the oscillator and with the negative sign if it flows into the oscillator). Now, consider the square contour consisting of four green wires, which form the upper face of the green cube. By KVL, the sum of voltages along the sides of this square must be zero, i.e., $r\tilde{J}_{12} + r\tilde{J}_{23} + r\tilde{J}_{34} - r\tilde{J}_{14} = 0$ (we put '+' when we go in the direction of the current and '−' otherwise). Hence,

$$\tilde{J}_{12} + \tilde{J}_{23} + \tilde{J}_{34} - \tilde{J}_{14} = 0. \tag{17}$$

Again, a similar equation is valid for each face of the green cube and for each face of the red cube, giving 12 new equations for the currents. Combining all the 20 equations (such as (16) and (17)) together, one can convert them to an equivalent system of fourteen equations in twelve variables $x_{ik} := J_{ik} + \tilde{J}_{ik}$. By direct verification, the rank of this system is equal to twelve, meaning that the system admits only the trivial solution $x_{ik} = 0$ for all the indices i, k. Thus,

$$\tilde{J}_{ik} = -J_{ik}. \tag{18}$$

The next step is to consider the rectangular contour formed by the green wire and the red wire connecting the oscillators No. 1 and No. 2 and by the vertical capacitance links of these two oscillators. Summing the voltages across the four sides of this contour, one obtains (see KVL)

$$u_1 - r\tilde{J}_{12} - u_2 + rJ_{12} = 0.$$

A similar relation holds for each edge of the cube, giving us 12 equations

$$\tilde{J}_{ik} - J_{ik} = \frac{u_i - u_k}{r}.$$

Combining these equations with relations (18) yields

$$-\tilde{J}_{ik} = J_{ik} = \frac{u_k - u_i}{2r}. \tag{19}$$

Finally, consider the circuit of the oscillator No. 1 (with the initial state η_0^1). It is described by the voltage equation and the current equation. The voltage equation is the same as for the decoupled oscillator (it says that the sum of the voltages across the LCR contour is zero):

$$L\frac{d}{dt}(j_1 + \Theta[\eta_0^1]j_1) + Rj_1 = u_1,$$

where j_1 is the current through the resistor R and the inductance and $\Theta[\eta_0^1]$ is the Preisach input-output operator defined by (14)–(15). The second equation says that the total current through the capacitance is equal to $-Cu_1'$:

$$Cu_1' = -j_1 + \sigma_1 u_1 - \sigma_2 u_1^3 + J_{15} + J_{14} + J_{12},$$

which, due to (19), is equivalent to

$$Cu_1' = -j_1 + \sigma_1 u_1 - \sigma_2 u_1^3 + \frac{1}{2r}(u_2 + u_4 + u_5 - 3u_1).$$

Writing down a similar pair of voltage-current equations for each oscillator and using the vector notation $j = (j_1, \ldots, j_8)$, $u = (u_1, \ldots, u_8)$, $u^3 = (u_1^3, \ldots, u_8^3)$ and the vector hysteresis operator $\Theta[\eta_0]j = (\Theta[\eta_0^1]j_1, \ldots, \Theta[\eta_0^8]j_8)$ with $\eta_0 = (\eta_0^1, \ldots, \eta_0^8)$, we arrive at the system

$$\begin{array}{rcl} \frac{d}{dt}(j + \Theta[\eta_0]j) & = & -\frac{R}{L}j + \frac{1}{L}u \\ u' & = & -\frac{1}{C}j + \frac{\sigma_1}{C}u - \frac{\sigma_2}{C}u^3 + \frac{1}{2rC}\mathcal{B}u \end{array} \tag{20}$$

with the interaction matrix

$$\mathcal{B} = \begin{pmatrix} -3 & 1 & 0 & 1 & 1 & 0 & 0 & 0 \\ 1 & -3 & 1 & 0 & 0 & 1 & 0 & 0 \\ 0 & 1 & -3 & 1 & 0 & 0 & 1 & 0 \\ 1 & 0 & 1 & -3 & 0 & 0 & 0 & 1 \\ 1 & 0 & 0 & 0 & -3 & 1 & 0 & 1 \\ 0 & 1 & 0 & 0 & 1 & -3 & 1 & 0 \\ 0 & 0 & 1 & 0 & 0 & 1 & -3 & 1 \\ 0 & 0 & 0 & 1 & 1 & 0 & 1 & -3 \end{pmatrix}.$$

4.2. Result

For all values of σ_1, system (20) has a trivial stationary solution with $j = u = 0$. Take $\alpha = \sigma_1$ as a bifurcation parameter. Before formulating the main result related to the S_4-equivariant Hopf bifurcation in (20) (see Theorem 1), some remarks are in order.

First, the periodicity of a solution to system (20) means that both the vector-valued functions $j = j(t)$, $u = u(t)$ are periodic and the evolution of the state $S_k(t)$ ($k = 1, ..., 8$) of each Preisach model involved is periodic with the same period (cf. figure 6). The periodicity of $S_k(t)$ implies periodicity of the output $\theta_k(t) = (\Theta[\eta_0^k]j_k)(t)$ of each Preisach model. The infinite dimensional parameters η_0^k describing the initial states $S_k(t_0)$ of the Preisach operators are a part of the solution as the choice of η_0^k should ensure the periodicity of $S(t) = (S_1(t), \dots, S_8(t))$. Observe also that system (20) is autonomous—every shift of time takes solutions to solutions. Hence, it is appropriate to talk about a cycle representing the time shifts of a periodic solution.

Second, every branch in question contains periodic solutions with an arbitrarily small norm $\sqrt{j_1^2 + \cdots + j_8^2 + u_1^2 + \cdots + u_8^2}$ of the u, v-components. For the rigorous definition of the branch of periodic solutions we refer to [8, 48, 49].

Next, the S_4-symmetry group of the graph presented in figure 1 determines in a standard way the S_4-representation on $\mathfrak{W} = \mathbb{R}^8$ and system (20) is equivariant with respect to $\mathfrak{W} \oplus \mathfrak{W}$. In order to characterize *symmetry* properties of bifurcating periodic solutions, we adopt the standard terminology from [6]. More precisely, given a periodic solution (j, u, S), the pair $(\gamma, e^{i\tau}) \in S_4 \times S^1$ is called a *spatio-temporal* symmetry of this solution if $\gamma j(t + \tau) = j(t)$, $\gamma u(t + \tau) = u(t)$, $\gamma S(t + \tau) = S(t)$ for all t (cf. [6], formula (3.2)). If the solution is *non-stationary*, then the set of all its spatio-temporal symmetries constitutes a *twisted* (one-folded) subgroup of $S_4 \times S^1$ (cf [6], formula (3.4)). Since S_4 is finite, the solution is the so-called *discrete rotating wave* (cf. [6], Definition 3.2 and Theorem 3.4). As usual (see, for example, [6], p. 93), $\mathfrak{W} \oplus \mathfrak{W}$ can be identified with the comlexification $\mathfrak{W}^c = \mathfrak{W} \otimes \mathbb{C}$, so the twisted subgroups can be viewed from the $S_4 \times S^1$-representation $(\mathfrak{W}^c, S_4 \times S^1)$ given by

$$(\gamma, e^{i\tau})(j \otimes z) := (\gamma(j) \otimes (e^{i\tau}z)) \ (j \in \mathfrak{W},\ z \in \mathbb{C},\ (\gamma, e^{i\tau}) \in S_4 \times S^1). \tag{21}$$

Clearly, any subgroup conjugate to a twisted subgroup is twisted as well (meaning that one can speak about "twisted orbit types"). Denote by $\Phi^t(\mathfrak{W}^c, S_4 \times S^1)$ the set of all twisted orbit types occurring in $(\mathfrak{W}^c, S_4 \times S^1)$. The set $\Phi^t(\mathfrak{W}^c, S_4 \times S^1)$ admits a partial order induced by the standard partial order on the set of all conjugacy classes of subgroups in $S_4 \times S^1$ (see, for example, [8], formula (2.1)).

If the Hopf bifurcation takes place for $\alpha = \overline{\alpha}$, then, following the scheme suggested in [8], Sections 9.2 and 9.3, one can associate to system (20) the equivariant bifurcation invariant $\omega_{S_4 \times S^1}(\overline{\alpha})$. This invariant is expressed in terms of the twisted equivariant degree (see [8], Section 5.2) having a direct link to twisted orbit types of branches of discrete rotating waves bifurcating from $\overline{\alpha}$ (in fact, $\omega_{S_4 \times S^1}(\overline{\alpha})$ takes its values in the free abelian $\mathbb{Z}$-module generated by conjugacy classes of twisted subgroups). We refer to Remark 3 where it is explained how to extract the information provided by $\omega_{S_4 \times S^1}(\overline{\alpha})$.

Finally, the following physical parameter

$$c = \frac{r}{R}\left(1 - \frac{R^2C}{L}\right), \tag{22}$$

plays an important role in our considerations. For simplicity, to prevent a steady-state/Hopf mode interaction, we assume that

$$c \neq 1, 2, 3. \tag{23}$$

Theorem 1 *Under the Assumption A, suppose that conditions (8) and (23) are satisfied. Then:*

(i) **(Occurrence)** *System (20) with the bifurcation parameter* $\alpha := \sigma_1$ *has four bifurcation points given by*

$$\alpha_i = \frac{RC}{L} + \frac{i}{r} \qquad (i = 0, 1, 2, 3);$$

(ii) **(Multiplicity and asymptotics)** *There are at least 1 branch of discrete rotating waves bifurcating from the zero equilibrium at each of the bifurcation points* $\alpha = \alpha_0, \alpha_3$, *and at least 27 branches of discrete rotating waves bifurcating from the zero equilibrium at each of the bifurcation points* $\alpha = \alpha_1, \alpha_2$*; the periods of all the bifurcating cycles are asymptotically close to* $2\pi\sqrt{LC}/\sqrt{1 - \frac{R^2C}{L}}$ *as* $\alpha \to \alpha_i$*;*

(iii) **(Equivaraint topology/physics interaction)** *Given values of the parameter c (see (22)), the equivariant bifurcation invariants* $\omega_{S_4 \times S^1}(\alpha_i)$, $i = 0, 1, 2, 3$ *(see [8], formulae (9.23), (9.30), (9.31) and Definition 9.30), are completely determined in Tables 1–4 (see also Remark 3);*

(iv) **(Symmetry properties)** *For each maximal orbit type* $(H) \in \Phi^t(\mathfrak{W}^c, S_4 \times S^1)$*, there exists a branch of discrete rotating waves with the orbit type* (H) *bifurcating from the zero equilibrium at some* α_i *(i=0,1,2,3) (in particular, for* $\alpha = \alpha_1, \alpha_2$*, symmetry groups of the 27 branches provided by (ii) are organized in five orbit types related to the maximal twisted types occurring in the corresponding "central spaces").*

4.3. Comments

Remark 3 How to use the invariant $\omega_{S_4 \times S^1}(\alpha_i)$**?** Let us explain how to "decode" the information provided by the invariants $\omega_{S_4 \times S^1}(\alpha_i)$ presented in Tables 1–4.

(a) Under the assumptions of Theorem 1, $\omega_{S_4 \times S^1}(\alpha_i) \neq 0$ for each α_i $(i = 0, 1, 2, 3)$. Therefore (cf. [8], Theorem 9.28) the Hopf bifurcation takes place for each α_i.

(b) To classify symmetry properties and minimal number of the bifurcating branches, observe that the maximal orbit types in $\Phi^t(\mathfrak{W}^c, S_4 \times S^1)$ are: (S_4), (S_4^-), $(\mathbb{Z}_4^c)$, (D_4^d), (D_2^d), (D_3), $(\mathbb{Z}_3^t)$, $(D_3^{\mathbb{Z}})$, $(D_4^{\mathbb{Z}})$ (see [8], Subsection 5.3.4, for the explicit description of the corresponding subgroups). Take, for example, $\omega_{S_4 \times S^1}(\alpha_1) = (D_4^d) + (D_3) + (D_2^d) + (Z_4^c) + (Z_3^t) - (D_1) - (Z_2^-)$ from Table 1. It follows from the proof of Theorem 9.28 in [8] and the definition of a twisted orbit type that there exist 3 branches (of discrete rotating waves) of type (D_4^d), 4 branches of type (D_3), 6 branches of type (D_2^d), 6 branches of type (Z_4^c), and 8 branches of type (Z_3^t) bifurcating from $(\alpha_1, 0)$ (altogether, $3 + 4 + 6 + 6 + 8 = 27$ branches).

(c) It is easy to see that statements (i), (ii) and (iv) of Theorem 1 are consequences of statement (iii), meaning that the invariant $\omega_{S_4 \times S^1}(\alpha_i)$ contains the essential information on bifurcating periodic solutions. Observe also that, in general, the invariant of type $\omega_{S_4 \times S^1}(\alpha_i)$ can be used to study twisted rotating waves with *non-maximal* symmetries (cf. [8], Remarks 10.15-16), however, this goes beyond the scope of the present paper.

Remark 4 (a) Recall that system (20) has a (non-smooth) hysteresis input-state-output operator leading, in particular, to an infinite dimensional phase space without a local linear structure. If the operator Θ in (20) is zero (i.e., one has a smooth non-hysteretic finite-dimensional system of ODEs), then conclusions (i) and (ii) of Theorem 1 as well as the existence of a branch of periodic solutions with symmetry H for any $\mathbb{C}$-axial isotropy H in the corresponding central subspace (that is parallel to conclusion (iv) of Theorem 1) follow from the Equivariant Hopf Theorem in [6, 7] (see also [22]).

(b) Conclusion (iii) of Theorem 1, on the one hand, shows that the twisted equivariant degree based method used in this paper is not sensitive to any smoothness/genericity requirements, and, on the other hand, connects directly the physics (value of c) with the equivariant topology (values of $\omega_{S_4 \times S^1}(\alpha_i)$) underlying the problem in question.

Table 1. Branches of periodic solutions at Hopf bifurcation points for $0 < c < 1$.

Bif. point	$\omega_{S_4\times S^1}(\alpha_i)$	# Branches
α_0	$-(S_4)$	1
α_1	$(D_4^d)+(D_3)+(D_2^d)+(Z_4^c)+(Z_3^t)-(D_1)-(Z_2^-)$	27
α_2	$(D_4^z)-(D_3^z)-(D_2^d)-(D_2^z)+(Z_4^c)-(Z_3^t)+3(D_1^z)$ $+(D_1)-(Z_2^-)-(Z_1)$	27
α_3	$(S_4^-)-2(D_3^z)-(D_2^z)-(Z_4^-)+(Z_3)+2(D_1^z)+(Z_2)$ $-(Z_1)$	1

Table 2. Branches of periodic solutions at Hopf bifurcation points for $1 < c < 2$.

Bif. point	$\omega_{S_4\times S^1}(\alpha_i)$	# Branches
α_0	$-(S_4)$	1
α_1	$-(D_4^d)-(D_3)-(D_2^d)-(Z_4^c)-(Z_3^t)+(D_1)+(Z_2^-)$	27
α_2	$(D_4^z)+(D_3^z)+(D_2^d)+(Z_4^c)+(Z_3^t)-(D_1^z)-(Z_2^-)$	27
α_3	$(S_4^-)-2(D_3^z)-(D_2^z)+3(D_1^z)-(Z_1)$	1

Table 3. Branches of periodic solutions at Hopf bifurcation points for $2 < c < 3$.

Bif. point	$\omega_{S^4\times S^1}(\alpha_i)$	# Branches
α_0	$-(S_4)$	1
α_1	$-(D_4^d)-(D_3)-(D_2^d)-(Z_4^c)-(Z_3^t)+(D_1)+(Z_2^-)$	27
α_2	$-(D_4^z)-(D_3^z)-(D_2^d)-(Z_4^c)-(Z_3^t)+(D_1^z)+(Z_2^-)$	27
α_3	(S_4^-)	1

Table 4. Branches of periodic solutions at Hopf bifurcation points for $c > 3$.

Bif. point	$\omega_{S^4\times S^1}(\alpha_i)$	# Branches
α_0	$-(S_4)$	1
α_1	$-(D_4^d)-(D_3)-(D_2^d)-(Z_4^c)-(Z_3^t)+(D_1)+(Z_2^-)$	27
α_2	$-(D_4^z)-(D_3^z)-(D_2^d)-(Z_4^c)-(Z_3^t)+(D_1^z)+(Z_2^-)$	27
α_3	$-(S_4^-)$	1

4.4. Sketch of the proof of Theorem 1

Step 1: Reduction to (j,u)-components. As a matter of fact, every periodic solution $(j(t),u(t),S(t))$ to system (20) belongs to the class of periodic solutions $\mathfrak{L} = (j(t),u(t),S_{\eta_o}(t))$ with the *same* (j,u)-components. From the physical viewpoint, variables (j,u) are observable, while the internal dynamics of the Preisach memory of the system (represented by the variable $S_{\eta_o}(t)$) is not. Using the *monocyclic* property of the hysteresis input-output operator Θ (see, for example, [14]) and choosing properly a representative in each class $\mathfrak{L}$, one can replace system (20) by a differential-operator system

$$\tfrac{d}{dt}(j+\hat{\Theta}^T j) = -\tfrac{R}{L}j + \tfrac{1}{L}u, \qquad u' = -\tfrac{1}{C}j + \tfrac{\sigma_1}{C}u - \tfrac{\sigma_2}{C}u^3 + \tfrac{1}{2rC}\mathcal{B}u \tag{24}$$

satisfying the following conditions: (i) (24) is a system in (j,u)-components only; (ii) any periodic solution to (24) canonically defines a class $\mathfrak{L}$ of periodic solutions to (20); (iii) (24) contains an operator $\hat{\Theta}^T$ (induced by Θ) which is globally Lipschitz continuous in the space $\mathfrak{C}$ of T-periodic continuous $\mathbb{R}$-valued functions, and satisfies the global estimate $\|\hat{\Theta}^T x\|_\infty \le K\|x\|_\infty^2$.

Step 2: Inversion of Preisach operator. Following the standard lines (cf. [15, 17]), in order to convert system (24) to the fixed-point operator equation in an appropriate functional space, one studies invertibility properties of the map $\mathrm{Id} + \hat{\Theta}^T$. Using (iii) from Step 1, one can show that the map $\mathrm{Id} + \hat{\Theta}^T$ takes $\mathfrak{C}$ onto itself, is invertible, and its inverse is globally Lipschitz continuous.

Step 3: Period normalization. As in the case of autonomous ODEs without hysteresis, one may consider the unknown period T as an additional parameter, and reduce studying T-periodic solutions of system (24) to looking for 2π-periodic solutions of an equivalent system, say (D). This transformation uses the rate-independence property of the input-output and input-state operators of the Preisach model (see, for example, [14]).

Step 4: Operator reformulation. Let $\mathbb{E}$ be the space of C^1-smooth 2π-periodic $\mathbb{R}^{16}$-valued functions equipped with the standard C^1-norm and $G := S_4 \times S^1$-action given by $((\gamma,\tau)x)(t) := \gamma x(t+\tau)$ ($\gamma \in S_4$, $e^{i\tau} \in S^1$, $t \in \mathbb{R}$ and $x \in \mathbb{E}$). Combining Steps 1 – 3 with the standard Sobolev compact embedding theorems one can replace system (D) with an equivalent operator equation

$$\mathfrak{F}(\alpha, T, x) = 0 \qquad (\alpha \in \mathbb{R},\ \ T \in (0,\infty),\ \ x \in \mathbb{E}), \tag{25}$$

where $\mathfrak{F}(\alpha, T, x) := x - \mathcal{F}(\alpha, T, x)$ and $\mathcal{F} : \mathbb{R} \times (0,\infty) \times \mathbb{E} \to \mathbb{E}$ is a completely continuous G-equivariant operator (cf. [8], pp. 263–264); here G acts trivially on $\mathbb{R} \times (0,\infty)$.

Step 5: Linearization. Let $L_\alpha : \mathbb{R}^{16} \to \mathbb{R}^{16}$ be the linearization of the right-hand side of system (20) with $\alpha := \sigma_1$. Then, $\frac{T}{2\pi} L_\alpha$ canonically determines the linearization of $\mathfrak{F}$ at $(\alpha, T, 0)$ (cf. (25)). Recall that $\mathfrak{F}$ has non-differentiability points in any neighborhood of $(\alpha, T, 0)$.

Step 6: Auxiliary function and invariant $\omega_{S_4 \times S^1}(\alpha)$. Take $\alpha_0 \in \mathbb{R}$ and $T_0 \in (0,\infty)$ such that $i\frac{T_0}{2\pi}$ is the purely imaginary eigenvalue of L_{α_0}. Combining Step 5 with the Implicit Function Theorem, one can construct a G-invariant neighborhood $\Omega \subset \mathbb{R} \times (0,\infty) \times \mathbb{E}$ of $(\alpha, T, 0)$ and an invariant continuous function $\eta : \overline{\Omega} \to \mathbb{E}$ (the so-called "auxiliary function", cf. [8], Definition 9.17) satisfying the following conditions: (a) any solution to the equation

$$\mathfrak{F}^\eta(\alpha, T, x) = 0 \qquad (\mathfrak{F}^\eta = (\eta, \mathfrak{F})) \tag{26}$$

is a solution to (25), and (b) $\mathfrak{F}^\eta$ does not have zeroes on $\partial\Omega$. In particular, the twisted G-equivariant degree of $\mathfrak{F}^\eta$ on Ω is correctly defined. Using this degree as the bifurcation invariant $\omega_{S_4 \times S^1}(\alpha)$ (cf. [8], Theorem 9.28) one can prove Theorem 1 (cf. Remark 3).

Acknowledgments

This publication has emanated from research conducted with the financial support of Federal Programme 'Scientists of Innovative Russia', grant 2009-1.5-507-007, and Russian Foundation for basic Research, grant 10-01-93112.

References

[1] Iudovick V I 1971 *Prikl. Mat. Mek.* **35** 638–655
[2] Turing A 1952 *Phil. Trans. Roy. Soc. B.* **237** 37–72
[3] Yoshida K 1982 *Hiroshima Math. J.* **12** 321–348
[4] Wu J 1998 *Trans. Amer. Math. Soc.* **350** 4799–4838
[5] Fiedler B 1988 *Global Bifurcation of Periodic Solutions with Symmetry* (New York: Springer)
[6] Golubitsky M and Stewart I N 2002 *The Symmetry Perspective* (Basel-Boston-Berlin: Birkhäuser)
[7] Golubitsky M, Schaeffer D G and Stewart I N 1988 *Singularities and Groups in Bifurcation Theory* vol 2 (New York: Springer)
[8] Balanov Z, Krawcewicz W and Steinlein H 2006 *Applied Equivariant Degree* (Springfield: AIMS)
[9] Field M J 2007 *Dynamics and Symmetry* (London: Imperial College Press)
[10] Rezaei-Zare A, Sanaye-Pasand M, Mohseni H, Farhangi Sh and Iravani R 2007 *IEEE Trans. Power Deliv.* **22** 919–929

[11] Colombo A, di Bernardo M, Hogan S J and Kowalczyk P 2007 *J. Nonlinear Sci.* **17** 85–108
[12] Logemann H, Ryan E P and Shvartsman I 2008 *Nonlinear Anal. Theory Methods & Appl.* **69** 363–391
[13] Krasnosel'skii M and Pokrovskii A 1989 *Systems with Hysteresis* (New York: Springer)
[14] Mayergoyz I D 1991 *Mathematical Models of Hysteresis* (Springer)
[15] Brokate M and Sprekels J 1996 *Hysteresis and Phase Transitions* (New York: Springer)
[16] Visintin A 1994 *Differential Models of Hysteresis* (Berlin: Springer)
[17] Krejči P 1996 *Hysteresis, Convexity and Dissipation in Hyperbolic Equations* (Tokyo: Gakkotosho)
[18] 2006 *The Science of Hysteresis*, three volume set, ed I Mayergoyz and G Bertotti (Elsevier, Academic Press)
[19] 2009 *Hysteresis*, ed R Iyer and X Tan (*IEEE Control Systems Magazine* **1**)
[20] Field M J and Swift J W 1994 *Nonlinearity* **7** 385–402
[21] Guo S and Lamb J S W 2008 *Proc. of the American Mathematical Society* **136** 2031-2041
[22] Ashwin P and Podvigina O 2003 *Proc. R. Soc. A* **459** 1801–1827
[23] Dias A P S and Rodrigues A 2009 *Nonlinearity* **27** 627–666
[24] Brokate M, Pokrovskii A, Rachinskii D and Rasskazov O 2006 *The Science of Hysteresis* vol 1, ed G Bertotti and I Mayergoyz (Academic Press) chapter 2 pp 127-291
[25] Brokate M, Collings I, Pokrovskii A and Stagnitti F 2000 *Z. Anal. Anw.* **19** 469–487
[26] Krejci P 2000 *Appl. Math.* **45** 439-468
[27] Krejci P, Sprekels J and Zheng S 2001 *J. Differential Equations* **175** 88–107
[28] Appelbe B, Flynn D, McNamara H, O'Kane P, Pimenov A, Pokrovskii A, Rachinskii D and Zhezherun A 2009 *IEEE Control Systems Magazine* **29** 44–69
[29] Krejci P, O'Kane P, Pokrovskii A and Rachinskii D 2010 *J. Physics: Conf. Series* to appear
[30] Diamond P, Kuznetsov N A and Rachinskii D I 2001 *J. Differential Equations* **175** 1–26
[31] Pokrovskii A, Power T, Rachinskii D and Zhezherun A 2006 *J. of Physics: Conf. Series* **55** 171–190
[32] Pimenov A and Rachinskii D 2009 *Discrete and Continuous Dynamical Systems B* **11** 997–1018
[33] Appelbe B, Rachinskii D and Zhezherun A 2008 *Physica B* **403** 301–304
[34] Kuznetsov N A, Rachinskii D and Zhezherun A 2008 *Doklady Math.* **78** 705–709
[35] Ize J and Vignoli A 2003 *Equivariant Degree Theory* (W. de Gruyter)
[36] Balanov Z and Krawcewicz W 2008 *Handbook of Differential Equations, Ordinary Differential Equations* vol. 4, ed F Battelli and M Feckan (Elsevier) pp 1–131
[37] Balanov Z, Krawcewicz W, Rybicki S and Steinlein H *J. of Fixed Point Theory Appl.* to appear
[38] Dancer E N 1985 *Ann. Inst. H. Poincaré Anal. Non Lineaire* **2** 1–18
[39] Dancer E N and Toland J F 1993 *Proc. London Math. Soc.* **66** 539–567
[40] Fuller F B 1967 *American Journal of Mathematics* **89** 133–148
[41] Balanov Z, Krawcewicz W and Ruan H 2006 *Discrete Contin. Dyn. Syst. Ser. A* **15** 983–1016
[42] Bredon G E 1972 *Introduction to Compact Transformation Groups* (New York-London: Academic Press)
[43] tom Dieck T 1987 *Transformation Groups* (Berlin: W. de Gruyter)
[44] Kirillov A A 1976 *Elements of the Theory of Representations* (Berlin-Heidelberg-New York: Springer)
[45] Kuznetsov Yu A 1995 *Elements of Applied Bifurcation Theory* (Berlin-New York: Springer)
[46] Krejci P 1989 *Appl. Math.* **34** 364-374
[47] Eleuteri M, Kopfova J and Krejči P 2008 *Physica B* **403** 448-450
[48] Krasnosel'skii M 1964 *Positive Solutions of Operator Equations* (Groningen: P. Noordhoff Ltd)
[49] Krasnosel'skii A M and Rachinskii D I 2002 *Doklady Math.* **65** 344–349

On transformation to the singularly perturbed system

V Bykov

Institute of Technical Thermodynamics
Karlsruhe Institute of Technology (KIT)
Engelbert-Arndold-Strasse 4, Geb. 10.91
D-76131 Karlsruhe, Germany

E-mail: `viatcheslav.bykov@kit.edu`

Abstract. A rapid progress in hard- and software development of computational facilities as well as in numerical methods has increased the role of numerical simulations in the quantitative system analysis of many engineering problems. At the same time, the system complexity (in terms of dimensionality and non-linearity) has grown considerably increasing demand for automatic methods of analysis of qualitative system behavior. For instance, nowadays, definition of key system parameters controlling the system dynamics and finding critical regimes automatically have become crucial issue of numerical system analysis. In the present paper a transformation to the Singularly Perturbed System (SPS) as a main theoretical framework to cope with the complexity and high dimensionality will be discussed in detail. Both simple but famous and meaningful model example of Van der Pol oscillator and an example of application to numerical analysis of chemical kinetics mechanisms will be used to show the potential of the suggested framework.

1. Introduction

Power of modern computational facilities increases rapidly over last decades, more and more detailed physical and mathematical models of various phenomena become treatable by numerical system analysis via numerical simulations. Numerical experiments are considered as a powerful tool of modern research taking a leading role in science, engineering and industry. In the last decade, for instance, if one has a look at chemical kinetics (see e.g. [1] in combustion, [2, 3] - atmospheric chemistry and [4, 5] in biochemistry etc.) one realizes that the dimension of the considered system of governing equations increases almost uncontrollable.

However, the increase of the system dimension and complexity might lead to loss of reliability of the results. Understanding and interpretation of results of simulations of such systems becomes problematic. It is not easy to elucidate key system parameters and their combinations governing distinguished dynamical regimes.

In many cases, however, it is quite often one does not need such complicated description of the phenomenon under consideration, most likely the detailed model has been developed to cover and to 'fit' all possible ranges of observations and experimental data whose amount constantly grows up with time. But, in particular simulations, only a small part a certain 'sub-model' can be efficiently used to reach required accuracy and level of prediction.

Thus, one has the detailed model, from one hand side, which reflects the current situation of the detailed model development for the whole range of observations, but from the other hand in numerical treatments/simulations only a 'small part' of this model is actually involved and important, while the rest is not relevant and redundant. In order to formalize this observation and develop a tool of automatic model reduction, i.e. the tool of automatic identification of such sub-model, the coordinate free singular perturbation approach has been developed [6, 7] last few years.

The basic idea behind the approach is in using low dimensional invariant (integral) manifolds embedded into the detailed system phase space, which reflect appropriate asymptotic limits for particular application of the detailed model for numerical simulation and analysis of system dynamics. In other words, the phase space of the detailed system is assumed to be split asymptotically into family of fast/slow invariant manifolds representing relatively slow and fast system motions.

This means the decomposition of motions becomes the main assumption of our approach. It is justified by the fact that in particular situations not all phase or state space of the original detailed system are accessed by the system solution trajectories. In a wide range of system's initial, boundary conditions as well as for a wide range of system parameters the system solution is often restricted to a subset, which is in many applications can be approximated by low-dimensional invariant manifolds (see e.g. [8, 9] for more details, where the decomposition is handled implicitly). Following this assumption one concludes that if there is a simplified (in terms of dimension) or reduced model satisfactory describing the detailed one, then it should correspond to the invariant manifold of low dimension. While the overall system dynamics can be restricted to this manifold.

Accordingly, the problem of model reduction or the problem of adequate modeling reduces to identification such manifolds in the system's phase space and study these properties. One is interested in an approach which for a given detailed model (system of governing equations, range of initial, boundary and system parameters are supposed to be known) performs

- decomposition of motions;
- singular versus regular perturbation analysis;
- definition appropriate manifolds for reduced model formulation;
- restriction or projection the overall system to these manifolds.

In the present paper, the transformation to the standard form of singularly perturbed system (SPS) by using singularly perturbed vector fields approach is represented and discussed as an ideal candidate for a theoretical framework that can be very efficient for model reduction. It is illustrated on both analytical and numerical levels by well known model of Van der Pol oscillator and by application to an auto-ignition problem of combustion theory.

2. SPVF: analytical approach, Van der Pol oscillator

The main and the key idea of the suggested approach is to find a transformation to the standard SPS system form, which allows straightforward application of powerful and well developed methods of perturbation analysis. Accordingly, a number of interesting questions arise, which are in the focus of the current study. Namely,

(i) Suppose one has such a system, which is overdetermined in the sense that more parameters and equations are used than needed for adequate description of particular phenomenon under investigation. How to check this? If the answer on the previous question is positive, then

(ii) How to estimate appropriate dimensions?

(iii) How to transform the system to the explicit decomposed form as the SPS system?

2.1. Van Der Pol oscillator, original model

In order to illustrate very important technical part of the suggested method and to address motivation of the paper, let us consider the well known system of Van der Pol oscillator in its original form given by

$$\ddot{x} - \mu\left(1 - x^2\right)\dot{x} + x = 0, \tag{1}$$

written in the phase coordinates (x, y), $y = \dot{x}$ as shown in figure 1 (left), equation (1) becomes equivalent to

$$\begin{array}{c} \frac{dx}{dt} = y \\ \frac{dy}{dt} = -x + \mu\left(1 - x^2\right)y \end{array} \tag{2}$$

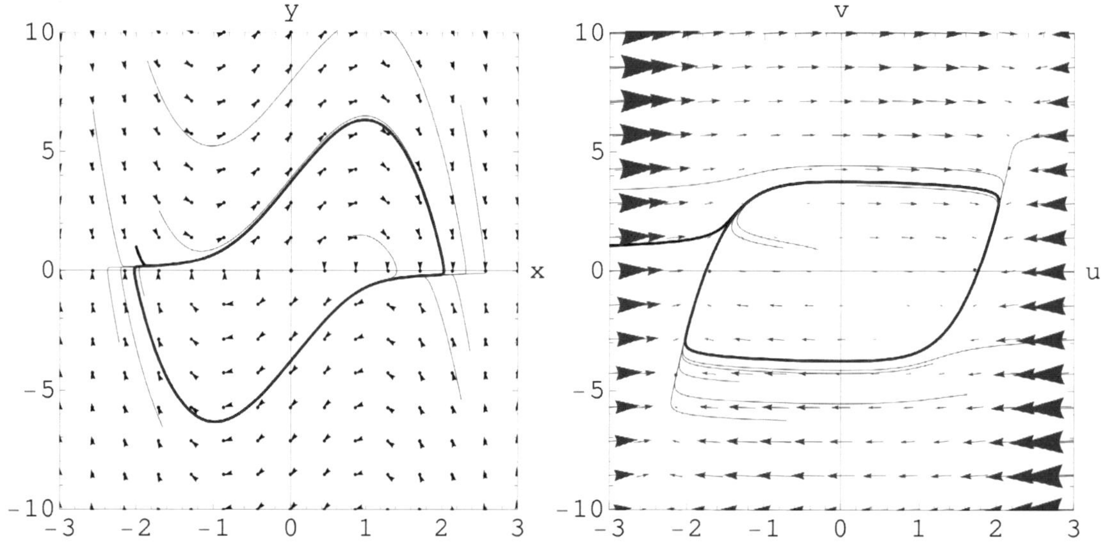

Figure 1. Phase space of the Van der Pol oscillator in the original coordinates (2) (left) and in the Lienard's coordinates (5) (right) with particular solutions trajectories and vector field.

This is an example of well known Lienard's system [10]

$$\ddot{x} + f(x)\dot{x} + g(x) = 0. \tag{3}$$

It can be considered as a well known example of one of the first successful transformation to the SPS standard form, i.e. in his study of (4) Lienard has suggested to transform the system coordinates in (4) according to

$$\begin{array}{c} u = x \\ v = \int f(x)\,dx + \dot{x} \end{array}, \tag{4}$$

which to be applied to (4) allowed to obtain well known simpler model of non-linear oscillations in the phase space coordinates, which is typically referenced as the Van der Pol oscillator (see figure 1 right)

$$\begin{array}{c} \frac{du}{dt} = v + \mu\left(1 - \frac{1}{3}u^2\right)u \\ \frac{dv}{dt} = -u \end{array}. \tag{5}$$

However, while looking at phase spaces of both systems one immediately realizes the difference between the models (just compare phase portraits shown in figure 1). The second representation in the Lienard's coordinates is more preferable, one can easily see two stages of the oscillatory behavior because the system is represented in the decomposed form, e.g. the first equation represents the fast system motions, while the second governs the slow system evolution. The transformation 4 decomposes of motions of Van der Pol model.

2.2. Van der Pol model with known decomposition

The main question of this subsection is the following: 'Is there a rigorous way to proceed and find transformation (4) and reconcile the meaning of the transformation?'. In order to answer this question and illustrate powerful technical skills of the suggested method let us rescale $\varepsilon\, d\tau = dt$ the time in (3)

$$\frac{d^2x}{d\tau^2} + \varepsilon\, f(x) \frac{dx}{d\tau} + \varepsilon^2 g(x) = 0. \tag{6}$$

Hence, in the phase space the original model of Van der Pol model becomes

$$\begin{array}{r} \frac{dx}{d\tau} = y \\ \frac{dy}{d\tau} = -\varepsilon^2 x + \varepsilon\mu\left(1 - x^2\right) y \end{array}. \tag{7}$$

Now, we apply the suggested SPVF similar to [6]. The proposed framework allows to redefine the global transformation on the basis of knowledge about multiple time scales $\varepsilon\mu \sim O(1)$ explicitly present in the scalled form (6). According to [6], initial step is a decomposition of the vector field

$$F = F_f + F_s, \tag{8}$$

where fast - F_f and slow - F_s sub-fields of the original vector field (6) are easily found due to the explicit small parameter ε as

$$\begin{array}{r} F_f = \begin{pmatrix} F_{f,1} \\ F_{f,2} \end{pmatrix} = \begin{pmatrix} y \\ \varepsilon\mu\left(1 - x^2\right) y \end{pmatrix}, \\ F_s = \begin{pmatrix} F_{s,1} \\ F_{s,2} \end{pmatrix} = \begin{pmatrix} 0 \\ -\varepsilon^2 x \end{pmatrix}. \end{array} \tag{9}$$

In order to represent the system as a SPS system one can proceed with the analysis of these sub-fields separately. This will lead to considerable simplification which is very important in an analytical study because it can reduce the system dimension and allows to work with much simple subsystems. For instance, if full system is considered, it is not a simple task to find the solution in any form, while the fast sub-field (subsystem) with $z = (x, y)$

$$\frac{dz}{d\tau} = F_f(z), \tag{10}$$

can be integrated easily and has the general solution in an implicit form given by

$$v = y - \varepsilon\mu\left(1 - \frac{1}{3}x^2\right)x + C_v = 0, \tag{11}$$

which means the fast motion is realized in the (x, y) plane on cubic parabola - $y - \varepsilon\mu\left(1 - \frac{1}{3}x^2\right)x + C_v = 0$ (see figure 2, left; where fast manifolds $\varepsilon = 1$, $\mu = 4$, $C_v = 0, \pm 2, \pm 4, \pm 6$ are shown by solid lines), this defines a family of fast manifolds - $\{M_{(x,y)}\}$. Moreover, the new variable v defined in this way, governs the slow system dynamics of the original system while

$$\frac{dv}{d\tau} = v_x \frac{dx}{d\tau} + v_y \frac{dy}{d\tau} = v_x \left(F_{f,1} + F_{s,1}\right) + v_y \left(F_{f,2} + F_{s,2}\right) = v_x F_{s,1} + v_y F_{s,2} \sim O(\varepsilon^2).$$

At the same time the slow subsystem

$$\frac{dz}{d\tau} = F_s(z),$$

obeys simple integral

$$u = x + C_u = 0, \tag{12}$$

and it is expected to represent fast motion (just compare to the previous system) according to

$$\frac{du}{d\tau} = u_x \frac{dx}{d\tau} + u_y \frac{dy}{d\tau} = u_x \left(F_{f,1} + F_{s,1}\right) + u_y \left(F_{f,2} + F_{s,2}\right) = u_x F_{f,1} + u_y F_{f,2} \sim O(1).$$

Hence, the standard SPS form of the system is redefined by using (11-12) as

$$\frac{du}{d\tau} = u_x F_{f,1} + u_y F_{f,2} = F_{f,1}\left(u, v\right) = v + \varepsilon\mu \left(1 - \frac{1}{3}u^2\right) u,$$

$$\frac{dv}{d\tau} = v_x F_{s,1} + v_y F_{s,2} = F_{s,2}\left(u, v\right) = -\varepsilon^2 u.$$

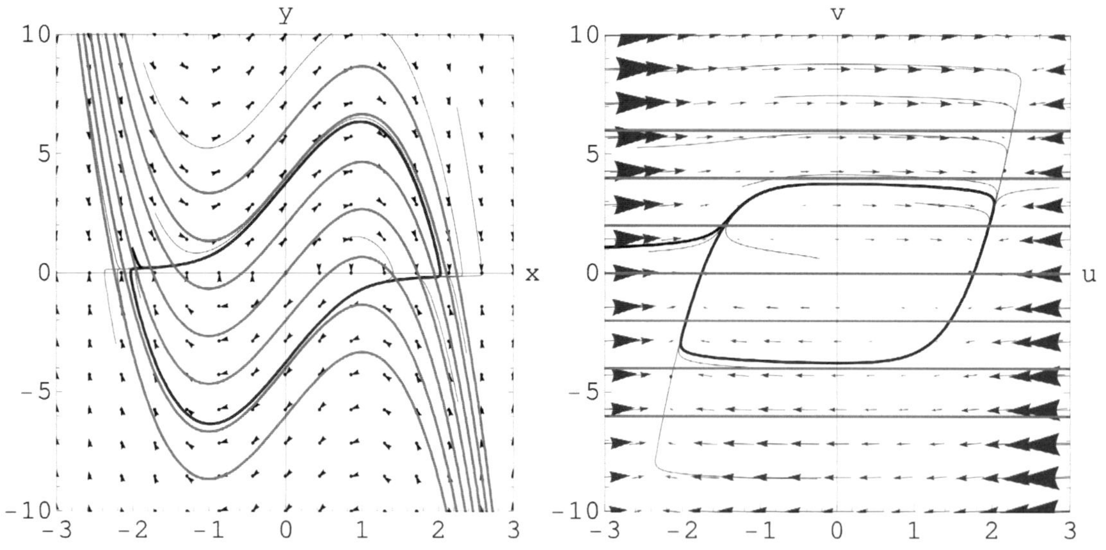

Figure 2. Phase space of the Van der Pol oscillator with fast manifolds (11) (red) as defined by the SPVF analysis in the original coordinates (2) (left) and in the Lienard's coordinates (5) (right) with particular solutions trajectories and vector field.

Here constants - C_u, C_v were chosen to be equal to zero in the coordinate change. Moreover, when the scaling factor is set to be one, the system purely coincides with the Lienard form of the original Van der Pol model. It means that the suggested procedure to check the system hierarchy and decompose the system fully reconciles the Lienard's change of the system variables to simplify the Van der Pol system representation.

3. SPVF: numerical approach, auto-ignition problem

In the previous section it was shown how the SPVF theory can be analytically applied if the decomposition is known explicitly. However, more interesting and complicated question concerns situations where there is no additional internal information available. Although our approach is very general, by now main applications are in reacting flow modelling.

3.1. Invariant manifolds method and connection to the decomposition

Consider a very simple model describing chemically reacting system, which is a pure homogeneous system. It is represented by a system of ordinary differential equations (ODEs) that describes the chemical kinetics mechanism by the system based on the mass action law. The equations of the system describe the evolution of the thermo-chemical state vector

$$\Psi = \left(h,\, p,\, \frac{w_1}{M_1}, \ldots, \frac{w_{n_s}}{M_{n_s}}\right)^T, \tag{13}$$

in time, where Ψ_i represents such quantities as the pressure of the mixture p, the enthalpy h, i.e. two thermodynamic quantities and specific mole numbers w_i/M_i (the mass fractions w_i divided by molar mass M_i) or mole fractions of chemical species that describe the mixture composition. In these vector notations the system of governing equations of a homogeneous system can be written in autonomous form as (see e.g. [11])

$$\frac{d\Psi}{dt} = F(\Psi), \quad \Psi \in \Omega \subset R^n. \tag{14}$$

Here the so-called chemical source term F represents the chemical kinetic mechanism comprises $A_1, \ldots, A_{n_s}$ chemical species ($n = n_s + 2$) participating in $i = 1, \ldots, n_r$ elementary chemical reactions

$$\alpha_{i,1}A_1 + \ldots + \alpha_{i,n_s}A_{n_s} \rightleftharpoons \beta_{i,1}A_1 + \ldots + \beta_{i,n_s}A_{n_s}, \tag{15}$$

The mass action low implies the polynomial form of elementary reaction rates (consider species concentrations only, i.e. $\Psi = ([A_1], \ldots, [A_{n_s}])$), then the elementary reactions rates are

$$R_i(\Psi) = k_i^+ \prod_{j=1}^{n_s} \Psi_j^{\alpha_{i,j}} - k_i^- \prod_{j=1}^{n_s} \Psi_j^{\beta_{i,j}}, \; i = 1, \ldots, n_r, \tag{16}$$

where k_i^+, k_i^- are the parameters (depending on the thermodynamic properties of the system $\Psi = ([A_1], \ldots, [A_{n_s}])$) of forward and inverse reactions rates typically modeled by the Arrhenius law [12]. Therefore, it yields for species evolution of $F(\Psi)$

$$F_j(\Psi) = \sum_{i=1}^{n_r} \gamma_{i,j} R_i(\Psi), \; j = 1, \ldots, n_s, \tag{17}$$

here $\gamma_{i,j} = \beta_{i,j} - \alpha_{i,j}$ components of the so-called stoichiometric vectors. Additionally, invariance of the domain Ω (composed by vectors with non-negative components) is assumed to ensure dissipative nature of chemical reaction kinetic mechanisms.

The main question 'What does it mean to reduce the system (14)?' is discussed in this subsection. One understands it as a reformulation of the system (14) in an appropriate form by introducing the so-called reduced state space $\theta = (\theta_1, \ldots, \theta_m)$, $m \ll n$ consisting of new variables set, such that any 'particular' solution of the system (14) will be 'represented sufficiently' accurate by the reduced model

$$\frac{d\theta}{dt} = \tilde{F}(\theta), \quad \theta \in R^m. \tag{18}$$

Clearly, the question of how the detailed and reduced spaces are connected to each other has to be in the focus of any model reduction strategy. In principle, one only needs to determine how the manifold is embedded into the phase space! This, in most cases not a trivial relation, should be the subject of any reduction methodology. If the reduced space can be accurately represented in an explicit form by

$$M = \{\Psi = \Psi(\theta) : \theta \in R^m\}, \tag{19}$$

a low-dimensional invariant manifold M embedded in the detailed system (14) state space, then model reduction means rewriting the system (14) on M. Substitution of the relation (19) to (14) yields

$$\frac{d\Psi(\theta)}{dt} = F(\Psi(\theta)),\ \Psi_\theta \frac{d\theta}{dt} = F(\Psi(\theta)) \rightarrow \frac{d\theta}{dt} = \tilde{F}(\theta) = \Psi_\theta^+ F(\Psi(\theta)). \tag{20}$$

here Ψ_θ^+ is the Moore-Penrose pseudo-inverse [13], which is a well defined function unless the local tangent space of M given by Ψ_θ degenerates. This simple observation makes clear the importance of definition of an appropriate manifold (19) in the state space and of studies of its properties. Obviously, not any manifold is appropriate for model reduction, only those which possess properties of

(i) invariance or approximate invariance, e.g. to be exact representation of a solution of (14) or to be an approximation to the exact solution;
(ii) attractiveness, i.e. small perturbations cannot drive the solution of (14) out of the manifold (19));
(iii) existence in the whole domain of interest, i.e. it has to approximate the detailed system (14) solutions for sufficiently wide range of initial conditions and system parameters.

These properties will automatically guarantee the closeness of the solutions of (14) $\Psi = \Psi(t)$ and of (18) through $M : \Psi = \Psi(\theta(t))$. Therefore, there is a strong need in methods of construction of appropriate invariant manifolds in the system (14) state/phase space and in approaches that can be applied to study their properties.

3.2. GQL

Because the definition of the invariant, attractive and low dimensional manifold, and the way it is embedded into the detailed system state space, is a central point of model reduction the reduction of a given arbitrary system with prescribed dimension and accuracy is almost impossible in general. Fortunately, systems governed by chemical reactions have certain properties that, nevertheless, make the model reduction feasible.

The existence of various scales or multiple scales is the well known feature of chemically reacting flows. Both in experiments and in detailed model simulations it is observed that not the entire possible range of the thermo-chemical state space is typically accessed by the system states (solutions), but only a part of it [14]. The standard mathematical framework to handle multi-scales phenomenon is the standard SPS theory (see e.g. [15, 16] for more detail). However, it is assumed that the system is represented as the SPS in the original coordinates, which restricts considerably the applicability range of such an approach. Therefore, as pointed out above, a coordinate free approach has been developed recently [17, 18]. The main advantage of the approach is in access to the decomposition structure without being restricted to particular coordinate frame.

It is shown above by the Van der Pol model that the approach is an efficient tool of asymptotic analysis when the decomposition is known. For cases of unknown a priory decomposition structure an algorithm of Global-Quasi Linearization (GQL) for identification of the system multi-scale hierarchy with its suitable representation has been emerged as well (see e.g. [18] for detailed description). Below a brief outline is only presented which is needed for further analysis and results presentation.

The central point of the approach is a global linear approximation or interpolation of the Right Hand Side (RHS) function of (14) given by $T\,\Psi \sim F\,(\Psi)$. It is assumed that if there are multiple scales of the reaction processes, then a gap between the eigenvalues of T can be used for decomposition. In this case two groups of eigenvalues have sufficiently different (by orders of magnitude) characteristic values. Their invariant subspaces (obtained by using the Schur decomposition and then by solving subsequently the Sylvester equation [19]) yield:

$$T = (Z_f\; Z_s) \cdot \begin{pmatrix} N_f & 0 \\ 0 & N_s \end{pmatrix} \cdot \begin{pmatrix} \tilde{Z}_f \\ \tilde{Z}_s \end{pmatrix}. \tag{21}$$

$$\tilde{Z} = Z^{-1} = (Z_f\; Z_s)^{-1} = \begin{pmatrix} \tilde{Z}_f \\ \tilde{Z}_s \end{pmatrix}, \tag{22}$$

here Z_s is the (n by m_s)-dimensional matrix belonging to the invariant subspace of the m_s eigenvalues having the smallest real parts (defined by N_s), while Z_f is the (n by m_f)-dimensional matrix related to the m_f eigenvalues (defined by N_f) having the largest real parts respectively, where m_s and m_f denotes the number of eigenvalues according to slow and fast processes. Thus, an invariant manifold describing the slow system dynamics is given as the manifold that annihilates the sub-processes spanned into the fast subspace, i.e. it defines the states where fastest processes are relaxed already

$$M_s^0 = \left\{\Psi : \tilde{Z}_f\, F(\Psi) = 0\right\}. \tag{23}$$

This equation system is an implicit definition of the correlation between the state variables of a reacting flow introduced by fast relaxing thermo-chemical processes. Whilst those split into slow subspace do not change during fast relaxation process, namely

$$M_f^0(\Psi_0) = \left\{\Psi : \tilde{Z}_s\, \Psi = \tilde{Z}_s\, \Psi_0\right\}. \tag{24}$$

Thus, new coordinates suitable for an explicit decomposition are given by

$$\begin{pmatrix} U \\ V \end{pmatrix} = \begin{pmatrix} \tilde{Z}_f \Psi \\ \tilde{Z}_s \Psi \end{pmatrix}, \quad \Psi = (Z_f\; Z_s) \begin{pmatrix} U \\ V \end{pmatrix}, \tag{25}$$

where and correspond to new fast and slow variables respectively. The decomposed form

$$\begin{cases} \frac{dU}{dt} = \tilde{Z}_f F\,(\Psi(U,V)) \\ \frac{dV}{dt} = \tilde{Z}_s F\,(\Psi(U,V)) \end{cases}, \tag{26}$$

and corresponding fast and slow subsystems are

$$\begin{cases} \frac{dU}{dt} = \tilde{Z}_f F\,(\Psi(U,V)) \\ \frac{dV}{dt} = 0 \end{cases}, \quad \begin{cases} 0 = \tilde{Z}_f F\,(\Psi(U,V)) \\ \frac{dV}{dt} = \tilde{Z}_s F\,(\Psi(U,V)) \end{cases}. \tag{27}$$

The small system parameter can be estimated by the gap between the smallest eigenvalue of the large group and the largest eigenvalue of the small group of eigenvalues [18]

$$\varepsilon = \frac{max\,|\lambda\,(N_s)|}{min\,|\lambda\,(N_f)|} \ll 1. \tag{28}$$

The reduced system dimension and the small system parameter are available due to the gap in (21) between limiting eigenvalues (28). In this way identified system's hierarchy, namely, approximations of slow, fast invariant subspaces (21), (25) and system small parameter (21) make the analysis of properties (existence, stability, invariance etc.) of the slow invariant manifold possible. For instance, the following eigenvalue problem of answers whether the points on the slow manifold (23), (24) are stable or not

$$Re\left(\lambda\left(\tilde{Z}_f\, D_\Psi F\left(\Psi^*\right)\, Z_f\right)\right)|_{\Psi^*\in M_s^0} < 0, \tag{29}$$

This condition means that all points of the manifold attractive to the system trajectory flow with respect to the fast manifolds. It is clear that without proper estimation of the invariant subspace of the fast motions this feature of the manifold can not be investigated. Additionally, the knowledge of the explicit decomposition (25) can be used not only to determine the slow, fast manifolds (23), (24), but it provides with possibility to improve the approximation of the slow manifold (23), for instance, an approximation of the first order of magnitude by (28) is given by

$$M_s^1 = \left\{\Psi : \tilde{Z}_f\, D_\Psi F\left(\Psi\right)\, F\left(\Psi\right) = 0\right\}, \tag{30}$$

where $D_\Psi F\left(\Psi\right)$ is the Jacobi matrix of (20). In general, the idea of the constant transformation is not new see e.g. [20], but principle point is in the developed algorithm that estimates this transformation, which produces the best decomposition with respect to existing multi-scales hierarchy.

3.3. Methane/air (C1) mechanism

In order to illustrate the analysis and main results a relatively small mechanism has been chosen [21]. It contains $n_s = 16$ species plus pressure and enthalpy, i.e. overall system dimension is 18, and $n_r = 92$ elementary reactions. By implementing the GQL at initial temperature of 1235 K (pressure 3.12 bar) one finds that there are two cases/gaps: the first gap between 8 and 9 in (21) i.e. $m_f^1 = 10$, $m_s^1 = 8$, and between 14 and 15 yielding $m_f^2 = 4$, $m_s^2 = 14$, with $\varepsilon^1 = 0.17$ and $\varepsilon^2 = 0.018$ correspondingly. Figure 3 shows the state space and profiles of the integration of these two subsystems (along with 2D GQL slow manifolds). First, the fast subsystem of 10 dimensions given by (27) in the left is implemented to represent the reduced model, and the second the slow subsystem of 14 dimensions is chosen to reduce the system (14), where 4 dimensional fast subspace is 'relaxed' (see (27) right).

Figure 4 shows the performance of two models as the fast and slow subsystem with respect to the delay times and relative errors of the reduced models. It is shown that the relative error is smaller for the slow invariant manifold approximation, which is clear because the gap condition is stronger for the slow subsystem (small parameters: $\varepsilon^1 = 0.17 \gg \varepsilon^2 = 0.018$), but both can be considered as appropriate reduced models because, typically, in applications logarithmic scales are compared for the delay times.

In particular, the performed analysis shows that the system (14) in this case can be described efficiently by the system of only 6 dimensions $n - m_s^1 - m_f^2 = 18 - 8 - 4 = 6$, because 8 linear combinations of original variables can be fixed and four processes are relaxed according to (23), (24).

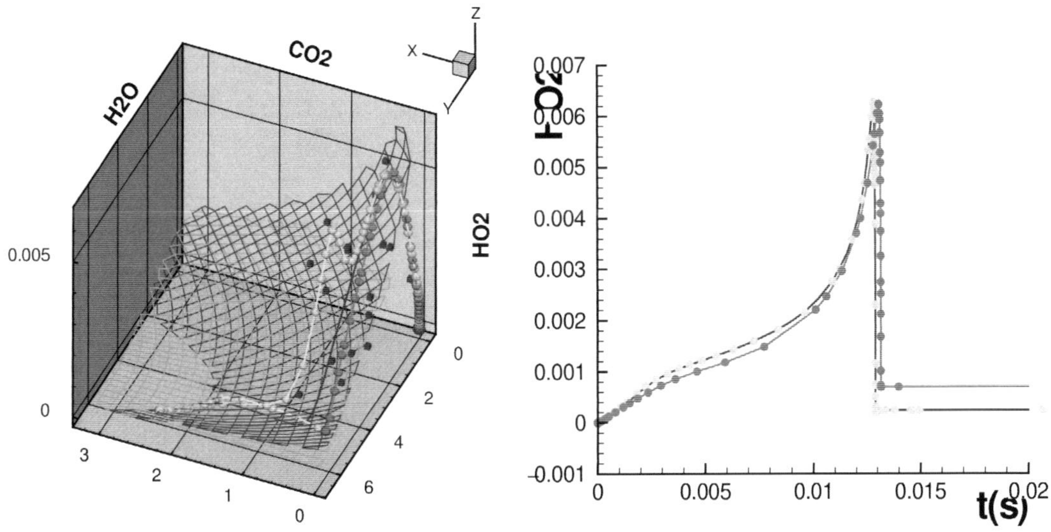

Figure 3. Results of GQL analysis of a methane/air homogeneous system shown in projection onto specific mole numbers and by solutions' profiles (right). Solid lines with symbols are fast subsystem solutions' trajectories and slow subsystem solutions, the mesh (left) is the 2D GQL slow surface. Solid line is the detailed system trajectory, cubes represent the set of point used for the GQL system analysis.

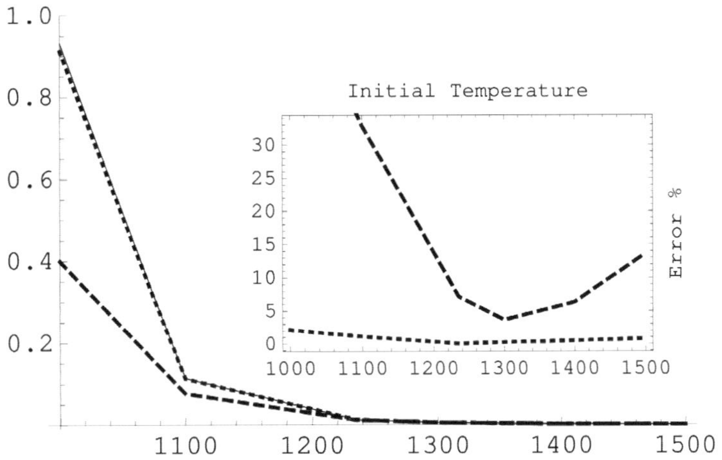

Figure 4. Delay times versus initial temperature and relative errors, the detailed model – black line, slow subsystem delay times – dotted line, fast subsystem delay times – dashed line.

3.4. Van der Pol oscillator, numerical treatment by GQL

It is interesting to compare analytical and results with numerical approach and to see what one can obtain in the case of not linear fast manifolds which, by now, is only the case when the GQL can be automatically applied. For this purpose we consider the original Van der Pol model with $\mu = 4$ and compare it with results for the transformed Van der Pol system where fast manifolds are linear (see figure 2 and 5 (right)). The results of the numerical analysis for $\Omega = [-3, 3] \times [-10, 10]$ are shown in figure 5 (left), where the red line is an approximation of the fast motion and blue line represents the slow curve. The result of the GQL analysis is

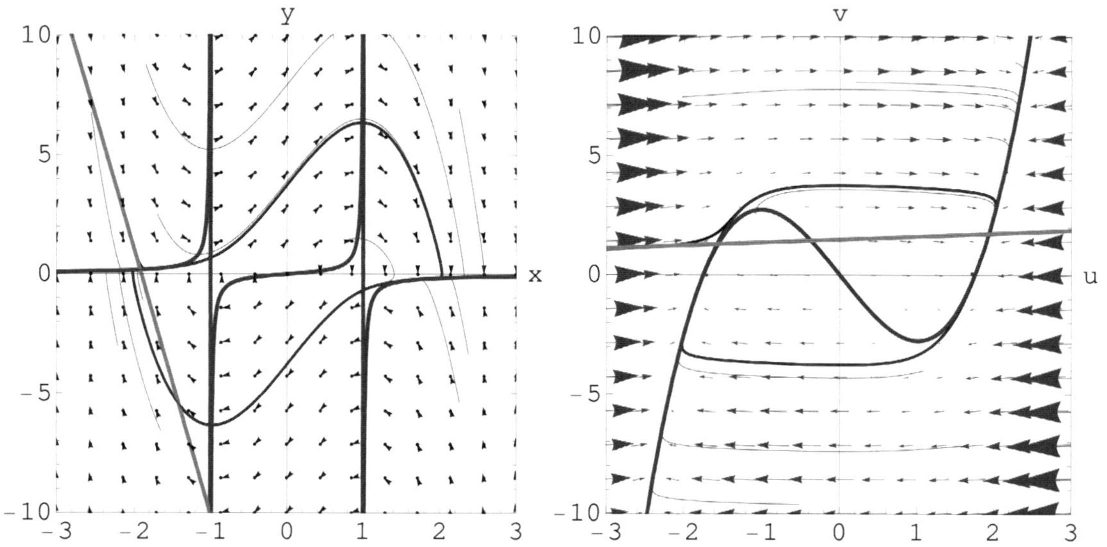

Figure 5. Phase space of the Van der Pol oscillator in the original coordinates (2) (left) and in the Lienard's coordinates (5) (right) with particular solutions trajectories and vector field. On both figures, the fast subspace of the GQL numerical analysis is shown by solid straight line (red), while the slow manifold is shown by bold solid curve (blue).

$$T = \begin{pmatrix} 0 & 1 \\ -0.96 & -8.76 \end{pmatrix}, \; Z = \begin{pmatrix} -0.11 & 0.99 \\ 0.99 & -0.11 \end{pmatrix}, \tag{31}$$

$$\lambda(T) = \{-8.64, -0.11\}.$$

The eigenvalues show that there is a decomposition in the considered system as it was expected (the small parameter is approximated by $\varepsilon = 0.1115$). Moreover, although the fast manifolds are not linear the slow curve, nevertheless, approximates quite well the invariant system manifold (see figure 5, where fast and slow manifolds are calculated for the Van der Pol original system (2) according to (23), (24) and (31)). Additionally, the direction of the linear fast motions, at least for a large part of the domain, given by the GQL analysis quite accurately approximate the fastest part of the non-linear fast manifolds (just compare figure 2 and 5, where fast direction represents family of non-linear fast manifolds).

4. Conclusions

In the present paper problem of model reduction and asymptotic system analysis is discussed. The problem of reduced description is represented from the invariant manifolds method point of view, which is a most suitable framework of model reduction. The natural assumption of multiple scales and the decomposition of motions allows to decouple the fast motions/processes and as a result reduce the system's dimension and stiffness. A method of coordinate free singular perturbations (SPVF) for model analysis and reduction has been discussed both at analytical and numerical levels.

The main feature of the approach in comparison to other methods is in approximating not only the slow system dynamics/manifolds, but the fast manifolds as well. Rigorous detailed analysis of system multi-scale hierarchy becomes possible. The approach can be successfully applied in both analytical and numerical aspects to yield the explicit decomposition of the system exhibiting the multi-scale behavior. The realization and implementation of the method lead to the general scheme of an automatic reduction procedure which is based on invariant system manifolds of low dimension evaluated by the developed GQL system decomposition.

The results show that the suggested method can be efficiently implemented to define of the fast/slow invariant manifold equations, decompose system (define the explicit SPS form) and study the properties of the invariant system manifold. The method improves considerably the model reduction concept of mechanisms of chemical kinetics, it represents a rigorous approach to treat systematically kinetics mechanisms and define the decomposition of detailed model.

Acknowledgments

I would like to thank Vladimir Gol'dshtein, my former supervisor, friend and close collaborator in the developing the approach for his remarks and suggestions. Financial support from the Deutsche Forschungsgemeinschaft (DFG) is gratefully acknowledged.

References

[1] Warnatz J, Maas M, Dibble R W 2004 *Combustion* 4 edn. (Springer-Verlag, Berlin Heidelberg)
[2] http://jpldataeval.jpl.nasa.gov/
[3] Atkinson et al. 2003 Summary of Evaluated Kinetic and Photochemical data for atmospheric chemistry, available at: http://www.iupac-kinetic.ch.cam.ac.uk/summary/IUPACsumm_web_latest.pdf
[4] Crampin E J, Schnell S and McSharry P E 2004 *Prog. Biophys. Mol. Biol.* **86** 177
[5] Maria G 2004 *Chem. Biochem. Eng. Q.* **18:3** 195
[6] Bykov V, Goldfarb I, Gol'dshtein V 2006 *J. Phys.: Conf. Series* **55** 28-44
[7] Bykov V, Gol'dshtein V 2008 *J. Phys.: Conf. Series* **138**, 012003
[8] Gorban A N, Karlin I V 2003 *Chem. Eng. Sci.* **58**, 4751
[9] Gorban A N, Karlin I V and Zinovyev A Yu 2004 *Physica A* **333**, 106
[10] Liénard A 1928 *Revue générale de l'électricité* **23** 901–912 and 946–954
[11] Williams F A 1985 *Combustion Theory*, 2nd ed. (Benjamin/Cummings, Menlo Park, California)
[12] Laidler K J 1997 *Chemical Kinetics*, 3rd ed. (Benjamin-Cummings)
[13] Golub G H and van Loan C F 1989 *Matrix Computation*, The Hopkins University Press (Baltimore, London)
[14] Maas U and Thévenin D 1998 *Proc. Combust. Inst.* **27** 1183–1189
[15] Fenichel N 1979 *J. Differential Equations* **31** 53–98
[16] Gol'dshtein V and Sobolev V 1992 *Amer. Math. Soc.*, Translations, ed.: S.G. Gindikin, s. 2, **153** 73–92
[17] Bykov V, Gol'dshtein V and Maas U 2008 *Combustion Theory and Modelling (CTM)* **12(2)** 389–405
[18] Bykov V and Maas U 2009 *Z. Phys. Chem.* **223(4-5)** 461–479
[19] Maas U 1993 *Habilitation Thesis*, Institut fuer Technische Verbrennung (Universitaet Stuttgart)
[20] Okino M S and Mavrovouniotis M L 1998 *Chem. Rev.* **98(2)** 391-408
[21] Smooke M D Puri I K and Seshadri K 1986 *Proc. Combust. Inst.* **21** 813–821

Journal of Physics: Conference Series **268** (2011) 012004 doi:10.1088/1742-6596/268/1/012004

Mechano-electric optoisolator transducer with hysteresis

I M Ciuruş, M Dimian and A Graur
Stefan cel Mare University/ Electrical Engineering and Computer Science,
13 University Street, 720 229, Suceava, Romania

E-mail: marcelciurus@usv.ro

Abstract. This article presents a theoretical and experimental study of designing a mechano-electric optoisolator transducer with hysteresis. Our research is centred upon designing transducers on the basis of optical sensors, as photoelectric conversions eliminate the influence of electromagnetic disturbances. Conversion of the rotation/translation motions into electric signals is performed with the help of a LED-photoresistor Polaroid optocoupler. The driver of the optocoupler's transmitter module is an independent current source. The signal conditioning circuit is a Schmitt trigger circuit. The device is designed to be applied in the field of automation and mechatronics.

Index Terms: Optoelectronics, Hysteresis, Polaroid optocoupler, Transducer.

1. Introduction

The applications of optoelectronics and photonics have extended far beyond the classical area of communications and have attracted a great amount of research interest from various disciplines over the last few years [1, 2]. Several areas aim at benefiting from various modalities of controlling signal output of an optocoupler by acting on the light fascicle that couples the emitter to the receiver [3-5].

Our recent work in this field has been focused on designing and testing the Polaroid optocoupler [6], which was realized by attaching Polaroid filters at the transmitter output and receiver input. As a result, the Polaroid optocoupler acts as a mechano-electric optoisolator transducer that transforms a mechanical rotational and/or translational motion into a variation of the electric signal. Since this transducer's sensor functions on the basis of photoelectric conversions, it is not influenced by the electromagnetic disturbances in the work area.

In the field of transducers' research, an important part is played by the study of hysteresis transducers [7, 8]. By using a Schmitt trigger (ST) circuit, the Polaroid optocoupler (PO) is transformed into a mechano-electrical optoisolator (MEO) transducer with hysteresis. Detailed analysis of it is presented in this paper along with suggested applications in the area of automation, communications and mechatronics.

2. Proposed scheme

MEO transducer with hysteresis is a device made up of the motion sensor PO, the driver of the PO's transmitter module (T_x) and the signal conditioning circuit, figure 1.

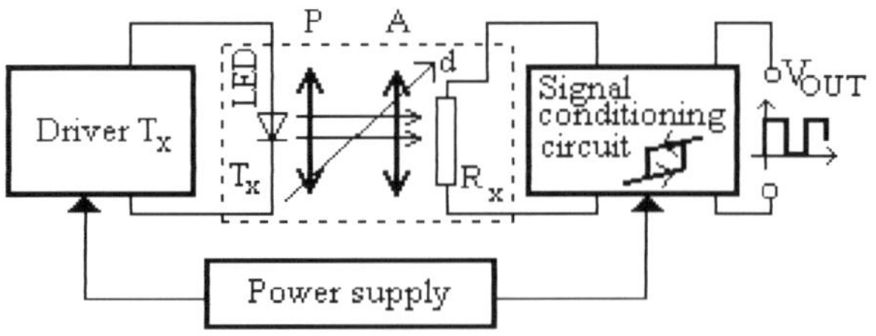

Figure 1. The block diagram of the mechanical-electrical optoisolator transducer with hysteresis

The motion sensor PO uses as light source a super bright white LED. This one's luminous intensity on the direction of the longitudinal axis is of (10365±226) mcd, for a current of 19mA. This current's value is kept constant by the driver of (T_x) module, figure 2.

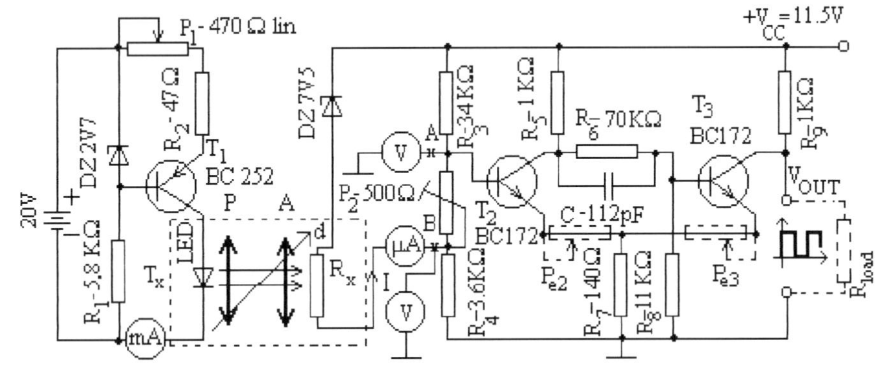

Figure 2. Circuitry of the mechanical-electrical optoisolator transducer with hysteresis

The thickness of the PO's Polaroid filters is 0.7mm. PO's photosensitive detector is a LDR07 photoresistor [9]. The signal conditioning circuit is a ST circuit with bipolar n-p-n BC172 transistors. This circuit is designed so as one of the two transistors is in saturation and the other one is close. The PO sensor is connected to the ST circuit by means of the resistive divisor R_4, P_2, R_3. This divisor is a current-tension converter (I, V_A).

3. Experimental results

The transducer's motion sensor from figure 2 is a PO LED-photoresistor with distance adjustment. The device is in the stage of prototype [10]. POs are optocouplers at which the user has access to the optical way of the light beam coupling the transmitter (T_x) module and the (R_x) receiver. At the optical output of (T_x) module is fixed a Polaroid polarizer filter (P) and at the optical input of (R_x) module is fixed a Polaroid analyser filter (A).

The optocoupler's mechanical system is designed so that the modules can perform axial translation and rotation motions one towards the other one. The main characteristic of this new type of optocoupler consists in the possibility to modify the intensity of the electric current at the output of the photosensitive element after the processes of the light beam's in-line polarization, re-orientation of its polarizing plan and translation motion of (T_x) module towards (R_x) module.

In table 1 are given the intensities of the electric current through photoresistor I, for different values of α angle between the polarizing plans of the two filters and distance d between (T_x) and (R_x) modules. These data are obtained for a supplying tension of photoresistor V=3V. By graphically representing data from table 1 are obtained the characteristics' families: $I=I(\alpha)_{d=const.}$ and $I=I(d)_{\alpha=const.}$ from figure 3.

The transducer's signal conditioning circuit was designed for the characteristic PO $I=I(\alpha)_{d=93\text{mm}}$. In figure 4, is represented the transfer characteristic of the motion sensor PO $I=I(\alpha)_{d=93\text{mm}}$ [curve(1)] and the transfer characteristic $V_{OUT}=V_{OUT}(\alpha)_{d=93\text{mm}}$ of the MEO transducer with hysteresis [curve (2)].

That device is a transducer with inverting hysteresis. Whenever the dihedral angle goes over the High Threshold Level $\alpha_H \cong 74°$ (when the output current of the PO sensor I_L=14μA), the output of the transducer is switched LOW ($V_{OUT,L}$ =1.63V). The output will remain in this state, as long as the dihedral angle is above the Low Threshold Level $\alpha_L \cong 19°$, when the output current of the PO sensor I_H=71μA. When the dihedral angle goes below this level, the output of the ST circuit will switch ($V_{OUT,H} \cong$ 11.5V).

Table 1. The output current intensities of the LED–photoresistor MEO transducer expressed in (μA), for different values of α and d

Angle	Distance (mm)							
(DEG)	78	83	88	93	98	103	108	113
0	105	95	87	78	71	65	60	56
10	102	93	84	75	70	64	59	54
20	95	85	77	70	65	58	54	50
30	85	75	67	60	56	51	47	44
40	70	63	55	51	46	42	39	37
50	54	47	43	40	36	33	31	29
60	35	32	30	27	25	23	21	19
70	22	19	18	17	15	14	12	12
80	12	10	10	9	8	7	7	6
90	7.2	6.5	5.9	5.3	4.9	4.5	4.1	3.8

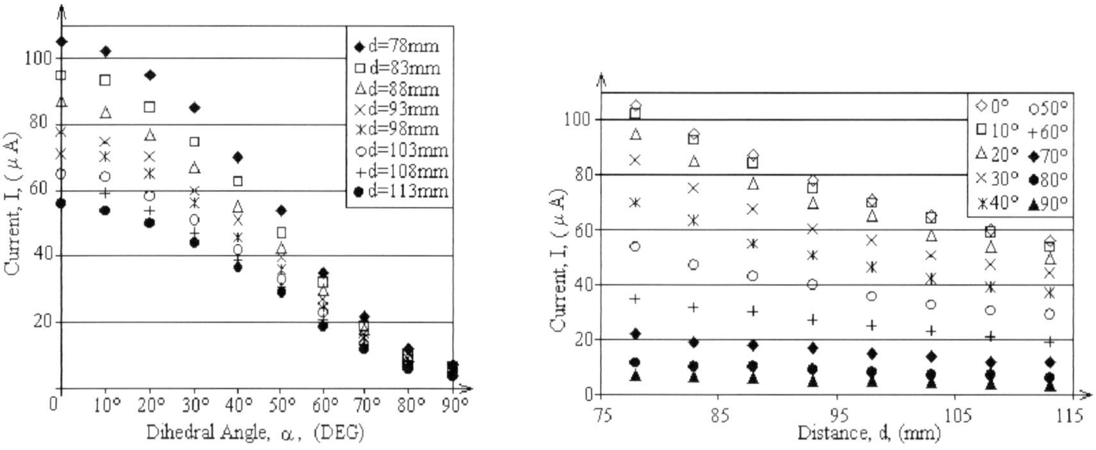

a) b)

Figure 3. The characteristics' family of the LED–photoresistor MEO sensors: a) $I=I(\alpha)_{d=const.}$ and b) $I=I(d)_{\alpha=const.}$.

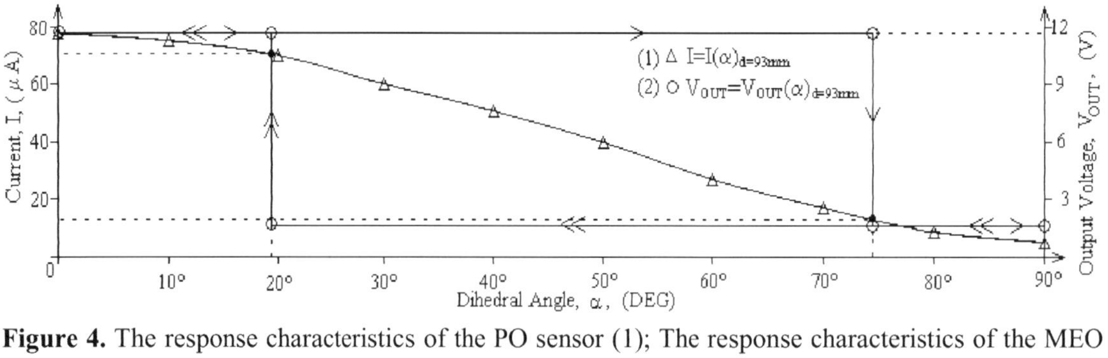

Figure 4. The response characteristics of the PO sensor (1); The response characteristics of the MEO transducer with hysteresis (2).

Journal of Physics: Conference Series **268** (2011) 012004 doi:10.1088/1742-6596/268/1/012004

4. Theoretical analysis of the device

If the intensity of the current through the LED in figure 5 is maintained constant, then the luminous intensity of the source $\mathcal{I}_v$ is constant in time. This radiation will produce a luminous illumination E_{vP} on the surface of a Polaroid polarizer filter which is at distance d_P from the LED.

$$E_{vP} = \frac{\mathcal{I}_v}{d_P^2} \tag{1}$$

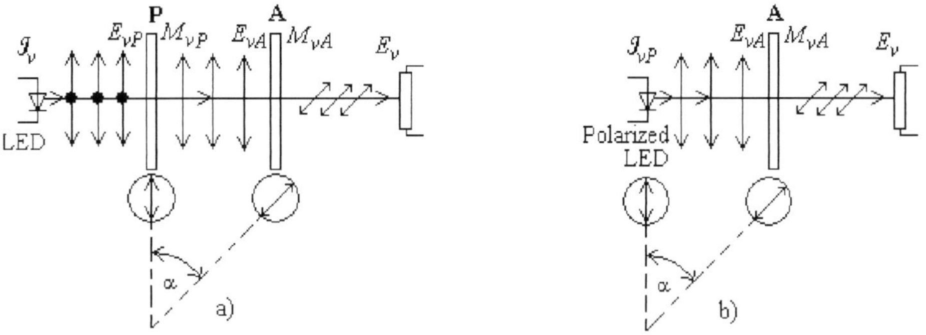

Figure 5. a) LED–photoresistor MEO transducer; b) Polarized LED–photoresistor MEO transducer

Equation (1) is true if the detector diameter is at least ten times smaller than the LED-photoresistor distance d – 'the rule of ten diameters' - and the light beam falls perpendicular to the detector surface. Knowing that the radiation's luminous illumination E_v and luminous exitance M_v are proportional with the square of the maximum value of the electrical field's intensity, according to Malus's law, one can show that radiation luminous exitance M_{vP} at the exit of the filter (P) is given by the following relation:

$$M_{vP} = \frac{E_{vP}}{2} = \frac{\mathcal{I}_v}{2 \cdot d_P^2} = \frac{\mathcal{I}_{vP}}{d_P^2} \tag{2}$$

Equation (2) considers ideal Polaroid filters (i.e. the transmission coefficient in the transmission axis is T'=1, the reflection coefficient is R'=0 and the absorption coefficient in the orthogonal axis is A'=1 and in any other situation, their values are: T=0, A=0, R=0). From relation (2) and figure 5 can be noticed that the system formed by a LED with luminous intensity $\mathcal{I}_v$ and the Polaroid filter (P) is equivalent with a polarized LED with luminous intensity $\mathcal{I}_{vP}$ [11].

In this case, according to Malus's law, the luminous illumination at the input of the Polaroid filter (A) E_{vA}, the luminous exitance at the output from the Polaroid filter (A) M_{vA} and the luminous illumination of the surface of the photoresistor E_v are given by the relations:

$$E_{vA} = \frac{\mathcal{I}_{vP}}{d_A^2} = \frac{\mathcal{I}_v}{2 \cdot d_A^2}; \quad M_{vA} = E_{vA} \cdot \cos^2\alpha = \frac{\mathcal{I}_{vP}}{d_A^2} \cdot \cos^2\alpha = \frac{\mathcal{I}_v}{2 \cdot d_A^2} \cdot \cos^2\alpha; \quad E_v = \frac{\mathcal{I}_v}{2 \cdot d^2} \cdot \cos^2\alpha \tag{3}$$

where d_A is the distance LED-Polaroid filter (A) and d is the LED-photoresistor distance.

When the photosensitive surface of the photoresistor from figure 2 is illuminated, through it passes a photocurrent given by the expression:

$$I = I_d + C_1 \cdot E_v^a = I_d + C_1 \cdot \left(\frac{\mathcal{I}_v}{2 \cdot d^2} \cdot \cos^2\alpha \right)^a \cong C \cdot \frac{\cos^{2a}\alpha}{d^{2a}} \tag{4}$$

where: $C = C_1 \cdot (\mathcal{J}_v / 2)^a$, I_d is the darkness current which is actually negligible as compared to photocurrent in the case of LDR07 photoresistor, a is a parameter depending on the level of the photoresistor's illumination and C_1 is a proportionality constant. It depends on the voltage (V) applied to the photoresistor, on the radiation frequency and the photoresistor type.

This photocurrent appears because of the photoconductivity's dependence on illumination.

Since when $\alpha = 90°$, the transmission coefficient in the transmission axis of real polarizer filter $T \neq 0$ (figure 3), an additional parameter T must be introduced in (4):

$$I(\alpha, d) = C \cdot \frac{\cos^{2a} \alpha + T}{d^{2a}} \tag{5}$$

Equation (5) represents the theoretical transfer characteristics' family $I=I(\alpha,d)$, for LED-photoresistor PO sensor. So as PO sensor can command the ST circuit, photocurrent I must be converted into the command tension V_A.

The transfer function of the current-to-voltage converter from figure 2 can be expressed as follows:

$$V_A(I) = \frac{V_{CC} \cdot (R_{P_2} + R_4) + I \cdot R_4 \cdot R_3}{R_3 + R_{P_2} + R_4} \tag{6}$$

From (5) and (6) is obtained the dependence function of the command voltage V_A' of ST circuit's on α angle.

$$V_A(\alpha) = \frac{V_{CC} \cdot (R_{P_2} + R_4) \cdot d^{2a} + C \cdot R_4 \cdot R_3 \cdot (\cos^{2a} \alpha + T)}{(R_3 + R_{P_2} + R_4) \cdot d^{2a}} \tag{7}$$

If $R_{Pe2}=R_{Pe3}=0$, the high trigger voltage V_{AH} and the low trigger voltage V_{AL} can be approximately calculated as:

$$V_{AH} = \frac{V_{CC} \cdot R_8}{R_5 + R_6 + R_8} - V_{BE_{SAT}} + V_{BE_{ON}} \cong \frac{V_{CC} \cdot R_8}{R_5 + R_6 + R_8} - 0.15 \cong 1.39\,\text{V} \tag{8}$$

$$V_{AL} = V_{BE_{SAT}} + R_7 \cdot \frac{\dfrac{V_{CC} \cdot R_8}{R_5 + R_6 + R_8} - V_{BE_{ON}}}{\dfrac{R_5 \cdot R_8}{R_5 + R_6 + R_8} + R_7} \cong 1.21\,\text{V} \tag{9}$$

where: $V_{BE_{SAT}} \cong 0.7\,\text{V}$ is the base emitter saturation voltage and $V_{BE_{ON}} \cong 0.55\,\text{V}$ is the base emitter opening voltage [12].

5. Results. Discussion

I for the extreme cases $\alpha=0°$ and $\alpha=90°$ when d=93mm, the values for the empirical constants T and C are obtained. Constants a are determined by superposing the characteristics given by (5) over the experimental characteristics from figure 3a. As a result, the theoretical transfer characteristic (10) for our prototype is as it follows:

$$I(\alpha, d) = 193.44\text{mA} \cdot (\text{mm})^{1.74} \cdot \frac{\cos^{1.74} \alpha + 0.073}{d^{1.74}} \tag{10}$$

In this case (7) becomes:

$$V_A(\alpha)=\frac{V_{CC}\cdot\left(R_{P_2}+R_4\right)\cdot d^{1.74}+193.44\text{mA}\cdot(\text{mm})^{1.74}\cdot R_4\cdot R_3\cdot\left(\cos^{1.74}\alpha+0.073\right)}{\left(R_3+R_{P_2}+R_4\right)\cdot d^{1.74}} \tag{11}$$

In figure 6 are rendered two-dimensional graphic representations $I=I(\alpha)_{d=const.}$ and $I=I(d)_{\alpha=const.}$ of (10).

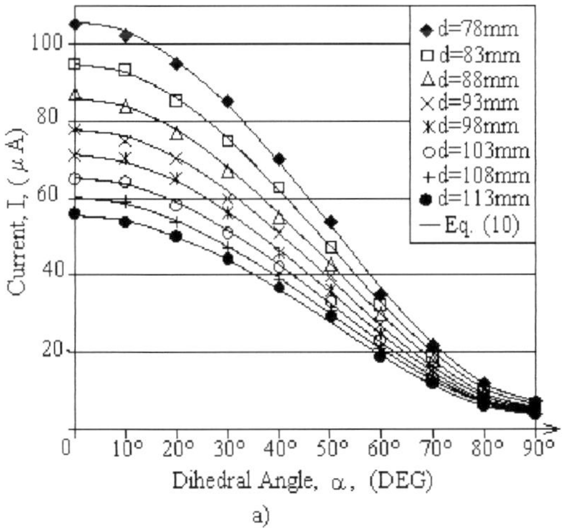

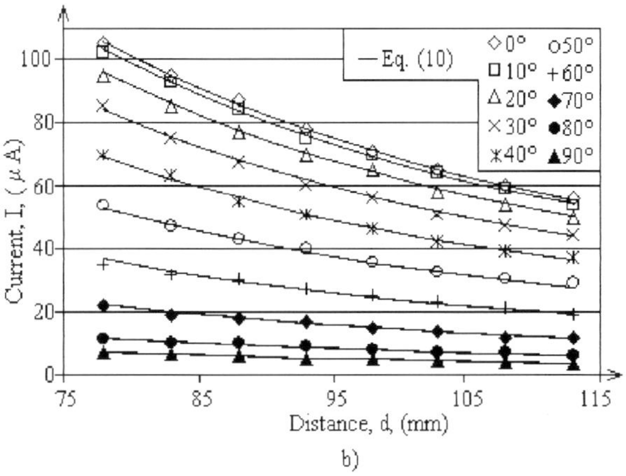

Figure 6. The families of the theoretical and experimental characteristics of the LED–photoresistor MEO sensors: a) $I=I(\alpha)_{d=const.}$ and b) $I=I(d)_{\alpha=const.}$

For comparison with the experimental data, in these graphics are also represented the points corresponding to the characteristics' family $I=I(\alpha)_{d=const.}$ *and* $I=I(d)_{\alpha=const.}$ from figure 3.

In figure 7 is rendered the (3-D) graphical representation $I=I(\alpha,d)$ of (10).

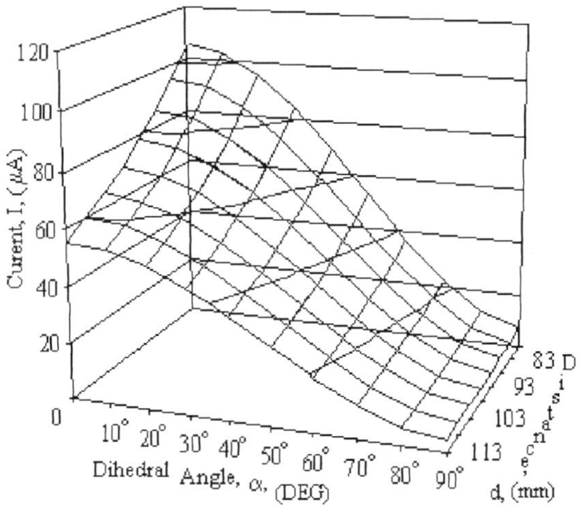

Figure 7. The 3D representation of the theoretical characteristics' family of the LED–photoresistor MEO sensors $I=I(\alpha,d)$

By replacing in (11) the dihedrals angle Threshold Level $\alpha_H \cong 74°$ and $\alpha_L \cong 19°$, the theoretical value of the voltage threshold level are obtained to change the state ST circuit: $V_{AH} \cong 1.39$V(8) and $V_{AL} \cong 1.21$V (9). These values have been calculated for a distance between (T_x) and (R_x), modules of d=93mm.

If in the circuit in figure 2, $R_{Pe1} \neq 0$ and $R_{Pe2} \neq 0$, the voltage threshold level V_{AH} and V_{AL} can be modified as:

$$V'_{AH} = V_{AH} \cdot \left(\frac{R_7 + R_{P_{e3}}}{R_7} \right) - V_{BE_{ON}} \cdot \frac{R_{P_{e3}}}{R_7} \tag{12}$$

$$V'_{AL} = V_{AL} \cdot \left(\frac{R_7 + R_{P_{e2}}}{R_7} \right) - V_{BE_{SAT}} \cdot \frac{R_{P_{e2}}}{R_7} \tag{13}$$

They can be also adjusted by modifying resistance R_{P2} (11).

The modification of the voltage threshold level V_{AH} and V_{AL} permits the adjustment by means of the electrical way of the dihedral angle Threshold Level α_H and α_L.

From figure 8 can be noticed that the adjustment of the dihedral angle's threshold level α_H and α_L can be achieved through optical way, too.

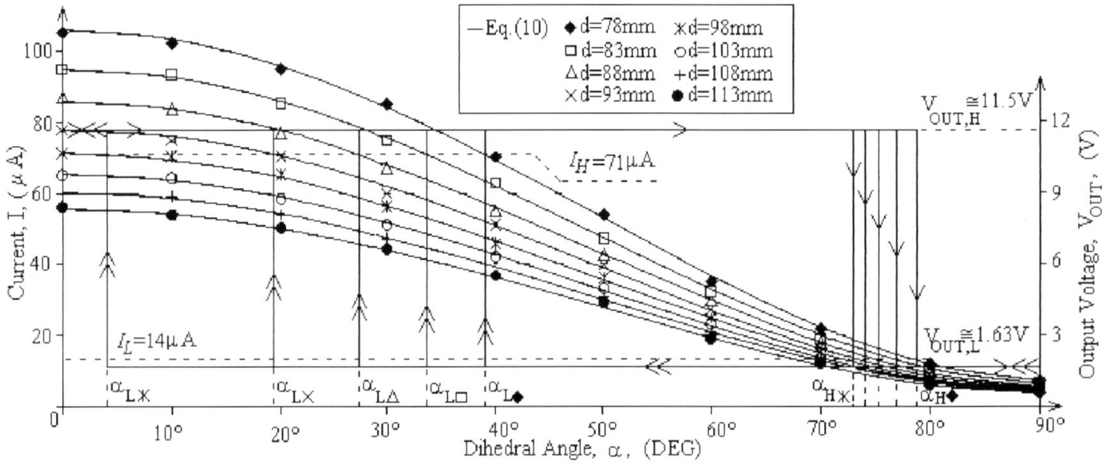

Figure 8. Mechano-optical adjstment of the dihedral angle's threshold level α_H and α_L

The increase in angle threshold level α_H and the decrease in angle threshold level α_L can be achieved by increasing distance d between (T_x) and (R_x) modules.

So as the voltage threshold level $V_{OUT,L}$ can be low, transistor T_3 must be in saturation state and the value of R_7 must be low. If applications in which this transistor is used need that the voltage threshold level $V_{OUT,L}$ to be low and the circuit's commutation to be rapid, T_2 and T_3 transistors must be Schottky transistors. These rapid transistors, fitted with Schottky limiting diodes between collector and base, allow the transistor to be wide opened without being deeply in saturation. In this case, the transistor's much lower switching time (t_{ON} and t_{OFF}), determine the increase of the work speed [13].

The modification of the distance between (T_x) and (R_x) can be also used to compensate in time the effects appearing as a consequence of the LED's ageing.

The applications' domain in which this new type can be used is much diversified.

If it is used as detector of the circuit voltage's presence figure 9, as compared to the classical optocouplers detectors, this device allows the mechano-optical adjustment of the level of detecting the circuit voltage which must be detected, as well as the domain of the voltages in which this one can fluctuate.

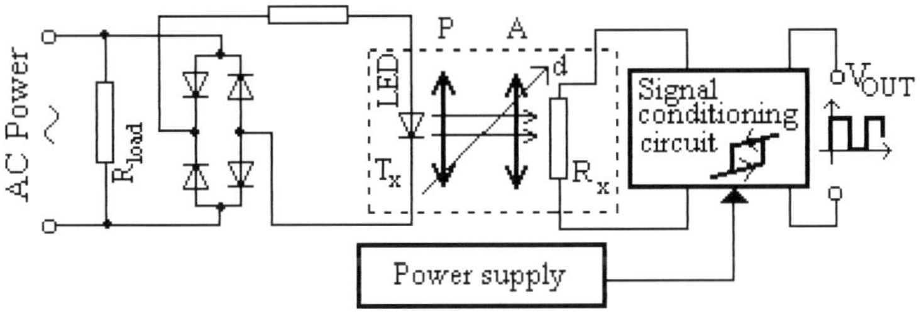

Figure 9. Line power detection circuitry

The adjustment of the nominal work voltage level is performed by modifying the Polaroid optocoupler's α parameter, and the settlement of the domain in which the circuit voltage can fluctuate is made by modifying parameter d, figure 8.

Mechano-optical adjustments allowed by this device offer it advantages as compared to classical optocouplers and also in the case when these ones are used for filtering digital signals accompanied by noise. If this transducer is used as tachometer, it isn't influenced by electromagnetic disturbances, and the modification of parameter d allows the modulation in width of the impulses formed by the device.

If α parameter is modified only in the interval (0°–90°), the transducer can be used for the automatic measurement of the water level in a dam, figure 10 [14]. The possibility to modify parameter d allows the transducer's re-calibration in time.

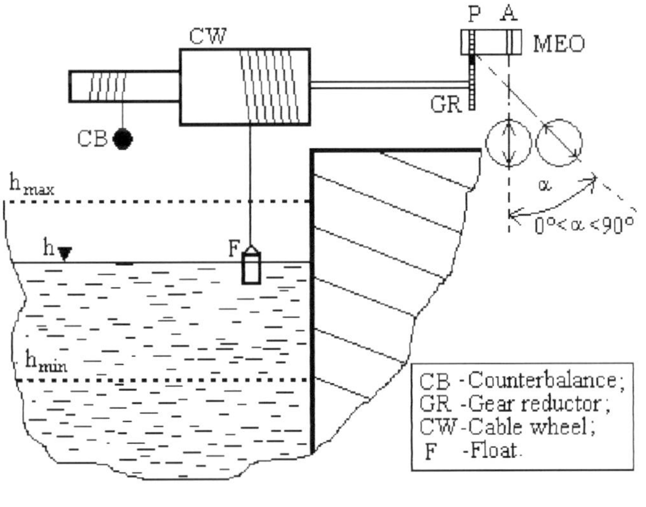

Figure 10. Measuring water level of dam with MEO transducer with hysteresis

6. Conclusions

The transducer proposed by us, using a Polaroid optocoupler sensor is not affected by electromagnetic disturbance. This characteristic advantages it when using it in areas with electromagnetic disturbances.

The use as a circuit conditioning a ST circuit, allowed the obtaining of a memory mechano-electric transducer. This characteristic determines its use in the applications where one needs the memorization of the machine-tools' extreme positions during motion.

The optocoupler's remote adjustment offers the advantage of its optimization or re-calibration by the mechano-optical way. This adjustment, unlike the electric adjustment (12, 13), doesn't depend on temperature.

The device designed by us is meant to be applied in the field of automation and mechatronics.

Acknowledgement

This work was partially supported by European Framework Program 7 under the contract no. PIRG02 GA-2007-224904.

References

[1] Kasap S, Ruda H and Boucher Y 2009 *Cambridge Illustrated Handbook of Optoelectronics and Photonics* (Cambridge: Cambridge University Press)

[2] Dakin J and Brown R G W (eds.) 2006 *Handbook of Optolectronics* (London: Taylor&Franics)

[3] Ouazzani K and Bentama J 2007 A promising optical technique to measure cake thickness of biological particles during a filtration process *Desalination* **206** 36-41

[4] Alcock R D and Emmony D C 2002 Sensitivity of reflection transducers *Journal of Applied Physics* **92** 1630-42

[5] Broistedt H, Petz M and Tutsch R 2009 Force transducer based on a geometric-optical displacement sensor *Proc. of IEEE Int. Symp. on Optomechatronic Technologies (Istanbul: TR, 21-23 Sept. 2009)* pp.164-168

[6] Ciuruş I M, Dimian M and Graur A 2010 Mechanical-Electrical Optoisolator Transducers *Proc. of IEEE 12th Int. Conf. Optimization of Electrical and Electronic Equipment (Brasov: RO, 20-22 May 2010)* pp. 740-745

[7] Garshelis I J and Cuseo J M 2009 "Negative" Hysteresis in Magnetoelastic Torque Transducers *IEEE Transactions on Magnetics* **45** 4471-74

[8] McNeill N, Iordanidis G, Holliday D and Mellor P 2006 Low-cost high-bandwidth current transducer for automotive applications *IEEE Transactions on Power Electronics* **21** 832-835

[9] http://www.datasheetarchive.com/ LDR07-datasheet.html#.

[10] Ciuruş I M 2009 *Optocuplor polaroid (patent pending a2009 00270)* (Bucureşti: OSIM).

[11] Wheatley J, Leatherdale C and Ouderkirk A 2005 *Polarized LED* PCT/US2005/034300.

[12] http://www.datasheetcatalog.org/datasheet/MicroElectronics/mXyzruur.pdf

[13] Jagadesh Kumar M and Venkatesh Rao D 2002 A new lateral PNM Schottky collector bipolar transistor (SCBT) on SOI for nonsaturating VLSI logic design *IEEE Transactions on Electron Devices* **49** 1070-72

[14] Guilin Zheng Hongyan Zong 2009 High accuracy surface perceiving water level gauge with self calibration *Proc. of IEEE Int. Conf. Mechatronics and Automation (Changchun: CN, 9-12 Aug. 2009)* pp. 3680 - 86

Compensation and control of two-inputs systems with hysteresis

D Davino[1], A Giustiniani[2] and C Visone[1]

[1] Department of Engineering, University of Sannio, P.zza Roma, 21 I-82100 Benevento, Italy
[2] DIIIE, Università degli Studi di Salerno I-84084 Fisciano (SA), Italy

E-mail: visone@unisannio.it

Abstract. The paper discuss the most assessed issues concerning the control of systems with memory and their application to realistic smart device, with particular focus to those applications where two-inputs fields (i.e. magnetic field and stress) are simultaneously concerned.

1. Introduction

In the last decade, the design of smart devices, based both on newly developed and known materials, has shown an enormous increase. Without any claim of exhaustivity we can recall the well known piezo-electric materials (PZT), magnetostrictives (GMM), or shape memory alloys (SMA), which are the oldest in the set of the so-called smart materials. As well known they can 'couple' electric, magnetic fields or temperature to mechanical stress or strain. Further, more recently, other solution have been proposed. For the sake of example we can recall ferro-fluids, where viscosity of the material is related to an applied magnetic field; Ni-Mn-Ga alloys (also called ferromagnetic shape memory), alloys (FSMA)) where a similar behavior, driven by the magnetic field, of classic SMAs is observed; other materials based on known phenomena, such as magneto-caloric effect, Peltier effect, etc. could also be mentioned and the list could be continued. All these alloys share the property that the constitutive relationship links together variable of different physical nature. A suitable name to describe this set could be that of *Multi-Functional Materials*, (MFM). Among them PZTs, SMAs, GMMs have been widely applied in actuators design in last years.

Mini-actuators based on SMA share the simplicity of mechanisms, cleanliness, noiseless actuation, remotability, sensing ability, low driving voltage. SMAs also show some weakness, such as low energy efficiency ($< 10\%$), quite a low response speed, nonlinear and irreversible behavior. They found wide applications in medical robotics, but also in other fields. The design of such devices as actuators should face with constructive and practical problems (clamping difficulties, low electrical resistance, etc.) and with very low efficiency, generally in the order of 4-5 per cent, [1], [2]. The recently developed FSMAs experience a huge deformation when subjected to a mechanical stress or to a magnetic field. Due to their fast response time (less than a millisecond) and to the high achievable strains (over 10%), FSMAs represent a promising candidate in the development of smart actuators, [3]. Electro-Rheological (ER) fluids are a further example of MF materials. Defined as a dispersion of particles in a insulated oil, they rapidly modifies the internal stress if driven by an external electric field. For this reason they could be a promising tool for the vibrations active control, [4].

Unlike the former, Magneto-Rheological (MR) fluids found successful applications in the development of new generation of car brakes, dampers, because of their accuracy, dynamics, fast responses and quite simple implementation. They can also find interesting applications in car or train suspension or vibration control of marine diesel engines, [5],[6]. Other interesting materials, controlled by the application of magnetic fields are MR foams and MR elastomers, where the controllable fluid, in the first case is contained into an absorbive matrix, while in the latter, the magnetic field controls the stiffness of the material (which unlike foams is a rubber-like solid material). Their possible application span from active vibration control to suspension devices, [7].

In the field of smart device, Piezo-materials are certainly the most known and widely applied. They show good efficiency, fast responses and quite high forces per unit volume. Applications range from mini-actuators for micro-positioning tasks, to ultrasonic sources and to linear drives. Moreover, due to their high integration capabilities, and again to their high force densities they also found an ever growing interest in Microelectromechanical systems (MEMS) where the successful integration of piezo-materials could yield to the innovation of inertial sensors for navigation, high-frequency crystal oscillators and filters, microactuators for RF applications, etc., [8],[9]. Even if more recently developed, Giant Magnetostrictive Materials (GMM) due to their promising (and to some extent complementary to Piezos) physical properties have focused a large interest for their potentialities. Applications could span from actuation, linear and rotational motors, sonar transducers, etc.,[10]. In particular, they found also a great interest in the design of sensors, such as load cells, torque meters, pressure gauges, tensile stress sensors, thinfilm thickness sensors, and dynamic stress sensors, strain sensing in civil engineering, [11],[12]. A review on the applications as sensor or actuators of the smart materials, mainly focused on piezoelectrics and SMAs is provided in [13]. There also GMM and Electrostrictive materials are considered as potential candidates for these applications. In this case however the need of an accurate modeling and a wide experimental characterization is outlined. In fact, almost all MFM share nonlinearities and rate independent memory effects (hysteresis) which has proved to be a drawback in several tasks. This undesired behavior could strongly limits the performances of smart actuator unless a suitable and accurate modelling and control strategy is adopted. In the paper, with the aim to improve the performances of smart devices, all the basic progresses in modelling and compensation of these phenomena are presented. A particular emphasis is given to the modeling of magnetostrictive materials.

2. Multi-variate systems with hysteresis

Multi-variate systems, such as magnetostrictive, PZTs or any device employing multifunctional materials share the property of linking together *more than two* physical quantities. As well known, these relationships show *rate independent* memory phenomena [14], and therefore a generalization of the standard hysteresis operators is required. So, accordingly to the definition of the Classical Preisach model, we can recall the Preisach magnetostrictive operator, as introduced in [14].

Let us consider the continous functions, u_k, with $k = 1, 2,\ \ u_k \in \mathcal{C}^1[t_0,\, T]$ of the kind

$$u_k : [t_0,\, T] \to \mathbb{R},$$

Definition 1 *Let be* $\hat{\gamma}_{\alpha,\beta}$ *the relay operator, and* $\mu_j(\alpha, \beta, x)$ *and* $\nu_j(\alpha, \beta, x)$ *with* $j = 1, 2$ *bounded measurable functions over the half-plane,* $T_\infty = \{(\alpha, \beta) \in \mathbb{R}^2 \mid \alpha \geq \beta,\, \forall x \in \mathbb{R}\}$.
The following correspondence:

$$y_j(t) = \iint\limits_{\alpha \geq \beta} \mu_j(\alpha, \beta; u_1(t))\, \hat{\gamma}_{\alpha\beta}\, u_2(t)\, d\alpha d\beta + \iint\limits_{\alpha \geq \beta} \nu_j(\alpha, \beta; u_2(t))\, \hat{\gamma}_{\alpha\beta}\, u_1(t)\, d\alpha d\beta \qquad (1)$$

maps the functions $u_k(t) \in \mathcal{C}^1[t_0, T]$, $k = 1, 2$, *into the functions* $y_j(t) \in \mathcal{C}^1[t_0, T]$, $j = 1, 2$, *and can be referred as* Multi-variate *Preisach operator.*

It should be noted that such definition slightly generalizes the definition provided in [14]. The two inputs could represent a mechanic and an electric or magnetic variable; in a similar way can be interpreted the output functions. The model defined above has been proposed in Mayergoyz's book and in [15], with only one output, since the attention was focused on magnetostrictive hysteresis. In that case it shows several important properties, such as *wiping-out* property, *equal vertical chords* and *path-independent* properties, [14]. The equation is an operator of Preisach-type linking a couple of input functions into a couple of output functions. An alternative approach, proposed by [16], exploits the structure of the vector Preisach models, proposed in [14], to describe the magnetostrictive coupling between mechanical and magnetic fields. The class of models defined in equation (1) has properties which make it quite interesting for the modelling of multi-functional (MF) devices. It can easily be employed to describe the behavior of giant magnetostrictive materials, based on Terfenol-D alloys, characterised by the following link between strain ε and flux density B on one side, and magnetic field H and mechanical stress σ on the other:

$$\varepsilon = \mathcal{G}_1(H, \sigma) \quad B = \mathcal{G}_2(H, \sigma). \tag{2}$$

The same machinery could also be employed to model the static constitutive characteristic of Piezo-electric materials, which could be stated as follows:

$$\varepsilon = \mathcal{F}_1(E, \sigma) \quad D = \mathcal{F}_2(E, \sigma), \tag{3}$$

being D and E the electric displacement and the electric field respectively, while ε and σ are the strain and the stress, respectively. A typical magnetostrictive behavior is shown in figures 1, and 2.

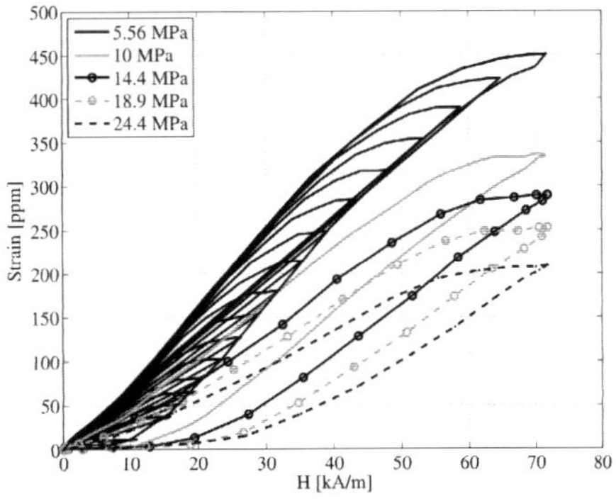

Figure 1. Typical measured magnetostrictive loops. It is shown the strong dependence of the magnetostrictive response on the applied stress.

Then, it should be stressed that these materials are employed for the design of smart devices where a robust control algorithms is required. The effects of rate independent memory yield the classical control strategies unsuitable and require different approaches to take hysteresis into account. To this end, the *model-based* control strategy [17]-[20] where a suitable inverse hysteresis operator is able to compensate memory effects, is the most applied. In fact, the availability of a compensator strongly simplifies the design of the controller. For these applications only the first component of the output field shown in equation (1) is considered.

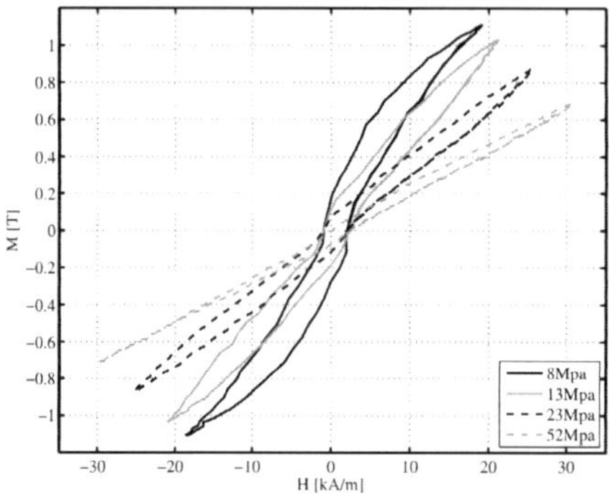

Figure 2. Typical experimental magnetic loops. It is evident the dependence on the applied stress.

Further, it should also be mentioned that almost all the proposed approaches considered the device operating under a constant applied stress. Hence, out of few exceptions, the model based control strategy employed the classical one-input one-output hysteresis operators, linking the input (magnetic or electric field) to the material strain. It should also be observed that, the compensation of Piezo-devices would be an easier task because of less evident saturation effects (cfr. figure 4.2 in [21]). Further, in this case, compensators in a closed form could maybe be exploited [22]. In summary, even if the model described in equation (1) could represent a suitable and complete approach to the multi-variable hysteresis modelling, when control applications are of concern simpler models should be applied. They can span from the classical one-input to one-output Preisach model, to a two-input/one-output operator, when, for example, also the applied stress has to be taken into account. The latter case is now of interest, since it allows to extend the limits of applicability of the approach. Unfortunately, the operator described in (1) has not an inverse in a closed form nor could be inverted with high computational efficiency, as the classical Preisach model, [23]. It is therefore such motivation which pushed to propose a new operator able to fully model the device but still holding the property to be inverted with a low computational cost [24]. This would guarantee applicability to real time control tasks.

The model we aim to discuss in the sequel, is a simplified version of the one proposed by Bergqvist and Engdahl in [25], and defined as follows:

Definition 2 *Let be $\mu(\alpha,\beta)$ a bounded measurable function over the half-plane, $T_\infty = \{(\alpha,\beta) \in \mathbb{R}^2 \mid \alpha \geq \beta\}$ while $u_k \in \mathcal{C}^1[t_0, T]$, with $k = 1, 2$, and Φ_e continuous funtions. The following equation defines a Preisach-type operator with two inputs.*

$$y(t) = \iint\limits_{\alpha\geq\beta} \mu(\alpha,\beta)\widehat{\gamma}_{\alpha\beta}\,\Phi_e(u_1, u_2, \alpha, \beta)\,d\alpha\, d\beta. \tag{4}$$

It is quite clear that the function Φ_e plays the role of what is commonly called *mean field* contribution. Here, the mean field depends on each point (α,β) in the Preisach plane. However, a general way to identify and model the material by this approach seemed to be not completely suitable for the defined control task. To this aim, therefore, a simplification to this model was proposed, with the aim to guarantee at the same time an easier model definition and implementation and the availability of a well suited compensation algorithm with a low computational cost.

Definition 3 *Let be $f : \Omega \subseteq \mathbb{R}^2 \to \mathbb{R}$; a continuous function in Ω, while $u_k \in \mathcal{C}^1[t_0, T]$, with $k = 1, 2$, $\mu(\alpha,\beta)$, a bounded measurable functions over the half-plane T_∞, and $\lambda \in \mathcal{C}^1(\Theta)$, with*

$\Theta \subseteq \mathbb{R}$. *The two-inputs operator of Preisach-type is defined as follows:*

$$y(t) = \iint_{\alpha \geq \beta} \mu(\alpha, \beta) \widehat{\gamma}_{\alpha\beta} f(u_1, u_2) d\alpha d\beta + \lambda(u_2). \tag{5}$$

The operator relates the input fields u_1 and u_2 to the variable of interest. It is important to observe that, unlike other attempts, [26], such a simplified approach guarantees the dependence of the state to both input fields, still admitting a compensator operator with low computational cost, as will be specified later. For the sake of example, and in order to particularize the functions introduced so far, let us recall some phenomena which can be observed in magnetostriction hysteresis, and reported in the figures 3 and 4, which clearly show the strong interaction of magnetic field and stress in affecting the global macroscopic behavior of the material, with particular emphasis to its deformation.

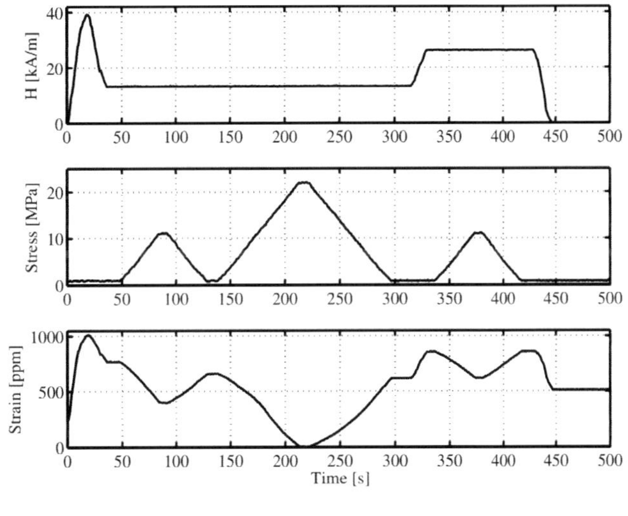

Figure 3. Experimental magnetic field, stress and strain vs. time. For the sake of clarity, the strain measurement has been normalized to zero by subtracting the sensor offset.

Both figures use experimental data over a magnetostrictive actuator and show the dependence of the output (strain) on both the magnetic field and stress and their linked action in affecting the internal state of the active material. In particular, figure 3 shows the time behavior of measured stress, strain and magnetic field of the sample; moreover, figure 4 displays the further dependence of the system internal state on the stress. In particular, observing the smaller loops, corresponding to the same stress variation but at different magnetic fields, we can infer that the lower the magnetic field applied to the sample, the smaller the loop width. This conclusion would be of help in specifying the parameters of the model.

When a magnetostrictive material is concerned, the model could be specified by assuming u_1 and u_2 as the magnetic field H and the stress σ, respectively, while the output, represents the displacement of the sample. The last term in equation, conversely, would represent the pure mechanical response, with zero applied magnetic field. In particular, assuming σ^* as the reference stress value (i.e. the pre-stress normally applied to the sample), the following conditions can be assumed:

$$\lambda(\sigma^*) = 0, \quad f(H, \sigma = \sigma^*) = H, \quad f(H = 0, \sigma) = 0.$$

The latter condition underline that in the working range stress values, no pure mechanical hysteresis is possible, without any applied magnetic field.

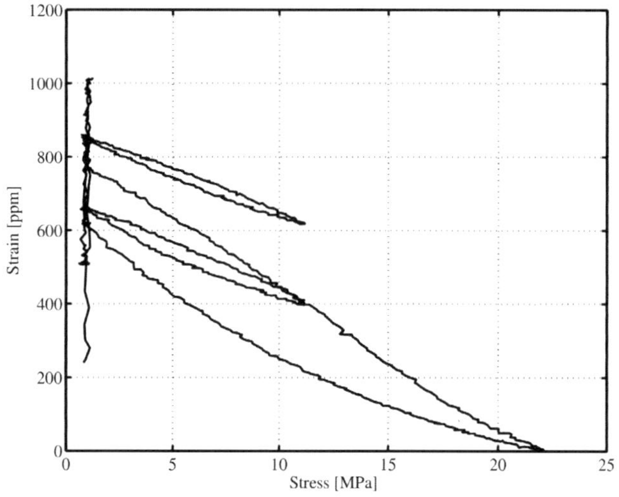

Figure 4. Measured mechanical loops. For the sake of clarity, the strain measurement has been normalized to zero by subtracting the sensor offset.

2.1. Identification of the model

The model defined so far, allows to exploit several useful results proposed for the classical Preisach operator. In particular, let us assume that the Preisach operator is invertible [27], [28]; moreover, let us assume the Preisach operator in the following fashion:

$$\Gamma[x] = \iint_{\alpha \geq \beta} \mu(\alpha, \beta) \widehat{\gamma}_{\alpha\beta}\, x \mathrm{d}\alpha \mathrm{d}\beta. \tag{6}$$

This allow to rewrite equation 5 as follows:

$$\varepsilon(t) = \Gamma[f(H, \sigma)] + \lambda(\sigma). \tag{7}$$

Preliminary, it should be stressed that the $\lambda(\cdot)$ function represents the pure elastic response of the sample and therefore it can be determined by a series of pure elastic measurements at zero magnetic field. Now recalling that if $\sigma = \sigma^*$ then $f(H, \sigma^*) = H$ and $\lambda(\sigma^*) = 0$, the operator takes the following classical form at $\sigma = \sigma^*$

$$\varepsilon(t) = \Gamma[H], \tag{8}$$

which could be easily identified [14]. Once the Preisach distribution function is known, the procedure to identify the $f(\cdot, \cdot)$ function can start. Recalling now that in suitable conditions Γ admits an inverse and that an efficient inversion algorithm is available, the equation (7) can easily be re-arranged as:

$$\Gamma^{-1}[\varepsilon - \lambda(\sigma)] = f(H, \sigma). \tag{9}$$

Let us preliminary define the interval of variation of the stress, let's say $[\sigma_{min}, \sigma_{max}]$. For the sake of example, in figure 5 the anhysteretic magnetostrictive responses at different mechanical load are shown. In that case we observe a monotonic decrease of the magnetostrictive response for increasing compressive stresses. Assume now that σ^* is the stress value corresponding to the maximum strain variation (which with reference to the figure, corresponds to a 1.05MPa applied mechanical stress). Moreover, let be $\sigma_{min} = \sigma^*$.

The interval $[\sigma_{min}, \sigma_{max}]$ can now be splitted into M elements, each of length $\Delta\sigma = \frac{\sigma_{max}-\sigma_{min}}{M}$, from $\sigma_0 = \sigma_{min} = \sigma^*$ to $\sigma_M = \sigma_{max}$. Now the following procedure for each sampled stress σ_k and $k = 1, .., M$ is applied.

(i) `Apply an AC demagnetization procedure to the sample;`

(ii) `Apply N times a Reduced Memory Sequence (RMS) for the H-field such that to obtain N points of the anhysteretic curve;let us call the output strain on such a curve` ε_i, $i = 1, ..N$[1];

(iii) `the following equation allows therefore to build up M X N samples of the unknown function f:`

$$\Gamma^{-1}\left[\varepsilon_i - \lambda(\sigma_k)\right] = f(H_i, \sigma_k).$$

Of course, any other value of the $f(\cdot,\cdot)$ can easily be computed by interpolation. Figure 6 shows the $f(\cdot,\cdot)$ identified for a sample of magnetostrictive material.

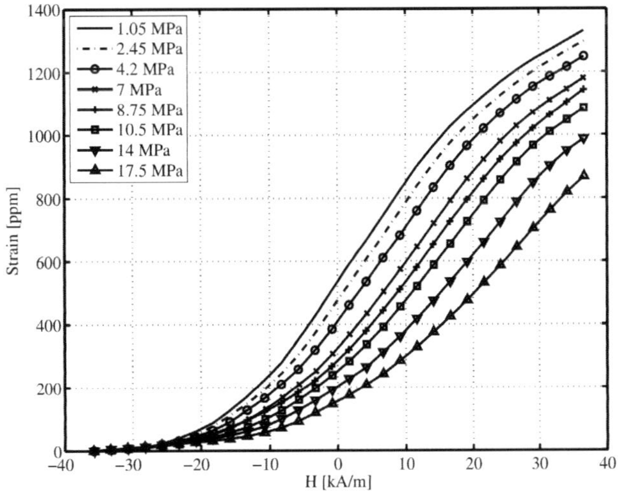

Figure 5. Anhysteretic magnetostrictive responses for varying applied stresses.

3. Model-based control of systems with hysteresis

The control of system with hysteresis requires an accurate updating of past input history. The most easy way to do this is to employ models of hysteresis which take memory into account in a natural way. This could be done by exploiting the state updating algorithm which characterizes any well-behaved hysteresis operator. In detail, starting from the device to be controlled, and defining the hysteresis operator able to describe the system, it is possible to build up the inverse of such operator (i.e. *compensator*) which allows the control system to operate as the device

[1] For the definition of Reduced Memory Sequence cfr. [29]

were linear. In figure 7 a typical *model-based* feedback control system is sketched. This of course strongly simplifies the controller design and is the reason why, in the last decade, a wide spread of contribution in the analysis and design of model-based control systems for smart devices (mostly based on Piezo-, SMA or magnetostrictive materials) were proposed, [30]-[35].

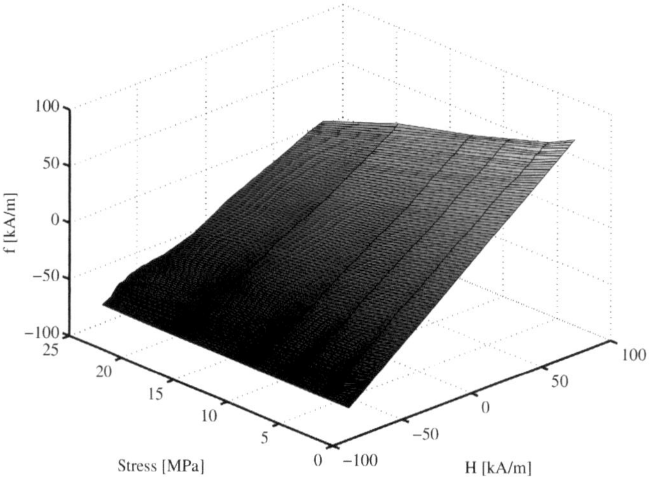

Figure 6. Identified f-function.

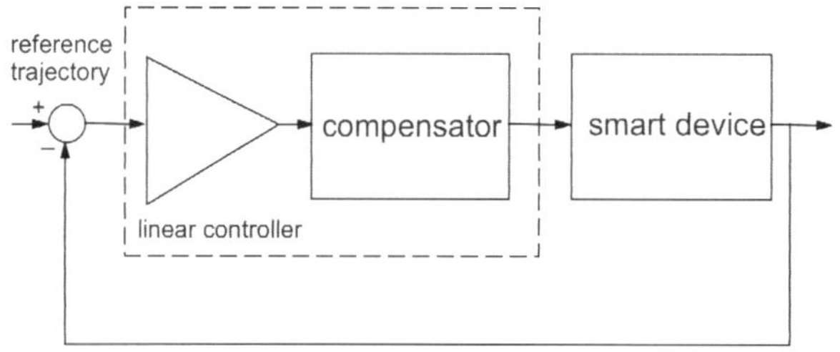

Figure 7. Scheme of the model-based control system.

Basically, the attention is focused on the definition of a reliable model of the device which allows the derivation of a suitable compensation algorithm with, hopefully, a low computational weight. The basic theory concerning the inversion of operators with hysteresis can be found in [36]-[39].

Several approaches have been proposed, to implement suitable compensators in control algorithms, spanning from the control of electromagnetic actuators, [40] or SMA-based devices, [41], [42], to magnetostrictive actuators [43]-[48]. A further contribution was also to take into account dynamic phenomena [20], [49] taking place in the device.
For the sake of example, in [20] the compensator has been exploited to identify the dynamic component of the process (supposed as linear), which allowed to improve the model performances. The design of the control device yielded to an improvement of the performances of the closed loop control system, both in terms of error tracking and stability, [45], [47]. The figures 8 and 9 show the improvements in the tracking of a reference signal when a compensator is employed. It is

evident that, with respect to the absence of a compensator, the tracking error is largely reduced, even with a complex reference signal. Moreover, in figure 10 it is shown how compensation could improve the stability of the system. In particular, it refers to a reference sine wave where the linear controller gain has been set to $k = 1000$ for both cases. The uncompensated system becomes unstable, while the model based strategy is still assuring stability to the system. It should be noted that this results have been achieved with a minumum increase in the system complexity.

The model-based control strategy adopted until now found its applicability on the assumption that only one of the input variable is varying, while the other (commonly the applied stress) is kept constant. This assumption works well as soon as micro-positioning tasks are concerned. Unfortunately, when both input variable are free to vary during the process, we assist to an evident worsening of the global performances of the device. Therefore, in any application where the stress is no longer a constant variable, a completely new approach has to be conceived. In the next section we will describe the new approach based on a two-variate modelling of the hysteretic device.

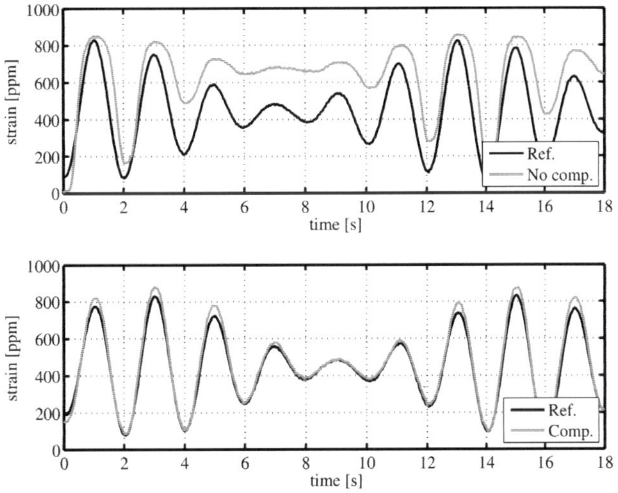

Figure 8. Tracking of a reference signal both for a uncompensated and compensated actuator.

4. Compensation algorithm for multi-variate systems with hysteresis

Before describing the procedure of model-based control for two-variate systems with hysteresis, it is useful to recall all those concepts regarding compensation procedures of classical (e.g. 1-input/1-output) systems with hysteresis. Let us recall the definition [29], [36], [38] as reported in [28]:

Definition 4 *A hysteresis operator $\mathcal{U}$ with initial state $\phi_{-1} \in \Psi_0$, is called a* compensator *(or inverse) of the operator $\mathcal{W}$, with initial state ψ_{-1} if, for any state $\phi \in \Psi_0$, there is a state $\psi \in \Psi_0$, such that $\mathcal{W}_\psi \circ \mathcal{U}_\phi\, x(t) = \mathcal{U}_\phi \circ \mathcal{W}_\psi\, x(t) = x(t)$, for every input function $x(t)$.*

Within the set of Preisach-kind operators, a sufficiently wide number of compensators for 1-input to 1-output operators with hysteresis is outlined, [28]. It is interesting to recall those which, starting from a result given in [39] proposed a generalized version of a compensator in a closed form, [20]. Moreover, other attempt were carried out, which driven to define a simple algorithm which is able to give the Preisach compensator with the same computational effort of the classical Preisach operator, [23].

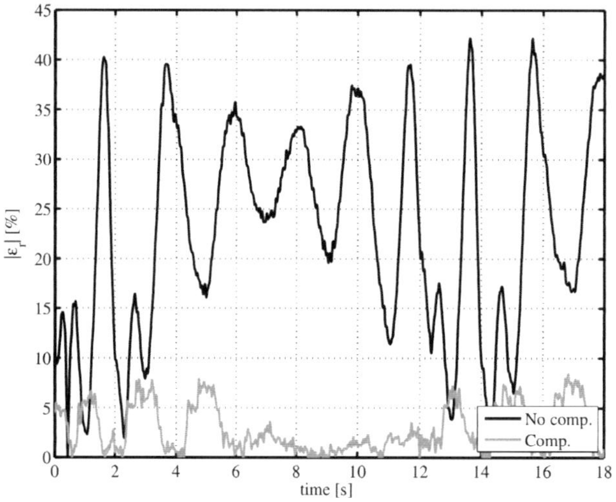

Figure 9. Tracking error of a reference signal both for a uncompensated and compensated actuator.

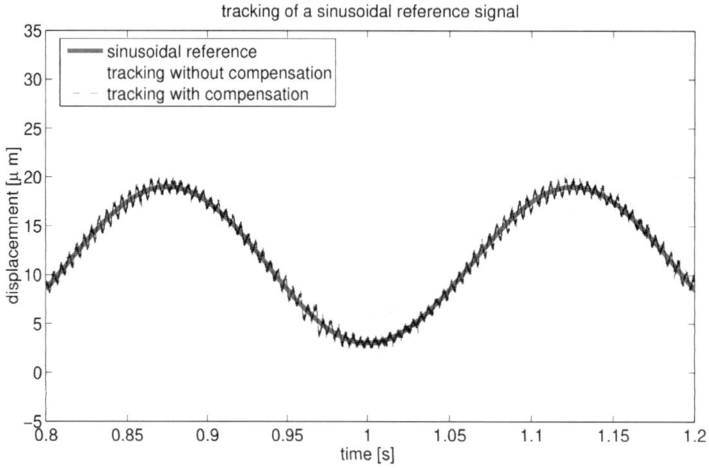

Figure 10. Example of tracking of a sinusoidal reference. The linear controller gain has been set to $k = 1000$ for both compensated and uncompensated control strategy.

Let us now consider a system with hysteresis (as those introduced above) having two input variables and one output variable of interest[2]. Let us assume that it is representable by the model given in equation (5) or, in the case example of magnetostrictive actuators, by equation (7). Let us also now assume that the operator starts from the initial state ψ_{-1}, [28], so for a

[2] It should be mentioned that all real systems have also two output variables. In the present case the interest is focused on only one of them

given output value y (or ε) and given input u_2 (or σ), the first input field is required, e.g. u_1 (or H). Let us state the following definition.

Definition 5 *Let be $u_k : [t_0, T] \to S' \subseteq \mathbb{R}$, with $k = 1, 2$ continuous functions, λ and f suitable continuous functions, and $z \equiv \Gamma^{-1}[y(t) - \lambda(u_2)]$, the equation*

$$f(u_1, u_2) = z \tag{10}$$

identifies the set

$$S \equiv \left\{(z, u_2) \subseteq \mathbb{R}^2 : \exists u_1 : f(u_1, u_2) \equiv z\right\}. \tag{11}$$

Definition 6 *Let be $u_k : [t_0, T] \to S' \subseteq \mathbb{R}$, with $k = 1, 2$ continuous functions, the function is defined:*

$$\Phi : (z, u_2) \in S \to u_1 \in S' \subseteq \mathbb{R}. \tag{12}$$

It is quite obvious that the following property holds:

Theorem 1 *The funtion Φ associates to every $(z, u_2) \in S$ only one value of $u_1 \in S'$, and vice-versa (e.g. Φ is invertible) if and only if equation (10) admits only one solution.*

This result guarantees that once the conditions of invertibility are stated, the definition of a compensation algorithm, for a two-inputs system with hysteresis, is well-posed.

Definition 7 *Let us now consider the following mapping $g : \mathbb{R}^2 \to \mathbb{R}^2$*

$$\begin{cases} v_1 = f(u_1, u_2); \\ v_2 = u_2 \end{cases} \tag{13}$$

the following domain $W \subseteq \Omega$ can be defined:

$$W = \{P_0 \in \Omega | J_g(P_0) \neq 0\}, \tag{14}$$

where

$$J_g(P_0) = \begin{vmatrix} \frac{\partial f}{\partial u_1} & \frac{\partial f}{\partial u_2} \\ 0 & 1 \end{vmatrix} = \frac{\partial f}{\partial u_1} \tag{15}$$

is the Jacobian of g in P_0.

Of course it is equal to the partial derivative of f with respect to u_1.

Theorem 2 *Given the mapping g defined above, for every $P_0 \in W \subseteq \Omega$, the function f is locally invertible in P_0.*

This theorem, according to the well-known theorem on implicit functions [50], [51], guarantees that it is possible to express for every point in W the variable u_1 as a function of z and u_2, i.e. $u_1 = \Phi(z, u_2)$. For the sake of example, figure 11 shows a case where the conditions specified above are fulfilled, and so it is possible to build up the function Φ, providing u_1 once z and u_2 are specified. In particular, for a given z it is possible to find the value of $\varepsilon - \lambda$ on the magnetostrictive anhysteretic response on the unperturbed curve (i.e. $u_2 = \sigma^*$). Now, once the actual value of u_2 is specified, it is possible to find the value of u_1 which guarantees the value of z for the new u_2. The procedure also gives the domain W within Φ is defined. In fact the same figure also shows z values having no u_1 image. Further, figure 12 give an example of the Φ function built up from suitable experimental data, picked from a magnetostrictive sample.

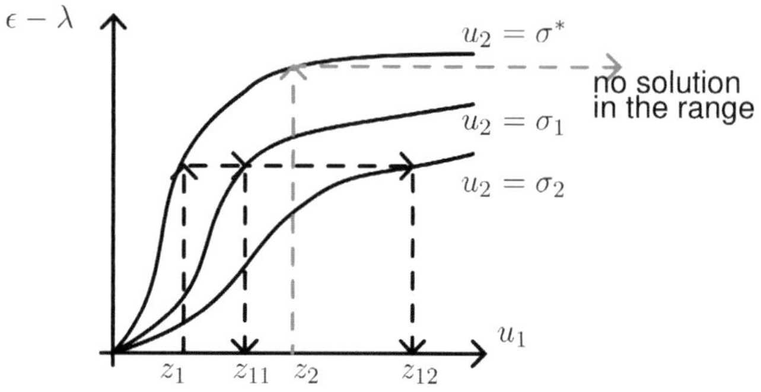

Figure 11. Sketch of the procedure to build up the Φ function, once the conditions of invertibility are specified.

5. Real time control algorithms and experimental verification

As mentioned above, in control systems applications where real time operations are of concern, the most important feature is the computation efficiency of the control algorithm. Indeed, the computation capabilities of the microcontrollers are usually limited by costs considerations. The two-inputs compensation algorithm presented above needs a computation effort that is comparable with the one of a one-input Preisach operator. At each computation time, t_k, the steps are the following:

- aquire the actual reference $y(t_k)$ and stress $u_2(t_k)$ values;
- compute the elastic displacement $\lambda[u_2(t_k)]$;
- compute the $z(t_k)$ as output of the one-input Preisach compensator, $z(t_k) = \Gamma^{-1}[y(t_k) - \lambda[u_2(t_k)]]$;
- compute the output as $u_1(t_k) = \Phi(z(t_k), u_2(t_k))$ as a double interpolation on a 2D look-up table;
- as reported in [23], the one-input Preisach compensator uses the same machinery of a standard Preisach operator plus a double interpolation.

Once the algorithm has been specified, it is important to use a versatile tool to implement and check the performances of the compensated control system in a real-time environment. The Matlab package xPCTarget has been chosen for this aim. Indeed, this toolbox allows to implement the control strategies in Simulink, a user-friendly environment, and then to compile and download the generated C-code to a real-time capable controller. A minimum hardware needed to implement the system should be composed by the following items:

- *desktop PC*: on this PC the simulink models are implemented. It allows to compile them in C-code and then to download onto the Target desktop PC by means of a LAN connection;
- *target desktop PC*: this PC runs a specific operating system and allows to run any algorithm in a real time environment. This PC has been equipped with a data acquisition/generation card to acquire and to generate the external signals;
- *bipolar amplifier*: this unit can operate as a voltage-controlled current generator, it is driven by the Target desktop PC and feed the actuator;
- *eddy current proximity sensor*: this sensor is used to measure the actuator displacement;
- *load cell*: it is used to measure the load applied to the actuator;
- *traction-compression test machine*: it is used to apply a time-varying mechanical load to the actuator.

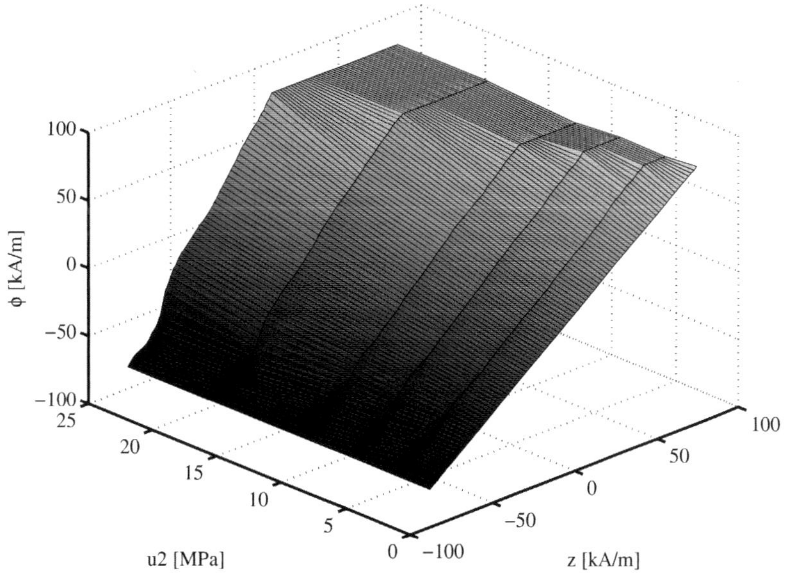

Figure 12. Sketch of the Φ function. It is worth noting that the flat region for high values of z represents the *no solution region*. The output is set to the maximum available value and a finite compensation error would be get.

In order to test the performance of the two-inputs compensator, two control systems have been implemented: a *open loop* one and a *closed loop* one. Both systems made use of a custom magnetostrictive actuator with a Terfenol-D active elements, preloaded with washer springs and with a permanent magnet. Typical values of the actuator in the performed tests were a current of ± 6 A and a maximum strain of 1350 ppm, equivalent to a displacement of 135 µm.

In the figure 13 it is presented the sketch of the implemented open loop control scheme. The inputs to the system are the desired displacement ϵ_d (generated by a reference signal generator) and the actual measured compressive stress σ_m. The pure elastic contribution ϵ_{el} is obtained from the memoryless function λ and is subtracted to the input. The displacement measurement, the load cell and the current amplifier are represented and modeled as constant gains k_l, k_m and k_a respectively. The gain k_g can be used to convert the scale of the generated signals to the maximum allowable input range of the compensator.

In the figure 14, it is reported the experimental results of a comparison between the stress-dependent compensator of figure 13 with a *standard* one-input compensator, i.e. without any stress-dependence. The reference signal was a triangular periodic wave (middle-plot) and the mechanical stress was increasing over time (upper-plot). Then, from the analysis of the relative error presented in the lower-plot, it is apparent that the stress-dependent compensator greatly improved the tracking of the reference strain.

In the figure 15, it is presented the sketch of the implemented closed loop control scheme. In this case, the feedback system is completed by a proportional-integral (PI) controller. The PI controller, if carefully tuned, helps to reduce the tracking error of the control system. The figure 16 shows the experimental result of a comparison between the system with the two-inputs compensator plus the PI controller with a closed loop system without the compensator. Both systems shared the same PI parameters. The reference signal was a sinc-like periodic wave (middle-plot) and the mechanical stress was increasing over time (upper-plot). Then, from the relative error presented in the lower-plot, it is apparent that both systems have a lower relative error with respect to the open loop system, because of the feedback effect. But, the system with the two-inputs compensator further reduce the tracking error of a 50% with respect to the closed loop system without compensation.

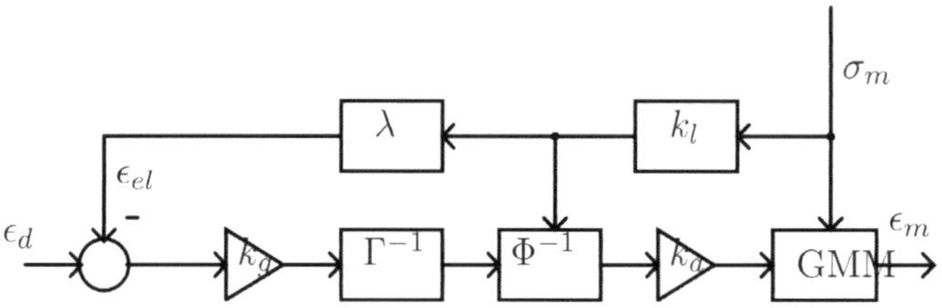

Figure 13. Sketch of the open loop control scheme.

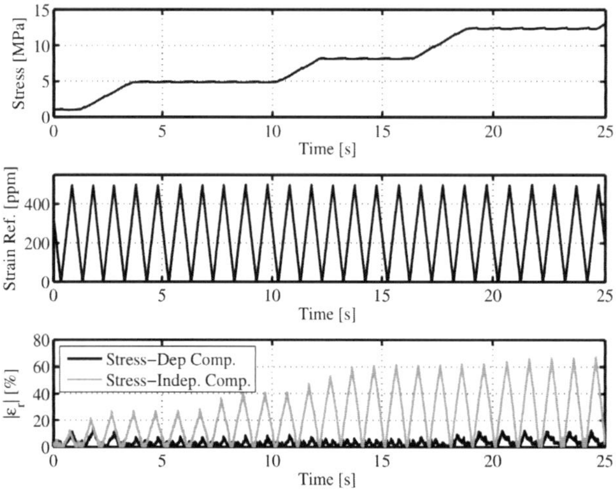

Figure 14. Comparison between the stress-dependent compensator and a standard one-input compensator with a triangular wave reference strain under a time increasing compressive stress.

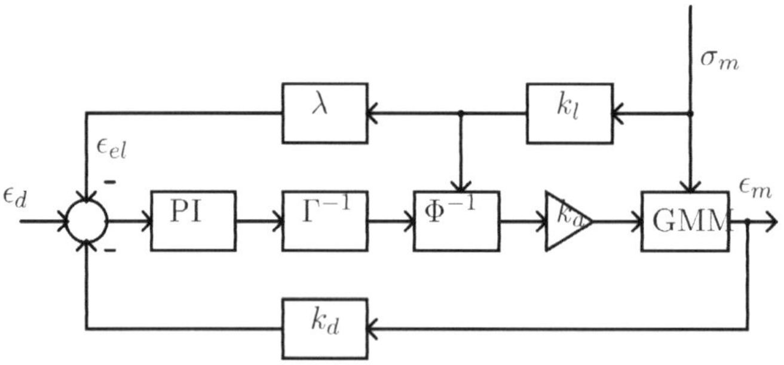

Figure 15. Sketch of the closed loop control scheme.

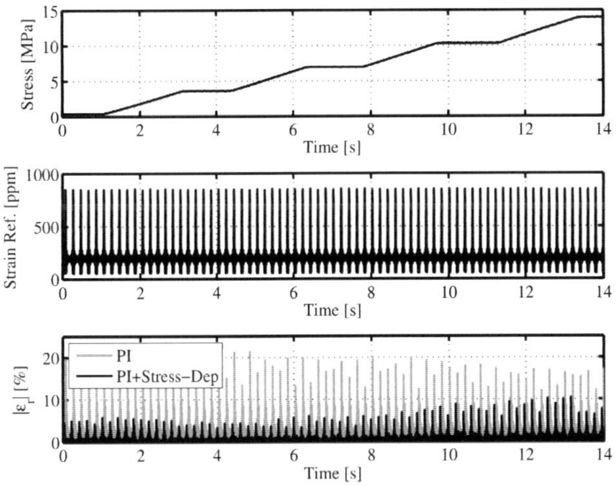

Figure 16. Comparison between the two closed loop control system with a sinc-like reference displacement under a time increasing compressive stress. One system has a stress-dependent compensator plus a PI controller while the other has only the PI controller.

6. Conclusion

In this paper it has been presented a review of the issues concerning the control of systems with memory, with particular emphasis to realistic smart device where, in many applications, two-inputs should be considered. In the latter case, a suitable two-inputs hysteresis model can be defined together with its inverse. The range of invertibility of this model is not restrictive.

The computational effort of this model is comparable with the one of a standard Preisach operator and, then, a real-time feedback control system can be built. Experimental results over a magnetostrictive actuator show that the improvement in the tracking error makes it suitable for an implementation in realistic feedback systems of smart devices.

References

[1] Bellouard Y 2008 *Materials Science and Engineering A* **481-482** 582
[2] Sreekumar M, Nagarajan T, Singaperumal M, Zoppi M and Molfino R 2007 *Industrial Robot: An International Journal* **34** 285
[3] Nespoli A, Besseghini S, Pittaccio S, Villa E and Viscuso S 2010 *Sensors and Actuators A* **158** 149
[4] Stanwayy R, Sprostonz J L and El-Wahedz A K 1996 *Smart Mater. Struct.* **5** 464
[5] Olabi A G and Grunwald A 2007 *Materials and Design* **28** 2658
[6] Muhammad A, Yao X and Deng Z 2006 *Journal of Marine Science and Application* **5** 17
[7] Carlson J D, Jolly M R, 2000 *Mechatronics* **10** 555
[8] Hemsel T and Wallaschek J 2000 *Ultrasonics* **38** (2000) 37
[9] Tadigadapa S and Mateti K 2009 *Meas. Sci. Technol.* **20** (2009) 092001.
[10] Olabi A G and Grunwald A 2008 *Materials and Design* **29** (2008) 469
[11] Hristoforou E and Ktena A 2007 *Journal of Magnetism and Magnetic Materials* **316** 372
[12] Pasquale M 2003 *Sensors and Actuators A* **106** 142
[13] Chopra I 2002 *AIAA Journal* **40** 2145
[14] Mayergoyz I D 1991 *Mathematical Models of Hysteresis* (Springer)
[15] Adly A A, Bergqvist A and Mayergoyz I D *J. Appl. Phys.* **69** 5777
[16] Adly A A and Mayergoyz I D 1996 *IEEE Trans. on Magnetics* **32** 4773
[17] Kuhnen K 2003 *Eu. J. of Control* **9** 407
[18] Tan X, Baras J S and Krishnaprasad P S 2005 *Syst. and Control Lett.* **54** 483
[19] Iyer R V, Tan X and Krishnaprasad P S 2005 *IEEE Trans. on Automatic Control* **50** 798
[20] Davino D, Natale C, Pirozzi S, Visone C 2004 *Physica-B* **343** 112

[21] Damjanovic D 2006 *Hysteresis in Piezoelectric and Ferroelectric Materials, The Science of Hysteresis* **3** I. Mayergoyz and G.Bertotti (Eds.) (Elsevier)
[22] Visone C and Sjöstrom M 2004 *Physica-B* **343** 148
[23] Davino D, Natale C, Pirozzi S and Visone C 2005 *J. of Mag. and Mag. Mat.* **290-291** 1351
[24] Davino D, Giustiniani A and Visone C 2009 *J. of Applied Physics*, **105** 07D512.
[25] Bergqvist A and Engdahl G 1991 *IEEE Trans. on Magn.* **27** 4796
[26] Cavallo A, Davino D et. al 2008 *Physica B* **403** 261
[27] Brokate M 1989 *IEEE Trans. on Magn.* **25** 2922
[28] Visone C 2008 *J. Phys.: Conf. Series* (2008) 012028
[29] Visintin A 1991 *Differential Models of Hysteresis* (Springer)
[30] Ge P and Jouaneh M 1996 *IEEE Trans. on Contr. Sys. Tech.* **4** 209
[31] Schäfer J and Janocha H 1995 *Sensors And Actuators A* **49** 97
[32] Webb G V, Lagoudas D C and Kurdila J 1998 *J. Of Int. Mat. Syst. And Struct.* **9** 432
[33] Krejci P and Kuhnen K 2001 *IEE Proc. on Control Theory and Appl.* **148** 185
[34] Tan X, Venkataraman R and Krishnaprasad P S 2001 *SPIE Modeling, Signal Processing, and Control in Smart Structures* **4326** 101
[35] Natale C, Velardi F and Visone C 2001 *Proc. of IEEE/ASME Int. Conf. on Advanced Intelligent Mechatronics, Como, Italy* 744
[36] Krasnoselskii M A and Pokrovskii A V 1989 *Systems with Hysteresis* (Springer-Verlag)
[37] Krejci P 1996 *Gakuto Int. Series Math. Sci. & Appl.* **8**
[38] Brokate M and Sprekels J 1996 *Hysteresis and Phase Transitions* (Springer)
[39] Krejci P 1986 *Mat. Zeit.* **193** 247
[40] Mittal S and Menq C H 2000 *IEEE/ASME Trans. on Mechatronics* **5** 394
[41] Majima S, Kodama K and Hasegawa T 2001 *IEEE Trans. on Control Sys. Tech.* **9**
[42] Hasegawa T and Majima S 1998 *Proc. IEEE Int. Symp. on Micromech. and Human Sci.* **171**
[43] Kuhnen K and Janocha H 2001 *Proc. of the 20th IASTED-Conf. on Modeling, Identification and Control* 375
[44] Natale C, Velardi C and Visone C 2001 *Physica-B* **306** 161
[45] Cavallo A, Natale C, Pirozzi S and Visone C 2003 *IEEE Trans. on Magn.* **39** 1389
[46] Tan X and Baras J S 2005 *IEEE Trans. on Automatic Control* **50** 827
[47] Cavallo A, Natale C, Pirozzi S, Visone C and Formisano A 2004 *IEEE Trans. on Magn.* **40** 876
[48] Davino D, Giustiniani A, Vacca V and Visone C 2006 *IEEE Trans. on Magn.* **42** 3443
[49] Ekanayake D B, Iyer R V and Dayawansa W P 2007 *Proc. of American Control Conference* 4321
[50] Munkres J R 1991 *Analysis on Manifolds* (Addison-Wesley)
[51] Weisstein E W "Implicit Function Theorem." From MathWorld–A Wolfram Web Resource. http://mathworld.wolfram.com/ImplicitFunctionTheorem.html

On coupled Lane-Emden equations arising in dusty fluid models

D Flockerzi[1] **and K Sundmacher** [1,2]

[1] Max Planck Institute for Dynamics of Complex Technical Systems, Sandtorstr. 1, D-39106 Magdeburg, Germany

[2] Otto-von-Guericke-University, Universitätsplatz 2, D-39106 Magdeburg, Germany

E-mail: flockerzi@mpi-magdeburg.mpg.de

Abstract. We investigate the existence of solutions for mixed boundary value problems in coupled Lane-Emden equations. Such problems arise e.g. in the study of multicomponent diffusion and reaction processes inside catalyst particles under spherical symmetry. In contrast to the frequently applied decomposition method of Adomian, we employ the geometric theory of ODEs to show that this boundary value problem can be transformed into a terminal value problem. To this end, we determine the integral manifold on which solutions of the boundary value problem necessarily lie. This will be done in suitable warped and blown-up coordinates. Moreover, we comment on the numerical implementation.

1. Introduction

1.1. Some history

In astrophysics, the Lane-Emden equation is Poisson's equation for the gravitational potential of a self-gravitating, spherically symmetric and polytropic fluid at hydrostatic equilibrium:

$$\frac{d}{dr}\left(r^2 \frac{d\rho}{dr}\right) + r^2 \rho^n = 0, \qquad \rho(0) = 1, \quad \frac{d\rho}{dr}(0+) = 0. \tag{1.1}$$

Here, the pressure P is assumed to be proportional to the power $\frac{n+1}{n}$ of the 'density' $\delta = const \cdot \rho^n$ via the polytropic relation $P = const \cdot \delta^{1+1/n} = const \cdot \rho^{n+1}$ of index n. So the solution of (1.1) provides the dynamics of a scaled density and hence of a scaled pressure.

Jonathan Homer Lane (1819-1880), an American astrophysicist, was the first to perform a mathematical analysis of the sun as a gaseous body. With his work *On the theoretical temperature of the sun under the hypothesis of a gaseous mass maintaining its volume by its internal heat and depending on the laws of gases known from terrestial experiments* from 1870 (cf. [1]) he initiated the theory of stellar evolution. Jacob Robert Emden (1862-1940), a Swiss astrophysicist and meteorologist, provided a mathematical model as a basis of stellar structure by his work *Gas balls: Applications of the mechanical heat theory to cosmological and meteorological problems* in 1907 (cf. [2]). In a series of papers, Ralph Howard Fowler (1889-1944) considered a generalization of (1.1) of the form

$$\frac{d}{dr}\left(r^\alpha \frac{d\rho}{dr}\right) + r^\sigma f(\rho) = 0 \tag{1.2}$$

in particular for $f(\rho)$ equal to ρ^n or $e^{\rho n}$ (see e.g. [3], [4]). A simple coordinate change $r \mapsto \theta$ entails $u_{\theta\theta} + \theta^{\gamma} f(u) = 0$ for some γ (cf. Chapter 7 of [5] or [6]). By now, there is a vast literature on the generalized Emden-Fowler equation

$$\frac{d}{dr}\Big(p(r)\frac{d\rho}{dr}\Big) + q(r)\, f(\rho) \;=\; 0\,, \tag{1.3}$$

The review article [7] of J.S.W.Wong offers a comprehensive survey for the years up to 1975. For more recent theoretical and numerical work on generalized Emden-Fowler equations we refer to [8], [9], [10], [11], [12], [13], [14], [15] and [16], [17], [18], [19], [20] [21], [22] and the references therein. Often, numerical computations are based on Adomian's decomposition method (see e.g. [16] or [20]).

1.2. Lane-Emden equation in chemical engineering applications

One of the important fields of application of the Lane-Emden equation is the analysis of the diffusive transport and chemical reaction of species inside a porous catalyst particle. Aris [23] and Metha & Aris [24] were among the first who solved this problem as boundary value problem, assuming power-law kinetics for a single chemical reaction taking place inside a spherical catalyst particle - the most relevant practical catalyst geometry:

$$\frac{1}{r^2}\frac{d}{dr}\Big(r^2\frac{dc}{dr}\Big) \;=\; \phi^2\, c^n\,, \qquad c(0) = 1, \quad \frac{dc}{dr}(0+) = 0 \tag{1.4a}$$

where the so called Thiele modulus ϕ is a constant parameter. The derivation of equation (1.4a) is based on the assumption that the diffusion of the considered species inside the porous catalyst obeys Fick's law with a constant, i.e. concentration-independent, diffusivity.

In multicomponent mixtures, the transport of species in porous catalysts should not be described by Fick's law, but by the Maxwell-Stefan equations. The application of the latter equations to mass transport phenomena in porous solids leads to the so called Dusty Gas Model [25]. In this model, the solid catalyst is treated as 'dust', i.e. an ensemble of 'huge' motionless molecules which undergo exchange of momentum with the fluid species moving inside the porous catalyst structure. This approach accounts for three different kinds of transport contributions: bulk and Knudsen diffusion, surface diffusion and viscous Poiseuille-type flow. The Dusty Gas Model was originally formulated for ideal gas mixtures. Later it was generalized (named Dusty Fluid Model) to make it also applicable to nonideal multicomponent fluid mixtures, see e.g. [26], [27].

Combining the Dusty Fluid Model with the mass balances of species being transported inside a porous spherical catalyst body yields [28]:

$$\frac{1}{r^2}\frac{d}{dr}\Big(r^2\, D(x)\,\frac{dx}{dr}\Big) = N\, R(x)\,, \tag{1.4b}$$

where x stands for the vector of mole fractions of chemical species and $D(x)$ for the concentration-dependent matrix of diffusion coefficients. Note that, based on the theory of irreversible thermodynamics, it can be shown that D is the product of two positive definite matrices (cf. [29]) and thus similar to a diagonal matrix with positive diagonal elements (cf. [30], Thm.7.6.3). On the right-hand side of eq. (1.4b), N represents the stoichiometry matrix of the considered reaction network, and $R(x)$ represents the vector of reaction kinetic expressions. Assuming a constant D-matrix, one gets the following simplified catalyst diffusion model:

$$\frac{1}{r^2}\frac{d}{dr}\Big(r^2\,\frac{dx}{dr}\Big) = f(x), \qquad f(x) = D^{-1}\, N\, R(x)\,. \tag{1.4c}$$

Among all possible rate expressions, linear, bilinear and quadratic functions $R(x)$ are of highest relevance in chemical kinetics, due to the fact that chemical reactions are mostly based on mono-molecular and/or bi-molecular events. Focusing on this class of reaction kinetics, one can identify a sub-vector z containing the concentrations of all species which only appear in linear and bilinear rate expressions, and another sub-vector y containing the concentrations of the species which also appear in the quadratic terms. A practical example for this situation is the acid-catalysed dimerization of C4-olefines, a complex reaction accompanied by the formation of butane trimers, tetramers and isomers [31]. In this example the main reactant isobutene is the only species which appears in quadratic kinetic terms.

The just discussed class of catalyst diffusion problems can be reformulated as quadratic Lane-Emden boundary value problem (with $' = \frac{d}{dr}$):

$$\begin{aligned} y'' + \frac{2}{r}y' &= L_1(y)\,[K_{11}y + K_{12}z]\,, \quad y'(0+) = 0, \quad y(1) = \beta_1\,, \\ z'' + \frac{2}{r}z' &= L_2(y)\,[K_{21}y + K_{22}z]\,, \quad z'(0+) = 0, \quad z(1) = \beta_2\,. \end{aligned} \tag{1.5a}$$

for $x = (y^{\mathrm{T}}, z^{\mathrm{T}})^{\mathrm{T}}$ and $r \in (0,1]$ where the $L_j(y)$ are matrices, linear in y, and the K_{ij} are constant matrices, all of appropriate sizes. In a more compact notation, the problem can be given as follows:

$$x'' + \frac{2}{r}x' = L(y)\,Kx\,, \quad x'(0+) = 0, \quad x(1) = \beta, \tag{1.5b}$$

for $x = (y^{\mathrm{T}}, z^{\mathrm{T}})^{\mathrm{T}}$ and $r \in (0,1]$.

1.3. Outlook and summary

In Section 2 we treat the scalar boundary value problem

$$(r^2x')' - k\,r^2\,x^n = 0\,, \tag{1.6a}$$

$$x'(0+) = 0\,, \quad x(1) = \beta > 0 \tag{1.6b}$$

of Lane-Emden-Fowler type (with $' = \frac{d}{dr}$). Based on the geometric theory of ordinary differential equations, we first determine the integral manifold $\mathcal{S}$ in (r, x, x')-space of the (1.6a)-solutions which satisfy the boundary condition $x'(0+) = 0$ and thus reveal the underlying geometric structure. In the scalar case, the incoming manifold $\mathcal{S}$ will be shown to be, globally, the graph of a smooth function $x' = y_s(r,x)$ so that, in a second step, the boundary value problem (1.6a)&(1.6b) can be converted to a terminal value problem (with $x'(1) = y_s(1,\beta)$). We establish in Section 3 that this geometric approach can also be taken for coupled Lane-Emden equations as (1.5a).

The idea of converting boundary value into initial value problems has a long history. Here, we just refer to [6], where W.F. Ames and E. Adams have used group methods for the Emden-Fowler equations with boundary conditions that are different from ours.

For a numerical implementation, the Taylor polynomials – near $x' = 0$ – for the incoming manifold $\mathcal{S}$ can be easily computed in a recursive algorithm. The backward flow of such a local approximation of the manifold then generates an approximation of the global incoming manifold. Its intersection with the terminal plane $\{(r, x, x') : r = 1, x = \beta\}$ gives an approximation for the initial value of the desired solution of (1.5a). In this 2-dimensional case, it is still open whether the incoming manifold $\mathcal{S}$ will be the graph of a smooth function $x' = f(r,x)$ over a sufficiently large domain containing $(r,x) = (1,\beta)$.

Generalizations to positive $K = K(x)$ or even to positive $K = K(r,x)$ are obvious since the constant matrix K in (1.5b) is not crucial for the existence of the above-mentioned incoming manifolds and thus not essential for the geometry inherent in these Lane-Emden type equations. One might as well consider differential operators $\frac{d}{dr}\,(r^p\,\frac{d}{dr})$ with $p \neq 2$.

2. Scalar Lane-Emden boundary value problems

2.1. The model for spherical porous catalyst pellets
We consider the boundary value problem

$$x'' + \frac{2}{r}\,x' \;=\; k\,x^n\,, \qquad x'(0+) = 0\,,\; x(1) = \beta > 0\,, \tag{2.1}$$

for $x \in \mathbb{R}$, $r \in (0,1]$, $k > 0$. Since the linear cases for $n = 0$ and $n = 1$ are readily solved – with the solutions $x(r) = \beta + k(r^2 - 1)/6$ and $x(r) = \beta\,sinh(\sqrt{k}\,r)/(r\,sinh\sqrt{k})$ resp. – we take $n \geq 2$ in the following discussions. Of course, the second order system (2.1) can equivalently be written as the first order non-autonomous system

$$x' = y, \quad y' = k\,x^n - \frac{2}{r}\,y\,, \qquad y(0+) = 0\,,\; x(1) = \beta \tag{2.2a}$$

in $\mathbb{R}^2$. System (2.1) was investigated in [32] and [33] by means of Adomian's decomposition method.

We introduce a 'time' variable t via $r = e^{-t}$ and employ the notation $\dot{} = \frac{d}{dt}$. So the bijection $t \in [0,\infty) \leftrightarrow r = e^{-t} \in [0,1]$ can be thought of as the solution of the initial value problem $\dot{r} = -r$, $r(0) = 1$. Hereby, the second order system (2.1) can equivalently be written as the first order autonomous system

$$\dot{r} = -r\,, \quad \dot{x} = -ry\,, \quad \dot{y} = 2y - krx^n\,, \qquad r(0) = 1\,,\; x(0) = \beta\,,\; y(\infty) = 0 \tag{2.2b}$$

in $\mathbb{R}^3$. The differential equation in (2.2b) possesses the x-axis $\mathcal{X} = \{(r,x,y) = (0,x,0) : x \in \mathbb{R}\}$ as a line of equilibria. Moreover, the r-axis is invariant for the differential equation in (2.2b). The linearizations at equilibria $(0,x_0,0)$ are given by

$$J(x_0) = \left(\begin{array}{cc|c} -1 & 0 & 0 \\ 0 & 0 & 0 \\ \hline -kx_0^n & 0 & 2 \end{array}\right) \tag{2.3}$$

possessing the eigenvalues -1, 0 and 2 with corresponding eigenspaces spanned by $(1, 0, \frac{k}{3}x_0^n)^{\mathrm{T}}$, $(0,1,0)^{\mathrm{T}}$ and $(0,0,1)^{\mathrm{T}}$ respectively. Thus, near such points $(0,x_0,0)$, there exist local center-stable manifolds – tangent to the center-stable eigenspaces – of the form $y = y_{cs}^{x_0}(r,x)$ with $y_{cs}^{x_0}(0,x) \equiv 0$ and $y_{cs}^{0}(r,0) \equiv 0$.

These local center-stable manifolds form the stable manifold $\mathcal{M}_s$ for compact parts $\mathcal{X}^0 = \{(r,x,y) = (0,x,0) : x \in [0,x^0]\}$. It allows – locally near $r = 0$ – the representation $y = y_s(r,x)$ with

$$y_s(r,0) = 0 \quad \text{and} \quad y_s(0,x) = 0\,. \tag{2.4a}$$

The flow on $\mathcal{M}_s$ is then determined by

$$\dot{r} = -r\,, \quad \dot{x} = -ry_s(r,x) \quad \text{or} \quad x' = y_s(r,x) \tag{2.4b}$$

with $(x, x') = (x_0, 0)$ for $r = 0+$ and $x_0 \in [0, x^0)$.

Figure 1 shows the stable manifold $\mathcal{M}_s$ in the phase portrait of (2.2b) for $n = 2$ and $k = 1$. The blue curve, connecting the vertical line $\mathcal{L} = \{(r, x, y) = (1, \beta, y) : y \in \mathbb{R}\}$ and the horizontal line $\mathcal{X}$, represents the sought boundary value solution within the stable manifold of the equilibrium manifold $\mathcal{X}^0$.

We will derive this stable manifold in the following subsection. To this end, we will define a suitable change of coordinates $(r, x, y) \mapsto (r, \xi, \eta)$ such that the origin of the (ξ, η)-system (cf. (2.9b) below) will have a unique smooth stable manifold $\eta = S(\xi)$ giving rise to the stable manifold of $\mathcal{X}^0$ for system (2.2b) (cf. in particular Remark 2.1).

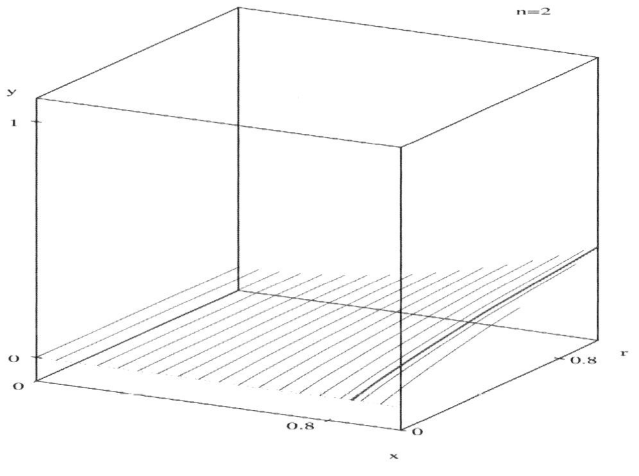

Figure 1. 'Geometry' of (2.2b) for $n = 2$.

2.2. Analysis of the model

Suppose $x(r)$ is the solution of (2.1) – satisfying (2.4b) for r near 0 – and define for positive σ the scalings

$$u(r) = r^\sigma x(r), \quad r = e^{-t}, \quad v(t) := u(r) \tag{2.5}$$

with the 'time' $t \geq 0$. We present the analysis of system (2.1) in these new coordinates.

(a) We compute the resulting boundary value problem for $v = v(t)$ and $w := \dot{v}(t)$ and note that

$$\sigma = \tfrac{2}{n-1} \in (0, 2] \tag{2.6}$$

generates the autonomous system

$$\begin{gathered} \dot{v} = w, \quad \dot{w} = \sigma(1-\sigma)\, v + (1-2\sigma)\, w + k\, v^n, \\ v(0) = \beta, \; v(\infty) = w(\infty) = 0 \end{gathered} \tag{2.7a}$$

with

$$x'(0+) = 0 \quad \Leftrightarrow \quad \lim_{t \to \infty} e^{(\sigma+1)t}[\sigma v(t) + w(t)] = 0\,. \tag{2.7b}$$

(b) The origin $(0,0)$ is the only equilibrium of (2.7a) in the right half-plane R for $n \geq 3$. For $n \in [2,3)$, there's a second equilibrium – of saddle type – in the right half-plane R. The eigenvalues of the linearization of (2.7a) at the origin are $-\sigma$ and $1-\sigma$. The eigenspace corresponding to the smaller eigenvalue $-\sigma$ is given by $\sigma v + w = 0$. The origin is exponentially stable for $n \in [2,3)$ and a saddle for $n > 3$. For $n = 3$, it is a weak saddle with a stable manifold and a 'weakly unstable' center manifold. The condition (2.7b) reveals that the desired solution of the boundary value problem has to lie on the strongly stable manifold of the origin.
Moreover, the case $n = 5$ is special in the sense that the vector field in (2.7a) will be divergence free. So, system (2.7a) will be integrable for $n = 5$. A first integral for (2.9b) will be given – in terms of new coordinates – in (2.9d) below.

(c) We employ the linear change

$$(v, w) \mapsto (v, z) = (v, \sigma v + w) \tag{2.8a}$$

leading to

$$\dot{v} = -\sigma v + z\,, \quad \dot{z} = (1-\sigma)z + kv^n\,, \tag{2.8b}$$

such that the (strong) stable eigenspace of the linearized system at the origin is given by the v-axis.

(d) Finally, to single out the strongly stable direction, we use the blow-up

$$(v, z) \mapsto (\xi, \eta) = (v^{1/\sigma}\,,\ z/v) \tag{2.9a}$$

and arrive at the following *autonomous blown-up quadratic planar system* on the right half-plane $H = \{(\xi, \eta) : \xi \geq 0\}$:

$$\dot{\xi} = \xi\,[\frac{\eta}{\sigma} - 1]\,, \quad \dot{\eta} = \eta - \eta^2 + k\,\xi^2\,. \tag{2.9b}$$

The origin $(\xi, \eta) = (0,0)$ now is a saddle for (2.9b) with a unique stable and a unique unstable manifold. The unstable manifold is contained in the η-axis. The local stable manifold $\eta = S(\xi)$ is tangent to the ξ-axis and thus of the form $S(\xi) = S_2\xi^2 + h.o.t.$ Moreover, the global stable manifold is a strictly decreasing function $S(\xi)$ over $\xi \geq 0$ with

$$0 > S(\xi) \geq -2k\xi\,[1 + \sqrt{1+4k\xi}\,]^{-1/2} \quad \text{for} \quad \xi > 0 \tag{2.9c}$$

where the lower bound comes from the nullcline of η. The reduced flow on $\eta = S(\xi)$ is determined by $\dot{\xi} = \xi\,[-1 + S(\xi)/\sigma]$.
For $n = 5$, the system (2.9b) possesses the first integral

$$I(\xi, \eta) = \tfrac{k}{3}\xi^3 + (\eta - \eta^2)\xi. \tag{2.9d}$$

We show the phase portrait of equation (2.9b) for $n = 2$, i.e. $\sigma = 2$, and $k = 1$ in figure 2 and for $n = 4 > 3$, i.e. $\sigma \leq 1$, and $k = 1$ in figure 3 – the blue curve $\eta = \eta(\xi)$ represents the stable manifold of the origin of equation (2.9b) and corresponds to the blue solution of the boundary value problem (2.2b) shown in figure 1.

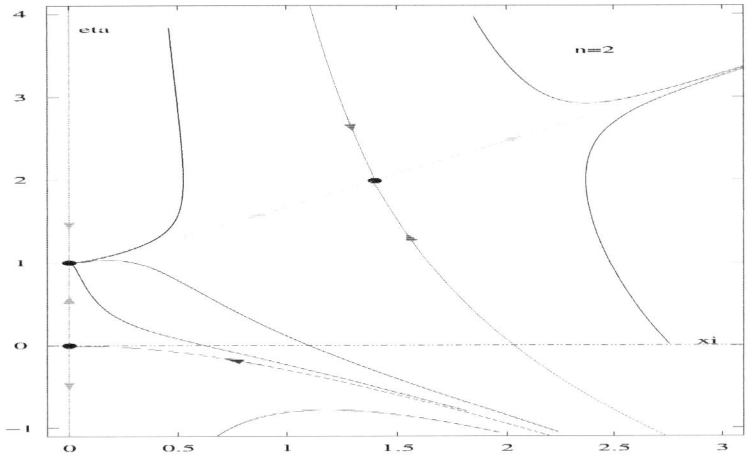

Figure 2. System (2.9b) for $n = 2$.

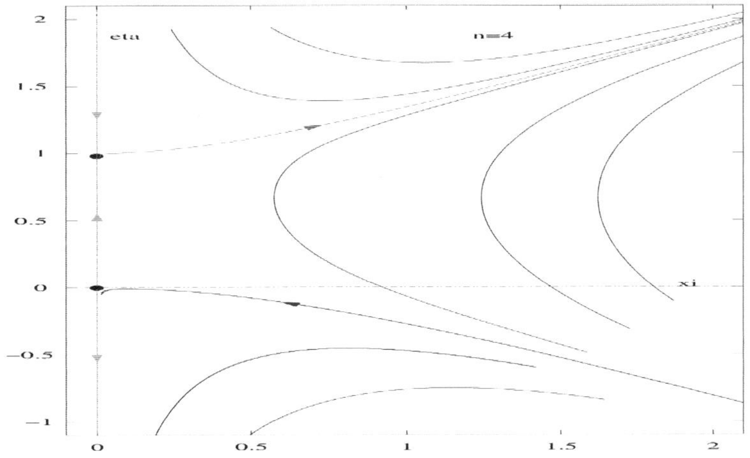

Figure 3. System (2.9b) for $n = 4$.

Remark 2.1 (Alternative approach)
One may compute the stable manifold $y = y_s(r, x)$ of the equilibrium line $\mathcal{X}^0$ of the 3D-system (2.2b) directly in a particular form. First note that the stable manifold $\eta = S(\xi) = S_2\xi^2 + h.o.t.$ of the origin in (2.9b) corresponds to an invariant manifold, the stable manifold of $\mathcal{X}^0$, for (2.2b) given by

$$r\,y \;+\; S(r x^{1/\sigma})\, x \;=\; 0\,.$$

Hence, the Ansatz

$$y = y_s(r, x) := -\;\frac{x}{r}\,S(\xi) \qquad with \qquad \xi := r\,x^{1/\sigma}\,, \tag{2.10a}$$

$S(\xi) = \mathcal{O}(\xi^2)$ and $1/\sigma = \frac{n-1}{2} \geq 1/2$ for the stable manifold $\mathcal{M}_s$ of the equilibrium line $\mathcal{X}^0$ will lead to a determining equation for S that is exactly the determining equation for $\eta = S(\xi)$ being the stable manifold of the origin with respect to equation (2.9b), i.e.

$$\xi\,[\tfrac{2S}{n-1} - 1]\,S_\xi \;=\; S - S^2 + k\xi^2\,, \quad S(0) = 0 \;\; S_\xi(0) = 0\,. \tag{2.10b}$$

Equation (2.10a) shows a special dependence of the invariant manifold on (r, x) and thus reveals the special 'symmetry' of the invariant manifold.

We would like to add that $\mathcal{M}_s$ restricted to $(r,x) \in [0,1] \times [0,x^0]$ is positive invariant with respect to (2.2b). This is a consequence of (2.4a) and (2.9c). □

We summarize the above analysis:

Theorem 2.2 (1D Lane-Emden boundary value problem)
The scalar Lane-Emden boundary value problem (2.1) has a unique positive solution $x = x(r)$. Its terminal value at $r = 1$ is given by

$$x(1) = \beta > 0, \qquad x'(1) = -\,\beta\, S(\beta^{(n-1)/2}) > 0\,, \tag{2.11}$$

where the function $\eta = S(\xi)$, $\xi \geq 0$, represents the unique stable manifold of the origin of (2.9b) within the right half-plane $H = \{(\xi,\eta) : \xi \geq 0\}$ (cf. (2.10b)). The solution of (2.1) can be computed from the reduced equation (cf. (2.4b))

$$x' = -\,\frac{x}{r}\,S(\,r x^{(n-1)/2}\,) \geq 0 \qquad x(1) = \beta\,. \tag{2.12}$$

We close this section with a few computational and technical remarks.

Remark 2.3

(a) The Taylor polynomials at $\xi = 0$ for Σ can be easily computed in a recursive algorithm. The backward flow of such a local approximation of the stable manifold then generates an approximation of the global stable manifold over $[0,\beta]$. Its intersection with the line $\{(\xi,\eta) : \xi = \beta\}$ determines the (approximate) initial value for the desired solution of (2.9b).

(b) Instead of the blow-up (2.9a) one could use

$$(v,z) \mapsto (\xi,\eta) = (v\,,\,z/v) \tag{2.13a}$$

and arrive at

$$\dot\xi = \xi\,[\,\eta - \sigma\,]\,, \quad \dot\eta = \eta - \eta^2 + k\,\xi^{n-1}\,. \tag{2.13b}$$

This system can be discussed in complete analogy to system (2.9b). Just note that the case $n = 2$ yields an additional linear term $k\xi$ so that the stable manifold of the origin is of the form $\eta = \Sigma(\xi) := -\frac{k}{3}\xi + h.o.t.$. In the original coordinates one then has the invariance of

$$ry + x\Sigma(r^\sigma x) = 0 \quad \text{with} \quad \Sigma(0) = 0\,. \tag{2.13c}$$

(c) Generalizations to positive $k = k(x)$ or even to positive $k = k(r,x)$ for $0 \leq r \leq 1$, $0 \leq x \leq x^0$ are obvious. One just has to ensure the existence of the stable integral manifold for the blown-up planar systems corresponding to (2.9b).

(d) Up to System (2.9b), the above analysis of System (2.1) is valid for negative k too. □

3. The 2D Lane-Emden boundary value problem from the dusty fluid approach

We consider the boundary value problem (1.5a)&(1.5b) in the form

$$\begin{aligned} x_1'' + \frac{2}{r} x_1' &= x_1 \left[K_{11} x_1 + K_{12} x_2 \right], \quad x_1'(0+) = 0, \quad x_1(1) = \beta_1 , \\ x_2'' + \frac{2}{r} x_2' &= x_1 \left[K_{21} x_1 + K_{22} x_2 \right], \quad x_2'(0+) = 0, \quad x_2(1) = \beta_2 . \end{aligned} \tag{3.1}$$

for $x = (x_1, x_2)^T := (y, z)^T \in \mathbb{R}^2$ and $r \in (0, 1]$ with positive constants β_i and a nonzero matrix K. We note that $x_1 \equiv 0$ is invariant for the differential equation in (3.1) whereas $x_2 \equiv 0$ is not in case $K_{21} \neq 0$.

We present the analysis of system (3.1) in new coordinates given by

$$u_i(r) = r^2 x_i(r), \quad r = e^{-t}, \quad v_i(t) := u_i(r) \tag{3.2}$$

with the 'time' $t \geq 0$ where the exponent 2 comes from the Fowler-transformation $u_i = r^\sigma x_i$ with $\sigma = \frac{2}{n-1}$ for the quadratic right-hand side of (3.1) ($n = 2$, cf. the scalings in (2.5)). Thereby we are let to the following *autonomous Lane-Emden boundary value problem* in $\mathbb{R}^4$ for $v_i(t) = u_i(r)$:

$$\begin{aligned} \dot v_1 &= w_1 \\ \dot w_1 &= -2\, v_1 - 3\, w_1 + \left[K_{11} v_1 + K_{12} v_2 \right] v_1 \\ \dot v_2 &= w_2 \\ \dot w_2 &= -2\, v_2 - 3\, w_2 + \left[K_{21} v_1 + K_{22} v_2 \right] v_1 \end{aligned} \tag{3.3a}$$

with

$$v(0) = \beta , \quad v(\infty) = 0 = w(\infty) \tag{3.3b}$$

and

$$x_i'(0+) = 0 \quad \Leftrightarrow \quad \lim_{t \to \infty} e^{3t} \left[2 v_i(t) + w_i(t) \right] = 0 . \tag{3.3c}$$

The Jacobian blocks $\begin{pmatrix} 0 & 1 \\ -2 & -3 \end{pmatrix}$ have the negative eigenvalues -1 and -2 with the eigenspaces $v + w = 0$ and $2v + w = 0$ respectively. As in Section 2, the boundary conditions $x_i'(0+) = 0$ ask for the determination of the strong stable manifolds. We will proceed in three steps.

(a) First we note that a blow-up of the fast eigen-directions via

$$\begin{aligned} &\begin{pmatrix} v_1 \\ w_1 \end{pmatrix} \to \begin{pmatrix} v_1 \\ z_1 = 2 v_1 + w_1 \end{pmatrix} \to \begin{pmatrix} \xi_1 = v_1 \\ \eta_1 = z_1 / v_1 \end{pmatrix}, \\ &\begin{pmatrix} v_2 \\ w_2 \end{pmatrix} \to \begin{pmatrix} v_2 \\ z_2 = 2 v_2 + w_2 \end{pmatrix} \to \begin{pmatrix} \xi_2 = v_2 \\ \eta_2 = z_2 / v_1 \end{pmatrix} \end{aligned} \tag{3.4a}$$

is adapted to the common factor x_1 on the right-hand side of (3.1) and hence to the common factor v_1 in the quadratic terms on the right-hand side of (3.3a). The transformation (3.4a) leads to the *autonomous blown-up Lane-Emden equations* in $\mathbb{R}^4$:

$$\begin{aligned} \dot\xi_1 &= -2\xi_1 && && +\xi_1 \eta_1 \\ \dot\xi_2 &= && -2\xi_2 && +\xi_1 \eta_2 \\ \dot\eta_1 &= K_{11}\xi_1 && +K_{12}\xi_2 + \eta_1 && -\eta_1^2 \\ \dot\eta_2 &= K_{21}\xi_1 && +K_{22}\xi_2 + \eta_2 && -\eta_1 \eta_2 . \end{aligned} \tag{3.4b}$$

We note that $\xi_1 = 0$ implies $\dot{\xi}_1 = 0$ so that

$$\{\, \xi_1 = 0 \,\} \text{ is invariant.} \tag{3.4c}$$

(b) This blown-up system (3.4b) has the origin as a saddle with Jacobian

$$\left(\begin{array}{c|c} -2\,I & 0 \\ \hline K & I \end{array}\right) = \left(\begin{array}{cc|cc} -2 & 0 & 0 & 0 \\ 0 & -2 & 0 & 0 \\ \hline K_{11} & K_{12} & 1 & 0 \\ K_{21} & K_{22} & 0 & 1 \end{array}\right) \tag{3.5a}$$

with eigenvalues $+1$ and -2. The corresponding (unique) 2D stable manifold of the origin for the blown-up system (3.4b) corresponds to the strong stable manifold of system (3.3a). The stable manifold of (3.4b) is locally of the form

$$\eta \;=\; \Sigma(\xi) \;=\; -\frac{1}{3}\,K\xi \;+\; h.o.t. \tag{3.5b}$$

So, for positive K_{ij}, $\Sigma(\xi)$ is negative for small positive ξ_1 and ξ_2 (cf. Remark 2.3(b)).

(c) The computations are done via two *reduced 2D-systems*. The first is describing the invariance of the stable manifold given by $\eta_i = \Sigma_i(\xi)$, i.e.

$$\dot{\eta} \;= \Sigma_\xi\, \dot{\xi} \quad \text{on} \quad \eta = \Sigma(\xi)\,,\; \Sigma(0) = 0\,. \tag{3.6a}$$

This 2-dimensional partial differential system (3.6a) reads

$$\begin{aligned} K_{11}\xi_1 + K_{12}\xi_2 + \Sigma_1 - \Sigma_1^2 \quad &= \;(\Sigma_1)_{\xi_1}\,[\Sigma_1\xi_1 - 2\xi_1] + (\Sigma_1)_{\xi_2}\,[\Sigma_2\xi_1 - 2\xi_2]\,, \\ K_{21}\xi_1 + K_{22}\xi_2 + \Sigma_2 - \Sigma_1\Sigma_2 \quad &= \;(\Sigma_2)_{\xi_1}\,[\Sigma_1\xi_1 - 2\xi_1] + (\Sigma_2)_{\xi_2}\,[\Sigma_2\xi_1 - 2\xi_2]\,. \end{aligned} \tag{3.6b}$$

The Taylor polynomials for the components Σ_j of Σ can be easily computed in a recursive algorithm. The backward flow of a local approximation of the stable manifold then generates an approximation of the global stable manifold. Its intersection, if a singleton, with the plane $\{(\xi,\eta) : \xi = \beta\}$ determines the (approximate) initial value for the desired solution. In case the global stable manifold is of the form $\eta = \Sigma_{glob}(\xi)$ over a sufficiently large domain in ξ-space (containing β), the second 2D-system determines the flow within the stable manifold via

$$\dot{\xi} \;= \xi_1\, \Sigma_{glob}(\xi) \;-\; 2\,\xi \quad \text{with} \quad \xi(0) = \beta \tag{3.6c}$$

with 2-dimensional ξ.

We summarize the above analysis:

Theorem 3.1 (2D Lane-Emden boundary value problem)
The solution of the two-dimensional Lane-Emden boundary value problem (3.1) satisfies for small $r > 0$

$$x'(r) = -\;\frac{x_1}{r}\,\Sigma(\,r^2 x) \tag{3.7}$$

where the right-hand side is given in terms of the function $\eta = \Sigma(\xi)$ representing the unique local stable manifold $\mathcal{M}_{loc}$ of system (3.4b) (cf. (3.5b)).
The intersection, if a singleton, of the corresponding global stable manifold $\mathcal{M}_{glob}$ of system (3.4b) with the plane $\{(\xi,\eta) : \xi = \beta\}$ determines the terminal value $x'(1)$. Moreover, the first solution component $x_1(r)$ is positive on $[0,1]$.

References

[1] Lane J H 1870 *Am. J. Science and Arts 2nd ser.* **50** 57
[2] Emden J R 1907 *Gaskugeln – Anwendungen der mechanischen Wärmetheorie auf Kosmologie und meteorologische Probleme* (Leipzig: B.G.Teubner)
[3] Fowler R H 1930 *Monthly Notices of the Royal Astronomical Society* **91** 63
[4] Fowler R H 193 *Quart. J. of Mathematics* **2** 259
[5] Bellman R 1953 *Stability theory of differential equations* (New York: McGraw-Hill)
[6] Ames W F and Adams E 1979 *Int. J. Nonlinear Mechanics* **14** 35
[7] Wong J S W 1975 *SIAM Review* **17** (2) 339
[8] Bandle C 1991 *ZAMM* **71** T565
[9] Goenner H 2000 *J. of Mathematical Physics* **41** (10) 7029
[10] Goenner H 2001 *General Relativity and Gravitation* **33** (5) 833
[11] Wong J S W 2002 *J. of Mathematical Analysis and Applications* **274** 746
[12] Berkovich L M 2007 *Applicable Analysis and Discrete Mathematics* **1** 122
[13] Govinder K S and Leach P G L 2007 *J. of Nonlinear Mathematical Physics* **14** (3) 443
[14] Khalique C M and Ntsime P 2008 *New Astronomy* **13** 476
[15] Cariñena J F, Leach P G L and de Lucas J 2009 *J. of Mathematical Physics* **50** 103515 1-21
[16] Wazwaz A M 2005 *Applied Mathematics and Computation* **166** 638
[17] Momoniat E and Harley C 2006 *New Astronomy* **11** 520
[18] Batiha K 2007 *American J. of Applied Sciences* **4** (7) 439
[19] Parand K and Taghavi A 2008 *Applied Mathematical Sciences* **2** (60) 2955
[20] Aslanov A 2009 *Int. Journal of Computer Mathematics* **86** (5) 807
[21] Geng F, Cui M and Zhang B 2010 *Nonlinear Analysis: Real World Applications* **11** 637
[22] Karimi S, Vanani A and Aminataei A 2010 *Computers and Mathematics with Applications* **59** 2815
[23] Aris R 1965 *Introduction to the analysis of chemical reactors* (Englewood Cliffs, N.J.: Prentice-Hall)
[24] Metha B N and Aris A 1971 *J. of Mathematical Analysis and Applications* **36** 611
[25] Jackson R 1977 *Transport in porous catalysts* (Amsterdam: Elsevier Scientific Publishing Company)
[26] Higler A, Krishna R and Taylor R 2000 *Ind. Eng. Chem. Res.* **39** 1596
[27] Krishna R and Wesselingh J A 1997 *Chem. Eng. Sci.* **52** (6) 861
[28] Katariya A, Freund H, Flockerzi D and Sundmacher K 2010 Approximated solution of multicomponent diffusion effects on multiple reactions in porous catalysts, *AIChE Journal* to be submitted
[29] Taylor R and Krishna R 1993 *Multicomponent mass transfer* (New York: John Wiley & Sons)
[30] Horn R A and Johnson C R 1985 *Matrix Analysis*, (Cambridge: Cambridge University Press)
[31] Talwalkar S, Mankar S, Kataiya A, Aghalayam A, Ivanoa M, Sundmacher K 2007 and Mahajani S *Ind. Eng. Chem. Res.* **46** 3024
[32] Shi-Bin L , Yan-Ping S and Scott K 2003 *Chem. Eng. Technology* **26** 87
[33] Yan-Ping S , Shi-Bin L and Scott K 2004 *Chem. Eng. Journal* **102** 1

Analysis of phase transitions in spin-crossover compounds by using atom – phonon coupling model

A Gîndulescu[1,2], A Rotaru[2] , J Linares[1], M Dimian[2], and J Nasser[3]

[1]GEMaC, CNRS-UMR 8635, Université de Versailles St. Quentin en Yvelines, 78035 Versailles, France
[2]Stefan cel Mare University, Electrical Engineering and Computer Science, Suceava, Romania
[3] Lab. LISV, Université de Versailles St. Quentin en Yvelines, 78035 Versailles, France

E-mail : jlinares@physique.uvsq.fr, aurelian.rotaru@gmail.com

Abstract. The spin – crossover compounds (SCO) have become of great interest recently due to their potential applications in memories, sensors, switches, and display devices. These materials are particularly interesting because upon application of heat, light, pressure or other physical stimulus, they feature a phase transition between a low-spin (LS) diamagnetic ground state and a high-spin (HS) paramagnetic state, accompanied in some cases by color change. The phase transition can be discontinuous (with hysteresis), in two steps or gradual. Our analysis is performed by using the atom – phonon coupling (APC) model which considers that neighboring molecules are connected through a spring characterized by an elastic constant depending on molecules electronic state. By associating a fictitious spin to each molecule that has -1 and +1 eigenvalues corresponding to LS and HS levels respectively, an Ising type model can be developed for the analysis of metastable states and phase transitions in spin-crossover compounds. This contribution is aimed at providing a review of our recent results in this area, as well as novel aspects related to SCO compounds behavior at low temperature. In the framework of the APC model, we will discuss about the existence of metastable and unstable states, phase transitions and hysteresis phenomena, as well as their dependence on sample size.

1. Introduction

The spin – crossover compounds (SCO) have become of great interest recently due to their potential applications in memories, sensors, switches, and display devices [1-3]. These materials are particularly interesting because upon application of heat, light, pressure or other physical stimulus, they feature a phase transition between a low-spin (LS) diamagnetic ground state and a high-spin (HS) paramagnetic state, accompanied by color change. The phase transition can take a variety of shapes depending on the material characteristics.

Spin crossover phenomena occurs in some octahedrically coordinated transition metal compounds with 4 to 7 electrons in the 3d orbital. As a consequence of the splitting of the energy of d orbitals into the t_{2g} and e_g sets in a ligand field, octahedral complexes may exist in high or low spin states. The transition takes place due to the competition between δ and Π , where δ is the splitting energy between the t_{2g} and e_g orbitals and Π is the spin pairing energy (see figure 1).

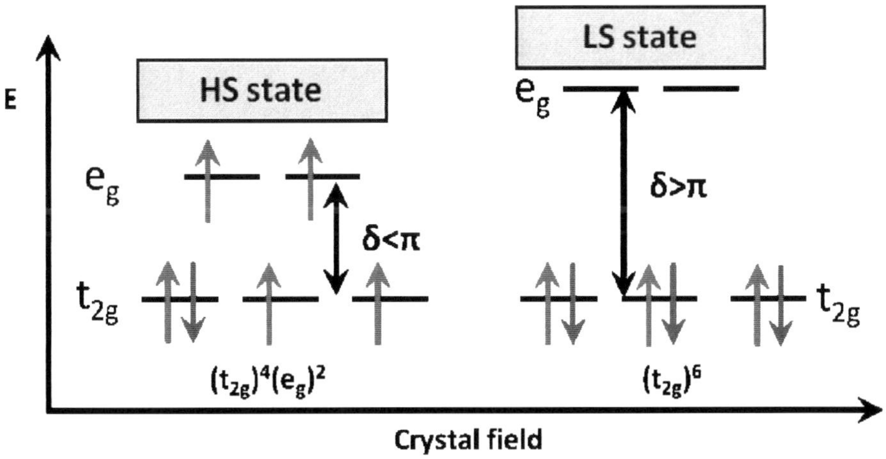

Figure 1. Electronic diagram of the HS and LS states for a Fe(II) ion in an octahedral ligand field.

The spin crossover or spin transition between the two states of a SCO leads to different physical properties of the systems such as magnetic, optic, structural and vibrational which can be determined using a very wide range of characterization techniques: structural characterization, magnetic susceptibility measurements, X-Ray diffraction, Raman spectroscopy, infrared and UV-visible spectroscopy, calorimetric, reflectance and ellipsometry measurements and optic microscopy [4-6]. While in the LS state we can talk about a diamagnetic material, in the HS state the compound is paramagnetic. The transition might be followed by a color change of the material. Most of the SCO found in LS state have a pink or violet color, while in the HS state they are colorless, pale yellow or white.

According to the type and the intensity of the interactions between the molecules, the SCO compounds present different types of spin transition, with or without hysteresis. The transitions with hysteresis present a considerable industrial interest, due to their applicability, for example, in information storage. The presence of hysteresis may be associated with the presence of very strong long - range interactions within the solid or with a change of crystallographic phase (in which case the system is called strongly cooperative).

The article is aimed at providing a review of our recent results in this area as well as novel aspects related to SCO compounds behavior at low temperature. The first part will introduce the Ising model and its applicability to describe SCO, followed by a short presentation of the atom-phonon coupling (APC) model. The third section presents the role of the parameters in the APC model. Next sections are dedicated to relaxation dynamics in SCO, the size effect on their behaviour and a low temperature analysis of obtained metastable and unstable states.

2. The Ising model

2.1. Configuration diagram

The configuration diagram representing the variation of the potential energy E in function of the Fe-ligand distance for the case when the compound is in the HS state, respectively in the LS state is shown in figure 2. The energy difference between the two states is given mainly by the ligand field force around Fe(II). The variation of the Fe-ligand field leads to a transition between the LS state and HS state.

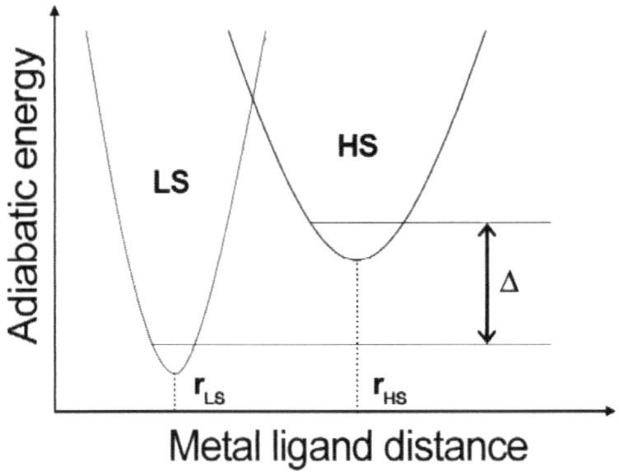

Figure 2. Representation of potential energy for the LS and HS state for a Fe(II) compound with spin conversion.

2.2. Thermodynamic aspects

As we mentioned earlier the interactions between the molecules (intermolecular interactions) play an important role in the phase transitions and influence the shape of the graphical representation of the high spin fraction with the variation of the temperature. In the case in which the hysteresis is present (figure 3) the system's response is going to follow two different paths when the sample is heated and cooled. In the figure $T_{1/2}\uparrow$ is the transition temperature in heating, $T_{1/2}\downarrow$while is the one in cooling. The hysteresis width is given by:

$$\Delta T = T_{1/2}\uparrow - T_{1/2}\downarrow \quad (1)$$

so for a given temperature that is comprised between the two temperatures defined earlier, the system can be either in the HS state, either in the LS state, depending on its previous state. If N_{HS} and N_{LS} are the number of molecules in HS state, respectively LS state, we can characterize the macroscopic state of our system using a parameter called the high spin fraction, defined as $n_{HS}= N_{HS}/(N_{HS}+N_{LS})$

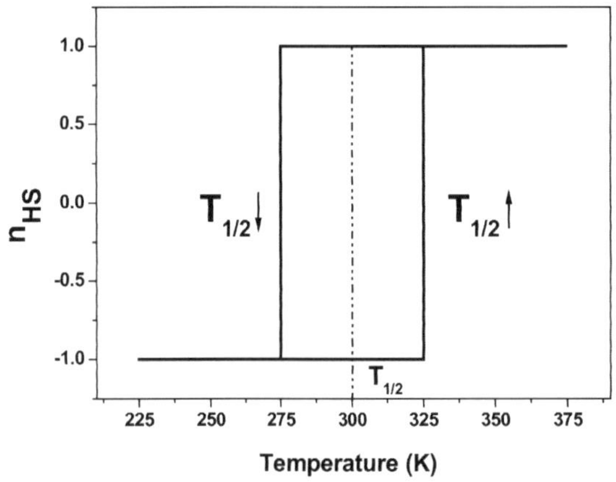

Figure 3. n_{HS} variation in function of temperature.

In order to study the behavior of SCO, a theoretical approach is used, in which to each spin we associate a fictitious operator <σ> with two eigenvalues: +1 for the HS state and -1 for the LS state.

If we consider that molecules do not interact with each other, the energy of a molecule is given by:

$$E = \pm\frac{\Delta}{2}, \tag{2}$$

where Δ is the energy difference between the minimal energies of the two states, the minus sign corresponding to the LS state, while the plus is for the HS state. Taking into account the Boltzmann law and the degenerescences for the two states g_{LS} and g_{HS} the "mean" state of a molecule at a certain temperature is given by:

$$<\sigma>=\frac{-g_{LS}\exp(\beta\times\Delta/2)+g_{HS}\exp(-\beta\times\Delta/2)}{g_{LS}\exp(\beta\times\Delta/2)+g_{HS}\exp(-\beta\times\Delta/2)}, \tag{3}$$

where $\beta=\frac{1}{k_B T}$

or

$$<\sigma>=\frac{r\exp(-\beta\times\Delta)-1}{r\exp(-\beta\times\Delta)+1} \tag{4}$$

if we consider $r=\frac{g_{HS}}{g_{LS}}$.

We consider N is the total number of molecules, N_{HS} the number of molecules in the HS state, we can define the high spin fraction n_{HS}:

$$<\sigma>=\frac{N_{HS}\times(1)+(N-N_{HS})\times(-1)}{N}=n_{HS}-(1-n_{HS}) \tag{5}$$

from which, n_{HS}, is given by:

$$n_{HS}=\frac{<\sigma>+1}{2} \tag{6}$$

If we consider that the molecules are interacting with each other, the term that describes in a phenomenological manner the interaction between the molecules (i) and (j) has the following form: $-J\sigma_i\sigma_j$. If the variable J is positive then the energy $-J\sigma_i\sigma_j$ is minimal because σ_i and σ_j have the same value being in the same state. Therefore the total energy of the system for a number of N molecules is given by:

$$\frac{\Delta}{2}\sum_{i=1}^{N}\sigma_i - \sum_{<i,k>}J\sigma_i\sigma_j \tag{7}$$

where the first sum takes into consideration all the molecules, while the second one refers to the pairs between neighboring molecules.

We apply the mean field approximation technique in order to find the average value of σ. For this we consider a particular molecule (i) isolating in equation (7) the part where this molecule interferes:

$$\frac{\Delta}{2}\sigma_i - \sigma_i\sum_{j\ neighboring\ molecules}J\sigma_j \tag{8}$$

$$\left[\frac{\Delta - 2\sum\limits_{j\ neighboring\ molecules}J\sigma_j}{2}\right]\sigma_i \tag{9}$$

From equation (9) we can consider that the molecule is under the influence of an "effective" energy gap:

$$\Delta_{eff} = \Delta - 2 \sum_{j\ neighboring\ molecules} J\sigma_j \qquad (10)$$

so the energy of the molecule (i) is given by the following formula:

$$\frac{\Delta_{eff}}{2}\hat{\sigma}_i \qquad (11)$$

The mean field approximation neglects the fluctuations in time of this Δ_{eff} replacing it with its average value:

$$\langle \Delta_{eff} \rangle = \Delta - 2 J p \langle \sigma \rangle, \qquad (12)$$

where p is the number of the nearest neighbors of a lattice site. In the mean field approximation each molecule of the crystal is submitted only to the action of the effective field, so we have again N independent molecules, placed in a uniform field, but this time it depends on the average value $\langle \sigma \rangle$ of the system. In the mean field approximation the energy of the molecule (i) can be calculated using the formula presented below:

$$E = \left[\frac{\Delta - 2Jp\langle \sigma \rangle}{2}\right]\hat{\sigma}_i \qquad (13)$$

Therefore the energy of a molecule found in the LS state is $E_{BS} = -\frac{\Delta - 2Jp\langle \sigma \rangle}{2}$, while for the one found in the HS state has the following form: $E_{HS} = \frac{\Delta - 2Jp\langle \sigma \rangle}{2}$ giving an average value of $\langle \sigma \rangle$:

$$\langle \sigma \rangle = \frac{r \exp\left(-\beta \times \left(\Delta - 2Jp\langle \sigma \rangle\right)\right) - 1}{r \exp\left(-\beta \times \left(\Delta - 2Jp\langle \sigma \rangle\right)\right) + 1} \qquad (14)$$

3. The atom-phonon coupling model

Although the Ising model offers an accurate description of the main characteristics of spin crossover phenomena, along with the models introduced by Wajnflasz and Pick [7] and Boussekou et al. [8] do not offer an explanation on the physical origin of the intermolecular interactions and the exchange like constants are just phenomenological parameters. In order to be able to describe the interactions as well as the exchange like constants, the atom – phonon coupling (APC) model [9, 10] was introduced, which considers that neighboring molecules are connected through a spring characterized by an elastic constant depending on molecules electronic state. By associating a fictitious spin to each molecule that has -1 and +1 eigenvalues corresponding to LS and HS levels respectively, an Ising type model can be developed for the analysis of metastable states and phase transitions in spin-crossover compounds.

This model considers that neighboring atoms are connected to each other with a spring characterized by an elastic constant which depends on the atoms electronics states: $C_{LL} = \lambda$ when both atoms are in the LS state, by $C_{HH} = \nu$ when both are in the HS state and by $C_{HL} = \mu$ when the atoms are in different states.

Using the Ising – like model approach in the APC model, we assume that each atom has associated a fictitious – spin operator $\hat{\sigma}$ which can take one of the two eigenvalues +1, respectively -1, corresponding to the electronic states of the atom (HS, respectively LS). The ratio between the degeneracy of the fundamental LS level g_{LS} and that of the excited HS level g_{HS} defines the degeneracy factor: $r = g_{HS} / g_{LS}$. The resulting Hamiltonian of the system formed by a number of N atoms disposed in chain is:

$$H = \sum_{i=1}^{N} \frac{\Delta}{2}\hat{\sigma}_i + \sum_{i=1}^{N} \frac{p_i^2}{2m_a} + \sum_{i=1}^{N} \frac{k_{i,i+1}}{2}\left(u_{i+1} - u_i\right)^2 \tag{15}$$

where: Δ is the energy gap between the LS and HS states, the second sum is the kinetic energy of the chain, u_i is the displacement of i – atom from its equilibrium position, m_a is the mass of the atom, $k_{i,i+1}$ is the elastic constant that can take one of the three values λ, ν or μ mentioned above, depending on the states of atoms i and $i+1$.

Therefore, using the Ising – like model, $k_{i,i+1}$ can be written as:

$$k_{i,i+1} = \frac{\lambda + 2\mu + \nu}{4} + \frac{\nu - \lambda}{4}\left(\hat{\sigma}_i + \hat{\sigma}_{i+1}\right) + \frac{\lambda - 2\mu + \nu}{4}\hat{\sigma}_i \cdot \hat{\sigma}_{i+1} \tag{16}$$

and the Hamiltonian (1) can be rewritten by using the following expression:

$$H = -\sum_{i=1}^{N} h_i\hat{\sigma}_i - \sum_{i=1}^{N} J_{i,i+1}\hat{\sigma}_i \cdot \hat{\sigma}_{i+1} + \sum_{i=1}^{N}\left[\frac{p_i^2}{2m_a} + \frac{\lambda + 2\mu + \nu}{8}\left(u_{i+1} - u_i\right)^2\right] \tag{17}$$

In the above equation, the first sum could be considered as a Zeeman – like interaction where each spin $\hat{\sigma}_i$ is under the influence of an effective field h_i, while the second sum resembles to an exchange – like interaction between the two neighboring spins $\hat{\sigma}_i$ and $\hat{\sigma}_{i+1}$. In this case the effective field will be given by: $h_i = -\{4\Delta + (\nu - \lambda)[(u_i - u_{i-1})^2 + (u_{i+1} - u_i)^2]/8$, while the interaction between the spins i and $i+1$ can be calculated using the formula $J_{i,i+1} = (\lambda - 2\mu + \nu)(u_{i+1} - u_i)^2/8$. One can observe now that the total Hamiltonian of our system is formed by a classical Ising Hamiltonian and a phonon Hamiltonian that can be solved using methods from the theory of magnetism.

By using the mean-field approximation in the APC model, we replace the elastic constants $k_{i,i+1}$ with one effective constant K which does not depend on the site. Therefore the effective field and the exchange coupling constant presented above are constant and we can note them with h and J. The total Hamiltonian can be studied now by using the transfer matrix so one arrives at the following two equations: $m = \langle\hat{\sigma}_i\rangle$, $s = \langle\hat{\sigma}_i\hat{\sigma}_{i+1}\rangle$, where $\langle\cdot\rangle$ are the thermal mean values.

Therefore, in the mean field approximation, the thermal mean value of the phonon Hamiltonian (H_{phon}) has the following expression (the formula was determined by using the calculations from [10]):

$$\langle H_{phon}(K)\rangle = \frac{\hbar\omega_M(\lambda)}{2}\sqrt{\frac{K}{\lambda}}\sum_{(i)}\left|\sin\frac{i\pi}{N}\right|\coth\left(\frac{\beta\hbar\omega_M(\lambda)}{2}\sqrt{\frac{K}{\lambda}}\sum_{(i)}\left|\sin\frac{i\pi}{N}\right|\right) \tag{18}$$

where $i = 0, \pm 1, \ldots \pm\left(\frac{N}{2} - 1\right), \frac{N}{2}$, $\beta = \frac{1}{k_B T}$ and K is calculated using the following formula:

$$K = \frac{\lambda + 2\mu + \nu}{4} + \frac{\nu - \lambda}{4}m + \frac{\lambda - 2\mu + \nu}{4}s \tag{19}$$

The maximum phonon angular frequency $\omega_M(\lambda)$ for a periodic chain with an elastic constant λ, is given by $\omega_M(\lambda) = 2\sqrt{\lambda/m_a}$.

The self consistent equations for the two parameters considered in this problem, m and s, were derived in [10] and they can be expressed as follows:

$$m=\frac{e^{\beta J}\sinh\left(\beta\left(-\frac{\Delta}{2}-\frac{\upsilon-\lambda}{4K}\frac{\left\langle H_{phon}(K)\right\rangle_T}{N}+k_B\frac{\ln r}{2}\right)\right)}{\sqrt{e^{2\beta J}\sinh^2(\beta h)+e^{-2\beta J}}} \tag{20}$$

$$s=1-\frac{2e^{-2\beta J}}{\left(e^{\beta J}\cosh(\beta h)+\sqrt{e^{2\beta J}\sinh^2(\beta h)+e^{-2\beta J}}\right)\sqrt{e^{2\beta J}\sinh^2(\beta h)+e^{-2\beta J}}} \tag{21}$$

where the effective field and effective exchange coupling constants are expressed by:

$$h=-\frac{\Delta}{2}-\frac{\nu-\lambda}{4K}\frac{\left\langle H_{phon}(K)\right\rangle_T}{N} \quad \text{and} \quad J=-\frac{\lambda-2\mu+\nu}{8K}\frac{\left\langle H_{phon}(K)\right\rangle_T}{N} \tag{22}$$

From the equations (20) – (22) for m and s one can find at low temperature the metastable states of a SCO compound. Using the equation for m one can also find easily the value of the high spin fraction by applying the following expression: $n_{HS} = (m+1)/2$.

Throughout this article we have used the following parameters [10]:

- The energy gap: $\delta=\frac{\Delta}{\hbar\omega_M(\lambda)}$;
- The reduced temperature: $t_r=\frac{k_BT}{\hbar\omega_M(\lambda)}$

4. The role of the atom – phonon coupling model parameter C_{LH}/C_{LL}

The parameters used in the APC model can be determined experimentally using calorimetric measurements (for the energy gap δ and the degenerescence r), Brillouin spectroscopy for the elastic constant ratios, inelastic diffusion of neutrons, inelastic diffusion of X-rays [11, 12]. The hysteretic properties of a system depend highly on C_{LH}/C_{LL} ratio, their influence can be observed in the figure 4.

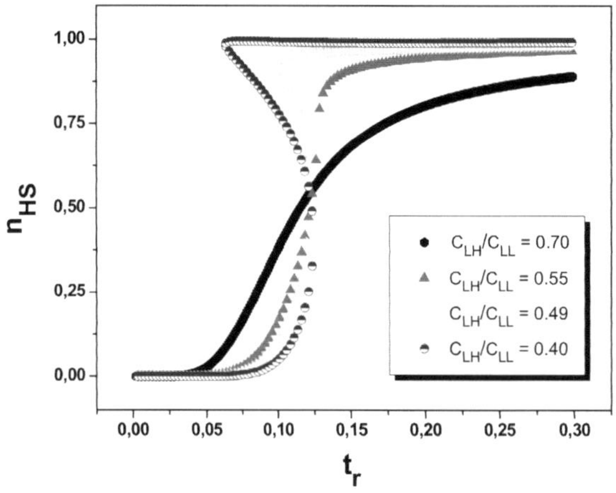

Figure 4. n_{HS} variation in function of reduced temperature for different values of C_{LH}/C_{LL} using $C_{HH}/C_{LL} = 0.2$, $r = 5$, $\delta = 0.55$, $N = 50$.

When two neighbouring atoms are in different states, the elastic constant C_{LH} between them can take values between C_{LL} and C_{HH}. The figure 4 presents the variation of the high spin fraction in function of temperature for different values of the ratio: C_{LH}/C_{LL}. The figure shows the importance of the parameter's value in obtaining results experimentally observed:

a. gradual spin conversion is obtained for $C_{LH} = (C_{LL} + C_{HH})/2$, in this case $C_{LH}/C_{LL} = 0.70$;
b. the limit value bellow which hysteresis is obtained is $C_{LH}/C_{LL} = 0.55$;
c. a typical hysteresis can be observed for $C_{LH}/C_{LL} = 0.49$;
d. a more abrupt and pronounced hysteresis can be achieved for values of $C_{LH}/C_{LL} = 0.40$.

From the figure we can also conclude that $T_{1/2}$ is slightly shifted towards smaller values of temperatures when the ratio C_{LH}/C_{LL} decreases. When C_{LH} takes values closer to C_{LL} the hysteresis becomes more abrupt. From an application point of view the last case presented is very important because the two electronic states HS and LS are very well defined.

5. Relaxation dynamics

A molecule can pass from the HS state to the LS state in two different ways, either by nonradiative relaxation from $^5T_2$ to $^1A_1$, either using the reverse LIESST effect by irradiation with light in the absorption band [13]. While the latter mechanism has a fast kinetics of the order of ps for a molecule, the direct relaxation from $^5T_2$ to $^1A_1$ is much slower (on the order of ms), being the subject of studies that try to determine the relaxation in time of the photo-excited states after application of a laser impulse.

The relaxation process is mainly determined by the tunnel effect at low temperatures ($T < 30K$) and thermally activated for higher temperatures ($T > 40K$). Many studies were undertaken by Hauser and his research group [13, 14] that rely on the properties of the HS state energy barrier in order to understand the properties of the cooperative relaxation. The relaxation rate k_{HL} on the thermal activation regime is:

$$K_{HL}(T,n) = k\ \exp\left(-\frac{E_a}{k_B T}\right) \tag{23}$$

where k is the relaxation rate at high temperature and E_a is the activation energy given by:

$$E_a = E_0 - E_{HS} \tag{24}$$

The energy difference between the two potential wells E_{HS} is given by:

$$E_{HS} = \frac{\Delta}{2} + \frac{E_{phonons}}{N} \tag{25}$$

so we can rewrite the equation :

$$K_{HL}(T,n) = k\exp\left(-\frac{E_a}{k_B T}\right) = k\exp\left(-\frac{A - \frac{\delta}{2} - \frac{\langle H_{ph}(K)\rangle_T}{N}}{t_r}\right) \tag{26}$$

Where :

$$A = \frac{E_0}{\hbar w_M(\lambda)} \tag{27}$$

The variation in time for the high spin fraction can be calculated using the following expression:

$$\frac{dn}{dt} = -nk\exp\left(-\frac{A - \frac{\delta}{2} - \frac{\langle H_{ph}(K)\rangle_T}{2}}{t_r}\right) \tag{28}$$

Using this equation, the relaxation curves that take into account the factors that can interfere in the real situations such as dilution, size or the temperature of the system [15, 16] are presented in the figure 5.

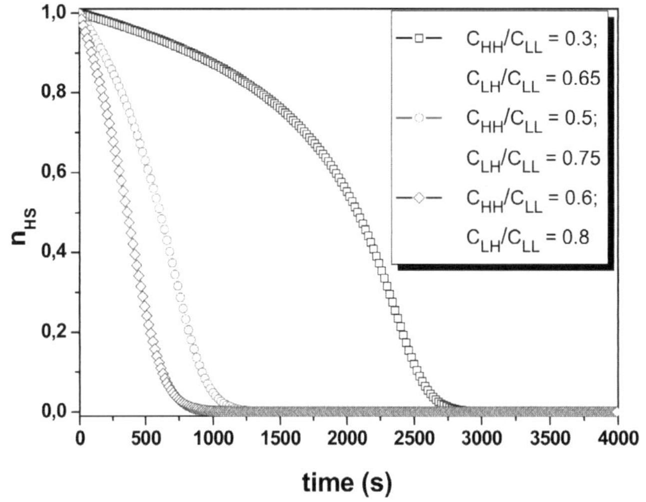

Figure 5. Relaxation curve without using light irradiation using: $\delta = 0.6$, $t = 0.03$, $N = 2000$, $r = 5$, $A = 0.8$, $k = 100$.

Another representation that allows us to analyze the relaxation mechanism is the Arrhenius representation in which: $\ln(k_{HL}) = f(n_{HS})$ (figure 6) [16].

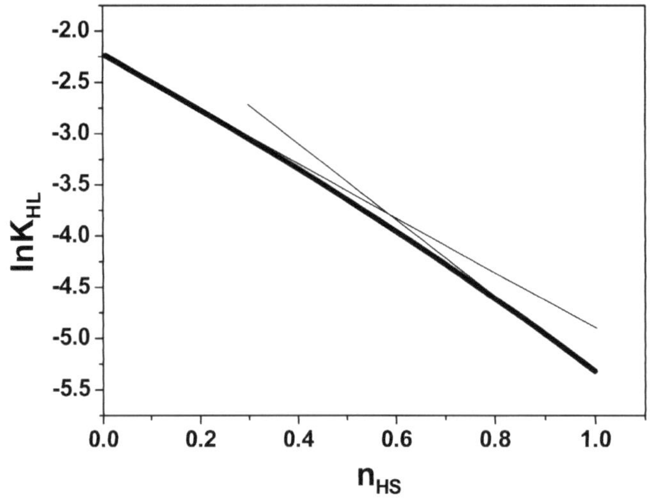

Figure 6. Arrhenius diagram using $\delta = 0.6$, $t = 0.03$, $N = 2000$, $r = 5$, $A = 0.8$, $k = 100$, $x = 0.5$, $y = 0$ [16].

In the figure 6 Arrhenius's diagram form presents two slopes and not a straight line like in the case of Hauser's model which comes in good agreement with the experimental data. Using the equation

(5.6) the theoretical results were superposed with the experimental data obtained for $[Fe_{0.5}Zn_{0.5}(btr)_2(NCS)_2]H_2O$ (figure 7).

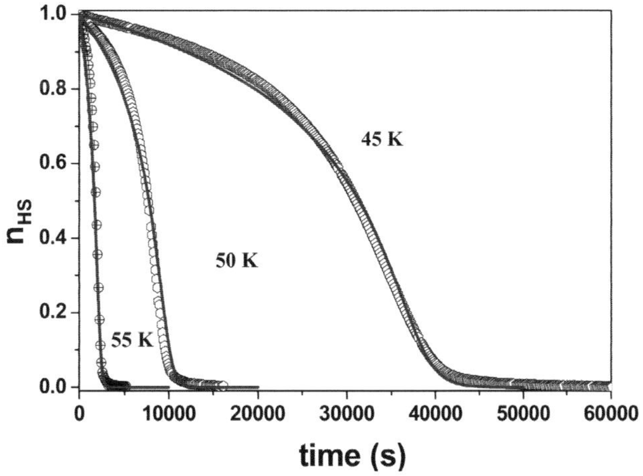

Figure 7. Relaxation curves at different temperatures obtained for $[Fe_{0.5}Zn_{0.5}(btr)_2(NCS)_2]H_2O$ after photo-excitation with light having the wavelength of 550nm wavelength. The continuous lines were calculated using the equation (28).

The final part of the curve is called 'relaxation queue' and it was explained by Hauser as being determined by the progressive installation of correlations created by the short range interactions. The figure clearly shows that the relaxation curves obtained theoretically are very well superposed on the one's obtained experimentally including the relaxation tail. In Hauser's model, the last part of the curve could be reproduced only if another term is included, accounting for the short range interactions [17-19]. The theoretical relaxation curves including this term are presented in the figure 8.

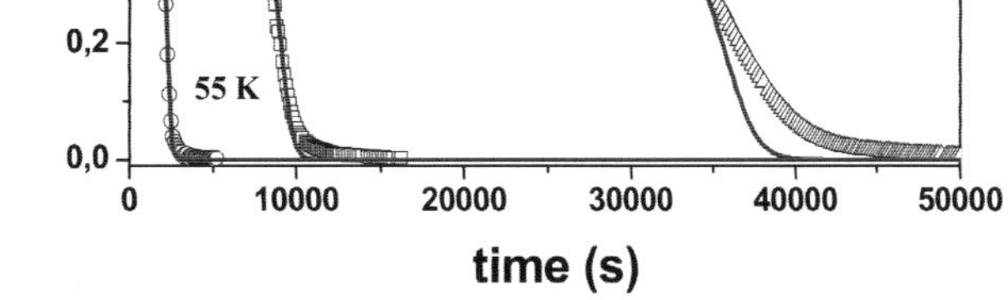

Figure 8. Relaxation curves at different temperatures obtained after photo-excitation with light having the wavelength of 550 nm wavelength. The continuous lines are the theoretical lines from Hauser's model.

6. Size effect

The size effect plays an important role from an applications point of view such as information storage. One of the most important analysis are based on finding the smallest size for which the system keeps its properties. In order to find the influence of the system's size on the width of the hysteresis cycle we have simulated the dependence of the n_{HS} fraction for different sizes of the chain of molecules.

Two important ratios that are parameters in the APC model influence highly the hysteretic properties of a system: the ratio between C_{HH} and C_{LL} and the ratio between C_{LH} and C_{LL} [20].

There is a strong dependence of the hysteresis width on the ratios C_{HH}/C_{LL} and C_{LH}/C_{LL} which offers an explanation for different results regarding the size effect in SCO. Following this idea we have simulated in a 1D SCO system the dependence of the hysteresis width in function of the number of molecules N for different values of C_{LH}/C_{LL} (figure 9).

The figure 9 shows that for both values of the ratio C_{LH}/C_{LL} the hysteresis loop disappears for a chain of 8 molecules. We can notice a variation of the hysteresis width with the chain's size is followed by saturation, which corresponds to an infinite system.

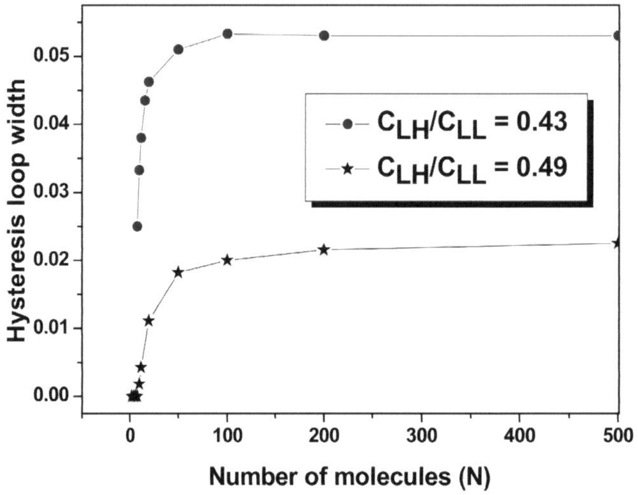

Figure 9. The dependence of the hysteresis loop width in 1D systems on the number of molecules for different values of C_{LH}/C_{LL} using $C_{HH}/C_{LL} = 0.4$, $r = 5$, $\delta = 0.55$.

7. Low temperature metastable states

An interesting phenomena observed in spin crossover compounds is the LIESST effect. Experimentally, it has been shown, first by Moussa et al [21] and recently by Neville et al [22] that in the case of a two-step thermal spin transition, by irradiating the sample at low temperature, the system can be photoexcited only 50% from the high spin fraction. By using the APC model we have been able to reproduce this kind of behavior [23, 24]. The results are displayed in figure 10.

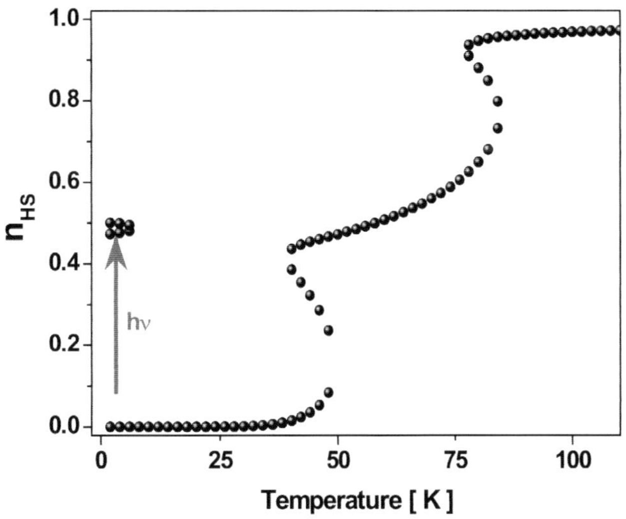

Figure 10. The variation of n_{HS} with reduced temperature for $C_{LH}/C_{LL} = 0.69$, $C_{HH}/C_{LL} = 0.226$, $r = 30$, $\delta = 0.606$, $N = 50$.

8. Conclusion

In this report the atom-phonon coupling model has been applied to a 1D spin crossover systems. Only the elastic force constant of the spring which links two atoms first neighbors is assumed to be depending on the electronic states of both atoms. As shown in the figure 4 the atom-phonon coupling model leads to spin conversion with or without first order phase transition depending on the values of the elastic constant between LS and HS atoms. On the bases of good agreement between simulated results and experimental one we can conclude the APC model can be applied to describe the relaxation process into a system with spin conversion. At low temperature we have obtained metastable states at $n_{HS} = 0.5$. These states can be associated to the "incomplete" photoconversion excited states obtained by irradiating the samples. We have presented a study on the size effect in the spin crossover compounds: our results point to the fact that to reduce the size of a system and keep the properties of its hysteresis loop, it is necessary to choose materials with C_{HH}/C_{LL} ratio as small as possible.

Acknowledgement: This work was partially supported by European Framework Program 7 under the contract no. PIRG02-GA-2007-224904 and by the Agence Universitaire de la Francophonie AUF (ref. PCSI 6313PS842). A.G. thanks the European Union program contract no. POSDRU/6/1.5/S/22.

References

[1] Kahn O and Martinez CJ 1998 *Science* **279** 44
[2] Hausser A, Gütlich P and Spiering H 1996 *Inorg. Chem.* **25** 4345
[3] Gütlich P and Goodwin HA 2004 *Spin Crossover in transition metal compounds I, II and III* vol 233-235 (Berlin: Springer)
[4] Guionneau P, Marchivie M, Bravic G, Letard JF and Chasseau D 2004 *Spin Crossover in Transition Metal Compounds II* vol 234 (Berlin: Springer-Verlag) p97
[5] Tuchagues JP, Bousseksou A, Molnar G, McGarvey JJ and Varret F 2004 *Spin Crossover in Transition Metal Compounds III* vol 235 (Berlin: Springer-Verlag) p85
[6] Rodriguez-Velamazan JA, Castro M, Palacios E, Burriel R, Costa JS and Letard JF 2007 *Chem. Phys. Lett.* **435/4-6** 358
[7] Wajnflasz J and Pick R 1971 *J. Phys. Colloq. France* **32** C1-91

[8] Bousseksou A, Nasser J., Linares J, Boukheddaden K and Varret F 1992 *J. Phys. I France* **2** 1381
[9] Nasser J 2001 *Eur. Phys. J. B.* **21** 3-10
[10] Nasser J, Boukheddaden K and Linares J 2004 *Eur. Phys. J. B.* **39** 219
[11] Jung J, Bruchhauser F, Feile R, Spiering H and Gutlich P 1996 *Zeitschrift Fur Physik B Condensed Matter* **100/4** 517
[12] Martin JP, Zarembowitch J, Bousseksou A, Dworkin A, Haasnoot JG and Varret F 1994 *Inorg. Chem.* **33/26** 6325
[13] Hausser A, Gütlich P and Spiering H 1986 *Inorg. Chem.* **25/23** 4245
[14] Hauser A 1992 *Chem. Phys. Lett.* **192/1** 65
[15] Rotaru A and Linares J 2007 *Journal of Optoelectronics and Advanced Materials* **9/9** 2724
[16] Rotaru A 2009 *Ph.D Thesis* University of Versailles Saint-Quentin-en-Yvelines
[17] Hoo B, Boukheddaden K and Varret F 2000 *Eur. Phys. J. B* **17/3** 449
[18] Boukheddaden K, Shteto I, Hoo B and Varret F 2000 *Phys. Rev. B* **62/22** 14796
[19] Boukheddaden K, Shteto I, Hoo B and Varret F 2000 *Phys. Rev. B* **62/22** 14806
[20] Nakamoto T, Tan ZC and Sorai M 2001 *Inorg. Chem.* **40/15** 3805
[21] Moussa NO, Molnár G, Bonhommeau S, Zwick A, Mouri S, Tanaka K, Real JA and Bousseksou A 2005 *Phys. Rev. Lett.* **94** 107205
[22] Neville SM, Etrillard C, Asthana S and Létard JF 2010 *Eur. J. Inorg. Chem.* **282** 282-288
[23] Rotaru A, Linares J, Mordelet S, Stancu A and Nasser J 2009 *J. Appl. Phys.* **106** 043507
[24] Gindulescu A, Rotaru A, Linares J, Dimian M and Nasser J 2010 *J. Appl. Phys.* **107** 1

Journal of Physics: Conference Series **268** (2011) 012008 doi:10.1088/1742-6596/268/1/012008

Combining CFD simulations with blockoriented heatflow-network model for prediction of photovoltaic energy-production

I E Haber[1] and I Farkas[2]

[1] University of Pecs/Department of Technical Informatics, Pecs, Hungary

[2] Szent Istvan University of Godollo/Department of Physics and Process Control, Godollo, Hungary

E-mail: ihaber@pmmk.pte.hu

Abstract. The exterior factors which influencing the working circumstances of photovoltaic modules are the irradiation, the optical air layer (Air Mass – AM), the irradiation angle, the environmental temperature and the cooling effect of the wind. The efficiency of photovoltaic (PV) devices is inversely proportional to the cell temperature and therefore the mounting of the PV modules can have a big affect on the cooling, due to wind flow-around and naturally convection. The construction of the modules could be described by a heatflow-network model, and that can define the equation which determines the cells temperature. An equation like this can be solved as a block oriented model with hybrid-analogue simulator such as Matlab-Simulink. In view of the flow field and the heat transfer, witch was calculated numerically, the heat transfer coefficients can be determined. Five inflow rates were set up for both pitched and flat roof cases, to let the trend of the heat transfer coefficient know, while these functions can be used for the Matlab/Simulink model. To model the free convection flows, the Boussinesq-approximation were used, integrated into the Navier-Stokes equations and the energy equation. It has been found that under a constant solar heat gain, the air velocity around the modules and behind the pitched-roof mounted module is increasing, proportionately to the wind velocities, and as result the heat transfer coefficient increases linearly, and can be described by a function in both cases. To the block based model the meteorological parameters and the results of the CFD simulations as single functions were attached. The final aim was to make a model that could be used for planning photovoltaic systems, and define their accurate performance for better sizing of an array of modules.

1. Introduction

There are mainly two factors, which influences the power output of the photovoltaic modules, they are the temperature of the solar cells and the irradiation, which comes from the Sun. The quantity of the radiation arriving from the Sun influences the first factor, what between the average Earth-Sun distance, outside the atmosphere of the Earth on a perpendicular surface is $1353 Wm^{-2}$, but a part of this in the atmosphere is absorbing or reflecting. The moisture content of the air and the clouding influences this parameter, while at clean weather, the radiation coming to the ground surface through the atmosphere is around $1000 Wm^{-2}$, till in cloudy time it could be only $200\ Wm^{-2}$ or less.

The other factor is the temperature of the solar cells. The performance of Si solar cells decrease with the increase of the temperature, averagely in a direct proportion with 0.2% (figure 1). The solar modules which can be bought in trade, are only tested on 25ºC, but the cells' temperature may attain 60-70ºC and in some cases above 100ºC in the practice.

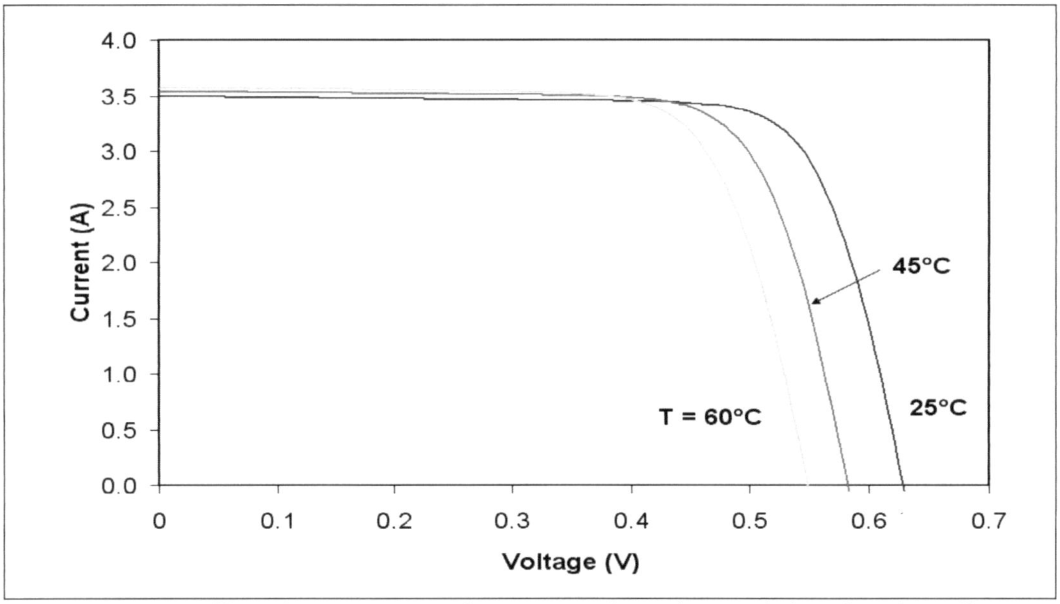

Figure 1. Effect of temperature on the current-voltage characteristics of a solar cell.

This temperature is beside certain circumstances artificially reducible, this could be made by many ways, starting with the extreme cases like fluent hydrogen refrigeration, ventilation, etc., but it was proven already, that the performance increase gained by these, is not in a proportion with the expenses. The repository of the passive refrigeration opportunities is wide. In this case the cooling medium is air, refrigeration happen due to convection, which may be natural (when the density difference causes the substance flow) or may be artificial.

Natural convection cooling is mainly used by pitched roofs or by PVs integrated in vertical facades, where the energy of the cooling medium will be used for heating the building.

The cooling effect at free standing PV modules is usually the wind and the buoyancy driven flows. In this work we examine the flow properties at free standing solar modules with the help of computational fluid dynamics (CFD).

The upwarming can be reduced with the suitable setup of the modules, taking the meteorological parameters into consideration: the prevailing wind direction and hereby the air circulation around the modules. To let us be able to calculate with this, it is necessary to examine the regional meteorological parameters, which will be the boundary conditions in the simulation case.

The performance of solar cell is normally evaluated under the standard test condition (STC), where an average solar spectrum at AM 1.5 is used, the irradiance is normalized to $1000Wm^{-2}$, and the cell temperature is defined as 25°C. To satisfy the requirement of temperature and insulation in STC, the test usually needs specified environment and some special testing equipment, such as an expensive solar simulator. In fact the modules temperature can reach 80-90°C between naturally conditions, where the efficiency looses either 20-30% from its originally given value. Some manufacturers give a number called NOCT (Nominal Operating Cell Temperature), which can better describe the operating conditions of a module. The NOCT number of a module will be determined at irradiation $800Wm^{-2}$, at temperature 20°C and wind velocity $1ms^{-1}$, while it is assumed that wind can flow around it.

$$T_{PVcell} = T_{air} + \frac{NOCT - 20}{80} I \quad (1)$$

Substituting the proper values in it, where T_{air} is the environmental temperature in °C and I is the irradiation in mWm^{-2} , we can get the actual operating temperature of our module.

The above formula is experimental and simplified, the convection and the heat conduction are taken into consideration, but presupposes it, that the heat conduction coefficient of the substance does not change powerfully with the temperature.

To calculate the modules accurate temperature, we have to know at least the surface heat transfer coefficients (α) at various flow rate, which defines the leaving quantity of heat from the modules.

Along an L surface, which has T_s temperature, the mediums temperature is T_w and a flow with v velocity, the upcoming heat flow hangs on many parameters. Either we have to solve the continuity equations of Fourier–Kirchoff, Navier–Stokes and Reynolds, or through calculating or measuring the Nusselt-number and expressing from the expression (2) [1].

$$Nu = \frac{\alpha L}{\lambda} \quad (2)$$

In this case, on the other hand, determining the Nusselt number is necessary, which is possible if the questionable flow space is known, while Nu = f(Gr, Pr, Re), so it depends on the Prandtl-, Reynolds- and Grashof criteria.

The problematic of the determination of the Nusselt number, was examined in several works already, in case of photovoltaic modules. One of the most accurate and most detailed models, which are based on measuring an artificially heated plate, is the work Sharples és Charlesworth (1997) [2].

$$Nu_L = 0.664 \, \mathrm{Re}_L^{0.5} \, \mathrm{Pr}^{0.33} \quad (3)$$

Into the above formula replacing the physical parameters of the air, at a surface area of $1.8 \times 0.9 m^2$ with 325K surface temperature, the modules heat transfer coefficient – at 300K environmental temperature – with the under mentioned context can be calculated, where v is the flow velocity of the air:

$$\alpha = 4.01 v_{\infty}^{0.5} \quad (4)$$

2. Methodolgy

2.1. Preparation

The Nusselt number and the heat transfer coefficient will be defined through CFD simulations, in our case. The reliability of CFD modelling was proven by Moshfegh and Sandberg. [3] They made a model, which was validated for buoyancy-induced flow and heat transfer in a tall open air cavity. The average difference between the predicted and measured velocities and for heat fluxes at and above $100 Wm^{-2}$ was less than 3%.

The Hungarian meteorological parameters were examined, principally the environment temperature and the wind, which has a strong influence on the power generation of the solar cells. We wanted to examine what happens at the highest levels of these parameters, this is because the irradiance was set automatically to 1000 Wm^{-2}, and at the other parameters we were searching for the highest averages. The average air temperature at our latitude is at a sunny summer day at about 21°C, but in Pécs were measured the highest low temperature in a day (in Hungary), which was 26.9°C. The wind velocities are mostly between 1 and $3 ms^{-1}$ but the wind has the biggest affect on the surface heat transfer coefficient so we made calculation from 0 to $7 ms^{-1}$.

The PV modules were defined as solid in the calculations with one domain, so we made a thickness weighted average on the density, heat conductivity(U) and on the specific heat capacity(C). A typical solar module stays from glass coverage, Ethylene-vinyl acetate (EVA) encapsulation for the cells and usually from a PVF, PVB or glass bracing on the backside. The thickness of the solar cells is usually 300μm, the glass has around 2mm thickness. The rest should be then about 7.7mm which are mainly plastics and have near the same conductivity and capacity for heat. Our calculations made out that the average values for our type of modules are: $\alpha = \sim 1500 kgm^{-3}$, $U = \sim 0.5 Wm^{-1}K^{-1}$ and $C = \sim 400 Jkg^{-1}K^{-1}$.

We do not want to specify the temperature of the modules, these numbers are to let us approach them in the course of calculations, because the buoyancy driven flows could be calculated more precisely this way. The buoyancy driven flows are important mostly in the lower or zero wind velocity calculations, where they are the only flow types near the wall of the modules, exactly in the wind-shaded areas.

The heat transfer coefficients, which will be served by the calculations, will be used at the heatflow-network model (figure 2) we presented in our previous works. [4] This network describes the material layer junctions of the photovoltaic module, represented with their heat resistance. This heatflow-network model was implemented in Matlab/Simulink, and this is a one-dimension steady state model which is able to describe the junction temperatures in the photovoltaic module.

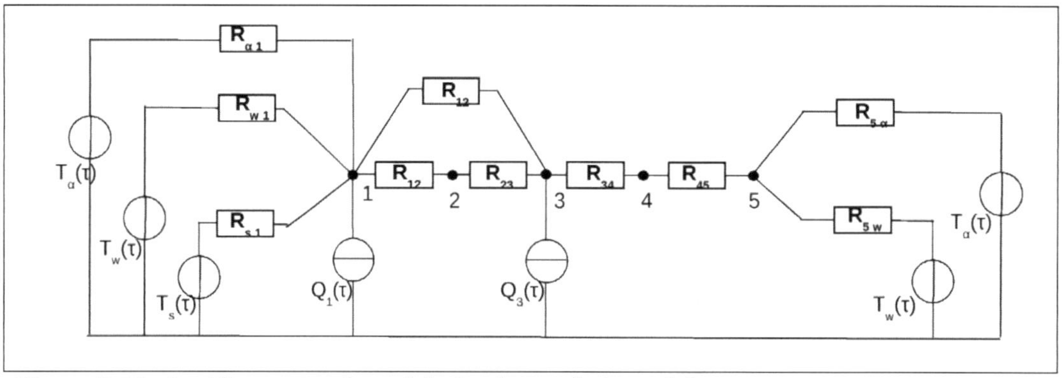

Figure 2. The heatflow-network model of a photovoltaic module.

2.2. Pre-processing the cases

The quality of the numerical grid is important to the numerical calculation, in this consideration on the wall of the solar module, prism elements were generated with growing side longitude (figure 3), like this the boundary layer is conformed for using the k-ε turbulence model 'Enhanced wall treatment' option, so we can take advantage of the opportunity of modelling the boundary layer detachments and the wall nearly flows adequately. In the rest of the parts of the space we applied triangle elements, on unstructured construction. Near the module the side length of the triangle elements is 4mm, and is growing gradually receding from the wall. The skewness – as the grids' qualitative indicator – comes to under a 0.6 value onto all of the grid which is relatively good for meshes containing triangular elements. The grid consist of 110 769 pieces of elements in the one module case and 233 730 pieces at the three modules. At the second case the finest smaller element region has a bigger extent, which is why it contains more elements.

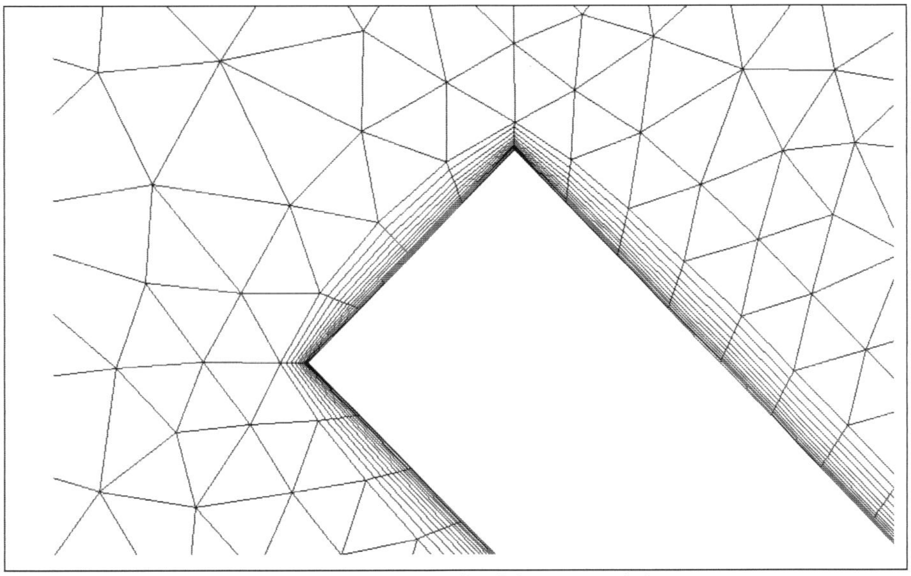

Figure 3. The mesh distribution at the wall of the pv modules.

The basis of the physical models is the continuity equation (5) the momentum equation (6)(7) and the energy equation (8), and we made use of the Boussinesq approximation (9) for the determination of the density change of the air. The form of the equations in case of two-dimensional stationary applications:

$$\frac{\partial u}{\partial x}+\frac{\partial v}{\partial y}=0\,, \tag{5}$$

$$u\frac{\partial u}{\partial x}+v\frac{\partial v}{\partial y}=g\beta\left(T-T_0\right)-\frac{1}{\rho}\frac{\partial p}{\partial x}+v\left(\frac{\partial^2 u}{\partial x^2}+\frac{\partial^2 v}{\partial y^2}\right), \tag{6}$$

$$u\frac{\partial u}{\partial x}+v\frac{\partial v}{\partial y}=g\beta\left(T-T_0\right)-\frac{1}{\rho}\frac{\partial p}{\partial y}+v\left(\frac{\partial^2 u}{\partial x^2}+\frac{\partial^2 v}{\partial y^2}\right), \tag{7}$$

$$\rho c_p\left(\frac{\partial T}{\partial x}+v\frac{\partial T}{\partial y}\right)=k\left(\frac{\partial^2 T}{\partial x^2}+\frac{\partial^2 T}{\partial y^2}\right), \tag{8}$$

$$\rho=\rho_0(1-\beta\Delta T)\,. \tag{9}$$

At the defining of the boundary conditions of the equations, the physical parameters of the air are constant, except the density (9). The environmental temperature is 300K, the temperature of the modules comes out from the radiation. At PV cells, most of the absorbed solar radiation is converted into heat, which increases the temperature of the cells. The solar heat gain by the PV modules was calculated from the insulation minus the reflection (5%) and the energy conversion into electricity (about 15%) [5]. A radiation model was used in the simulation case too, to calculate the radiation heat losses.

In all cases five different wind speeds were examined for the heat transmission of the modules, in order v_{wind} = {0, 1, 3, 5, 7}ms^{-1}, based on the wind speeds occurring in Hungary mostly.

3. Results

An average of the values were made from the calculated heat transfer coefficients, hereby the biggest deflection was 2.54%. The linear function of the average values may be used in the heatflow network model, while its ingoing parameter is the wind velocity, which can be delivered from one of the initial values of the block oriented model.

- Pitched roof
  - $\alpha_{avg\text{-}pitched\text{-}north}(v) = 1.85x + 2.93$;
  - $\alpha_{avg\text{-}pitched\text{-}south}(v) = 3.62x + 2.93$;
- Flat roof
  - $\alpha_{avg\text{-}flat\text{-}north}(v) = 1.93x + 2.90$;
  - $\alpha_{avg\text{-}flat\text{-}south}(v) = 2.17x + 2.90$;
- Free standing modules
  - $\alpha_{avg\text{-}array\text{-}north}(v) = 4.188x + 2.90$;
  - $\alpha_{avg\text{-}array\text{-}south}(v) = 3.128x + 2.90$.

In figure (figure 4) of the schematic connections of the heatflow-network model we can see that other parameters are still missing like the wind speed, solar irradiation and the environmental temperature.

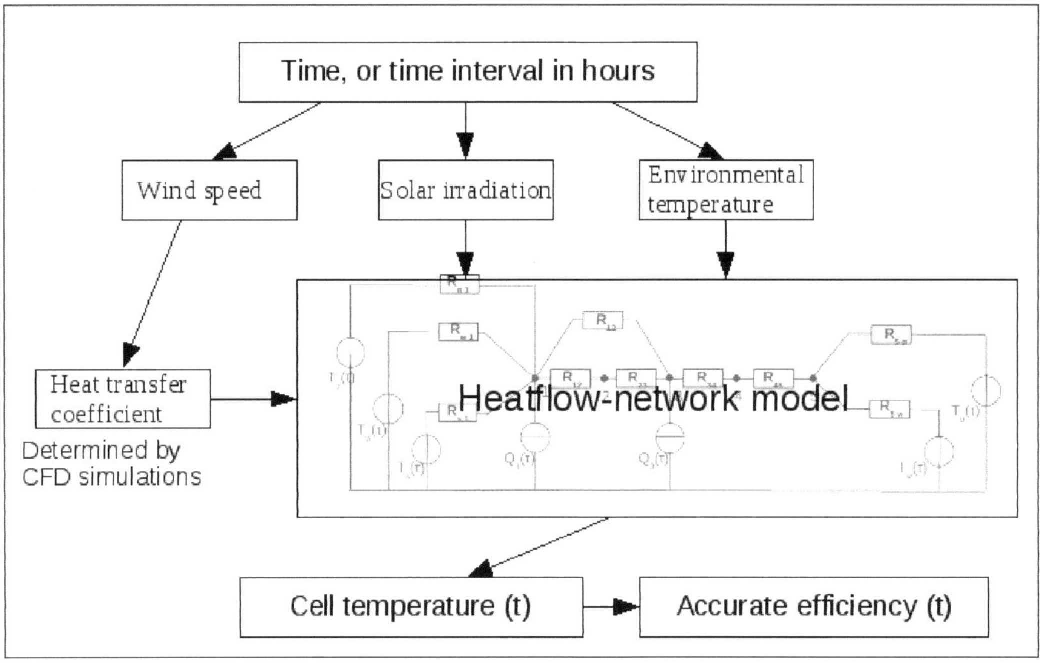

Figure 4. Schematic model.

Functions of the meteorological parameters were generated from the data's of the Hungarian Meteorological Service (OMSZ). With these functions the energy production of the photovoltaic modules can be predicted.

$$t_w(\tau) = t_m \cos[\alpha(\tau - \tau_0) + t_a] \quad (10)$$

$$I(\tau)=\begin{cases} I_m \cos[\beta(\tau-\tau_0)]\exp\left[-\frac{\tau-\tau_0}{\tau_h}\right]^2, & \text{if } |\tau-\tau_0|\le\tau_h \\ 0, & \text{if } |\tau-\tau_0|>\tau_h \end{cases} \tag{11}$$

$$w(\tau_h)=e^{a(N_{day})(\tau_h-b(N_{day}))^2}+c(N_{day}). \tag{12}$$

4. Conclusions

A method, with which the heat heat transfer coefficient of the modules can be defined with a good approach, got developed and implemented in the present work, that in the course of the later ones with measurements can be supported, but the calculated values appear suitable, based on the literatures. These values will be used in the mentioned Matlab/Simulik model [4]. Important, that the calculation can be parameterized, changes can be made in the temperature and examine the effect of the convective flow, or in case of the wind cooling. Similarly can be calculated other typical placement manners, like varying the distance from the ground, that may improve on the circulating quality.

5. References

[1] Brinkworth B J 2000 Solar Energy, 69 413-420
[2] Sharples S, Charlesworth P S 1998 Solar Energy 62 69-77
[3] Moshfegh B, Sandberg M 1998 Energy Reviews 2 287-301
[4] Háber I, Farkas I 2008 Conference Review PhD Symposium Pécs
[5] Gan G 2008 Solar Energy 83 1253–1273

5th International Workshop on Multi-Rate Processes and Hysteresis (MURPHYS 2010) IOP Publishing
Journal of Physics: Conference Series **268** (2011) 012009 doi:10.1088/1742-6596/268/1/012009

Mass action realizations of reaction kinetic system models on various time scales

K M Hangos[1,2] and G Szederkényi[1]

[1] Process Control Research Group, Computer and Automation Reseach Institute, Kende u. 13-17, H-1111 Budapest, Hungary

[2] Department of Electrical Engineering and Information Systems, University of Pannonia, Egyetem u. 10, Veszprém, Hungary

E-mail: hangos,szeder@scl.sztaki.hu

Abstract. Complex chemical reaction networks often exhibit different dynamic behaviour on different time scales. A combined approach is proposed in this work for determining physically meaningful mass action realizations of complex chemical reaction networks that describe its dynamic behaviour on different time scales. This is achieved by appropriately reducing the detailed overall mass action kinetic scheme using quasi steady state assumptions fit to the particular time scale, and then searching for an optimal realization using mixed integer linear programing.

Furthermore, the relationship between the properties (reversibility, deficiency, stability) of the obtained realizations of the same system on different time scales are also investigated and related to the same properties of the detailed overall model. It is shown that the reduced models obtained by quasi steady state assumptions may show exotic nonlinear behaviour, such as oscillations, when the original detailed is globally asymptotically stable. The proposed methods are illustrated by using a simple Michaelis-Menten type reaction kinetic example. The simplified versions of the well known Brusselator model have also been investigated and presented as a case study.

1. Introduction

Chemical Reaction Networks (CRNs) form an important and wide class of positive (or non-negative) systems attracting significant attention not only among chemists but in numerous other fields such as physics, or even pure and applied mathematics where nonlinear dynamical systems are considered [1]. Beside pure chemical reactions, CRNs are often used to model the dynamics of intracellular processes, metabolic or cell signalling pathways [2]. The increasing interest towards reaction networks as a well defined special class of positive nonlinear systems is clearly shown by recent tutorial and survey papers [3], [4], [5].

Most of the complex CRNs exhibit dynamic behaviour that spans a wide range in the time scale, i.e. they are multi scale by nature. A model that describes the behaviour at a particular time scale is often obtained from a detailed overall model by model reduction. There are various popular engineering model reduction transformations, such as the quasi steady state and quasi equilibrium assumptions or the variable lumping that are widely applied for this purpose [6] but their effect on the dynamic properties of reaction kinetic models is not very well known. Some earlier investigations about biochemical reaction networks indicate that approximate reduced dynamic models can only be applied if certain parametric and/or state-dependent conditions are met, see e.g. [7], [8]. At the same time, quite different dynamic properties (e.g. special nonlinear effects like oscillatory or chaotic behaviour) have been detected for different simplified versions of the same model. A recent paper [9] even warns against applying "inconsistent assumptions" in mathematical models of protein interaction networks.

Therefore, the aim of this paper is to identify the reason for the above spurious dynamical properties with a view to later propose ways to overcome them.

2. Chemical reaction networks and their properties

A brief summary of the most important properties of reaction kinetic networks obeying the mass action law is given in this section based largely on [10].

2.1. Mass action law, physical and chemical background

The origin of the reaction kinetic models is found in the molecular collision picture of gas phase chemical reactions, where the reaction rate is proportional to the number of collisions, i.e. to the concentration of the reactants. Here each elementary reaction step is *irreversible* in the form of

$$\sum_{i=1}^{n} \alpha_{ij} \mathbf{X}_i \rightarrow \sum_{i=1}^{n} \beta_{ij} \mathbf{X}_i \quad j = 1, ..., r \tag{1}$$

where $\mathbf{X}_i$ are the components, and the stoichiometric coefficients α_{ij} and β_{ij} are always non-negative integers, while n is the number of components, and r is the number of reactions.

It is also assumed in classical reaction kinetics that *the system is closed and evolves under isotherm conditions.*

The so called *complexes* $C_k = \sum_{i=1}^{n} \alpha_{ij} \mathbf{X}_i$ or $C_k = \sum_{i=1}^{n} \beta_{ij} \mathbf{X}_i$ $(k = 1, ..., m)$ are associated to the left- and right-hand sides of the reaction steps.

When the reaction kinetic system obeys the mass action law, then the *reaction rate* of the jth irreversible reaction step has the following form

$$\rho_j = k_j \prod_{i=1}^{n} [\mathbf{X}_i]^{\alpha_{ij}} = k_j \prod_{i=1}^{n} x_i^{\alpha_{ij}} \quad , \quad j = 1, ..., r \tag{2}$$

where $[\mathbf{X}_i] = x_i$ is the concentration of the ith component, and $k_j > 0$ is the *reaction rate constant* of the jth reaction, that is always positive.

Note that reversible reactions

$$\sum_{i=1}^{n} \alpha'_{ij} \mathbf{X}_i \rightleftarrows \sum_{i=1}^{n} \beta'_{ij} \mathbf{X}_i \quad j = 1, ..., r'$$

are regarded as two irreversible steps that share the two complexes but have different reaction rates.

2.2. Algebraic characterization

A set of irreversible reaction steps with MAL kinetics can be characterized by the following model elements. The stoichiometric matrix Y and the reaction monomials $\varphi_j(x)$ are in the form of

$$[Y]_{ij} = y_{ij} \quad , \quad \varphi_j(x) = \prod_{i=1}^{n} x_i^{y_{ij}} \quad , \; j = 1, .., m; \; i = 1, ..., n \tag{3}$$

where the matrix elements are the stoichiometric coefficients of the reactans, i.e. $y_{ij} = \alpha_{ij}$.

Algebraically, the so-called *Kirchhoff matrix of the reaction graph* $A_k \in \mathbb{R}^{m\times m}$ characterizes the reaction structure

$$[A_k]_{ij} = \begin{cases} -\sum_{l=1}^{m} k_{il} & \text{if} \quad i=j \\ k_{ji} & \text{if} \quad i\neq j \end{cases} \tag{4}$$

that is a column-convservation matrix with non-positive diagonal and non-negative off-diagonal entries.

The dynamic model equations The dynamic state equation for the ith component is in the form

$$\frac{dx_i}{dt} = -\sum_{j=1}^{r} \alpha_{ij}\varphi_j(x) + \sum_{i=1}^{n} \beta_{ij}\varphi_j(x) \tag{5}$$

The above equations for all components can be descibed in vector-matrix form

$$\frac{dx}{dt} = Y \cdot A_k \cdot \varphi(x) = \mathcal{M} \cdot \varphi(x) \tag{6}$$

that forms the dynamic model of the reaction kinetic system with $\mathcal{M} = YA_k$.

2.3. Open systems and the zero complex

The *zero complex* will play an important role in analyzing the dynamical implications of model reduction. It is a special complex with zero stoichiometric coefficients, i.e. $C_0 = 0$. This appears in the reaction schemes where one considers either an external source or sink of some components, i.e. elementary reaction steps of the form

$$\emptyset \to \sum_{i=1}^{n} \beta_{ij}\mathbf{X}_i \quad \text{or} \quad \sum_{i=1}^{n} \alpha_{ij}\mathbf{X}_i \to \emptyset$$

In both cases *the systems is thermodynamically* ***not*** *closed*, therefore the conservation of the total mass in the system generally does not hold.

2.4. Structural and dynamic properties

Positive (nonnegative) systems The concepts and basic results in this paragraph are mostly taken from [5]. A function $f = [f_1 \ \dots \ f_n]^T : [0,\infty)^n \to \mathbb{R}^n$ is called *essentially nonnegative* if, for all $i = 1,\dots,n$, $f_i(x) \geq 0$ for all $x \in [0,\infty)^n$, whenever $x_i = 0$. In the linear case, when $f(x) = Ax$, the necessary and sufficient condition for essential nonnegativity is that the off-diagonal entries of $A \in \mathbb{R}^{n\times n}$ are nonnegative (such a matrix is often called a *Metzler-matrix*).

Consider an autonomous nonlinear system

$$\dot{x} = f(x), \ x(0) = x_0 \tag{7}$$

where $f : \mathcal{X} \to \mathbb{R}^n$ is locally Lipschitz, $\mathcal{X}$ is an open subset of $\mathbb{R}^n$ and $x_0 \in \mathcal{X}$. Suppose that the nonnegative orthant $[0,\infty)^n = \bar{\mathbb{R}}^n_+ \subset \mathcal{X}$. Then the nonnegative orthant is invariant for the dynamics (7) if and only if f is essentially nonnegative.

Kinetic realizability of nonnegative polynomial systems An autonomous polynomial nonlinear system of the form (7) is called *kinetically realizable* or simply *kinetic*, if a mass action reaction mechanism given by equation (6) can be associated to it that exactly realizes its dynamics, i.e. $f(x) = Y \cdot A_k \cdot \varphi(x)$ where φ contains the monomials, matrix Y has nonnegative integer elements and A_k is a valid Kirchhoff matrix. In such a case, the pair (Y, A_k) will be called a *realization* of the polynomial system (7) (note that Y contains all information about the composition of the monomials in φ in the case of mass-action dynamics). As it is expected from linear algebra, the same polynomial system may have many parametrically and/or structurally different realizations, and this is particularly true for kinetic systems coming from application domains other than chemistry. The problem of kinetic realizability of polynomial vector fields was first examined and solved in [11] where the constructive proof contains a realization algorithm that produces the directed graph of a possible associated mass action mechanism. According to [11], the necessary and sufficient condition for kinetic realizability is that all coordinates functions of the right hand side of (7) must have the form

$$f_i(x) = -x_i g_i(x) + h_i(x), \; i = 1, \ldots, n \tag{8}$$

where g_i and h_i are polynomials with nonnegative coefficients. The very short description of the realization algorithm presented in [11] is the following. Let us write the polynomial coordinate functions of the right hand side of a kinetic system (7) as

$$f_i(x) = \sum_{j=1}^{r_i} m_{ij} \prod_{k=1}^{n} x^{b_{jk}} \tag{9}$$

where r_i is the number of monomial terms in f_i. Let us denote the transpose of the ith standard basis vector in $\mathbb{R}^n$ as e_i and let $B_j = [b_{j1} \; \ldots \; b_{jn}]$.

Algorithm 1 from [11]
For each $i = 1, \ldots, n$ and for each $j = 1, \ldots, r_i$ do:

(i) $C_j = B_j + \text{sign}(m_{ij}) \cdot e_i$
(ii) Add the following reaction to the graph of the realization

$$\sum_{k=1}^{n} b_{jk}\mathbf{X}_k \longrightarrow \sum_{k=1}^{n} c_{jk}\mathbf{X}_k$$

with reaction rate coefficient $|m_{ij}|$, where $C_j = [c_{j1} \; \ldots \; c_{jn}]$.

Roughly speaking, condition (8) means that *kinetic systems cannot contain negative cross-effects.* From this, it is easy to see that all nonnegative linear systems are kinetic, since a linear system characterized by a Metzler matrix where only the diagonal elements can have negative coefficients is obviously in the form of (8). Moreover, classical Lotka-Volterra (LV) systems are always kinetic. However, there are many essentially nonnegative polynomial systems that are not directly kinetic, since some of the monomials in f_i that do not contain x_i may have negative coefficients. To circumvent this problem, one possible way is the application of a state-dependent time-rescaling [12] with a monomial function (see subsection 3.1).

Deficiency and stability Similarly to [13], we can assign the following directed graph (see, e.g. [14]) to the reaction network (1) in a straightforward way. The directed graph $D = (V_d, E_d)$ of a reaction network consists of a finite nonempty set V_d of vertices and a finite set E_d of ordered pairs of distinct vertices called directed edges. The vertices correspond to the complexes, i.e.

Journal of Physics: Conference Series **268** (2011) 012009 doi:10.1088/1742-6596/268/1/012009

$V_d = \{C_1, C_2, \dots C_m\}$, while the directed edges represent the reactions, i.e. $(C_i, C_j) \in E_d$ if complex C_i is transformed to C_j in the reaction network. The reaction rate coefficients k_j for $j = 1, \dots, r$ in (2) are assigned as positive weights to the corresponding directed edges in the graph. Where it is more convenient, the notation k'_{ij} will be used for denoting the reaction rate coefficient corresponding to the reaction $C_i \to C_j$. A set of complexes $\{C_1, C_2, \dots, C_k\}$ is a *linkage class* of a reaction network if the complexes of the set are linked to each other in the reaction graph but not to any other complex [15].

We can associate an n-dimensional vector with each reaction in the following way. For the reaction $C_i \to C_j$, the corresponding *reaction vector* denoted by h_k is given by

$$h_k = [Y]_{\cdot,j} - [Y]_{\cdot,i} \tag{10}$$

where $[Y]_{\cdot,i}$ denotes the ith column of Y. Similarly to reaction rate coefficients, whenever it is more practical, h'_{ij} denotes the reaction vector corresponding to the reaction $C_i \to C_j$.

The *rank* of a reaction network denoted by s is defined as the rank of the vector set $\{h_1, h_2 \dots, h_r\}$ where r is the number of reactions.

The positive *stoichiometric compatibility class* containing a concentration x_0 is the following set [15]:

$$(x_0 + S) \cap \mathbb{R}^n_+$$

where $\mathbb{R}^n_+$ denotes the positive orthant in $\mathbb{R}^n$, and $S = \text{span}\{h_1, h_2 \dots, h_r\}$.

The deficiency d of a reaction network is defined as [13], [15]

$$d = m - l - s \tag{11}$$

where m is the number of complexes in the network, l is the number of linkage classes and s is the rank of the reaction network.

A reaction network is called *reversible*, if each of its reactions is a reversible reaction. A reaction network is called *weakly reversible*, if each complex in the reaction graph lies on at least one directed cycle (i.e. if complex C_j is reachable from complex C_i on a directed path in the reaction graph, then C_i is reachable from C_j on a directed path).

The well-known *Deficiency Zero Theorem* [15] says that the ODEs of a weakly reversible deficiency zero CRN are globally stable with a known logarithmic Lyapunov function for all positive values of the reaction rate coefficients. Furthermore, these CRNs have a unique steady-state point in each positive stoichiometric compatibility class. Therefore (among other realization problems) it is of interest whether we can find a weakly reversible deficiency zero kinetic realization of a nonnegative polynomial system.

Computing preferred realizations of CRNs using mixed integer programming Mixed integer linear programming (MILP) is a powerful tool for combining discrete (boolean or integer) and continuous variables within an optmization framework where the objective function, equality and inequality constraints are linear [16]. It was shown that qualitative knowledge about the modeled system that can be expressed in propositional logic form has an equivalent representation as linear equations and inequalities [17]. Using this computational approach, it is possible to rewrite and solve certain CRN synthesis problems as MILP problems [10]. So far, the following problems have been solved: (a) computing the realizations containing the minimal/maximal number of reactions (called sparse and dense realizations, respectively), (b) computing the realization containing the minimal/maximal number of complexes (from within a given set), (c) computing reversible realizations (if they exist) [18]. These tools will be used to compute different realizations of kinetic systems in sections 3 and 4.

Conservation equations Assuming irreversible reaction equations (1) the conservation of component masses associated to the jth reaction is

$$\sum_{i=1}^{n} \alpha_{ij} M_i = \sum_{i=1}^{n} \beta_{ij} M_i$$

where M_i is the molecular weight of the ith component. This can be written in the following homogeneous form

$$\sum_{i=1}^{n} (\alpha_{ij} - \beta_{ij}) M_i = 0 \tag{12}$$

with the matrix $[\mathcal{N}]_{ij} = \alpha_{ij} - \beta_{ij}$ and vector M composed of the molecular weights. In order to have valid component mass conservation it should be required that the homogeneous equation $\mathcal{N}M = 0$ has at least one positive solution.

In order to develop a conservation equation for the component masses, one should take the reaction equation for the ith component in equation (5) and multiply it by its molar weight M_i, because the reaction equations refer to concentrations in molar units, e.g. in mol/m^3:

$$M_i \frac{dx_i}{dt} = \sum_{j=1}^{r} (\alpha_{ij} - \beta_{ij}) M_i \varphi_j(x)$$

Summing up the above reaction equations (both of their left and right hand sides), and performing some algebraic manipulations we obtain:

$$\frac{d}{dt} \sum_{i=1}^{n} (x_i M_i) = \sum_{j=1}^{r} \left(\sum_{i=1}^{n} (\alpha_{ij} - \beta_{ij}) M_i \right) \varphi_j(x) \tag{13}$$

Because of the mass conservation of the jth reaction in equation (12), the factor in the bracket on the right hand side in equation (13) is zero, therefore we obtain $\frac{d}{dt} \sum_{i=1}^{n} (x_i M_i) = 0$. This proves the lemma below.

Lemma 1 *A closed reaction kinetic system with MAL kinetics and component mass conservation for every reaction has a linear first integral corresponding to the total component mass conservation in the form* $\sum_{i=1}^{n} M_i x_i = const$ *if the equation* $\mathcal{N}M = 0$ *has at least one positive solution.*

2.5. An example: The Michaelis-Menten reaction scheme

The Michaelis-Menten reaction scheme describes a simple enzyme-kinetic reaction with the following reaction and dynamic model equations

$$E + S \rightleftarrows ES \qquad ES \rightleftarrows E + P \qquad ES + S \rightleftarrows ESS$$

$$\begin{aligned}
E &: \frac{dx_1}{dt} = -k_1^+ x_1 x_2 + k_1^- x_3 - k_2^- x_1 x_5 + k_2^+ x_3 \\
S &: \frac{dx_2}{dt} = -k_1^+ x_1 x_2 + k_1^- x_3 - k_3^+ x_2 x_3 + k_3^- x_4 \\
ES &: \frac{dx_3}{dt} = +k_1^+ x_1 x_2 - k_1^- x_3 + k_2^- x_1 x_5 - k_2^+ x_3 - \\
&\quad -k_3^+ x_2 x_3 + k_3^- x_4 \\
ESS &: \frac{dx_4}{dt} = +k_3^+ x_2 x_3 - k_3^- x_4 \\
P &: \frac{dx_5}{dt} = -k_2^- x_1 x_5 + k_2^+ x_3
\end{aligned}$$

where the concentration state vector x is formed as

$$x_1 = [E],\ x_2 = [S],\ x_3 = [ES],\ x_4 = [ESS],\ x_5 = [P]$$

from the components E for the enzyme, S for the substrate, P for the product, and ES and ESS being intermediate complexes.

The dynamical behaviour of the concentration variables and the reaction graph that consists of two reversible connected components are seen in figure 1, with the following model parameters used for the simulations:

$$k_1^+ = 1,\ k_1^- = 0.8,\ k_2^+ = 1.1,\ k_2^- = 0.75,\ k_3^+ = 2,\ k_3^- = 1.8$$

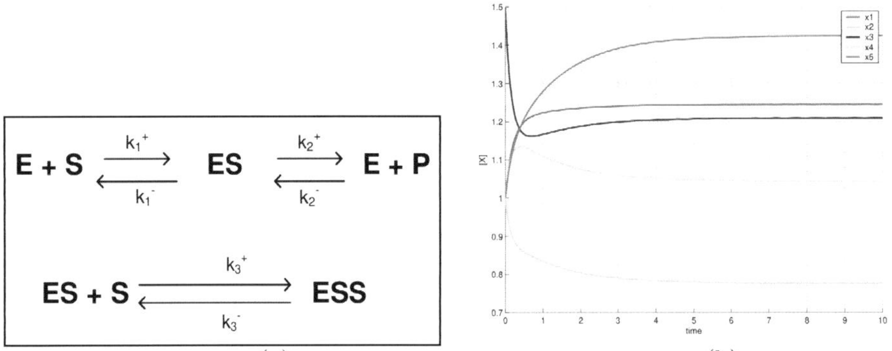

(a) (b)

Figure 1. The Michaelis-Menten reaction scheme (a) structure graph (b) system dynamics

The deficiency of this reversible model is zero, therefore its equilibrium points are globally stable because of the deficiency zero theorem.

3. Transformation of reaction kinetic models

The different transformations that are used here are described as formal mathematical transformations applied to the reaction kinetic model equations with MAL.

3.1. Time-reparametrization transformation

The time reparametrization transformation is an equivalence transformation where only the algebraic form but not the structural properties of the model are changed. It is performed by introducing a new time variable $dt = x_1^{\nu_1} \cdot \ldots \cdot x_n^{\nu_n} \cdot dt'$ with $\nu_i \in \{0, 1\}$. The transformed model equations are

$$\frac{dx}{dt'} = \mathcal{M}\varphi'(x)$$

where only the reaction monomials change to

$$\varphi_j'(x) = \prod_{i=1}^{n} x_i^{y_{ij}+\nu_i}$$

It immediately follows from the definition that *the transformed model remains in the MAL reaction kinetic class with the* ***same structural properties****.*

The time reparametrization transformation will be used later for the *elimination of negative cross-effects* that may arise during the model reduction.

3.2. Model simplification by steady state assumption transformation

The model reduction using quasi steady state (QSS) assumption(s) is the most popular method of obtaining dynamic models with lower dimension (i.e. with less differential state variables). If the jth component is considered to be in quasi steady state, formally this is written as

$$\frac{dx_j}{dt} = F_j(x) = 0 \quad , \quad x_j = x_j^* = \text{const} \tag{14}$$

The *formal model reduction transformation* is performed in two consecutive formal steps:

SS1 eliminating the component $\mathbf{X}_j$ by setting its concentration x_j to a constant x_j^* whenever it appears,

SS2 substituting the resulting algebraic equation from equation (14), e.g. $F_j(x) = 0$ into the differential ones.

The first step decreases the number of state variables by one, and then step SS2 reduces it further by one.

It is important to note that the algebraic equation from equation (14) is not considered further, i.e. step SS2 is not performed, if the component assumed to be in QSS has formed a complex in itself. In this case this complex is replaced by the zero complex in the reduced model indicating that the reduced system became open. Then the resulting constant terms in the individual reaction kinetic equations may serve as input variables with a help of which the algebraic equation can be enforced.

The steady state assumption transformation is a projection in mathematical sense, therefore the structural properties of the model may change. It is important to note that even the reaction kinetic form may be lost during this transformation. For a more detailed analysis of this transformation we refer to [6].

3.3. Simplified Michaelis-Menten reaction schemes and their realizations

This section shows how the steady state assumption transformation changes the structural properties of a reaction kinetic system model and identifies conditions when the dynamic properties of the reduced model may be different from the original one.

The constant values x_1^* and x_3^* in the following subsections were chosen from the steady state simulation results of the original model described in section 2.5.

3.3.1. Quasi steady state for the component ES (x_3) Here only the first SS1 step of the steady state transformation is performed, because the component ES forms a complex in itself.

$$\begin{aligned}
E &: \quad \frac{dx_1}{dt} = -k_1^+ x_1 x_2 + \overline{k}_1^- - k_2^- x_1 x_5 + \overline{k}_2^+ \\
S &: \quad \frac{dx_2}{dt} = -k_1^+ x_1 x_2 + \overline{k}_1^- - \overline{k}_3^+ x_2 + k_3^- x_4 \\
ESS &: \quad \frac{dx_4}{dt} = \overline{k}_3^+ x_2 - k_3^- x_4 \\
P &: \quad \frac{dx_5}{dt} = -k_2^- x_1 x_5 + \overline{k}_2^+
\end{aligned}$$

where the new reaction rate constants are $\overline{k}_i^{\pm} = k_i^{\pm} x_3^*$.

The model *remains reaction kinetic (in **generalized** sense)* with the reaction graph seen in figure 2(a).

We can eliminate the resulting zero complex through a time reparametrization transformation by introducing a new time variable $dt = x_1 dt'$ to obtain

$$
\begin{aligned}
E &: \frac{dx_1}{dt'} = -k_1^+ x_1^2 x_2 - k_2^- x_1^2 x_5 + (k_1^- + k_2^+) x_3^* x_1 \\
S &: \frac{dx_2}{dt'} = -k_1^+ x_1^2 x_2 - k_3^+ x_3^* x_1 x_2 + k_3^- x_1 x_4 + k_1^- x_3^* x_1 \\
ESS &: \frac{dx_4}{dt'} = k_3^+ x_3^* x_1 x_2 - k_3^- x_1 x_4 \\
P &: \frac{dx_5}{dt'} = -k_2^- x_1^2 x_5 + k_2^+ x_3^* x_1
\end{aligned}
$$

The resulting model equations are kinetic with the corresponding reaction structure seen in figure 2(b). This structure is fully reversible and of deficiency 0, thus the model is stable.

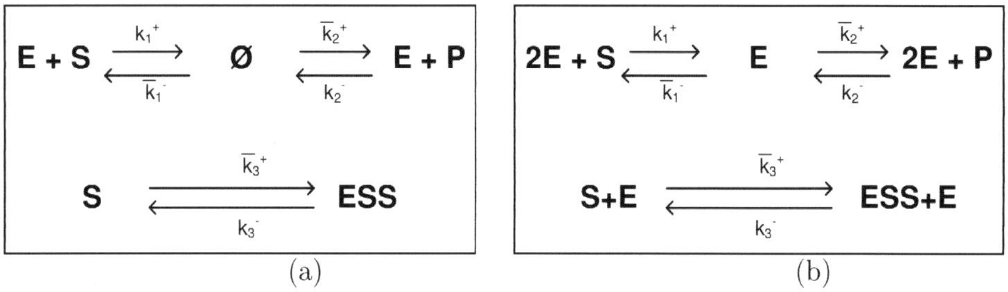

Figure 2. The simplified Michaelis-Menten reaction scheme with ES in QSS (a) reduced reaction graph (b) after time reparametrization

The dynamics of the reduced model is seen in figure 3 with $x_3^* = 1.21$. By comparing figures 1(b) and 3 we can say that even the details of the dynamics are fairly similar.

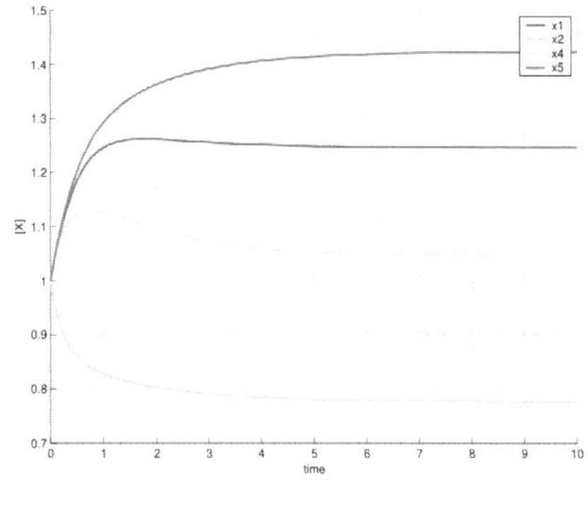

Figure 3. The dynamics of the simplified Michaelis-Menten reaction scheme with ES in QSS

3.3.2. Quasi steady state for the component E

The first step SS1 of this transformation is to set $x_1 = x_1^*$ without substituting the resulting first integral that results in a model

$$\begin{aligned}
S : \frac{dx_2}{dt} &= -k_1^+ x_1^* x_2 + k_1^- x_3 - k_3^+ x_2 x_3 + k_3^- x_4 \\
ES : \frac{dx_3}{dt} &= k_1^+ x_1^* x_2 + (-k_1^- - k_2^+) x_3 + k_2^- x_1^* x_5 - k_3^+ x_2 x_3 + k_3^- x_4 \\
ESS : \frac{dx_4}{dt} &= k_3^+ x_2 x_3 - k_3^- x_4 \\
P : \frac{dx_5}{dt} &= k_2^+ x_3 - k_2^- x_1^* x_5
\end{aligned}$$

The obtained reaction graph is seen in figure 4(a), which is a reversible deficiency 0 structure, and thus structurally stable. The dynamics of the state variables is shown in figure 4(b), that is fairly similar to the original dynamics for the remaining state variables with $x_1^* = 1.246$.

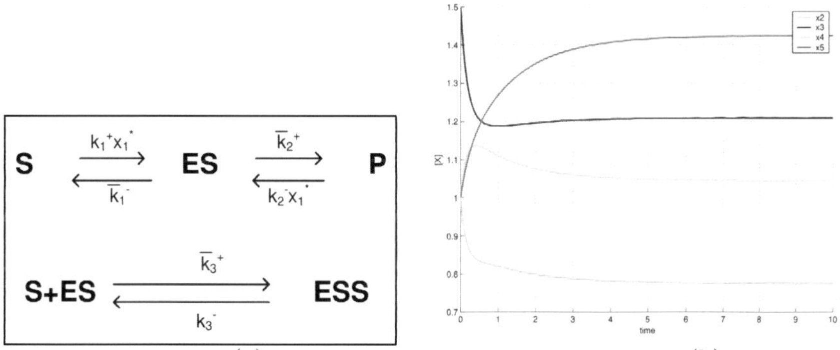

(a) (b)

Figure 4. The simplified Michaelis-Menten reaction scheme E in QSS without substituting the algebraic equation (a) structure graph (b) system dynamics

The second step SS2 requires to substitute the resulting algebraic equation from the first differential equation with $x_1 = x_1^*$, i.e.

$$-k_1^+ x_1^* x_2 + (k_1^- + k_2^+) x_3 - k_2^- x_1^* x_5 = 0$$

into the model. This can be performed by expressing x_5 and substituting it to all of the remaining equations to obtain

$$\begin{aligned}
S : \frac{dx_2}{dt} &= -k_1^+ x_1^* x_2 + k_1^- x_3 - k_3^+ x_2 x_3 + k_3^- x_4 \\
ES : \frac{dx_3}{dt} &= \underbrace{(k_1^+ x_1^* - k_2^- x_1^* l_2)}_{0} x_2 + \underbrace{(k_2^- x_1^* l_3 - k_1^- - k_2^+)}_{0} x_3 - k_3^+ x_2 x_3 + k_3^- x_4 \\
ESS : \frac{dx_4}{dt} &= k_3^+ x_2 x_3 - k_3^- x_4
\end{aligned}$$

where $l_2 = \frac{k_1^+}{k_2^-}$ and $l_3 = \frac{k_1^- + k_2^+}{k_2^- x_1^*}$. It is important to note that the polynomial form of the equations remained.

We can now search for a reaction kinetic realization of the above reduced model using MILP. It is obtained that *there exists no reversible realization*, therefore a realization with the minimum number of reactions has been searched for. The resulting irreversible reaction kinetic network is shown in figure 5(a) that does not fulfill the conditions of the deficiency zero theorem. Moreover, the dynamics of the remaining state variables (see figure 5(b)) is rather different from the original ones.

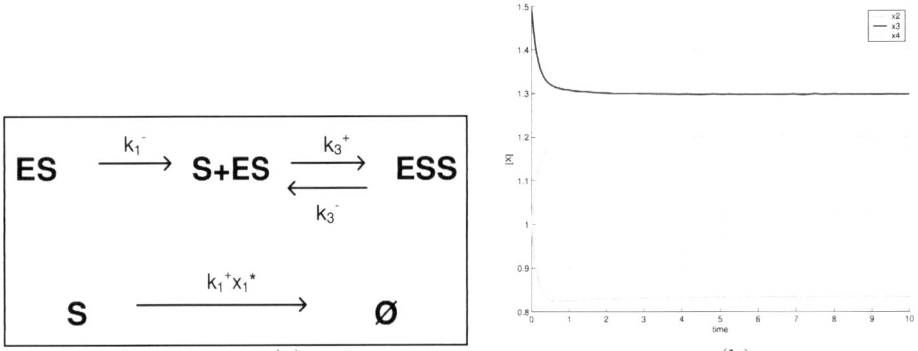

(a) (b)

Figure 5. The simplified Michaelis-Menten reaction scheme E in QSS with substituting the algebraic equation (a) structure graph (b) system dynamics

In conclusion one can observe that the second step SS2 is the one that may drastically modify the structure and the dynamics of the system, and this is the one that may bring out the model from the polynomial system class.

Without this step, however, there is no guarantee, that the component assumed to be in QSS is indeed constant, this implies that the pseudo reaction rate coefficients will vary in time. At the same time it is known, that as long as the structure of a deficiency zero CRN remains such that the deficiency zero weakly reversible property is preserved in each time instant, the time-variation of the reaction rate coefficients cannot destroy stability [19].

4. The Brusselator case study

In this section, the effect of QSS assumptions will be shown on the reversible and irreversible versions of the well-known Brusselator example. In order to have polynomial reduced models, only the first step SS1 of the QSS assumption transformation is performed in this case. Thereafter the latest results of optimization based reaction mechanism generation [10, 18] will be used to compute kinetic realizations of the simplified models.

4.1. The irreversible and reversible full models and their properties

The starting point is the most frequently used reaction scheme for the irreversible and reversible Brusselator model that are shown in figure 6 (a) and (b), respectively. The kinetic differential equations of the irreversible model are the following:

$$\dot{x}_1 = -k_{12}x_1 \tag{15}$$
$$\dot{x}_2 = k_{12}x_1 - k_{23}x_2 + k_{45}x_2^2x_4 - k_{67}x_2x_5 \tag{16}$$
$$\dot{x}_3 = k_{23}x_2 \tag{17}$$
$$\dot{x}_4 = -k_{45}x_2^2x_4 + k_{67}x_2x_5 \tag{18}$$
$$\dot{x}_5 = -k_{67}x_2x_5 \tag{19}$$
$$\dot{x}_6 = k_{67}x_2x_5 \tag{20}$$

The parameter values used for the numerical computations (resulting in oscillatory behaviour in the reduced model) were the following:

$$k_{12} = 1.2,\ k_{23} = 2.1,\ k_{45} = 4.2,\ k_{67} = 6.5 \tag{21}$$

The equations corresponding to the reversible Brusselator model version can be written as

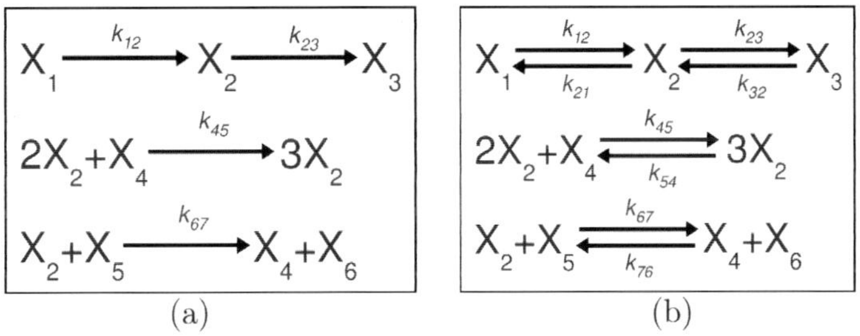

Figure 6. Brusselator reaction schemes: (a) irreversible case, (b) reversible version

$$\dot{x}_1 = -k_{12}x_1 + k_{21}x_2 \tag{22}$$
$$\dot{x}_2 = k_{12}x_1 - (k_{21} + k_{23})x_2 + k_{32}x_3 + k_{45}x_2^2x_4 - k_{54}x_2^3 - k_{67}x_2x_5 + k_{76}x_4x_6 \tag{23}$$
$$\dot{x}_3 = k_{23}x_2 - k_{32}x_3 \tag{24}$$
$$\dot{x}_4 = -k_{45}x_2^2x_4 + k_{54}x_2^3 + k_{67}x_2x_5 - k_{76}x_4x_6 \tag{25}$$
$$\dot{x}_5 = -k_{67}x_2x_5 + k_{76}x_4x_6 \tag{26}$$
$$\dot{x}_6 = k_{67}x_2x_5 - k_{76}x_4x_6 \tag{27}$$

The parameter values (also causing oscillatory behaviour in the reduced model) were the following:

$$k_{ij} = 1,\ \forall i,j \tag{28}$$

It can be easily computed that the deficiency is zero for both of the above models. This means that there is no equilibrium point in the interior of the positive orthant in the case of the irreversible Brusselator, and there is exactly one equilibrium point in each stoichiometric compatibility class of the reversible model. Moreover, the dynamics of the reversible model is globally stable with a known Lyapunov function, and it cannot produce complex behaviour such as oscillations in its original closed form.

4.2. Simplified versions of the irreversible variant

To produce oscillations, two concentrations, namely x_1 and x_5, must be kept constant. This practically means opening the originally closed network and feeding in the necessary amount of these species. From the modeling point of view, this is a QSS assumption where $x_1 = x_1^*$ and $x_5 = x_5^*$ is assumed. The resulting dynamical equations are then the following:

$$\dot{x}_2 = k_{12}x_1^* - (k_{23} + k_{67}x_5^*)x_2 + k_{45}x_2^2x_4 \tag{29}$$
$$\dot{x}_3 = k_{23}x_2 \tag{30}$$
$$\dot{x}_4 = -k_{45}x_2^2x_4 + k_{67}x_5^*x_2 \tag{31}$$
$$\dot{x}_6 = k_{67}x_5^*x_2 \tag{32}$$

The numerical values of the chosen steady states were $x_1^* = x_5^* = 1$. It is easy to check that the polynomial system (29) - (32) is kinetic. Using the methodology described in [10], the sparse realization of the model can be determined which is shown in figure 7. Equations (30) and (32)

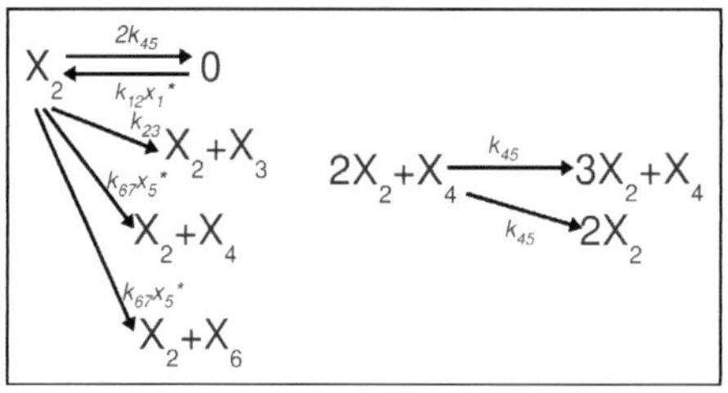

Figure 7. Sparse reaction scheme corresponding to the kinetic system (29) - (32)

can be omitted from the model if we are not interested in the dynamics of product composition. In this case, the remaining equations are

$$\dot{x}_2 = k_{12}x_1^* - (k_{23} + k_{67}x_5^*)x_2 + k_{45}x_2^2x_4 \tag{33}$$
$$\dot{x}_4 = -k_{45}x_2^2x_4 + k_{67}x_5^*x_2 \tag{34}$$

Naturally, the above equations are still kinetic, and the corresponding sparse reaction network can also be determined. This structure can be seen in figure 8. It is immediately visible that the network in figure 8 is a proper subgraph of the one shown in figure 7. The deficiency of both CRNs are 2, and with proper parametrization (for example the one in equation (21)), both are able to produce stable limit cycles. Moreover, it can also be shown through optimization, that no reversible CRN realization exists either for equations (29)-(32) or for (33)-(34) which is not very surprising, since the orignal closed CRN shown in figure 6 (a) was completely irreversible (although one reversible reaction pair appears between the zero complex and X_2).

In summary, we can observe that the deficiency and even the minimal number of reactions in a kinetic system may increase as a result of a QSS assumption. This might be contrary to our expectations, since the dynamics is seemingly simplified by the elimination of certain differential equations. Additionally, the physical opening of the system (needed to keep the appropriate concentrations constant) appears in the violation of component mass conservation (zero complex and autocatalytic reactions). Then, these factors can result in complex dynamical behaviour such as a limit cycle that is not observable in the original closed system.

4.3. Simplified versions of the reversible model

Let us consider now the reversible Brusselator variant given by equations (22)-(27). The following QSS assumptions have to be made for oscillating behaviour:

$$x_i = x_i^*, \quad i = 1, 3, 5, 6 \tag{35}$$

The remaining dynamics is then given by

$$\dot{x}_2 = (k_{12}x_1^* + k_{32}x_3^*) + (-k_{21} - k_{23} - k_{67}x_5^*)x_2 - k_{54}x_2^3 + k_{45}x_2^2x_4 + (k_{76}x_6^*)x_4 \tag{36}$$

$$\dot{x}_4 = -k_{45}x_2^2x_4 + k_{54}x_2^3 + k_{67}x_5^*x_2 - k_{76}x_6^*x_4 \tag{37}$$

The values of the steady states were chosen as follows: $x_1^* = 1$, $x_3^* = 1$, $x_5^* = 16$, $x_6^* = 0.5$. It is immediately visible that equations (36) - (36) are kinetic, therefore we can run **Algorithm 1** to compute a CRN realization which is shown in figure 9. It can be seen from the figure that this realization of deficiency 4 contains 8 reactions and none of the two linkage classes is (at least weakly) reversible.

However, an attempt to compute a fully reversible realization is successful in this case and gives the CRN that is depicted in figure 10. We remark that this CRN is also a sparse realization of the dynamics (36) - (37), since the MILP optimization method tells us that the minimal number of reactions needed to realize that dynamics is 6. The deficiency of the obtained reversible network is 1, therefore we can apply the *Deficiency one theorem* [15] which says that the differential equations of the CRN admit precisely one steady-state in each positive stoichiometric compatibility class. This fact, however, does not exclude the possibility of oscillating behaviour, that is to have a stable limit cycle around this steady-state. The stable limit cycle produced by the CRN with parameters in equation (28) started from the initial state $x(0) = [2\ 2.5]^T$ can be seen in figure 11.

5. Conclusion and future work

The models relevant on a given time scale of a CNR are constructed by using QSS assumption(s) in this paper, that are defined as formal model transformations consisting of two steps. A MILP

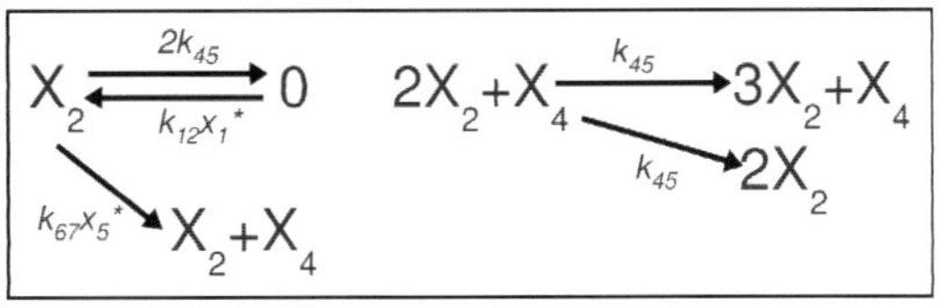

Figure 8. Sparse reaction scheme corresponding to the kinetic system (33) - (34)

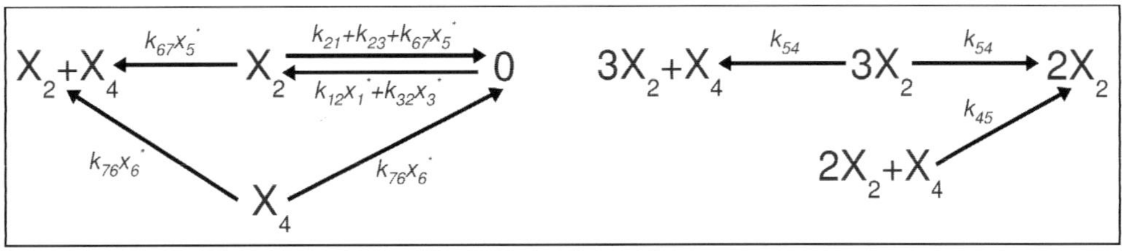

Figure 9. Reaction network produced by **Algorithm 1** corresponding to the kinetic system (36) - (37)

approach is proposed further for determining physically meaningful mass action realizations of these reduced models.

The relationship between the properties (reversibility, deficiency, stability) of the obtained realizations of the same system on different time scales were also investigated and related to the same properties of the detailed overall model. It has been shown that the QSS assumption can cause the inrease of the deficiency and that of the minimal number of reaction in the realization, as well as the appearence of the zero complex. This implies that spurious exotic nonlinear behaviour, such as multiple steady-states and oscillations can appear in the reduced models that were not present in the oroginal detailed ones.

The proposed methods are illustrated by using a simple Michaelis-Menten type reaction kinetic example. The simplified versions of the well known Brusselator model have also been investigated and presented as a case study.

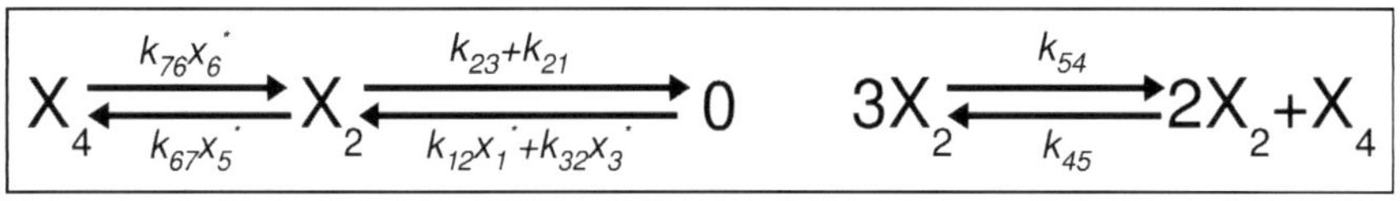

Figure 10. Sparse reversible realization corresponding to the kinetic system (36) - (37)

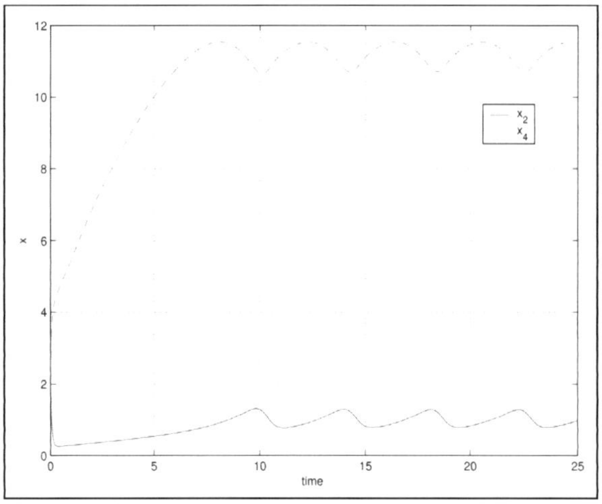

Figure 11. Time-domain behaviour of the CRN shown in Fig 10

Acknowledgments
This research was partially supported by the Hungarian Scientific Research Fund and the National Office for Research and Technology through grant no. K67625 and by the Control Engineering Research Group of HAS at Budapest University of Technology and Economics.

References
[1] Érdi and Tóth J 1989 *Mathematical Models of Chemical Reactions. Theory and Applications of Deterministic and Stochastic Models* (Manchester, Princeton: Manchester University Press, Princeton University Press)
[2] Haag J, Wouver A and Bogaerts P 2005 *Mathematical Biosciences* **193** 25–49
[3] Sontag E 2001 *IEEE Trans. Autom. Control* **46** 1028–1047

[4] Angeli D 2009 *European Journal of Control* **15** 398–406
[5] Chellaboina V, Bhat S P, Haddad W M and Bernstein D S 2009 *IEEE Control Systems Magazine* **29** 60–78
[6] Hangos K 2010 *Entropy* **12** 772–797
[7] Tzafriri A R and Edelman E R 2004 *Journal of Theoretical Biology* **226** 303–313
[8] Millat T, Bullinger E, Rohwer J and Wolkenhauer O 2007 *Mathematical Biosciences* **47** 40–57
[9] Sabouri-Ghomi M, Ciliberto A, Kar S, Novak B and Tyson J J 2008 *Journal of Theoretical Biology* **250** 209–218
[10] Szederkényi G 2010 *Journal of Mathematical Chemistry* **47** 551–568 dOI: 10.1007/s10910-009-9525-5 URL `http://www.springerlink.com`
[11] Hárs V and Tóth J 1981 *Qualitative theory of differential equations* (North Holland) chap On the inverse problem of reaction kinetics, pp 363–379
[12] Szederkényi G, Hangos K and Magyar A 2005 *Physics Letters A* **334** 288–294
[13] Feinberg M 1979 *Lectures on chemical reaction networks* (Wiscosin: Notes of lectures given at the Mathematics Research Center, University of Wisconsin)
[14] Bang-Jensen J and Gutin G 2001 *Digraphs: Theory, Algorithms and Applications* (Springer)
[15] Feinberg M 1987 *Chemical Engineering Science* **42** 2229 – 2268
[16] Floudas C 1995 *Nonlinear and mixed-integer optimization* (Oxford University Press)
[17] Raman R and Grossmann I 1991 *Computers and Chemical Engineering* **15** 73–84
[18] Szederkényi G, Hangos K M and Péni T 2011 *MATCH Commun. Math. Comput. Chem.* **65** 309–332
[19] Chaves M 2005 *SIAM J. Control Optim.* **44** 704–727

Journal of Physics: Conference Series **268** (2011) 012010 doi:10.1088/1742-6596/268/1/012010

Full-scale tests on steel frames under cyclic loading

M Ivanyi[1], P Ivanyi [1] and M M Ivanyi[2]

[1] University of Pécs, Pollack Mihály Faculty of Engineering, Pécs, Hungary
[2] UVATERV Ltd. Budapest, Hungary

E-mail: steelivanyi@epito.bme.hu

Abstract. An experimental investigation on full-scale frames aimed at determining the effect of flexibly in the beam-column connections with cycling loading is described. The paper presents an overview of the experimental set-up and briefly explains how the complexities arising from the three-dimensional nature of the tests were addressed. The quality of the collected experimental data has justified the effect, which was expended in addressing the experimental complexities.

1. Introduction

The structure of a building can be supplied by different types of loads and environmental impacts. In practice the loading of a structure is modeled as a continuously increasing, quasi-static system. On the other hand temporary aspects of different environmental impacts and loads should also be investigated, since meteorological loads (snow, wind, and temperature change) cannot be modeled as continuously increasing system. In the experiments beside the increasing vertical force, an alternating horizontal force or pulsating force was also applied. Due to the effect of these different loading types the behavior and response of structures can also be quite different. For example plastic states develop which in the case of repetitive loading will result in a hysteretic behavior in structures.

Semi-continuous framing, i.e. framing in which partial strength and semi-rigid joints according to the Eurocode 3 [1] definitions are applied, is gaining importance in building structures due to the economy offered and the possibilities for modeling and calculation provided by a new set of design codes [2]. The connection models presented in these design standards and in the technical literature in general are based on the enormous number of tests, experimental and numerical, which were performed in the last decade or two (see e.g. [3]), most of them examining joints isolated from the whole structure in which they are to be incorporated. However, it is not evident how these joints behave when they become part of a real frame; it is recognized that this problem needs experimental investigations performed on full-scale frames and realistic solutions for connection behavior.

A series of tests in 1993/94 consisting of four, two-storey and two-bay frames [4], in which partial strength and semi-rigid joints were applied for column bases as well as beam-to-column connections. However, these joints were not designed to accommodate plastic hinges; therefore the failure of the frames in all instances was due either to loss of member stability or to connection failure in a brittle manner.

The new test series presented in this paper consists of frames prepared with semi-rigid beam-to-column joints where special attention was devoted to the detailing of these joints in order to ensure a rotation capacity sufficient for a plastic mechanism to occur within the frame. Furthermore, the tests concentrated mostly on the behavior of these frames under quasi-static cyclic loading with different types of load histories. These way two objectives were set at the start of the tests: (i) to examine the behavior of the frame itself under the circumstances covered, and (ii) to look at the behavior of the partial strength and semi-rigid connections within these frames.

2. Testing program

2.1. Structure

The experimental research project was carried out in the Laboratory of the Department of Bridges and Structures, Budapest University of Technology and Economics. In figure 1 there is the general view of the experimental test frame.

Figure 1. General view of experimental test frame.

An overall view of the testing arrangement is shown in figure 2. The frames examined are two-storey single-bay ones. Both the columns and the beams are welded I sections, see table 1. Pairs of test frames are identical. Columns are connected to a rigid steel base element by two bolts through an end plate (layout generally regarded as pinned joint in the practice).

Beams and columns are connected with flush end plate joints (figure 3). In OTKA frames 1 to 5, the connections are strengthened with single-sided additional web plates. These were found to be necessary on the basis of an analysis of joint behavior according to Eurocode 3 Part 1.8. [5] According to the predictions, the joints were to fail in their tension zone, by Mode 1 bending of the end plate, a failure mode generally recognized as one ensuring sufficient rotation capacity to accommodate a plastic hinge within the connection. These predictions also showed that omitting the additional web plates would have caused column web buckling to become the governing failure mode, which is regarded to be ineffective in terms of plastic hinge behavior. Despite these predictions, joints in

OTKA frame 6, in which these plates were indeed omitted, showed similarly ductile behavior as those in OTKA frames 1 to 5.

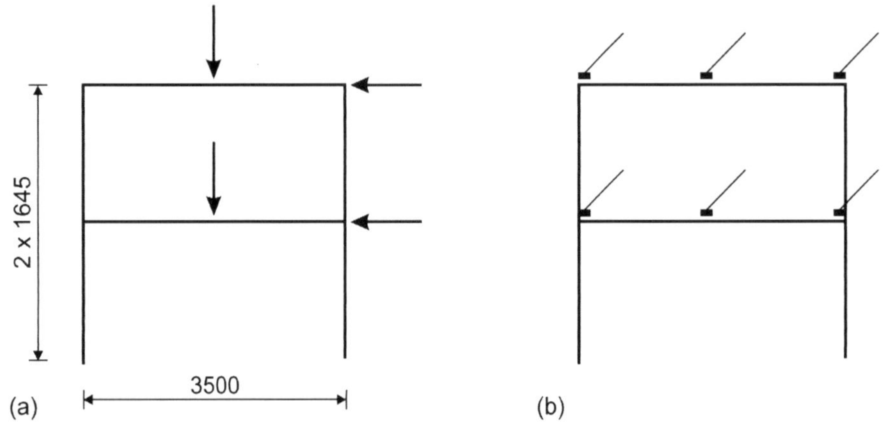

Figure 2. The tested frames: (a) main geometry and (b) lateral restrains.

Table 1. Beam and column sections applied in the tests.

Frame	Beam (flange/web)	Column (flange/web)
1, 2	130-8/260-8	200-12/260-8
3, 4	160-10/260-8	200-12/260-8
5, 6	130-8/260-8	160-10/260-8

In addition, these joints showed a type of asymmetric behavior which, according to the best knowledge of the authors, is not treated in the technical literature, and which is not covered by design code regulations. This behavior, characterized by excessive yielding of the column flange at one side and similarly excessive yielding of the end plate at the other, is due to the asymmetry of the joint because of the single-sided additional web plate, and suggests that these web plates play a more sophisticated role in the joint behavior than as implied in current European regulations. This issue is discussed in more detail elsewhere [6].

In order to avoid lateral-torsional buckling, lateral restraints are applied to the frame at the beam-to-column joint locations and at the mid-spans of the beams, see figure 2.

2.2. Loading

The frame is loaded by two vertical concentrated loads at the mid-spans of the beams, and two horizontal loads applied at one side of the frame in the levels of the beams (figure 2).

The two vertical loads are increased and decreased proportionally using three hydraulic jacks (one larger to the lower beam and two smaller and identical to the upper) connected into one oil circuit. Because of the slight difference between the pressure surfaces of the larger jack on one hand and the two smaller jacks on the other, the lower beam was loaded by a concentrated load 89% in magnitude of the load on the upper beam. The vertical loads are applied through so-called gravity load simulators [7], devices, which ensure the verticality of the loads within certain limits of lateral displacements of the points of application of the loads.

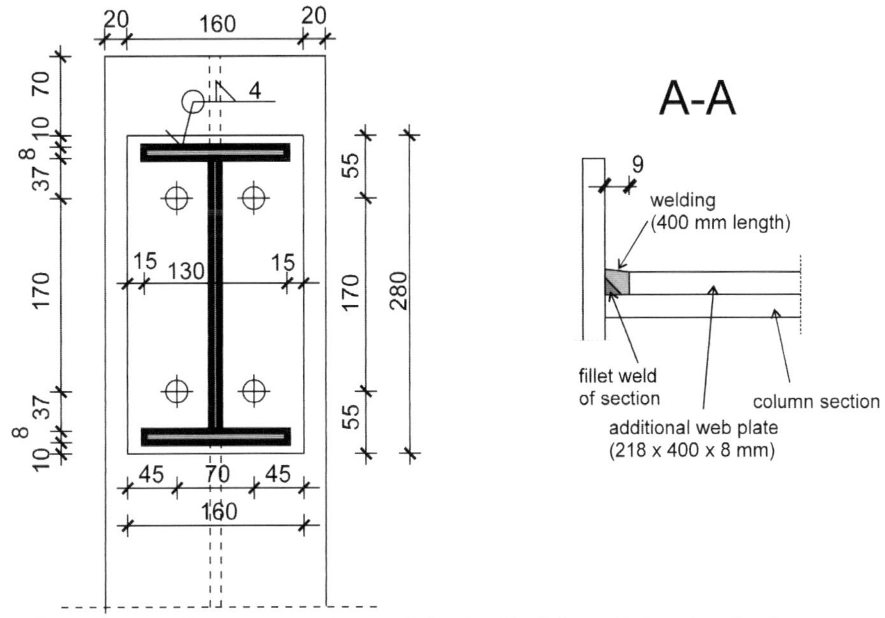

Figure 3. Beam-to-column connection and the detail of the additional web plate.

The horizontal loads are applied using one hydraulic jack through a simply supported vertical beam, which ensures the applied load, to be equally distributed between the two beam levels. The direction of these horizontal loads is reversible.

The loading histories under which test OTKA frames 1 to 6 were examined are shown in table 2 and figure 4. OTKA frame 1 was loaded by proportionally increasing loads; OTKA frames 2 to 6 were loaded by increasing vertical loading and cyclic horizontal loading at the levels of the vertical loads. In the case of OTKA frames 2, 3, 5 and 6, the amplitude of the horizontal loads was set as a ratio of the vertical loads, while in the case of frame 4, this amplitude was set to a constant value. In the case of all frames except OTKA frame 4, the ratio of the vertical loads and horizontal loads was set to 1:6 (value of one horizontal force as compared to the vertical load acting on the upper beam); for OTKA frame 4, the amplitude of the horizontal load cycles was set to a constant value of 36 kN (value of one horizontal force).

Table 2. Loading program.

Frame	Program description			Highest vertical load (kN)
	vertical load	horizontal load	amplitude of load cycles	
1	monotonic	monotonic		193.4
2	monotonic	cyclic	fixed ratio	208
3	monotonic	cyclic	fixed ratio	240
4	monotonic	cyclic	fixed value	240
5	monotonic	cyclic	fixed ratio	196
6	monotonic	cyclic	fixed ratio	200

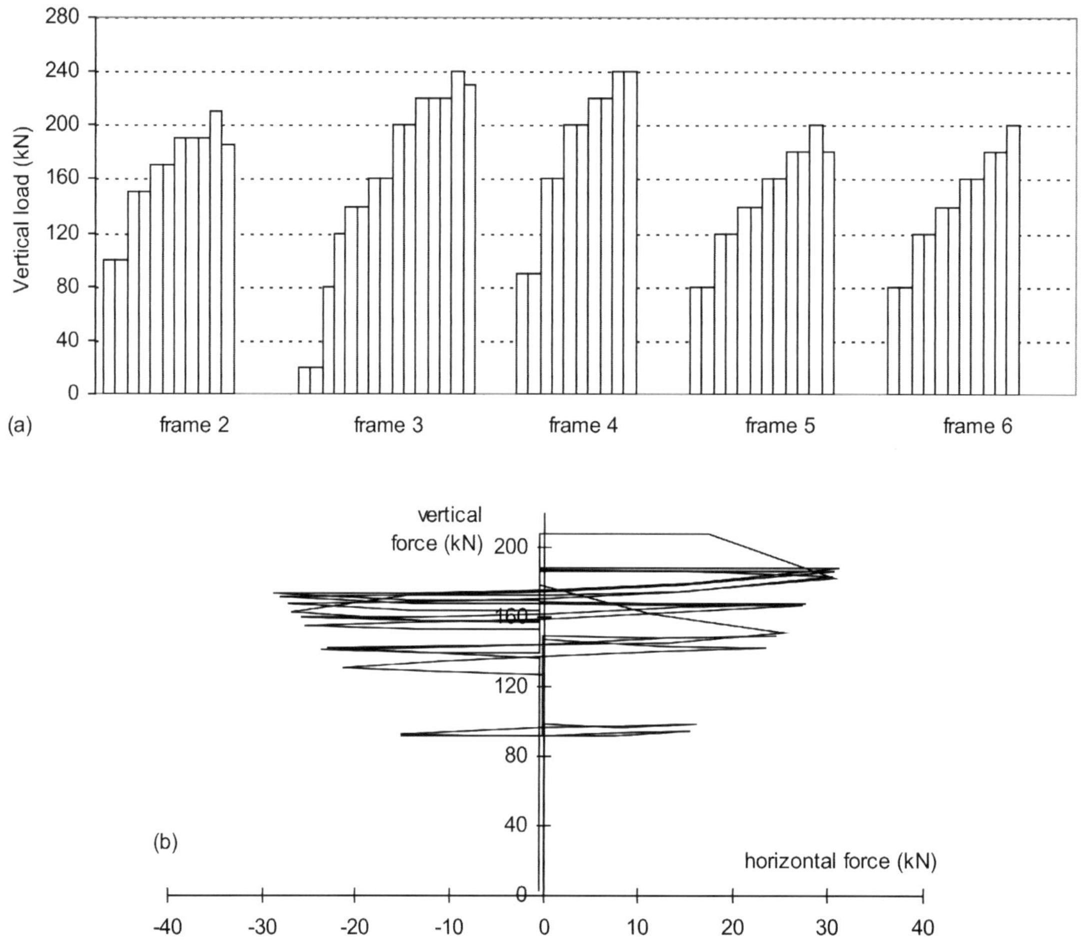

Figure 4. Loading program, (a) general program in terms of vertical forces for OTKA frames 2 to 6 (columns represent cycles of horizontal loads); (b) vertical force versus horizontal force for OTKA frame 2 as measured.

3. Measurements

During the tests we measured beam mid-span deflections and lateral storey displacements by using inductive transducers, forces by pressure transducers built into the oil circuit of the hydraulic system, strains by strain gauges and cross-section rotations by purpose-designed devices (figure 5).

The strain measurement was taken at the end cross-sections of beams and columns, by using ten gauges for each cross-section. The arrangement we applied was first used by [8], and makes possible to derive the four cross-sectional internal forces (the axial force, the two bending moments and the bimoment) even in the case of plasticity occurring within a part of the cross-section (figure 6). We did not, however, expect excessive plasticity in these cross-sections due to the fact that the beams and columns were connected by partial strength connections.

We applied a special purpose-designed device referred to as rotation indicator for the measurement of the absolute rotations of cross-sections (figure 7). This device consists of spring steel connected rigidly on its top to the cross-section being examined. While the cross-section and the top of the spring element rotates, a weight connected to the bottom of this spring element tries to remain in place, which causes bending within the spring steel element.

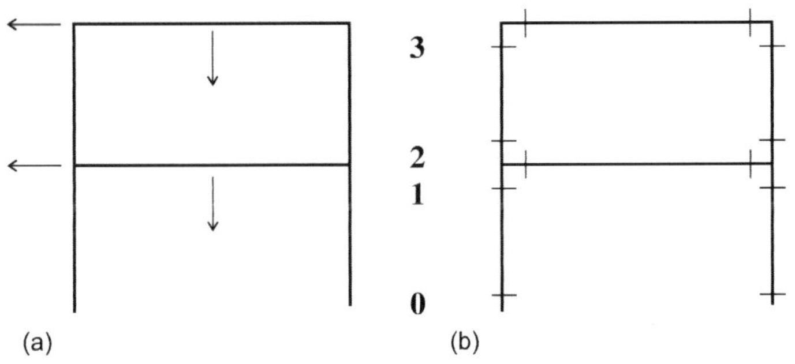

Figure 5. Locations of measurements: (a) measurement of displacements; (b) cross-sections where rotations and strains were measured.

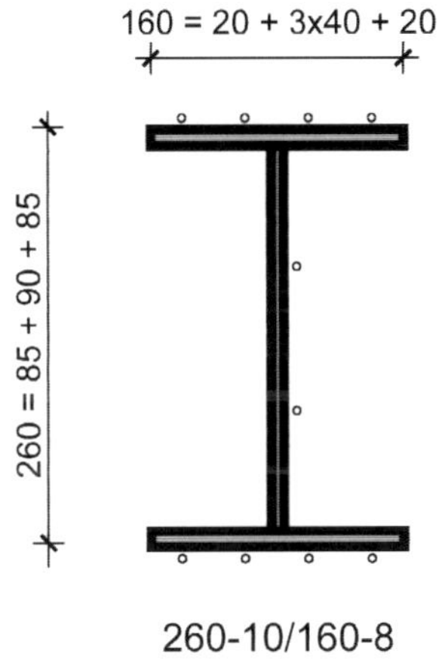

Figure 6. The arrangement of strain gauges within a cross-section.

The deformations depend on the rotation to be measured, and are assessed by two strain gauges on the two surfaces of the spring steel element. The device is suitable for the measurement of rotations within the range of ±10 degrees. The device was calibrated for this range and was found to behave linearly; its coefficient, i.e. the ratio between the rotation and the difference of strains as measured by the two strain gauges, was in the range of 7.8 to 9.0 × 10^{-3} degrees per micro-strains (the results come from the calibration of the fifteen rotation indicators constructed).

Material properties of the applied steel was determined on the basis of tension tests, see table 3.

Table 3. Material prosperities.

Plate element		Yield strength (MPa)	Ultimate strength (MPa)
Thickness	Purpose		
8 mm	web, flange	300	429
10 mm	flange	276	384
10 mm	end plate	320	444
12 mm	flange	274	382

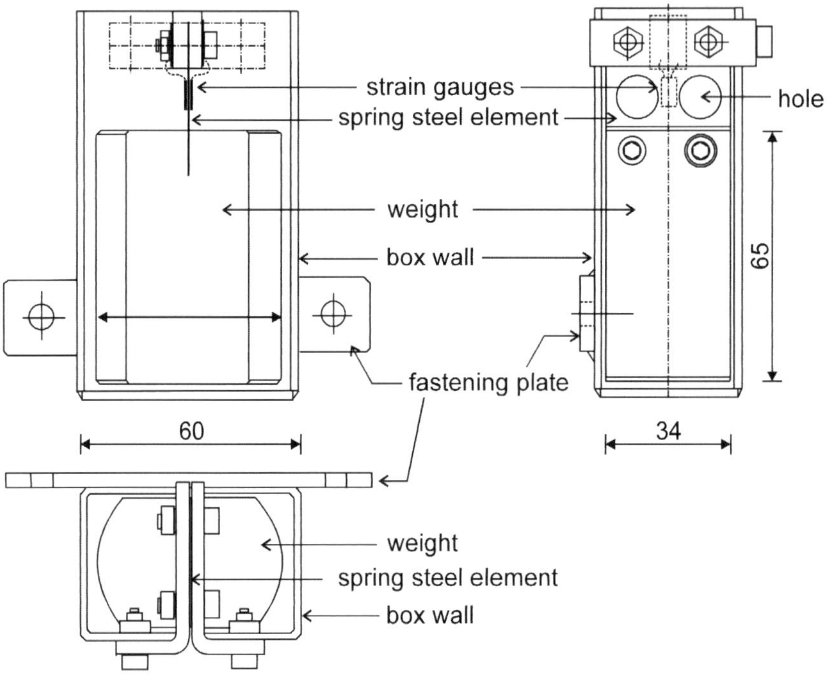

Figure 7. The rotation indicator. The 'weight' element is a ϕ50×67 solid element; the 'spring' element is a 0.5×34×40 steel element, with a cross-section weakened to 7 mm^2 at the strain gauges.

4. Results of OTKA-1 with proportional loads

The frames examined are two-storey single-bay ones. Both the columns and the beams are welded I sections. Pairs of test frames are identical. Columns are connected to a rigid steel base element by two bolts through an end plate (layout generally regarded as pinned joint in the practice).

Beams and columns are connected with flush end plate joints. In frame OTKA-1, the connections are strengthened with single-sided additional web plates. These were found to be necessary on the basis of an analysis of joint behavior according to [5].

The frame is loaded by two vertical concentrated loads at the mid-spans of the beams, and two horizontal loads applied at one side of the frame in the levels of the beams. The two vertical loads are increased and decreased proportionally using three hydraulic jacks (one larger to the lower beam and two smaller and identical to the upper) connected into one oil circuit. Because of the slight difference between the pressure surfaces of the larger jack on one hand and the two smaller jacks on the other, the lower beam was loaded by a concentrated load 89% in magnitude of the load on the upper beam. The vertical loads are applied through so-called gravity load simulators, devices which ensure the verticality of the loads within certain limits of lateral displacements of the points of application of the loads. The horizontal loads are applied using one hydraulic jack through a simply supported vertical beam, which ensures the applied load to be equally distributed between the two beam levels. The direction of these horizontal loads is reversible.

Concerning the experimental frame OTKA-1, the relation of load-deflection curve develops according to figure 8. The Approximate Engineering Method is presented on test frame OTKA-1 (figure 8) [9]. The comparison shows that the Approximate Engineering Method gives satisfactory results for the maximum loads and the descending state path of whole structure as well; and at the same time the analysis can be done at the 'desk of the designer'.

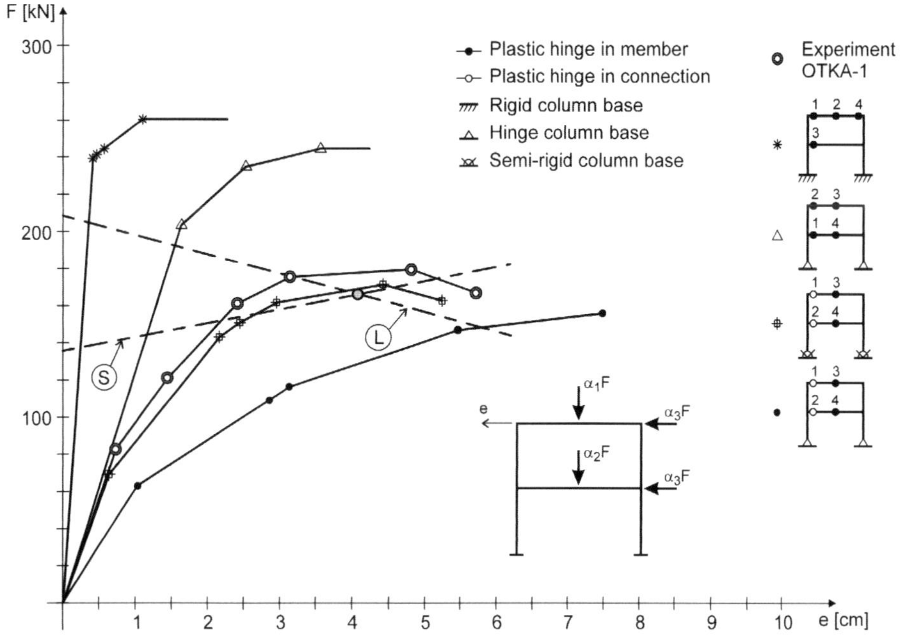

Figure 8. Load–displacement curve of experiment OTKA frame 1.

5. Results of OTKA 2-6 with cyclic loadings

The ultimate behavior of the OTKA frames was characterized by excessive yielding within the joint zones (plastic mechanism), and in the beams below the vertical loads, and the failure was in nearly all instances due to the buckling of the compression flanges of the top beams below the vertical loads, and then by the initiation of lateral-torsional buckling in the top beams.

In OTKA frames 2 to 6 the cross sections for beams and columns were different, see table 1, The loading program were different too, see table 2. Figures 9-13 have shown the similar load vs. displacement and rotation curves for OTKA frames 2 to 6. In the figures there are some details of experimental results:

(a1) Vertical load of upper beam versus top storey drift
(a2) Total horizontal load versus top storey drift
(b1) Vertical load of upper beam versus vertical deflection
(b2) Total horizontal load versus vertical deflection
(c1) Vertical load of upper beam versus rotation at '0'
(c2) Total horizontal load versus rotation of section '0'
(d1) Vertical load of upper beam versus normal force at '2'
(d2) Total horizontal load versus normal force at section '2'.

6. Conclusion

The paper sets the objective to present the testing methods applied during a test series consisting of six frames with similar geometry and, in five cases, under quasi-static cyclic loading. The testing methods presented in the paper prove to be suitable for the given purpose. Special attention is paid to a device, a new development, referred to as rotation indicator, which has been constructed to measure the rotations of cross-sections.

Further work related to these tests will concentrate on the detailed analysis of the results collected. The focus will be to look at the connection behavior, which in one sense is peculiar due to its asymmetry relative to the plane of the frame, and which will also be interesting from the point of view

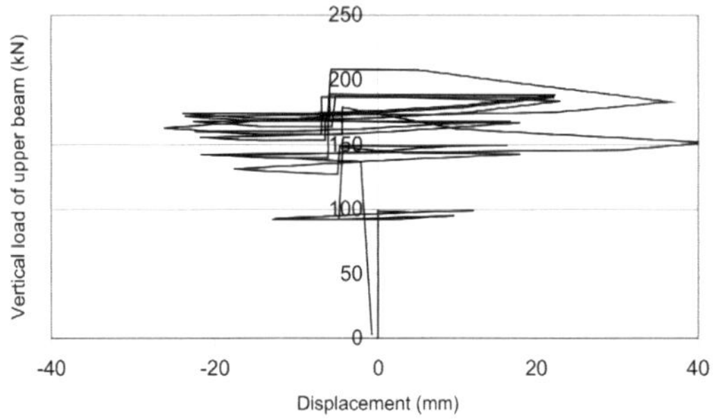

(a1) Vertical load of upper beam versus top storey drift

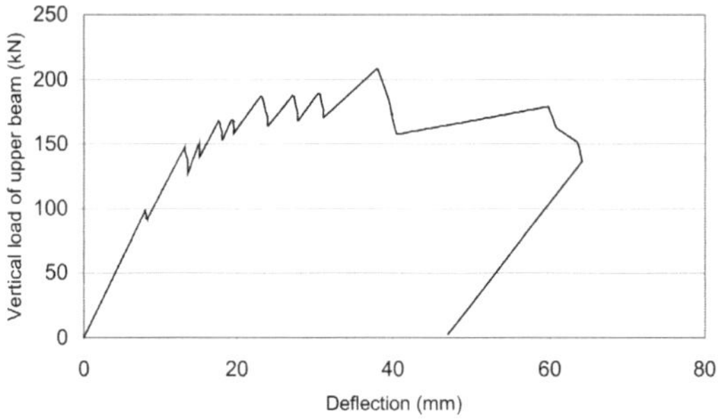

(b1) Vertical load of upper beam versus vertical deflection

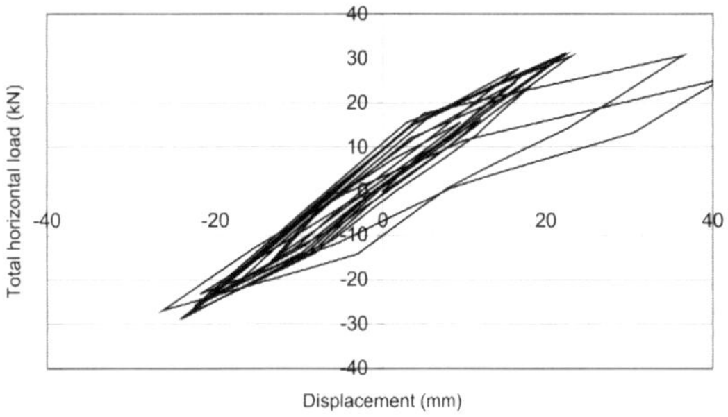

(a2) Total horizontal load versus top storey drift

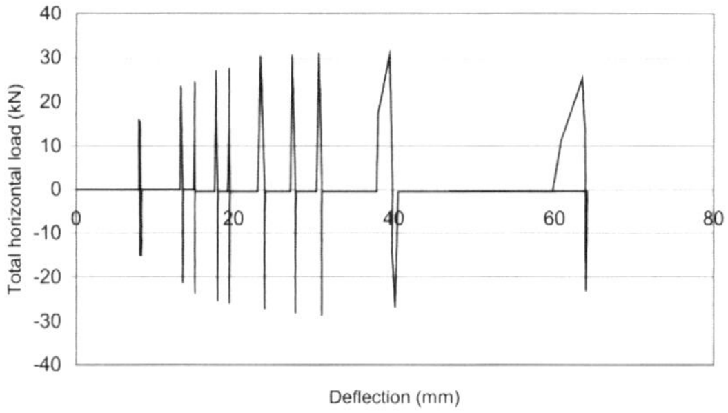

(b2) Total horizontal load versus vertical deflection

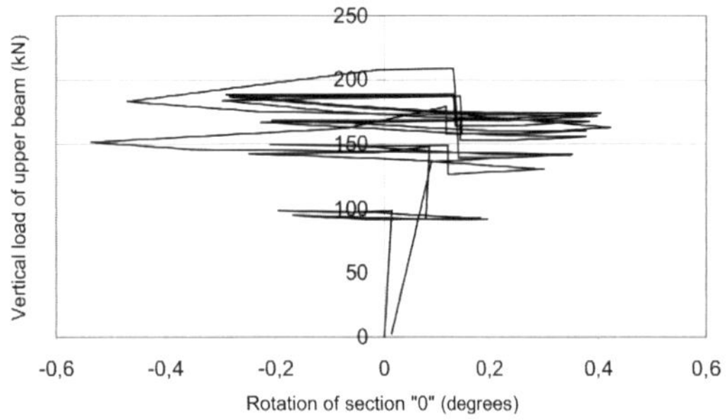

(c1) Vertical load of upper beam versus rotation at '0'

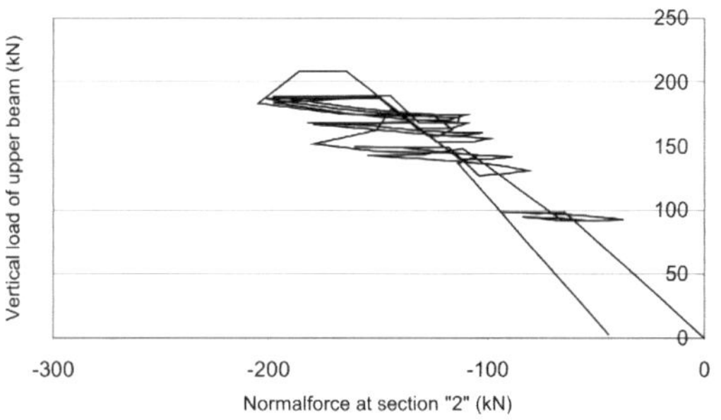

(d1) Vertical load of upper beam versus normal force at '2'

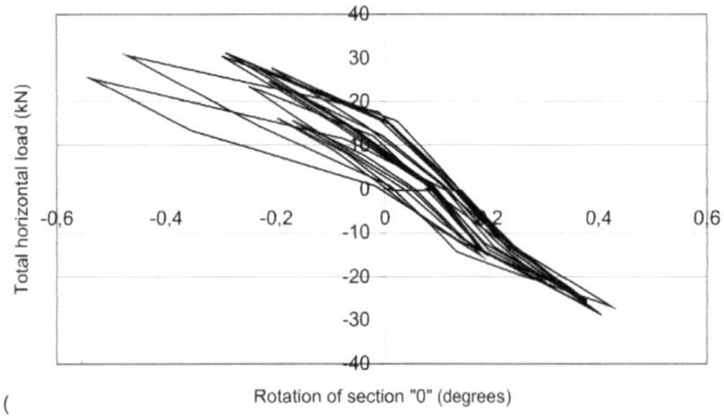

(c2) Total horizontal load versus rotation of section "0"

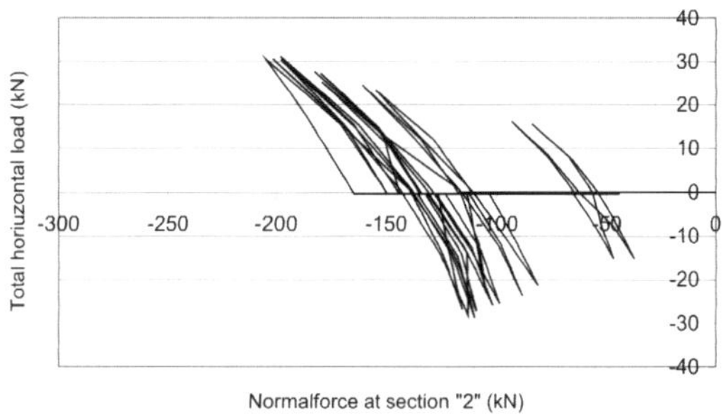

(d2) Total horizontal load versus normal force at section "2"

Figure 9. Load vs. displacement and rotation curves for OTKA frame 2.

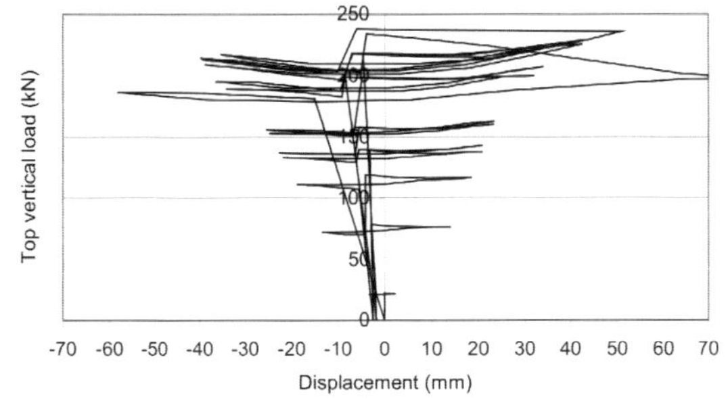

(a1) Vertical load of upper beam versus top storey drift

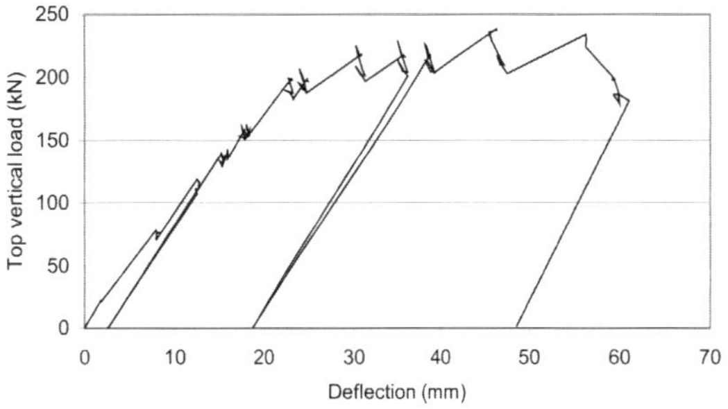

(b1) Vertical load of upper beam versus vertical deflection

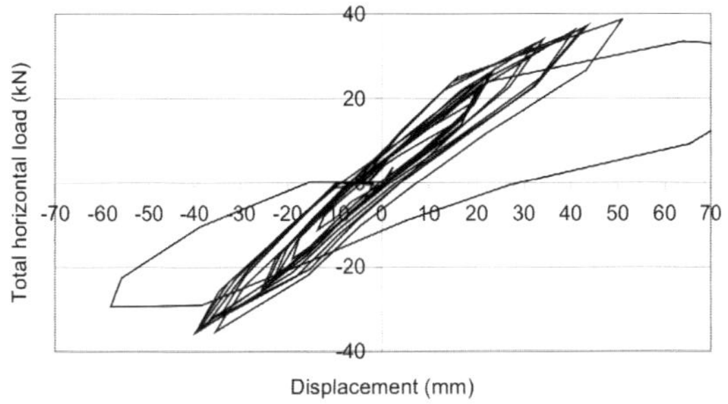

(a2) Total horizontal load versus top storey drift

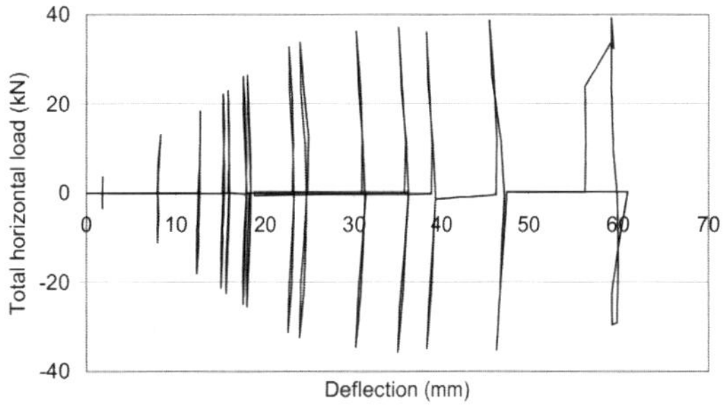

(b2) Total horizontal load versus vertical deflection

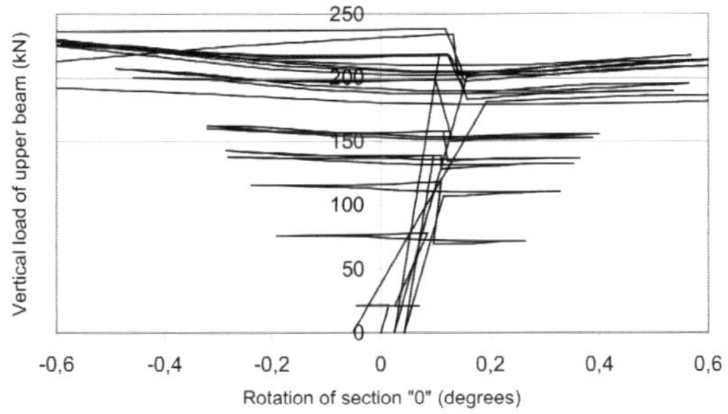

(c1) Vertical load of upper beam versus rotation at '0'

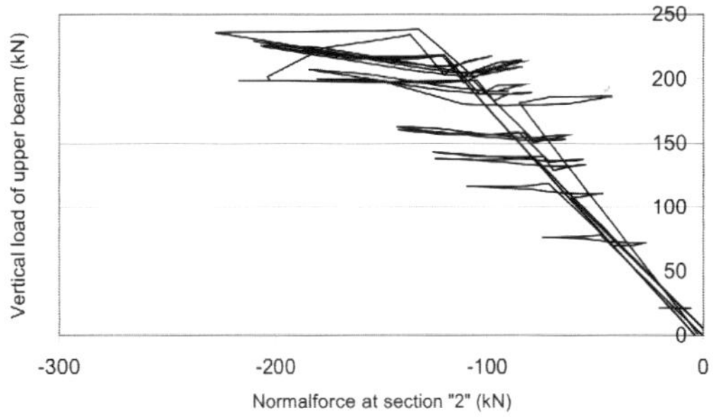

(d1) Vertical load of upper beam versus normal force at '2'

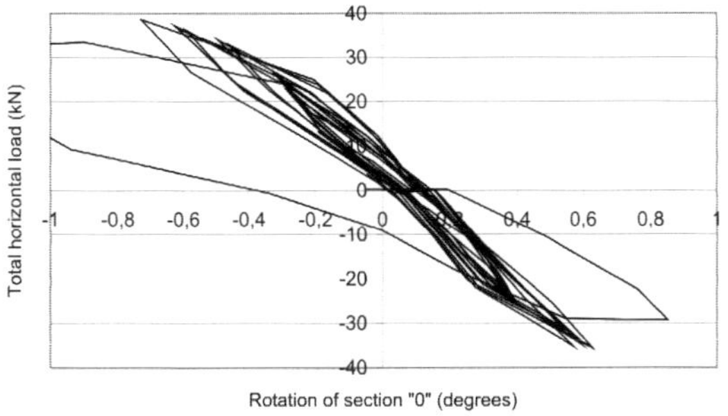

(c2) Total horizontal load versus rotation of section '0'

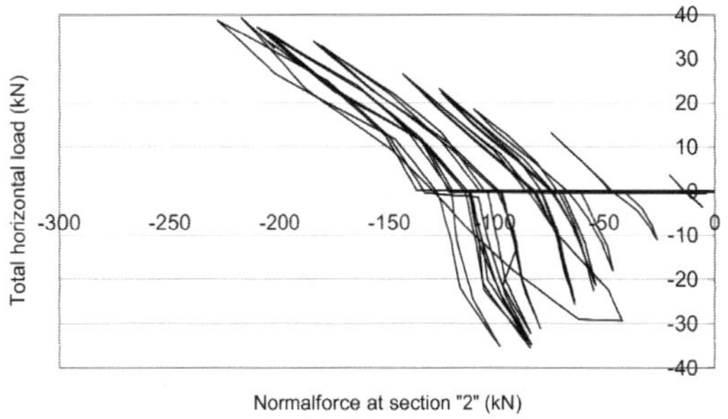

(d2) Total horizontal load versus normal force at section '2'

Figure 10. Load vs. displacement and rotation curves for OTKA frame 3.

Journal of Physics: Conference Series **268** (2011) 012010 doi:10.1088/1742-6596/268/1/012010

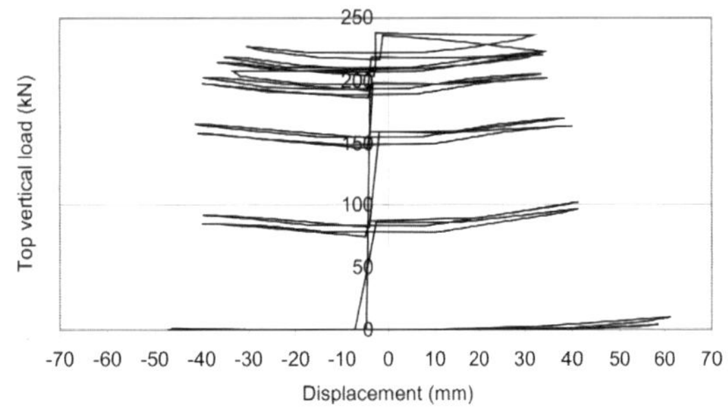

(a1) Vertical load of upper beam versus top storey drift

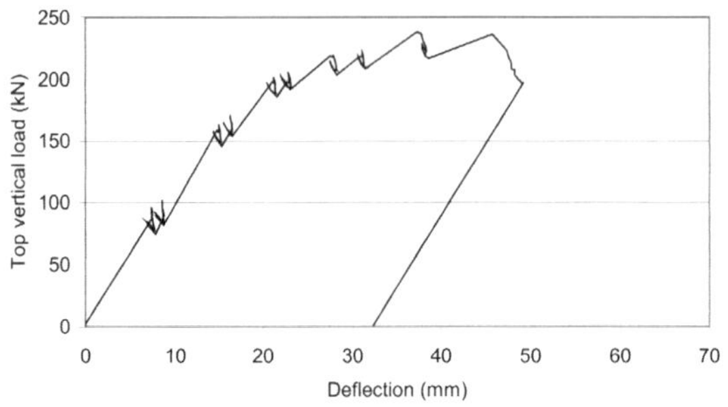

(b1) Vertical load of upper beam versus vertical deflection

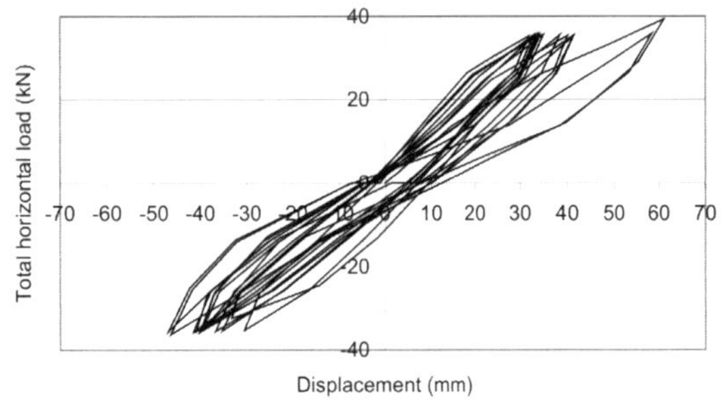

(a2) Total horizontal load versus top storey drift

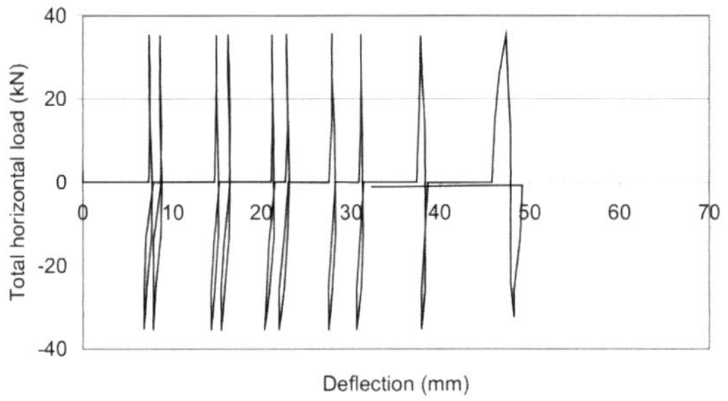

(b2) Total horizontal load versus vertical deflection

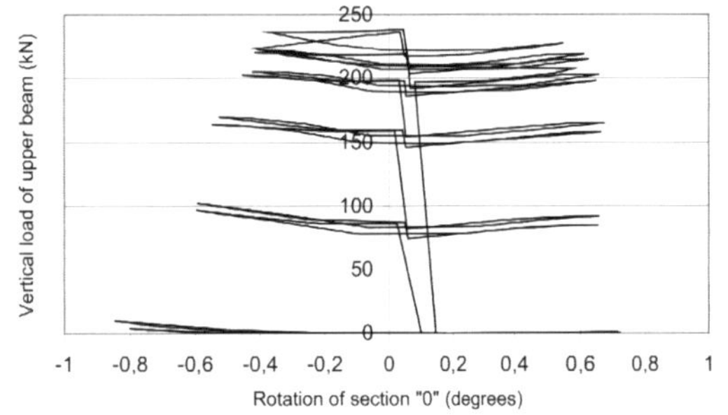

(c1) Vertical load of upper beam versus rotation at '0'

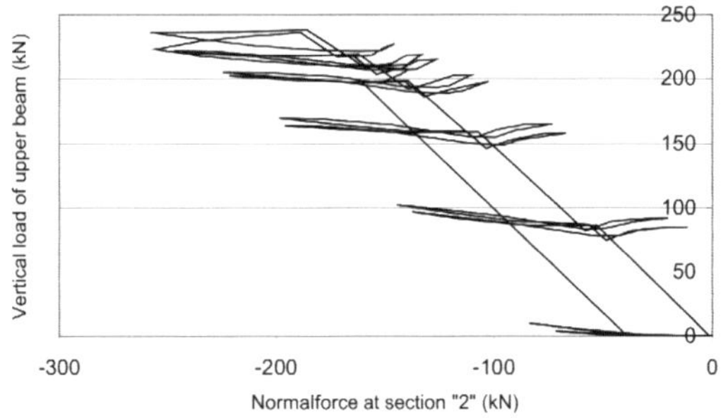

(d1) Vertical load of upper beam versus normal force at '2'

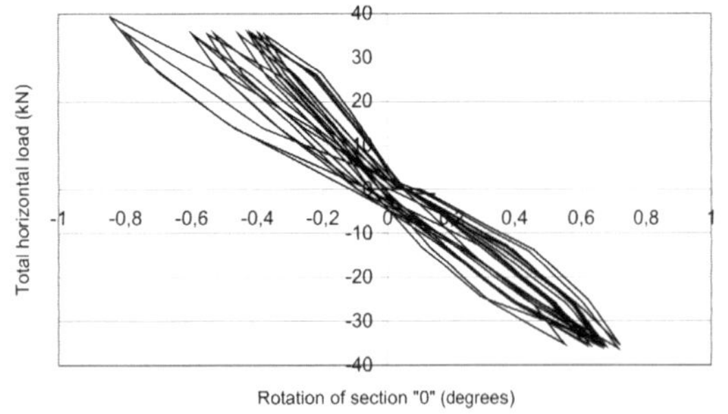

(c2) Total horizontal load versus rotation of section '0'

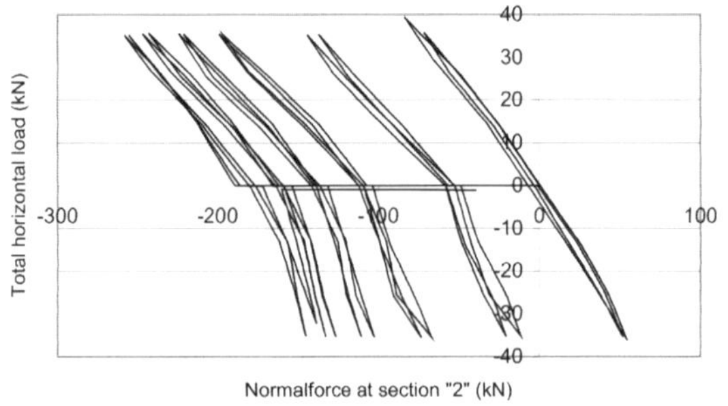

(d2) Total horizontal load versus normal force at section '2'

Figure 11. Load vs. displacement and rotation curves for OTKA frame 4.

Journal of Physics: Conference Series **268** (2011) 012010 doi:10.1088/1742-6596/268/1/012010

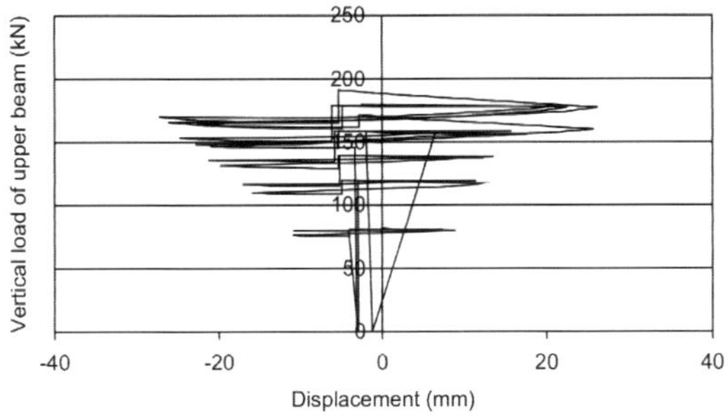

(a1) Vertical load of upper beam versus top storey drift

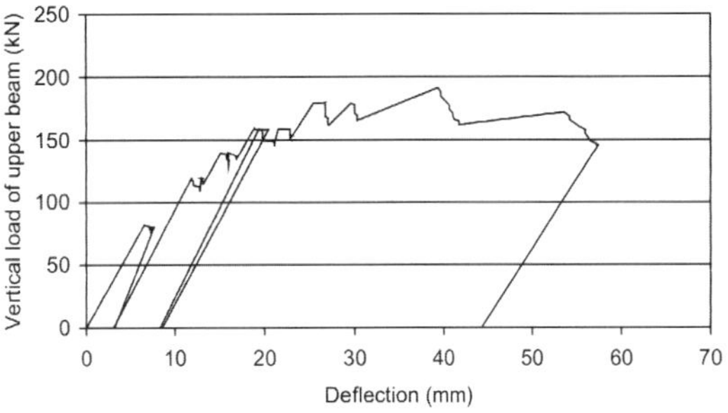

(b1) Vertical load of upper beam versus vertical deflection

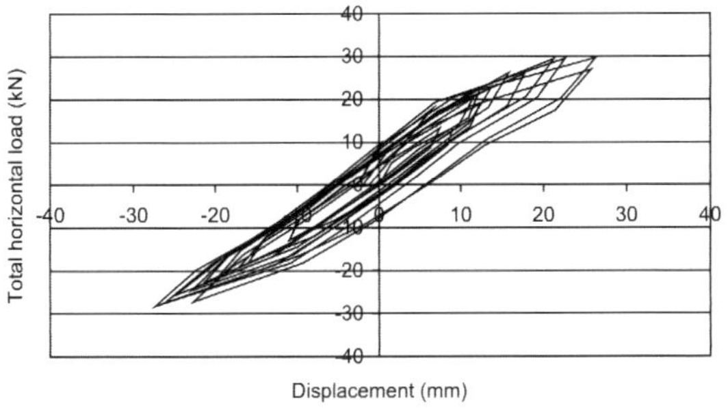

(a2) Total horizontal load versus top storey drift

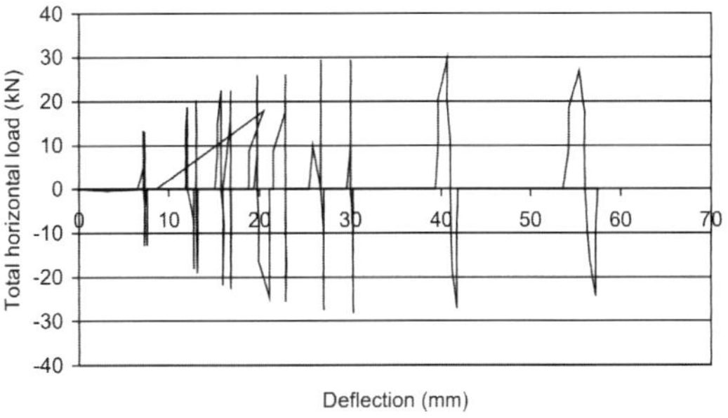

(b2) Total horizontal load versus vertical deflection

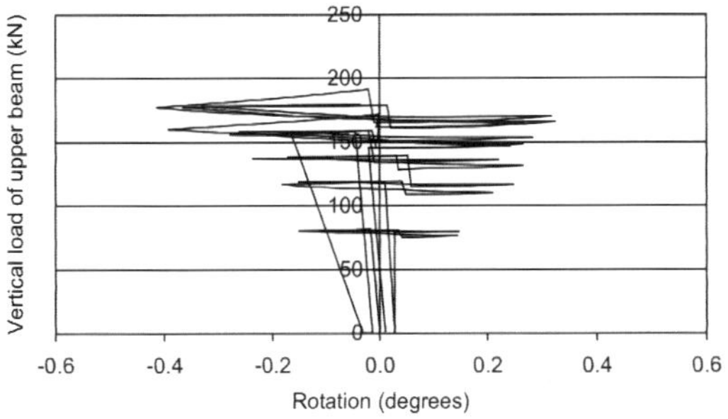

(c1) Vertical load of upper beam versus rotation at '0'

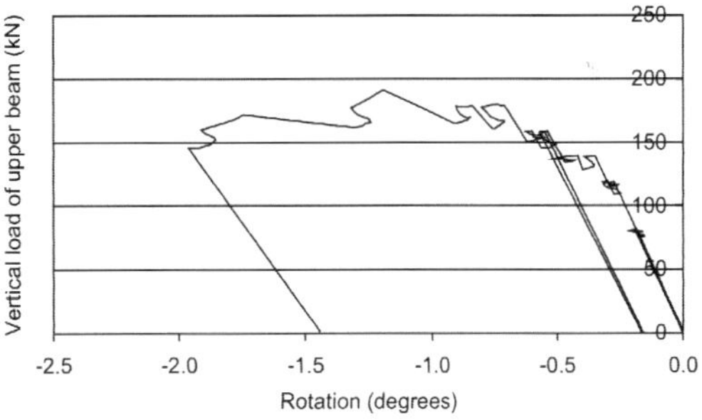

(d1) Vertical load of upper beam versus normal force at '2'

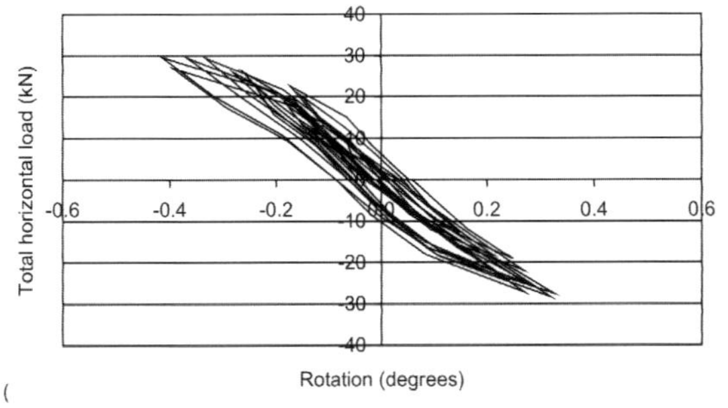

(c2) Total horizontal load versus rotation of section '0'

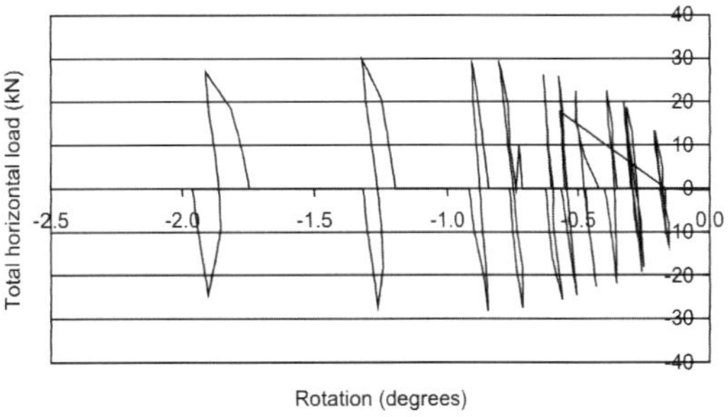

(d2) Total horizontal load versus normal force at section '2'

Figure 12. Load vs. displacement and rotation curves for OTKA frame 5

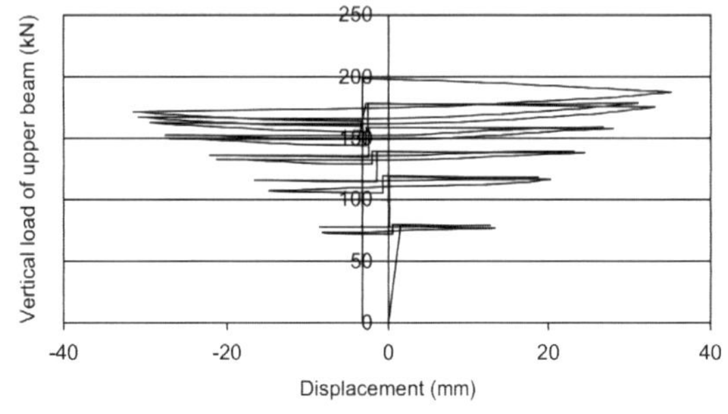

(a1) Vertical load of upper beam versus top storey drift

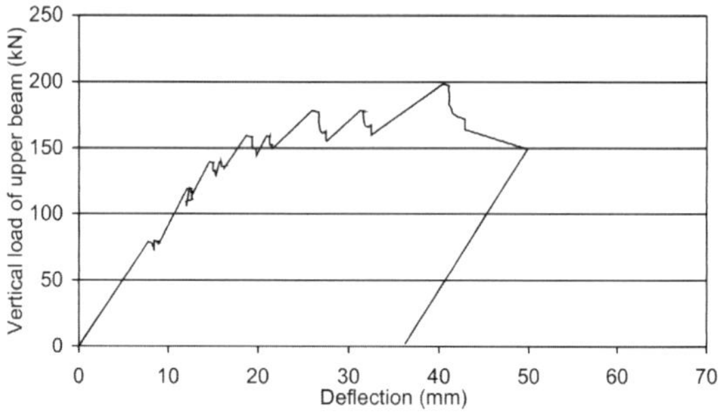

(b1) Vertical load of upper beam versus vertical deflection

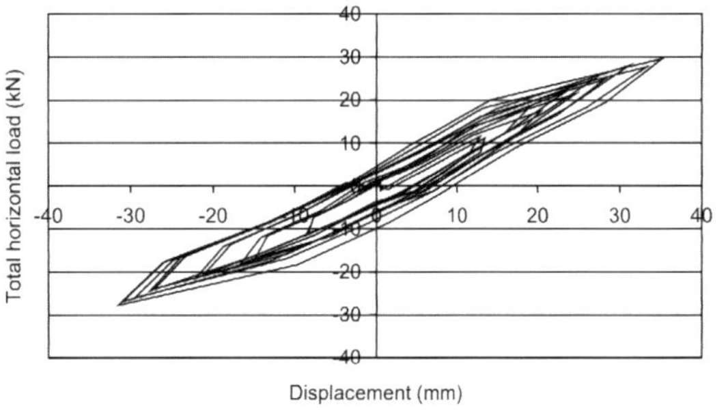

(a2) Total horizontal load versus top storey drift

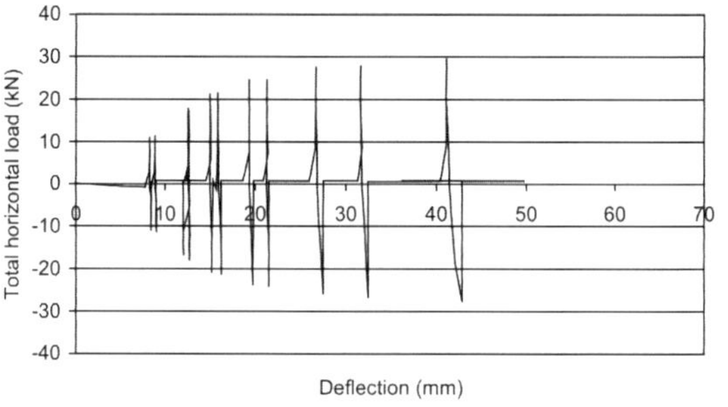

(b2) Total horizontal load versus vertical deflection

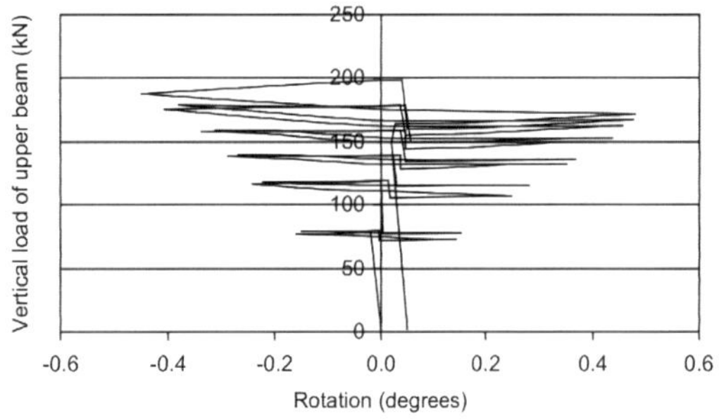

(c1) Vertical load of upper beam versus rotation at '0'

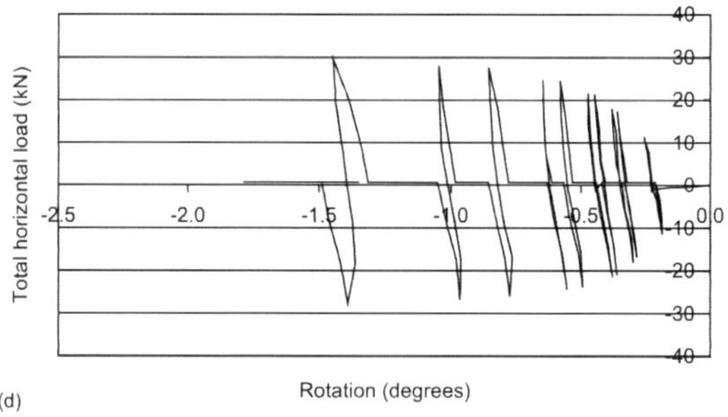

(d1) Vertical load of upper beam versus normal force at '2'

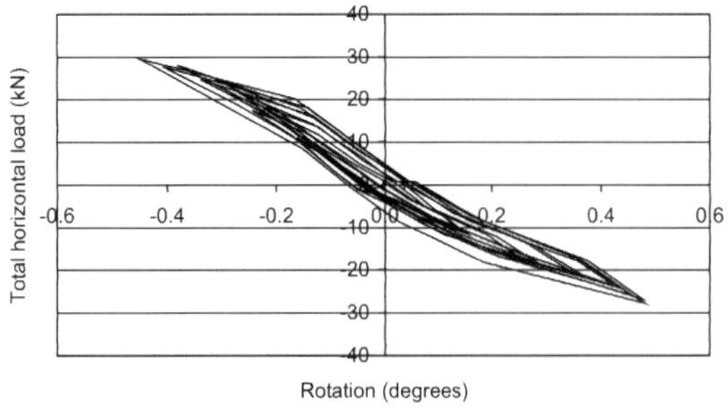

(c2) Total horizontal load versus rotation of section '0'

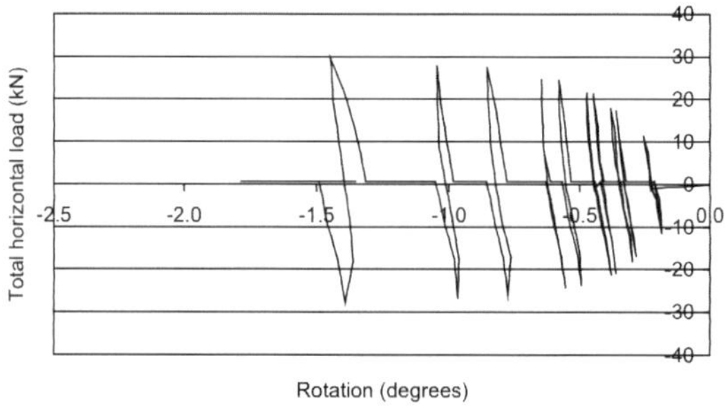

(d2) Total horizontal load versus normal force at section '2'

Figure 13. Load vs. displacement and rotation curves for OTKA frame 6.

of the information available about behavior of connections tested off-frame. Another main direction of the further assessment of results will concentrate on the behavior of the frame as a whole.

In the traditional (standard) load bearing capacity analysis the ultimate limit state is investigated under continuously increasing loading. The shakedown state is also usually analyzed under alternating and pulsating horizontal loading. Although these analyses correspond to the load bearing capacity, it is also important to investigate the process, which leads to this ultimate state, therefore it is necessary to investigate the hysteretic behavior of structures. The presented experimental results can provide good bases for further theoretical and numerical investigations.

References

[1] EN 1993-1-1:2005 *Eurocode 3*

[2] Davies J M 1996 *Stability of Steel Structures,* (Ed. M. Iványi, *Akadémiai Kiadó, Budapest, Hungary*) **1** 377

[3] Mazzolani F M 1992, *Stability Problems of Steel Structures* (Eds. M. Iványi and M. Skaloud, *Springer-Verlag*) 357

[4] Iványi M 2000 *Engineering Structures* **22** 168

[5] EN 1993-1-8:2005 *Eurocode 3*

[6] Iványi M and Varga G 1999 *Eurosteel 99 Conference* (Prague, Czech Republic)

[7] Halász O and Iványi M 1979 *Periodica Polytechnica Civil Engineering* (Budapest) **23** 3-4 151

[8] Gibbons C, Kirby P A and Nethercot D A 1991 *Journal of Strain Analysis*, **26** 1, 31

[9] Iványi M 2000 *Semi-rigid connections in structural steelwork* (Eds. Iványi M. Baniotopoulos C.C. *Springer-Verlag*) 87

Micromagnetics with eddy currents

R Iyer, J Millhollon and K Long

Room 201, Department of Mathematics and Statistics, Texas Tech University, Lubbock, TX 79409.

E-mail: `ram.iyer@ttu.edu`

Abstract. In this paper, we study the modified Landau-Lifshitz-Gilbert (LLG) equation for of a conducting, magnetic body. The modified LLG equations include the magnetic field due to eddy currents in the total effective magnetic field. We derive an expression for the magnetic field due to eddy current losses and show that it is well defined. We then show that the work done by the eddy currents in opposing the change of magnetization is a Rayleigh type dissipation function, and derive the modified LLG equations using the calculus of variations. Finally, we show that the modified LLG equations lead to a decrease in the Gibbs energy. This implies that the LLG equations describes a dynamic process proceeding spontaneously forward in time.

1. Introduction

Core losses in a magnetic material are typically considered to be due to three sources: hysteresis losses, classical eddy current losses, and excess losses. For obvious reasons, eddy and excess losses are important in induction motors [1], magnetostrictive actuators [2], amorphous magnetic ribbons, guided spin wave devices [3]. The frequency ranges at which the eddy and excess losses have been measured is about 0 - 1 MHz [2, 4, 5]. An interesting question is whether the theory of micromagnetics can explain the observed excess losses. Below microwave frequencies, the classical LLG equations could be used to study the excess losses (at microwave frequencies, the LLG equation needs to be modified [6]). However, the magnetic field due to eddy currents needs to be accounted for in the LLG equations, which is not done in the vast majority of the literature on micromagnetics [7, 8, 9, 10, 11, 12, 13]. The only references that do include the eddy currents [14, 15, 16] simply state them without derivation. As eddy current losses are a form of dissipation, one anticipates that a Rayleigh type dissipation function could be found, and the classical LLG equations modified to include eddy currents. That is the contribution of this paper. The key result that is used is Corollary 2.1, where it is shown that the eddy current field at a point in the magnetic body does not depend on the rate of change of magnetization at that point.

1.1. The Landau-Lifshitz equation

In 1935, Landau and Lifshitz [17] proposed a new theory with which they were able to compute the thickness of walls between magnetic domains, and also explain domain formation in ferromagnetic materials. This theory which now goes by the name of *micromagnetics* [8] has been instrumental in the understanding and development of magnetic memories. Landau and Lifshitz considered the Gibbs energy $\mathcal{G}$ of a magnetic material to be composed of three terms – exchange, anisotropy and Zeeman energies (due to the external magnetic field) [17], and postulated that the observed magnetization per unit volume $\boldsymbol{M}$ field would correspond to a local minimum of the Gibbs energy. Later researchers added other terms to $\mathcal{G}$ such as magnetoelastic energy and demagnetization energy [7, 11]. They also derived the Landau-Lifshitz (LL) equation using physical arguments (but not using the calculus of variations):

$$\dot{\boldsymbol{M}} = \gamma_0\, \boldsymbol{M} \times \boldsymbol{H}_{eff} - \frac{\alpha_0}{M_s^2}\, \boldsymbol{M} \times (\boldsymbol{M} \times \boldsymbol{H}_{eff}), \tag{1}$$

where $\gamma_0 = \frac{e}{m\,c}$ with m the electronic mass, e the electronic charge, c the speed of light; α_0 is a positive constant, $\|\boldsymbol{M}\| = M_s$ is assumed to be constant over the material. The effective field $\boldsymbol{H}_{eff} = -\frac{\partial \mathcal{G}}{\partial \boldsymbol{M}}$ is the negative gradient of the Gibbs energy with respect to the magnetization, and can be expressed as: $\boldsymbol{H}_{eff} = \boldsymbol{H}_0 + \boldsymbol{H}_{demag} + \boldsymbol{H}_{anis} + \boldsymbol{H}_{exch} + \boldsymbol{H}_{me}$ where $\boldsymbol{H}_0$ is the external magnetic field, $\boldsymbol{H}_{demag}$ is the demagnetization field, $\boldsymbol{H}_{anis}$ is the field due to the anisotropy of the sample, $\boldsymbol{H}_{exch}$ is the field due to exchange interactions, and $\boldsymbol{H}_{me}$ is the field due to magnetoelastic interactions.

1.2. The Landau-Lifshitz-Gilbert equation and variational formulation

Equation (1) was modified by Gilbert [7] who derived the equation:

$$\dot{\boldsymbol{M}} = \gamma_0\, \boldsymbol{M} \times (\boldsymbol{H}_{eff} - \eta\, \dot{\boldsymbol{M}}), \tag{2}$$

where $\eta > 0$, and which showed a better match with certain experimental results [18]. He also showed that the equation:

$$\dot{\boldsymbol{M}} = \gamma'\, \boldsymbol{M} \times \boldsymbol{H}_{eff} - \frac{\alpha'}{M_s^2}\, \boldsymbol{M} \times (\boldsymbol{M} \times \boldsymbol{H}_{eff}), \tag{3}$$

can be obtained from the LLG equation (2), if one chose $\gamma' = \frac{\gamma_0}{1+\gamma_0^2\,\eta^2\,M_s^2}$ and $\alpha' = \frac{\eta\,\gamma_0^2\,M_s^2}{1+\eta^2\,\gamma_0^2\,M_s^2}$. Gilbert also showed that the second term on the right hand side of (2) could be obtained from a Rayleigh dissipation function:

$$R_G(\dot{\boldsymbol{M}}) = \frac{\eta}{2} \int_\Omega \dot{\boldsymbol{M}} \cdot \dot{\boldsymbol{M}},$$

if one interpreted (2) as Euler-Lagrange equations with a Lagrangian of the type: Kinetic energy minus Potential energy. His contribution were limited by two issues: (i) the kinetic energy was not described, and (ii) eddy currents were neglected. The magnetic field due to eddy currents $\boldsymbol{H}_{eddy}$ is missing as eddy currents were neglected. It should be noted that as the eddy currents do not contribute to the Gibbs energy, it is not straight forward to see how they could be added to the total magnetic field H_{eff} described earlier.

Brown [8] showed the existence of a kinetic energy - a limiting form of the classical rigid body kinetic energy - which along with the Gibbs energy $\mathcal{G}$, and Gilbert's damping function $R_G(\dot{\boldsymbol{M}})$ could be used to derive the LLG equation (2). Brown did not include eddy currents in his theory either.

1.3. The LLG equations with eddy currents

A suggestion to add a magnetic field term $\boldsymbol{H}_{eddy}$ to $\boldsymbol{H}_{eff}$ is found in Bertotti [14], though a derivation was not presented. Following [14], Torres et al. [15] included the eddy current field in their computations, but ignored it in their numerical computations. Serpico, Mayergoyz and Bertotti [16] include the eddy current field in their study of a thin magnetic film, again without providing the justification for modifying the LLG equation.

In this paper, we derive an expression for the magnetic field due to eddy current losses and show that it is well defined. We then show that the modified LLG equations lead to a decrease in the Gibbs energy. This implies that the LLG equations describes a dynamic process proceeding spontaneously forward in time. Finally, we show that the work done by the eddy currents in opposing the change of magnetization is a Rayleigh type dissipation function, and derive the modified LLG equations using the calculus of variations.

2. Eddy Currents in a conducting magnetic material

Consider a magnetic material with magnetization per unit volume $\boldsymbol{M}$, occupying a compact, connected set Ω in $\mathbb{R}^3$, with a Lipschitz continuous boundary. We assume that an external magnetic field $\boldsymbol{H}_0$ is imposed by an external source. The free charge in the material is assumed to be zero.

Maxwell's equations are given by (neglecting $\frac{\partial \boldsymbol{D}}{\partial t}$) :

$$\begin{array}{rclrcl} \nabla\cdot\boldsymbol{B} &=& 0 & \boldsymbol{B}_n^+ - \boldsymbol{B}_n^- &=& 0 \\ \nabla\times\boldsymbol{H} &=& \boldsymbol{J} & \boldsymbol{H}_t^+ - \boldsymbol{H}_t^- &=& \boldsymbol{J}_s\times\boldsymbol{n} \\ \nabla\times\boldsymbol{E} &=& -\frac{\partial B}{\partial t} & \boldsymbol{E}_t^+ - \boldsymbol{E}_t^- &=& \boldsymbol{0} \\ \nabla\cdot\boldsymbol{D} &=& 0, & \boldsymbol{D}_n^+ - \boldsymbol{D}_n^- &=& \boldsymbol{0}, \end{array} \tag{4}$$

The relation between the fields are: $\boldsymbol{B} = \mu_0\,(\boldsymbol{H}+\boldsymbol{M})$, $\boldsymbol{D} = \epsilon_0\,\boldsymbol{E}$, $\boldsymbol{J} = \sigma\boldsymbol{E}$, where σ is the conductivity of the material. σ is required to be a measurable and bounded function. To make physical sense, it is also required to be non-negative.

The (change in the) magnetization $\boldsymbol{M}$ has to be related to the applied field using the Landau-Lifshitz theory of micromagnetics [17, 8]. To this end, let G be the Gibbs free energy given by:

$$\mathcal{G} = E_{demag} + E_{anis} + E_{exch} + E_{me} - \int_\Omega \boldsymbol{H}_0\cdot\boldsymbol{M}\,dv,$$

where E_{demag} is the demagnetization energy, E_{exch} is the exchange energy, E_{anis} is the anisotropy energy, and E_{me} is the magnetoelastic energy. The expressions for these energies can be found in Brown [8] (page 38).

In what follows, we employ CGS units as in Brown [8]. Consider Figure 1. Let P be a point in Ω. Suppose $\epsilon > 0$ is any given number sufficiently small, and consider a small (relative) neighborhood S that is removed from Ω. S is contained in a small ball of radius ϵ with center at P, that is, $S \subset B_\epsilon(P)$. Consider an infinitesimal tube of surface area dA and length dl located at the point Q in $\Omega\cap S^c$ (S^c is the complement set of S with S^c containing points in space that are not in S), such that the current $\boldsymbol{J}(Q)\,dA(Q)$ in the tube is along the axis of the tube. Let $\boldsymbol{r}_{PQ}$ be the vector centered at Q and Then, the work done by the current $\boldsymbol{J}(Q)\,dA(Q)$ in opposing the change in the magnetization $\boldsymbol{M}(P)$ at the point P is given by [8] (page 27):

$$\frac{dR}{dt}(P;S) = \int_{\Omega\setminus S}\frac{\mu_0}{4\pi}\,\frac{\boldsymbol{J}(Q)\,dv(Q)\cdot\left(\dot{\boldsymbol{M}}(P)\times\boldsymbol{r}_{PQ}\right)}{\|r_{QP}\|^3} = \int_{\Omega\setminus S}\frac{\mu_0}{4\pi}\,\frac{(\boldsymbol{r}_{PQ}\times\boldsymbol{J}(Q))}{\|r_{PQ}\|^3}\,dv(Q)\;\cdot\;\dot{\boldsymbol{M}}(P), \tag{5}$$

where $dv(Q) = dA\,dl$ is the volume of the infinitesimal tube located at the point Q.

Definition 2.1 *(Absolute convergence of the eddy current integral)*

Let $P\in\Omega$. If the last integral in (5) exists for all $\epsilon > 0$ and $S\subset B_\epsilon(P)$, converges absolutely as $\epsilon\to 0$, and if the limit is independent of the shape of S, then we say that the eddy current field $\mathbf{H}_{eddy}$ exists at P, and

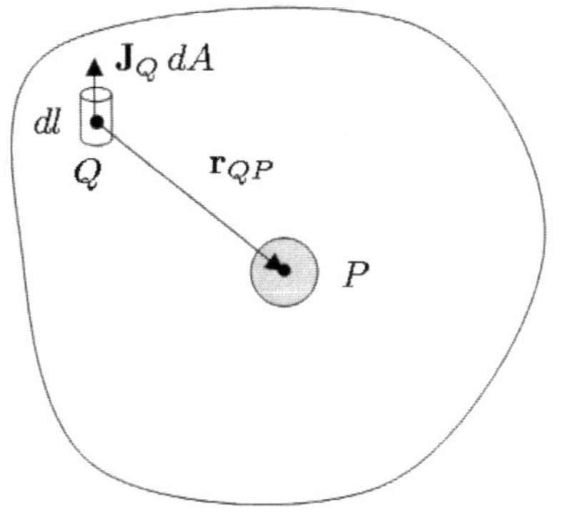

Figure 1. Computation of the loss due to eddy currents.

$$\boldsymbol{H}_{eddy}(P) = \int_{\Omega} \frac{\mu_0}{4\pi} \frac{(\boldsymbol{r}_{PQ} \times \boldsymbol{J}(Q))}{\|r_{PQ}\|^3}\, dv(Q) = \lim_{\epsilon \to 0} \int_{\Omega \setminus S} \frac{\mu_0}{4\pi} \frac{(\boldsymbol{r}_{PQ} \times \boldsymbol{J}(Q))}{\|r_{PQ}\|^3}\, dv(Q). \tag{6}$$

Remark: In the SI system, the left hand side would be called $\boldsymbol{B}_{eddy}$ the effective magnetic flux density, and $\boldsymbol{H}_{eddy}$ is defined according to: $\boldsymbol{B}_{eddy} = \mu_0\, \boldsymbol{H}_{eddy}$. □

We can define:

$$\frac{dR}{dt}(P) = \lim_{\epsilon \to 0} \frac{dR}{dt}(P; S) = \lim_{\epsilon \to 0} \int_{\Omega \setminus S} \frac{\mu_0}{4\pi} \frac{(\boldsymbol{r}_{PQ} \times \boldsymbol{J}(Q))}{\|r_{PQ}\|^3}\, dv(Q) \;\cdot\; \dot{\boldsymbol{M}}(P), \tag{7}$$

and the work done by the eddy current field $\boldsymbol{J}$ in opposing the change in the magnetization in the material is given by:

$$\int_{\Omega} \frac{dR}{dt}(P)\, dv(P) = \int_{\Omega} \boldsymbol{H}_{eddy} \cdot \dot{\boldsymbol{M}}\, dv. \tag{8}$$

This is work done by the "system" consisting of the magnetic material on its environment, and hence it is non-positive. Hence, the Rayleigh dissipation function R_E defined by:

$$R_E = -\int_{\Omega} \boldsymbol{H}_{eddy} \cdot \dot{\boldsymbol{M}}\, dv \geq 0, \tag{9}$$

and it represents the work done by the battery in overcoming the demagnetizing effect of eddy currents.

Remark: In the SI system, we have $R_E = -\mu_0 \int_{\Omega} \boldsymbol{H}_{eddy} \cdot \dot{\boldsymbol{M}}\, dv$ □

Below is the main lemma of this paper.

Lemma 2.1 *The limit in (6) converges absolutely irrespective of the shape of the open set S, which is assumed to contain P and be a subset of $\mathbf{B}_\epsilon(P)$.*

Proof: Consider the integral:

$$\mathbf{I} = \int_{S} \frac{\mu_0}{4\pi} \frac{(\boldsymbol{r}_{PQ} \times \boldsymbol{J}(Q))}{\|r_{PQ}\|^3}\, dv(Q)$$

where S is a small neighborhood $\subset B_\epsilon(P)$ that is a subset of Ω.

The proof of absolute convergence of $\mathbf{I}$ as $\epsilon \to 0$, depends on showing the boundedness of $\boldsymbol{J}$, which is implied by the boundedness of $\boldsymbol{E}$ over all of Ω. As the material is without polarization (that is, $\boldsymbol{P} = \mathbf{0}$), we have $\boldsymbol{D} = \epsilon_0\,\boldsymbol{E}$, and the boundedness of $\boldsymbol{E}$ is equivalent to that of $\boldsymbol{D}$. From the Maxwell's equation:

$$\int_V \nabla \cdot \boldsymbol{D}\,dv = \int_{\partial V} \boldsymbol{D} \cdot \boldsymbol{n}\,dS = 0,$$

where $V \subset D$ is any region with a Lipschitz continuous boundary. If $\boldsymbol{D}$ is unbounded or undefined at any point on ∂V, then the third equality in the above equation cannot hold. Hence, $\boldsymbol{D}$ must be defined everywhere on the boundary of an arbitrary region $V \subset \Omega$, which implies that $\boldsymbol{D}$ is defined everywhere on Ω. As Ω is a compact set, $\boldsymbol{D}$ is bounded everywhere on Ω, and there exists $C > 0$ such that $\|\boldsymbol{D}\| \leq C$ on Ω.

The rest of the proof is an adaptation of a method in Leathem [19] (page 15). Let T be a point on the boundary of S that is at the shortest distance from P. As the boundary of S is a compact set, there exists at least one such point. Let $\|\boldsymbol{r}_{PT}\| = \eta$. By construction, $\eta < \epsilon$, and:

$$\begin{aligned}
|I| &\leq \frac{\mu_0}{4\pi}\int_S \left|\frac{(\boldsymbol{r}_{PQ} \times \boldsymbol{J}(Q))}{\|r_{PQ}\|^3}\right| dv(Q) \\
&\leq \frac{\mu_0}{4\pi}\int_S \frac{\|\boldsymbol{J}(Q)\|}{\|r_{PQ}\|^2}\,dv(Q) \\
&\leq \frac{\mu_0}{4\pi}\int_\eta^\epsilon \frac{\sigma_{\max}\,C}{r^2}\,4\,\pi\,\epsilon_0\,r^2\,dr \\
&= \frac{\mu_0}{\epsilon_0}\,C\,\sigma_{\max}\,(\epsilon - \eta) \\
&\leq C\,\sigma_{\max}\,Z_0^2\,\epsilon,
\end{aligned}$$

where $\sigma_{\max}$ is the maximum value of the conductivity σ on Ω, and $Z_0 = \sqrt{\frac{\mu_0}{\epsilon_0}}$ is the characteristic impedance of free space. The inequality above shows that irrespective of S, we have $\lim_{\epsilon\to 0}|I| = 0$. Now, $\int_{\Omega\setminus S}\left|\frac{(\boldsymbol{r}_{PQ}\times\boldsymbol{J}(Q))}{\|r_{PQ}\|^3}\right| dv(Q)$ is finite because $\boldsymbol{J}$ is bounded on Ω, and the denominator never approaches 0. Hence, we have:

$$\frac{\mu_0}{4\pi}\int_\Omega \frac{(\boldsymbol{r}_{PQ}\times\boldsymbol{J}(Q))}{\|r_{PQ}\|^3}\,dv(Q) = \frac{\mu_0}{4\pi}\int_{\Omega\setminus S} \frac{(\boldsymbol{r}_{PQ}\times\boldsymbol{J}(Q))}{\|r_{PQ}\|^3}\,dv(Q) + \frac{\mu_0}{4\pi}\int_S \frac{(\boldsymbol{r}_{PQ}\times\boldsymbol{J}(Q))}{\|r_{PQ}\|^3}\,dv(Q),$$

as both the integrals on the right side exist. At this point, the left hand side seems to be function of S. However,

$$\left|\frac{\mu_0}{4\pi}\int_\Omega \frac{(\boldsymbol{r}_{PQ}\times\boldsymbol{J}(Q))}{\|r_{PQ}\|^3}\,dv(Q) - \frac{\mu_0}{4\pi}\int_{\Omega\setminus S} \frac{(\boldsymbol{r}_{PQ}\times\boldsymbol{J}(Q))}{\|r_{PQ}\|^3}\,dv(Q)\right| \leq C\,\sigma_{\max}\,Z_0^2\,\epsilon, \qquad (10)$$

which shows that $\boldsymbol{H}_{eddy}(P)$ does not depend on S. □

Corollary 2.1 *Let $P \in \Omega$. The eddy current magnetic field $\mathbf{H}_{eddy}(P)$ does not depend on $\dot{\mathbf{M}}(P)$, and $\frac{\partial R_E}{\partial \dot{\mathbf{M}}}(P) = \mathbf{H}_{eddy}(P)$.*

Proof: The claim follows from the fact that $\frac{\mu_0}{4\pi}\int_{\Omega\setminus S}\frac{(\boldsymbol{r}_{PQ}\times\boldsymbol{J}(Q))}{\|r_{PQ}\|^3}\,dv(Q)$ converges absolutely to $\boldsymbol{H}_{eddy}$, and the contribution of $\frac{\mu_0}{4\pi}\int_S\frac{(\boldsymbol{r}_{PQ}\times\boldsymbol{J}(Q))}{\|r_{PQ}\|^3}\,dv(Q)$ to $\boldsymbol{H}_{eddy}$ can be made as small as desired.

As $\boldsymbol{H}_{eddy}(P)$ is independent of $\dot{M}(P)$, we get from (9), the second claim in the corollary. □

Remark: In SI units, the corollary above reads $\frac{\partial R_E}{\partial \dot{\boldsymbol{M}}}(P) = \mu_0\,\boldsymbol{H}_{eddy}(P)$. □

3. The modified Landau-Lifshitz-Gilbert equation with eddy currents

Brown [8] (pages 31 - 43) showed the existence of a Routhian function $\mathcal{T}$ such that with the Lagrangian $\mathcal{L} = \mathcal{T} - \mathcal{G}$ such that:

$$\left(\frac{d}{dt}\frac{\partial \mathcal{L}}{\partial \dot{M}} - \frac{\partial \mathcal{L}}{\partial \boldsymbol{M}}\right) \cdot \delta \boldsymbol{M} = \frac{\partial \mathcal{R}_G}{\partial \dot{M}} \cdot \delta \boldsymbol{M}, \tag{11}$$

where $\delta \boldsymbol{M} = \boldsymbol{M} \times \mathbf{w}$ is a virtual variation, with $\mathbf{w}$ an arbitrary vector field on Ω, which is $\mathbf{0}$ at the initial and final times, yields the LLG equation (2):

$$\frac{1}{\gamma_0}\dot{\boldsymbol{M}} - \boldsymbol{M} \times \boldsymbol{H}_{eff} = \eta\,(\dot{\boldsymbol{M}} \times \boldsymbol{M}). \tag{12}$$

We consider the dissipation function $\mathcal{R} = \mathcal{R}_G + \mathcal{R}_E$ where $\mathcal{R}_E$ is given by (9), which leads to the Euler-Lagrange equations:

$$\left(\frac{d}{dt}\frac{\partial \mathcal{L}}{\partial \dot{M}} - \frac{\partial \mathcal{L}}{\partial \boldsymbol{M}}\right) \cdot \delta \boldsymbol{M} = \frac{\partial \mathcal{R}_G}{\partial \dot{M}} \cdot \delta \boldsymbol{M} + \frac{\partial \mathcal{R}_E}{\partial \dot{M}} \cdot \delta \boldsymbol{M}. \tag{13}$$

By Corollary 2.1, we get:

$$\frac{1}{\gamma_0}\dot{\boldsymbol{M}} - \boldsymbol{M} \times \boldsymbol{H}_{eff} = \eta\,(\dot{\boldsymbol{M}} \times \boldsymbol{M}) - \boldsymbol{H}_{eddy} \times \boldsymbol{M}.$$

Therefore, modified LLG equation is given by:

$$\dot{\boldsymbol{M}} = \gamma_0\, \boldsymbol{M} \times \boldsymbol{H}_{total} - \gamma_0\, \eta\, \boldsymbol{M} \times \dot{\boldsymbol{M}} = \gamma_0\, \boldsymbol{M} \times (\boldsymbol{H}_{total} - \eta\, \dot{\boldsymbol{M}}), \tag{14}$$

where $\eta > 0$ and $\gamma_0 = \frac{e}{mc} < 0$. The total magnetic field is given by:

$$\begin{aligned} \boldsymbol{H}_{total} &= \boldsymbol{H}_{eff} + \boldsymbol{H}_{eddy} \\ &= \boldsymbol{H}_0 + \boldsymbol{H}_{demag} + \boldsymbol{H}_{anis} + \boldsymbol{H}_{exch} + \boldsymbol{H}_{me} + \boldsymbol{H}_{eddy} \\ &= \boldsymbol{H} + \boldsymbol{H}_{anis} + \boldsymbol{H}_{exch} + \boldsymbol{H}_{me}, \end{aligned}$$

as the magnetic field $\boldsymbol{H}$ in Maxwell's equations is given by:

$$\boldsymbol{H} = \boldsymbol{H}_0 + \boldsymbol{H}_{demag} + \boldsymbol{H}_{eddy}.$$

From (14),

$$\dot{\boldsymbol{M}} \cdot (\boldsymbol{H}_{total} - \eta\, \dot{\boldsymbol{M}}) = \mathbf{0},$$

which implies:

$$\dot{\boldsymbol{M}} \cdot \boldsymbol{H}_{eff} = \eta \dot{\boldsymbol{M}} \cdot \dot{\boldsymbol{M}} - \dot{\boldsymbol{M}} \cdot \boldsymbol{H}_{eddy}. \tag{15}$$

Now,

$$\frac{dG}{dt}(\boldsymbol{M}, \boldsymbol{H}_0)dt = \int_\Omega \frac{\partial G}{\partial \boldsymbol{M}}\dot{\boldsymbol{M}} + \frac{\partial G}{\partial \boldsymbol{H}_0}\dot{\boldsymbol{H}}_0\, dv = \int_\Omega (-\boldsymbol{H}_{eff} \cdot \dot{\boldsymbol{M}} - \boldsymbol{M} \cdot \dot{\boldsymbol{H}}_0)\, dv. \tag{16}$$

Using the reciprocity relation for magnetostatics (page 37, [8]),

$$\int_\Omega \boldsymbol{M} \cdot \dot{\boldsymbol{H}}_0\, dv = \int_\Omega \dot{\boldsymbol{M}} \cdot \boldsymbol{H}_0\, dv,$$

we get,

$$\begin{aligned} \frac{dG}{dt}(\boldsymbol{M}, \boldsymbol{H}_0)dt &= \int_\Omega (-\boldsymbol{H}_{eff} \cdot \dot{\boldsymbol{M}} - \dot{\boldsymbol{M}} \cdot \boldsymbol{H}_0)\, dv & (17) \\ &= \int_\Omega (-\eta\, \dot{\boldsymbol{M}} \cdot \dot{\boldsymbol{M}} + \dot{\boldsymbol{M}} \cdot \boldsymbol{H}_{eddy} - \dot{\boldsymbol{M}} \cdot \boldsymbol{H}_0)\, dv. & (18) \end{aligned}$$

Each of the three terms on the right hand side is non-negative. The second term is negative as it is work done by the magnetic material on its environment, while the term $\int_\Omega \dot{\boldsymbol{M}} \cdot \boldsymbol{H}_0\, dv$ is positive as it is the work done by the battery on changing the magnetization of the material. Hence, Equation (14) describes a physically valid, irreversible phenomenon.

4. Conclusion

In this paper, we studied the modified Landau-Lifshitz-Gilbert (LLG) equation for of a conducting, magnetic body. The modified LLG equations include the magnetic field due to eddy currents in the total effective magnetic field. We derived an expression for the magnetic field due to eddy current losses and show that it is well defined. We showed that the work done by the eddy currents in opposing the change of magnetization is a Rayleigh type dissipation function, and derive the modified LLG equations using the calculus of variations. Finally, we showed that the modified LLG equations lead to a decrease in the Gibbs energy. This implies that the LLG equations describes a dynamic process proceeding spontaneously forward in time.

References

[1] Diaz G, Arboleya P, Gonzalez-Moran C and Gomez-Aleixandre J 2007 *IET Electr. Power Appl.* **1** 75
[2] Ekanayake D, Iyer R and Dayawansa W 2007 *Proc. American Control Conf. (July 2007 New York)* 4321
[3] Hertel R 2007 Guided Spin Waves *Handbook of Magnetism and Advanced Magnetic Materials* vol 2, ed Kronmüller H and Parkin S (Wiley) pp 1003-20
[4] Willard M, Francavilla T and Harris V 2005 *J. Appl. Phys.* **97** 10F502 001
[5] Fiorillo F and Novikov A 1990 *IEEE Trans. Mag.* **26** 2904
[6] Thomas L and Parkin S 2007 Current Induced Domain-wall Motion in Magnetic Nanowires, *Handbook of Magnetism and Advanced Magnetic Materials* vol 2, ed Kronmüller H and Parkin S (Wiley) 942
[7] Gilbert T L 2004 *IEEE Trans. Mag.* **40** 3443
[8] Brown Jr W F 1962 *Micromagnetics* (New York: John Wiley & Sons)
[9] D'Aquino M (Dissertation: Dept. of Electrical Engineering Universitá degli Studi Di Napoli "Federico II", December 2004)
[10] Bertotti G, Mayergoyz I and Serpico C 2009 *Nonlinear Magnetization Dynamics in Nanosystems* (Amsterdam: Elsevier) 21
[11] Kronmüller H 2007 *Handbook of Magnetism and Advanced Magnetic Materials* vol 2, ed Kronmüller H and Parkin S (Wiley) 703
[12] Miltat J and Donahue M 2007 *Handbook of Magnetism and Advanced Magnetic Materials* vol 2, ed Kronmüller H and Parkin S (Wiley) 742
[13] Schrefl T, Hrkac G,Bance S, Suess D, Ertl O and Fidler J 2007 *Handbook of Magnetism and Advanced Magnetic Materials* vol 2, ed Kronmüller H and Parkin S (Wiley) 765
[14] Bertotti G 1998 *Hysteresis in Magnetism* (Boston: Academic Press)
[15] Torres L, Martinez E, Lopez-Diaz L and Alejos O 2004 *Phys. B* **343** 257
[16] Serpico C, Mayergoyz I D and Bertotti G 2001 *IEEE Trans. Magn.* **37** 3546
[17] Landau L and Lifshitz E 1965 *Collected Papers of L. D. Landau* (New York: Gordon and Breach) 101
[18] Gilbert T and Kelly J 1955 *Proc. Conf. Magnetism and Magnetic Materials*, (American Institute of Electrical Engineers, New York) 253
[19] Leathem J G 1960 *Volume and Surface Integrals used in Physics* second edition (New York: Hafner Publishing)

Nonlinear multigrid diffusion

I Jancskar

Department of Information Technology, Pollack Mihály Faculty of Engineering, University of Pecs, Rokus u. 2, H-7624 Pecs, Hungary

E-mail: jai@morpheus.pte.hu

Abstract. In this paper the feasibility of multigrid techniques in the solution of nonlinear heat equations is investigated. The solution of the heat equation with hysteretic diffusivity requires improvements to the presently applied numerical procedures. Not only the diffusivity hysteresis must be modelled but the nonlinear heat equation requires development of a unique solver. In this study the so-called coarse level iterating multigrid method is proposed to reduce the solution time and memory demands. When the exact solution of the problem is sufficiently smooth, which in diffusion processes is a reasonable assumption, this type of problem reduction does not cause considerable deviations in results.

1. Introduction

The analyses of transient thermal responses in composite materials, for instance in chalcogenide glass, could be of great importance in various engineering and scientific fields. These materials have received a lot of attention over the past several years due to their wide range of applications. Experiments prove that thermal diffusivity of composite materials shows hysteretic behavior [1], [2]. For example hysteresis of thermal diffusivity of silica based materials used in coke oven linings is reported in [1]. The thermal conductivity of chalcogenide semiconductor glass has been studied in [3] and hysteresis properties appearing in certain compositions of Se-Te-Cu alloys have been proved. Measured hysteresis values of thermal conductivity under an applied electromagnetic field can be found in [4]. The hysteresis is related to microstructure changes such as phase transitions, micro-cracking induced by thermal expansion anisotropy, interfacial decohesion between particles and so forth.

Several numerical methods have been developed for the analysis of heat transfer problems in solids [5], [6], however solutions of transient problems with variable thermal properties in more than one dimension are very resource demanding and time-consuming, therefore most existing models have been developed under constant coefficient conditions. Efforts to take into account the temperature dependence of thermal diffusivity tend towards polynomial or exponential approximations [7], [8], [9]. Simulation of heat diffusion with hysteresis requires improvements to the presently applied numerical procedures.

The parabolic heat equation with hysteretic diffusivity can only be solved numerically. The finite difference method with high spatial resolution leads to a large system of equations. In the last few years, the multigrid technique has become one of the most popular numerical solvers of heat conduction problems [10], [11], [12], [13], [14]. The multigrid technique can produce very good convergence rates. It can solve a problem with N unknowns in $O(N)$ time [15]. However in case of

Journal of Physics: Conference Series **268** (2011) 012012 doi:10.1088/1742-6596/268/1/012012

hysteretic diffusivity, the so-called inner iteration can dramatically increase the computational cost as it should handle the local hysteresis memory at each grid point.

2. Model definition with hysteresis

In this study a 2D transient heat transfer problem is assumed with temperature-dependent thermal diffusivity κ. (It is assumed that the temperature dependence of the volumetric heat capacity c_v can be neglected and the expression of thermal diffusivity $\kappa = \lambda / c_v$ is valid, where λ is the thermal conductivity). The heat equation is written in the form

$$\frac{\partial \theta}{\partial t} = \frac{\partial}{\partial x}\left[\kappa(\theta)\frac{\partial \theta}{\partial x}\right] + \frac{\partial}{\partial y}\left[\kappa(\theta)\frac{\partial \theta}{\partial y}\right] + s(\theta, t) \; , \tag{1}$$

where θ is the dimensionless temperature, the thermal diffusivity $\kappa(\theta)$ is temperature dependent, $x, y \in \Omega$, $\Omega = [0, L] \times [0, L]$ are spatial coordinates. The source term $s(\theta, t)$ will be zero in numerical examples below. The initial condition of $\theta(x, y, 0) = \theta_{init}$, $x, y \in \Omega$ and the first order boundary conditions $\theta(x, y, t) = \theta_w$, $x, y \in \partial\Omega$ are assumed. The nonlinearity of the system originates from the composite material properties, since dependence of on temperature is hysteretic. The thermal diffusivity as a function of the temperature is shown in figure 1 and corresponds to the measured data presented in [1]. Santos et al. [1] observed that during a heating cycle, in the range of lower temperatures, there is a sudden change in the thermal diffusivity, associated with the microstructure modification, in the referred case, the phase transition of $\alpha \to \beta$ cristobalite. The thermal diffusivity upon heating remains relatively unaltered until a higher temperature range, where it presents another steep increase. During cooling a reverse process takes place, but higher diffusivity values can be expected. Furthermore, experiments showed that thermal cycling can reduce the hysteresis of diffusivity in composites.

In this study we assumed that the temperature dependent diffusivity hysteresis could be modeled with a temperature dependent hysteresis operator [16], [17], namely, with an appropriate parameterized Preisach-type hysteresis model [18].

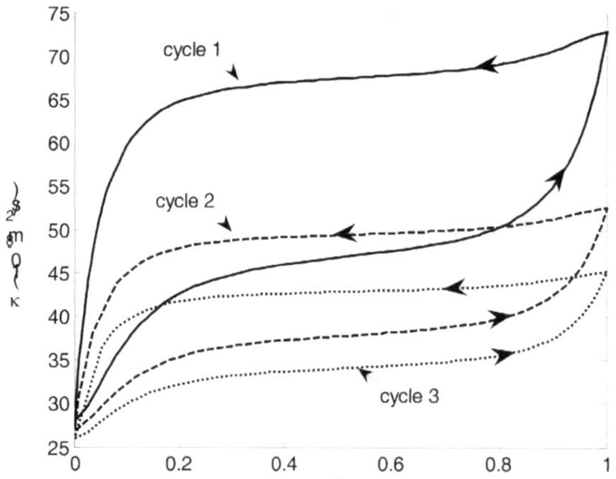

Figure 1. Hysteresis of thermal diffusivity with experimentally observed damping effects of thermal cycling [1].

Repeated heating-cooling cycles can decrease the magnitude of the thermal diffusivity and decrease the area of the hysteresis loops as well. This can be introduced into the model through a cycle factor, ν. The numerical implementation of the Preisach scalar model is based on the Preisach triangle of elementary hysteresis relays [18]. The diffusivity at grid points $\kappa(\theta^{k+1})$ is calculated by a recursive formula provided by discretization of the Preisach operator [17]

$$\kappa(\theta^{k+1}) = H(\theta^{k+1}) = \kappa(\theta^k) \pm \left(b^\nu \frac{T_P}{m} \sum_j \sum_i \mu(\alpha_i, \beta_j) \right), \; ij = m, \; (i,j) \in P, \tag{2}$$

where T_P is the area of the domain P over which in the last input step the hysteresis relays have changed their state and m is the number of relays in the domain P, ν is the heating-cooling cycle number. The sign of the second part of the right-hand side depends on the sign of the temperature change. The normalized distribution function $\mu(\alpha,\beta) = 1/N_{2D}(m_1, m_2, \sigma_1^2, \sigma_2^2)$ is discretized on a 500×500 grid, N_{2D} represents the normalized density function of 2D Gaussian distribution with mean values m_1, m_2 and variances σ_1^2, σ_2^2.

The hysteresis cycles in figure 2 can be provided by the following parameters: $m_1 = m_2 = 0.525$, $\sigma_1 = \sigma_2 = 0.18$, in first thermal cycle $\kappa_0^1 = 28 \cdot 10^{-8}\ \mathrm{m^2/s}$, $b^1 = 1$, in the second cycle $\kappa_0^2 = 27 \cdot 10^{-8}\ \mathrm{m^2/s}$, $b^2 = 0.57$ and in the third and all higher cycles $\kappa_0^\nu = 26 \cdot 10^{-8}\ \mathrm{m^2/s}$, $b^\nu = 0.43$, $\nu = 3,4,\cdots$.

3. Discretization of the nonlinear model

The solution domain Ω has been discretized by a Cartesian mesh with a uniform mesh size $h = L/n$. Time differentiation is discretized by the Crank-Nicholson second-order method [15]. The right hand side of (1) is discretized in two steps. First the thermal flux differences are calculated, considering cell interfaces at the distance of $\pm h/2$ from grid points. For example between points (i,j) and $(i+1,j)$, the assumed flux is $J_{i+1/2,j} = \kappa_{i+1/2,j}(\theta_{i+1,j} - \theta_{ij})/h$, where $\kappa_{i+1/2,j}$ is a space averaged diffusivity, $\kappa_{i+1/2,j} = 0.5(\kappa_{i,j} + \kappa_{i+1,j})$.

Depending on the time averaging of the diffusivity, various approximations to (1) can be obtained. The Crank-Nicholson method with space averaged, time-instantaneous diffusivity has the form of

$$\frac{\theta_{ij}^{k+1,l+1} - \theta_{ij}^k}{\Delta t} = \frac{1}{h}\left(\sigma\left(\Delta J_{i\pm1/2,j}^{k+1,l+1} + \Delta J_{i,j\pm1/2}^{k+1,l+1}\right) + (1-\sigma)\left(\Delta J_{i\pm1/2,j}^{k} + \Delta J_{i,j\pm1/2}^{k}\right)\right) + s^{k+1/2}, \tag{3}$$

where the first superscript k refers to the time step and the second superscript l refers to the iteration cycle. The weighting factor is $\sigma = 0.5$ for the Crank-Nicholson method, J is the heat flux and ΔJ denotes the flux differences. Each flux difference has been computed in the same manner, for example

$$\Delta J_{i\pm1/2,j}^{k+1,l+1} = J_{i+1/2,j}^{k+1,l+1} - J_{i-1/2,j}^{k+1,l+1}. \tag{4}$$

The thermal fluxes in the iteration step $l+1$ and at time step $k+1$ are determined by the diffusivity hysteresis model $\kappa_{ij}^{k+1,l+1} = H(\theta_{ij}^{k+1,l+1})$. The hysteretic diffusivity $\kappa_{ij} = H(\theta_{ij}^k)$ is not involved in the inner iteration marked by l.

Another approximation scheme uses time-averaged temperatures to determine the diffusivity hysteresis. In this case the discretized form of (1) is

$$\frac{\theta_{ij}^{k+1,l+1}-\theta_{ij}^{k}}{\Delta t}=\frac{1}{h}\left(\sigma\left(\Delta J_{i\pm 1/2,j}^{k+1,l+1}+\Delta J_{i,j\pm 1/2}^{k+1,l+1}\right)+(1-\sigma)\left(\Delta J_{i\pm 1/2,j}^{k,l+1}+\Delta J_{i,j\pm 1/2}^{k,l+1}\right)\right)+s^{k+1/2}. \tag{5}$$

Grid points at times k and $k+1$ have the same values of diffusivity, $\kappa_{ij}^{k+1/2,l+1}=H\left(0.5\left(\theta_{ij}^{k+1,l}+\theta_{ij}^{k}\right)\right)$. Diffusivity at cell interfaces is interpolated linearly by

$$\kappa_{i+1/2,j}^{k+1/2,l+1}=0.5\left(\kappa_{i+1,j}^{k+1/2,l+1}+\kappa_{ij}^{k+1/2,l+1}\right). \tag{6}$$

The thermal flux across the cell interfaces at time $k+1$ and in iteration cycle $l+1$ is calculated according to

$$J_{i+1/2,j}^{k+1,l+1}=\kappa_{i+1/2,j}^{k+1/2,l+1}\left(\theta_{i+1,j}^{k+1,l+1}-\theta_{ij}^{k+1,l+1}\right)/h. \tag{7}$$

At time k the thermal flux is $J_{i+1/2,j}^{k,l+1}=\kappa_{i+1/2,j}^{k+1/2,l+1}\left(\theta_{i+1,j}^{k}-\theta_{ij}^{k}\right)/h$. All the other thermal fluxes can be derived in the same manner. Substituting thermal fluxes (7) into (5) produces the so-called time-centered discretization scheme

$$\begin{aligned}&(1+p)\theta_{ij}^{k+1,l+1}-\frac{\Delta t}{2h^2}\left(\kappa_{i\pm 1/2,j}^{k+1/2,l+1}\theta_{i\pm 1,j}^{k+1,l+1}+\kappa_{i,j\pm 1/2}^{k+1/2,l+1}\theta_{i,j\pm 1}^{k+1,l+1}\right)=\\&(1-p)\theta_{ij}^{k}+\frac{\Delta t}{2h^2}\left(\kappa_{i\pm 1/2,j}^{k+1/2,l+1}\theta_{i\pm 1,j}^{k}+\kappa_{i,j\pm 1/2}^{k+1/2,l+1}\theta_{i,j\pm 1}^{k}\right)+s_{ij}^{k+1/2}\Delta t,\end{aligned} \tag{8}$$

where $\kappa_{i\pm 1/2,j}^{k+1/2,l+1}\theta_{i\pm 1,j}^{k+1,l+1}$ is an abbreviation of $\kappa_{i+1/2,j}^{k+1/2,l+1}\theta_{i+1,j}^{k+1,l+1}+\kappa_{i-1/2,j}^{k+1/2,l+1}\theta_{i-1,j}^{k+1,l+1}$ and

$$p=\frac{\Delta t}{2h^2}\left(\kappa_{i-1/2,j}^{k+1/2,l+1}+\kappa_{i+1/2,j}^{k+1/2,l+1}+\kappa_{i,j-1/2}^{k+1/2,l+1}+\kappa_{i,j+1/2}^{k+1/2,l+1}\right). \tag{9}$$

Assembling (8) for each interior node yields a system of equations

$$\mathbf{A}_h^{k+1,l+1}\boldsymbol{\theta}_h^{k+1,l+1}=\mathbf{B}_h^{k+1,l+1}\boldsymbol{\theta}_h^{k}+\mathbf{S}_h^{k+1/2}, \tag{10}$$

where $\mathbf{A}_h^{k+1,l+1}$ is the matrix of the coefficients on the left-hand side of (8), $\boldsymbol{\theta}_h^{k+1,l+1}$ is the vector of unknown temperatures, $\mathbf{B}_h^{k+1,l+1}$ is the matrix of coefficients on the right-hand side, $\boldsymbol{\theta}_h^{k}$ is the solution temperature vector at time step k, $\mathbf{S}_h^{k+1/2}$ is the vector of the source terms multiplied by Δt. Subscript h refers to the level of the spatial resolution.

The semi-implicit finite difference method (8) resembles the 2D variation of the DuFort-Frankel scheme, which is consistent with the original PDE only if $\Delta t\kappa_{max}/h^2 \ll 1$, where $\kappa_{max}=max(\kappa_{ij})$, $i,j=1,\cdots,n-1$ [15]. To avoid solutions that do not satisfy the physical nature of the modeled system, the upper limit to the time step can be evaluated from the inequality $\Delta t\le h^2/(2(1-\sigma)\kappa_{max})$.

4. The coarse level iterating multigrid solver

The nonlinearity of κ in method (8) needs iteration at every time step. The Newton iteration is very expensive in CPU time because of the large set of unknown variables [15]. Hysteresis-like temperature dependence of the diffusivity increases dramatically the computational cost as it requires to manage the local memory at each grid point (10). Treatment of this memory is very resource and time consuming. The finer the resolution, the more hysteresis memory has to be treated. Using coarser resolution decreases the computational costs, but produces a coarser solution as well. In this study we propose a coarse level iterating multigrid algorithm, in which the fine resolution of the temperature field is preserved, but the hysteresis of diffusivity is reduced to a coarser grid. Thus, this multigrid algorithm works on several resolutions [16], [19].

In a general case, the multigrid solver assumes an initial guess on the fine resolution for the diffusivity $\boldsymbol{\kappa}_h^{k+1/2,0} = H(\boldsymbol{\theta}_h^k)$ and for the temperature field $\boldsymbol{\theta}_h^{k+1,0} = \boldsymbol{\theta}_h^k$. In every iteration cycle the diffusivity coefficients have to be re-calculated for all fine grid points. To reduce the solution time, before the iteration, problem (10) should be reduced to the next coarser grid, with mesh size $H = 2h$, in the form of

$$\mathbf{A}_H^{k+1,l+1}\boldsymbol{\theta}_H^{k+1,l+1} = \mathbf{B}_H^{k+1,l+1}\boldsymbol{\theta}_H^k + \mathbf{S}_H^{k+1/2}, \tag{11}$$

where the initial temperature $\boldsymbol{\theta}_H^{k+1,0}$ of time step $k+1$ has been determined by the restriction of the fine grid solution at time step k, $\boldsymbol{\theta}_H^{k+1,0} = \boldsymbol{\theta}_H^k = \mathbf{R}\boldsymbol{\theta}_h^k$. The initial diffusion coefficient field has been defined with the hysteresis operator as $\boldsymbol{\kappa}_H^0 = H(\boldsymbol{\theta}_H^k)$. The source term $\mathbf{S}_h^{k+1/2}$ has been restricted to $\mathbf{S}_H^{k+1/2}$, or $\mathbf{S}_H^{k+1/2}$ could be calculated directly as well. The temperature field approximation for time step $k+1$ and mesh size H could be evaluated by iteration (11) with the new values of diffusivity $\boldsymbol{\kappa}_H^{k+1/2,l+1} = H\left(0.5\left(\boldsymbol{\theta}_H^{k+1,l} + \boldsymbol{\theta}_H^{k+1,0}\right)\right)$, until the absolute difference decreases below a preliminarily defined ε limit, $\left\|\boldsymbol{\theta}_H^{k+1,l+1} - \boldsymbol{\theta}_H^{k+1,l}\right\| < \varepsilon$. After this so-called inner-iteration, the diffusivity and the temperature are interpolated to the fine grid. By denoting $\mathbf{P}$ as the prolongation operator and $\mathbf{I}_H^h$ as the interpolation operator, the fine grid diffusivity is $\boldsymbol{\kappa}_h^{k+1/2} = \mathbf{S}_{GS}\mathbf{P}\boldsymbol{\kappa}_H^{k+1/2,l+1}$, the initial temperature is $\boldsymbol{\theta}_h^{k+1,0} = \mathbf{I}_H^h\boldsymbol{\theta}_H^{k+1,l+1}$. The prolongated diffusivity needs smoothing, $\mathbf{S}_{GS}$ denotes the Gauss-Seidel smoothing operator [19][20]. The interpolated temperature could be seen as a good initial solution to the grid level h. To evaluate the fine grid solution $\boldsymbol{\theta}_h^{k+1}$, multigrid *V*-cycles are applied, until the convergence criterion is reached. During the fine grid *V*-cycles the diffusivity field remains unaltered. The graphical scheme of the iteration method can be seen in figure 2.

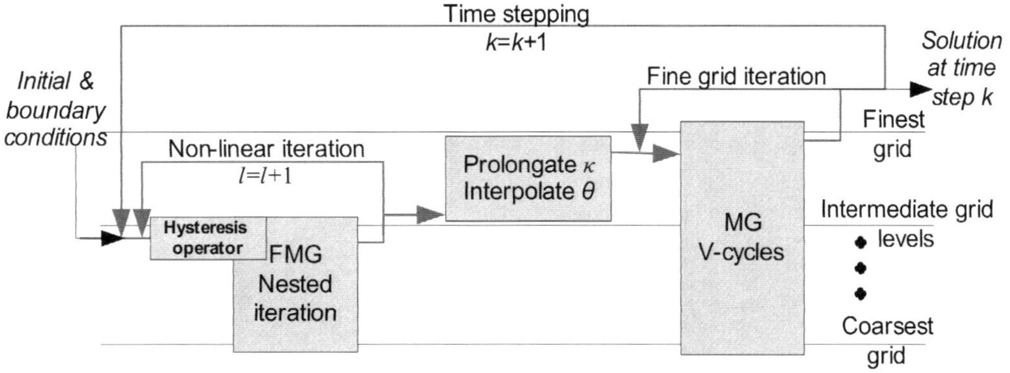

Figure 2. Schematic representation of the coarse level iterating multigrid method.

This proposed coarse level iterating hierarchical algorithm consists of two types of multigrid methods. The inner-iteration is on the coarser grid, the solver is the so-called full multigrid algorithm. An advantage of this so called Full Multigrid (FMG) method is the exact initial solutions on the coarsest grid, and therefore a good convergence. The nonlinear iteration has a reduced set of variables and the amount of hysteresis memory data also decreases on the coarser grid. The stencils of $\mathbf{A}_H^{k+1,l+1}$

and $\mathbf{B}_H^{k+1,l+1}$ are changed point-by-point, depending on the diffusivity. The stencils of $\mathbf{A}_H^{k+1,l+1}$ and $\mathbf{B}_H^{k+1,l+1}$ are

$$\mathbf{A}_H^{k+1,l+1} = \begin{bmatrix} 0 & -\beta\kappa_{i,j-1/2_H}^{k+1/2,l+1} & 0 \\ -\beta\kappa_{i-1/2,j_H}^{k+1/2,l+1} & (1+p)_H & -\beta\kappa_{i+1/2,j_H}^{k+1/2,l+1} \\ 0 & -\beta\kappa_{i,j+1/2_H}^{k+1/2,l+1} & 0 \end{bmatrix}, \tag{12}$$

and

$$\mathbf{B}_H^{k+1,l+1} = \begin{bmatrix} 0 & -\beta\kappa_{i,j-1/2_H}^{k+1/2,l+1} & 0 \\ -\beta\kappa_{i-1/2,j_H}^{k+1/2,l+1} & (1-p)_H & -\beta\kappa_{i+1/2,j_H}^{k+1/2,l+1} \\ 0 & -\beta\kappa_{i,j+1/2_H}^{k+1/2,l+1} & 0 \end{bmatrix}, \tag{13}$$

respectively, where $\beta = \Delta t / 2H^2$. In the proposed algorithm the diffusivity fields $\mathbf{\kappa}$, the temperature fields $\mathbf{\theta}$ and the source terms $\mathbf{S}$ are stored at every grid level. The FMG determines the solutions at each grid level, so the approximated truncation error can be the stop criterion of the FMG cycles. The hysteresis memory is treated and stored on the coarser grid level.

To refine the temperature field after the inner iteration, some simple multigrid V-cycles run on the finest grid. The temperature field is iterated on the coarser grid because of nonlinearity, and after interpolation it is iterated on the fine grid to increase the accuracy. The diffusivity is iterated on the coarser grid as well, but diffusivity values remain unchanged on the fine grid. The high level V-cycles (MG-V) coarsening technique and stencil determination are similar to the techniques used in FMG, but the temperature field is calculated only on the finest grid. On the coarse levels MG-V works with the corrections of the temperatures. The V-cycles iteration could be stopped when the residual norm has been reduced below the η times of the initial norm, i.e. $\left\|\mathbf{r}_h^{k+1}\right\| < \eta \left\|\mathbf{A}_h^{k+1,l+1}\mathbf{I}_H^h\mathbf{\theta}_H^{k+1,l+1} - \mathbf{B}_h^{k+1,l+1}\mathbf{\theta}_h^k - \mathbf{S}_h^{k+1/2}\right\|$, $\eta < 1$. The coarsest problem has been solved in both multigrid algorithms by the successive over-relaxations method (SOR). The schematic flowchart of the proposed coarse level iterating multigrid algorithm is shown in figure 3.

5. Numerical results

In this section the proposed method and algorithm are analysed. In all the test problems the source term is considered to be zero. The test parameters are $L = 0.5\,\mathrm{m}$, $\kappa_0 = 2.5\cdot10^{-7}\,\mathrm{m^2/s}$, $\Delta t = 250\,\mathrm{s}$, the finest level consists of 33×33 mesh points. In all tested FMG and also in MG-*V*-cycles, full weighting restriction and bilinear interpolation have been applied, except for the diffusivity prolongation at the end of the coarse level iteration. SOR was the coarsest grid solver. The number of pre- and post-relaxation is equal to 3, the coarsest grid resolution has 5×5 points, for nonlinear iteration $\varepsilon = 1\cdot10^{-4}$ and the initial norm reduction in MG-*V* is $\eta = 0.1$.

5.1. Verifying the coarse level iterating algorithm

The finite difference methods (3) and (5) are compared without reduction of the diffusivity hysteresis, i.e. the diffusivity has been iterated in both cases at the finest resolution. Initial conditions are $\mathbf{\theta}^0 = 1$ with constant boundary conditions $\theta(x,y,0) = 0$, $x,y\in\partial\Omega$. Figure 4*a* shows the norm of differences between the solutions obtained by two distinct difference schemes versus time. The highest norm value is less than $2.5\cdot10^{-3}$.

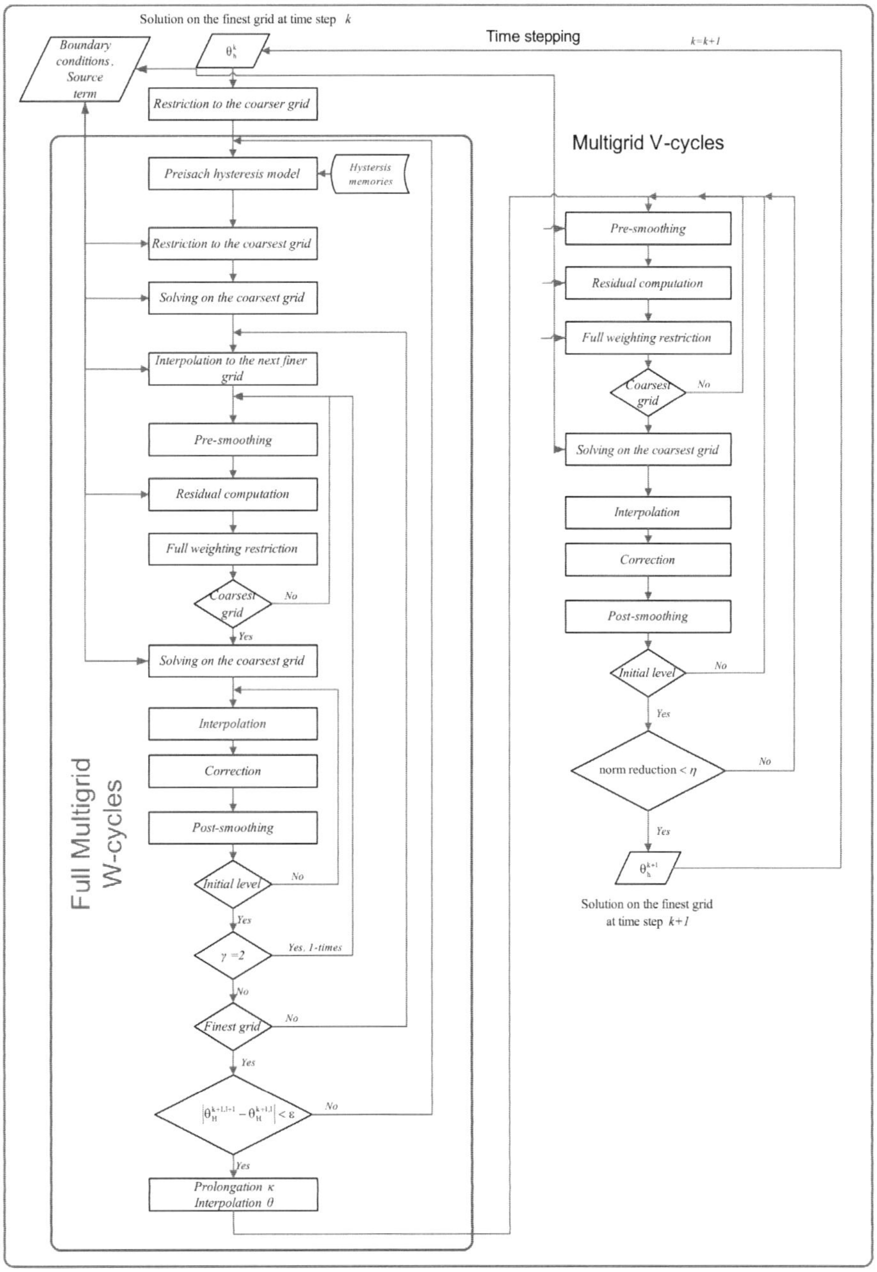

Figure 3. Flow chart of the proposed coarse level iterating multigrid algorithm.

Comparing the numbers of iteration cycles, it can be stated that the proposed scheme (5) generally needs two iterations, however formulation (3) at larger temperature differences needs considerably more iteration cycles [16], [17][19]. When approaching the steady-state, the cycle number could also decrease. In the proposed algorithm, after the coarse level iteration, the diffusivity remains unchanged. The crucial question is how the diffusivity should be prolongated to the finest grid. When dealing with two-dimensional cases the bilinear interpolation is the most popular technique. Another possibility is the simple prolongation. According to the required rapid iteration, higher order interpolation is not suitable. Figure $4b$ and figure $4c$ show the norms of differences between the fine grid iteration and

Journal of Physics: Conference Series **268** (2011) 012012 doi:10.1088/1742-6596/268/1/012012

coarse level iteration with two distinct diffusivity projections. The deviation in the norm with the prolongation algorithm (curve c) all the time remains beneath the norm of bilinear interpolation (curve b). Comparing the results, it can be concluded, that the algorithm with prolongation approximates the results obtained by using the fine grid iteration algorithm better, therefore we propose that the temperature field can be interpolated but the diffusivity field should be prolongated to the finest grid.

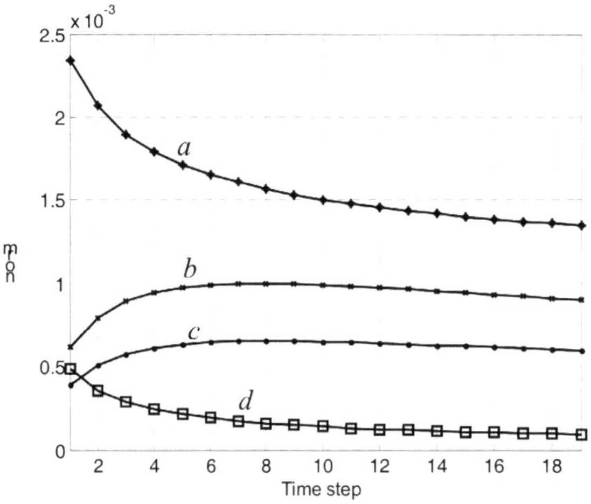

Figure 4. Norm tendencies differences *a)* between schemes (3) and (5), *b)* between CLI-MG by bilinear interpolation and fine level iteration, *c)* CLI-MG by prolongation and fine level iteration, *d)* CLI-MG with $\Delta t = 500\text{s}$ and $\Delta t = 50\text{s}$.

The correctness of the time-centered diffusivity has been tested by employing different time steps using the same resolutions in space. The reference time step was a tenth of the original time step. figure $4d$ shows the norm of differences between two steps in time. Apart from the first few steps, there is no difference between the two calculations, so time averaging is suitable, as long as regularity and consistency are ensured.

Comparing the solution time, we measured twenty time steps. Assuming that the solution time for the coarse level iteration is one unit, the solution time with fine grid iteration is four units. Analysing the memory usage, the memory requirement of the multigrid part of the two methods are roughly the same. The discrete Preisach algorithm in the proposed form has one large distribution matrix. The memory character of the Preisach model requires to store the previous input values of all grid points in a matrix of size $(n+1)\times(n+1)$ and the return points of the characteristic of $\theta_{ij}(k\Delta t)$ at each grid point, which means that about 30 values have to be stored and treated per cycle and per grid point. If the Preisach algorithm works on the next coarser grid, the memory demand can be calculated in the same way, replacing n by $N = n/2$. The saved memory size if $n >> 1$ is about $\approx 23n^2$ data. With large problems, significant differences in memory usage can be accompanied by reduced CPU-time as well [17][19].

5.2. Hysteresis induced heating-cooling asymmetry

To investigate the results of simulations using the proposed algorithm, the heating-cooling thermal cycles with constant and temperature dependent hysteretic thermal diffusivity are compared. Numerical simulations represent well the thermal cycling asymmetry induced by hysteresis of diffusivity. In the first test the thermal diffusivity changes on the main hysteresis curves. It can be reached by monotone heating or cooling of the boundaries. In the heating process the initial and boundary conditions are $\boldsymbol{\theta}(x, y, 0) = \mathbf{0}$ at $t = 0$, and $\theta(x, y, t) = 1$, $x, y \in \partial\Omega$ at $t > 0$. The initial and boundary conditions in the cooling process are $\boldsymbol{\theta}(x, y, 0) = \mathbf{1}$ at $t = 0$; $\theta(x, y, t) = 0$, $x, y \in \partial\Omega$ and $t > 0$. In the tests presented, the constant diffusivity is considered to be the base diffusivity of the hysteretic diffusion κ_0. Temperature fields and diffusivity fields at four different time steps are shown in figure 5. Temperature transients at the middle of the domain are shown in figure 6.

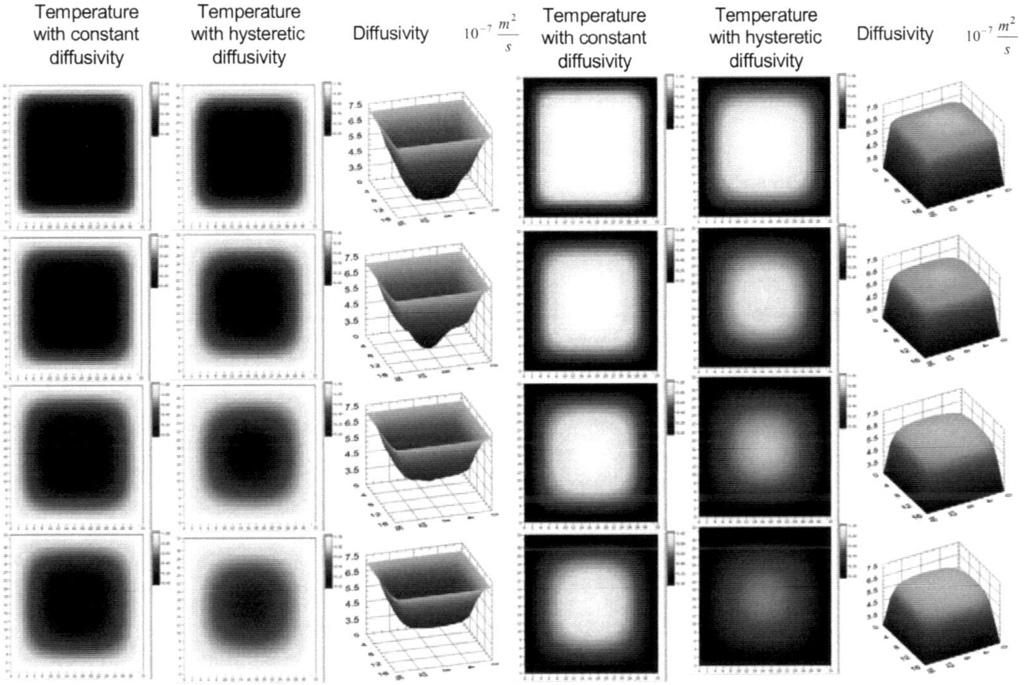

Figure 5. Heating and cooling a rectangular 2D sample with constant and with hysteretic diffusivity.

Heating-cooling transients with constant thermal diffusivity are symmetrical. In hysteretic diffusion, the heating-cooling transients differ from each other. The thermal process speeds up with increased diffusivity. The cooling process begins at higher diffusivity [17]. Differences between constant and hysteretic diffusion processes are expressed by the Euclidean norm. Figure 7 shows the norm of differences for heating and cooling processes. The norm first increases in time and after reaching a maximum slowly decreases. In the transient time domain, the difference between cooling processes is higher than the difference between heating processes. The cooling process with hysteretic diffusivity approaches the steady-state more rapidly.

5.3. Periodically changing boundary temperatures

The hysteresis minor loops have been avoided using the presented monotone heating or cooling models. However with periodically heating-cooling of the boundaries, it has to be assumed, that diffusivity could vary along minor loops. In this test, a rectangular domain with sinusoidal varying temperatures at two neighboring boundaries, zero temperature boundary conditions on the other two sides of the domain are considered. The transient thermal characteristic of a material with diffusivity hysteresis is compared to the characteristic of a material with constant diffusivity. Hundred time steps have been evaluated. The temperature fields and diffusivities at time steps $k = 10, 20, 50, 100$ can be seen in figure 8. In figure 8*a-d* the linear, in figure 8*e-h* the hysteresis heat diffusion time series and in figure 8*i-l* the appropriate diffusivity fields can be seen. Diffusivity minor loops are shown in figure 9.

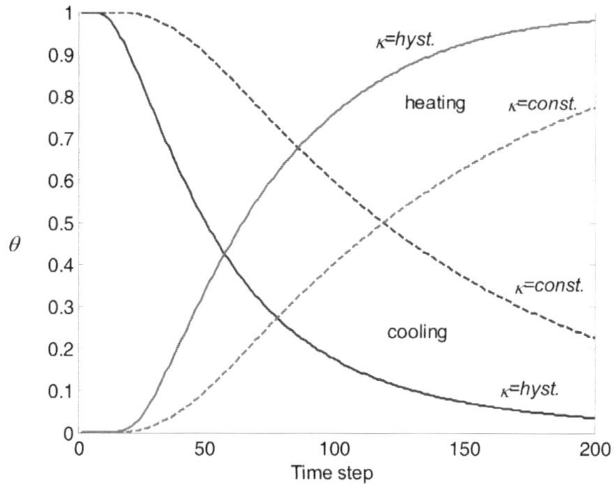

Figure 6. Temperature transients in the middle of the domain with hysteretic and with constant diffusivity.

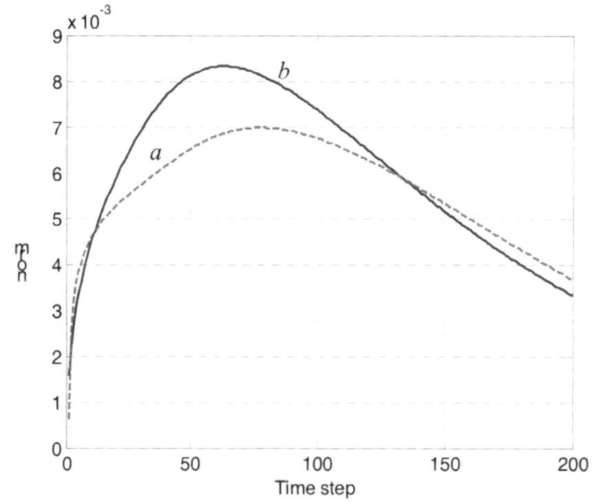

Figure 7. Norm of differences between hysteretic and normal heat diffusion *a)* heating and *b)* cooling along the main hysteresis curves.

The periodical changing of boundary temperatures can be seen in figure 10. From determining the norm of differences of the nonlinear and the linear temperature fields in time, it can be concluded, that the norm is increasing rapidly when the boundary temperatures are increasing and is decreasing slowly when the boundary temperatures are decreasing. The characteristic of the curve approaches a quasi-static state. The asymmetry in the norm cycles represents well the nonlinear effects of the hysteresis.

5.4. Nonlinear diffusion with inhomogeneous initial conditions

Special problems arise from inhomogeneous initial conditions. In these cases the initial diffusivity field depends on how the initial temperature distribution has been reached. A hypothetical initial temperature distribution can be seen in figure 11*a*. To have a correct solution, the initial diffusivity field has to be determined as well. If the initial diffusivity field can be calculated on the main hysteresis curves, different initial diffusivity fields can be found on the increasing and decreasing main hysteresis curves, see figure 11*b* and figure 11*c*, respectively. Considering zero temperature boundary conditions, each initial diffusivity field induces different transients in the cooling process that can be seen in figure 12 for constant diffusivity (figure 12*a-c*), for hysteretic diffusivities starting from the increasing curve (figure 12*d-f*) and for hysteretic diffusivities starting from the decreasing curve (figure 12*g-i*) at three time steps, *k=6,20,40*. The temperature fields are homogenized faster in the initially high temperature domain with nonlinear diffusivity than with constant low values of diffusivity. After $k = 40$ time steps, the temperature fields are well smoothed in all cases, but the effect of initial conditions has been retained in the diffusivity fields, see figure 13*a* and figure 13*b*.

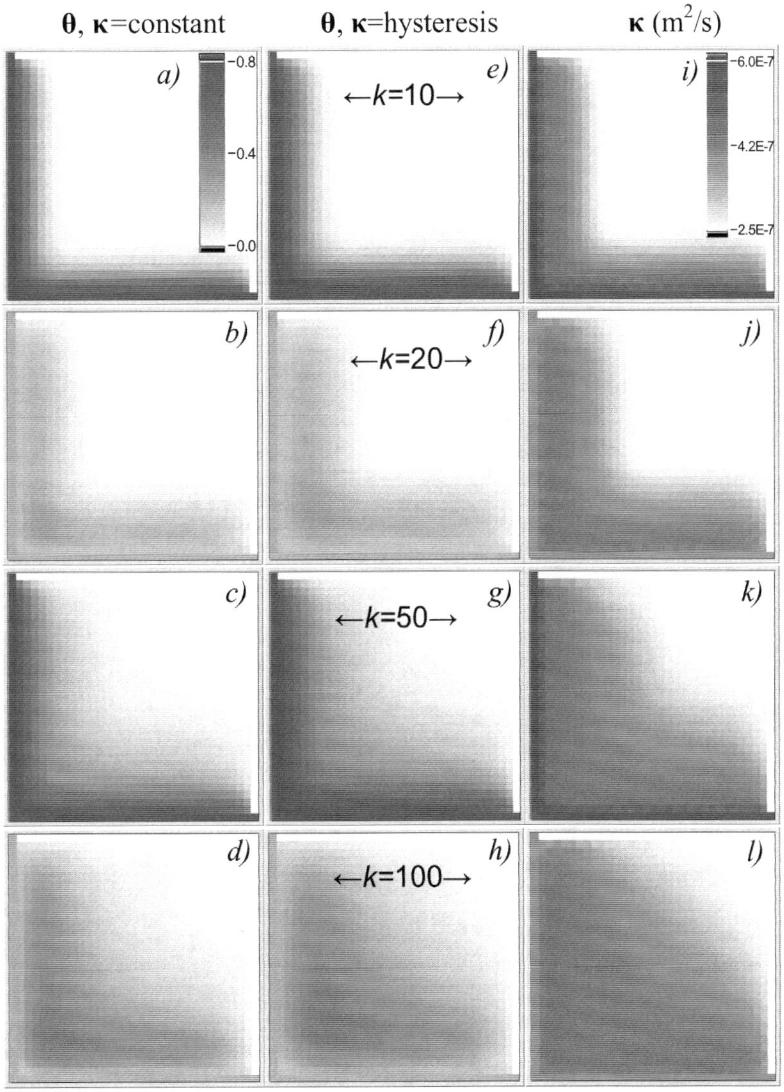

Figure 8. Temperature fields with *a)-d)* linear, *e)-h)* nonlinear heat diffusion and *i)-l)* diffusivity at four different time steps.

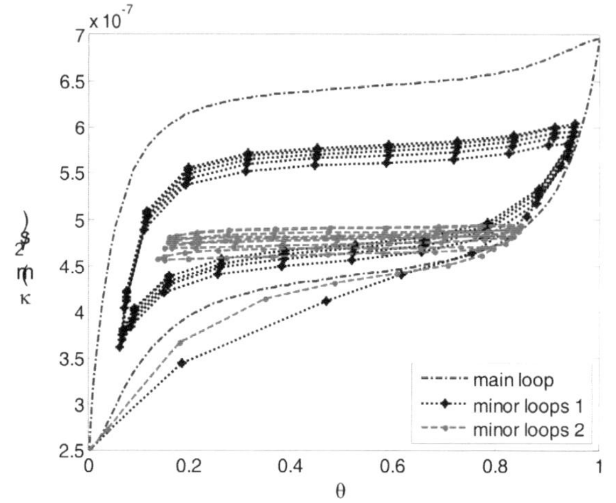

Figure 9. Minor loops of diffusivity near the boundaries.

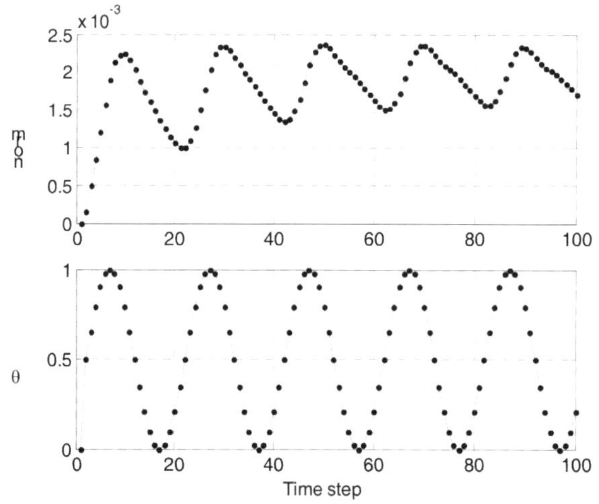

Figure 10. Top: Norm of differences between linear and nonlinear heat diffusions. bottom: Periodically changing boundary temperatures.

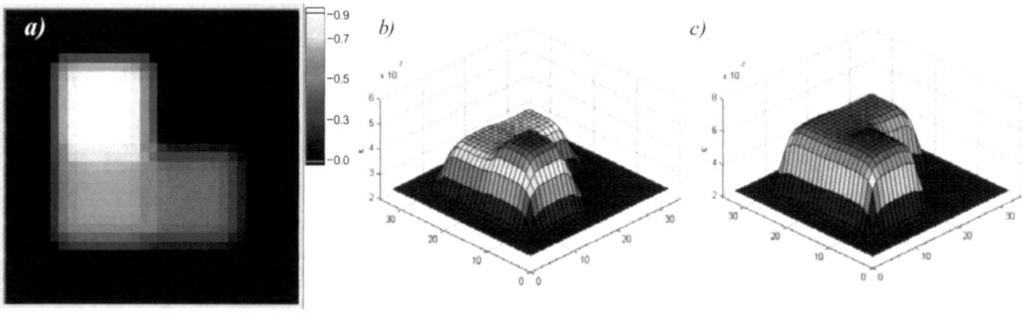

Figure 11. *a)* Hypothetical initial temperature field, *b)* initial diffusivity field determined on the main increasing curve and *c)* initial diffusivity field determined on the main decreasing curve.

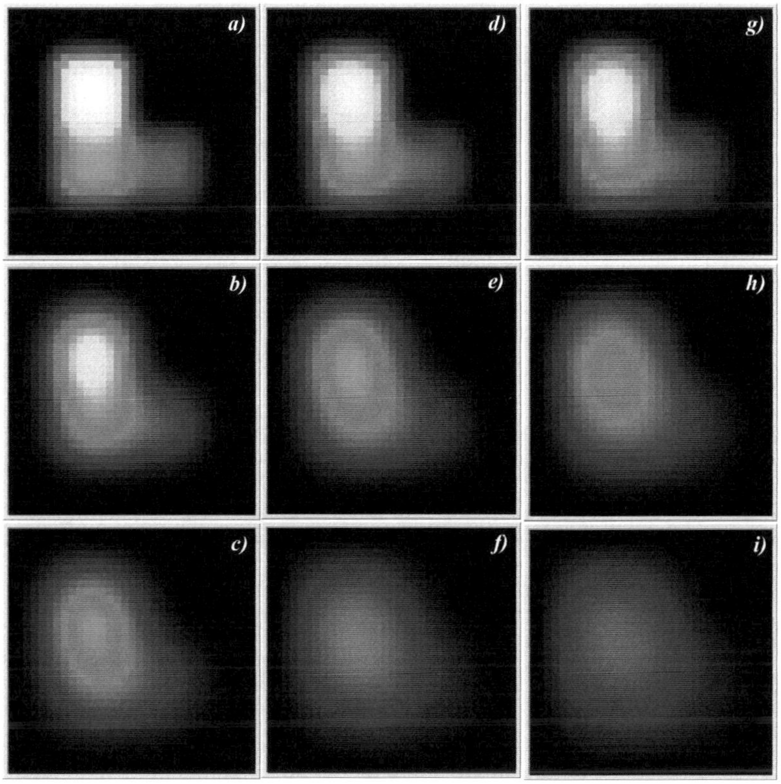

Figure 12. Temperature fields at three different time steps, top to bottom, k=6,20,40, with constant diffusivity (a-c), with initial diffusivity of figure 11b (d-f), and with initial diffusivity of figure 11c (g-i).

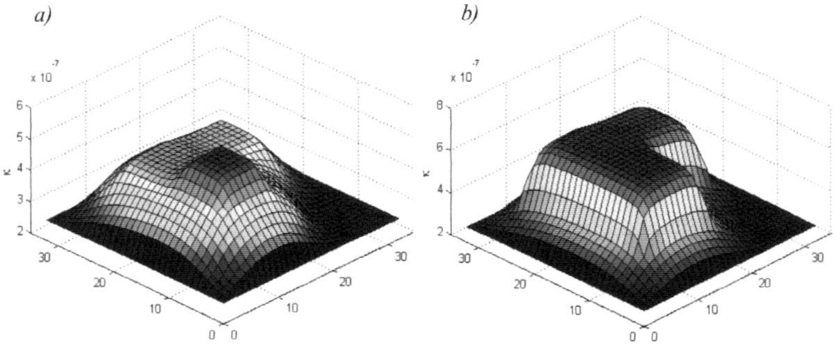

Figure 13. Diffusivity fields after k=40 time steps.

6. Conclusion

The experimentally proved hysteresis of diffusivity for composite materials is not yet incorporated in numerical models of systems applied in common engineering applications. Hysteresis of diffusivity can be approximated from a macroscopic point of view by a hysteresis operator. This operator describes the temperature dependence of diffusivity. Numerical solution of nonlinear heat equations in more than one dimension is very demanding in computational resources and time consuming, even more so with hysteresis memory incorporated at every discretization unit. This study has proved that the execution time and the memory demand can be reduced by the proposed coarse level iterating multigrid technique. Even with the lower computational costs, good convergence properties have been achieved by applying the full multigrid method in the inner iteration. When the exact solution of the problem is sufficiently smooth, which in diffusion processes is a reasonable assumption, this type of problem reduction does not cause considerable deviations in the results.

The presented transient heat diffusion examples indicate that due to the hysteretic character of diffusivity, heat transfer processes become non-symmetric which can be different from processes with constant diffusivity. The transient process depends not only on the initial and boundary conditions but on the prehistory of the sample, i.e. the correct solution requires in addition to the initial distribution of temperature the initial diffusivity distribution and the prehistory of diffusivity at the initial moment. We can conclude that for the proper analysis of heat transfer processes over a broad temperature range inside composite materials, the hysteretic temperature dependence of the diffusivity should not be neglected.

References

[1] Santos W N, Baldo J B and Taylor R 2000 *Materials Research Bulletin* **35** 2091
[2] Singh K, Singh A K and Saxena N S 2008 *Current Appl. Phys.* **8** 159
[3] Wawryk R, Marucha Cz, Balcerek K, Terzijska B M and Ivanova Z 2000 *Cyrogenics* **40** 749
[4] Kashiwada Y, Fujishiro H and Ibeke M 2003 *Phys. B* **329-333** 924
[5] Imre L 1983 *Heat transfer in complex structures,* (in Hungarian) (Budapest: Akadémiai Kiadó)
[6] Jaeger J C and Carslaw H S 1959 *Conduction of heat in solids*, 2nd ed., (Oxford: Oxford University Press)
[7] Dai W and Woodward P R 1998 *J. Comp. Phys.* **139** 58
[8] G-Fernández C F, Alhama F, Alarcón M and L-Sánchez J F 1998 *Comp. Phys. Comm.* **111** 53
[9] Goto T and Suzuki M 1996 *Int. J. Heat Mass Trans.* **39** (4) 823
[10] Brandt A 1994 *SIAM J. Numer. Anal.* **31** 1695
[11] de Lemos M J S and Rabi J A 2001 *Appl. Math. Comp.* **124** 215
[12] Hackbusch W 1985 *Multigrid methods and applications* (Berlin: Springer)
[13] Hackbush W 1994 *Iterative solution of large sparse system of equations* (New York: Springer)
[14] Zhang J, Gupta M M and Kouatchou J 1997 *J. Comp. Phys.* **132** 123
[15] Stojan G and Takó G 1995 *Numerikus módszerek* (Budapest: ELTE)
[16] Jancskár I and Iványi A 2006 *Acta Agraria Kaposváriensis* **10** (1) 143
[17] Jancskar I. and Ivanyi A 2006 *Phys. B*, **372** 222
[18] Iványi A 1997 *Hysteresis models in electromagnetic computation* (Budapest: Akadémiai Kiadó)
[19] Jancskar I and Ivanyi A 2005 *Proc. The Tenth Int. Conf. Civil, Struct. Env. Eng. Comp.* (Rome, Elsevier) paper 138
[20] Jancskar I and Ivanyi A 2005 *Przeglad Elektrotechniczny* **6** 37

Journal of Physics: Conference Series **268** (2011) 012013 doi:10.1088/1742-6596/268/1/012013

Bichromatic field propagation in a resonant medium: Floquet analysis

Z Kis[1], S Guérin[2] and HR Jauslin[2]

[1] H.A.S. Research Institute for Solid State Physics and Optics, H-1525 Budapest, P.O.Box 49, Hungary

[2] Laboratoire Interdisciplinaire Carnot de Bourgogne (ICB), UMR 5209 CNRS-Université de Bourgogne, 9 Av. A. Savary, BP 47 870, F-21078 Dijon Cedex, France

E-mail: zsolt@szfki.hu

Abstract. We study the propagation of a bichromatic field in a resonant medium. The two modes of the incoming fields are pulsed and delayed with respect to each other. It is shown that in the course of the propagation, new Raman sidebands will be generated. The achievable frequency spacing between the sidebands is determined from experimental data. A numerical example is shown for realistic physical parameters.

1. Introduction

Resonant nonlinear optics (RNLO) is an excellent concept to control the electric susceptibility of a medium by light. In RNLO the light fields propagating in a medium are resonant or nearly resonant with some atomic transitions. The atoms of the medium not only mediate the interaction between the field modes, but form a compound quantum system with the photons, hence a previously unimaginable range of phenomena emerge. In the heart of this new class of processes there is the quantum coherence and quantum interference, which can be used to control the propagation of light pulses in the underlying media. Moreover, the efficiencies of the processes are increased dramatically compared to the conventional nonlinear optical processes with far off-resonant fields. The pioneer work in the domain of RNLO was electromagnetically induced transparency (EIT), which relies on the destructive quantum interference between two atomic transition paths: an initially absorbing medium becomes transparent for a weak probe field due to the interaction with a strong control field [1, 2, 3]. Based on EIT a number of interesting and important applications have been developed, our present work is related to non-linear field generation [4, 5, 6, 7, 8, 9, 10, 11, 12, 13, 14], and pulse propagation in multistate systems such as in a double-lambda [15, 16]. For recent review on the EIT process and related applications see [17].

The first experiments have been realized in hot/cold gases, and in BEC. However, doped single crystals proved to be as good as or even superior medium to gases for RNLO applications. The doping atoms play the role of the active media, which are embedded into some host material. The host material can be either a glass or a crystalline structure. As for the doping atoms, one of the most often used group of elements are the rare-earths. Their relevant properties are the following: the atomic density in the host crystal may reach $10^{18} - 10^{20}\,\mathrm{cm}^{-3}$; the radiative lifetime of the atomic levels is of the order of a milli-second, while the homogeneous linewidth is in the kHz range; the inhomogeneous linewidth may be several GHz; the power of the control lasers should be chosen much smaller than in gases because of the low damage threshold of the host media. However, by means of sophisticated pumping techniques, a subset of doping atoms with the same transition frequency can be selected, and excellent experimental results can be obtained, e.g. photon storage times greater than a second [18]. Several RNLO processes have already been successfully realized in rare-earth doped single crystals, such as EIT [19]; coherent Raman beats [20]; slow down of light pulses in a solid [21, 22, 23]; quantum storage of light pulses [18, 21, 24].

In a series of works [25, 26, 27], Raman-sideband generation has been developed and experimentally realized, where a strong, bichromatic field interacted in an off-resonant Raman-configuration with the atoms of the medium. As a result, several Raman-sidebands have been generated, leading to a sequence of short laser pulses.

In this paper we consider the propagation of a strong, bichromatic laser pulse which interacts resonantly with the ground and excited states of the atoms of a medium. The dynamics of the light-atom interaction is studied with the aid of the Floquet-analysis. The condition of Raman-sideband generation is presented for delayed incoming pulses. The derived Maxwell-Scrödinger equations are solved numerically and an example is shown for Raman-sideband generation in a system with realistic physical parameters.

The paper is organized as follows: in Section 2 we present the model system and derive the equations which govern the time evolution of the atoms and the propagation of the fields. The summary of the Floquet analysis of the atomic dynamics is given in Section 3. The propagation of the field modes is discussed in Section 4, where a numerical example is shown as well. The results are summarized in Section 5.

2. The model system

We consider the propagation of an initially bichromatic field interacting with an ensemble of effective two-state atoms. The coupling configuration is shown in figure 1. It is assumed that the field components are nearly resonant with the atomic transition. Furthermore, the fields are pulsed, with pulse length much shorter than the energy and coherence relaxation times of the atoms. Incoherent line broadening is not taken into account. Under these circumstances, the time evolution of the atoms is described by the Schrödinger equation

$$i\hbar\frac{\partial|\psi\rangle}{\partial t} = \mathsf{H}|\psi\rangle\,, \tag{1}$$

where the Hamiltonian using the rotating wave approximation with respect to the atomic transition frequency $\omega_{eg} = (E_e - E_g)/\hbar$, is given by

$$\mathsf{H} = \hbar\omega_{eg}|e\rangle\langle e| - \frac{1}{2}(d_{eg}E^{(+)}(t,z)|e\rangle\langle g| + \text{H.c.})\,, \tag{2}$$

where H.c. denotes the Hermitian-conjugate of the previous term, $d_{eg} = \langle e|d|g\rangle$ is the matrix element of the dipole operator between the excited and ground states, and $E^{(+)}(t,z)$ denotes the positive-frequency part of the electric component of the applied oscillating electromagnetic field, $E^{(-)} = (E^{(+)})^*$. Initially, the field consists of two frequency components

$$E^{(+)}(t,z) = \sum_{l=-1,0} \mathcal{E}_l(t,z)e^{i(k_l z - \omega_l t)}\,, \tag{3}$$

where ω_l and $k_l = \omega_l/c_n$ are the carrier frequency and the wave number of the field components, respectively. Here c_n is the speed of light in the medium $c_n = c_{\text{vac}}/n$, with n being the refractive

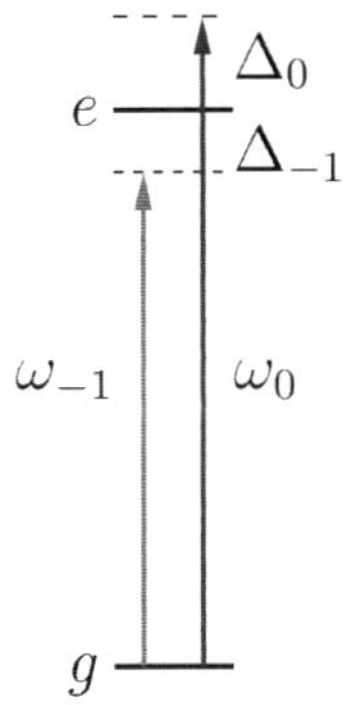

Figure 1. The interaction scheme between the bichromatic field and the atoms.

index of the host medium. We ignore the small variation of n which results from dispersion. In vacuum $n = 1$, whereas in a crystal n is the refractive index of the crystal without the atoms participating in the RNLO process. The slowly varying envelope of the field components is denoted by $\mathcal{E}_l(t,z)$.

The field is fed at $z = 0$ into the medium. Then it propagates a distance Z, while it interacts nearly resonantly with the atoms comprising the medium. The wave propagation is described by the equation

$$\sum_l k_l \left(\frac{\partial}{\partial z} + \frac{1}{c_n} \frac{\partial}{\partial t} \right) \mathcal{E}_l(t,z) e^{i(k_l z - \omega_l t)} = -i \frac{\mu_0}{2} \frac{\partial^2}{\partial t^2} P^{(+)}(t,z) , \tag{4}$$

which is valid in the slowly varying envelope approximation [28], i.e. $|\partial^2 \mathcal{E}_l(z,t)/\partial z^2| \ll |k_l \partial \mathcal{E}_l(z,t)/\partial z|$ and $|\partial^2 \mathcal{E}_l(z,t)/\partial t^2| \ll |\omega_l \partial \mathcal{E}_l(z,t)/\partial t|$. Here and in the following we exploit the relation $[\partial/\partial z + (1/c_n)\partial/\partial t] \exp(i[k_l z - \omega_l t]) = 0$. The positive frequency part $P^{(+)}(t,z)$ of the microscopic polarization is given by

$$P^{(+)}(t,z) = 2\mathcal{N} d_{ge} \varrho_{eg}(t,z) , \tag{5}$$

where $\mathcal{N}$ is the density of the atoms participating in the process, $\varrho_{eg}(t,z)$ is the coherence between the excited and ground states of the atoms.

We assume that the polarization of the medium can be expanded into a harmonic series

$$P^{(+)}(t,z) = \sum_l \mathcal{P}_l(t,z) e^{i(k_l z - \omega_l t)} . \tag{6}$$

Then the time-derivative in the r.h.s. of equation (4) can be easily evaluated. In leading order in ω_l one finds

$$\sum_l k_l \left(\frac{\partial}{\partial z} + \frac{1}{c_n} \frac{\partial}{\partial t} \right) \mathcal{E}_l(t,z) e^{i(k_l z - \omega_l t)} = i \sum_l k_l^2 \frac{\mathcal{P}_l(t,z) e^{i(k_l z - \omega_l t)}}{2\varepsilon_0 \varepsilon_r} , \tag{7}$$

where $\varepsilon_r = n^2$ is the relative permittivity. In order to obtain a simpler expression we introduce the frequency differences Δ_l through the relation

$$\omega_l = \omega - \Delta_l , \tag{8}$$

Journal of Physics: Conference Series **268** (2011) 012013 doi:10.1088/1742-6596/268/1/012013

where ω is a reference frequency, its choice depends on convenience. Let's assume that $\Delta k_l = \Delta_l / c_n$ is several orders of magnitude smaller than $k = \omega / c_n$. Then k_l can be replaced by k in the sums of equation (7), but in the exponents we keep k_l. Multiplying both sides of the equation by $e^{i(\omega t - kz)}$ one obtains

$$\sum_l \left(\frac{\partial}{\partial z} + \frac{1}{c_n} \frac{\partial}{\partial t} \right) \mathcal{E}_l(t,z) e^{i\Delta_l(t - z/c_n)} = ik \sum_l \frac{\mathcal{P}_l(t,z) e^{i\Delta_l(t-z/c_n)}}{2\varepsilon_0 \varepsilon_r} . \tag{9}$$

In this equation we can identify the slowly varying electric field $\mathcal{E}(t,z)$ and polarization $\mathcal{P}(t,z)$

$$\mathcal{E}(t,z) = \sum_l \mathcal{E}_l(t,z) e^{i\Delta_l(t-z/c_n)} , \tag{10a}$$

$$\mathcal{P}(t,z) = \sum_l \mathcal{P}_l(t,z) e^{i\Delta_l(t-z/c_n)} . \tag{10b}$$

In terms of these quantities equation (9) takes the form

$$\left(\frac{\partial}{\partial z} + \frac{1}{c_n} \frac{\partial}{\partial t} \right) \mathcal{E}(t,z) = ik \frac{\mathcal{P}(t,z)}{2\varepsilon_0 \varepsilon_r} , \tag{11}$$

as in the single mode case.

We define a rotating frame for the atomic systems by the unitary transformation

$$\mathsf{U} = \exp(-i(\omega t - kz)|e\rangle\langle e|) . \tag{12}$$

By means of U we transform the Shrödinger equation (1) into the rotating wave picture

$$i\hbar \frac{\partial |\psi_R\rangle}{\partial t} = \mathsf{H}_R |\psi_R\rangle , \tag{13}$$

where $|\psi_R\rangle = \mathsf{U}^\dagger |\psi\rangle$, and the rotating wave Hamiltonian is given by

$$\mathsf{H}_R = i \frac{\partial \mathsf{U}^\dagger}{\partial t} \mathsf{U} + \mathsf{U}^\dagger \mathsf{H} \mathsf{U} = \hbar \Delta |e\rangle\langle e| - \frac{1}{2}(d_{eg} \mathcal{E}(t,z)|e\rangle\langle g| + \text{H.c.}) , \tag{14}$$

where $\Delta = \omega_{eg} - \omega$.

Now our task is to solve the Maxwell-Schrödinger equations (11) and (13) simultaneously. It is assumed that the atoms are initially prepared in the ground state $|g\rangle$, the bichromatic field is fed into the medium at $z = 0$, and the propagation takes place in the domain $[0, Z]$.

3. Floquet analysis of the atomic state evolution

It turns to be convenient to introduce a retarded frame $\tau = t - z/c_n$ and $\zeta = z$. Then in equations (10) the exponents are transformed as $\exp(i\Delta_l[t - z/c_n]) \to \exp(i[\Delta_l \tau + \theta_l])$, where θ_l defines an initial phase. Hence for frozen field amplitudes the Hamiltonian in equation (14) is time periodic with two frequencies, the two field-components has a beat frequency (frequency difference) $\delta = \omega_0 - \omega_{-1} = \Delta_{-1} - \Delta_0$. Later we will see that useful results can be obtained when the beat frequency has the same magnitude as the Rabi frequencies of the input fields, $\delta \approx \max_t\{|\Omega_l(t,z)|\} \ll \omega_{eg}$, $(l = 0, -1)$, where the Rabi frequencies are defined as $\Omega_l(t,z) = -d_{eg}\mathcal{E}_l(t,z)/\hbar$. In such a case it is natural to introduce the strong-field dressed Hamiltonian (the so called Floquet or quasienergy Hamiltonian) [29]. Following the derivation in [30], we treat the initial phases θ_l as dynamical variables acting on the Hilbert space of 2π

periodic functions $\mathcal{L}_2(\theta_l/2\pi)$. In the rotating wave approximation, an effective reduced Hilbert space $\mathcal{K}_{\text{eff}}$ can be defined as a tensor product of the Hilbert space of the atoms $\mathcal{H}_{\text{a}} \equiv \mathbb{C}^2$ and the Hilbert space associated with the classical fields, $\mathcal{K}_{\text{eff}} = \mathcal{H}_{\text{a}} \otimes \mathcal{L}_2(\eta/2\pi)$, where $\eta = \theta_0 - \theta_{-1}$. The effective Floquet Hamiltonian K_R acting on $\mathcal{K}_{\text{eff}}$ is given by

$$\mathsf{K}_R = -i\hbar\delta\frac{\partial}{\partial\eta} + \frac{\hbar}{2}\begin{bmatrix} 0 & \sum_{l=-1,0}\Omega_l^* e^{il\eta} \\ \sum_{l=-1,0}\Omega_l e^{-il\eta} & 2\Delta \end{bmatrix}, \tag{15}$$

where we choose $\omega = \omega_0$ for the reference frequency, corresponding to $\Delta = \Delta_0$. This effective model is valid only if the two frequencies ω_0 and ω_{-1} are different. The derivative term represents the relative number of photon pairs, one ω_0 photon minus one ω_{-1} photon. Thus the absorption of one "effective photon" of frequency δ corresponds to the absorption of one photon of frequency ω_0 and emission of an other photon of frequency ω_{-1}.

The basic properties of the effective Hamiltonian of equation (15) can be understood by determining numerically its quasienergy (eigenenergy) surfaces as a function of Ω_{-1} and Ω_0. A section of the hierarchy of the computed quasienergy surfaces is shown in figure 2. There are two families of periodic eigenvalues and eigenvectors for frozen Ω_{-1} and Ω_0: $\lambda_{n;k,-k}(\vec{\Omega}) = \lambda_{n;0,0}(\vec{\Omega}) + \hbar k\delta$, and $|n;\vec{\Omega},k,-k\rangle = |n;\vec{\Omega},0,0\rangle\exp(ik\eta)$, $(n = e,g)$ with $k \in \mathbb{Z}$, and $\vec{\Omega} = [\Omega_{-1},\Omega_0]$. The label $n = e,g$ indicates that the dressed eigenstate families converge to either the state $|g;\vec{0},k,-k\rangle \equiv |g\rangle\exp(ik\eta)$ or $|e;\vec{0},k,-k\rangle \equiv |e\rangle\exp(ik\eta)$ in the limit of $\vec{\Omega} \to \vec{0}$. Any two neighboring surfaces have points of contact that are conical intersections. All intersections are located either along the line $\Omega_{-1} = 0$ or $\Omega_0 = 0$, corresponding to the situation where only one of the laser fields is interacting with the atom.

For time dependent Rabi-frequencies $\vec{\Omega}(t)$ the system evolves along a trajectory in the quasienergy space. Let us assume that the system is prepared initially in the state $|g;\vec{0},0,0\rangle$. If $\vec{\Omega}(t)$ is a smooth, slowly varying function of time, then the system evolves adiabatically, which is manifested in the quasienergy space as a trajectory which lays in the quasienergy surfaces. For sufficiently large amplitude of one of the components of $\vec{\Omega}(t)$ while the other one is kept zero, the trajectory will cross the intersections of the quasienergy surfaces. In figure 2, we draw a trajectory which results from the interaction with two delayed, sine-square shaped pulses. The peak Rabi frequencies are so large that the system crosses two intersections and evolves to the state $|g;\vec{0},2,-2\rangle$. Using the Landau-Zener analysis it can be shown that the intersections of the quasienergy surfaces are true crossings, i.e. in the adiabatic limit the system evolves from one quasienergy surface to the other one. As a result, the system may traverse several quasienergy surfaces leading to the absorption/emission of several "effective photons".

If adiabaticity is violated in the course of the dynamics then the trajectory in the quasienergy space will not lay in the quasienergy surfaces. Consequently, the state vector of the system is not a single eigenstate of the effective Hamiltonian, rather a superposition of several eigenstates.

4. Propagation of the bichromatic field

In order to consider the propagation of an incoming bichromatic pulse in a medium consisting of an ensemble of two-state atoms, we have to evaluate the induced polarization of the atoms defined in equation (5). We have shown in the previous section that when the interaction between a bichromatic light pulse and the atoms is in the strong-field regime, the atomic dynamics can be described conveniently in an effective Floquet space. The position-dependence of the atomic excitation is taken into account by assigning position-dependence to the atomic state vector. The state vector in the retarded frame is defined as

$$|\psi(\tau,\zeta)\rangle_R = \sum_k (c_{g;k}(\tau,\zeta)|g\rangle + c_{e;k}(\tau,\zeta)|e\rangle)e^{ik\eta}, \tag{16}$$

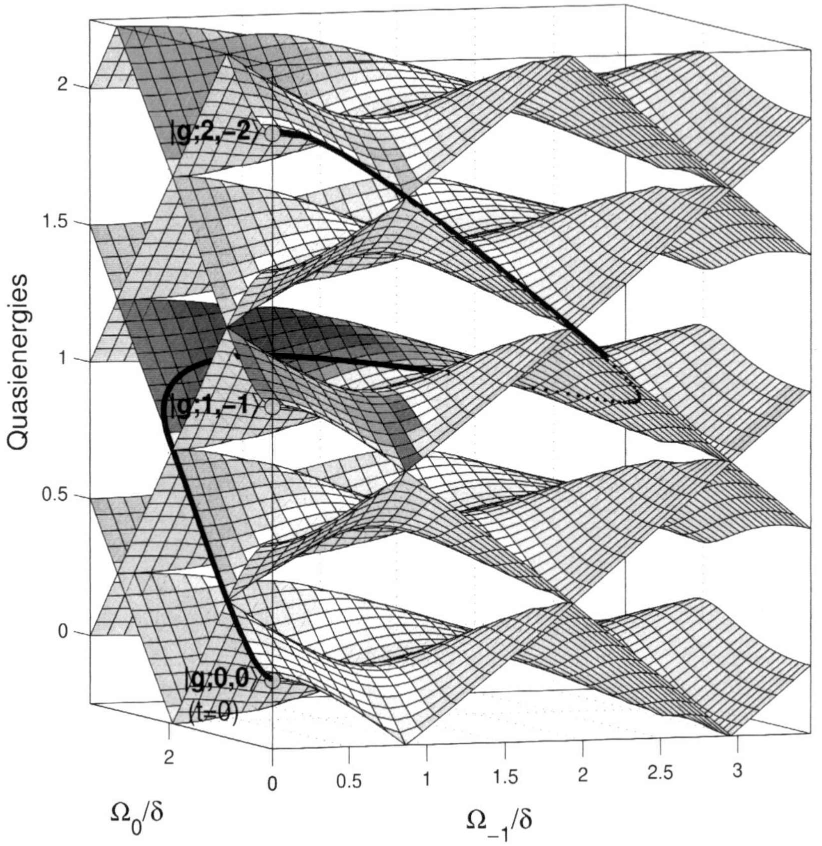

Figure 2. (color online) Quasienergy surfaces of the effective Floquet-Hamiltonian. The trajectory starts from the state $|g;\vec{0},0,0\rangle$ and terminates in the state $|g;\vec{0},2,-2\rangle$.

in the Floquet space, where the state vector is expanded in the basis $\{|n\rangle \exp(ik\eta)\}_{n=g,e;\,k\in\mathbb{Z}}$. It is transformed back to the atomic Hilbert space by replacing the variable η with $\delta t + \eta_0$, the resulting state is denoted by $|\psi(\tau,\zeta)\rangle$. Then the required $\varrho_{eg}(\tau,\zeta)$ is given by

$$\varrho_{eg}(\tau,\zeta) = \langle e|\psi(\tau,\zeta)\rangle\langle\psi(\tau,\zeta)|g\rangle = \sum_l \left(\sum_k c_{e;k}(\tau,\zeta) c_{g;k+l}(\tau,\zeta)^*\right) e^{-il\delta\tau} \equiv \sum_l \varrho_{eg}^{(l)}(\tau,\zeta)\, e^{-il\delta\tau}. \tag{17}$$

For smooth probability amplitudes $c_{e;k}(\tau,\zeta)$ and $c_{g;l}(\tau,\zeta)$, the coherence $\varrho_{eg}(\tau,\zeta)$ has discrete frequency components with amplitudes $\varrho_{eg}^{(l)}(\tau,\zeta)$, that we call partial coherences, which are separated by δ. The probability amplitudes are smooth if the characteristic time of their variation is much longer than δ^{-1}. Inserting equation (17) into equation (5), we can observe that the polarization induced by a bichromatic field may involve several higher harmonics of the fundamental frequency δ. In fact, the partial coherence $\varrho_{eg}^{(l)}(\tau,\zeta)$ in equation (17) is proportional to the polarization component $\mathcal{P}_l$ of equation (10b). Therefore, in the course of the propagation of the bichromatic field, new field components may be generated, which are higher harmonics of the fundamental frequency δ. The generated field is of the form of equation (10a), where the summation for l goes from minus infinity to plus infinity in principle. Then we have to change the effective Floquet Hamiltonian in equation (15) since due to propagation, there emerge several field components with frequency separation of δ, which interact with the atoms.

Let us consider a multimode field given by

$$\mathcal{E}(\underline{\theta}+\underline{\omega}t) = \sum_{l=-L}^{L} \mathcal{E}_l \cos(\omega_l t + \theta_l)\,, \tag{18}$$

where $\omega_l = \omega_0 + l\delta$, $(l \in [-L, L])$, the limits for the summation $-L, L$ are chosen for later convenience, they can be arbitrary. The analogous Hamiltonian to the one in equation (2) for a two-level system interacting with the multimode field of equation (18), reads

$$\mathsf{H}(\underline{\theta}+\underline{\omega}t) = \begin{bmatrix} 0 & V(\underline{\theta}+\underline{\omega}t)^* \\ V(\underline{\theta}+\underline{\omega}t) & \hbar\omega_{eg} \end{bmatrix}, \tag{19}$$

where

$$V(\underline{\theta}+\underline{\omega}t) = -d_{eg}\mathcal{E}(\underline{\theta}+\underline{\omega}t)\,, \tag{20}$$

where the rotating wave approximation has not been applied yet. The Floquet Hamiltonian associated with the Hamiltonian of equation (19) is given by

$$\mathsf{K} = -i\hbar\underline{\omega}\frac{\partial}{\partial\underline{\theta}} + \begin{bmatrix} 0 & V(\underline{\theta})^* \\ V(\underline{\theta}) & \hbar\omega_{eg} \end{bmatrix}. \tag{21}$$

We define a rotating wave coordinate system by means of the unitary matrix

$$\mathsf{R} = \begin{bmatrix} 1 & 0 \\ 0 & e^{-i\theta_0} \end{bmatrix}. \tag{22}$$

Then the transformed Floquet Hamiltonian $R^{\dagger}K(\underline{\theta})R$ takes the form

$$\mathsf{K}_R = -i\hbar\underline{\omega}\frac{\partial}{\partial\underline{\theta}} + \begin{bmatrix} 0 & V(\underline{\theta})^* e^{-i\theta_0} \\ V(\underline{\theta})e^{i\theta_0} & \hbar(\omega_{eg}-\omega_0) \end{bmatrix}. \tag{23}$$

The coupling terms are given by

$$V(\underline{\theta})e^{i\theta_0} = \frac{1}{2}d_{eq}\sum_{l=-L}^{L}\mathcal{E}_l\left(e^{i(\theta_l+\theta_0)} + e^{i\eta_l}\right), \tag{24}$$

where $\eta_l = \theta_0 - \theta_l$. In terms of the variables $\{\eta_l\}$ the rotating wave multimode Floquet Hamiltonian equation (23) is given by

$$\mathsf{K}_R \;=\; -i\hbar\omega_0\frac{\partial}{\partial\eta_a} - i\hbar\delta\sum_l(-l)\frac{\partial}{\partial\eta_l} + \frac{\hbar}{2}\begin{bmatrix} 0 & \sum\Omega_l^* e^{-i\eta_l} \\ \sum\Omega_l e^{i\eta_l} & 2\Delta \end{bmatrix},$$

where $\eta_a = \theta_0$, $\Omega_l = -d_{eg}\mathcal{E}_l/\hbar$, and $\Delta = \omega_{eg} - \omega_0$. The first derivative term commutes with all the other terms, producing only a global phase and can thus be omitted. The second one looks like a directional derivative: the gradient $\partial_{\vec{\eta}}$ is taken in the direction $-\vec{l}$. We can define the new variable η through the relation

$$\frac{\partial}{\partial\eta} = -\sum_l l\frac{\partial}{\partial\eta_l}\,. \tag{25}$$

The derivative of η_l with respect to η is given by

$$\frac{\partial}{\partial \eta}\eta_l = -\sum_j j \frac{\partial}{\partial \eta_j}\eta_l = -l\,. \tag{26}$$

It follows that

$$\eta_l = -l \cdot \eta + c_l\,, \tag{27}$$

where c_l are irrelevant constants, hence can be omitted. Then the rotating wave Floquet Hamiltonian in the RWA reads

$$\mathsf{K}_R = -i\hbar\delta\frac{\partial}{\partial \eta} + \frac{\hbar}{2}\begin{bmatrix} 0 & \sum \Omega_l^* e^{il\eta} \\ \sum \Omega_l e^{-il\eta} & 2\Delta \end{bmatrix}. \tag{28}$$

This Hamiltonian has exactly the same form as that of equation (15), the only difference is that now the summation goes from $-L$ to L. The state vector of equation (16) is equivalent with $\vec{c}(\tau,\zeta) = [c_g(\tau,\zeta), c_e(\tau,\zeta)]^T$ in the basis $\{|n\rangle \exp(ik\eta)\}_{n=g,e;\, k\in\mathbb{Z}}$. The equation of motion for $\vec{c}(\tau,\zeta)$ is given by

$$i\hbar\frac{\partial}{\partial \tau}\vec{c}(\tau,\zeta) = \mathsf{K}(\tau,\zeta)\vec{c}(\tau,\zeta)\,, \tag{29}$$

where $\mathsf{K}(\tau,\zeta)$ is the matrix representation of $K_R(\eta)$ in equation (28).

It follows from equations (5), (6), (10b), and (17) that the polarization components $\mathcal{P}_l(\tau,\zeta)$ are given by

$$\mathcal{P}_l(\tau,\zeta) = 2\mathcal{N} d_{ge}\varrho_{eg}^{(l)}(\tau,\zeta)\,. \tag{30}$$

Using this result, the Maxwell equation (11) can be decomposed as

$$\frac{\partial}{\partial \zeta}\Omega_l(\tau,\zeta) = -i\,\alpha\,\varrho_{eg}^{(l)}(\tau,\zeta)\,, \tag{31}$$

where $\alpha = \mathcal{N}k|d_{eg}|^2/\hbar\varepsilon_0\varepsilon_r$. The propagation of a bichromatic pulse is obtained by solving simultaneously the equations (29) and (31). If the partial coherences $\varrho_{eg}^{(l)}(\tau,\zeta)$ on the rhs of equation (31) has an appreciable value for $l \neq -1, 0$, then new field modes will be generated. Initially the atoms are in the ground state $c_{g,0}(0,\zeta) = 1$, all other amplitudes $c_{q,k}(0,\zeta)$ are zero. Therefore, for a large value of the partial coherence $\varrho_{eg}^{(l)}(\tau,\zeta)$ it is necessary that the products $c_{e,k}(\tau,\zeta)c_{g,k-l}(\tau,\zeta)$, $l \neq 0$ become non-zero, and interfere constructively. Our purpose is to show that it is possible to generate new field modes in this setup, we made several numerical experiments to point this out.

Initially only the Rabi frequencies $\Omega_l(\tau,0)$ associated with the field modes $\mathcal{E}_{-1}(\tau,0)$ and $\mathcal{E}_0(\tau,0)$ are non-zero. At the input of the medium the envelope of the Rabi frequencies are chosen as

$$\Omega_{-1}(\tau,0) = \Omega_{-1}\sin^2\left(\frac{\pi(\tau-\tau_{12})}{T}\right), \tag{32a}$$

$$\Omega_0(\tau,0) = \Omega_0\sin^2\left(\frac{\pi\tau}{T}\right), \tag{32b}$$

where τ_{12} describes a time delay between the two peaks of the two frequency components. In our numerical examples we choose the parameters of the praseodymium doped crystal $Y_2SiO_5{:}Pr^{3+}$. The dipole moment for the strongest transition ${}^3H_4(\pm\frac{5}{2}) \leftrightarrow {}^1D_2(\pm\frac{5}{2})$ of the Pr^{3+} ions is $d_{eg} = 2.6\times10^{-26}\,C\,\mu m$. The population relaxation time is $T_1 = 165\mu s$, whereas the coherence relaxation time is $T_2 = 111\mu$s, [31]. We ignore hyperfine splitting and inhomogeneous broadening in our model calculations, and treat the Pr^{3+} ions as ideal two-state systems. In figure 2 we used the pulse-shapes of equation (32) to compute the trajectory. In order to have two crossings in the course of the evolution, we have chosen $\Omega_l \equiv \max_t\{\Omega_l(t)\} = 2.5\,\delta$, $(l = -1, 0)$ for the input fields. As for the other parameters, we have used $\mathcal{N} = 5 \times 10^{18}\mathrm{cm}^{-3}$, $\lambda_{\mathrm{vac}} = 0.605\mu$m, and $n = 1.8$, $\Delta = -\delta/2$.

To determine the possible range of the peak input field Rabi frequencies Ω_l, $(l = -1, 0)$, and δ we have to take into account the following factors: Ω_l should be a few times larger than δ in order to have some crossings in the quasienergy space. The upper bound of Ω_l ($\mathcal{E}_l$) is determined by the damage threshold of the medium and the available laser power. We use $\Omega_l = 25$MHz $(l = -1, 0)$ in our model calculations, which are achievable values in Y_2SiO_5:Pr^{3+}. We select $\delta = 10$MHz for the beat-frequency. The pulse-length T should be so large that the product $N = \delta \times T$ is of the order of a few hundreds in order to have adiabatic evolution [30]. Adiabaticity ensures that the coefficients $c_{q,k}(\tau, \zeta)$ do not oscillate more rapidly than δ. On the other hand we want to have a coherent evolution. This requires that the pulse-length should be much shorter than the T_2 time of the transition under consideration, $T \ll T_2$. From these considerations it follows that a coherent evolution requires that the relation $N \ll \delta \times T_2$ should be fulfilled. For the parameters of our system we find that N could be a few hundreds, in agreement with the condition for adiabaticity we have seen earlier. In practice it proved to be a reasonable choice for the time delay τ_{12} between the two components of the incoming bichromatic field the one-third of the sine-square pulse length, $\tau_{12} = T/3$. In the numerical simulations we chose $T = 21\mu$s, so that the total duration of the excitation process was 28μs at the input of the medium.

There is only one free parameter left: the length Z of the sample. In the numerical evaluation of equation (31) it is useful to introduce the phase-shift $\xi = \alpha\zeta$. In this way the equation becomes more universal. In turn, the length of the sample should also be scaled, hence the domain of integration is given by $\xi \in [0, \alpha Z]$. We conclude that the propagation of the bichromatic field leads to the same result if we vary together α and Z in such a way that $\alpha Z = \mathrm{const}$. For the parameters given above $\alpha = 2.08855 \times 10^5\,\mathrm{cm}^{-1}$. We choose $Z = 0.25\,\mathrm{cm}$ as the length of the sample.

We have solved numerically the pair of equations (29) and (31) using the parameters defined above. In figure 3 we plot the normalized Rabi frequencies $|\Omega_l(\tau, Z)/\Omega_0|$, $(l = -8 \ldots 7)$ as a function of the scaled time $\delta \times \tau$. It can be seen that the initial modes become significantly distorted and new modes are generated. The zeroth order mode is switched on at $\delta \times \tau = 0$. This pulse has suffered some delay. The fringes starting around $\delta \times \tau \approx 40$ results from Rabi oscillations between the states $|g; \vec{\Omega}, 0, -0\rangle$ and $|e; \vec{\Omega}, 0, -0\rangle$. The second incoming pulse is switched on at $\delta \times \tau = 70$. After this time instant, new field modes, Raman sidebands are generated as a result of the strong modulation of the coherence $\rho_{eg}(\tau, Z)$. The spectrum of the total outgoing field is shown in figure 4, whereas the time evolution of the coherence $\rho_{eg}(\tau, \zeta)$ is shown in figure 5. These latter two plots are obtained from a direct time domain simulation, i.e. by solving the equations (11) and (13) in the retarded frame. It can be seen that due to the reshaping of the pulses (see figure 3), a small amount of coherence is left in the system at $\zeta = Z$, which means that the atoms continue to radiate. The emitted field is damped, the damping rate is determined by the T_1 and T_2 times of the atoms.

5. Summary

We have studied the propagation of a pair of bichromatic pulses in a resonant medium. We have shown using the Floquet-technique, that if the two incoming pulses are delayed and the Rabi-frequencies are a few times larger than the beat frequency between the incoming modes, then new Raman sidebands will be generated in the course of the propagation. The coherence time of the atomic system limits the timescale of the process, which in turn determines the maximal

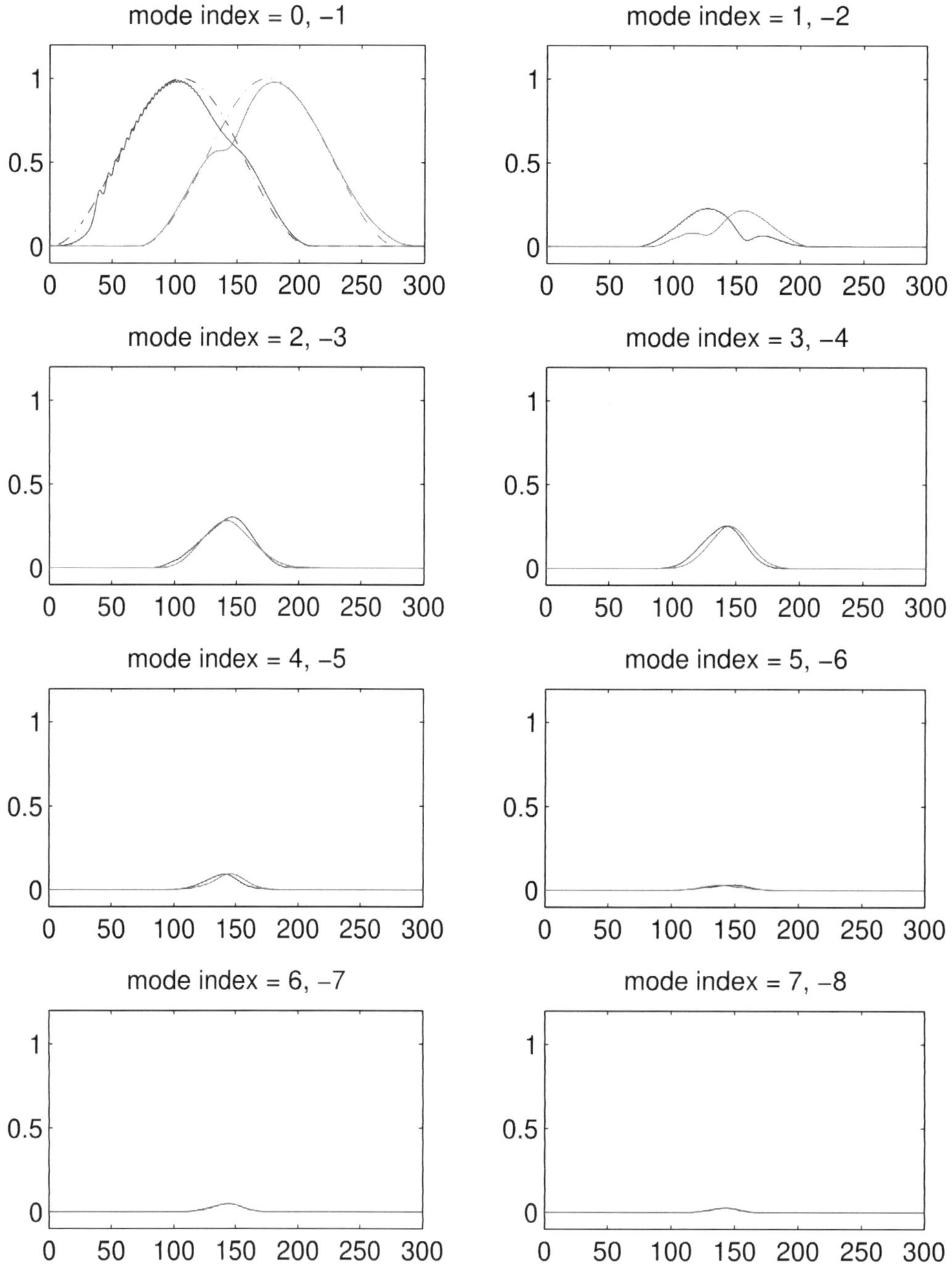

Figure 3. (color online) The time evolution of the normalized Rabi frequencies $|\Omega_l(\tau, Z)/\Omega_0|$, as a function of the scaled time $\delta \times \tau$. For reference, in the top-left panel we plot the initial envelopes of $\Omega_{-1}(\tau, 0)$ and $\Omega_0(\tau, 0)$ with dash-dot line.

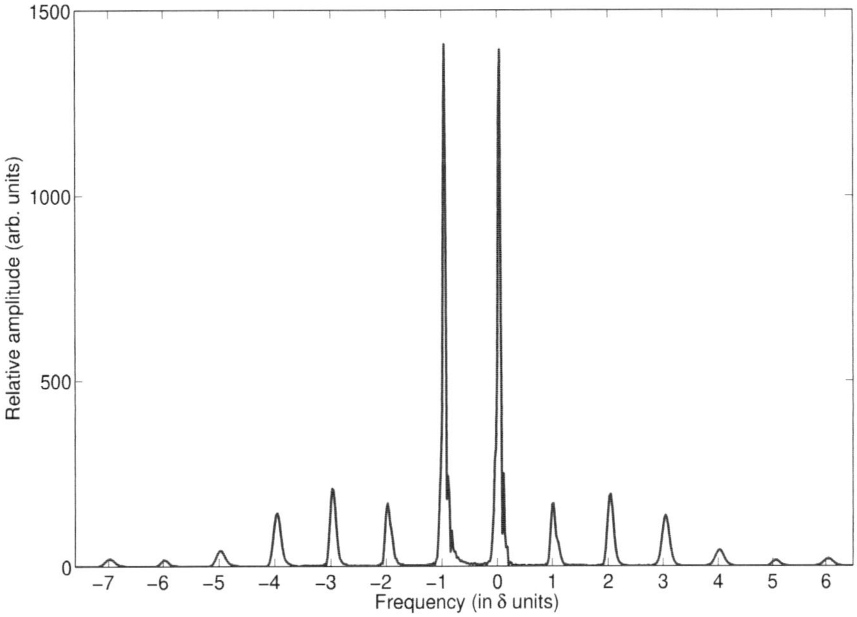

Figure 4. (color online) The frequency spectrum of the total field at the output of the medium.

beat frequency between the incoming pulses, $T_2\delta \gg T\delta =$ few hundred. In the numerical example we have chosen $\delta = 10\,\mathrm{MHz}$, however, the previous relation suggests that all results remain valid for scaled total time T/q and beat frequency $\delta \times q$. Therefore, if in a system the Rabi frequency can be as large as a few gigahertz, then the beat frequency can also be in the gigahertz range. As a result, the spacing between the generated Raman sidebands will be in the gigahertz domain.

Acknowledgments

We acknowledge the support from the French Agence Nationale de la Recherche (Project CoMoC) and from the European Marie Curie Initial Training Network Grant No. CA-ITN-214962-FASTQUAST. Z.K. is grateful for the hospitality at the University Bourgogne. Z.K. acknowledges the support of the Bolyai Program of the Hungarian Academy of Sciences.

References

[1] S.E. Harris, J.E. Field, and A. Imamoğlu 1990 *Phys. Rev. Lett.* **64** 1107
[2] K.-J. Boller, A. Imamoğlu, and S.E. Harris 1991 *Phys. Rev.* A **66** 2593
[3] M.M. Kash, V.A. Sautenkov, A.S. Zibrov, L. Hollberg, G.R. Welch, M.D. Lukin, Y. Rostovtsev, E.S. Fry, and M.O. Scully 1999 *Phys. Rev. Lett.* **82** 5229
[4] M. Jain, H. Xia, G.Y. Yin, A.J. Merriam, and S.E. Harris 1996 *Phys. Rev. Lett.* **77** 4326
[5] S.E. Harris and M. Jain 1997 *Opt. Lett.* **22** 636
[6] W. Harshawardhan and G.S. Agarwal 1998 *Phys. Rev.* A **58** 598
[7] L. Deng, M.G. Payne, and R.W. Garrett 1998 *Phys. Rev.* A **58**, 707
[8] E. Paspalakis, N.J. Kylstra, and P.L. Knight 2002 Phys. Rev. A **65** 053808
[9] E. Paspalakis and Z. Kis 2002 *Phys. Rev.* A **66** 025802
[10] E. Paspalakis and Z. Kis 2002 *Optics Letters* **27** 1836
[11] Z. Kis and E. Paspalakis 2003 *Phys. Rev.* A **68** 043817

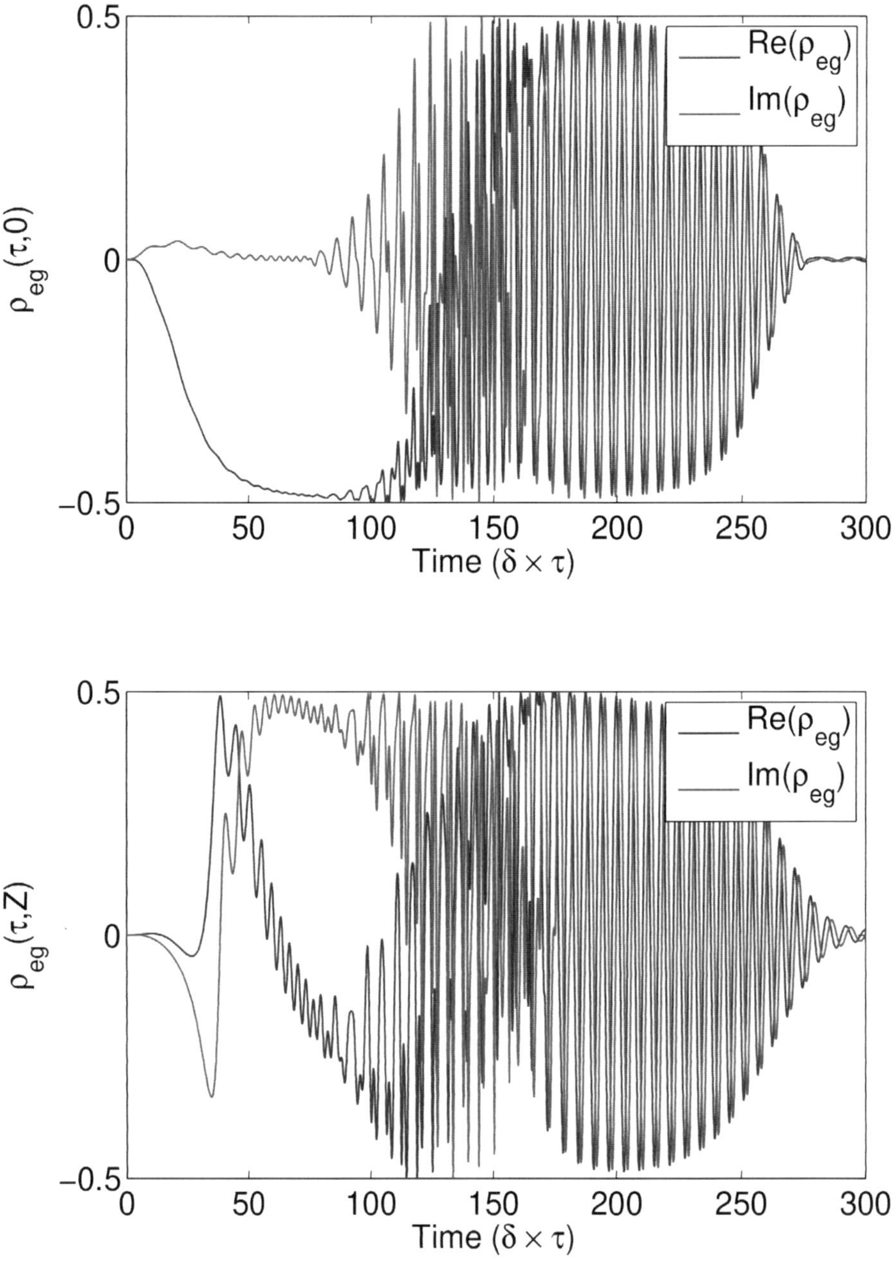

Figure 5. (color online) The time evolution of the atomic coherence $\varrho_{eg}(\tau, \zeta)$ as a function of the scaled time $\delta \times \tau$. In the upper panel $\zeta = 0$, whereas in the lower one $\zeta = Z$.

[12] V.A. Sautenkov, C.Y. Ye, Y.V. Rostovtsev, G.R. Welch, and M.O. Scully 2004 *Phys. Rev.* A **70** 033406
[13] A. Eilam, A.D. Wilson-Gordon, and H. Friedmann 2006 *Phys. Rev.* A **73** 053805
[14] A. Eilam, A.D. Wilson-Gordon, and H. Friedmann 2007 *Opt. Commun.* **277** 186
[15] M.D. Lukin, P.R. Hemmer, M. Löffler, and M.O. Scully 1998 *Phys. Rev. Lett.* **81** 2675
[16] E.A. Korsunsky and D.V. Kosachiov 1999 *Phys. Rev.* A **60** 4996
[17] M. Fleischhauer, A. Imamoğlu, and J.P. Marangos 2005 *Rev. Mod. Phys.* **77** 633
[18] J.J. Longdell, E. Fraval, M.J. Sellars, and N.B. Manson 2005 *Phys. Rev. Lett.* **95** 063601
[19] K. Ichimura, K. Yamamoto, and N. Gemma 1998 *Phys. Rev.* A **58** 4116
[20] A. Louchet, J.S. Habib, F. Bretenaker, F. Goldfarb, I. Lorgeré, and J.-L. Le Gouët 2007 *J. Luminescence* **127** 89
[21] A.V. Turukhin, V.S. Sudarshanam, M.S. Shahriar, J.A. Musser, B.S. Ham, and P.R. Hemmer 2002 *Phys. Rev Lett.* **88** 023602
[22] E. Kuznetsova, O. Kocharovskaya, P. Hemmer, and M. O. Scully 2005 *Phys. Rev.* A **66** 063802
[23] Q. Sun, Y.V. Rostovtsev, J.P. Dowling, M.O. Scully, and M.S. Zubairy 2005 *Phys. Rev.* A **72** 031802(R)
[24] S.A. Moiseev, V.F. Tarasov, and B.S. Ham 2003 *J. Opt. B: Quantum Semiclass. Opt.* **5** S497
[25] S.E. Harris and A.V. Sokolov 1997 *Phys. Rev.* A **55** R4019
[26] S.E. Harris and A.V. Sokolov 1998 *Phys. Rev. Lett.* **81** 2894
[27] A.V. Sokolov, D.R. Walker, D.D. Yavuz, G.Y. Yin, and S.E. Harris 2000 *Phys. Rev. Lett.* **85** 562
[28] L. Allen and J.H. Eberly 1987 *Optical Resonance and Two-Level Atoms* (New York: Dower Publications, INC.)
[29] J.H. Shirley 1965 *Phys. Rev.* **138** B979
[30] S. Guérin, L.P. Yatsenko, and H.R. Jauslin 2001 *Phys. Rev.* A **63** R031403
[31] R.W. Equall, R.L. Cone, and R.M. Macfarlane 1995 *Phys. Rev.* B **52** 3963

Journal of Physics: Conference Series **268** (2011) 012014 doi:10.1088/1742-6596/268/1/012014

Hysteresis in flow through porous media

P Kordulová

Mathematical Institute, Silesian University in Opava, 746 01 Opava, Czech Republic

E-mail: Petra.Kordulova@math.slu.cz

Abstract. Porous media flow is described by a class of PDEs whose model equation is represented by the Richards equation with soil-moisture hysteresis term. We assume that the hysteresis is represented by the Preisach hysteresis operator. Under suitable assumptions on the hysteresis operator we investigate the existence of the solution of our problem.

1. Introduction

This paper deals with hysteresis in the unsaturated case of the diffusive form of Richards equation. The equation is coupled with initial-boundary data. We investigate the existence of solution of the problem coupled with Preisach operator under the time derivative under appropriate conditions.

The important role of hysteresis has been observed in hydrology and soil physics for a long time. A soil is a collection of grains forming the soil matrix, with interconnected pore spaces forming the channels available for water transport. Water provided by rainfalls or irrigations moves through unsaturated soil, gets into pores and passes through infiltration, drainage, evaporation and absorption by plant roots. The pores in unsaturated soil are filled partly with water and partly with air. The cycle of soil wetting-drying process exhibits hysteresis in porous media, roughly explained by the fact that the volumetric water content has different profiles with respect to the wetting and drying processes. The soil water hysteresis is relation between the volumetric water content and the total potential.

Seventy-nine years ago, Haines [1] observed the presence of hysteresis in the relationship between the water content in the soil and the water potential. For further information about the history and descriptions of soil-moisture hysteresis, see [2, 3, 4, 5]. We refer to Poulovassilis and Childs [6], Mualem [7] and the further references therein for experimental studies of the hysteresis relation.

A model equation with no hysteretic relation between the volumetric water content and the total potential, has been studied by Alt, Luckhaus and Visintin [8], Otto [9]. We refer to the monographs of Brokate and Sprekels [10], Krasnosel'skii and Pokrovskii [11], Krejčí [12], Mayergoyz [13] and Visintin [14] for details about the hysteresis theory. Various studies on the porous media flow with hysteretic behaviour have been published up to now and we mention here the papers [15, 16, 17, 18] and the references given there. In [15] Richards equation is coupled with memory effect in the constitutive law consisting of hysteresis operator of play type and a rate dependent component. The existence result applied also for a Preisach hysteresis operator is presented. Paper [16] describes nonlinear diffusion equation with hysteresis coupled with boundary conditions of Neumann and Signorini type. Existence and asymptotic stability of a solution for the initial-boundary-value problem are established. In paper [17] the problem of filtration model with hysteresis is formulated in the form of a quasi-variational inequality and a time global solution is proved. In paper [18], well-posedness of the hysteretic model coupled with the potential form of Richards equation with hysteretic hydraulic laws for a wetting-drying cycles is shown. The abstract semigroup theory is used.

The paper is organized as follows. In Section 2 we briefly introduce the concept of model equation. In Section 3, we recall some basic definitions and properties about play and Preisach hysteresis operators. In Section 4 the main result is postulated. The proof of the existence result is led through three steps: approximation, a priori estimates and limit procedure.

2. Model formulation

We assume that the porous medium is rigid, homogeneous and isotropic, that the fluid is inviscid and incompressible.

Richards equation that describes the dynamic flow in an unsaturated zone was deduced by combining generalised Darcy's law for the flow

$$q = -k \operatorname{grad} \psi$$

with equation of continuity

$$\frac{\partial \theta}{\partial t} = -\nabla \cdot q,$$

where q is the flux density, k is the hydraulic conductivity, ψ is the total potential and θ is the volumetric moisture content. The potential comprises several component potentials. The most important are the matric potential ψ_m and the gravitational potential ψ_g. Hence, $\psi = \psi_m + \psi_g, \quad \psi \leq 0$. The gravitational potential ψ_g is negative when the position coordinate is taken as positive downwards from the surface, i.e., $\psi_g = -gz$.

Hence, Richards equation representing transient water flow is

$$\frac{\partial \theta}{\partial t} = \nabla \cdot (k \nabla \psi_m - kg). \tag{1}$$

θ_s is the volumetric water content at natural saturation and θ is the volumetric moisture content of the medium. When the soil matrix is perfectly dryed then $\theta = 0$, when the matrix is fully saturated with water, then $\theta = \theta_s < 1$ and finally for in-between states the matrix contains both air and water. Hence θ is bounded between 0 and θ_s, i.e., $0 \leq \theta \leq \theta_s < 1$.

The constitutive relation between the volumetric water content θ and the matric potential ψ_m is typically represented by a relation of the form

$$\theta(x,t) \in h(\psi_m)(x,t),$$

where $h : \mathbb{R} \to [0,1]$ is a maximal monotone graph as in figure 1.

The value θ_r represents a residual water content, the quantity that remains in a soil after any drainage imposed by the gravitational forced has ceased.

Here we assume hysteretic relation between volumetric water content θ and the matric potential ψ_m models by the Preisach operator, i.e., $\theta(x,t) = \mathcal{W}(\psi_m)(x,t)$, see figure 2.

Heuristically speaking, since θ is a hysteretic function of ψ_m, it follows that the hysteretic behaviour will be involved in all functions depending on θ.

In our case we assume that the hydraulic conductivity depends nonlinearly on matric potential. Moreover, the hydraulic conductivity is single-valued, positive, twice differentiable,

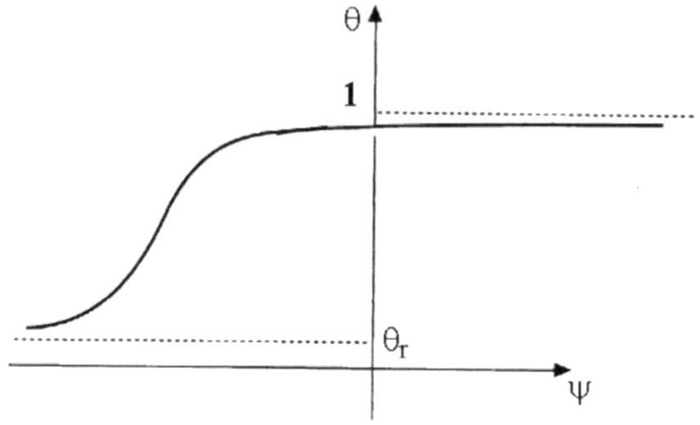

Figure 1. Water content versus matric potential constitutive relation without hysteresis

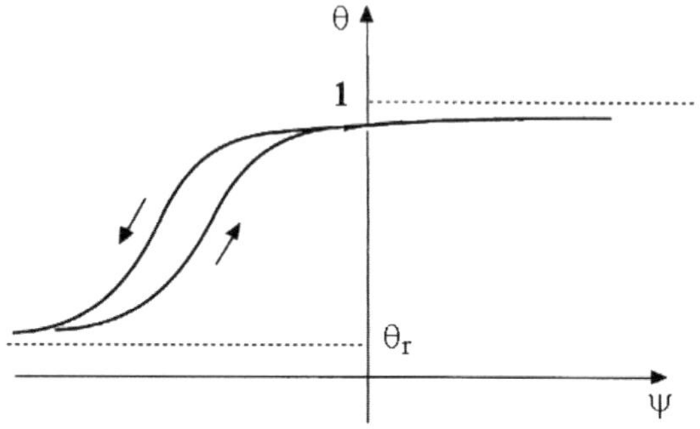

Figure 2. Water content versus matric potential constitutive relation with hysteresis

monotonically increasing function of matric potential.
Here we can apply the *Kirchhoff transformation*:

$$\mathcal{K} : \psi_m \mapsto u := \int_0^{\psi_m} k(s)\mathrm{d}s.$$

Since $k(s)$ is positive, this transformation can be inverted and equation (1) can be rewritten in terms of a new variable, $u := \mathcal{K}(\psi_m)$. Defining now

$$\nabla u = \nabla \mathcal{K}(\psi_m) = k(\psi_m)\nabla\psi_m,$$

$$\tilde{k}(u) = k(\mathcal{K}^{-1}(u))$$

and letting e_z denote the vertical unit vector, equation (1) becomes

$$\frac{\partial\theta}{\partial t} = \nabla\cdot(\nabla u - \tilde{k}(u)e_z). \tag{2}$$

By [19], Theorem 4.17, the mapping $u \mapsto \tilde{\mathcal{W}}[\lambda, \mathcal{K}^{-1}(u)]$ is again a Preisach operator, hence we obtain (2) with $\theta = \mathcal{W}[\lambda,\cdot] = \tilde{\mathcal{W}}[\lambda, \mathcal{K}^{-1}(\cdot)]$.

This procedure was used in [8] to prove existence of solution for the problem without hysteresis.

On the other side, if the hydraulic conductivity is a nonlinear function of water content, $k(\theta)$, then hysteresis in θ versus ψ_m dependence entails occurrence of hysteresis in the k versus ψ_m relation. In this case, it is not clear how the Kirchhoff transformation might be extended to this setting, see [15, 16].

Remark 2.1 *We give a general view to the model. In this paper we study model equation only in two dimensions, i.e., equation (2) without gravity term. Three-dimensional Richards equation is studied (paper is in preparation).*

3. Hysteresis operators

Some remarks concerning hysteresis operators.

3.1. The play operator

Firstly we briefly mention definition and properties of the classical play operator. Definition is given in this way: for a given input function $u \in C([0,T])$ and initial condition $x_r^0 \in [-r,r]$, we define the output $\xi := \mathcal{P}_r[x_r^0, u] \in C([0,T]) \cap BV(0,T)$ of the play operator

$$\mathcal{P}_r : [-r,r] \times C([0,T]) \to C([0,T]) \cap BV(0,T)$$

as the solution of the variational inequality

$$\begin{aligned} &\int_0^T [u(t) - \xi(t) - y(t)]\,\mathrm{d}\xi(t) \geq 0, \qquad \forall y \in C([0,T]),\ \max_{0\leq t\leq T} |y(t)| \leq r, \\ &|u(t) - \xi(t)| \leq r, \quad \forall t \in [0,T], \\ &\xi(0) = u(0) - x_r^0. \end{aligned} \tag{3}$$

The concept of memory in connection to hysteresis operators is related to the fact that at any instant time t, the output $v(t)$ may depend not only on the input $u(t)$ and the initial condition but also on the previous evolution of input value $u(t)$.

We notice that we can associate to any $r \in \mathbb{R}$ the corresponding value x_r^0; this suggests the idea of making the initial configuration of the play system independent of the initial conditions $\{x_r^0\}_{r>0}$ for the output function by the introduction of some suitable function of r. More precisely, following [12], Section II.2, let us consider any function $\lambda \in \Lambda$ where

$$\Lambda := \left\{ \lambda \in W^{1,\infty}(0,\infty); \left|\frac{d\lambda(r)}{dr}\right| \leq 1 \quad \text{a.e. in } [-r,r] \right\}.$$

We also introduce subspaces of Λ, i.e.,

$$\Lambda_K := \{\lambda \in \Lambda; \lambda(r) = 0 \quad \text{for } r \geq K\}, \quad \Lambda_0 := \bigcup_{K>0} \Lambda_K. \tag{4}$$

Λ is called *configuration space* and the functions λ are called *memory configurations.*
If $Q_r : \mathbb{R} \to [-r,r]$ is the projection

$$Q_r(x) := \operatorname{sign}(x) \min\{r, |x|\} = \min\{r, \max\{-r, x\}\},$$

then we set

$$x_r^0 := Q_r(u(0) - \lambda(r)).$$

This implies that the initial configuration of the play system depends on λ and on $u(0)$. So we can introduce the following more convenient notation

$$p_r[\lambda, u] := \mathcal{P}_r[x_r^0, u], \tag{5}$$

for any $\lambda \in \Lambda$, $u \in C([0,T])$ and $r > 0$.
For the sake of completeness $p_0[\lambda, u] = u$. Moreover, the operator $p_r : \Lambda \times C([0,T]) \to C([0,T])$ is Lipschitz continuous in the following sense (see [12], Section II.2, Lemma 2.3).

Lemma 3.1 *For every $u, w \in C([0,T])$, $\lambda, \mu \in \Lambda$ and $r > 0$ we have*

$$|p_r[\lambda, u] - p_r[\mu, w]|_\infty \leq \max\{|\lambda(r) - \mu(r)|, ||u - w||_\infty\}.$$

The introduction of the function λ plays an important role in the characterization of the memory of the play system, in the sense that, for any given λ, we can construct the play operator $p_r[\lambda, u]$ starting from λ and from a sequence of values (t_j, r_j) which is so called *memory sequence* (see [12], Section II.2 or [14], Section III.6) of any input u at a certain instant t with respect to the initial configuration λ. These values are what one simply has to know in order to evaluate the output of the play operator.

In [20], the play operator is definied in the space $G_R(0,T)$ of right-continuous regulated functions. This is the space of functions $u : [0,T] \to \mathbb{R}$ which admits the left limit $u(t_-)$ at each point $t > 0$ and the right limit $u(t_+)$ exists and coincides with $u(t)$ at each point $t \geq 0$. The space $G_R(0,T)$ is endowed with the norm

$$\|u\|_{[0,T]} = \sup\{|u(\tau)|; \tau \in [0,T]\} \quad \text{for } u \in G_R(0,T) \tag{6}$$

so the $G_R(0,T)$ is a Banach space. By Theorem 2.1 and Proposition 2.4 of [20], this is Lipschitz continuous in the sense that

$$|p_r[\lambda, u](t) - p_r[\mu, w](t)| \leq \max\{|\lambda(r) - \mu(r)|, \|u - w\|_{[0,T]}\}, \tag{7}$$

for any $\lambda, \mu \in \Lambda$, $u, w \in G_R(0,T)$ and $t \in [0,T]$. For step functions $u \in G_R(0,T)$ of the form

$$u(t) = \sum_{k=1}^{m} u_{k-1}\chi_{[t_{k-1,t_k})}(t) + u_m\chi_{\{T\}}(t), \tag{8}$$

where $0 = t_0 < t_1 < \ldots t_m = T$ is a given division of $[0,T]$, we have in particular

$$p_r[\lambda, u](t) = \sum_{k=1}^{m} \xi_{k-1}(r)\chi_{[t_{k-1,t_k})}(t) + \xi_m(r)\chi_{\{T\}}(t), \tag{9}$$

where $\chi_\omega(t)$ is the characteristic function of a set $\omega \subset [0,T]$, and

$$\xi_0(r) = \mathcal{P}[\lambda, u_0](r), \quad \xi_k(r) = \mathcal{P}[\xi_{k-1}, u_k](r), \tag{10}$$

with $\mathcal{P} : \Lambda \times \mathbb{R} \to \Lambda$ defined as

$$\mathcal{P}[\lambda, w](r) = \max\{w - r, \min\{w + r, \lambda(r)\}\}. \tag{11}$$

3.2. The Preisach operator

Now we briefly recall definition and some basic properties of the Preisach operator. More information about the Preisach operator can be found in [10, 11, 12, 13, 14, 21].
Let us introduce the Preisach half-plane, defines as

$$\mathbb{R}^2_+ := \{(r, w) \in \mathbb{R}^2 : r > 0\} \tag{12}$$

and assume that a function $\psi \in L^1_{loc}(\mathbb{R}^2_+)$ (the Preisach density) is given with the following property.

Assumption 3.1 *There exist $\beta_1 \in L^1_{loc}(0,\infty)$, such that*

$$0 \leq \psi(r,w) \leq \beta_1(r) \quad \textit{for a.e. } (r,w) \in \mathbb{R}^2_+.$$

We put

$$b_1(K) := \int_0^K \beta_1(r)\,\mathrm{d}r \quad \textit{for } K > 0 \tag{13}$$

and

$$g(r,w) := \int_0^w \psi(r,z)\,\mathrm{d}z \quad \textit{for } (r,w) \in \mathbb{R}^2_+. \tag{14}$$

We define the Preisach operator as follows.

Definition 3.1 *Let $\psi \in L^1_{loc}(\mathbb{R}^2_+)$ be given and let g be as in (14). Then the Preisach operator $\mathcal{W} : \Lambda_0 \times G_R(0,T) \to G_R(0,T)$ generated by the function g is defined by the formula*

$$\mathcal{W}[\lambda, u](t) := \int_0^\infty g(r, p_r[\lambda, u](t))\,\mathrm{d}r = \int_0^\infty \int_0^{p_r[\lambda,u](t)} \psi(r,z)\,\mathrm{d}z\,\mathrm{d}r \tag{15}$$

for $\lambda \in \Lambda_0$, $u \in G_R(0,T)$ and $t \in [0,T]$.

Then we have the following result (see [12], Section II.3, Proposition 3.11).

Proposition 3.1 *Let Assumption 3.1 be satisfied and let $K > 0$ be given. Then for every $\lambda, \mu \in \Lambda_K$ and $u, w \in G_R(0,T)$ such that $\|u\|_{[0,T]}, \|w\|_{[0,T]} \leq K$, the Preisach operator (15) satisfies*

$$\|\mathcal{W}[\lambda, u] - \mathcal{W}[\mu, w]\| \leq \int_0^K |\lambda(r) - \mu(r)|\beta_1(r)\,\mathrm{d}r + b_1(K)\|u - w\|_{[0,T]} \quad \forall t \in [0,T].$$

We introduce the Preisach potential energy $\mathcal{U}$ as

$$\mathcal{U}[\lambda, u] := \int_0^\infty G(r, p_r[\lambda, u])\,\mathrm{d}r, \tag{16}$$

where

$$G(r, w) := wg(r, w) - \int_0^w g(r, z)\,\mathrm{d}z = \int_0^w z\psi(r,z)\,\mathrm{d}z, \tag{17}$$

and the dissipation operator is given by

$$\mathcal{D}[\lambda, u] := \int_0^\infty rg(r, p_r[\lambda, u])\,\mathrm{d}r. \tag{18}$$

We recover the following result (see [12], Section II.4, Theorem 4.3).

Proposition 3.2 *Let the Preisach operator $\mathcal{W}$ satisfy Assumptions 3.1 and let $K > 0$ be given. For arbitrary $\lambda \in \Lambda_K$ and $u \in W^{1,1}(0,T)$ such that $\|u\|_{C([0,T])} \leq K$, we put*

$$v := \mathcal{W}[\lambda, u] \quad U := \mathcal{U}[\lambda, u] \quad D := \mathcal{D}[\lambda, u].$$

Then we have

(i) $U(t) \geq \frac{1}{2b_1(K)} v^2(t) \quad \forall t \in [0,T]$,

(ii) $\dot{v}(t)u(t) - \dot{U}(t) = |\dot{D}(t)|$ *a.e.*

The folowing result can be found in [12], Section II.4, Proposition 4.8.

Proposition 3.3 *Let Assumptions 3.1 be satisfied and let $K > 0$ be given. Suppose moreover $b \geq 0$, $\lambda \in \Lambda_K$, and $u \in W^{1,1}(0,T)$ be given such that $\|u\|_{C([0,T])} \leq K$. Put $v := bu + \mathcal{W}[\lambda, u]$. Then*

$$b\dot{u}^2(t) \leq \dot{v}(t)\dot{u}(t) \leq (b + b_1(K))\dot{u}^2(t). \tag{19}$$

In our equation, both the input function and the initial memory configuration depend on the space variable x. If $\lambda(x,\cdot)$ belongs to Λ_0 and $u(x,\cdot)$ belongs to $C([0,T])$ for (almost) every x, then we can define

$$\mathcal{W}[\lambda,u](x,t) := \int_0^\infty g(r, p_r[\lambda(x,\cdot),u(x,\cdot)](t))\,\mathrm{d}r. \tag{20}$$

In the following we will often write $\mathcal{W}(u)$ instead of $\mathcal{W}[\lambda,u]$ for brevity or when λ is clear from the context.

We conclude this subsection with the convexification of the Preisach operator, i.e., that in a certain region, the convexity of the loops is satisfied (see [12], Section II.4, Proposition 4.22). Let $R > 0$ be fixed, set

$$\mathcal{D}_R := \{(r,w) \in \mathbb{R}^2_+ : |w| + r \le R\}.$$

In addition to Assumption 3.1 we prescribe the following conditions.

Assumption 3.2

(i) $\frac{\partial\psi}{\partial w} \in L^\infty_{loc}(\mathbb{R}^2_+)$;
(ii) $A_R := \inf\{\psi(r,w); (r,w) \in \mathcal{D}_R\} > 0$.

Furthermore, denote

$$C_R := \sup\left\{\left|\frac{\partial}{\partial w}\psi(r,w)\right|; (r,w) \in \mathcal{D}_R\right\}.$$

Taking possibly a smaller $R > 0$, if necessary, we may assume that

$$K_R := \frac{1}{2}A_R - RC_R > 0. \tag{21}$$

We modify the density ψ outside $\mathcal{D}_R$ and set

$$\psi_R(r,w) = \begin{cases} \psi(r,w) & \text{if} \quad (r,w) \in \mathcal{D}_R, \\ \psi(r,-R+r) & \text{if} \quad w < -R+r,\ r \le R, \\ \psi(r,R-r) & \text{if} \quad w > R-r,\ r \le R, \\ \psi(R,0) & \text{if} \quad r > R. \end{cases} \tag{22}$$

We define the convexified Preisach operator $\mathcal{W}_R$ by the formula

$$\mathcal{W}_R[\lambda,u](t) = \int_0^\infty \int_0^{p_r[\lambda,u](t)} \psi_R(r,w)\,\mathrm{d}w\,\mathrm{d}r \tag{23}$$

for $\lambda \in \Lambda_0$ and $u \in W^{1,1}(0,T)$. The convex character of $\mathcal{W}_R$ will be exploited in Section 4.

4. Main result

Consider an open bounded set of Lipschitz class $\Omega \subset \mathbb{R}^2$ and set $Q := \Omega \times (0,T)$. We set $V := H^1(\Omega)$, $H := L^2(\Omega)$ and introduce the space of functions with null average in Ω

$$V_* := \{u \in V : \int_\Omega u = 0\}.$$

We formally have the equation

$$v_t - \Delta u = 0, \tag{24}$$

which we couple with the boundary condition

$$\nabla u \cdot \nu = E, \tag{25}$$

on $\partial\Omega$, where ν is the unit outward normal vector, $E \in L^\infty(\partial\Omega \times (0,T))$ is a given outer matric potential. We set $v := \mathcal{W}[\lambda, u]$. We assume that $u_0(x), v_0(x) \in L^2(\Omega)$ are given initial conditions. The compatibility condition

$$\int_\Omega \nabla u_0(x)\nabla\phi(x)\mathrm{d}x - \int_{\partial\Omega} E(x,0)\phi(x)\mathrm{d}\sigma = 0 \tag{26}$$

holds for every $\phi \in V$.
We want to solve the following problem.

Problem 4.1 *Let us consider a Preisach operator $\mathcal{W}$ and let $u_0 \in L^2(\Omega)$, $\lambda : \Omega \to \Lambda$ be given initial data. We search for a function u such that $u(x,0) = u_0(x)$ a.e. in Ω and for any $\phi \in V$, and for a.e. $t \in (0,T)$ we have*

$$\int_\Omega \frac{\partial v}{\partial t}\phi\,\mathrm{d}x + \int_\Omega \nabla u\nabla\phi\,\mathrm{d}x = \int_{\partial\Omega} E\phi\,\mathrm{d}\sigma \tag{27}$$

The main result can be stated as follows.

Theorem 4.1 *(Existence) Let us assume operator $\mathcal{W}$ be the Preisach operator introduced in (20) and satisfying Assumptions 3.1 and 3.2. And let $R > 0$ be fixed as in Subsection 3.2. Let $K \in [0,R]$ and $\lambda : \Omega \to \Lambda_K$ be given. Moreover $E \in L^\infty(\partial\Omega \times (0,T))$, $E_t \in L^2(\partial\Omega \times (0,T))$, $u_0 \in V$, $v_0 \in L^2(\Omega)$ and (26) holds. Set $\alpha := \max\{\|u_0\|_V, \|E\|_{L^\infty(\partial\Omega\times(0,T))}, \|E_t\|_{L^2(\partial\Omega\times(0,T))}\}$. Then there exists a constant $\beta > 0$ such that if $\alpha \le \beta$, then Problem 4.1 has at least one solution such that*

$$u \in C^0(Q),$$
$$u_t \in L^2(0,T;V_*).$$

Proof
(i) **Approximation.** Let us fix a time step $\tau := \frac{T}{m}$, for some $m \in N$. Furthermore we consider for $k = 1,\dots,m$ and for any $\phi \in V$ the following reccurent system

$$\frac{1}{\tau}\int_\Omega (v_k - v_{k-1})\phi\,\mathrm{d}x + \int_\Omega \nabla u_k\nabla\phi\,\mathrm{d}x = \int_{\partial\Omega} E_k\phi\,\mathrm{d}x, \tag{28}$$

where E_k is defined by $E_k(x) = \frac{1}{\tau}\int_{(k-1)\tau}^{k\tau} E(x,t)\,\mathrm{d}t$.
Here we set

$$v_k(x) = \int_0^\infty g_R(r, \xi_k(x,r))\,\mathrm{d}r \tag{29}$$

with

$$g_R(r,w) = \int_0^w \psi_R(r,w')\,\mathrm{d}w', \tag{30}$$

where ψ_R is function introduced in (22), i.e., we are working with the convexified Preisach operator $\mathcal{W}_R(u)$. The sequence ξ_k is defined recursively by

$$\xi_0(x,r) := P[\lambda(x,\cdot), u_0(x)](r), \quad \xi_k(x,r) := P[\xi_{k-1}(x,\cdot), u_k(x)](r). \tag{31}$$

Setting

$$\bar{u}^{(\tau)}(x,t) = \sum_{k=1}^m u_{k-1}(x)\chi_{[(k-1)\tau,k\tau)}(t) + u_m(x)\chi_{\{T\}}(t), \tag{32}$$

and

$$\bar{\xi}_r^{(\tau)}(x,t) = \sum_{k=1}^{m} \xi_{k-1}(x,r)\chi_{[(k-1)\tau,k\tau)}(t) + \xi_m(x,r)\chi_{\{T\}}(t), \tag{33}$$

we thus have, in the view of (8) - (11)

$$\bar{\xi}_r^{(\tau)}(x,t) = p_r[\lambda, \bar{u}^{(\tau)}](x,t). \tag{34}$$

The solution to (28) can be constructed by induction over k, then using the Browder-Minty fixed point theorem, the monotonicity of the mapping $g(r,\cdot)$ and $P[\lambda,\cdot]$.

(ii) **A discrete first order energy inequality.** We established here a discrete counterpart of Proposition 3.2. We set $\xi_k^r(x) := \xi_k(x,r)$, where $\xi_k(x,r)$ was introduced in (31). Let ψ be an arbitrary function satisfying Assumption 3.1. We define a discrete versions of the Preisach potential energy $\mathcal{U}$ and dissipation operator $\mathcal{D}$, introduced in (16) and (18) as

$$U_k(x) = \int_0^\infty G(r, \xi_k^r(x))\mathrm{d}r,$$

$$D_k(x) = \int_0^\infty r g(r, \xi_k^r(x))\mathrm{d}r,$$

with G given by (17). The discrete version of the first order energy inequality can be stated as follows:

$$(v_k - v_{k-1})u_k - (U_k - U_{k-1}) \geq \int_0^\infty \int_{\xi_{k-1}^r}^{\xi_k^r} r\psi(r,w)\mathrm{d}w\mathrm{d}r = |D_k - D_{k-1}|. \tag{35}$$

(iii) **A discrete second order energy inequality.** The discrete version of the second order energy inequality can be stated as follows: For every $k = 2,\ldots,n$, $n \in \{1,\ldots,m\}$ and a.e. $x \in \Omega$

$$(z_k - z_{k-1})q_k \geq \frac{1}{2}(z_k q_k - z_{k-1}q_{k-1}), \tag{36}$$

where we set

$$q_k := \frac{u_k - u_{k-1}}{\tau}, \quad z_k := \frac{v_k - v_{k-1}}{\tau} \quad \text{for } k = 1,\ldots,m;.$$

The time continuous case is treated in detail in ([12], Section II.3 and II.4).

(iv) **First a priori estimate.** In the estimates below, let C denote a constant independent of α and τ. Indeed, the value of C may vary from one formula to another.
We choose $\phi = \tau u_k$ in (28), we obtain

$$\int_\Omega (v_k - v_{k-1})u_k \,\mathrm{d}x + \tau\int_\Omega |\nabla u_k|^2 \,\mathrm{d}x = \tau\int_{\partial\Omega} E_k\, u_k \,\mathrm{d}\sigma.$$

Therefore, using (35) and summing up for $k = 1,\ldots,n$, for every $n \in \{1,\ldots,m\}$, we obtain

$$\tau\sum_{k=1}^n \int_\Omega |\nabla u_k|^2 \,\mathrm{d}x \leq \tau\sum_{k=1}^n \int_{\partial\Omega} |E_k|^2 \,\mathrm{d}\sigma + \frac{1}{4}\tau\sum_{k=1}^n \int_{\partial\Omega} |u_k|^2 \,\mathrm{d}\sigma + \int_\Omega U_0 \,\mathrm{d}x.$$

Using the regularity of the initial data and trace operator theorem, it follows that

$$\tau \sum_{k=1}^{n} \|u_k\|_{V_*}^2 \leq C\alpha^2 \tag{37}$$

(v) **Second a priori estimate.** Due to the monotonicity and local Lipschitz continuity of the functions $g(r,\cdot)$ and $\mathcal{P}[\lambda,\cdot](r)$, we have the pointwise inequality

$$q_k(x) z_k(x) \geq 0 \quad \text{a.e. in } \Omega,\ k = 1,\ldots,m. \tag{38}$$

We set for brevity

$$E_{tk} := \frac{E_k - E_{k-1}}{\tau}.$$

We take the incremental ratio in time in (28) and then test by τq_k, we get

$$\int_{\Omega} (z_k - z_{k-1}) q_k \,\mathrm{d}x + \tau \int_{\Omega} |\nabla q_k|^2 \,\mathrm{d}x = \tau \int_{\partial\Omega} E_{tk}\, q_k \,\mathrm{d}\sigma. \tag{39}$$

Then let us sum for $k = 1,\ldots,n$, for every $n \in \{1,\ldots,m\}$, use (36) for $k \geq 2$ and Hölder inequality, we obtain

$$\int_{\Omega} (q_1 z_1 - q_1 z_0)\,\mathrm{d}x + \frac{1}{2}\sum_{k=2}^{n} \int_{\Omega} (q_k z_k - q_{k-1} z_{k-1})\,\mathrm{d}x + \tau \sum_{k=1}^{n} \int_{\Omega} |\nabla q_k|^2 \,\mathrm{d}x$$
$$\leq \tau \sum_{k=1}^{n} \int_{\partial\Omega} |E_{tk}|^2 \,\mathrm{d}\sigma + \frac{1}{4}\tau \sum_{k=1}^{n} \int_{\partial\Omega} |q_k|^2 \,\mathrm{d}\sigma.$$

Futhermore we use (38), compatibility condition (26), the estimates on the initial data and apply trace operator theorem to get

$$\tau \sum_{k=1}^{n} \|q_k\|_{V_*}^2 \leq C\alpha^2. \tag{40}$$

(vi) **Third a priori estimate.** Finally we need L^∞ bound for our solution. We prove by induction over $k = 1,\ldots,m$ that there exists $B > 0$ independent of k and m such that

$$\|u_k\|_{L^\infty(\Omega)} \leq B\alpha \quad \text{for all } k = 0,\ldots,m. \tag{41}$$

For $k = 0, 1, \ldots, m$ we set

$$B_k^{(m)} = \frac{1}{\alpha} \max\{\|u_j\|_{L^\infty(\Omega)}; j = 0, 1, \ldots, k\}.$$

We have $B_0^{(m)} \leq C$ independently of m by hypothesis on initial data. Let now $1 \leq k_0 \leq m$ be fixed and assume that $B_{k_0-1}^{(m)} < \infty$. By direct comparison in equation (28), we derive for $k = 1,\ldots,k_0$ the estimate

$$\|\triangle u_k\|_H \leq \|z_k\|_H. \tag{42}$$

This yields that $B_{k_0}^{(m)} < \infty$. Acording to (29) - (31), (22) and by definition of α and by (40), we obtain from (42) that

$$\|\triangle u_{k_0}\|_H \leq C\alpha(1 + \max\{B_{k_0-1}^{(m)}\alpha, \|u_{k_0}\|_{L^\infty(\Omega)}\}). \tag{43}$$

A combined use of the Poincare inequality, the embedding theorem, Gagliardo-Nirenberg inequality and (37) yield $\|u_{k_0}\|_{L^\infty(\Omega)} \le C\alpha(1+\max\{B_{k_0-1}^{(m)}\alpha, \|u_{k_0}\|_{L^\infty(\Omega)}\})^{\frac{2}{3}}$. Assume that $B_{k_0}^{(m)} > B_{k_0-1}^{(m)}$. Then

$$B_{k_0}^{(m)} = \frac{1}{\alpha}\max\{\|u_{k_0}\|_{L^\infty(\Omega)}\} \le C(1+\alpha_0 B_{k_0}^{(m)})^{\frac{2}{3}},$$

hence $B_{k_0}^{(m)} \le \max\{C, B_{k_0-1}^{(m)}\}$ with a constant C independent of k and m, and the desired estimate (41) follows. Inequality (43) implies in particular that

$$\|\triangle u_k\|_{L^2(\Omega)} \le C\alpha \quad \text{for all } k=1,\dots,m. \tag{44}$$

(v) **Limit procedure.** With the sequence $\{u_k\}$ constructed above (see (32) - (34)) we define for each fixed time step τ their piecewise linear and piecewise constant time interpolates according to the following scheme:

$$\begin{cases} \bar{u}_+^{(\tau)}(x,t) = u_k(x), & \bar{u}_-^{(\tau)}(x,t) = u_{k-1}(x), \\ \bar{v}_+^{(\tau)}(x,t) = v_k(x), & \bar{E}_+^{(\tau)}(x,t) = E_k(x), \end{cases} \tag{45}$$

and

$$\begin{cases} \hat{u}^{(\tau)}(x,t) = u_{k-1}(x) + \frac{t-(k-1)\tau}{\tau}(u_k(x) - u_{k-1}(x)) \\ \hat{v}^{(\tau)}(x,t) = v_{k-1}(x) + \frac{t-(k-1)\tau}{\tau}(v_k(x) - v_{k-1}(x)) \end{cases} \tag{46}$$

for every $x \in \Omega$ and $t \in [(k-1)\tau, k\tau)$, $k = 1, 2, \dots, m$, continuously extended to $t = T$. We have

$$\bar{v}_+^{(\tau)} = \mathcal{W}_R[\lambda, \bar{u}_+^{(\tau)}]. \tag{47}$$

As a consequence of the estimates (40), (44) and by comparison of the terms of the equation (28) we get that $u \in L^\infty(0,T; W^{2,2}(\Omega)\cap V_*)$, $u_t \in L^2(0,T;V_*)$. The a priori estimates we found allow us to conclude that there exists u, u_t such that, along a subsequence as $\tau \to 0$, we have

$$\begin{array}{lcl} \hat{u}^{(\tau)} \to u & \text{weakly star in} & L^\infty(0,T;W^{2,2}(\Omega)), \\ \hat{u}_t^{(\tau)} \to u_t & \text{weakly in} & L^2(0,T;V_*), \end{array} \tag{48}$$

By compact embedding, we have, passing again to a subsequence, if necessary

$$\begin{array}{lcl} \nabla\hat{u}^{(\tau)} \to \nabla u & \text{strongly in} & L^2(Q;\mathbb{R}^2), \\ \hat{u}^{(\tau)} \to u & \text{uniformly in} & C^0(\bar{Q}). \end{array} \tag{49}$$

We further have for every τ and every $(x,t) \in Q$ that

$$|\hat{u}^{(\tau)}(x,t) - \bar{u}_\pm^{(\tau)}(x,t)|^2 \le \max_k |u_k(x) - u_{k-1}(x)|^2 \le \sum_{k=1}^m |u_k(x) - u_{k-1}(x)|^2,$$

$$|\hat{v}^{(\tau)}(x,t) - \bar{v}_+^{(\tau)}(x,t)|^2 \le \max_k |v_k(x) - v_{k-1}(x)|^2 \le C\sum_{k=1}^m |u_k(x) - u_{k-1}(x)|^2.$$

From (40) it follows that

$$\|\hat{u}^{(\tau)} - \bar{u}_{\pm}^{(\tau)}\|_{L^2(\Omega;G_R(0,T))} + \|\hat{v}^{(\tau)} - \bar{v}_{+}^{(\tau)}\|_{L^2(\Omega;G_R(0,T))} \le C\sqrt{\tau}, \tag{50}$$

$$\|\nabla\hat{u}^{(\tau)} - \nabla\bar{u}_{\pm}^{(\tau)}\|_{L^2(Q;\mathbb{R}^2)} \le C\sqrt{\tau}, \tag{51}$$

Hence $\bar{u}_{\pm}^{(\tau)}$ converge strongly to u in $L^2(\Omega; G_R(0,T))$ as $\tau \to 0$. By Proposition 3.1 we may pass to the limit in (47) and obtain

$$\bar{v}_{+}^{(\tau)} \to v = \mathcal{W}_R[\lambda, u] \quad \text{strongly in} \quad L^2(\Omega; G_R(0,T)). \tag{52}$$

This and (50) yield

$$\hat{v}^{(\tau)} \to v \quad \text{strongly in} \quad L^2(\Omega; G_R(0,T)). \tag{53}$$

The convergences (48)-(49), (52)-(53) and inequality (51) enable us to pass to the limit as $\tau \to 0$ and obtain

$$\int_{\Omega} (v_t\phi + \nabla u \nabla\phi)\, \mathrm{d}x = \int_{\partial\Omega} E\phi\, \mathrm{d}\sigma. \tag{54}$$

At this point we can deduce that there exists a constant β depending on R such that if $\alpha \le \beta$ then $|u(x,t)| \le R$ a.e. in Q. This implies that $\mathcal{W}(u) = \mathcal{W}_R(u)$ (see Lemma II.2.4 in [12] and the definition of ψ_R (22)). This completes the proof of Theorem. □

Remark 4.1 *We assume equation (24) with homogenous Dirichlet or Neumann boundary condition and compatibility condition (26) is then without the second term. The system is isolated and starts at the steady state, i.e. compatibility condition holds. Because there are no outer sources on the boundary then the solution stays at the steady state trivially. We can see this from the proof of Theorem 4.1 where the estimate for time derivative is equal to zero.*

Acknowledgments

The research was supported, in part, by the Grant Agency of the Czech Republic, grant No. 201/09/P163 and by project MSM4781305904 from the Czech Ministry of Education.

References

[1] Haines W B 1930 *J. Agric. Sci.* **20** 97-116
[2] Everett D H 1958 *Some Problems in the Investigation of Porosity by Adsorption Methods, The Structure and Properties of Porous Materials* (London: Butterworth)
[3] Everett D H 1967 *The Solid-Gas Interface* (New York: Dekker) p 1005
[4] Poulovassilis A 1962 *Soil Sci.* **93** 405
[5] Flynn D, McNamara H, O'Kane J P and Pokrovskii A 2005 *Sci. Hyst.* **3** 698-744
[6] Poulovassilis A and Childs E C 1971 *Soil Sci.* **112** 301-312
[7] Mualem Y 1984 *Soil Sci.* **137** 283-291
[8] Alt H W, Luckhaus S and Visintin A 1984 *Ann. Mat. Pura Appl.* **136** 303-316
[9] Otto F 1997 *Adv. Math. Sci. Appl.* **7** 537-553
[10] Brokate M and Sprekels J 1996 *Hysteresis and Phase Transitions* (Berlin: Springer)
[11] Krasnosel'skii M and Pokrovskii A 1989 *Systems with Hysteresis* (New York: Springer)
[12] Krejčí P 1996 *Hysteresis, Convexity and Dissipation in Hyperbolic Equations* (Tokyo: Gakkotosho)
[13] Mayergoyz I D 1991 *Mathematical Models of Hysteresis* (New York: Springer)
[14] Visintin A 1994 *Differential Models of Hysteresis* (Heidelberg: Springer)
[15] Bagagiolo F and Visintin A 2000 *J. An. Appl.* **19** 977-997
[16] Bagagiolo F and Visintin A 2004 *Advan. Math. Sci. Appl.* **14** 379-403
[17] Kubo M 2004 *J. Diff. equation* **201** 75-98
[18] Marinoschi G 2008 *Nonlin. An.: Real World Appl.* **9** 518-535
[19] Krejčí P 1991 *Apl. Mat.* **36** 305-326
[20] Krejčí P 2006 *J. Phys.: Conf. Series* **55** 144-154
[21] Preisach F 1983 *Z. Physik* **94** 277-302

Journal of Physics: Conference Series **268** (2011) 012015 doi:10.1088/1742-6596/268/1/012015

Interactive multimedia and tele-presence in production

G L Kovács[1,2] and G Haidegger[2]

[1]University of Pécs, PMMK, H-7624 Pécs, Rókus u. 2. Hungary
[2]Computer and Automation Research Institute, Hungarian Academy of Sciences, H-1111 Budapest, Kende u. 13-17. Hungary

E-mail: gkovacs@sztaki.hu, haidegger@sztaki.hu

Abstract: The authors of this paper give a short summary on the research and development actions carried out and on results achieved in the CIM Research Laboratory in the past years, partly during a Hungarian National Research and Development Program addressed as 'Digital Factory'. This paper outlines some tasks solved in relation to interactive multimedia and tele-presence based industrial applications. These solutions may be helpful for test, diagnoses and quality control of the production, or sometimes they just help the operator's work. The basic research phase of the work was followed by applied research and development, while the concluding phase enabled industrial applications and test scenarios. The authors explain some details of their achievements including aspects of virtual reality as well.

1. Introduction

This paper gives some details of the R&D works devoted to interactive multimedia and tele-presence applications for Digital Enterprises and Production Networks (Digital Factory in short). Information technology developments have given faith to system designers to plan for a new generation of manufacturing environment. Some experts tend to call this new era as 'digital revolution' in manufacturing, and foresee a time in the near future, when the 'old idea' of the 'paper-less office' will be replacing the present workplaces. In a similar process as the design and planning, also the execution and control of the manufacturing processes in a factory will (or could) run in a paper-less style, and this vision has led to the birth of a new topic: 'running a digital factory' [1].

2. Main goals of the research

There are several fields within a factory environment, where it is essential to involve human experts to observe, to test, to control and to feel the details of the production processes, both in the design and the planning stages, and during the manufacturing, testing and verification processes. As soon as digital technology reaches a complexity within the environment, appropriate digital communication and information processing subsystems can meet the above listed user requirements. These new features raise the quality level, the efficiency, the observability and controllability of the production, but also allow the customer and end-user to actively take part in the processes of manufacturing. Implementing interactive multimedia services for tele-presence, advances the technology of the given manufacturing facility into a new, higher level. Dreams were envisaged to introduce tele-presence oriented multimedia services within specific productions scenarios at several industrial factory sites. The foreseen applications and dreams were giving us an inspiration, and helped us to achieve the basic goals.

3. Tele-presence from different viewpoints

When this research was started we could realize that the terms we were using were not at all definite for us, and no wonder it was deeply confusing for others, who just started to deal with these phrases. Even after a course of basic technology presentations, newcomers could heavily hinder the progress of our work in defining the detailed work. As it will be shown in the next section, our work had started with a wide range of 'fundamental underlying technology' familiarisation, and only after that could we start to focus on a narrow gap for selecting our experimental domains.

As a beginning 3 viewpoints were identified for understanding tele-presence, each of them is in respect to multimedia.

3.1. The historical approach of tele-presence

It is an ancient ambition of mankind is to overcome our time- and dimensional-limitations. We have gained several great achievements derived from this ambition like writing, telephony, radio, TV and movie. However a new media, the Internet made the real solution possible. The opening of the direct tele-presence research was the cybernetic tele-operation in the sixties of the twentieth century.

That time the cybernetic model of human behaviour has been worked out. The first definition was formulated by Csíkszentmihályi in 1975 [2]: tele-operation is concentrated on some tasks as to render one unconscious of stimuli outside of the task, including even awareness of self and the passage of time. In the beginning of the nineties the tele-presence research changed its focus, and instead of the remote control of an object the sensing of an object became the goal of the investigation [3]. Tele-presence is called 'as a compelling impression of being at the location occupied by the slave device' and identified it with the phenomenon called distal attribution, in which a person 'externalizes', creating an identification of self to include the external world.

One of the most general definitions of tele-presence was drawn as the ultimate goal of these efforts as to produce a transparent link from human to machine; a user interface through which information is passed so naturally between operator and environment that the user achieves a complete sense of presence within the remote site. Considering the up-to-date intermediate link, the tele-presence is the ability to operate remotely in a computer-mediated world. On the other hand a virtual presence can also be defined in the frame of tele-presence as experience by a person when sensory information is generated only by and within a computer compels a feeling of being present in an environment other than one is actually in [4]. The virtual presence involves some new definitions, too, such as virtual reality and augmented reality. Virtual reality is a modelled environment generally generated by high capacity computers. This environment illustrates a real environment however it has not got any connections with the real environment. Augmented reality works using graphical objects mounted onto the pictures of the real world, and shows an alternative reality of the environment.

3.2. The multimedia approach

All the previously mentioned definitions contain the time and dimensional independence of the connection of a subjective part and an objective part. We can define the tele-presence as a general and natural set connection of a subjective activity with an object that is independent of the time and the distance. The earlier definitions however did not analyse this connection sufficiently.

The connections in tele-presence can be demonstrated using a 3-dimensional diagram. The 3 dimensions represent the time, the distance or dimensional and the functional connections respectively (figure 1).

The time variable of this diagram shows how the tele-presence is realized: as real time connection or as a delayed one. A more interesting meaning of the time domain analysis considers if the previously sensed information can be gained in parallel ('storytelling function'), and - as a new approach - how the information is used in forecasting. In this aspect the scale of the time axis demonstrates the perspective of the forecasting and storytelling ability.

The dimensional variable of the diagram is not a simple distance measure, it means mainly how the object in the sensing environment of the subject or the sensor data of the remote environment are concentrated (positive direction) either in a control room, or in the intranet of an institution or under internet-connected realization. Another conceptual realization of tele-presence is, when not the senses but the virtual realization of the object can be reached in a control room, or in the intranet or in a less concentrated way (negative direction).

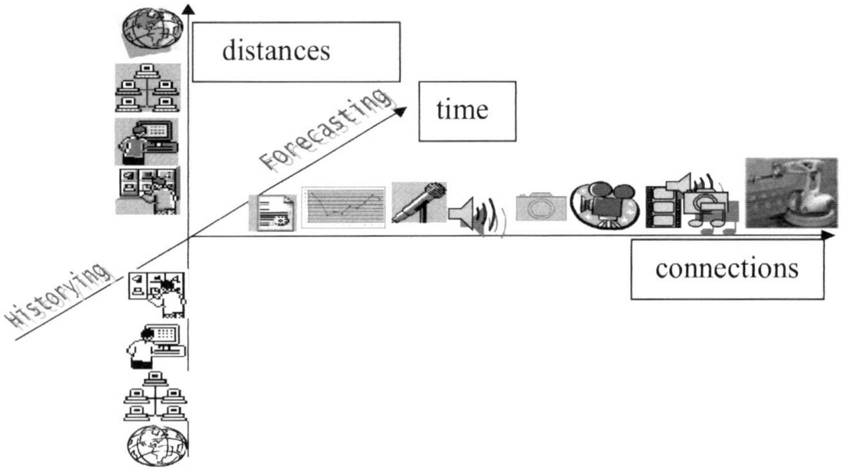

Figure 1. 3-dimensional demonstration of tele-presence.

The third dimension in the function diagram demonstrates the less analysed connection set. In this axis only the positive half is valid and the scale of the dimension measures how rich is the transmitted information between the sensing subject and the object. The different information forms start from the simple binary codes through the numerical or audio files till the video information or the combination of these. The highest level of the tele-presence function is the physical remote presence using mechanical activities as tele-robotics for example.

3.3. The manufacturing approach

The integration of the appropriate software solutions (CAD, CAM, CAPP, etc.) with the up-to-date manufacturing systems and the computer networks resulted in the paradigm of the computer integrated manufacturing (CIM). The simplified definition of CIM contains the parallel realization of the material and information flow that is necessary for product manufacturing and provides a more efficient production infrastructure. In the information flow we need to follow the manufacturing of parts (this dimension represents the life cycle of the product) including assembly activities.

The information must be provided for all the participants of the production such as the designer, quality team, manufacturing staff, accounting and resource management or the general management (this dimension represents the production tasks), and the information must be present in the full hierarchy of the production from the process level to the factory level (hierarchy dimension). A possible information model of the production is demonstrated in figure 2.

4. Research directions and goals to achieve

At the beginning of the project we prepared assessment reports, trend analysis and requirement specifications. Based on the RUP design technology [5], new concept was introduced and learned. The following specific topics were under heavy investigation.

Reliability and safety - a trend analysis was produced to give an up-to-date status report on the available techniques and methods. A survey was prepared to detail the most well known industrial and consumer applications for tele-presence operations.

Since tele-presence and interactive multimedia involve the integration of human persons, the research work also dealt with man-machine interaction means and with defining VMD (Virtual Manufacturing Device) devoted for human entities within the production area. A survey on wearable computers analyses the presently available technology for applying computer platforms around the human body.

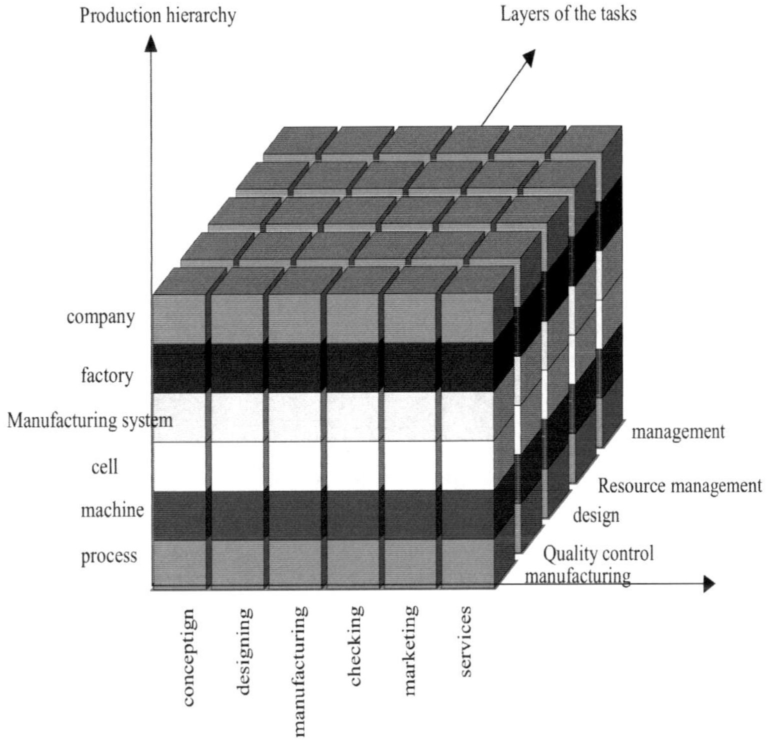

Figure 2. Information model of a production system.

4.1. Wearable Computers, interactive multimedia and tele-presence

A market trend report was prepared based on the available eye-glass-integrated or head-mounted displays (HMD) and on the technology and architecture of wearable computer platforms. Figure 3 shows the summary view of the analysed HMD-s.

The focus of the research work was preparing the concept of software development for generalized software that enables tele-presence operations with interactive multimedia features. The work that is now called as IMUTA stands for Interactive Multimedia for Tele-presence Applications.

Major effort was devoted to the planning of pilot demonstrator implementations. It was required to find such demonstration sites, which could well inspire the hosts and potential further customers to implement similar functions - and service-intensive applications. Models were developed for the specific application scenarios to derive a generalized framework. Such business models are becoming more and more common and important in the search for gaining market share. One of the steps in business modelling is the modelling of dynamic behaviour of the business. The activity diagram provides high-level (macro activity) view in the process of modelling the dynamic behaviour of the business.

5. Virtual and augmented reality

By integrating real-life vision and computer-animated entities, virtual and augmented reality services can be offered for specific user applications. For this purpose we developed a distributed simulation

Journal of Physics: Conference Series **268** (2011) 012015 doi:10.1088/1742-6596/268/1/012015

and visualization environment for flexible manufacturing systems [6]. It enables its users to visualize the results of a simulation of a Flexible Manufacturing System (FMS) with 3D graphics, or visualize the actual data of a real system (see figure 4). Figure 5 shows a typical process model of providing MM services.

Figure 3. Analysed versions for Head-mounted Displays (HMDs).

Figure 4. The real and the simulated manufacturing cell.

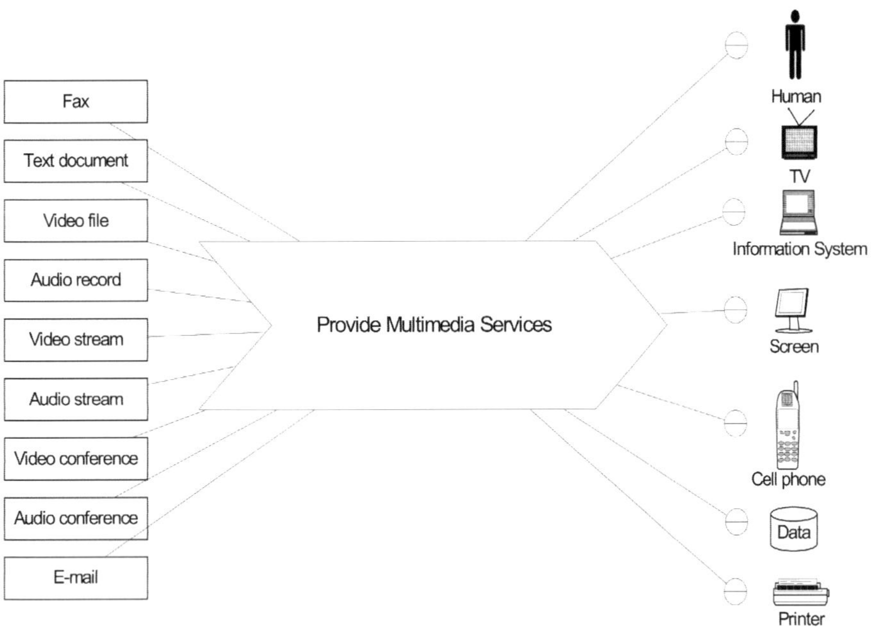

Figure 5. Process model for providing MM services.

On the client side we use a Virtual Reality Modeling Language (VRML) and Java, so we need only a web browser with the appropriate plug-ins to use it. The users can view the scene provided by the simulation system through an Internet access, they can see it from the desired point of view and interact with it in the simulation phase.

The IMUTA concept supports the integration of augmented reality [7].

The international MPEG (Moving Picture Experts Group) standardization body takes responsibility to define and harmonize the new definitions, rules and methodologies for multimedia elements, tools and services. Our involvement was to get a deeper understanding of the present processes and design the IMUTA to allow future evolvement of MPEG standards.

The authors investigated and evaluated several computer platforms for applicability on humans. Special interfaces for mouse operations, mobile functions etc. have been studied to suggest versions for separate and independent IMUTA functions.

An experimental wearable computer set up (figure 6) was built in our laboratory with the simplest parts available. The main aims were to give the feeling of the wearable computer to the user and to have it with low cost commercial parts.

During the early experiments there were no equipment to determine the local position and the orientation, so the experimental wearable computer was not yet good for augmented reality applications.

6. Enterprise resources and multimedia

IMUTA is a Framework of Resource Integration and Multimedia. In this framework we extended Enterprise Resources with multimedia information and functionality. The Extended Resource model provides a richer and more human oriented view of resources and processes, for example:

Machining might visually be observed real-time and remotely; relevant process data, product drawings, technology plans, quality reports can be chosen and shown according to the needs.

Documentation of products might be extended with multimedia attachments to ensure the customer satisfaction.

Journal of Physics: Conference Series **268** (2011) 012015 doi:10.1088/1742-6596/268/1/012015

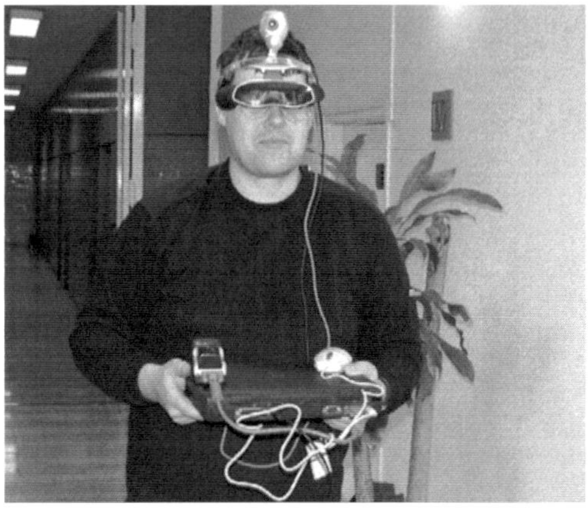

Figure 6. Our experimental wearable computer.

Customers might observe the production flow of a product.

The framework features as follows:

- Multimedia Document Management;
- Media Stream and Process Management;
- Semantic Resource Model;
- Resource Adapter Architecture.

We considered semantic information management as a key issue of the resource integration. In our approach, semantic features make possible to identify the nature and behaviour of the entity.

The applied semantic model is flexible and extendible, so we may feed more and more 'knowledge' into the system. Since enterprise resources do not form a well-defined and closed set of entities, the framework provides adapter architecture easy to implement for different types of resources.

The current version supports some key domains (figure 7):

Actor: active participants outside of the system, users, software agents, etc.

Media: media processing entities like streams, cameras, processes, etc.

Content: entities of the content management system, like storages, files, documents, etc.

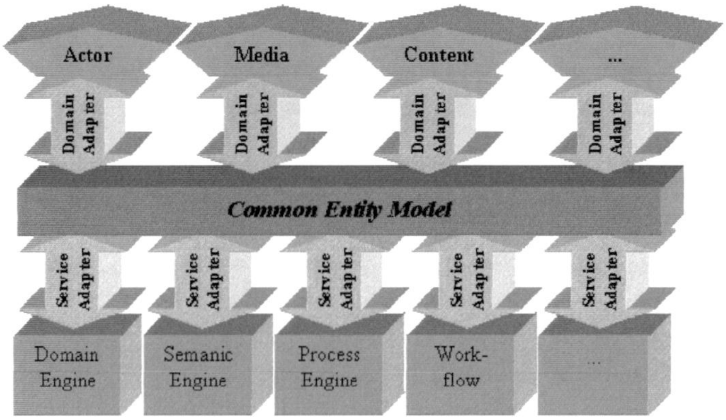

Figure 7. IMUTA architecture,

7. Intelligent document management

We have to mention, as a part of our work, the intelligent documents' management, which replaces traditional documents with new ones with highly sophisticated new functions for almost all user categories. Besides applying all the multimedia offered data element types, it offers intelligence embedded into the user-interface, it supports tele-teaching, tele-maintenance with guided measurement tasks, self-learning, verification of learnt materials, etc.

8. Multi-user multi-task information management

A multi-user, multi-task distributed audio and video capturing, processing and recording system (MUMUS) was developed (see [8]). The system is implemented in Java using Sun Microsystems, Inc.'s Java Media Framework (JMF) to assure portability and easy extension. It runs in Microsoft Windows, Linux and Sun Solaris environment. It uses the OS' standard media management interface, so a wide range of capturing devices can be used. The software includes fault tolerance routines to limit effects of communication, configuration and resource faults and keep smooth operation of intact tasks and parts. Control is decentralized with different embeddable interfaces and provides flexible management of different tasks.

Developing MUMUS had two aims:

- versatile media playback/transmit/recording system, which could be used in the following examples:
  general-purpose surveillance system with archiving; videoconference system; production process monitoring and event inspecting with circular buffer, etc.;
- real-time distributed media processing system. This would require an efficient and loss-less video codec for Real-time Transport Protocol (RTP) transmission. Motion-JPEG could be used with care (high-quality settings). The processing algorithms depend on the task, so they are missing, but easy to add.

MUMUS consists of:

- one or more skeletons, to run the processing part on each participating computer;
- tasks spreading to a subset of skeletons of MUMUS meaning a definite task to perform (can be one of the above examples);
- modules, atomic parts of tasks providing functions like capture, file playback or recorder, transmitter, receiver, circular buffer for delay, etc.;
- internal media streams flowing from one module to an other;
- sources and sinks which send or receive external media streams (to or from other application);
- one or more controllers, which are embeddable in other application and send commands to skeletons and receive answers; and
- messages to enable communication of skeletons and controllers.

The following example (figure 8) consists of three computers. The first one captures audio and video and transmits them to the second one, which displays both. The video stream is split in the first one and in turn transmitted to the last one in low resolution and is archived.

The following figure shows the participating computers according to their roles in this task (one skeleton and the interested controller runs on the same computer).

9. Use cases - Business modelling

By analysing use cases and applying business modelling with Rational Unified Process (RUP) methodology, business process models were developed in a sequence of joint work with the end-users [9]. The derived object models could be gained from the systematic approach and industry managers could also consult that intermediate level.

The final part of the technology enabled the preparations of the System Use case models. Figure 9 shows a derived model that upgraded the process within a manufacturing plant.

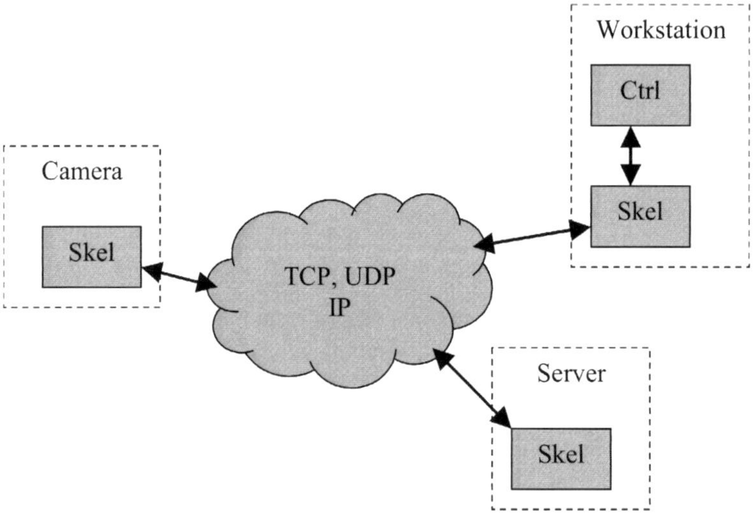

Figure 8. Editing the project file.

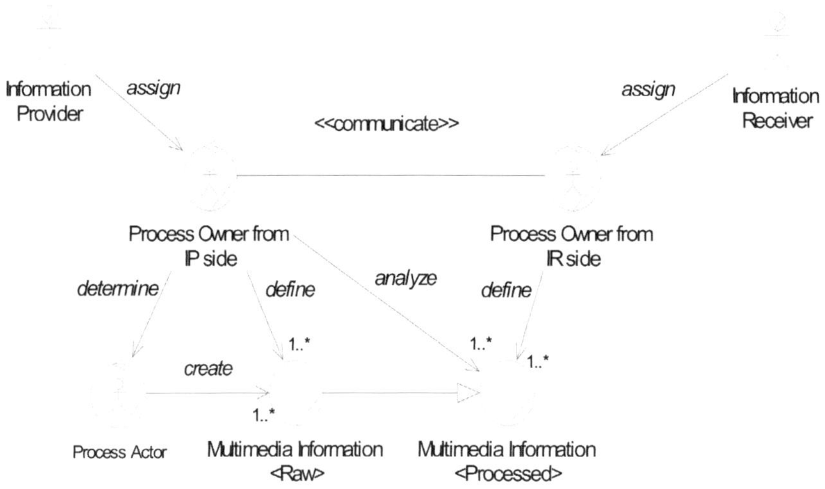

Figure 9. Example of Business Class Diagram.

10. Open system – further possibilities

IMUTA contains hooks and existing Application Programming Interfaces (APIs) that allow new and proprietary applications (e.g. home) to be built around remote access protocol. For example, third-party monitoring and repair services for home appliances with proactive notifications of device failures, 'health and wealth' monitoring for vacation homes, medical applications for senior citizens, and many others - all from the same back end installation. IMUTA serves as a tool to allow the partners to be creative as they look to design products and services that add a new stream of revenue, increase efficiency, and differentiate themselves from their competition.

Some of the applications that we offer today are following:

Energy management
Internet enabled wireless thermostats; Load controls (sprinklers, pool heaters, water heaters); Device control; Temperature scheduling (Local and Remote); Load shedding, curtailment, Diagnostics; Usage history

Security
Remote arm/disarm security system; Video verification; Streaming video (Local & Remote); Sensor notification (via e-mail, page, or cell phone). Power on/off controls; Full home automation functionality.

Healthcare
Fall detectors, Video verification for falls, Motioned detectors, Breathing Monitors

Appliance control and diagnostic
Remote Diagnostics, Remote Control, Remote Software Upgrades

Home automation
Streaming video (Local & Remote); Sensor notification (via e-mail, page, or cell phone); Power on/off controls; Door locks; Lights

11. Application scenarios

Based on the modules prepared, and based on the business case evaluations, test scenarios or 'IMUTA IMPLEMENTATIONS' were set up at our industrial partners sites'. Complex structures of media-elements, streams of video and sound - with text and graphics, documentations and PowerPoint presentation slides are managed harmoniously under the IMUTA, to enhance important management functions within production control, supervision or quality control.

11.1. Test scenario No. 1: 'Customer Witness'.

When large-scale expensive products are manufactured, the customer has an obligation to pay for its order in instalments, according to the status of the product being prepared, and tested, verified. The visits to the production sites are very time-consuming and rather expensive, but due to legal obligations, most often they cannot be neglected. A long list of actions and tasks are devoted to these verification phases.

By applying IMUTA services, interactive multimedia solutions can help in avoiding many of the problems identified and modelled in the definition stage. Not necessarily all and every, but predictably most of the factory visits can be substituted by tele-presence-oriented interactive multimedia. The IMUTA-based multimedia solution can provide the customer with all necessary information almost online. With a flexibility to run and schedule the test operations, the customer-representative can see, hear and witness the testing procedure at the very same acceptance-level as through real physical presence. All documents, reports, calibration and measurement devices' output data can be recorded by the measuring devices, while being connected to both the commanding computer node, and also to the video and sound-documenting subsystem.

11.2. Test scenario No. 2: 'Enhanced trading with internally produced expensive parts.'

Where large-scale expensive products are manufactured, the parts not reaching the tiniest tolerance or the highest quality grade can still be valuable for another level of customer. A quality inspection point inserted into the production line or sequence will monitor the products, usually one by one. The inspector, who might use high-precision measuring machine, will be able to check for the basic metrical parameters, but not all the problems can be directly seen by the human body's limitations.

When second quality grade products appear, the shop-floor engineer must take the decision on that part, to be reworked, or to be disposed, or to accept it as top-quality product. The tested scenario has evaluated in details the actions, activities, information flow and decision making processes, and with business modelling steps, redesigned the complete process. The IMUTA software system helps in the generation of mixed media document items, like text, drawings, video, still pictures with high resolution, and also for voice, all that involves information on the given product piece. The reselling of such 'second-grade' products is a god financial activity, compared to the 'lost' value, when the piece is just discarded or thrown to the scrap.

11.3. Test scenario No. 3

When products are manufactured in a 'fast running technological site', - e.g. at a conveyor belt - some activities' effects and real results can only be measured, or tested after certain time has elapsed. When the measurement points out the scrap products or injured modules, already several dozens (hundreds) of products had already been produced on the line. The feedback time might not be shortened, when the technology cannot support it, but the IMUTA software can still help, by storing all production oriented and monitored values, or sights. At our test environment, each minute 30 products are awaited at the end of the transfer line. When problems occur, the dispatcher or service person receives an automatic notice, and the video and other relevant production-related information elements are recalled from the multimedia real-time storage subsystem. The time-stamps, the digital measurements are stored and documented. In a control room environment, the dispatcher or the service personal will have full access to the details of all previous operations, and will be eager to 'run back in time'. By archiving the key production processes through video and sound and text files, tele-maintenance operations can also be implemented. In this case, the service people - without being trained for a specific action list - will be directed on-line by replaying the canned courses and training sessions.

All test cases proved valuable for the end-user company, and also for the developers to get some feedback on applicability.

12. Conclusions

We prepared a development work on interactive multimedia for tele-presence by research on several underlying domains. The focus of the work was to develop generalized software (IMUTA) to allow the easy integration and management of multimedia information flows. The main results are the following:

- a coherent definition on term and terminology of tele-presence, & interactive multimedia (IMM);
- basic research in the filed of security and safety in MM application environment;
- specific and general models of business cases;
- a better understanding of the emerging MPEG standards;
- integrating virtual environment interfaces for augmented reality;
- systematic approach for the preparation of industrial implementations.

13. Acknowledgments

The authors express thanks to some active members of university and industrial partners, who helped in reaching valuable experiences and real-life results. The first phases of our work were supported by the Hungarian Ministry of Education under the Széchenyi NKFP research program 2/040/2001.

References

[1] Monostori L, Haidegger G, Váncza J and Viharos Zs J 2003 *Proc. of the Third Conf. on Mechanical Engineering* 540
[2] www-cdr.standfor.edu/telepresence

[3] www.ornl.gov/rpsd/humfac/page01d07.html
[4] Sheridan T 2001 *A Progress Report Control Eng. Practice* **3** (2) 205
[5] www.rational.com
[6] Kopácsi S, Sárközy F and Kovács G L 2002 *Proc. of the 11th Int. Workshop on Robotics in Alpe-Adria-Danube region (RAAD 2002)* 289
[7] Samet H 1990 *Applications of Spatial Data Structure* (Reading, MA: Addison-Wesley)
[8] Haidegger G, Kovács G L et al 2004 *Proc. of the 37th CIRP Int. Seminar on Manufacturing Systems* (Ed. L. Monostori Budapest, HU, 19-21 May 2004) 449
[9] Anufriev A 2003 *CIM Res.Lab. Digital Factory Working Document*

Journal of Physics: Conference Series **268** (2011) 012016 doi:10.1088/1742-6596/268/1/012016

Stability results for a soil model with singular hysteretic hydrology

P Krejčí[1], J P O'Kane[2], A Pokrovskii[3] and D Rachinskii[3]

[1] Mathematical Institute, Academy of Sciences of the Czech Republic, Žitná 25, CZ-11567 Praha 1, Czech Republic
[2] Department of Civil & Environmental Engineering, University College Cork, Ireland
[3] Department of Applied Mathematics, University College Cork, Ireland

E-mail: d.rachinskii@ucc.ie

Abstract. We consider a differential equation describing the mass balance in a soil hydrology model with noninvertible Preisach-type hysteresis. We approximate the singular Preisach operator by regular ones and show, as main result, that the solutions of the regularized problem converge to a solution of the original one as the regularization parameter tends to zero. For monotone right hand sides, we prove that the solution is unique. If in addition the external water sources are time periodic, then we find sufficient conditions for the existence, uniqueness, and global asymptotic stability of periodic solutions.

Introduction

Large highly structured systems often admit a large number of locally stable equilibrium configurations. This is in particular the case of solid mechanics, where dislocations at the crystal level lead to a multitude of different equilibria for the same stress distribution. The instantaneous state at some time $t_0 > 0$ thus may depend on the previous history $t \leq t_0$ of the process. Similar phenomena occur in the complex dynamics of electro-magneto-mechanical processes in ferromagnetic, piezoelectric, and magnetostrictive materials.

In engineering applications, the goal of modelling is to predict the behaviour of a system at the *macroscopic* level, where the knowledge of the exact complex time and space distribution of the microstates is of minor importance with respect to the necessity of having a reliable and robust numerical method for a global simulation.

The mathematical theory of hysteresis operators, introduced by M. A. Krasnosel'skii and his collaborators in the 1970', see [1], seems to provide an efficient tool for such a macroscopic description of internal microstructure evolution in the situation, where the structure changes are much faster than the observer's time scale. Then the process can be considered as rate-independent which is, besides causality, the main feature of hysteresis. Among more recent publications devoted to different aspects of modelling and analysis of systems with hysteresis we may cite e. g. [2–8].

We focus here on a model of soil hydrology proposed in [9–11]. It is based on the same idea of describing the complicated mass exchange dynamics in the soil, where the microstructure is due to solid grains, pores, plant roots, animal activity etc., by an input-output hysteresis system. The spatial dependence is neglected, and the dynamics is driven by the mass balance between the soil water potential and the volumetric moisture content. The experiments described in [9] show that the hysteresis relation between the potential u and the moisture content w exhibits the so-called *return point memory* (or *wiping-out property*), that is, every minor loop returns back to its starting point, see figure 1. Furthermore, it is expected from experiments that periodic processes on the same potential level may take place with different amounts of the water content. Assuming that the mass balance is given by the system of equations (1) below, this observation leads to the conclusion that all possible responses w corresponding to the same process $u(t)$ differ only by an additive constant independent of t. In the (u, w) phase plane, this means that all single closed loops with endpoints $(u_1, w_1), (u_2, w_2)$ and $(u_1, w_1'), (u_2, w_2')$ have the same shape, see figure 1. This property is called the *congruency* and a classical result by Mayergoyz in [5] states that every hysteresis relation with return point memory and congruent loops can be represented by the Preisach model.

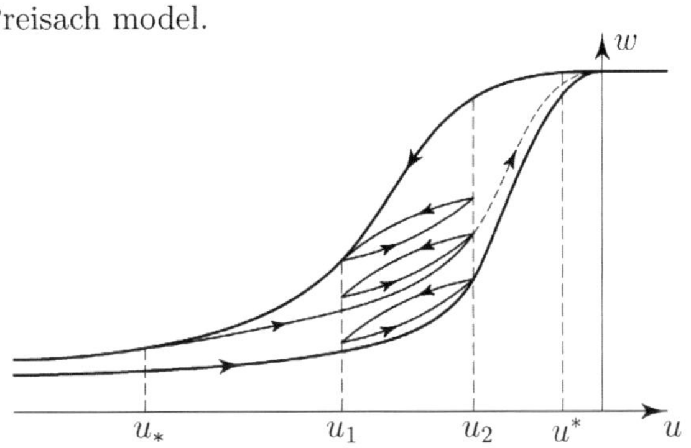

Figure 1. Return point memory and congruency.

A detailed discussion about the technical aspects of the model as well as the parameter identification is done in [9, 11]. The resulting system governing the process is of the form

$$\left.\begin{array}{rcl} \dot{w}(t) & = & f(t, u(t)) \\ w(t) & = & \mathcal{F}[\lambda, u](t) \\ u(0) & = & u^0 \end{array}\right\} \tag{1}$$

for $t \geq 0$, where $u^0 \in \mathbb{R}$ is a given initial condition, $f : [0, \infty) \times \mathbb{R} \to \mathbb{R}$ is a given function, and $\mathcal{F}[\lambda, \cdot]$ is a Preisach hysteresis operator with initial memory configuration λ, see Section 1.

Equation (1) is not autonomous, as the water-soil system is not physically closed. Water is coming in and out. The t-dependence in the right-hand side describes the external water exchange (rain, evaporation, drainage, etc.).

A derivation of system (1) and constructive methods for its solutions can be found in [12]. Here, we pursue these investigations and develop a general existence and stability theory for this problem. In Section 3, we show that solutions u to (1) can be locally uniformly approximated as $\varepsilon \to 0+$ by solutions u_ε to the regularized problem

$$\left.\begin{array}{rcl} \dot{w}_\varepsilon(t) & = & f_\varepsilon(t, u_\varepsilon(t)) \\ w_\varepsilon(t) & = & \varepsilon u_\varepsilon(t) + \mathcal{F}_\varepsilon[\lambda^\varepsilon, u_\varepsilon](t) \\ u_\varepsilon(0) & = & u_\varepsilon^0, \end{array}\right\} \tag{2}$$

where $f_\varepsilon, \mathcal{F}_\varepsilon, \lambda^\varepsilon, u_\varepsilon^0$ are suitable approximations of $f, \mathcal{F}, \lambda, u^0$, respectively.

This is indeed a regularization. The mappings $\mathcal{F}_\varepsilon^* : u \mapsto \varepsilon u + \mathcal{F}_\varepsilon[\lambda^\varepsilon, u]$ are Lipschitz continuous and admit Lipschitz continuous inverses for all $\varepsilon > 0$ in the Banach space $C[0, T]$ of all continuous functions on $[0, T]$ for every $T > 0$; hence, the existence and uniqueness of solutions to (2) follows from the contraction principle for every $\varepsilon > 0$. The convergence result is based on the observation that small amplitude Preisach hysteresis loops are convex. We prove here in Proposition 2.3

a refined form of an integral inequality (called *convexity inequality*, or *second order energy inequality*), which has been originally established in [13] in connection with hysteresis wave propagation and then used in [14] for solving a singular oscillation problem.

The solutions to (1) may in general be nonunique (see Example 3.3 (ii)). On the other hand, we show that the solution u to (1) is unique as long as $u(t)$ does not leave an interval $[u_*, u^*]$, where f is nonincreasing. If the interval $[u_*, u^*]$ contains 0, then the solution remains in this interval for all times. If moreover f is periodic in t and decreasing in u, then every solution trajectory converges to a unique periodic solution of the problem. The argument is based in a substantial way on the Hilpert inequality established in [15].

In the hydrological context, a nonincreasing function f is relevant to a common situation where water flows into the system are driven by negative feedback loops. Examples include infiltration of rain, drainage of water under a slab of soil, transpiration from plants leaves for vegetated soil and combinations of such flows [10, 11]. The results below guarantee the global stability of the corresponding models (1). However, a positive feedback loop can destabilize real soil-water systems. For instance, a destabilizing effect of the runoff flow from the soil surface, leading to a bifurcation, which has strong implications for the environment, has been shown in [16]. Equations (1) resulting from modelling hydrological systems with both negative and positive feedbacks can have non-monotone functions f.

The following text is divided into four sections. Section 1 is a survey of mathematical properties of the Preisach operator. For our purposes, it is convenient to use the alternative description of the Preisach model as a nonlinear one-parametric combination of continuous play operators, which goes back to [13], instead of the traditional one proposed in [17] and further developed in [1, 5], which defines the model as a linear superposition of two-parametric discontinuous relay operators. The equivalence of the two definitions was proved in [13, Section 1], see also [2, Section 2.4] or [3, Section II.3]. The variational approach based on the system of play operators makes in our situation the analysis simpler and more transparent. A detailed proof of the convexity inequality for the Preisach operator is carried out in Section 2. Solutions to Problem (2) for $\varepsilon > 0$ are investigated in Section 3. On a fixed time interval $[0, T]$, we derive a uniform bound for u_ε in $W^{1,2q}(0, T)$ for some $q > 1$ independent of ε, which enables us to prove the convergence along a subsequence to a solution of (1) as $\varepsilon \to 0+$. Global existence, uniqueness, periodicity, and asymptotic stability of solutions to (1) are discussed in Section 4.

1. Preisach operator

Let us denote by $\mathbb{R}_+$ the interval $[0, \infty)$. We work in the space $C(\mathbb{R}_+)$ of continuous functions $u : \mathbb{R}_+ \to \mathbb{R}$ endowed with a system of seminorms $\|u\|_{[0,t]} = \max\{|u(\tau)|\,;\, \tau \in [0, t]\}$ for $t \geq 0$. The metric

$$d(u, v) = \sup_{t \geq 0} \left(\frac{\|u - v\|_{[0,t]}}{1 + \|u - v\|_{[0,t]}} \right) \tag{3}$$

transforms $C(\mathbb{R}_+)$ into a Fréchet space. We similarly denote by $AC(\mathbb{R}_+)$ the set of absolutely continuous functions $u : \mathbb{R}_+ \to \mathbb{R}$ endowed with a system of seminorms $\|u\|_{AC[0,t]} = |u(0)| + \int_0^t |\dot{u}(\tau)| \,\mathrm{d}\tau$ for $t \geq 0$ and a metric analogous to (3).

We first introduce the Preisach state space Λ as the set of all functions $\lambda : \mathbb{R}_+ \to \mathbb{R}$ such that $|\lambda(r_1) - \lambda(r_2)| \leq |r_1 - r_2|$ for all $r_1, r_2 \in \mathbb{R}_+$ and

$$\exists R > 0 : \quad \lambda(r) = 0 \quad \forall r \geq R\,. \tag{4}$$

The set of all $\lambda \in \Lambda$ satisfying (4) will be denoted Λ_R in the sequel. Note that this set is compact with respect to the sup-norm.

For each given $r > 0$ we define the *play operator* $\mathfrak{p}_r : \Lambda \times AC(\mathbb{R}_+) \to AC(\mathbb{R}_+) : (\lambda, u) \mapsto \xi_r$, which with each $\lambda \in \Lambda$ and $u \in AC(\mathbb{R}_+)$ associates the unique solution ξ_r of the variational inequality (see figures 2, 3)

$$|u(t) - \xi_r(t)| \le r \qquad \forall t \in \mathbb{R}_+ , \tag{5}$$

$$\dot\xi_r(t)\,\big(u(t) - \xi_r(t) - z\big) \ge 0 \quad \text{a. e.} \qquad \forall z \in [-r, r] , \tag{6}$$

$$\xi_r(0) = \min\{u(0) + r, \max\{u(0) - r, \lambda(r)\}\} . \tag{7}$$

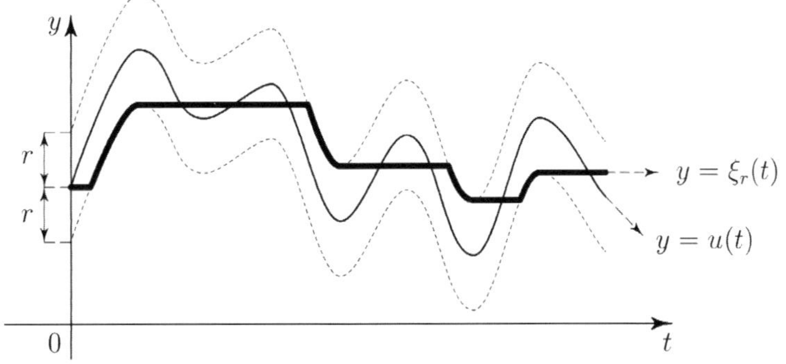

Figure 2. Time evolution of the play operator.

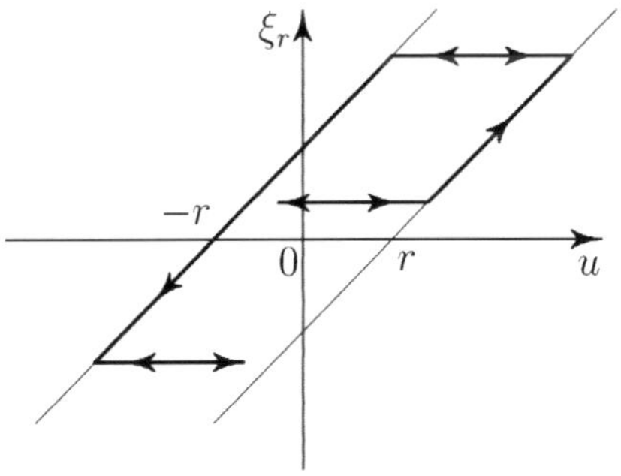

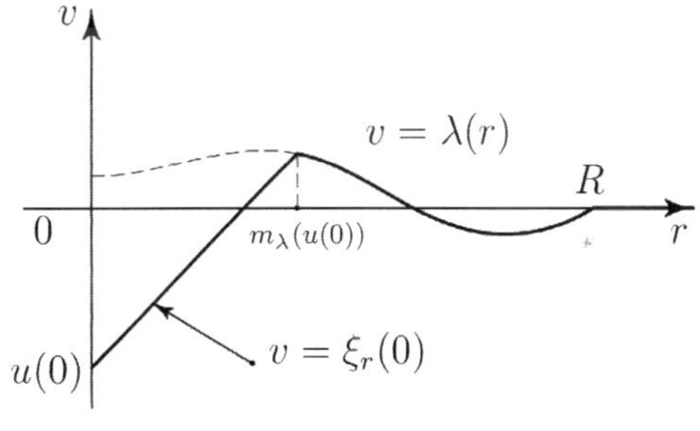

Figure 3. A phase diagram of the play (left) and the initial memory curve $v = \xi_r(0)$ (right).

More about the relationship between the variational definition of the play and a constructive approach in [1, 2] can be found e. g. in [3, 7].

As an immediate consequence of the definition, we first note that for all $h > 0$ sufficiently small we have $\dot\xi_r(t)\,\big(u(t) - u(t \pm h) - \xi_r(t) + \xi_r(t \pm h)\big) \ge 0$ a. e., hence

$$\dot\xi_r(t)\,\dot u(t) = \dot\xi_r^2(t) \quad \text{a. e.} \tag{8}$$

In other words, we have the implication $\dot\xi_r(t) \ne 0 \Rightarrow \dot\xi_r(t) = \dot u(t) = \frac{1}{r}\,|\dot\xi_r(t)|\,(u(t) - \xi(t))$.

We now recall an important inequality established originally by Hilpert in [15]. For the reader's convenience, we give an elementary proof.

Lemma 1.1 *For $u_1, u_2 \in AC(\mathbb{R}_+)$ and $\lambda_1, \lambda_2 \in \Lambda$ put $\xi_r^i(t) = \mathfrak{p}_r[\lambda_i, u_i](t)$, $i = 1, 2$. Then for every locally Lipschitz continuous non-decreasing function $g : \mathbb{R} \to \mathbb{R}$ we have*

$$\frac{\mathrm{d}}{\mathrm{d}t}\left(g(\xi_r^1) - g(\xi_r^2)\right)^+(t) \le H(u_1(t) - u_2(t))\,\frac{\mathrm{d}}{\mathrm{d}t}\left(g(\xi_r^1) - g(\xi_r^2)\right)(t) \quad a.\,e., \tag{9}$$

where $x^+ = \max\{x, 0\}$ for $x \in \mathbb{R}$, and H is the left continuous Heaviside function

$$H(x) = 0 \text{ for } x \le 0 , \quad H(x) = 1 \text{ for } x > 0 .$$

Proof. From (5) – (6) it follows that (we omit the argument t for simplicity)

$$g'(\xi_r^1)\,\dot\xi_r^1\left((u_1-\xi_r^1)-(u_2-\xi_r^2)\right)\ge 0,\quad -g'(\xi_r^2)\,\dot\xi_r^2\left((u_1-\xi_r^1)-(u_2-\xi_r^2)\right)\ge 0\quad \text{a.e.}\,,$$

hence

$$\frac{\mathrm{d}}{\mathrm{d}t}\left(g(\xi_r^1)-g(\xi_r^2)\right)\left((u_1-\xi_r^1)-(u_2-\xi_r^2)\right)\;\ge\;0\qquad\text{a.e.}\tag{10}$$

Using the implication $c(a-b)\ge 0\;\Rightarrow\;c(H(a)-H(b))\ge 0$ for $a,b,c\in\mathbb{R}$, we obtain from (10)

$$H(\xi_r^1-\xi_r^2)\,\frac{\mathrm{d}}{\mathrm{d}t}\left(g(\xi_r^1)-g(\xi_r^2)\right)\;\le\;H(u_1-u_2)\,\frac{\mathrm{d}}{\mathrm{d}t}\left(g(\xi_r^1)-g(\xi_r^2)\right)\qquad\text{a.e.}\tag{11}$$

For a.e. $t>0$ we have

$$\begin{aligned} g(\xi_r^1(t))=g(\xi_r^2(t)) \quad&\Rightarrow\quad \frac{\mathrm{d}}{\mathrm{d}t}\left(g(\xi_r^1)-g(\xi_r^2)\right)^+(t)=\frac{\mathrm{d}}{\mathrm{d}t}(g(\xi_r^1)-g(\xi_r^2))(t)=0\,,\\ g(\xi_r^1(t))\ne g(\xi_r^2(t)) \quad&\Rightarrow\quad H(\xi_r^1(t)-\xi_r^2(t))=H(g(\xi_r^1(t))-g(\xi_r^2(t)))\,,\end{aligned}$$

and (9) follows from the identity $\frac{\mathrm{d}}{\mathrm{d}t}x^+(t)=H(x(t))\dot x(t)$ a.e. for each $x\in AC(\mathbb{R}_+)$. ■

For completeness, we summarize the classical continuity properties of the play $\mathfrak{p}_r$. For a proof, see [2, Sect. 2.3] or [3, Proposition II.1.1].

Lemma 1.2 *Under the hypotheses of Lemma 1.1 we have*

$$\begin{aligned} &|\dot\xi_r^1-\dot\xi_r^2|(t)+\tfrac{\mathrm{d}}{\mathrm{d}t}|(u_1-u_2)-(\xi_r^1-\xi_r^2)|(t)\;\le\;|\dot u_1-\dot u_2|(t)\qquad \textit{a. e.}\,,\\ &\left|\xi_r^1-\xi_r^2\right|(t)\;\le\;\max\{|\lambda_1(r)-\lambda_2(r)|\,,\,\|u_1-u_2\|_{[0,t]}\}\quad\forall t\in\mathbb{R}_+\,.\end{aligned}$$

Lemma 1.2 states that $\mathfrak{p}_r:\Lambda\times AC(\mathbb{R}_+)\to AC(\mathbb{R}_+)$ is continuous and admits a continuous extension to $\Lambda\times C(\mathbb{R}_+)\to C(\mathbb{R}_+)$, and its restriction to any bounded interval $[0,T]$ is Lipschitz continuous in both cases.

We now continue with some finer memory properties of the play. Their proofs can be found e. g. in [2, Sect. 2.3] or [3, Sect. II.2].

Lemma 1.3 *For given $R>0$, $u\in C(\mathbb{R}_+)$, $\lambda\in\Lambda_R$, and $t\ge 0$ put*

$$\lambda_t(r)=\mathfrak{p}_r[\lambda,u](t)\qquad\textit{for}\;\;r>0\,,\quad\lambda_t(0)=u(t)\,.\tag{12}$$

Then

(i) $\lambda_t\in\Lambda_{R(t)}$, *where* $R(t)=\max\{R,\|u\|_{[0,t]}\}$*;*

(ii) *for all $s,t\in\mathbb{R}_+$ the following semigroup property holds:*

$$\mathfrak{p}_r[\lambda,u](s+t)\;=\;\mathfrak{p}_r[\lambda_s,u(s+\cdot)](t).\tag{13}$$

The play operator thus generates for every $t\ge 0$ a continuous *state mapping* $\Pi_t:\Lambda\times C(\mathbb{R}_+)\to\Lambda$ which with each $(\lambda,u)\in\Lambda\times C(\mathbb{R}_+)$ associates the state $\lambda_t\in\Lambda$ at time t. In other words, $\mathfrak{p}_r[\lambda,u](t)$ can be considered in a dual way: either as a function ξ_r of the time variable t for fixed r, or as a function λ_t of the memory variable r at fixed time t.

Let us recall another property of the play as a special case of [18, Lemma 3.1.2].

Lemma 1.4 *Let $u \in C(\mathbb{R}_+)$ and $t \geq 0$ be given. Set $u_{\max}(t) = \sup_{\tau\in[0,t]} u(\tau)$, $u_{\min}(t) = \inf_{\tau\in[0,t]} u(\tau)$. Then for all $\lambda \in \Lambda$ and $r > 0$ we have*

$$\mathfrak{p}_r[\lambda, u](\tau) \leq \max\{\lambda(r),\, u_{\max}(t) - r\}, \quad \mathfrak{p}_r[\lambda, u](\tau) \geq \min\{\lambda(r),\, u_{\min}(t) + r\} \quad \text{for } \tau \in [0, t]$$

and $\mathfrak{p}_r[\lambda, u](t) = \lambda(r)$ for $r > \|m_\lambda \circ u\|_{[0,t]}$, where for $v \in \mathbb{R}$ we put

$$m_\lambda(v) = \inf\{r \geq 0\,;\, |\lambda(r) - v| = r\}\,. \tag{14}$$

The meaning of m_λ can be illustrated on figures 3, 4. For a function u which is monotone in an interval $[t_0, t_1]$ and $u(t_0) = \lambda(0)$, the interval $[0, (m_\lambda \circ u)(t)]$ determines the active moving part of the memory curve $r \mapsto \lambda_t(r)$. More specifically, in the identity (8), we have in this case

$$\dot{u}(t) \neq 0 \implies \dot{\xi}_r(t) = \dot{u}(t) \ \text{ for } \ r < m_\lambda(u(t)) \ \text{ and } \ \dot{\xi}_r(t) = 0 \ \text{ for } \ r > m_\lambda(u(t))\,.$$

Let us derive some consequences of Lemma 1.4. Assume that $m_\lambda(u(\cdot))$ attains at a point $\bar{t} \geq 0$ its maximum over $[0, \bar{t}]$, that is,

$$\bar{r} := m_\lambda(u(\bar{t})) = \|m_\lambda(u(\cdot))\|_{[0,\bar{t}]}\,. \tag{15}$$

The case $\bar{r} = 0$ is trivial, as it implies $u(t) = \lambda(0)$ for all $t \in [0, \bar{t}]$. For $\bar{r} > 0$ we distinguish the cases: (i) $u(\bar{t}) = \lambda(\bar{r}) + \bar{r}$, and (ii) $u(\bar{t}) = \lambda(\bar{r}) - \bar{r}$. If (i) holds and $u(t) > u(\bar{t})$ for some $t \in [0, \bar{t}]$, then $\lambda(\bar{r}) + \bar{r} < u(t)$, hence $m_\lambda(u(t)) > \bar{r}$ in contradiction with (15). We thus have $u(\bar{t}) = u_{\max}(\bar{t})$. From (5) it follows that $\mathfrak{p}_r[\lambda, u](\bar{t}) \geq u(\bar{t}) - r$. On the other hand, (14) yields $u(\bar{t}) - r < \lambda(r) \Rightarrow m_\lambda(u(\bar{t})) < r$; hence, by Lemma 1.4 we have

$$\mathfrak{p}_r[\lambda, u](\bar{t}) = \max\{\lambda(r), u(\bar{t}) - r\}\,. \tag{16}$$

Similarly, in the case (ii) we have $u(\bar{t}) = u_{\min}(\bar{t})$ and $\mathfrak{p}_r[\lambda, u](\bar{t}) = \min\{\lambda(r), u(\bar{t}) + r\}$.

The above considerations imply the following well-known result on periodic inputs, cf. [1, Chap. 1, Sect. 2.8].

Corollary 1.5 *Let $u \in C(\mathbb{R}_+)$ be T-periodic, that is, $u(t+T) = u(t)$ for all $t \geq 0$, with a fixed period $T > 0$. Then $\mathfrak{p}_r[\lambda, u]$ is T-periodic for $t \geq T$ for all $\lambda \in \Lambda$.*

For a function $u \in C(\mathbb{R}_+)$ which is monotone (non-decreasing or non-increasing) in an interval $[t_0, t_1]$, we easily deduce from the semigroup property (13) and Lemma 1.4 the representation

$$\mathfrak{p}_r[\lambda, u](t) = \max\{u(t) - r, \min\{u(t) + r, \lambda_{t_0}(r)\}\} \tag{17}$$

for $t \in [t_0, t_1]$, see figures 2, 3. It is perhaps interesting to note that (17) has originally been used in [1] as alternative definition of the play on continuous piecewise monotone inputs, extended afterwards by density and continuity to the whole space of continuous functions.

The evolution of the graph of λ_t in dependence on t is depicted on figure 4. We now pass to the alternative definition of the Preisach operator as suggested in [13], see also [2, 3].

Definition 1.6 *Let $R > 0$ be given. Let $\mu : \mathbb{R}_+ \times \mathbb{R} \to \mathbb{R}_+$ be a locally bounded measurable function. For $(r, v) \in \mathbb{R}_+ \times \mathbb{R}$ put*

$$g(r, v) = \int_0^v \mu(r, z)\,\mathrm{d}z\,. \tag{18}$$

Then the mapping $\mathcal{F} : \Lambda_R \times C(\mathbb{R}_+) \to C(\mathbb{R}_+)$ defined as

$$\mathcal{F}[\lambda, u](t) = \int_0^\infty g(r, \mathfrak{p}_r[\lambda, u](t))\,\mathrm{d}r \quad \textit{for} \ (\lambda, u) \in \Lambda_R \times C(\mathbb{R}_+)\,,\ t \geq 0 \tag{19}$$

is called a Preisach operator.

The integral in (19) is finite due to the fact that $g(r,0) = 0$ for all $r > 0$ and $\mathfrak{p}_r[\lambda, u](t) = 0$ for r sufficiently large by virtue of Lemma 1.3 (i). We now list some basic properties of the Preisach operator. For the proofs see [2, Section 2.11] or [3, Section II.3].

Hypothesis 1.7 *Let μ be as in Definition 1.6. We assume in addition that*

(i) *For every $B > 0$ there exists a constant $\gamma_B > 0$ such that*

$$0 \leq \mu(r,v) \leq \gamma_B \quad a.\,e. \quad in \ \{(r,v) \in \mathbb{R}_+ \times \mathbb{R}\,,\ r + |v| \leq B\}\,.$$

(ii) *The function $\mu(r,\cdot)$ is locally Lipschitz continuous in $\mathbb{R}$ for almost all $r > 0$, and there exist constants $\alpha, \beta > 0$ and a continuous function $\varrho : \mathbb{R} \to (0,\infty)$ such that*

$$\mu(r,v) \geq \alpha, \quad \left|\frac{\partial \mu}{\partial v}(r,v)\right| \leq \beta \quad a.\,e. \ in \ \mathcal{C} := \{(r,v) \in \mathbb{R}_+ \times \mathbb{R}\,, 0 < r < \varrho(v)\}\,. \tag{20}$$

Taking a smaller $\varrho(v) > 0$, if necessary, we may assume that

$$\alpha \ \geq \ 4\beta\,\varrho(v) \quad \forall v \in \mathbb{R}\,. \tag{21}$$

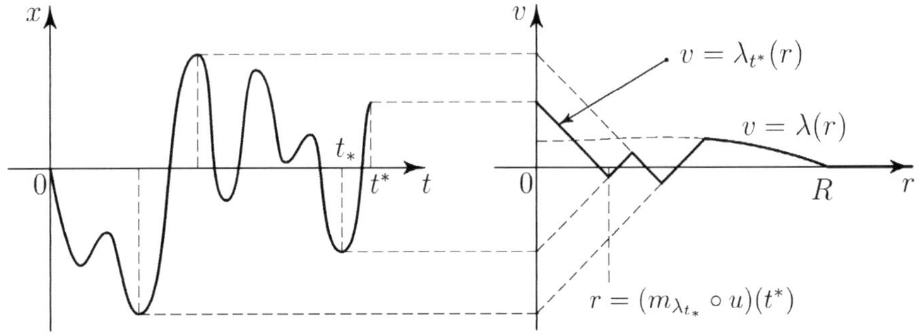

Figure 4. The memory curve $v = \lambda_t(r)$ at $t = t^*$.

We will see that $\mathcal{C}$ is the *convexity domain* of the Preisach operator in the sense that *closed hysteresis loops are convex and counterclockwise oriented as long as the active memory stays in $\mathcal{C}$*. The generalized convexity inequality below in Proposition 2.3 makes use of this fact. Its energetic interpretation is discussed in detail in [3, Section II.4]. We start with Lipschitz continuity properties of the Preisach and the inverse Preisach operator.

Proposition 1.8 *Let Hypothesis 1.7 (i) hold.*

(i) *Then for every $T > 0$, $B \geq R$, $\lambda_1, \lambda_2 \in \Lambda_R$ and $u_1, u_2 \in C(\mathbb{R}_+)$ such that $\|u_i\|_{[0,T]} \leq B$ for $i = 1,2$ and every $t \in [0,T]$ we have*

$$|\mathcal{F}[\lambda_1, u_1] - \mathcal{F}[\lambda_2, u_2]|(t) \ \leq \ \gamma_B \left(\int_0^R |\lambda_1 - \lambda_2|(r)\,\mathrm{d}r + B\,\|u_1 - u_2\|_{[0,t]}\right). \tag{22}$$

(ii) *For every $\alpha > 0$, $\lambda \in \Lambda$, the operator $\alpha\,I + \mathcal{F}[\lambda,\cdot] : C(\mathbb{R}_+) \to C(\mathbb{R}_+)$, where I is the identity mapping, is invertible and its inverse $\mathcal{G} = (\alpha\,I + \mathcal{F}[\lambda,\cdot])^{-1} : C(\mathbb{R}_+) \to C(\mathbb{R}_+)$ satisfies for every $u_1, u_2 \in C(\mathbb{R}_+)$ and every $T > 0$ the inequality*

$$\|\mathcal{G}[u_1] - \mathcal{G}[u_2]\|_{[0,T]} \ \leq \ 2\alpha^{-1}\,\|u_1 - u_2\|_{[0,T]}\,.$$

As an easy consequence of (9), we obtain the Hilpert inequality for the Preisach operator (19) in the following form.

Proposition 1.9 *Let μ and g be as in Definition 1.6, and let $\lambda_1, \lambda_2 \in \Lambda_R$ and $u_1, u_2 \in AC(\mathbb{R}_+)$ be given. Then for a. e. $t > 0$ we have*

$$
\begin{aligned}
& H(u_1(t) - u_2(t)) \frac{\mathrm{d}}{\mathrm{d}t} \left(\mathcal{F}[\lambda_1, u_1] - \mathcal{F}[\lambda_2, u_2]\right)(t) \\
& \quad \geq \frac{\mathrm{d}}{\mathrm{d}t} \int_0^\infty \left(g(r, \mathfrak{p}_r[\lambda_1, u_1]) - g(r, \mathfrak{p}_r[\lambda_2, u_2])\right)^+(t)\, \mathrm{d}r\,. \qquad (23)
\end{aligned}
$$

Interchanging the roles of u_1 and u_2 we obtain as a consequence of Proposition 1.9 that

$$
\begin{aligned}
& \operatorname{sign}(u_1(t) - u_2(t)) \frac{\mathrm{d}}{\mathrm{d}t} \left(\mathcal{F}[\lambda_1, u_1] - \mathcal{F}[\lambda_2, u_2]\right)(t) \\
& \quad \geq \frac{\mathrm{d}}{\mathrm{d}t} \int_0^\infty \left|g(r, \mathfrak{p}_r[\lambda_1, u_1]) - g(r, \mathfrak{p}_r[\lambda_2, u_2])\right|(t)\, \mathrm{d}r\,. \qquad (24)
\end{aligned}
$$

In the special cases $u_1 = u$, $\lambda_1 = \lambda$, $u_2 \equiv 0$, $\lambda_2 \equiv 0$, or $u_1 \equiv 0$, $\lambda_1 \equiv 0$, $u_2 = u$, $\lambda_2 = \lambda$, we obtain for a. e. t, as a special case of inequality (23), that

$$
H(u(t)) \frac{\mathrm{d}}{\mathrm{d}t} \mathcal{F}[\lambda, u](t) \;\geq\; \frac{\mathrm{d}}{\mathrm{d}t} \int_0^\infty \left(g(r, \mathfrak{p}_r[\lambda, u](t))\right)^+ \mathrm{d}r\,, \qquad (25)
$$

$$
-H(-u(t)) \frac{\mathrm{d}}{\mathrm{d}t} \mathcal{F}[\lambda, u](t) \;\geq\; \frac{\mathrm{d}}{\mathrm{d}t} \int_0^\infty \left(g(r, \mathfrak{p}_r[\lambda, u](t))\right)^- \mathrm{d}r\,, \qquad (26)
$$

where we denote $x^- = \max\{-x, 0\}$ for $x \in \mathbb{R}$.

2. Generalized convexity inequality

This section is devoted to the convexity inequality as mentioned in the Introduction. We first investigate in detail the behavior of $\mathcal{F}[\lambda, u]$ on monotone inputs u.

Lemma 2.1 *Let $R > 0$, $\lambda \in \Lambda_R$, $u \in C(\mathbb{R}_+)$ and $t_0 \geq 0$ be given. Let Hypothesis 1.7 hold, and let $\mathcal{F}$ be the Preisach operator (19). Set $\lambda_{t_0}(r) = \mathfrak{p}_r[\lambda, u](t_0)$ for $r > 0$ and*

$$
\begin{aligned}
\Phi_+(v) &= \int_{\lambda_{t_0}(0)}^{v} \int_0^{m_{\lambda_{t_0}}(z)} \mu(r, z - r)\, \mathrm{d}r\, \mathrm{d}z \qquad \text{for } v \geq \lambda_{t_0}(0)\,, \\
\Phi_-(v) &= -\int_{v}^{\lambda_{t_0}(0)} \int_0^{m_{\lambda_{t_0}}(z)} \mu(r, z + r)\, \mathrm{d}r\, \mathrm{d}z \qquad \text{for } v \leq \lambda_{t_0}(0)\,,
\end{aligned}
$$

where $m_{\lambda_{t_0}}$ is as in (14). Let u be monotone (nondecreasing or nonincreasing) in an interval $[t_0, t_1]$. Then for all $t \in [t_0, t_1]$ we have

$$
\mathcal{F}[\lambda, u](t) - \mathcal{F}[\lambda, u](t_0) = \begin{cases} \Phi_+(u(t)) & \text{if } u(t_1) \geq u(t_0)\,, \\ \Phi_-(u(t)) & \text{if } u(t_1) \leq u(t_0)\,. \end{cases} \qquad (27)
$$

Equation (27) is an explicit formula for the *Preisach loading curves* Φ_+, Φ_- associated with monotone loadings. In particular, the *initial loading curves* are given by the equations

$$
\hat{\Phi}_+(v) = \int_{\lambda(0)}^{v} \int_0^{m_\lambda(z)} \mu(r, z - r)\, \mathrm{d}r\, \mathrm{d}z \qquad \text{for } v \geq \lambda(0)\,, \qquad (28)
$$

$$
\hat{\Phi}_-(v) = -\int_{v}^{\lambda(0)} \int_0^{m_\lambda(z)} \mu(r, z + r)\, \mathrm{d}r\, \mathrm{d}z \qquad \text{for } v \leq \lambda(0)\,. \qquad (29)
$$

Condition (21) implies that the function ϱ is bounded and we denote by $\hat{\varrho}$ its smallest upper bound. We further introduce the notation

$$\varrho_+ = \int_{\lambda(0)+\hat{\varrho}}^{\infty} \varrho(v)\ \mathrm{d}v \ \in\ (0,+\infty]\,, \qquad \varrho_- = \int_{-\infty}^{\lambda(0)-\hat{\varrho}} \varrho(v)\ \mathrm{d}v \ \in\ (0,+\infty]\,.$$

We easily check that under Hypothesis 1.7, we have

$$\lim_{v\to\infty} \hat{\Phi}_+(v) \geq \alpha\varrho_+\,, \quad \lim_{v\to-\infty} \hat{\Phi}_-(v) \leq -\alpha\varrho_-\,. \tag{30}$$

Indeed, an application of Fubini's Theorem yields

$$\begin{aligned}\hat{\Phi}_+(v) &= \int_0^{m_\lambda(v)} \int_{\lambda(r)}^{v-r} \mu(r,z)\ \mathrm{d}z\ \mathrm{d}r \geq \int_0^{(v-\lambda(0))/2} \int_{\lambda(0)+r}^{v-r} \mu(r,z)\ \mathrm{d}z\ \mathrm{d}r \\ &= \int_{\lambda(0)}^{v} \int_0^{\min\{v-z,z-\lambda(0)\}} \mu(r,z)\ \mathrm{d}r\ \mathrm{d}z \geq \alpha \int_{\lambda(0)}^{v} \min\{v-z, z-\lambda(0), \varrho(z)\}\ \mathrm{d}z\,.\end{aligned}$$

The computation for $\hat{\Phi}_-$ is similar, and (30) follows.

Proof of Lemma 2.1. Assume first that u in nondecreasing in $[t_0,t_1]$. Then, by (16),

$$\mathfrak{p}_r[\lambda,u](t) \ = \ \max\{\lambda_{t_0}(r), u(t)-r\} \quad \text{for } \ r>0,\ t\in[t_0,t_1]\,.$$

In particular, $\mathfrak{p}_r[\lambda,u](t) = \lambda_{t_0}(r)$ for $r > m_{\lambda_{t_0}}(u(t))$. Hence, by Fubini's Theorem,

$$\mathcal{F}[\lambda,u](t) - \mathcal{F}[\lambda,u](t_0) = \int_0^\infty (g(r,\mathfrak{p}_r[\lambda,u](t)) - g(r,\lambda_{t_0}(r)))\ \mathrm{d}r$$

$$= \int_0^{m_{\lambda_{t_0}}(u(t))} \int_{\lambda_{t_0}(r)+r}^{u(t)} \mu(r,z-r)\ \mathrm{d}z\ \mathrm{d}r = \int_{\lambda_{t_0}(0)}^{u(t)} \int_0^{m_{\lambda_{t_0}}(z)} \mu(r,z-r)\ \mathrm{d}r\ \mathrm{d}z = \Phi_+(u(t))\,.$$

Similarly, if u in nonincreasing, $\mathfrak{p}_r[\lambda,u](t) \ = \ \min\{\lambda_{t_0}(r), u(t)+r\}$ for $r>0$, $t\in[t_0,t_1]$, hence,

$$\begin{aligned}&\mathcal{F}[\lambda,u](t_0) - \mathcal{F}[\lambda,u](t) = \int_0^\infty (g(r,\lambda_{t_0}(r)) - g(r,\mathfrak{p}_r[\lambda,u](t)))\ \mathrm{d}r \\ &= \int_0^{m_{\lambda_{t_0}}(u(t))} \int_{u(t)}^{\lambda_{t_0}(r)-r} \mu(r,z+r)\,\mathrm{d}z\,\mathrm{d}r = \int_{u(t)}^{\lambda_{t_0}(0)} \int_0^{m_{\lambda_{t_0}}(z)} \mu(r,z+r)\,\mathrm{d}r\,\mathrm{d}z = -\Phi_-(u(t)).\end{aligned} \tag{31}$$

■

In the situation of Lemma 2.1, set for $s\geq 0$

$$\varphi_+(s) = \int_0^s \mu(r,\lambda_{t_0}(s)+s-r)\ \mathrm{d}r\,, \qquad \varphi_-(s) = \int_0^s \mu(r,\lambda_{t_0}(s)-s+r)\ \mathrm{d}r\,. \tag{32}$$

Then

$$\Phi_+'(v) = \varphi_+(m_{\lambda_{t_0}}(v)) \quad \text{for a.e.} \quad v>\lambda_{t_0}(0); \quad \Phi_-'(v) = \varphi_-(m_{\lambda_{t_0}}(v)) \quad \text{for a.e.} \quad v<\lambda_{t_0}(0). \tag{33}$$

Note that

$$m_{\lambda_{t_0}}(v_1) - m_{\lambda_{t_0}}(v_2) \ \geq\ (v_1-v_2)/2 \quad \text{for} \quad v_1>v_2>\lambda_{t_0}(0), \tag{34}$$

$$m_{\lambda_{t_0}}(v_1) - m_{\lambda_{t_0}}(v_2) \ \geq\ (v_2-v_1)/2 \quad \text{for} \quad v_1<v_2<\lambda_{t_0}(0). \tag{35}$$

Referring to the notation of Lemma 2.1, we fix some $B \ge \max\{R, \|u\|_{[0,t_1]}\}$. By Lemma 1.4, we have $|\lambda_{t_0}(s)| \le \max\{B - s, 0\}$ for all $s \ge 0$. In particular, this yields for $0 < r < s \le B$

$$|\lambda_{t_0}(s) + s - r| + r \le B\,, \qquad |\lambda_{t_0}(s) - s + r| + r \le B\,.$$

Let ϱ_B be any number such that

$$0 < \varrho_B \le \min\{\varrho(v); |v| \le B\}\,. \tag{36}$$

Making use of Hypothesis 1.7 (ii), we obtain from (32) that

$$\varphi_\pm(s) \ge \alpha\varrho_B \ \text{ for } \ s \in [\varrho_B, B]\,; \qquad \varphi'_\pm(s) \ge \alpha - 2\beta\varrho_B \ge \alpha/2 \ \text{ for a.e. } \ s \in (0, \varrho_B)\,. \tag{37}$$

Combined with (33)–(35), these relations imply

$$(\Phi'_+(v_1) - \Phi'_+(v_2))/(v_1 - v_2) \ge \alpha/4 \ \text{ if } \ v_1 > v_2 > \lambda_{t_0}(0),\ 0 \le m_{\lambda_{t_0}}(v_2) < m_{\lambda_{t_0}}(v_1) \le \varrho_B \tag{38}$$

$$(\Phi'_-(v_2) - \Phi'_-(v_1))/(v_2 - v_1) \le -\alpha/4 \ \text{ if } \ v_1 < v_2 < \lambda_{t_0}(0),\ 0 \le m_{\lambda_{t_0}}(v_2) < m_{\lambda_{t_0}}(v_1) \le \varrho_B\,. \tag{39}$$

This is precisely what we had in mind when we introduced the convexity domain $\mathcal{C}$ in Hypothesis 1.7. By virtue of (27) and (38)–(39), Φ_+ is a convex function describing increasing branches and Φ_- is a concave function describing decreasing branches as long as the evolution takes place in $\mathcal{C}$, hence small closed hysteresis loops are counterclockwise convex.

We now state an auxiliary result, which is a basis for the generalized convexity inequality for the Preisach operator we give below.

Lemma 2.2 *Let $[a, b] \subset \mathbb{R}$, $\delta > 0$ and let $u, v \in W^{1,1}(a, b)$ be given functions such that $v(t) \ge 0$, and u is monotone (nondecreasing or nonincreasing) in $[a, b]$. Let $\psi : \mathrm{Conv}\{u(a), u(b)\} \to (0, \infty)$, where $\mathrm{Conv}\, A$ denotes the convex hull of a set A, be a function such that*

$$\psi(z_1) - \psi(z_2) \ \ge \ \delta(z_1 - z_2) \qquad \text{if } \ u(a) \le z_2 < z_1 \le u(b)\,, \tag{40}$$

$$\psi(z_1) - \psi(z_2) \ \le \ -\delta(z_1 - z_2) \qquad \text{if } \ u(b) \le z_2 < z_1 \le u(a)\,. \tag{41}$$

Then for an arbitrary $p > 0$

$$\int_a^b \frac{\dot v(t)}{\psi^p(u(t))}\,\mathrm{d}t \ \ge \ \left[\frac{v(t)}{\psi^p(u(t))}\right]_a^b + p\delta \int_a^b \frac{|\dot u(t)|\, v(t)}{\psi^{p+1}(u(t))}\,\mathrm{d}t\,. \tag{42}$$

Proof. We only treat the case that u is nondecreasing. Then ψ is positive and increasing; hence, $\psi \circ u$ is nondecreasing and the integrals are meaningful. We may approximate ψ by continuously differentiable functions ψ_n satisfying (40) and such that $\psi_n(z) \to \psi(z)$ for each $z \in [u(a), u(b)]$. Inequality (42) holds for each ψ_n as a direct consequence of an integration by parts. Passing to the limit in n we obtain the assertion. ∎

Consider now some $\varepsilon_0 > 0$ and a family $\{\mathcal{F}_\varepsilon\,;\, \varepsilon \in (0, \varepsilon_0)\}$ of Preisach operators of the form (18)–(19), associated with generating functions μ_ε and initial states $\lambda^\varepsilon \in \Lambda_R$ for some fixed $R > 0$, assuming that Hypothesis 1.7 is satisfied independently of ε. For $\varepsilon \in (0, \varepsilon_0)$ we define the operator $\mathcal{G}_\varepsilon : C(\mathbb{R}_+) \to C(\mathbb{R}_+)$ by the formula

$$\mathcal{G}_\varepsilon \ = \ (\varepsilon I + \mathcal{F}_\varepsilon[\lambda^\varepsilon, \cdot])^{-1}\,. \tag{43}$$

By Proposition 1.8, $\mathcal{G}_\varepsilon$ is Lipschitz continuous with Lipschitz constant $2/\varepsilon$. As the main result of this section, we derive an estimate for $\mathcal{G}_\varepsilon$ *independent of* ε and similar to [14, Theorem 2.1] as a generalization of [3, Proposition II.4.22]. Consistently with (12), we denote $\lambda^\varepsilon_t(r) = \mathfrak{p}_r[\lambda^\varepsilon, u_\varepsilon](t)$ for $t \ge 0$ and $r \ge 0$.

Proposition 2.3 (Generalized convexity inequality) *Let $\mathcal{G}_\varepsilon$ be as in (43) and let $q > 1$ be given. Let $\{w_\varepsilon\}_{\varepsilon\in(0,\varepsilon_0)}$ be a system of functions in $W^{2,q}_{\mathrm{loc}}(\mathbb{R}_+)$ parameterized by $\varepsilon > 0$, and set*

$$u_\varepsilon \;=\; \mathcal{G}_\varepsilon[w_\varepsilon]\,. \tag{44}$$

Let $B > R$ be given, and let ϱ_B, α be as in (36). Assume that for every $\varepsilon \in (0, \varepsilon_0)$ we have

$$\text{either }\; \dot w_\varepsilon(0) = 0\,, \quad \text{or }\; (m_{\lambda_0^\varepsilon} \circ u_\varepsilon)(0+) \ge \varrho_B\,. \tag{45}$$

Then there exists a constant $C_{q,B} > 0$ depending only on q and B such that for every $T > 0$ and every $\varepsilon \in (0, \varepsilon_0)$ with the property

$$\|u_\varepsilon\|_{[0,T]} \;<\; B \tag{46}$$

we have

$$\int_0^T |\dot u_\varepsilon(t)|^{2q}\,\mathrm{d}t \;\le\; C_{q,B}\int_0^T \left(|\ddot w_\varepsilon(t)|^q + |\dot w_\varepsilon(t)|^{2q}\right)\mathrm{d}t\,. \tag{47}$$

Relation (45) is a condition of compatibility between the initial memory configuration λ^ε and the initial velocity. Note that for every monotone function u on an interval $[t_0, t_1]$, every $\lambda \in \Lambda$, and every $t \in (t_0, t_1)$, we have the implication $\exists \dot u(t) \ne 0 \implies (m_{\lambda_{t_0}} \circ u)(t) > 0$. Indeed, $(m_{\lambda_{t_0}} \circ u)(t) = 0$ would imply $u(t) = \lambda_{t_0}(0) = u(t_0)$, which is a contradiction. The following Example 2.4 shows that Proposition 2.3 does not hold if condition (45) is violated.

Example 2.4 Consider the case $\lambda^\varepsilon(r) = \lambda(r) \equiv 0$, $\mu_\varepsilon(r,v) = \mu(r,v) \equiv 1$, $w_\varepsilon(0) = w(0) = 0$. In particular, $m_\lambda(v) = |v|$ for all $v \in \mathbb{R}$. By (7) and (19), we have

$$\mathcal{F}[\lambda, u_\varepsilon](0) \;=\; \int_0^\infty g(r, \xi_r^\varepsilon(0))\,\mathrm{d}r$$

with $\xi_r^\varepsilon(0) \;=\; \min\{u_\varepsilon(0) + r, \max\{u_\varepsilon(0) - r, 0\}\}$, hence $\mathcal{F}[\lambda, u_\varepsilon](0) = \frac{1}{2}|u_\varepsilon(0)|u_\varepsilon(0)$. From (44) it follows that $\varepsilon u_\varepsilon(0) + \frac{1}{2}|u_\varepsilon(0)|u_\varepsilon(0) = 0$, hence $u_\varepsilon(0) = 0$. Let now $w_\varepsilon = w$ be increasing in $[0, T]$. Then the functions u_ε are increasing as well by virtue of (8), and from Lemma 2.1 we obtain $\mathcal{F}[\lambda, u_\varepsilon](t) = \frac{1}{2}u_\varepsilon^2(t)$, hence $u_\varepsilon(t)$ is the solution of the equation

$$\varepsilon u_\varepsilon(t) + \frac{1}{2}u_\varepsilon^2(t) = w(t)\,.$$

If, for example, $w(t) = t$ for $t \in [0, T]$, then $u_\varepsilon(t) = \sqrt{\varepsilon^2 + 2t} - \varepsilon$. We see that $\int_0^T |\dot u_\varepsilon(t)|^{2q}\,dt$ is not bounded independently of ε for any $q \ge 1$, although the right hand side of (47) is bounded.

Proof of Proposition 2.3. We fix $T > 0$ and $\varepsilon > 0$ such that (46) holds, and assume first that w_ε has only finitely many monotonicity intervals in $[0, T]$. Let $[a, b] \subset [0, T]$ be a monotonicity interval of w_ε. Set

$$c \;=\; \sup\{t \in [a, b];\, (m_{\lambda_a^\varepsilon} \circ u_\varepsilon)(t+) \le \varrho_B\}$$

with the convention $c = a$ if $(m_{\lambda_a^\varepsilon} \circ u_\varepsilon)(a+) \ge \varrho_B$. The function $t \mapsto m_{\lambda_a^\varepsilon}(u_\varepsilon(t))$ is left-continuous and nondecreasing in $[a, b]$. By (31)–(37) we have

$$|\dot w_\varepsilon(t)| \;\ge\; (\varepsilon + \varrho_B\alpha)\,|\dot u_\varepsilon(t)| \qquad \text{for a.e. } \; t \in (c, b)\,. \tag{48}$$

In (a, c), we distinguish the cases that w_ε increases or decreases. First assume $\dot w_\varepsilon(t) \ge 0$ for $t \in (a, c)$. Let Φ_+ be the loading curve on $[u_\varepsilon(a), u_\varepsilon(c)]$ for $\mathcal{F}_\varepsilon$ as in Lemma 2.1. Then

$$\dot w_\varepsilon(t) \;=\; \left(\varepsilon + \Phi_+'(u_\varepsilon(t))\right)\dot u_\varepsilon(t) \qquad \text{for a.e. } \; t \in (a, c)\,. \tag{49}$$

Set $p = 2q - 2$. Then

$$\ddot{w}_\varepsilon(t)\,\dot{u}_\varepsilon^p(t) \;=\; \frac{\frac{\mathrm{d}}{\mathrm{d}t}\left(\dot{w}_\varepsilon^{p+1}(t)\right)}{(p+1)\left(\varepsilon + \Phi'_+(u_\varepsilon(t))\right)^p} \qquad \text{for a.e.} \quad t \in (a,c)\,. \tag{50}$$

By (38), we may use Lemma 2.2 with $v(t) = \dot{w}_\varepsilon^{p+1}(t)$, $\psi(z) = \varepsilon + \Phi'_+(z)$, $u(t) = u_\varepsilon(t)$, and obtain from (49) that

$$\int_a^c \ddot{w}_\varepsilon(t)\,\dot{u}_\varepsilon^p(t)\;\mathrm{d}t \geq -\frac{\dot{w}_\varepsilon^{p+1}(a)}{(p+1)\left(\varepsilon + \Phi'_+(u_\varepsilon(a))\right)^p} + \frac{\alpha p}{4(p+1)} \int_a^c \frac{\dot{u}_\varepsilon(t)\,\dot{w}_\varepsilon^{p+1}(t)}{\left(\varepsilon + \Phi'_+(u_\varepsilon(t))\right)^{p+1}}\,\mathrm{d}t$$

$$= -\frac{1}{p+1}\dot{u}_\varepsilon^p(a)\,\dot{w}_\varepsilon(a) + \frac{\alpha p}{4(p+1)}\int_a^c \dot{u}_\varepsilon^{p+2}(t)\,\mathrm{d}t = -\frac{1}{p+1}\dot{u}_\varepsilon^p(a)\,\dot{w}_\varepsilon(a) + \frac{\alpha p}{4(p+1)}\int_a^c \dot{u}_\varepsilon^{2q}(t)\,\mathrm{d}t\,.$$

Hence, by Hölder's inequality,

$$\frac{\alpha p}{4(p+1)}\int_a^c \dot{u}_\varepsilon^{2q}(t)\;\mathrm{d}t \leq \frac{1}{p+1}\dot{u}_\varepsilon^p(a)\,\dot{w}_\varepsilon(a) + \left(\int_a^c |\ddot{w}_\varepsilon(t)|^q\;\mathrm{d}t\right)^{2/(2+p)} \left(\int_a^c \dot{u}_\varepsilon^{2q}(t)\;\mathrm{d}t\right)^{p/(2+p)}.$$

We see that there exists a constant $C > 0$ depending only on p and α such that

$$\int_a^c \dot{u}_\varepsilon^{2q}(t)\;\mathrm{d}t \;\leq\; C\left(\dot{u}_\varepsilon^p(a)\,\dot{w}_\varepsilon(a) + \int_a^c |\ddot{w}_\varepsilon(t)|^q\;\mathrm{d}t\right).$$

Taking into account inequality (48), we thus have, with a possibly different constant depending only on q, ϱ_B, and α,

$$\int_a^b \dot{u}_\varepsilon^{2q}(t)\;\mathrm{d}t \;\leq\; C\left(\dot{u}_\varepsilon^p(a)\,\dot{w}_\varepsilon(a) + \int_a^c |\ddot{w}_\varepsilon(t)|^q\;\mathrm{d}t + \int_c^b |\dot{w}_\varepsilon(t)|^{2q}\;\mathrm{d}t\right).$$

Let now $\dot{w}_\varepsilon(t) \leq 0$ in (a,b), hence also $\dot{u}_\varepsilon(t) \leq 0$ a.e. in (a,b). Then (48) still holds, and, as a counterpart of (49)–(50), we have $\dot{w}_\varepsilon(t) \;=\; \left(\varepsilon + \Phi'_-(u_\varepsilon(t))\right)\dot{u}_\varepsilon(t)$ for a.e. $t \in (a,c)$ and

$$-\ddot{w}_\varepsilon(t)\,|\dot{u}_\varepsilon(t)|^p \;=\; \frac{\frac{\mathrm{d}}{\mathrm{d}t}\left((-\dot{w}_\varepsilon)^{p+1}(t)\right)}{(p+1)\left(\varepsilon + \Phi'_-(u_\varepsilon(t))\right)^p} \qquad \text{for a.e.} \quad t \in (a,c)\,.$$

We now argue as above using Lemma 2.2, and obtain again

$$\int_a^b |\dot{u}_\varepsilon|^{2q}(t)\;\mathrm{d}t \;\leq\; C\left(|\dot{u}_\varepsilon(a)|^p\,|\dot{w}_\varepsilon(a)| + \int_a^c |\ddot{w}_\varepsilon(t)|^q\;\mathrm{d}t + \int_c^b |\dot{w}_\varepsilon(t)|^{2q}\;\mathrm{d}t\right).$$

We conclude that there exists $C_{q,B}$ with the desired properties such that in every monotonicity interval $[a,b] \subset [0,T]$ we have

$$\int_a^b |\dot{u}_\varepsilon|^{2q}(t)\;\mathrm{d}t \;\leq\; C_{q,B}\left(|\dot{u}_\varepsilon(a)|^p\,|\dot{w}_\varepsilon(a)| + \int_a^b \left(|\ddot{w}_\varepsilon(t)|^q + |\dot{w}_\varepsilon(t)|^{2q}\right)\mathrm{d}t\right). \tag{51}$$

Let now $0 = t_0 < t_1 < \cdots < t_m = T$ be a minimal partition of $[0,T]$ such that w_ε is monotone in each interval $[t_{k-1}, t_k]$, $k = 1, \ldots, m$. We now apply formula (51) with $a = t_{k-1}$, $b = t_k$. For all $k = 2, \ldots, m$ we have $\dot{w}_\varepsilon(t_{k-1}) = 0$ (note that w_ε is continuously differentiable). Moreover, in the case $a = 0$, $b = t_1$, we have by hypothesis (45) that either $\dot{w}_\varepsilon(0) = 0$, or $|\dot{w}_\varepsilon(t)| \geq (\varepsilon + \varrho_B \alpha)\,|\dot{u}_\varepsilon(t)|$ for a. e. $t \in (0, t_1)$. In both cases we conclude that

$$\int_{t_{k-1}}^{t_k} |\dot{u}_\varepsilon|^{2q}(t)\,\mathrm{d}t \;\leq\; C_{q,B} \int_{t_{k-1}}^{t_k} \left(|\ddot{w}_\varepsilon(t)|^q + |\dot{w}_\varepsilon(t)|^{2q}\right) \mathrm{d}t, \qquad k = 1, \ldots, m.$$

Summing up over k we obtain the assertion for every piecewise monotone function w_ε. Let now $w_\varepsilon \in W^{2,q}_{\mathrm{loc}}(\mathbb{R}_+)$ and $T > 0$ be arbitrary. We approximate w_ε on $[0, T]$ by piecewise monotone functions $w_\varepsilon^{(k)}$ converging strongly to w_ε in $W^{2,q}(0, T)$. In particular, $w_\varepsilon^{(k)}$ converge to w_ε uniformly in $C[0, T]$, hence, by continuity of $\mathcal{G}_\varepsilon$, $u_\varepsilon^{(k)} = \mathcal{G}_\varepsilon[w_\varepsilon^{(k)}]$ converge uniformly to $u_\varepsilon = \mathcal{G}_\varepsilon[w_\varepsilon]$. For sufficiently large k, condition (46) is preserved, and by the above considerations,

$$\int_0^T |\dot{u}_\varepsilon^{(k)}|^{2q}(t)\,\mathrm{d}t \;\leq\; C_{q,B} \int_0^T \left(|\ddot{w}_\varepsilon^{(k)}(t)|^q + |\dot{w}_\varepsilon^{(k)}(t)|^{2q}\right) \mathrm{d}t\,,$$

hence, $\{\dot{u}_\varepsilon^{(k)}\}$ is a bounded sequence in $L^{2q}(0, T)$. Consequently, $\dot{u}_\varepsilon^{(k)}$ converge weakly to $\dot{u}_\varepsilon$ in $L^{2q}(0, T)$, and the assertion follows. ■

3. Regularization

In this section, we show that solutions to problem (2) converge locally uniformly as $\varepsilon \to 0+$ to solutions to (1) under the following hypotheses.

Hypothesis 3.1 *Let $q > 1$, $\varepsilon_0 > 0$, $R > 0$ be fixed, let $f_\varepsilon : \mathbb{R}_+ \times \mathbb{R} \to \mathbb{R}$ be continuous functions in both variables, and let $\mathcal{F}_\varepsilon$ be Preisach operators of the form (18)–(19), associated with generating functions μ_ε and initial states $\lambda^\varepsilon \in \Lambda_R$ for $\varepsilon \in (0, \varepsilon_0)$. The following conditions are assumed to hold.*

(i) *Hypothesis 1.7 holds for each μ_ε independently of $\varepsilon \in (0, \varepsilon_0)$.*

(ii) *There exists $f^0 \in L^{2q}_{\mathrm{loc}}(\mathbb{R}_+)$ such that $f^0(t) \geq 0$, and for all $(\varepsilon, t, v) \in (0, \varepsilon_0) \times \mathbb{R}_+ \times \mathbb{R}$*

$$v\, f_\varepsilon(t, v) \leq f^0(t)|v|. \tag{52}$$

(iii) *For every $B > 0$ there exist functions $\ell_B \in L^{2q}_{\mathrm{loc}}(\mathbb{R}_+)$, $f_B^1 \in L^q_{\mathrm{loc}}(\mathbb{R}_+)$, such that for almost all $(t, v) \in \mathbb{R}_+ \times (-B, B)$ and all $\varepsilon \in (0, \varepsilon_0)$ we have*

$$\left|\tfrac{\partial f_\varepsilon}{\partial t}(t, v)\right| \leq f_B^1(t)\,, \qquad \left|\tfrac{\partial f_\varepsilon}{\partial v}(t, v)\right| \leq \ell_B(t)\,. \tag{53}$$

(iv) *There exists $\varrho_0 > 0$ such that the initial compatibility conditions*

$$f_\varepsilon(0, u_\varepsilon^0) > 0 \implies m_{\lambda_0^\varepsilon}(u_\varepsilon^0 +) \geq \varrho_0\,, \qquad f_\varepsilon(0, u_\varepsilon^0) < 0 \implies m_{\lambda_0^\varepsilon}(u_\varepsilon^0 -) \geq \varrho_0\,,$$

hold for every $\varepsilon \in (0, \varepsilon_0)$, where $m_{\lambda_0^\varepsilon}$ is the mapping defined in (14).

(v) *There exists $U_0 \geq R$ such that $|u_\varepsilon^0| \leq U_0$ for every $\varepsilon \in (0, \varepsilon_0)$.*

(vi) *There exist $u^0 \in \mathbb{R}$, a locally bounded measurable function $\mu : \mathbb{R}_+ \times \mathbb{R} \to \mathbb{R}_+$, a continuous function $f : \mathbb{R}_+ \times \mathbb{R} \to \mathbb{R}$, and a function $\lambda \in \Lambda_R$ such that*

$$\lim_{\varepsilon \to 0+} u_\varepsilon^0 = u^0; \;\; \lim_{\varepsilon \to 0+} f_\varepsilon(t, v) = f(t, v) \text{ for } (t, v) \in \mathbb{R}_+ \times \mathbb{R}; \;\; \lim_{\varepsilon \to 0+} \lambda^\varepsilon(r) = \lambda(r) \text{ for } r \geq 0;$$

$$\lim_{\varepsilon \to 0+} \int_\Omega (\mu_\varepsilon - \mu)(r, v)\,\mathrm{d}r\,\mathrm{d}v = 0 \text{ for every open bounded set } \Omega \subset \mathbb{R}_+ \times \mathbb{R}\,.$$

We denote by $\hat{\Phi}^\varepsilon_\pm$ the initial loading curves associated with the Preisach operator $\mathcal{F}_\varepsilon$ according to formulas (28)–(29). A similar computation as in (30) shows, as a consequence of Hypotheses 1.7 and 3.1, that they have the property

$$\hat{\Phi}^\varepsilon_+(v) \ge \int_{U_0+\hat{\varrho}}^{v-\hat{\varrho}} \varrho(z)\,\mathrm{d}z \quad \text{for } v > U_0 + 2\hat{\varrho}; \quad \hat{\Phi}^\varepsilon_-(v) \le -\int_{v+\hat{\varrho}}^{U_0-\hat{\varrho}} \varrho(z)\,\mathrm{d}z \quad \text{for } v < U_0 - 2\hat{\varrho}. \tag{54}$$

We prove the following result.

Theorem 3.2 *Let Hypothesis 3.1 hold. Then for every $\varepsilon \in (0, \varepsilon_0)$ there exists a unique solution $u_\varepsilon \in W^{1,2q}_{\mathrm{loc}}(\mathbb{R}_+)$ to (2). If moreover $u^\varepsilon_1, u^\varepsilon_2 : \mathbb{R}_+ \to \mathbb{R}$ are two solutions to (2) corresponding to two initial conditions $u^{01}_\varepsilon \le u^{02}_\varepsilon$ and $\lambda^\varepsilon_1(r) \le \lambda^\varepsilon_2(r)$ for all $r \ge 0$, then $u^\varepsilon_1(t) \le u^\varepsilon_2(t)$ for every $t \ge 0$. If, in addition, there exist $B > R$ and $T > 0$ such that*

$$|u_\varepsilon(t)| \le B \quad \forall \varepsilon \in (0, \varepsilon_0) \ \ \forall t \in [0,T], \tag{55}$$

then there exists a constant $C_B > 0$ independent of ε such that

$$\int_0^T |\dot{u}_\varepsilon(t)|^{2q}\,\mathrm{d}t \le C_B \int_0^T \left(|f^1_B(t)|^q + |f^0(t)|^{2q} + |\ell_B(t)|^{2q}\right) \mathrm{d}t\,, \tag{56}$$

and a sequence $\varepsilon_k \to 0+$ such that u_{ε_k} converge uniformly in $C[0,T]$ and weakly in $W^{1,2q}(0,T)$ to a solution u of the limit system (1).

Proof. For the sake of completeness, we give an existence proof for solutions to (2) for every $\varepsilon \in (0, \varepsilon_0)$, which is fairly standard. Notice first that there is a one-to-one correspondence between the initial conditions for u_ε and w_ε in (2), which has the form

$$w_\varepsilon(0) = w^0_\varepsilon := \varepsilon u^0_\varepsilon + \int_0^\infty g(r, \xi^0_r)\,\mathrm{d}r\,, \quad \xi^0_r = \min\{u^0_\varepsilon + r, \max\{u^0_\varepsilon - r, \lambda(r)\}\}$$

by virtue of (7). Keeping $\varepsilon \in (0, \varepsilon_0)$ and $T > 0$ fixed, we define

$$\hat{w}_\varepsilon = \varepsilon |u^0_\varepsilon| + \int_0^\infty |g(r, \xi^0_r)|\,\mathrm{d}r\,, \quad B^* = \frac{1}{\varepsilon}\left(\hat{w}_\varepsilon + \int_0^T f^0(t)\,\mathrm{d}t\right).$$

Relation (52) implies no lower bound for f_ε if $v > 0$ and no upper bound if $v < 0$. Hence we put

$$f^*_\varepsilon(t,v) = \begin{cases} f_\varepsilon(t,v) & \text{for } (t,v) \in (0,T) \times [-B^*, B^*]\,, \\ f_\varepsilon(t,B^*) & \text{for } (t,v) \in (0,T) \times (B^*, +\infty)\,, \\ f_\varepsilon(t,-B^*) & \text{for } (t,v) \in (0,T) \times (-\infty, -B^*)\,. \end{cases}$$

By Hypotheses 3.1 (ii), (iii) we have $|f_\varepsilon(t,0)| \le f^0(t)$ for all t, therefore

$$|f^*_\varepsilon(t,v)| \le f^0(t) + B^* \ell_{B^*}(t) \quad \forall (t,v) \in (0,T) \times \mathbb{R}\,. \tag{57}$$

We introduce the set

$$K = \left\{\tilde{u} \in C[0,T] : \tilde{u}(0) = u^0_\varepsilon,\ \|\tilde{u}\|_{[0,T]} \le B^* + \frac{B^*}{\varepsilon}\int_0^T \ell_{B^*}(t)\,\mathrm{d}t\right\}.$$

We now check that the mapping $\mathcal{S} : K \to C[0,T]$, which with each $\tilde{u} \in K$ associates the solution u of the problem

$$\left.\begin{array}{rcl} \dot{w}(t) & = & f_\varepsilon^*(t, \tilde{u}(t)) \\ w(t) & = & \varepsilon u(t) + \mathcal{F}_\varepsilon[\lambda^\varepsilon, u](t) \\ u(0) & = & u_\varepsilon^0 , \end{array}\right\} \tag{58}$$

is a contraction on K, endowed with the norm

$$\|\tilde{u}\|_* = \max\left\{\left|\mathrm{e}^{-L^*(t)}\tilde{u}(t)\right| : t \in [0,T]\right\} \qquad \text{with} \qquad L^*(t) = \frac{1}{\varepsilon}\int_0^t \ell_{B^*}(\tau)\,\mathrm{d}\tau\ .$$

The existence and uniqueness of the solution u to (58) follows immediately from the Lipschitz continuity property in Proposition 1.8 (ii). To see that $u \in K$, we test equation (58) consecutively by $H(u(t))$ and $-H(-u(t))$, which yields, by virtue of (25)–(26), that $\varepsilon|u(t)| \le \hat{w}_\varepsilon + \int_0^t |f_\varepsilon^*(\tau, \tilde{u}(\tau))|\,\mathrm{d}\tau$, and it suffices to use inequality (57) to conclude that $u \in K$. The contracting property of $\mathcal{S}$ can be proved via Hilpert's inequality (24). For $\tilde{u}_1, \tilde{u}_2 \in K$, we test the identity

$$\frac{\mathrm{d}}{\mathrm{d}t}\big(\varepsilon(u_1(t) - u_2(t)) + \mathcal{F}_\varepsilon[\lambda^\varepsilon, u_1](t) - \mathcal{F}_\varepsilon[\lambda^\varepsilon, u_2](t)\big) = f_\varepsilon^*(t, \tilde{u}_1(t)) - f_\varepsilon^*(t, \tilde{u}_2(t))$$

by $\operatorname{sign}(u_1 - u_2)$ to obtain, using also the second relation in (53), that $\varepsilon\frac{\mathrm{d}}{\mathrm{d}t}|u_1(t) - u_2(t)| \le \ell_{B^*}(t)|\tilde{u}_1(t) - \tilde{u}_2(t)|$. For $t \in [0,T]$ we then have

$$\mathrm{e}^{-L^*(t)}|u_1(t) - u_2(t)| \;\le\; \frac{1}{\varepsilon}\|\tilde{u}_1 - \tilde{u}_2\|_* \int_0^t \mathrm{e}^{L^*(\tau)-L^*(t)}\ell_{B^*}(\tau)\,\mathrm{d}\tau \;\le\; \left(1 - \mathrm{e}^{-L^*(T)}\right)\|\tilde{u}_1 - \tilde{u}_2\|_*\,,$$

which yields the desired contraction. To see that the fixed point u of the mapping $\mathcal{S}$ satisfies system (2), we test the equation $\frac{\mathrm{d}}{\mathrm{d}t}\big(\varepsilon u(t) + \mathcal{F}_\varepsilon[\lambda^\varepsilon, u](t)\big) = f_\varepsilon^*(t, u(t))$ by $\operatorname{sign}(u(t))$ and obtain by (25), (26), and Hypothesis 3.1 (ii) that $\varepsilon|u(t)| \le \hat{w}_\varepsilon + \int_0^t f^0(\tau)\,\mathrm{d}\tau$. We see that $\|u\|_{[0,T]} \le B^*$, hence $f_\varepsilon^*(t, u(t)) = f_\varepsilon(t, u(t))$, and the existence proof is complete.

Let now $u_1^\varepsilon, u_2^\varepsilon : \mathbb{R}_+ \to \mathbb{R}$ be two solutions corresponding to two different initial conditions $u_\varepsilon^{01} \le u_\varepsilon^{02}$ and $\lambda_1^\varepsilon(r) \le \lambda_2^\varepsilon(r)$ for all $r \ge 0$. We fix some $T > 0$ and $B_T > \max\{R, \|u_1^\varepsilon\|_{[0,T]}, \|u_2^\varepsilon\|_{[0,T]}\}$. Then for all $t \in [0,T]$ and $r > 0$ we have by Lemma 1.4 that $|\mathfrak{p}_r[\lambda_i^\varepsilon, u_i^\varepsilon](t)| \le B_T$, and $\mathfrak{p}_r[\lambda_i^\varepsilon, u_i^\varepsilon](t) = 0$ for $r > B_T$, $i = 1,2$. Testing the difference of the equations for $u_1^\varepsilon, u_2^\varepsilon$ by $H(u_1^\varepsilon - u_2^\varepsilon)$ and using Proposition 1.9, we obtain a.e.

$$\frac{\mathrm{d}}{\mathrm{d}t}\left(\varepsilon(u_1^\varepsilon(t) - u_2^\varepsilon(t))^+ + \int_0^{B_T}(g(r, \mathfrak{p}_r[\lambda_1^\varepsilon, u_1^\varepsilon]) - g(r, \mathfrak{p}_r[\lambda_2^\varepsilon, u_2^\varepsilon]))^+(t)\,\mathrm{d}r\right) \le \ell_{B_T}(t)(u_1^\varepsilon(t) - u_2^\varepsilon(t))^+.$$

We have $\mathfrak{p}_r[\lambda_1^\varepsilon, u_1^\varepsilon](0) \le \mathfrak{p}_r[\lambda_2^\varepsilon, u_2^\varepsilon](0)$, and using the classical Gronwall argument we obtain $u_1^\varepsilon(t) \le u_2^\varepsilon(t)$ for all $t \in [0,T]$. Since $T > 0$ has been chosen arbitrarily, the solution exists and is unique globally in time for all $\varepsilon \in (0, \varepsilon_0)$.

To derive the upper bound (56) independent of ε, we use Proposition 2.3 in a substantial way. It follows from Hypothesis 3.1 (iv) that (45) holds for all $\varepsilon \in (0, \varepsilon_0)$, provided we set $\varrho_B = \min\{\varrho_0, \min\{\varrho(v); |v| \le B\}\}$. Proposition 2.3, Hypotheses 3.1 (ii)–(iii), and Hölder's inequality then yield

$$\begin{aligned}\int_0^T |\dot{u}_\varepsilon(t)|^{2q}\,\mathrm{d}t \;&\le\; C\int_0^T \left(\left(f_B^1(t) + \ell_B(t)\,|\dot{u}_\varepsilon(t)|\right)^q + |f^0(t)|^{2q}\right)\mathrm{d}t \\ &\le\; \hat{C}\int_0^T \left(|f_B^1(t)|^q + |f^0(t)|^{2q}\right)\mathrm{d}t + \hat{C}\left(\int_0^T |\ell_B(t)|^{2q}\,\mathrm{d}t\right)^{\frac{1}{2}}\left(\int_0^T |\dot{u}_\varepsilon(t)|^{2q}\,\mathrm{d}t\right)^{\frac{1}{2}}\end{aligned}$$

with a constant $\hat{C}$ independent of ε; hence, (56) holds. In particular, the functions $\dot{u}_\varepsilon$ are uniformly bounded in $L^{2q}(0,T)$. By compact embedding, there exists a uniformly convergent subsequence $u_{\varepsilon_k} \to u$ as $\varepsilon_k \to 0+$. We have $\mathfrak{p}_r[\lambda^\varepsilon, u_\varepsilon](t) = 0$ for all $r \geq B$ and $\varepsilon \in (0, \varepsilon_0)$ by virtue of Lemma 1.3. Hence, $\mathcal{F}_\varepsilon[\lambda^\varepsilon, u_\varepsilon](t) - \mathcal{F}_\varepsilon[\lambda, u](t) \to 0$ as $\varepsilon \to 0+$, as a consequence of (22), and the formula

$$\mathcal{F}_\varepsilon[\lambda, u](t) - \mathcal{F}[\lambda, u](t) = \int_0^B \int_0^{\mathfrak{p}_r[\lambda,u](t)} (\mu_\varepsilon - \mu)(r, v)\, \mathrm{d}v\, \mathrm{d}r$$

together with Hypothesis 3.1 (vi) enable us to pass to the limit as $\varepsilon \to 0+$ in (2) and check that u is a solution to the limit problem (1). ■

Example 3.3 Consider again the case $\lambda^\varepsilon(r) = \lambda(r) \equiv 0$, $\mu_\varepsilon(r,v) = \mu(r,v) \equiv 1$, and $u^0_\varepsilon = u^0 = 0$. The computation in Example 2.4 shows that $\mathcal{F}[\lambda, u](t) = \frac{1}{2}u^2(t)$ if u increases, and $\mathcal{F}[\lambda, u](t) = -\frac{1}{2}u^2(t)$ if u decreases in $[0,T]$.

(i) To show that Hypothesis 3.1 (iv) is necessary, set $f_\varepsilon(t,v) = f(t,v) \equiv 1$. We are in the situation of Example 2.4, and u_ε converge uniformly to the function $u(t) = \sqrt{2t}$, which does not belong to $W^{1,2q}(0,T)$ for any $q \geq 1$.

(ii) Uniqueness of solutions to the limit problem (1) cannot be expected in general even if Hypothesis 3.1 (iv) holds. For the same λ, μ, u^0 as in (i), and for $f_\varepsilon(t,v) = f(t,v) \equiv v$, all hypotheses of Theorem 3.2 are fulfilled, but $u_1(t) = 0$, $u_2(t) = t$, $u_3(t) = -t$ are three distinct solutions of (1). However, ε-approximations u_ε in Theorem 3.2 all converge to u_1.

The following theorem ensures uniqueness of a solution to problem (1) with a non-increasing f.

Theorem 3.4 *Let Hypothesis 3.1 hold, and let there exist $T > 0$ and $u_* < u^*$ such that*

$$\frac{\partial f}{\partial v}(t,v) \leq 0 \quad a.\,e.\ in\ \ (0,T) \times (u_*, u^*)\,.$$

Let there exist two solutions $u_1, u_2 : [0,T] \to [u_*, u^*]$ to (1) corresponding to two initial conditions $u^0_1 \leq u^0_2$ and $\lambda_1, \lambda_2 \in \Lambda_R$, $\lambda_1(r) \leq \lambda_2(r)$ for all $r \geq 0$. Then we have $u_1(t) \leq u_2(t)$ for every $t \in [0,T]$. In particular, the solution to (1) is unique for every initial condition $u^0 \in [u_*, u^*]$.

Proof. Set $B = \max\{R, |u_*|, |u^*|\}$. For all $t \in [0,T]$ and $r > 0$ we have by Lemma 1.4 that $|\mathfrak{p}_r[\lambda_i, u_i](t)| \leq B$, and $\mathfrak{p}_r[\lambda_i, u_i](t) = 0$ for $r > B$, $i = 1,2$. Proposition 1.9 yields

$$\frac{\mathrm{d}}{\mathrm{d}t}\int_0^B (g(r, \mathfrak{p}_r[\lambda_1, u_1]) - g(r, \mathfrak{p}_r[\lambda_2, u_2]))^+(t)\, \mathrm{d}r \leq 0 \ \text{ a.e. in } \ (0,T)\,.$$

We have $\mathfrak{p}_r[\lambda_1, u_1](0) \leq \mathfrak{p}_r[\lambda_2, u_2](0)$, hence $g(r, \mathfrak{p}_r[\lambda_1, u_1](t)) \leq g(r, \mathfrak{p}_r[\lambda_2, u_2](t))$ for all $t \in [0,T]$ and $r \in (0,B)$. By Hypothesis (20), there exists $\varrho_B > 0$ such that for $r < \varrho_B$ we have $g(r,v_1) - g(r,v_2) \geq \alpha(v_1 - v_2)$ for all $B > v_1 > v_2 > -B$. Consequently, $\mathfrak{p}_r[\lambda_1, u_1](t) \leq \mathfrak{p}_r[\lambda_2, u_2](t)$ for $t \in [0,T]$ and $r \in (0, \varrho_B)$. Since $|\mathfrak{p}_r[\lambda_i, u_i](t) - u_i(t)| \leq r$ by definition of the play, we may let r tend to zero and obtain $u_1(t) \leq u_2(t)$ for all $t \geq 0$. ■

We now state a sufficient condition for the estimate (55).

Proposition 3.5 *Let Hypothesis 3.1 hold. Let there exist U_0 such that $|u^0_\varepsilon| \leq U_0$ for every $\varepsilon \in (0, \varepsilon_0)$, and let $T > 0$ be such that*

$$-\int_{-\infty}^{U_0 - \hat{\varrho}} \varrho(z)\, \mathrm{d}z < \int_0^T f^0(t)\, \mathrm{d}t < \int_{U_0 + \hat{\varrho}}^{\infty} \varrho(z)\, \mathrm{d}z\,, \tag{59}$$

where $\hat{\varrho}$ is the least upper bound of the function ϱ introduced in (20). Then there exists a positive constant $B > R$ such that (55) holds.

Proof. Consider the system

$$\dot{w}^\sharp_\varepsilon(t) = f^0(t); \quad w^\sharp_\varepsilon(t) = \varepsilon u^\sharp_\varepsilon(t) + \mathcal{F}_\varepsilon[\lambda^\varepsilon, u^\sharp_\varepsilon](t); \quad u^\sharp_\varepsilon(0) = (u^0_\varepsilon)^+ . \tag{60}$$

We have $\dot{u}^\sharp_\varepsilon \geq 0$ a.e. and $u^\sharp_\varepsilon(0) \geq 0$, hence $u^\sharp_\varepsilon$ is positive and nondecreasing, and we have the representation formula $\mathcal{F}_\varepsilon[\lambda^\varepsilon, u^\sharp_\varepsilon](t) = \mathcal{F}_\varepsilon[\lambda^\varepsilon, u^\sharp_\varepsilon](0) + \hat{\Phi}^\varepsilon_+(u^\sharp_\varepsilon(t))$ for $t \geq 0$, where $\hat{\Phi}^\varepsilon_+$ is the positive initial loading curve associated with $\mathcal{F}_\varepsilon$. We integrate the first equation in (60) in time and use the first inequality in (54) to obtain

$$\int_0^t f^0(\tau)\, \mathrm{d}\tau \geq \varepsilon \left(u^\sharp_\varepsilon(t) - u^\sharp_\varepsilon(0) \right) + \int_{U_0+\hat{\varrho}}^{u^\sharp_\varepsilon(t)-\hat{\varrho}} \varrho(z)\, \mathrm{d}z \,.$$

Condition (59) entails that $u^\sharp_\varepsilon(t)$ are uniformly bounded from above in $[0, T]$ by some $B > 0$. We test the difference of equations (1) and (60) by $H(u_\varepsilon(t) - u^\sharp_\varepsilon(t))$, and obtain from Proposition 1.9 and Hypothesis 3.1 (ii) that

$$\frac{\mathrm{d}}{\mathrm{d}t} \left(\varepsilon(u_\varepsilon - u^\sharp_\varepsilon)^+ + \int_0^\infty (g_\varepsilon(r, \mathfrak{p}_r[\lambda^\varepsilon, u_\varepsilon]) - g_\varepsilon(r, \mathfrak{p}_r[\lambda^\varepsilon, u^\sharp_\varepsilon]))^+ \, \mathrm{d}r \right)(t) \leq 0 \ \text{ a.e.},$$

hence $u_\varepsilon(t) \leq u^\sharp_\varepsilon(t) \leq B$ for all $t \in [0, T]$. The lower bound is obtained similarly by considering the system

$$\dot{w}^\flat_\varepsilon(t) = -f^0(t); \quad w^\flat_\varepsilon(t) = \varepsilon u^\flat_\varepsilon(t) + \mathcal{F}_\varepsilon[\lambda^\varepsilon, u^\flat_\varepsilon](t); \quad u^\flat_\varepsilon(0) = -(u^0_\varepsilon)^- ,$$

and the assertion follows. ∎

Theorem 3.2 and Proposition 3.5 imply local existence of solutions to problem (1) if Hypothesis 3.1 and relation (59) hold.

4. Global and periodic solutions

In this section we give sufficient conditions for the existence, uniqueness, and asymptotic stability of global and time periodic solutions to (1).

Hypothesis 4.1 *An interval $[u_*, u^*]$, an initial condition $u^0 \in [u_*, u^*]$, and a continuous function $f : \mathbb{R}_+ \times [u_*, u^*]$ are given such that*

(i) $f(t, u^*) \leq 0$, $f(t, u_*) \geq 0$ *for all* $t \geq 0$;

(ii) *There exist functions $\ell \in L^{2q}_{\mathrm{loc}}(\mathbb{R}_+)$, $f^1 \in L^q_{\mathrm{loc}}(\mathbb{R}_+)$, such that for almost all $(t, v) \in \mathbb{R}_+ \times (u_*, u^*)$ we have*

$$\left| \frac{\partial f}{\partial t}(t, v) \right| \leq f^1(t) , \qquad \left| \frac{\partial f}{\partial v}(t, v) \right| \leq \ell(t) .$$

(iii) *There exists $\varrho_0 > 0$ such that the following initial compatibility conditions hold:*

$$f(0, u^0) > 0 \implies m_{\lambda_0}(u^0+) \geq \varrho_0 , \qquad f(0, u^0) < 0 \implies m_{\lambda_0}(u^0-) \geq \varrho_0 .$$

The main results of this section read as follows.

Theorem 4.2 *Let Hypotheses 1.7 and 4.1 hold, and let* $R > \max\{|u_*|, |u^*|\}$ *and* $\lambda \in \Lambda_R$ *be given. Then there exists a solution* $u \in W^{1,2q}_{\mathrm{loc}}(\mathbb{R}_+)$ *to Problem (1) such that* $u(t) \in [u_*, u^*]$ *for all* $t \geq 0$*. If moreover* $f(t, \cdot)$ *is nonincreasing in* $[u_*, u^*]$*, and* $u_1, u_2 : \mathbb{R}_+ \to [u_*, u^*]$ *are two solutions to (1) corresponding to two initial conditions* $u_1^0 \leq u_2^0$*, then* $u_1(t) \leq u_2(t)$ *for every* $t \geq 0$*. In particular, the solution to (1) is unique.*

Theorem 4.3 *Let the hypotheses of Theorem 4.2 hold, let* $f(t, \cdot)$ *be decreasing in* $[u_*, u^*]$*, and let there exist* $T > 0$ *such that*

$$f(t, v) = f(t + T, v) \tag{61}$$

for all $t \geq 0$ *and* $v \in [u_*, u^*]$*. Let* u *be a solution to Problem (1). Then there exists* $\lambda^T \in \Lambda_R$ *and a solution* $u^T \in W^{1,2q}_{\mathrm{loc}}(\mathbb{R}_+)$ *to the problem*

$$\left.\begin{array}{rcl} \dot{w}^T(t) & = & f(t, u^T(t)) \\ w^T(t) & = & \mathcal{F}[\lambda^T, u^T](t) \\ u^T(t) & = & u^T(t+T) \end{array}\right\} \quad \forall t > 0 \tag{62}$$

such that $\lim_{t\to\infty} |u(t) - u^T(t)| = 0$*. Moreover, if* (λ_1^T, u_1^T)*,* (λ_2^T, u_2^T) *are two solutions of (62), then* $u_1^T = u_2^T$*.*

In the last statement of Theorem 4.3, the case $\lambda_1^T \neq \lambda_2^T$ and $w_1^T - w_2^T \equiv \mathit{const} \neq 0$ cannot be excluded.

Proof of Theorem 4.2. We define the extended function $\tilde{f}$ by

$$\tilde{f}(t, v) = f(t, v) \text{ for } v \in [u_*, u^*]; \quad \tilde{f}(t, v) = f(t, u_*) \text{ for } v < u_*; \quad \tilde{f}(t, v) = f(t, u^*) \text{ for } v > u^*$$

and consider for $\varepsilon > 0$ the problem

$$\dot{w}_\varepsilon(t) = \tilde{f}(t, u_\varepsilon(t)); \quad w_\varepsilon(t) = \varepsilon u_\varepsilon(t) + \mathcal{F}[\lambda, u_\varepsilon](t); \quad u_\varepsilon(0) = u^0\,. \tag{63}$$

We first prove that the solution u_ε to (63) satisfies the bounds $u_\varepsilon(t) \in [u_*, u^*]$ for all $\varepsilon > 0$ and $t \geq 0$. We test the first equation of (63) by $H(u_\varepsilon(t) - u^*)$ and obtain $\dot{w}_\varepsilon(t)\, H(u_\varepsilon(t) - u^*) \leq 0$ for all $t > 0$. Using the fact that $\mathcal{F}[\lambda, u^*]$ is a constant, we may apply Proposition 1.9, which yields

$$\frac{\mathrm{d}}{\mathrm{d}t}\left(\varepsilon(u_\varepsilon - u^*)^+ + \int_0^\infty (g(r, \mathfrak{p}_r[\lambda, u_\varepsilon]) - g(r, \mathfrak{p}_r[\lambda, u^*]))^+ \,\mathrm{d}r\right)(t) \leq 0 \ \text{ a.e.}$$

Hence, $u_\varepsilon(t) \leq u^*$ for all $\varepsilon > 0$ and $t \geq 0$. The lower bound $u_\varepsilon(t) \geq u_*$ is derived similarly by testing with $H(u_* - u_\varepsilon(t))$. The existence of a global solution to (1) now follows from Theorem 3.2; the comparison result for two solutions u_1, u_2 is a direct consequence of Theorem 3.4. ∎

We now pass to the periodic case.

Proof of Theorem 4.3. Let u be the solution to (1) from Theorem 4.2. For $n \in \mathbb{N}$ put $\lambda^{(n)}(r) = \mathfrak{p}_r[\lambda, u](nT)$, $u^{(n)}(t) = u(t + nT)$, $w^{(n)}(t) = w(t + nT)$ for $t \geq 0$. By Lemma 1.3 (ii) and (61), we have $w^{(n)}(t) = \mathcal{F}[\lambda^{(n)}, u^{(n)}](t)$ for all $t \in [0, T]$.

We define an auxiliary function

$$D(u_1, u_2, \lambda_1, \lambda_2)(t) = \int_0^\infty |g(r, \mathfrak{p}_r[\lambda_1, u_1]) - g(r, \mathfrak{p}_r[\lambda_2, u_2])|(t)\,\mathrm{d}r$$

for $u_1, u_2 \in AC(\mathbb{R}_+)$, $\lambda_1, \lambda_2 \in \Lambda_R$, and $t \geq 0$. Whenever (u_i, w_i), $w_i = \mathcal{F}[\lambda_i, u_i]$ for $i = 1, 2$ are two solutions of (1), then, as a consequence of (24), we have

$$\frac{d}{dt} D(u_1, u_2, \lambda_1, \lambda_2)(t) + |f(t, u_1(t)) - f(t, u_2(t))| \ \leq\ 0 \quad \text{a.e.}\ , \tag{64}$$

hence, by virtue of Lemma 1.3 (ii), $\frac{d}{dt}D(u^{(n)}, u^{(m)}, \lambda^{(n)}, \lambda^{(m)})(t) + |f(t, u^{(n)}(t)) - f(t, u^{(m)}(t))| \leq 0$ a.e. $\forall n, m \in \mathbb{N}$. We have in particular $\frac{d}{dt}D(u^{(1)}, u, \lambda^{(1)}, \lambda)(t) \leq 0$, hence there exists the limit

$$\Delta = \lim_{t\to\infty} D(u^{(1)}, u, \lambda^{(1)}, \lambda)(t) \geq 0. \tag{65}$$

The set Λ_R is compact with respect to the sup-norm, hence we may find a sequence $\{n_k\}$ in $\mathbb{N}$ such that $u^{(n_k)}(0) \to \bar{u}_0 \in [u_*, u^*]$, $\lambda^{(n_k)} \to \bar{\lambda} \in \Lambda_R$ uniformly. By Theorem 3.2, the sequence $\{u^{(n)}\}$ is bounded in $W^{1,2q}(0,T)$. Hence, by compact embedding of $W^{1,2q}(0,T)$ in $C[0,T]$, n_k can be chosen in such a way that $u^{(n_k)} \to \bar{u} \in C[0,T]$ uniformly.

Set $\bar{w} = \mathcal{F}[\bar{\lambda}, \bar{u}]$. By continuity of $\mathcal{F}$ we have that $(\bar{u}, \bar{w})$ is a solution of (1). To see that $\bar{u}$ is T-periodic, put $\hat{u}(t) = \bar{u}(t+T)$ for $t \geq 0$, $\hat{\lambda}(r) = \mathcal{F}[\bar{\lambda}, \bar{u}](T)$. Using the identity $\mathcal{F}[\lambda^{(n_k)}, u^{(n_k)}](t+T) = \mathfrak{p}_r[\lambda^{(1)}, u^{(1)}](t + n_k T)$ we obtain from (65) that

$$D(\hat{u}, \bar{u}, \hat{\lambda}, \bar{\lambda})(t) = \lim_{k\to\infty} D(u^{(1)}, u, \lambda^{(1)}, \lambda)(t + n_k T) = \Delta \qquad \forall t \geq 0,$$

hence $\frac{d}{dt}D(\hat{u}, \bar{u}, \hat{\lambda}, \bar{\lambda})(t) \equiv 0$ and (64) yields that $f(t, \hat{u}(t)) = f(t, \bar{u}(t))$ a. e. Since f is decreasing, we have $\hat{u}(t) = \bar{u}(t)$ for all $t \geq 0$. This enables us to use (64) and conclude for all $n \in \mathbb{N}$ that

$$D(u^{(n+1)}, \bar{u}, \lambda^{(n+1)}, \bar{\lambda})(0) = D(u^{(n)}, \bar{u}, \lambda^{(n)}, \bar{\lambda})(T) \leq D(u^{(n)}, \bar{u}, \lambda^{(n)}, \bar{\lambda})(0).$$

We see that $\{D(u^{(n)}, \bar{u}, \lambda^{(n)}, \bar{\lambda})(0)\}$ is a nonincreasing sequence with a subsequence converging to 0. Hence, it converges to 0, and by virtue of (64), $\lim_{n\to\infty} \int_0^T |f(t, u^{(n)}(t)) - f(t, \bar{u}(t))|\, dt = 0$. Using again the compactness of the sequence $\{u^{(n)}\}$ in $C[0,T]$ and the fact that f is injective, we conclude that $u^{(n)}$ converge to $\bar{u}$ uniformly, and the assertion follows. The uniqueness of $\bar{u}$ is also a direct consequence of (64). ■

Acknowledgments

This publication has emanated from research conducted with the financial support of Federal Programme 'Scientists of Innovative Russia', grant 2009-1.5-507-007, Russian Foundation for basic Research, grant 10-01-93112, and GAČR grant P201/10/2315.

References

[1] Krasnosel'skii M A and Pokrovskii A V 1989 *Systems with Hysteresis* (Springer)
[2] Brokate M and Sprekels J 1996 *Hysteresis and Phase Transitions* (New York: Springer)
[3] Krejčí P 1996 *Hysteresis, Convexity and Dissipation in Hyperbolic Equations* (Tokyo: Gakkōtosho)
[4] Kuhnen K 2001 *Inverse Steuerung piezoelektrischer Aktoren mit Hysterese-, Kriech- und Superpositionsoperatoren* (Aachen: Shaker Verlag)
[5] Mayergoyz I D 1991 *Mathematical Models for Hysteresis* (New York: Springer)
[6] Mayergoyz I D 2003 *Mathematical Models of Hysteresis and Their Applications* (New York: Elsevier Science)
[7] Visintin A 1994 *Differential Models of Hysteresis* (New York: Springer)
[8] 2006 *The Science of Hysteresis*, three volume set, ed I Mayergoyz and G Bertotti (Elsevier, Academic Press)
[9] Flynn D, McNamara H, O'Kane P J and Pokrovskii A 2006 *The Science of Hysteresis* vol III, ed I Mayergoyz and G Bertotti (Elsevier, Academic Press) pp 689–744
[10] O'Kane J P 2005 *J. Phys.: Conf. Series* **22** 148–63
[11] O'Kane J P 2005 *Acta Geophys. Pol.* **53** 373-83
[12] Appelbe B, Flynn D, McNamara H, O'Kane P, Pimenov A, Pokrovskii A, Rachinskii D and Zhezherun A 2009 *IEEE Control Systems Magazine* **29** 44–69
[13] Krejčí P 1989 *Apl. Mat.* **34** 364–74
[14] Krejčí P 2008 *Physica B: Condensed Matter* **403** 297–300
[15] Hilpert M 1989 *Mathematical Models for Phase Change Problems*, ed J F Rodrigues (Basel: Birkhäuser) pp 377–88
[16] O'Kane J P, Pokrovskii A and Flynn D 2004 *Geophys. Res. Abstr.* **6** 07303
[17] Preisach F 1935 *Z. Phys.* **94** 277–302
[18] Krejčí P 2005 *Handbook of Evolution Equations: Evolutionary Equations* vol 2, ed C M Dafermos and E Feireisl (Elsevier) pp 303–70

Journal of Physics: Conference Series **268** (2011) 012017 doi:10.1088/1742-6596/268/1/012017

Development of a resistive model for overhead line poles

Z Kvasznicza and G Elmer

University of Pécs, Pollack Mihály Faculty of Engineering, Pécs, Hungary

E-mail: kvasznicza@vili.pmmf.hu, elmer@vili.pmmf.hu

Abstract. Quality of electricity is of increasing importance within conditions enabling to maintain it to an even less extent. Now consumers neither tolerate outages with duration of seconds. This fact forces the electric utility companies to maintain network operation during single-phase earth faults. This paper presents the results of a joint research investigating the potential of overhead line poles during faults with a newly developed pole model.

1. Introduction

Quality of electric power distributed by electric utility companies is of increasing importance, however the circumstances under which this good quality should be maintained enable this to a continuously decreasing extent because of the increasing number of non-linear loads. In our days consumers do not tolerate outages with duration of as low as one or two seconds. After these short time outages electronic devices have to be readjusted that can last hours in offices or even in residential areas having no uninterruptible power sources.

Measures characterizing the reliability of electric power networks are the CAIFI, i.e. Customer Average Interruption Frequency Index and the CAIDI, i.e. Customer Average Interruption Duration Index [1]. These indexes have quite low values for cable networks operated in urban areas with high consumption density. However in rural areas supplied by overhead lines with a voltage of 20 kV in Hungary this overhead network is exposed to weather conditions causing interruptions in the supply. The values characterizing the reliability of the supply of electric energy has to be reduced in rural areas as well.

This expectation forces the electric utility companies to attempt maintaining the operation of their medium voltage distribution networks during single phase earth faults as well. 20 kV medium voltage distribution networks are mainly operated compensated – with a coil (Petersen coil) connected between the star point and the earth – in Hungary. In case of these networks there is a possibility maintaining the supply with single-phase earth fault as well.

However in this case if somebody touches the pole of the faulty network at the location of the fault, he/she can be exposed to a fatal electric shock. This paper presents the results of a joint research of the electric network operating company E.ON acting also in the region of South-Transdanubia, Hungary and of the University of Pécs, Pollack Mihály Faculty of Engineering, Institute for Information Technology and Electrical Engineering.

Within the frame of this research a resistive quasi-distributed parameter model of reinforced concrete poles used for low and medium voltage overhead lines has been developed. With the help of this model and the software TINALab II [2] simulation sessions have been performed for different circumstances, different grounding resistance values and for different pole conditions to find out the potential along the pole in different cases. To find out the parameters of the model measurement series have been conducted on poles and networks.

Main objective of the research is to develop a data group of value ranges of the several factors influencing the pole potential resulting a pole potential of 50 VAC or less. Along with the knowledge of actual values on the network and simulating earth faults on network circuits with the novel pole model this data will then enable to predict pole potentials. Then the results will enable to make a decision if the operation of a network with earth fault can be maintained or not. Developing medium voltage networks with earth fault pole potentials below 50 V will result in a more reliable electric energy supply of the consumers.

2. Earth faults on compensated networks

As a first task the current resulting potentials more than zero on the poles has to be determined. This current is the earth fault current. The earthing method applied on medium voltage networks resulting minimum earth fault current is that compensated by placing a Petersen coil between the star point of the network and earth [3]. Figure 1 shows a compensated network taking into account the conductance to the earth and the resistance of the Petersen coil. In case of a theoretic circuit with no resistances zero current would flow to the earth.

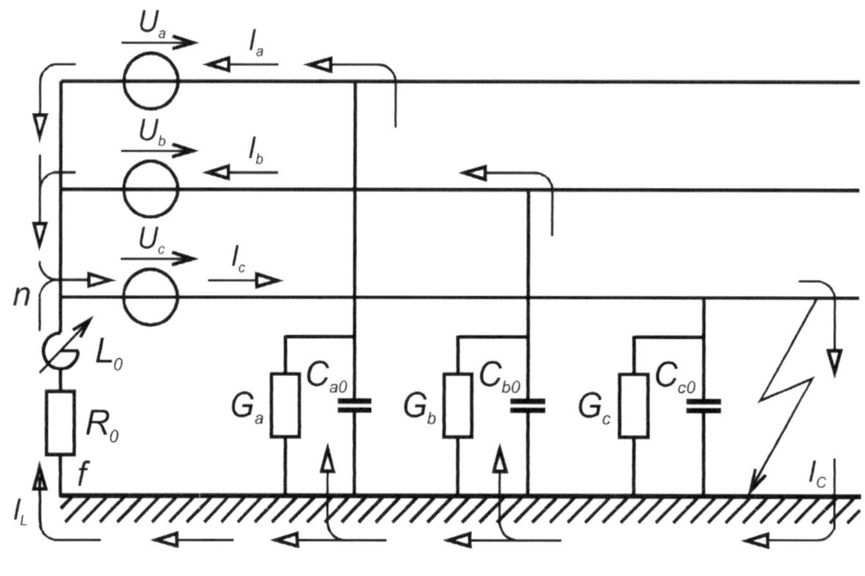

Figure 1. Earth fault on a compensated network

Figure 2 shows the theoretical circuit and the vector diagram of the same case. The residual current I_c is mainly determined by the line to earth conductance (G_a, G_b and G_c) and by the resistance of the Petersen coil. R_0, contains the coil resistance plus the earthing resistance of the neutral point.

This residual current has to be maintained at low values to realize the main task of the coil, i.e. extinguishing the arc at the location of the earth fault on one side and to maintain the potential of the pole at a maximum value of 50 V at the same location on the other side. Thus the residual current should not exceed 10 A.

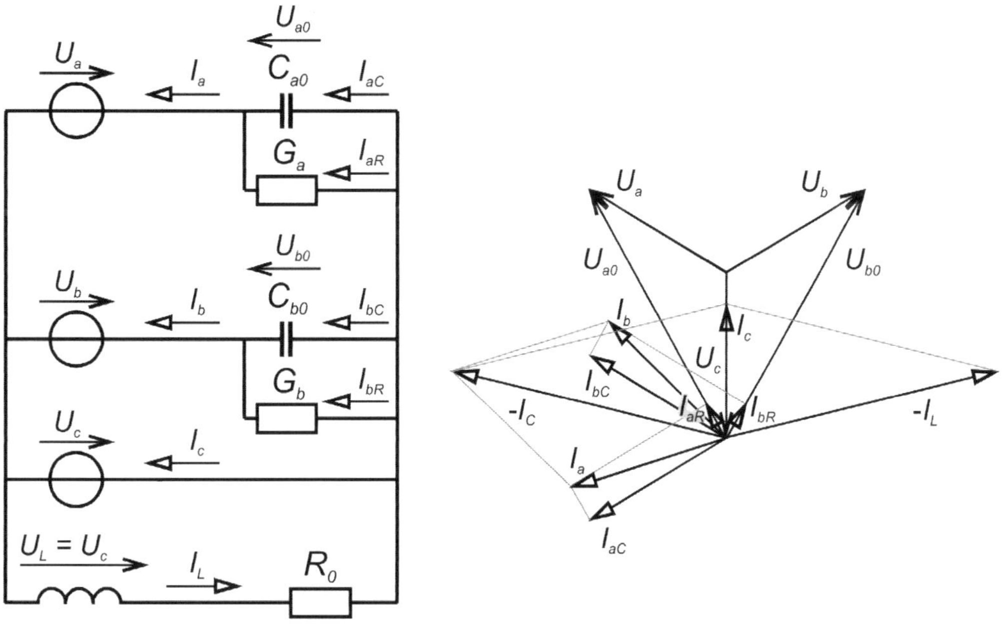

Figure 2. Theoretical circuit and vector diagram

Potential of the pole touched by a human is one factor to be taken into account, but the potential of the ground around the pole where the human stands is of interest as well. Considering a hemisphere earthing probe the *dR* resistance of the earth of a *dr* thickness around it is

$$dR = \rho \frac{l}{A(r)} = \rho \frac{dr}{2 \cdot r^2 \cdot \pi}, \tag{1}$$

where ρ is the specific resistance of the earth, *A* is the surface of the hemisphere and *r* is its radius. *R* total resistance of the earth between the distances r_1 and r_2 is

$$R = \int_{r_1}^{r_2} dR = \int_{r_1}^{r_2} \rho \frac{dr}{2 \cdot r^2 \cdot \pi}. \tag{2}$$

Then the subsequent resistance values of the earth with distances *i* an *j* can be calculated with

$$R_{i,j} = \frac{\rho}{2 \cdot \pi} \cdot \int_{r_i}^{r_j} \frac{1}{r^2} dr = \frac{\rho}{2 \cdot \pi} \cdot \left[-\frac{1}{r} \right]_{r_i}^{r_j} =$$
$$= \frac{\rho}{2 \cdot \pi} \cdot \left(-\frac{1}{r_j} + \frac{1}{r_i} \right) = \frac{\rho}{2 \cdot \pi} \cdot \left(\frac{1}{r_i} - \frac{1}{r_j} \right). \tag{3}$$

Considering a specific earth resistance of $\rho = \pi\ \Omega\text{m}$, then the values of the subsequent half meters going away from the pole are listed in table 1. Step length is always half meter. The values shown by the table are not typical, demonstrate only the proportions of the step values. Typical values for normal support poles are of several ten ohms.

Table 1.
Earth resistance around the pole

Distance from pole (m)	Resistance values (Ω)
0.5 – 1.0	0.500
1.0 – 1.5	1.6667
1.5 – 2.0	0.0833
2.5	0.0500

3. Development of the model

A decision for a resistive model have been made by the development team in spite of the fact that there are transient phenomena when earth faults occur. The reason of this decision is that, the objective of the model is to discover pole potentials during operation with earth fault for hours. There is a very low possibility that somebody touches the pole during the seconds of the transient period. If the fault is caused by a human e.g. by a service personnel of the utility company, then there will not be an operation with maintained because in cases of repair works a special operation mode is chosen with a definitive switching of the line without any delay.

Specific resistance of concrete can vary through orders of magnitude depending on its water content. To find out the resistance values of the reinforced concrete poles used by the electric utility company in our region and the typical grounding values a series of measurements have been performed on new and old poles and in sub-regions with different arts of soils. Bottom view of the measured pole is shown by Figure 3. In this figure the cross section of the iron wires reinforcing the pole can be seen. There are 12 steel wires with a diameter of 6 mm and a separate wire with a diameter of 10 mm integrated into the pole along its whole length.

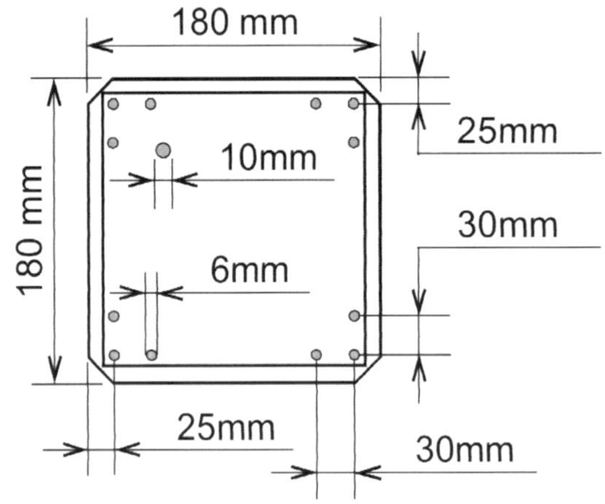

Figure 3. Bottom view of the reinforced concrete pole

The steel wire with the diameter of 10 mm is used for connecting the cross-arm and other fittings to the grounding. Resistance of the concrete has been taken into account as well. Measurement layout used for determining concrete resistance is shown by Figure 4.

A cross section of the pole and aluminium cables with a length identical with the cross section have been used for the measurement. The measured value of the specific resistance of concrete is $\rho_C = 1.26 \cdot 10^9\ \Omega mm^2/m = 1260\ \Omega m$.

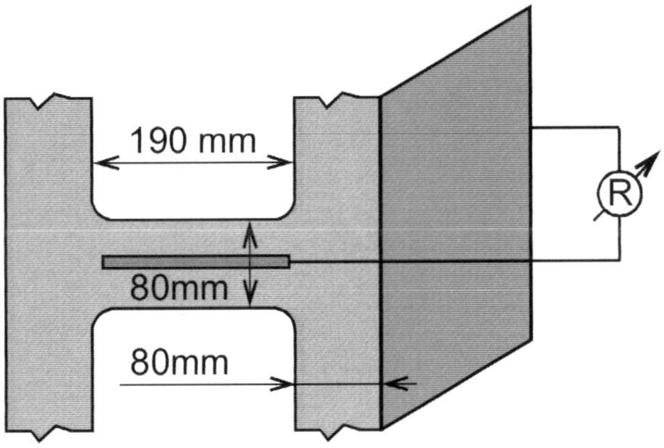

Figure 4. Measurement of concrete resistance

This specific resistance value of concrete is nearer to the minimum order of magnitude given by the Portland Cement Association R & D Lab Bull for wet concrete, i.e. to 100 Ωm [4].

Since the investigated pole is produced as a structure of 15 nearly identical parts and two end parts the model has been developed as a quasi-distributed parameter model made up of 17 units. Resistors in the model taking into account the steel wires are proportional to the lengths of the units. The value of the different resistors taking into account the concrete parts have been calculated with the equation

$$R_{Ci} = \rho_C \frac{0.6}{\left(188 + i \cdot d''\right)^2} = 1.26 \cdot 10^9 \frac{0.6}{\left(188 + i \cdot 8.5\right)^2}, \tag{4}$$

where i is the number of the pole unit, R_{Ci} is the resistance of the pole unit number i, ρ_C is the measured specific resistance of concrete and d" is the difference between the width values between two neighboring units.

Apart from pole parameters resistance values of cross-arm parts and of other fittings and of the grounding of the pole have been considered. As the supplying network resistance, inductance, main and ground capacitance values of a 20 kV network have been taken into account. The basic circuit used for simulation is shown by Figure 5.

In Figure 5 U_{L1}, U_{L2} and U_{L3} stand for the three phase-to-ground voltages of the rural overhead line networks in Hungary with nominal phase-to-phase voltage of 20 kV. L_T is the inductance of the transformer taken into account. Two transformer sizes have been considered: 25 MVA and 40 MVA typically applied for supplying 20 kV networks.

R_{OL} and L_{OL} are the resistance and inductance of the line between the transformer and the investigated pole. R_{ISO} is the resistance of the insulators between the cross-arm and the cable and R_{ISO} is the resistance of the cross-arm parts being asymmetrical.

Switches shown by Figure 3 above the insulators are used for simulating earth faults. Since in case of single earth faults a high risk of occurrence of two-phase earth faults arise, then the opportunity of simulating two-phase earth faults has to be given.

Since 20 kV overhead lines are mainly operated as networks with star points grounded through Petersen coils however with the possibility of switching to grounding through a resistor there are switches integrated into the model for switching these devices. As also low voltage networks have been investigated with direct grounding, this kind of grounding is available in the model a well.

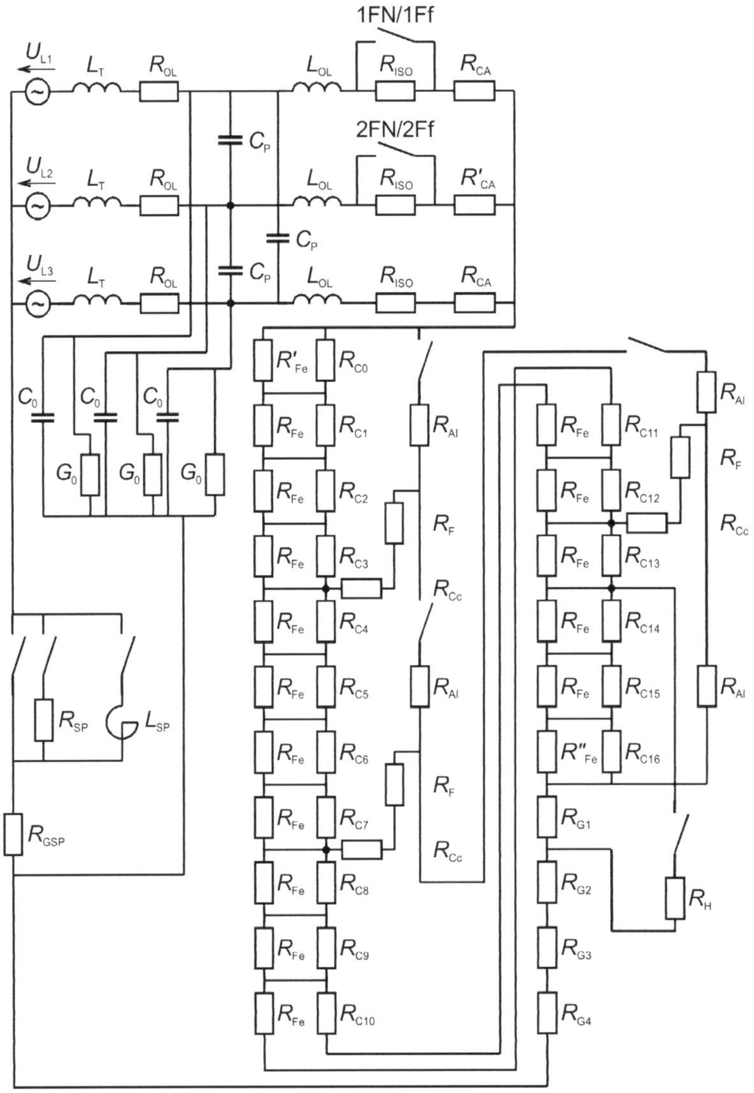

Figure 5. Simulation circuit

R_{Fe} values are resistance values of the steel wires in the reinforced concrete in which R'_{Fe} is the resistance of the upper most part of the pole, R_{Fe} that of the 15 nearly identical parts and R''_{Fe} is the resistance of the upper most part of the pole. The same is valid for the R_C values, however having slight differences from each-other because of the decreasing cross section of the pole up to its top.

Fittings are taken into account along the pole (R_F) e.g. for operating pole switches or supporting the transformer, distribution cabinet, etc. this conductive parts mounted onto the pole are connected to each-other and to the ground with a thick aluminium cable (R_{Al}).

R_{G1} to R_{G4} are the parts of the grounding resistance with a soil thickness of 0.5 m. Simple line supporting poles, operating poles with pole switches and those supporting transformers have different arts of grounding [5], [6]. Specific soil resistance (ρ) and typical grounding resistance values (R_G) have been measured [7] with the results listed in table 2.

Table 2.
Soil resitance values

#	Art of soil	ρ (Ωmm^2/m)	R_G (Ω)
1	Rock	116	12
2	Clay	25	2.5
3	Sand	260	35
4	Mixed (town)	24	3.2

Values in table 2 have been measured after a slight rain following a long dry period. Grounding conditions of the poles of different ages were different to a high extent to a perfect condition in case of newly installed poles to completely broken grounding in case of some very old poles.

4. Simulation sessions and results

During simulation sessions the following parameters have been varied [8]:

- Nominal power of the transformer (25 MVA and 40 MVA);
- art of star point handling (direct grounding, grounding through a resistor or coil);
- value of the grounding resistance of the star point;
- value of the grounding resistance of the pole;
- art of the earth fault.

Finally the resistance of human being with a value of $R_H = 1000\ \Omega$ has been taken into account standing 0.5 m far from the pole and touching a fitting at a height of about 1.8 m.

Results of simulation sessions are as follows:

- Overhead line poles made of reinforced concrete can be considered as equipotential elements because of the high cross section of the steel wires integrated into them.
- Potential of a pole is determined mainly by the grounding resistance of the pole and by the art of the fault.
- Low voltage pole potential remains under 50 V in case of a pole grounding resistance of about 10 Ω or less influenced by other parameters as matter of course.
- Increasing the length of the overhead line to the pole or increasing the nominal power of the transformer increases the potential of a pole.
- If fittings mounted onto the pole are connected to each-other and to the grounding or not does not have influence on the potential of the pole if the grounding of the pole is connected to the steel wires in the pole.
- If the connection between the grounding of the pole and the steel wires in the concrete is broken then a significant increase occurs in the potential of the pole.
- If the connection between the grounding of the pole and the steel wires in the concrete is broken then a connecting cable between the fittings and the grounding decreases the potential to a value identical with that with existing connection to the steel wires.
- In case of 20 kV lines only a Petersen coil between the star point and the ground assures pole potentials under 50 V.
- Two-phase earth faults result always in a dangerous potential of the pole.

Because of the equipotential nature of the concrete pole resulted by the high total cross section of the reinforcement irons wires inside it the model of the pole can be simplified to much less parts. A decision has been made by the development team to maintain as many parts as many fitting on the pole are placed. Figure 6 shows the actual simulation circuit used during the simulation sessions performed with TIBALab.

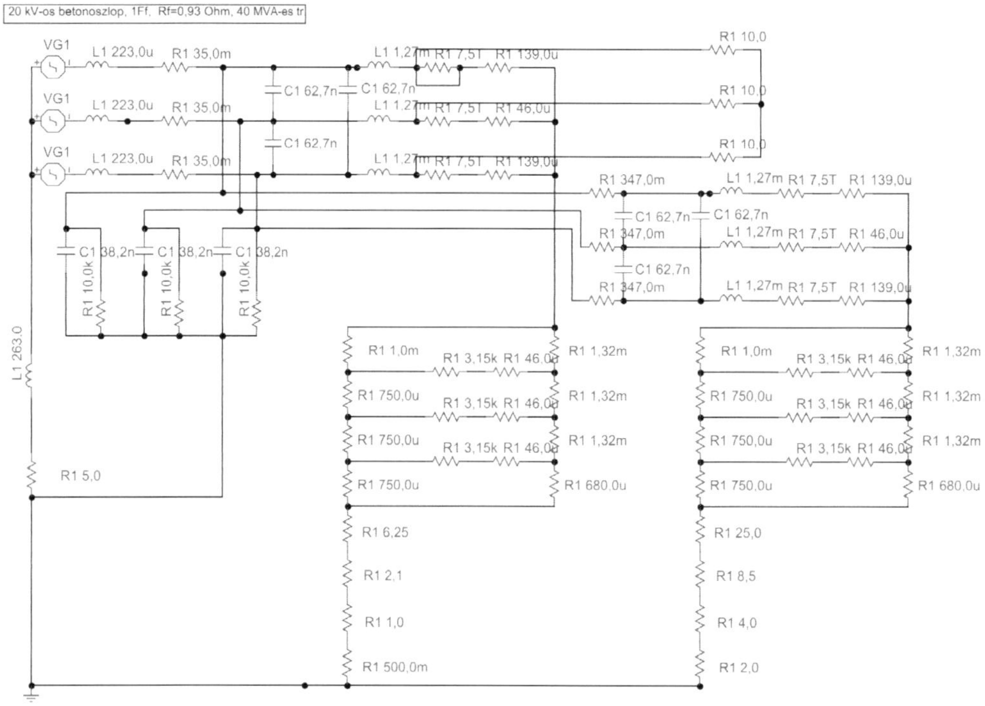

Figue 6. Simulation circuit used with TINALab

Circuit in Figure 6 contains two pole models enabling the simulation of 2Ff faults and contains a symmetrical load as well. Pole potential values have then been calculated for several earthing resistance values and line to earth conductance values – given as resistors as well.

Simulated pole potential values are shown in table 3 and table 4. Earthing resistance values vary from 40 Ω to 10 Ω as typical values measured on the actual network and line to earth conductance values vary from 10 kΩ to 1 MΩ supposed to be present on the network.

Table 3.
Pole potential values (10 km)

R_{lc} (kΩ)	10	20	50	100	200	500	1000
R_f (Ω)							
40	92.21	46.82	19.98	11.88	8.74	7.64	7.47
35	81.24	41.22	17.58	10.45	7.69	6.72	6.57
20	69.42	35.19	15.01	8.92	6.56	5.73	5.61
25	57.80	29.28	12.48	7.42	5.46	4.77	4.66
20	46.38	23.48	10.00	5.94	4.37	3.82	3.73
15	34.92	17.67	7.52	4.47	3.29	2.87	2.81
10	23.20	11.73	4.99	2.93	2.18	1.91	1.86

Table 4.
Pole potential values (1 km)

R_{le} (kΩ)	10	20	50	100	200	500	1000
R_f (Ω)							
40	95.51	48.49	20.69	12.83	9.06	7.91	7.73
35	82.23	41.71	17.79	10.58	7.78	6.80	6.65
20	71.91	36.45	15.54	9.24	6.80	5.94	5.81
25	62.28	31.55	13.45	7.99	5.88	5.74	5.02
20	48.04	24.32	10.36	6.96	4.53	3.96	3.87
15	36.17	18.13	7.79	4.63	3.41	2.97	2.91
10	24.03	12.15	5.17	3.07	2.26	1.97	1.93

The difference between the pole potential values given in table 3 and table 4 is the distance of the faulty pole from the 20 kV bus of the transformer station. Table 3 shows the values for a pole 10 km far from the station and table 4 shows them for a pole 1 km far from the station.

Potential values given in table 3 and Table 4 are shown in Figure 7 and Figure 8 as well. In case of compensated networks it is rather easy to maintain the pole potential under 50 V however neither the earthing resistance, nor the line to earth conductance can be controlled on an easy way.

In Figure 9 pole potential values can be seen in case of a 100 Ω high voltage resistance connected between the neutral point of the network and the earth. The order magnitude of the pole potential is the kV. Minimum potential – 700 V – is achieved with 10 Ω pole earthing resistance and 10,000 Ω resistance between lines and earth.

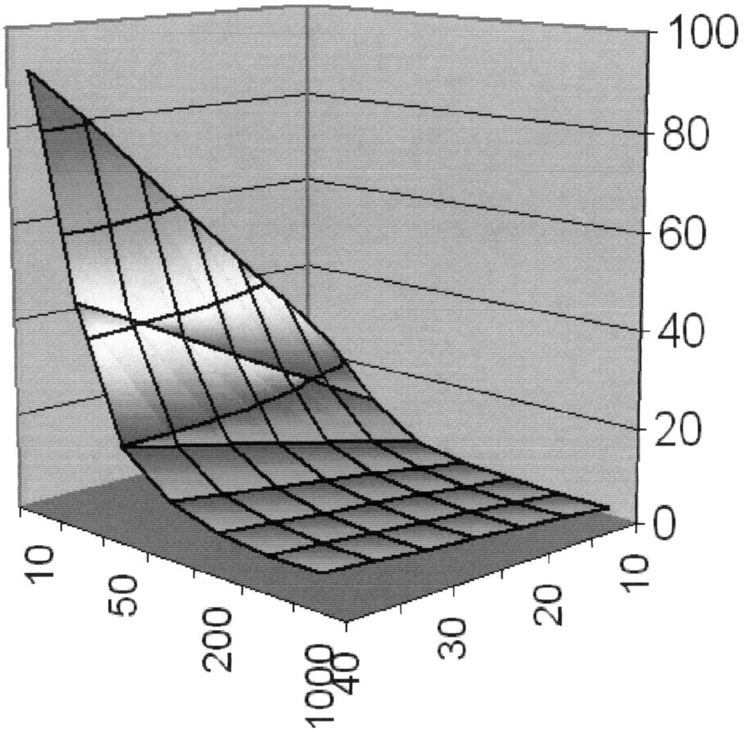

Figure 7. Pole potential in a distance of 10 km

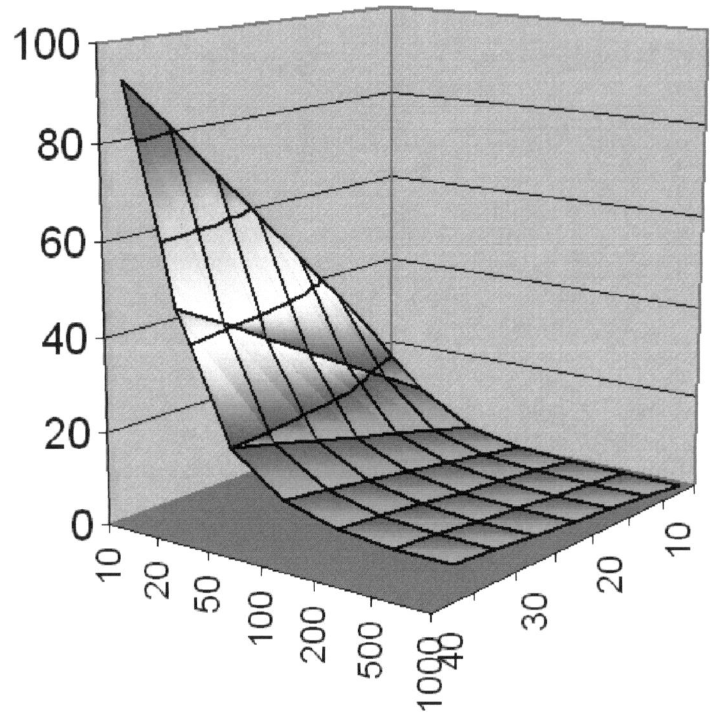

Figure 8. Pole potential in a distance of 1 km

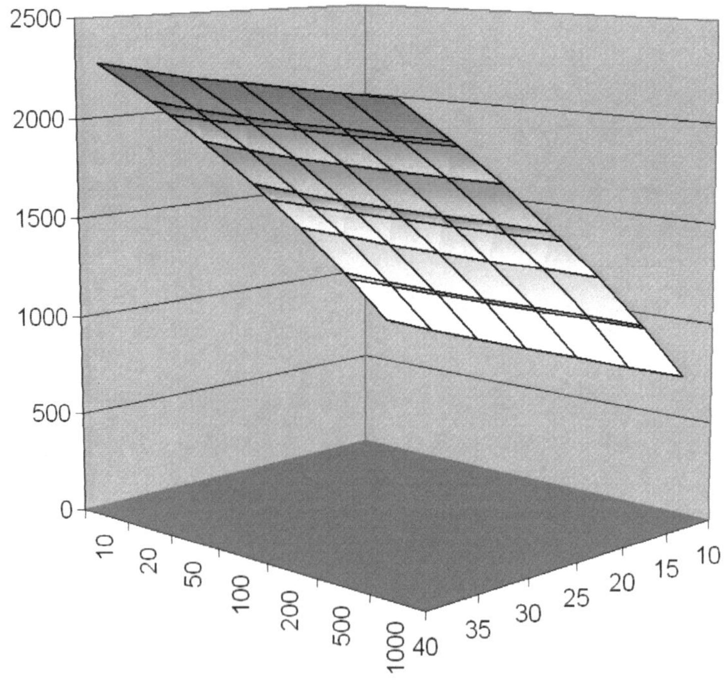

Figure 9. Pole potential with resistance as earthing

An other characteristic of this case that can be seen in Figure 9 is the much less dependence of pole potential of the parasite conductance between lines and earth.

In Figure 10 single earth fault pole potential values can be seen with the same parameters, however in case of three different dimension of the network, i.e. in case of three different values of earthing capacitances. Highest potential values have been given in case of the network with highest extension. These values are shown by thick red lines. Values of the network with medium extension are shown by magenta lines of medium thickness. Lowest potential values have been given by the network with least extension and shown by thin green lines.

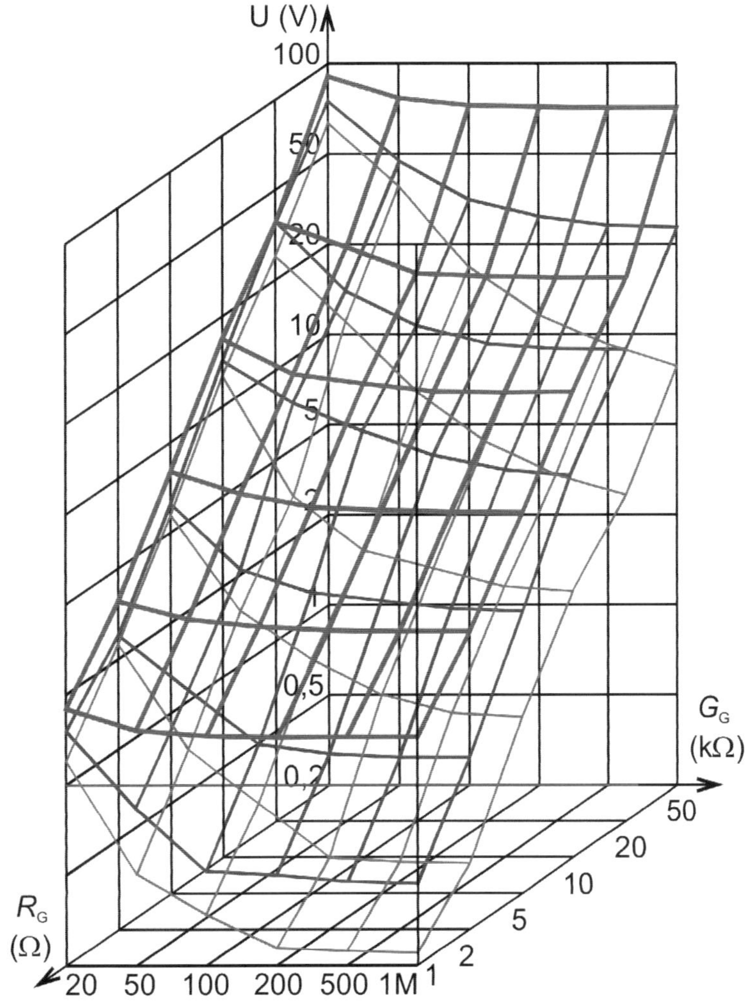

Figure 10. Pole potential values of lines of different extensions

An other phenomenon that can be seen in the figure is the decreasing dependence of the potential on the conductance to the earth with increasing network dimension.

5. Further developments

Since the residual current in case of a single line to earth fault is mainly determined by the current flowing from the line to the earth through parasite conductance, a most possible proper knowledge of this conductance is inevitable for a further investigation.

More simulation sessions have to be performed with model considering different values of also other parameters influencing the pole potential, like magnitude, harmonic content and symmetry of the consumer current, Petersen coil resistance value, etc.

An other case to be simulated is the period when the high voltage resistance of 100 Ω is connected between neutral point and earth since then more rigorous conditions have to be met for reducing pole potential to a desired value.

With this knowledge network sizes and conditions can be determined under which the operation of the medium voltage network can be maintained also in case of single phase earth faults.

References

[1] Nikolovski S, Maric P, Slipac G 2009 The Reliability of the Croatian Electric Power System for the Planned Construction Period of New Generation and Transmission Capacities, *27th International Conference Science in Practice, Pécs, Hungary*

[2] 2002 TINALab II High Speed Multifunction PC Instrument, User Manual, 2002 *DesignSoft, Budapest, Hungary*

[3] Bayliss C, Hardy B 2007 Transmission and Distribution Electrical Engineering, *ELSEVIER, London*

[4] 2010 Portland Cement Association R & D Lab Bull, The Electrical Resistivity of Concrete, *Etris.trb.org*

[5] ERŐTERV 2000, Grounding equipment of overhead line poles, *ERŐTERV, Budapest, Hungary*, 2000

[6] Horváth T 2007, Lightning protection, *Hungarian Electro-technical Association, Budapest, Hungary*

[7] MSZ EN 4851-2:1990, Measurement of specific soil resistance and grounding resistance

[8] Kvasznicza Z, Elmer G 2010, Simulating with a novel resistive model of overhead line poles, *28th Conference Science in Practice, Subotica, Serbia*, 2-4 June

Journal of Physics: Conference Series **268** (2011) 012018 doi:10.1088/1742-6596/268/1/012018

Modelling computer networks[1]

G Max
Budapest University of Technology and Economics,
Department of Automation and Applied Informatics
H-1521, Budapest, Po. Box. 91. Hungary

E-mail: max@aut.bme.hu

Abstract. Traffic models in computer networks can be described as a complicated system. These systems show non-linear features and to simulate behaviours of these systems are also difficult. Before implementing network equipments users wants to know capability of their computer network. They do not want the servers to be overloaded during temporary traffic peaks when more requests arrive than the server is designed for. As a starting point for our study a non-linear system model of network traffic is established to exam behaviour of the network planned. The paper presents setting up a non-linear simulation model that helps us to observe dataflow problems of the networks. This simple model captures the relationship between the competing traffic and the input and output dataflow. In this paper, we also focus on measuring the bottleneck of the network, which was defined as the difference between the link capacity and the competing traffic volume on the link that limits end-to-end throughput. We validate the model using measurements on a working network. The results show that the initial model estimates well main behaviours and critical parameters of the network. Based on this study, we propose to develop a new algorithm, which experimentally determines and predict the available parameters of the network modelled.

1. Introduction

In a large-scale computer network data packets may choose among many routes to transfer their destination address. These selections are based on knowledge of recent nodes of the computer networks. In most cases parts of the computer networks do not know all levels of the networks. Though almost all nodes take part in the process of forwarding packets, nodes of the computer network know more on neighbours than far ones. When users want to follow data transfers between starting and destination nodes, they need more knowledge on data transfers. To get know characteristics of computer networks, the simulation is a widely used method. After the introduction the existing works are summarized. In third session our non-linear model and its behaviours are established. The third session shows main elements of the computer networks. Next session explains the estimation process of model parameters and criteria. The fifth session presents main elements of the communication function which were implemented in our model, while in the sixth session a case study is given. The study compares an existing network to our model and observes system parameters. These system parameters and criteria are used to analyze quantitatively behaviour of the system as

[1] This work is connected to the scientific program of the "Development of quality-oriented and harmonized R+D+I strategy and functional model at BME" project. This project is supported by the New Hungary Development Plan (Project ID: TAMOP-4.2.1/B-09/1/KMR-2010-0002).

well as to investigate the impact of system parameters, such as data density, load level, link capacity on system stability. The congestion problems are followed with special attention. Last session shows tendency of future works.

2. Related works

To study network behaviour, traffic parameters are needed to measure and analyse, and to find their statistical laws, such as was done by Bhole and Popescu [1]. After having known the statistical rules, some traffic models are built, such as the establishment of a model for Novel network traffic by Jiang and Papavassiliou in [2] or we try to understand why buffers are overflowed like Sidiroglou et al. did in [3].

If the time-scale of network traffic is considered, the network traffic behaviour will be different in different time-scale. Paxson and Floyd [4] showed that the traffic behaviour of millisecond-time scale is not self-similar by the influence of network protocol. Due to the influence of environment, the traffic behaviour whose time-scale is larger than ten minutes is not also self-similar and is a non-linear time-series. Only the traffic behaviour in second-time scale is self-similar. This problem becomes known when transactions of the computer networks have to be described.

Insufficient operations of the network trace back to congestion problems. Model of Haifeng et al. [5] performs an effective network congestion control method for multilayer network.

If operational cost is important read Change's article [6] on Pricing Multicast Communication. .

3. Network elements

We may divide the elements of a real computer network into two big groups. On the one hand, the nodes imply the storage units; on the other hand, the transfer mediums connect the nodes to each other.

Forming a computer network several geometry features set conditions which are not implemented in our model:

- Geometrical conditions:
    - Distance of nodes.
    - Materials of the transfer medium.
    - Number of users.

- Seasonal effects:
    - The number of the users using the network actually moves on a very wide scale.

- Diverse external factors:
    - Weather.
    - Electromagnetic storm.

3.1. Nodes

That statement appears unambiguous, that in the network actually the nodes, the active elements of the network communicate with each other, and the nodes form the peaks of the network graph in the graph theory model of the computer network.

3.2. Transfer mediums

The process of the communication is implemented through the transfer medium. The communication happens with almost speed of light. This hilltop speed makes the transfer of a big amount of data possible already under short time. In order to avoid data collisions in the computer networks, the one-way communication is forced always along the network. Naturally, two-way communication may happen when dataflow is separated in time on the same medium. This means when a message splits into the packets to transfer them through the medium, in the same time only one data bit of the split message is crossed the transfer medium. For example, the message may represent a very long train crossing on an extraordinary short railway rail. Therefore, while the backside of the train still stays on the starting railway station and the engine has already arrived to the destination station only one railway carriage stays always on the examined railway rail. Similarly, although the network model's graph may imply one-way directed edges, in most case these mediums are used as two-ways channels since their communications are divided in time.

3.3. The communication

Communication between the nodes of the computer network is always dynamic. Peak hours are usual between the nodes, but no broadcast periods also exist. The temporal changes of the dynamic message transmissions are modelled with the help of a communicational matrix. The task of the communicational matrix is to take into consideration all elements of the network as well as their features, which regulate the data traffic between the nodes. These rules prescribe the conditions for the usage of the data transfer between the nodes and the data transfer medium. During the communication of the network's active elements, the model examines full connection system of the network. To the faultless simulation of the communication, the model has to imply some important parameters yet namely, the lengths and numbers of data transfer sections, the physical parameters of the medium, the sizes of storage units in the nodes, the degree of the utilisation, etc. .

3.4. Some special characteristics of the computer network in case of faultless data transfer

- In the course of data transfer, although transfer velocity of one data bit is constant in time, transmission times of packets with the same longitude may be different.
- In the course of data transfer, the data transfer sections running in parallel with each other do not affect each other directly, but in the nodes, for example the appearance of multiplied packets makes disturbing effects.
- Two-way traffic does not exist.
- The intensity of inner communication changes in time between the nodes directly connected to each other. In case of wrongly chosen parameters for example this inner communication capable to create peak-load on the examined network without traffic arriving from the external network.
- To control affecting message transmission, an inner communicational system works between the nodes connected to each other. For example, the receiver can receive a message vainly if the transmitter does not have a message to be sent on.

4. Traffic simulation model

Transactions in computer networks show non-linear features. For example, sometimes no data transfer happens between two nodes, because buffer of receiver node is full, though sender node is able to transfer data. It means no tools to describe transactions by mathematics of linear control systems. In this paper, a non-linear traffic behaviour model is set up. By the aid of this model, the non-linear features of traffic can be described. We also present a method to recognise congestions.

The computer network models known by the literature trace the data traffic back to the description of a communication happening between edges and nodes of network. The models handle the nodes as important elements in this description method. The outcome of a model is a communication graph that faithfully imitates the physical arrangement of the computer network, where the nodes, the active elements of the system are the peaks of the graph, which are connected to each other by the transfer mediums called edges. This statement is important, because the nodes, the active elements of the network communicate with each other, and the nodes form the peaks of the network graph in the graph theory model of the computer network. Figure 1 shows a plain example how nodes communicate to each other.

This description gives back that view on a natural manner. In this model the central place are occupied by nodes as the peaks of the network graph and the edges of the network graph show transactions of the traffic in which the peaks communicate with each other along the data lines connecting them.

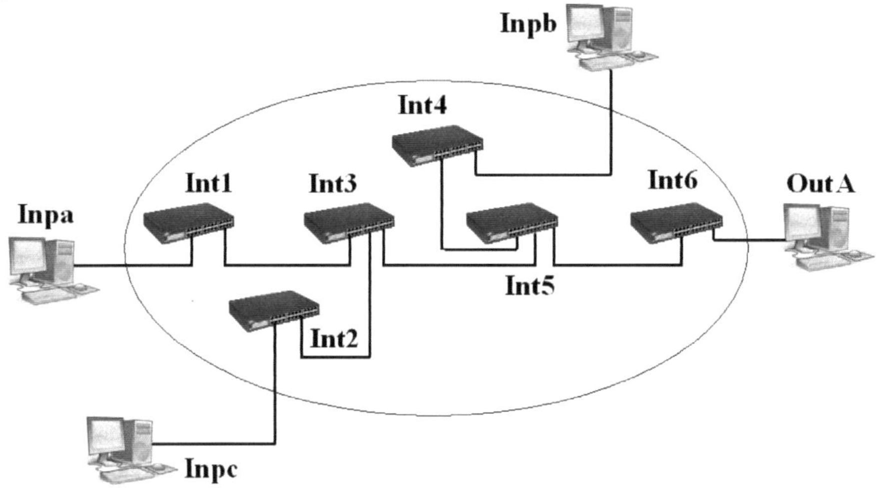

Figure 1. Internal and external elements of a network measured and simulated.

This process separates our internal nodes, where switches or bridges are located with a closed curve from their input and output workstation nodes, as it is shown in figure 1. In the following the internal nodes are numbered, input nodes are marked by small letters and output nodes are marked by capitals. Consider the network is known in a time t and numbers of bits stored in the nodes are marked with $\mathbf{N}(t)$. Each node has some buffer for their messages. This capacity is measured by data density (1) that shows rate of recent data quantity in time and the maximal data quantity in node i.

$$b_i(t) = \frac{\text{number of recent databits in buffer of node i}}{\text{buffer size of node i}} = \frac{N_i(t)}{A_i} \tag{1}$$

In our model the data density of node i is a number without dimension between $0 \leq b_i \leq 1$. Traditionally, the numeration of data density shows the difference of outgoing and incoming bits in time t or in other words it can show the maximum number of data bits that can be transferred in the next time unit. Now data density is introduced as a rate of the length of messages stored in node i and the maximal message length. This property shows non-linearity, because no chance is to receive more bits than the maximal size of the data buffer or node i can not receive any bit, if node j has no data to send. In our network data density of node i is marked with b_i. In our model $\mathbf{N}(t)$ and $\mathbf{b}(t)$ are vectors of n elements (2). From (1)

$$N_i(t) = A_i \cdot b_i(t) \tag{2}$$

shows the amount of data stored in node i, where A_i represents the buffer size of node i. Since no buffer is between node i and node j, $\mathbf{A}$ is a diagonal matrix (3), that is

$$\mathbf{N}(t)=\begin{bmatrix} A_1 & 0 & \cdots & 0 \\ 0 & A_2 & \cdots & 0 \\ \cdots & \cdots & \cdots & \cdots \\ 0 & 0 & \cdots & A_n \end{bmatrix}\cdot\begin{bmatrix} b_1(t) \\ b_2(t) \\ \cdots \\ b_n(t) \end{bmatrix}=\mathbf{A}\cdot\mathbf{b}(t) \tag{3}$$

If elements of $\mathbf{N}(t)$ are known, total number of databits stored in the observed area is calculated (4) as a vector scalar product:

$$N(t)=\sum_{i=1}^{n} A_i\cdot b_i(t)=diag(\mathbf{A})\cdot\mathbf{b}(t) \tag{4}$$

Now let us examine status of the network in time $t+\Delta t$. The amount of data stored in node i changes as (5):

$$N_{\mathrm{i}}(t+\varDelta t)=N_{\mathrm{i}}(t)+\varDelta N_{\mathrm{i}}^{\text{internal}}+\varDelta N_{\mathrm{i}}^{\text{input}}\text{-}\Delta N_{\mathrm{i}}^{\text{output}} \tag{5}$$

In the network the transmission speed means how many bits can be transported in one seconds between two nodes. Transmission speed may be different in some parts of the network, but from node j to node i the same value is supposed and marked with v_{ij}. If transfer speeds are known between the nodes, data changes can be expressed as (6)

$$\varDelta N_{\mathrm{i}}^{\text{internal}}=\sum_{\substack{j=1\\ i\neq j}}^{n} c_{\mathrm{ij}}\cdot v_{\mathrm{ij}}\cdot\varDelta t,\ \ \varDelta N_{\mathrm{I}}^{\text{input}}=\sum_{I=a}^{k} c_{\mathrm{iI}}\cdot v_{\mathrm{iI}}\cdot\varDelta t,\ \ \varDelta N_{\mathrm{O}}^{\text{output}}=\sum_{O=A}^{M} c_{\mathrm{Oi}}\cdot v_{\mathrm{Oi}}\cdot\varDelta t. \tag{6}$$

where c is communication function of the node. The communication function describes the quality of transmissions. Indexes of the communication function show the direction of the data transfer. The first index shows the receiver node and the second one represents the sender node. Input and output nodes are marked with I and O respectively. After summarizing data changes in all nodes, we get total data change (7) in our network:

$$\begin{bmatrix} N_1(t+\Delta t) \\ N_2(t+\Delta t) \\ \cdots \\ \cdots \\ N_n(t+\Delta t) \end{bmatrix}=\begin{bmatrix} N_1(t) \\ N_2(t) \\ \cdots \\ \cdots \\ N_n(t) \end{bmatrix}+\begin{bmatrix} \sum_{j=1;\, i\neq 1}^{n} c_{1\mathrm{j}}\cdot v_{1\mathrm{j}}+\sum_{I=1}^{k} c_{1\mathrm{I}}\cdot v_{1\mathrm{I}}-\sum_{O=1}^{m} c_{\mathrm{O}1}\cdot v_{\mathrm{O}1} \\ \sum_{j=1;\, i\neq 2}^{n} c_{2\mathrm{j}}\cdot v_{2\mathrm{j}}+\sum_{I=1}^{k} c_{21}\cdot v_{21}-\sum_{O=1}^{m} c_{\mathrm{O}2}\cdot v_{\mathrm{O}2} \\ \cdots \\ \cdots \\ \sum_{j=1;\, i\neq n}^{n} c_{\mathrm{nj}}\cdot v_{\mathrm{nj}}+\sum_{I=1}^{k} c_{\mathrm{nI}}\cdot v_{\mathrm{nI}}-\sum_{O=1}^{m} c_{\mathrm{On}}\cdot v_{\mathrm{On}} \end{bmatrix}\cdot\Delta t \tag{7}$$

Forming (7) and do $\Delta t\rightarrow 0$, the result is shown by (8):

$$\lim_{\varDelta t\to 0}\frac{N(t+\varDelta t)-N(t)}{\varDelta t}=N'(t)=\sum_{i=1}^{n}\Big(\sum_{j=1;\, i\neq n}^{n} c_{\mathrm{ij}}\cdot v_{\mathrm{ij}}+\sum_{I=1}^{k} c_{\mathrm{iI}}\cdot v_{\mathrm{iI}}-\sum_{O=1}^{m} c_{\mathrm{Oi}}\cdot v_{\mathrm{Oi}}\Big) \tag{8}$$

(8) says to add up all bits in transactions. In the observed area between two nodes a transaction transports the same amount of data. Only their signs are different, that is the total amount of data transported is equal to zero. (8) becomes more simple (9).

$$N'(t) = \sum_{i=1}^{n}\left(\sum_{I=1}^{k} c_{iI} \cdot v_{iI} - \sum_{O=1}^{m} c_{Oi} \cdot v_{Oi}\right) = \sum_{i=1}^{n}\sum_{I=1}^{k} c_{iI} \cdot v_{iI} - \sum_{i=1}^{n}\sum_{O=1}^{m} c_{Oi} \cdot v_{Oi} \tag{9}$$

(9) represents that the change of numbers of databits independents of change of internal dataflow in the observed area. (9) shows the difference of numbers of incoming and outgoing databits in a time unit or in other words (10)

$$\int_{t}^{t+\Delta t} N'(t) \cdot dt = \left(\sum_{i=1}^{n}\sum_{I=1}^{k} c_{iI} \cdot v_{iI} - \sum_{i=1}^{n}\sum_{O=1}^{m} c_{Oi} \cdot v_{Oi}\right) \cdot \Delta t = \text{Inputs} - \text{Outputs} \tag{10}$$

5. Communication function

The communication function plays big role in (6) – (10). There are some subfunctions implemented in c_{ij}. As we could see before $c_{ij}=c_{ij}(b_i,b_j,t)$. Let us examine parts of this function. $C_{ij}(b_i,b_j,t)$ must contain at least the following:

- Internal connections of the network.
- Environmental connections of the network.
- Size of internal buffers in nodes.
- Capabilities of nodes to send and receive data.

The subfunction of internal or environmental connection, marked with k grants the connection when node j can communicate with node i. The connection subfunction is product of two components. This function is created node by node. Each node is examined and if a physical connection exists between nodes j and node i, f_{ij} is set to 1, otherwise it is zero (11). Elements of f_{ij} creates a symmetrical matrix, because if node i is connected to node j, than node j is connected to node i, too. If there is the opportunity of the permanent communication between two nodes, and the node j works for node i, then $k_{ij}(t)= f_{ij} \cdot p_{ij}(t)$, where $p_{ij}(t)$ is the probability of data transmission from node j to node i. The probability of data transmission, $p_{ij}(t)$ presents the distribution proportion of given routes belonging to a node with relative weighting, where $0 < p_{ij}(t) < 1$, if node j works for more than one node and $p_{ij}(t)$ may differ from $p_{ji}(t)$. The relative weightings of node j are shown in column j of the matrix k and $\Sigma p_{ij}(t) = 1$ for each column. If there is no physical connection between these nodes, then $k_{ij}=0$.

$$k_{ij} = \begin{cases} f_{ij} \cdot p_{ij}, & \text{if connection exists between nodes } j \text{ and node } i \\ 0, & \text{if no connection is between nodes } j \text{ and node } i \end{cases} \tag{11}$$

where $i \neq j$ and $1 \leq i,j \leq n$.

As we could see before the size of the buffers is included in data density of the node.

The third subfunction presents the properties of the connection. We know that data link mechanisms depend on density of the traffic. In our model, two internal subfunctions are used. $S_j(t)$ defines properties of sender node, while $R_i(t)$ declares features of the receiver.

$S_j(t)$ is an internal subfunction (12) of transmitter node with values of 1 or 0. It shows whether node j has any message to send or not. The connection is disable if data density of node j ($b_j(t)$) is equal to 0, anyway 1.

$$S_j(t) = \begin{cases} 1, & b_j(t) > 0 \\ 0, & b_j(t) = 0 \end{cases} \tag{12}$$

$R_i(t)$ is another internal subfunction (13) of the receiver node with values of 1 or 0. The connection is enable if $b_i(t)$, the data density of node i smaller, than 1, anyway it is 0. Zero means that buffer of node i has been overloaded, so node i closes its communication port to direction of node j, therefore node i does not receive any message from node j.

$$R_i(t) = \begin{cases} 1, & b_i(t) < 1 \\ 0, & b_i(t) = 1 \end{cases} \tag{13}$$

Finally, the communication function between node j and node i is a product of three subfunctions (14):

$$c_{ij} = k_{ij}(t) \cdot S_j(b_j(t), t) \cdot R_i(b_i(t), t) \tag{14}$$

Suppose that dataflow is working normally. Input nodes can always send data to the internal nodes, no any trouble is among the internal nodes and output nodes can receive all data from our network. In this case both $S_j(b_j,t)$ and $R_i(b_i,t)$ are set to 1, that is $v_{ij} \cdot \Delta t \leq (1-b_i(t)) \cdot A_i$ as well as $v_{ij} \cdot \Delta t \leq b_j(t) \cdot A_j$ and the average transfer speed for the total time period is $e_{ij}(t) \cdot v_{ij}$, where $e_{ij}(t)$ is the effectiveness of the transmission channel that shows the time rate of real data transfers and the total working time.

6. Simulation and results

6.1. Test environment

An existing network which can be seen in Figure 1 was measured and simulated at transfer speed of 100Mb/sec. The first file of 89 MB was transported from Inpa and the second one of 91 MB started from Inpb. First channel used TCP protocol, while second file was transported by FTP protocol. The second file was started in 8 seconds after the first one. The third input channel named Inpc did not work. The result was measured at channel of OutA. Measured figures were observed by Wireshark. (Wireshark is registered trademarks of the Wireshark Foundation.) The transfer was repeated 60 times. The calculated efficiency rate was 72% in the 100 Mb/sec channels, because an average transfer took 20 sec.

6.2. Simulation unit

The simulation takes too much time if dataflow is simulated by bit by bit. We wanted to make a unit for our simulation. First the lengths of the frames were measured. Data transfers were made in different time period (early morning, afternoon, midnight), with different data lengths (from some hundred bytes to hundred megabytes) and different transmission speed (from 24 Kbits/s up to 100 Mbits/s). Measurements were made by the Wireshark Network Protocol Analyzer. After having collected more hundred thousands messages, the incoming and the outgoing frames were analyzed. The result showed two significant figures among the values. One of them represented the length of confirmation messages, while the other one was the typical length of data messages. The most frequented values are shown in Table 1. Taking into consideration figures of Table 1, we chose 55 bytes as unit of frame length. Using this frame length an average data length is 26 units.

Table 1. Number of pieces of the most frequented frame length [in bytes].

Length	Pieces	Total length
56	165 535	8 938 890
66	11 773	777 018
1082	10 730	11 609 860
1434	271 383	389 163 222
1514	43 127	65 294 278

6.3. Simulation

The simulation was written in MatLab. The same files were transferred through our model like we used in the real network. We are mainly interested in transfer systems of large number of messages. Our work is motivated by the following questions:

- How to implement update rules?
- How to implement update rules in different systems (I.e.: mono or multiplied channels)?
- How to reduce or avoid collision?
- How to measure effectiveness?

Main features of our update rules in our system are the following:

- No multichannel mode is implemented in nodes.
- The effects of collisions take into consideration in efficiency rate.
- Messages always move forwards or stay in nodes.

Our system uses parallel update rule. The internal nodes are numbered from 1 to n while input nodes are numbered from 1 to m. At each time step, each node is updated at least once. Each time step is divided into three stages. Update is performed as follows:

- Stage 1: If there is at least message in a node to transfer its move is based on communication function (transfer probabilities as well as own and neighbour status).
- Stage 2: The conflicts of transfers are resolved. The conflicts arise because more than one messages want to move to the same node at the same time step. If at least two messages want to move to the same node, one message is randomly chosen to move to that node and the other conflicting messages remain at their current node during the current time step.
- Stage 3: the actual message transfer bases on the transfer decisions and conflict resolutions of the first two stages. These three stages are used to perform parallel update at each time step.

Initially, the messages are located in their nodes with density b_i where $0 \leq b_i \leq 1$ and $1 \leq i \leq n$ or $a \leq i \leq m$. Each transfer medium can be occupied by at most one message at any time. The data density of the internal nodes is measured and calculated because the messages are either added or removed after the initialization of the network, so the densities of the nodes change in time. The update rules for nodes and messages are the following:

- The transfer probabilities depend on the probability of data transmission from node j to node i ($p_{ij}(t)$).
- Messages can move forward but not backwards.
- Using different internal buffers, bottleneck effects can also be tested.

6.4. Results

Figure 2 and figure 3 show two simulations with different buffer sizes. In figure 2 5 MByte internal buffer was used, while in figure 3 only 1 MByte buffer was allocated. In Figure 2 inputs work linearly, while figure 3 shows non-linear elements. Figure 3 helps us to find bottlenecks of the network. Inpb was blocked by node Int1, because there were no enough room for incoming frames.

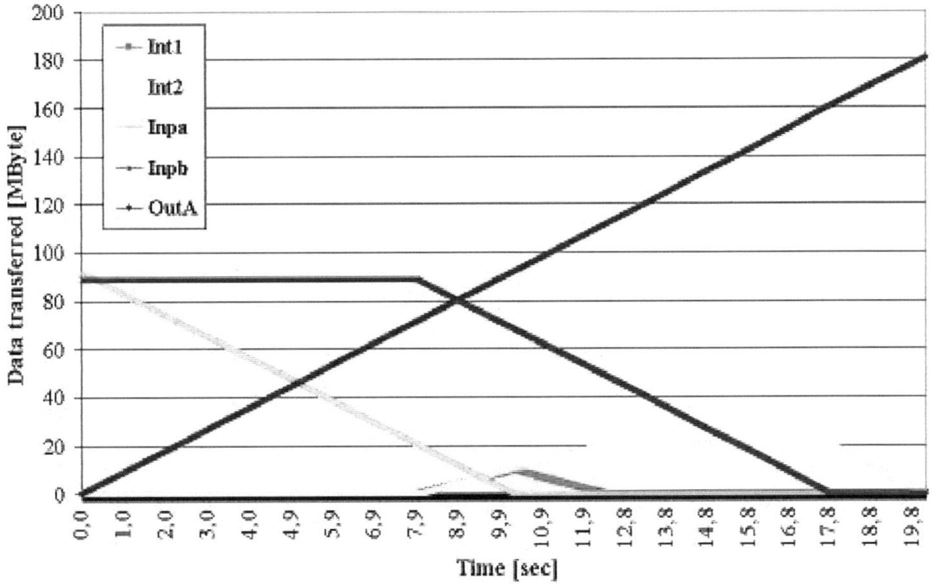

Figure 2. Data transfer using 5 MB internal buffers.

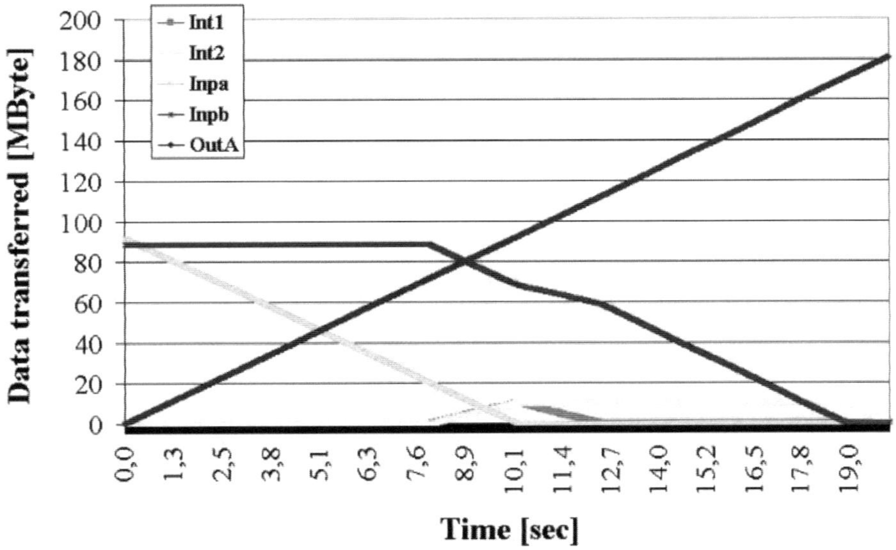

Figure 3. Data transfer using 1 MB internal buffers.

7. Conclusion and future work

This paper looks for possibilities to describe dataflow model of large-scale computer networks. A model was presented that was applicable to simulation, planning and regulation of computer network's traffic. At the time of the model's establishment, the partial differential equations were avoided in the mathematical model because of the specially chosen state variables. The nodes have honoured roles in this non-linear model because storage capacities of the transmission medium are practically zero. Nodes either communicate to each other or not. In our model, the mean of the data density is the proportion of the size of data stored in the single node and the data quantity, which can be stored maximally. Our model examines change of data density occurred by data flow among the nodes in a region demarcated by close curve. Input and output data densities are regarded as known. At first sight, these processes are the inputs and outputs of the model. Effectively, these processes together form the actual inputs of the mathematical model. State variables present data densities arising in the internal nodes of the system. Our system applies a data traffic model involving n internal and m external nodes. To create the mathematical model the communication matrices defining the network has fundamental importance. Our model applies four communication matrices. During simulation we supposed that speed of transmission equals between nodes. It must be changed in the future. We have to work out a quick test to calculate possible congestion points. Finally, a simply example was presented to demonstrate effectiveness of the model and how this model is applicable to the simulation, planning or regulation of computer networks.

References

[1] Bhole Y and Popescu A 2005 Measurement and Analysis of HTTP Traffic *Journal of Network and Systems Management* **13** 357-371

[2] Jiang J and Papavassiliou S 2004 Detecting Network Attacks in the Internet via Statistical Network Traffic Normality Prediction *Journal of Network and Systems Management* **12** 51-72

[3] Sidiroglou S, Giovanidis G and Keromytis A 2005 A Dynamic Mechanism for Recovering from Buffer Overflow Attacks *ISC 2005, LNCS 3650* (J. Zhou et al.) 1–15

[4] Paxson V and Floyd S 1995 Wide area traffic: The failure of Poisson modelling *IEEE/ACM Transactions on Networking* **3** 226–244

[5] Haifeng D, Yang X and Lingyun L 2008 An Effective Network Congestion Control Method For Multilayer Network *Journal Of Electronics (China)* **25**

[6] Change J and Sirbu M 2001 Pricing Multicast Communication: A Cost-Based Approach, *Telecommunication Systems* **17** 281–297

Attempts at a numerical realisation of stochastic differential equations containing Preisach operator

S McCarthy[1] **and D Rachinskii**[1]

[1]Department of Applied Mathematics, University College Cork, Cork, Ireland

E-mail: `stephen.p.mccarthy@umail.ucc.ie`

Abstract. We describe two Euler type numerical schemes obtained by discretisation of a stochastic differential equation which contains the Preisach memory operator. Equations of this type are of interest in areas such as macroeconomics and terrestrial hydrology where deterministic models containing the Preisach operator have been developed but do not fully encapsulate stochastic aspects of the area. A simple price dynamics model is presented as one motivating example for our studies. Some numerical evidence is given that the two numerical schemes converge to the same limit as the time step decreases. We show that the Preisach term introduces a damping effect which increases on the parts of the trajectory demonstrating a stronger upwards or downwards trend. The results are preliminary to a broader programme of research of stochastic differential equations with the Preisach hysteresis operator.

1. Introduction

A good body of work has been done of developing, studying and solving deterministic differential models which contain the Preisach operator. The simplest models consist of one scalar differential equation coupled with the Preisach operator input-output relationship. In particular, these types of models have been used in the context of terrestrial hydrology of homogeneous soil water systems with the Preisach operator introducing a hysteretic constitutive relationship between the moisture content and the pressure (matric potential) to the ordinary differential equation obtained by averaging out the spatial variation from the Philip-Richards balance equation [1–3]. Another example is provided by models of macroeconomic systems and multiagent market models where the Preisach operator is used to incorporate shock type memory into the system [4,5]. Both of these areas contain a stochastic nature which is not fully captured by deterministic models.

Stochastic aspects of the input-output relationship defined by the Preisach operator have been fairly well understood in the case of open loop systems where the properties of the stochastic input are known or measurable *a priori* [6]. The formalism of the phenomenological Preisach model presents a system as a weighted superposition of many two-state non-ideal relays that are individually and independently driven by the same input [7,8]. As this formalism is rather general in modelling rate-independent systems exhibiting hysteresis and shock type memory, the theory of a stochastically driven Preisach operator has been motivated by, and successfully applied to, a variety of problems such as modelling thermal relaxation and viscosity (after-effect) in ferromagnetic materials [9,10], creep in superconductors [11], signal processing and passage of noise through hysteretic systems [12], effect of noise on data storage technologies and data collapse [13] and others. Differential models of closed loop systems involving the Preisach operator have received less attention. The existing body of results refers mainly to systems with piecewise smooth trajectories such as models of stochastically driven mechanical systems, shape memory alloys and hysteretic oscillators [14]; estimation of damage and fatigue [15]; and, models of terrestrial hydrology [16].

As a natural stochastic extension of the deterministic equations with the Preisach operator considered in [3–5] in the context of modelling hydrological systems and economics systems, we propose the following formal stochastic differential equation

$$\xi dx_t + d(Px)_t = a(x_t, t)dt + b(x_t, t)dW_t \tag{1}$$

where P denotes the Preisach operator, *i.e.*, $(Px)_t = (Px)(t)$ denotes output of the Preisach model with continuous input $x_t = x(t)$; ξ is a positive parameter; W_t is the Wiener process; and, $a(\cdot,\cdot)$, $b(\cdot,\cdot)$ are continuous functions. This extension may be of interest in the areas of hydrology, economics and finance where stochastic effects are prevalent. As a prototype example, we will consider below a simple price dynamics model leading to a discrete time counterpart of equation (1) with constant a and b. This price dynamics model adopts, yet in an essentially simplified form, the philosophy of the modelling approach proposed in [17–21] which leads to a hierarchy of powerful multiagent models of economics systems.

Equation (1) can also be used for examining the effect of feedback on various open loop systems modelled by the Preisach operator and by the inverse Preisach operator with a stochastic input, as well as for testing the effect of noise on deterministic differential models involving the shock type memory.

We note that the simplest example of the shock type memory is the running maximum (running minimum) of the input which is one of the standard market indicators in financial setting. Stochastic differential equations involving running extrema have been studied, for instance, in [22]. From this perspective, equation (1) can be viewed as a similar model incorporating a more complex memory of the past shock type events (called the main extrema of the input in [23]). The role of such shocks in economics systems is discussed in [4]. Equation (1) is thus the simplest model incorporating both the Preisach memory operator and the stochastic component.

We stress that the mathematical formalism of the continuous time system (1) is not the subject of this paper. We will consider and compare two natural Euler type formal discretisation of equation (1). Some numerical evidence will be given that the two discrete time schemes converge as the time step becomes smaller. We will then discuss a damping effect introduced by the Preisach memory term to the system. A characteristic feature of this effect is that damping is stronger on the parts of a trajectory that demonstrate a more pronounced upward or downward trend.

The paper is organized as follows. In the next section, we briefly remind the construction of the Preisach operator introducing the necessary notation. In Section 3, we discuss a price dynamics model leading to a discrete counterpart of equation (1). In Section 4, two formal discretisations of equation (1) are presented. Section 5 contains results of implementation of the two proposed numerical schemes and discussion. The last section presents conclusions.

The results presented here are preliminary to a programme of future work which will aim at (a) developing mathematical formalism of the continuous time equation (1); (b) developing numerical methods of solving equation (1) and analysis of convergence of the discrete time schemes to the continuous limit; (c) improving the efficiency and accuracy of the numerical schemes (using, in particular, the Brownian bridge techniques); (d) adapting analytical tools of the theory of stochastic differential equations, such as the Fokker-Planck equation etc., to equation (1) wherever possible; and, (e) examining the long term effect of stochastic inputs on the Preisach memory configuration in closed loop systems. This work will be done elsewhere.

2. Preisach model

Here we detail some of the properties of the Preisach operator as they pertain to the work examined in this paper.

2.1. *Non-ideal relay*

The non-ideal relay is a basic hysteresis element. The dynamics of the relay are shown in figure 1. The state of the relay can be either 0 (the relay is switched off) or 1 (the relay is switched on) at any instant (sometimes, the states ± 1 are used instead with a straightforward change in all the definitions). The dynamics of the state is defined by the equation

$$y(t) = R_{\alpha,\beta}[\eta_0]x(t) = \begin{cases} 1 & \text{if } \exists\, t_1 \in [t_0, t] \text{ such that } x(t_1) \geq \beta, x(\tau) > \alpha\ \forall\, \tau \in [t_1, t] \\ 0 & \text{if } \exists\, t_1 \in [t_0, t] \text{ such that } x(t_1) \leq \alpha, x(\tau) < \beta\ \forall\, \tau \in [t_1, t] \\ \eta_0 & \text{if } \alpha < x(\tau) < \beta\ \forall\, \tau \in [t_0, t] \end{cases} \tag{2}$$

for $t \geq t_0$, where $\eta_0 = y(t_0)$ is the initial state of the relay at time t_0. Hence, the state of the relay is uniquely determined after the moment t_0 for a known continuous input $x(t)$, $t \geq t_0$, if the initial state η_0 is given.

2.2. *The Preisach Operator*

The Preisach operator can be thought of as a weighted sum of independent non-ideal relays with a common input. The threshold values of the relays are represented by points in the Preisach half-plane, $\Pi = \{(\alpha, \beta) : \beta > \alpha\}$. The operator maps a continuous input $x(t)$, $t \geq t_0$, to the continuous output

$$y(t) = P[\eta_0(\alpha,\beta)]x(t) = \iint\limits_{\alpha<\beta} \mu(\alpha,\beta) R_{\alpha,\beta}[\eta_0(\alpha,\beta)]x(t)\,d\alpha\, d\beta \qquad t \geq t_0 \tag{3}$$

of the Preisach model. Here the integrable nonnegative function $\mu(\alpha, \beta)$ called the Preisach weight function describes the weighting of the non-ideal relays; the binary function $\eta_0 = \eta_0(\cdot, \cdot)$ called the initial state of the Preisach operator is an infinite-dimensional parameter describing the states of all the relays at the initial moment t_0. We will also adopt a shorter notation $y(t) = (Px)(t)$ for the Preisach operator omitting the reference to the initial state parameter $\eta_0 = \eta_0(\alpha, \beta)$ wherever it does not lead to a confusion.

Similarly, a discrete version of the Preisach operator can be defined, where the number of relays is finite, the integral in (3) is replaced with summation, and the Preisach weight function is replaced with a finite set of nonnegative weights μ_i.

2.3. *Geometrical interpretation*

According to equation (3) the output of the Preisach operator is given by an integral over the Preisach half-plane. An efficient method of calculating the output is based on the following geometrical interpretation of the states of the Preisach operator, see, for example, [23] for details. Let us consider the area on the Preisach half-plane for which the relays are switched on at a given moment t. In figure 2, this area lies to the left of the staircase line $S(t)$, while the relays to the right of this line are switched off. Hence, the output value $y(t)$ equals the integral of the Preisach weight function $\mu(\alpha, \beta)$ over the area to the left of the line $S(t)$. The evolution of the input results in the evolution of the line $S(t)$ and thus in the evolution of the output $y(t)$. The evolution of $S(t)$ is defined by a few simple rules. We note that all the segments of the staircase line $S(t)$ are parallel to the coordinate axes at every time instant; for systems with piecewise monotone inputs or discrete time inputs, which can be interpolated to piecewise

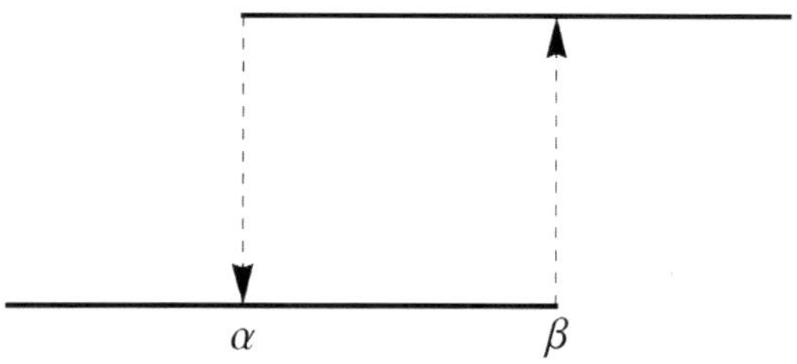

Figure 1. Non-ideal relay. Representation of the dynamics of the non-ideal relay with thresholds α and β, with $\beta > \alpha$.

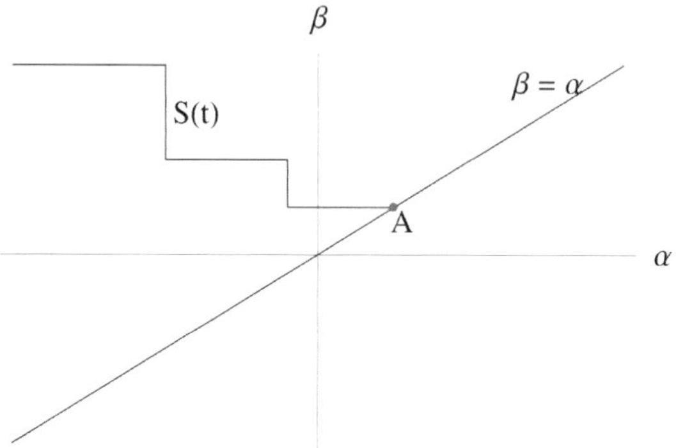

Figure 2. The Preisach half-plane. The boundary line $S(t)$ separates the domain where the points (α, β) represent the relays which are switched on (below the line $S(t)$) from the domain where the relays are switched off (above the line $S(t)$) at a given moment t. The coordinates of the point A equal $x(t)$, the input value.

monotone inputs and which are sufficient for the purposes of this paper, it suffices to consider the lines $S(t)$ with a finite (possibly, varying) number of corners.

The coordinates of the point of intersection between the lines $S(t)$ and $\beta = \alpha$, marked A in figure 2, equal the current value of the input $x(t)$. We can convert changes in $x(t)$ to movement of the point A along the line $\beta = \alpha$, where increases in $x(t)$ correspond to a movement of A from left to right, similarly decreases are movements from right to left. The movements of A result in changes in $S(t)$. For example, if figure 2 represents a moment t_1 when the input $x(t)$ reaches a local maximum value and then $x(t)$ decreases until a later moment t_2, then figure 3 shows the line $S(t)$ at the moment t_2. Note that a new corner in the line $S(t)$ has been created. Assuming further that t_2 is a local minimum point for the input and $x(t)$ increases after the moment t_2, figure 4 shows the line $S(t)$ at the later moment $t_3 > t_2$ when the value of the input has become equal to the ordinate of one of the corners of the line $S(t_2)$ shown in figure 3. To be noted here is that $S(t)$ has again changed by the removal of any corners that have a lower ordinate than the current value $x(t_3)$ of the input. The complete set or rules which define the dynamics of $S(t)$ can be found in [23] and will be detailed in Section 4 below. We also refer to the interactive applet available at http://euclid.ucc.ie/hysteresis/

Figure 5 illustrates how the changes in $S(t)$ define the changes in the output (3) of the Preisach operator.

2.4. *Preisach memory state*

The boundary line $S(t)$ is known as the state of the Preisach operator. Its position on the Preisach half-plane is defined by the location of its corners described by the coordinate pairs (m_k, M_k). Figure 6 shows the projection of these corners onto the bisector $\beta = \alpha$. The dynamics of the input results in the deletion of certain corners and the creation of new ones and is thereby translated to the dynamics of the state and the output of the Preisach operator as described in the previous subsection. At a given moment t, the abscissas $m_0, m_1, \ldots, m_k$ of the corners of $S(t)$ form a subset of the set of all the local minima of the input $x(t)$ which have been achieved prior to the moment t. They are called the running main minima of the input, see [23] for the definition and discussion. Likewise the ordinates $M_0, M_1, \ldots, M_{k'}$ of the corners of $S(t)$ are a specific selection of the local maxima of the input which $x(t)$ has achieved prior to the moment t; they are called the running main maxima. The running main extrema (also known as past shock values of the input) at every instant are ordered as

$$\cdots < m_k < \cdots < m_1 < m_0 < x(t) < M_0 < M_1 < \cdots < M_k < \cdots$$

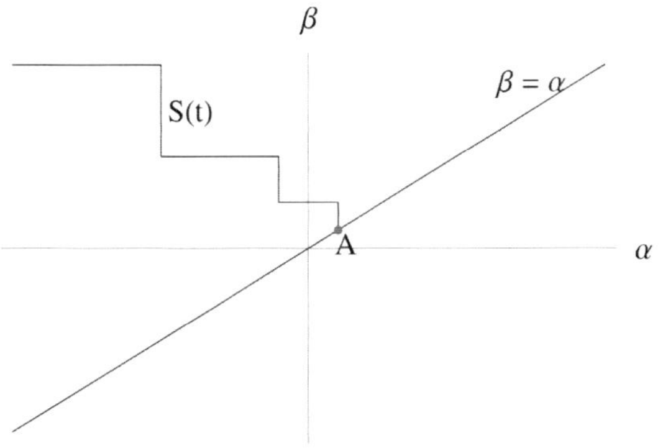

Figure 3. An example of creating a new corner in the line $S(t)$ at an extremum (turning) point of the input $x(t)$.

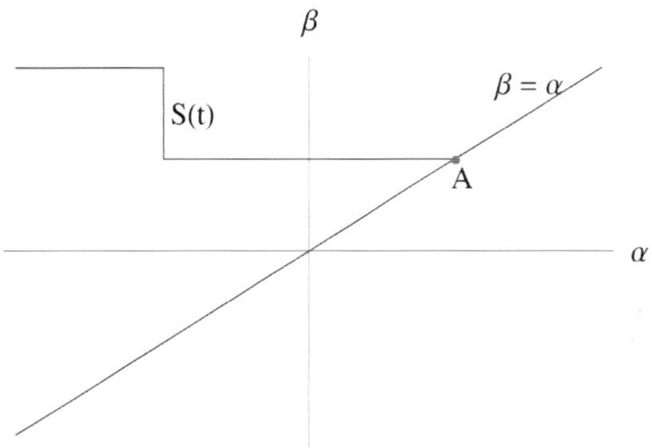

Figure 4. An example of erasing (wiping out) corners from the line $S(t)$.

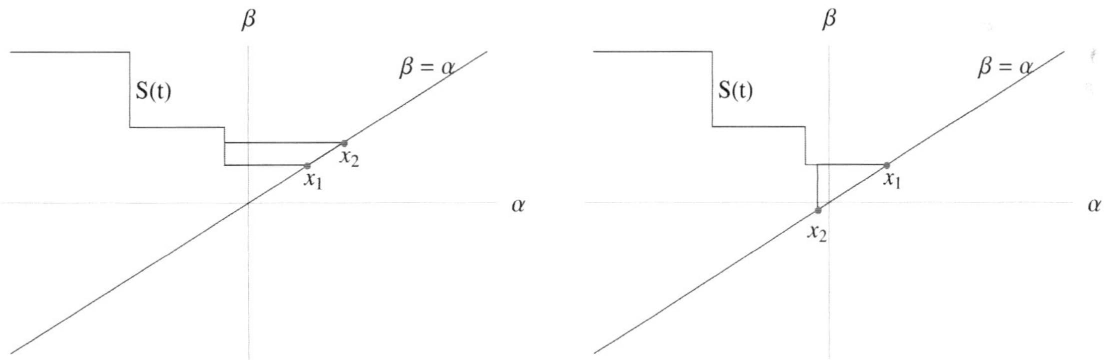

(a) (b)

Figure 5. Changes in the output of the Preisach operator. (a) The input increases from the value x_1 to the value x_2. As a result the relays in the shaded area switch on; other relays do not change their state. The output (3) increases by the integral of the Preisach weight function $\mu = \mu(\alpha, \beta)$ over the shaded area. (b) The input decreases from the value x_1 to x_2. The relays in the shaded area switch off. The output decreases by the integral of μ over the shaded area.

Figure 6 shows that given the same sequences of the main maxima and minima the state $S(t)$ can be arranged in two possible ways, one where the last link of $S(t)$ connecting the corner point (m_0, M_0) with the point $(x(t), x(t))$ on the bisector $\alpha = \beta$ is horizontal and the other where the last link is vertical. In order to uniquely describe the state, we use an additional variable ζ which can have either the value 1 or -1. The value of 1 corresponds to the arrangement of the state for which the last link is horizontal and the -1 value means that the last link is vertical. Thus, given the input value $x(t)$ at a moment t, the value of the variable ζ and the two arrays $\omega_m = \{m_0, m_1, \ldots\}$, $\omega_M = \{M_0, M_1, \ldots\}$ containing the values of the running main extrema at this moment completely define the state $S(t)$ (and thus the memory of the system at the moment t). This notation is used in Section 4 below where updating the state and output of the Preisach operator in response to changes of the input is implemented as part of the numerical schemes for solving equation (1).

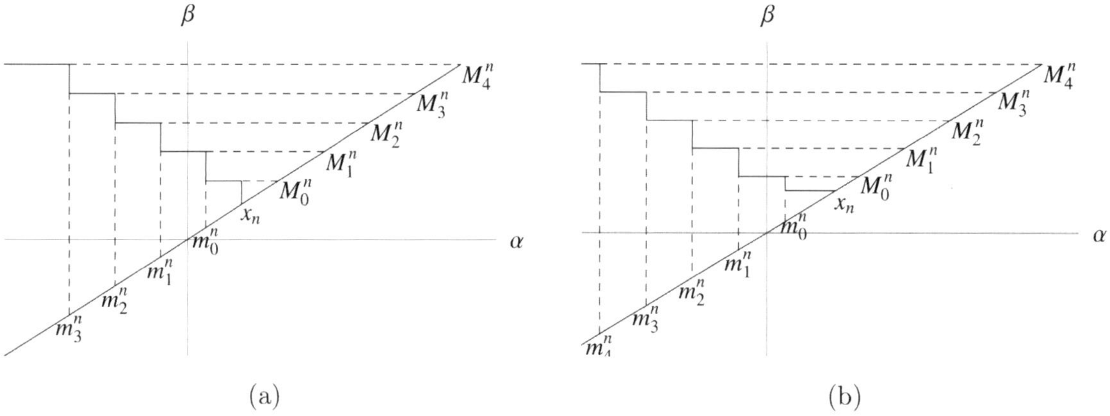

Figure 6. The Preisach operator state with the corners $(m_k, M_k) = (m_k^n, M_k^n)$ at a moment $t = n$ where the last link is (a) vertical, $\zeta = -1$; (b) horizontal, $\zeta = 1$. The arrays of the main extrema are $\omega_m = \{m_0^n, m_1^n, \ldots\}$, $\omega_M = \{M_0^n, M_1^n, \ldots\}$.

3. A motivating example: A Price dynamics model

Let us consider a model of a market of many traders. We model each single trader by a non-ideal relay with thresholds α, β where $\alpha < \beta$; the thresholds are different for different traders. Hence, a single trader has two possible states $s = \pm 1$. Let $s = -1$ be the 'long' state, *i.e.*, the trader has purchased an asset at a price lower than the value of his β threshold and waits until the price goes up to the value of β before selling it with profit and switchs to the state $s = 1$. The state $s = 1$ is 'short', the trader has not purchased an asset and waits for the price to fall to the level of his α threshold at which he wants to buy the asset at this point switching to the state $s = -1$.

Now we consider a pool of M traders, with the i-th trader having thresholds α_i, β_i, and construct a discrete version of the Preisach model of them. If we take the input to the system to be the price p of the asset, then, following [20], the output

$$\sigma = \frac{1}{M} \sum_{i=1}^{M} \mu_i s_i \tag{4}$$

measures the sentiment of the market. Here s_i is the state of the i-th trader; $\mu_i > 0$ is his weight in forming public opinion about the value of the asset (possibly related to the amount of money or asset he has, etc.).

We assume the dynamics of the price of the asset to follow the equation from [20]. That is, the price evolves in discrete time as

$$p_{t+h} = p_t \exp\left[\eta_t \sqrt{h} + \left(a - \frac{1}{2}\right) h - \kappa \Delta\sigma_t\right] \tag{5}$$

where h is the time step; the sequence of independent Gaussian random quantities $\eta_t \sim N(0,1)$ models the external random information stream; a is the trend term; $-1/2$ correction term is the same as in the classical setting where it comes from Ito's formula; $\kappa > 0$ is a parameter measuring the effect of the sentiment; and, $\Delta\sigma_t = \sigma_{t+h} - \sigma_t$ is the change in sentiment which creates a pressure on price. The minus sign in front of the hysteresis term $\kappa\Delta\sigma_t$ accounts for the fact that when more traders switch to the 'long' state this creates a positive pressure pushing the price up. Similarly, switching of more traders to the 'short' state creates a pressure on the price to move down. That is, there is a positive feedback loop. Note that setting $\kappa = 0$ transforms (5) to the discrete form of the standard geometric Brownian motion.

If we perform a logarithmic scaling of the price by introducing the variable

$$r = \ln \frac{p}{p_0}$$

instead of p, then equation (5) can be rewritten as

$$\Delta r_t = \eta_t\sqrt{h} + \left(a - \frac{1}{2}\right)h - \kappa\Delta\sigma_t \tag{6}$$

with $\Delta r_t = r_{t+h} - r_t$. It should be noted that according to (4), σ_t is the output of the discrete Preisach model with the input p_t. However, we can use r_t as the input instead of p_t if we also rescale the thresholds of the relays (or, equivalently, the Preisach measure density function). This is achieved by scaling the threshold values α_i, β_i of the i-th relay to instead be $\ln(\alpha_i/p_0), \ln(\beta_i/p_0)$. Let us denote by P the discrete Preisach operator obtained with this scaling. Then equation (6) becomes

$$\Delta r_t + \kappa\Delta(Pr)_t = \eta_t\sqrt{h} + \left(a - \frac{1}{2}\right)h \tag{7}$$

where $\Delta(Pr)_t = (Pr)_{t+h} - (Pr)_t$. We will consider some simulations of this price dynamics model in Section 5. The formal continuous time counterpart of equation (7),

$$dr_t + \kappa d(Pr)_t = \left(a - \frac{1}{2}\right)dt + dW_t \tag{8}$$

has the form of equation (1) with constant coefficients.

A few remarks are in order. First, equation (8) can be integrated explicitly in terms of the inverse Preisach operator, namely equation (8) is equivalent to

$$r_t - r_0 = (I + \kappa P)^{-1}\left[(a - 1/2)(t - t_0) + W_t - W_0\right] \tag{9}$$

where I is the identity operator and t_0, r_0, W_0 are the initial conditions (for details regarding the properties of the inverse operator $(I + \kappa P)^{-1}$, see [8]). This formula provides a rigorous interpretation of solutions to continuous time equation (8) and can be used to prove the convergence of the discretisations (7) to the continuous time limit (8) as $h \to 0$. A similar formula and the same approach can be used when the trend term $a = a(t)$ is a function of time (for example, accounting for seasonality in the trend). However, if a depends on the price, then (8) becomes a closed loop system not amenable to the explicit integration. In this case, a mathematical formalism for the continuous time equation (8) is an open problem as it is for the more general equation (1).

As the second remark, equation (8), when rewritten in the form (1), has the parameter $\xi = 1/\kappa$. This quantity can be assumed to be of order 1. However, in models of economics systems proposed in [5] (as well as in the hydrological context, see [3]) the cases of interest are $\xi \ll 1$ and $\xi = 0$. The case $\xi = 0$ can be singular and requires special treatment (as can be seen, for example, from equations (11), (10) below where $l_n = 0$ at every turning point of a solution to the discrete time counterpart of (1)). For example, in the deterministic setting, it has been shown that solutions of equation (1) with $b = 0$ have better regularity for $\xi > 0$ than for $\xi = 0$. Moreover, non-uniqueness of solutions to the initial value problem is possible for $\xi = 0$, see [3]. Nevertheless, natural generic conditions ensure the uniqueness property and the convergence of the solution of the deterministic equation with $\xi > 0$ to the solution of the equation with $\xi = 0$ as $\xi \to 0$. Numerical schemes for solving deterministic equations with $\xi = 0$ have been developed in [24–26]; the limit $\xi \to 0$ have been used in [16] to solve equations with a stochastic component that have piecewise smooth trajectories.

4. Numerical schemes

In this section we outline two Euler type time discretisations of equation (1). The two schemes differ in the method we use to evaluate the change in the output of the Preisach operator on each time step. The first scheme we call the rectangular method. Here the increment Δy_n of the output is evaluated in the same way as proposed in the deterministic setting in [24–26]. It is assumed that the time step is sufficiently small so that (i) the shaded area $S(t_{n+1}) \setminus S(t_n)$ shown in figure 5 (a) is well approximated by the rectangle which has the width equal to the increment Δx_n of the input and the length equal to the last link of the staircase line $S(t_n)$; (ii) a typical Δx_n is much smaller than ξ and hence the increment Δy_n of the output at the turning points of x_n is close to $\xi \Delta x_n$; and, (iii) for most other (non-turning) points of x_n the set $S(t_{n+1}) \setminus S(t_n)$ is trapezoid as in figure 5 (a).

Alternatively, we calculate Δy_n using the exact expression for the shaded area shown in figure 5 and call the corresponding numerical scheme the triangular method.

To simplify the notation we consider the Preisach operator which has the constant weight function $\mu(\alpha, \beta) = 1$ in the area of variation of x_n.

4.1. Rectangular method

We replace the right hand side of equation (1) with the finite difference

$$d_n = a(x_n, t_0 + nh)h + b(x_n, t_0 + nh)\sqrt{h}\Delta W_n$$

where h is the time step and ΔW_n are independent Gaussian random variables

$$\Delta W_n \sim N(0, 1)$$

In the rectangular method, we approximate the increment of the output of the Preisach operator on the n-th time step by the product of the increment $\Delta x_n = x_{n+1} - x_n$ of the input and the length l_n of the last link of the state of the Preisach operator at the moment n. This leads to the discretisation of equation (1) of the form

$$(l_n + \xi)\Delta x_n = d_n \tag{10}$$

where

$$l_n = \begin{cases} x_n - m_0^n & \text{if } d_n > 0, d_{n-1} > 0 \\ 0 & \text{if } d_n d_{n-1} < 0 \\ M_0^n - x_n & \text{if } d_n < 0, d_{n-1} < 0 \end{cases} \tag{11}$$

Here m_0^n and M_0^n are the first elements of the arrays ω_m and ω_M of the main minima and maxima at the moment n; d_{-1} equals the value ζ_0 of the variable ζ at the initial moment. We note that the sign of the increment Δx_n coincides with the sign of d_n; l_n is zero at the turning points of x_n, that is $l_n = 0$ whenever $\Delta x_n \Delta x_{n-1} < 0$ (equivalently, $d_n d_{n-1} < 0$), while $l_n > 0$ whenever $\Delta x_n \Delta x_{n-1} > 0$ (equivalently, $d_n d_{n-1} > 0$).

The numerical scheme has been implemented in the following steps to account for updating the memory arrays ω_m and ω_M, the variable ζ_n and the value of x_n. For $t = t_0 + nh$:

(i) Calculate the value of d_n;

(ii) Evaluate for turning points as follows:
- If $d_n d_{n-1} < 0$ (*i.e.*, x_n is a turning point), then:
  (a) If $d_n > 0$ add the element x_n to the memory array ω_m and set $\zeta_{n+1} = 1$;
  (b) If $d_n < 0$ add the element x_n to the array ω_M and set $\zeta_{n+1} = -1$;
- If $d_n d_{n-1} > 0$ then the memory arrays ω_m, ω_M remain the same and $\zeta_{n+1} = \zeta_n$;

(iii) Calculate the new x value as $x_{n+1} = x_n + \frac{d_n}{l_n+\xi}$;

(iv) Update the memory arrays for the new position of x by performing deletions of their elements if either of the following conditions are met:
  (a) For $d_n > 0$ remove all the elements $M_0^n, \ldots, M_k^n$ satisfying $M_i^n \leq x_{n+1}$ from the array ω_M; remove the same number of elements $m_0^n, \ldots, m_k^n$ from the array ω_m;
  (b) For $d_n < 0$ remove all the elements $m_0^n, \ldots, m_k^n$ satisfying $m_i^n \geq x_{n+1}$ from the array ω_m and the same number of elements $M_0^n, \ldots, M_k^n$ from ω_M.

4.2. Triangular method

In this alternative method, we use the exact expression for the increment Δy_n of the output of the Preiash operator corresponding to the increment Δx_n of the input, *i.e.*,

$$\Delta y_n = (Px)(t_0 + (n+1)h) - (Px)(t_0 + nh)$$

where $x(t)$ is a monotone interpolation of the input between the discrete values $x_n = x(t_0 + nh)$ and $x_{n+1} = x(t_0 + (n+1)h)$. This leads to the discretisation

$$\xi \Delta x_n + \Delta y_n = d_n \tag{12}$$

of equation (1) with the same d_n as above. This formula defines x_{n+1} implicitly given the value x_n and the state of the Preisach operator (*i.e.*, the arrays ω_m and ω_M and the variable ζ) at the moment n; again, the sign of the increment Δx_n equals the sign of d_n and Δy_n has the same sign. Solving equation (12) for x_{n+1} is equivalent to explicit inversion of the operator $\xi I + P$ and leads to four cases depending on the signs of d_n and ζ_n. The geometrical interpretation considered in Section 2 makes this inversion straightforward (cf. Figure 5).

Let us consider in some detail the case where $d_n > 0$ and $\zeta_n = 1$, *i.e.*, the last link of the state is horizontal at the moment n. An iterative procedure is used to find x_{n+1} with the number of iterations depending on how many corners should be erased from the state of the Preisach operator when the input increases from x_n to x_{n+1}. We introduce the value

$$\Sigma_0 = \xi(M_0^n - x_n) + \frac{(M_0^n - m_0^n)^2}{2} - \frac{(x_n - m_0^n)^2}{2}$$

If $\Sigma_0 \geq d_n$ then $x_{n+1} \leq M_0^n$ (*i.e.*, the value of x_{n+1} lies below the first corner of the staircase state $S(t_0 + nh)$) and equation (12) is equivalent to

$$\xi(x_{n+1} - x_n) + \frac{(x_{n+1} - m_0^n)^2}{2} - \frac{(x_n - m_0^n)^2}{2} = d_n$$

hence

$$x_{n+1} = m_0^n - \xi + \sqrt{(m_0^n - \xi)^2 - \left((m_0^n)^2 - (x_n - m_0^n)^2 - 2(\xi x_n + d_n)\right)}$$

where the positive root is chosen to ensure that $x_{n+1} > x_n$. If however $\Sigma_0 < d_n$ then we compare d_n consecutively with the elements $\Sigma_1, \Sigma_2, \ldots$ of the sequence

$$\Sigma_k = \Sigma_{k-1} + \frac{(M_k^n - m_k^n)^2}{2} - \frac{(M_{k-1}^n - m_k^n)^2}{2} + \xi(M_k^n - M_{k-1}^n)$$

until a Σ_k is found such that $\Sigma_k \geq d_n > \Sigma_{k-1}$. These relations ensure that $M_{k-1}^n < x_{n+1} \leq M_k^n$ and equation (12) is equivalent to

$$\Sigma_{k-1} + \xi(x_{n+1} - M_{k-1}^n) + \frac{(x_{n+1} - m_k^n)^2}{2} - \frac{(M_{k-1}^n - m_k^n)^2}{2} = d_n$$

which implies

$$x_{n+1} = m_k^n - \xi + \sqrt{(m_k^n - \xi)^2 - \left((m_k^n)^2 - (M_{k-1}^n - m_k^n)^2 - 2(\xi M_{k-1}^n + d_n - \Sigma_{k-1})\right)}$$

A similar algorithm for solving equation (12) applies in the three other cases. For $d_n < 0$, $\zeta_n = 1$, we set

$$\Sigma_{-1} = 0; \quad \Sigma_k = \Sigma_{k-1} - \frac{(M_{k-1}^n - m_k^n)^2}{2} + \frac{(M_{k-1}^n - m_{k-1}^n)^2}{2} + \xi(m_k^n - m_{k-1}^n), \quad k = 0, 1, 2, \ldots$$

then relations $\Sigma_k \leq d_n < \Sigma_{k-1}$ imply

$$x_{n+1} = M_{k-1}^n + \xi - \sqrt{(M_{k-1}^n + \xi)^2 - \left((M_{k-1}^n)^2 - (M_{k-1}^n - m_{k-1}^n)^2 + 2(\xi m_{k-1}^n + d_n - \Sigma_{k-1})\right)}$$

where we use the convention $M_{-1}^n = m_{-1}^n = x_n$. For $d_n > 0$, $\zeta_n = -1$, the relations $\Sigma_{k-1} < d_n \leq \Sigma_k$ with

$$\Sigma_{-1} = 0; \quad \Sigma_k = \Sigma_{k-1} + \frac{(M_k^n - m_{k-1}^n)^2}{2} - \frac{(M_{k-1}^n - m_{k-1}^n)^2}{2} + \xi(M_k^n - M_{k-1}^n), \quad k = 0, 1, 2, \ldots$$

imply

$$x_{n+1} = m_{k-1}^n - \xi + \sqrt{(m_{k-1}^n - \xi)^2 - \left((m_{k-1}^n)^2 - (M_{k-1}^n - m_{k-1}^n)^2 - 2(\xi M_{k-1}^n + d_n - \Sigma_{k-1})\right)}$$

For $d_n < 0$, $\zeta_n = -1$, the relations $\Sigma_k \leq d_n < \Sigma_{k-1}$ with

$$\Sigma_{-1} = 0; \quad \Sigma_k = \Sigma_{k-1} - \frac{(M_k^n - m_k^n)^2}{2} + \frac{(M_k^n - m_{k-1}^n)^2}{2} + \xi(m_k^n - m_{k-1}^n), \quad k = 0, 1, 2, \ldots$$

imply

$$x_{n+1} = M_k^n + \xi - \sqrt{(M_k^n + \xi)^2 - \left((M_k^n)^2 - (M_k^n - m_{k-1}^n)^2 + 2(\xi m_{k-1}^n + d_n - \Sigma_{k-1})\right)}$$

The implementation of the triangular method has the same steps (i), (ii) and (iv) as described in the previous subsection for the rectangular method. Step (iii) is modified and x_n is updated to obtain the next iteration x_{n+1} according to the algorithm presented in this subsection.

Table 1.

ξ	h	x_0
0.1	0.01	2.0

5. Results

As the first model example, we consider discretisations of the formal equation

$$\xi dx_t + d(Px)_t = -x_t dt + dW_t \tag{13}$$

obtained by the rectangular and triangular methods using different time step size h. We compare trajectories obtained by the two methods for the same realisation of the discretised Wiener process (random walk) W_t. We also compare the trajectories of equation (13) with the trajectories of the Ornstein-Uhlenbeck equation

$$\xi dx_t = -x_t dt + dW_t \tag{14}$$

obtained by the Euler method.

As the second example, the rectangular and triangular discretisations of the price dynamics model (8) are considered.

In all the examples, the Preisach weight function is set to $\mu(\alpha, \beta) \equiv 1$ for the whole domain of the Preisach plane where relays experience at least one switch during the simulation. The initial state of the Preisach operator is the staircase with all the steps of size 1.

5.1. Rectangular method

The values of parameters used in the simulations shown in this subsection are given in table 1. A numerical realisation of the rectangular method for equation (13) and the Euler method for equation (14) using these parameters is shown in figure 7; x_0 is the initial value of the solution.

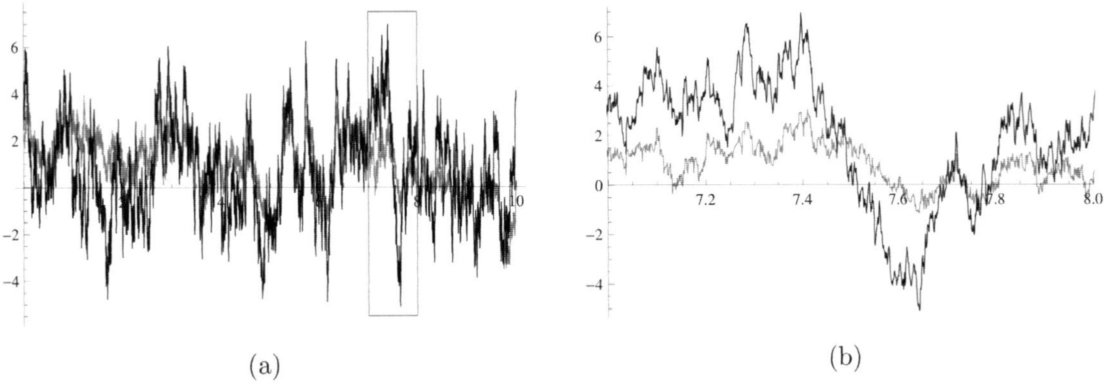

Figure 7. Data generated from the implementation of the rectangular method. (a) The red line is the trace of solution x_t to equation (13) including the Preisach operator. The black line denotes the solution of equation (14) without the Preisach operator for the same realisation of W_t. (b) A zoom of the region indicated in (a).

A look at figure 7 (b) which presents a zoomed in part of the full trajectory shows that while solutions of discretisations of equations (13) and (14) experience ups and downs at roughly

Journal of Physics: Conference Series **268** (2011) 012019 doi:10.1088/1742-6596/268/1/012019

Table 2.

ξ	h	x_0
0.1	0.1	2.0

the same time moments for the same realisation of the random walk, the inclusion of the Preisach operator has limited the effect of the jumps. It appears that the introduction of the Preisach operator in the stochastic equation (14) acts as a damping factor to the trajectory. It is characteristic of the damping that it is stronger on the parts of the trajectory demonstrating a stronger upwards or downwards trend. This feature can be explained by formulas (11) and (10). If the solution satisfies $x_n \geq x_{n_0}$ on some time interval $n_0 \leq n \leq N$, then at every moment n from this interval such that $x_n > x_k$ for all $n_0 \leq k \leq n$, the quantity l_n also satisfies $l_n > l_k$ for all $n_0 \leq k \leq n$. In this sense, the upward trend of the solution leads to the upward trend of l_n (and, similarly, the downward trend of x_n leads to the upward trend of l_n). At the same time, relation (10) implies stronger damping of the increment $\Delta x_n = x_{n+1} - x_n$ for larger l_n.

5.2. Triangular method

The parameters taken for the implementation of the triangular method for equation (13) and the Euler method for equation (14) are shown in table 2. Figure 8 presents solutions obtained with these parameters for the same realisation of W_t. The Preisach operator term seems to be a limiting factor on the size of jumps, as is the case with the implementation of the rectangular method. Figure 8b shows that while the two trajectories undergo up and down movements at the same time the inclusion of the Preisach operator has caused the changes to be less severe.

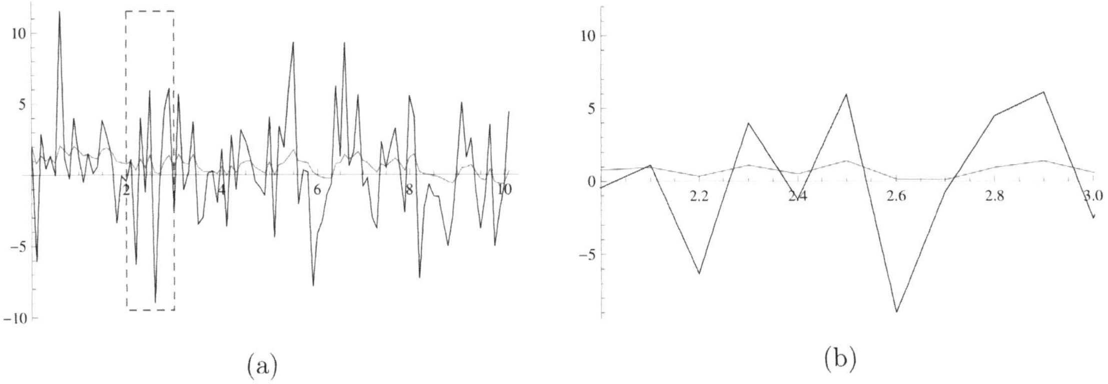

Figure 8. (a) Data generated from the implementation of the triangular method for equation (13) (red) and the Euler method for equation (14) (black), (b) A zoom of the region indicated in (a).

5.3. Comparison of triangular and rectangular methods

Figures 9 and 10 show a realisation of the rectangular and triangular methods for equation (13). The parameters used for the generation of these results are $\xi = 0.1$ and $x_0 = 2$; the value of h for each plot is stated in the caption. We see that the agreement between the two methods improves for smaller time step h. In particular, for the given parameters, a good agreement

Table 3.

κ	h	x_0	a
1	0.0001	2.0	0.5

between the two methods is achieved at $h = 0.00001$, see figure 10. Here the trajectories are close apart from a few discrepancies at some turning points where the rectangular method tends to produce larger jumps for the chosen small value of the parameter ξ.

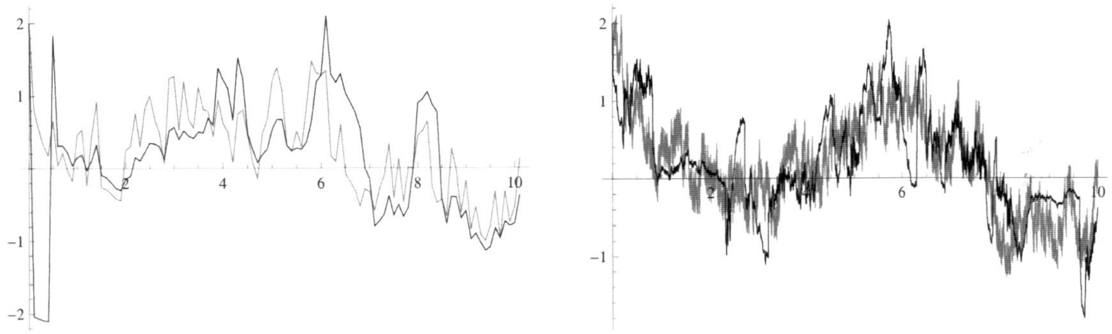

(a) (b)

Figure 9. Solution of the discretisations of equation (13) as generated from the implementation of both the rectangular method (black) and the triangular method (red) for different time step size: (a) $h = 0.1$, (b) $h = 0.001$.

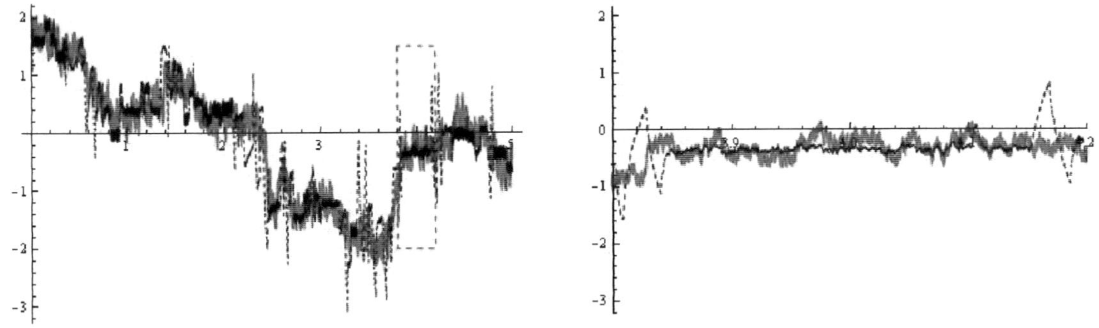

(a) (b)

Figure 10. Same as Figure 9 for $h = 0.00001$, (b) A zoom of the region marked in (a).

5.4. Price dynamics model

We now consider the price dynamics model (8) presented in Section 3. Figure 11 compares the results of implementation of each of the rectangular and triangular numerical methods in solving equation (8) with positive κ with the solution of the same equation for $\kappa = 0$ (*i.e.*, equation

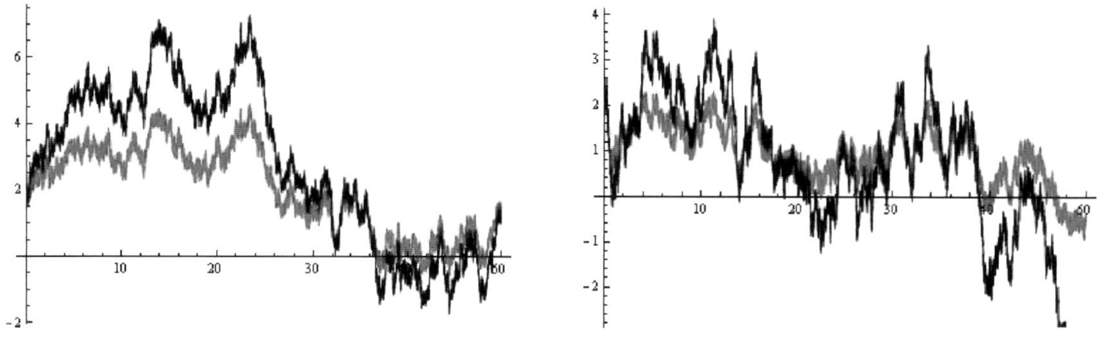

(a) (b)

Figure 11. The red line shows the plot of a solution $r(t)$ of equation (8) as generated from the implementation of (a) the rectangular method, and (b) the triangular method. Parameters are presented in table 3. The black line in each panel is the solution of equation (8) with $\kappa = 0$, which coincides with the realisation of the random walk W_t used in the red plot for the parameter values from the table.

$dr_t = (a - 1/2)dt + dW_t$ without the Preisach operator term) obtained by the Euler method. The values of the parameters used for these plots are shown in table 3. The plots demonstrate the same features as we observed in the other examples above. We see that solutions of the model including the Preisach operator ($\kappa = 1$) and the model without the Preisach operator ($\kappa = 0$) have similar shape, however the inclusion of the Preisach operator leads to a damping effect. In the model including the Preisach operator the price does not go as high, as well as falls in the price are limited and less severe, as in the model with $\kappa = 0$.

6. Conclusions

We have proposed and tested two discrete time counterparts of the formal continuous time stochastic differential equation (1) involving the Preisach operator. These two numerical schemes have been applied to an Ornstein-Uhlenbeck type equation and to a simple price dynamics model involving the Preisach memory term. We have shown that there is a good agreement between the two methods once the time step is sufficiently small. It has also been shown that the inclusion of the Preisach operator acts as a limiting damping effect when compared with an equivalent form of a regular stochastic differential equation. The damping is more pronounced on the parts of the trajectory demonstrating a stronger upwards or downwards trend. This effect can be explained by the fact that both upwards or downwards trend of the trajectory result in an upwards trend of the quantity (11). The results of this paper are preliminary to a broader programme of research of stochastic equations with the Preisach operator outlined in the Introduction.

Acknowledgments

The authors are thankful to Prof. Harbir Lamba for valuable discussions of the results of the paper and to Prof. Mihai Dimian and Prof. Ram Iyer for instructive discussions of numerical algorithms for solving discrete and stochastic equations with the Preisach operator. The work was partially supported by Russian Foundation for Basic Research, grant 10-01-93112.

References

[1] O'Kane J P 2005 *J. Phys.: Conf. Ser.* **22** 148
[2] Flynn D, McNamara H, O'Kane J P and Pokrovskii A 2006 *The Science of Hysteresis* vol III, ed I Mayergoyz and G Bertotti (Elsevier, Academic Press) 689

[3] Appelbe B, Flynn D, McNamara H, O'Kane P, Pimenov A, Pokrovskii A, Rachinskii D and Zhezherun A 2009 *IEEE Control Systems Magazine* **29** 44
[4] Cross R, Grinfled M and Lamba H 2009 *IEEE Control Systems Magazine* **29** 30
[5] Cross R, McNamara H and Pokrovskii A V 2008 *Physica B* **403** 451
[6] Mayergoyz I D 2006 *The science of hysteresis* vol I, ed I Mayergoyz and G Bertotti (Elsevier, Academic Press) 529
[7] Krasnosel'skii M A and Pokrovskii A V 1989 *Systems with hysteresis* (Springer)
[8] Brokate M and Sprekels J 1996 *Hysteresis and phase transitions* (Springer)
[9] Mayergoyz I D and Korman C E 1991 *J. Appl. Phys.* **69** 2128
[10] Rugkwamsook P, Korman C, Bertotti G and Pasquale M 1999 *J. Appl. Phys.* **85** 4361
[11] Mayergoyz I D, Adly A, Huang M W and Kraft C 2000 *IEEE Trans. Magn.* **36** 3208
[12] Dimian M, Coca E and Popa V 2009 *J. Appl. Phys.* **105** 07D515
[13] Dimian M, Adedoyin A, Gindulescu A and Andrei P 2009 *IEEE Trans. Magn.* **45** 5231
[14] Spanos P D, Cacciola P and Redhorse J 2004 *Nonlinear Dynamics* **36** 405
[15] Rychlik I and Gupta S 2007 *Int. J. of Fatigue* **29** 406
[16] Brokate M, McCarthy S, Pimenov A, Pokrovskii A and Rachinskii D 2010 *Environ. Model. Assess.* in press
[17] Lamba H and Seaman T 2008 *Int. J. of Theoretical and Applied Finance* **11** 717
[18] Cross R, Grinfeld M, Lamba H and Seaman T 2007 *The European Phys. Journal B* **57** 213
[19] Cross R, Grinfeld M, Lamba H and Seaman T 2005 *Physica A* **354** 463
[20] Lamba H and Seaman T 2009 arXiv:0709.4242, *Physica A* in press
[21] Lamba H 2009 arXiv:0908.0949v1
[22] Appleby J A D and Wu H 2008 *J. Phys.: Conf. Ser.* **138** 012002
[23] Mayergoyz I 2003 *Mathematical models of hysteresis and their applications* (New York: Elsvier)
[24] Flynn D and Rasskazov O 2005 *J. Phys.: Conf. Ser.* **22** 43
[25] Flynn D, O'Kane J P and Zhezherun A 2006 *J. Phys.: Conf. Ser.* **55** 63
[26] Flynn D, McNamara H, Pokrovskii A and O'Kane J P 2008 *Physica B* **403** 440

Journal of Physics: Conference Series **268** (2011) 012020 doi:10.1088/1742-6596/268/1/012020

Magnetization curve modelling of soft magnetic alloys

I Mészáros

Department of Materials Science and Engineering, Budapest University of Technology and Economics, Bertalan L. street 7., Budapest, H-1111, Hungary

E-mail: meszaros@eik.bme.hu

Abstract. In this paper we present an application of the so called hyperbolic model of magnetization. The model was modified and it was applied for nine different soft magnetic alloys. The tested samples were electro-technical steels (FeSi alloys) and a permalloy (FeNi alloy) with strongly different magnetic properties. Among them there are top, medium and definitely poor quality soft magnetic materials as well. Their minor hysteresis loops and normal magnetization curves were measured by alternating current measurement. The hyperbolic model of magnetization was applied for the experimental normal magnetization curves. It was proved that the applied model is excellent for describing mathematically the experimental magnetization curves.

1. Introduction

In this work a novel phenomenological model of magnetization was used and validated. The applied model is called hyperbolic model of magnetization (HMM). The predecessors of this model were the Flatley-Henretty model [1] and the so called T(x) model developed by J. Takacs in [2]. The hyperbolic model of magnetization provides an easy analytical solution to the complex phenomena of magnetic hysteresis. It is known that the thermodynamically equilibrium state of a magnetic material can be characterized by its anhysteretic magnetization curve [3]. In the hyperbolic model the following equation (1) was used for describing the anhysteretic magnetization curve:

$$\frac{M}{M_0} = \frac{\sum_{i=1}^{N}\left[A_i \tanh\left(N_i H\right)\right]}{\sum_{i=1}^{N} A_i}, \tag{1}$$

where: N_i are scaling parameters, A_i are the amplitudes of magnetization components ($A_1 = 1$).

As it can be seen the anhysteretic magnetization curve is composed as a linear combination of N pieces of tangent hyperbolic functions. They will be referred as hyperbolic components of magnetization in the following.

Following the way of Takacs's calculation [2] the whole hysteresis loop can be constructed. The components of the anhysteretic curve (1) shifted horizontally by different α_i as well as vertically in symmetrical way. The loci of the crossover points of the up-going and down-going parts of the symmetrical minor loops is referred as normal magnetization curve [4]. From the hyperbolic model the following mathematical function can be obtained for the normal magnetization curve:

$$\frac{M}{M_0} = \frac{\sum_{i=1}^{N} \{ A_i \tanh[N_i(H+\alpha_i)] + A_i \tanh[N_i(H-\alpha_i)] \}}{2\sum_{i=1}^{N} A_i}, \quad (2)$$

where: α_i. is the shifting factor which is associated with the coercivity.

In this approach the model was applied for $N = 2$ therefore two hyperbolic components were supposed. The theoretical normal magnetization curve (2) was fitted to the measured normal magnetization curves. As it can be seen the applied form of the hyperbolic model has five fitting parameters.

In this work eight different so-called electro-technical steels and a permalloy were taken as test samples. They were chosen intentionally to have strongly different magnetic properties. Among them there are top, medium and definitely poor quality products as well. Their minor hysteresis loops and normal magnetization curves were measured by alternating current measurement.

2. Experimental arrangement and samples

A permeameter-type magnetic property analyzer developed at our department was used for measuring the magnetization curves. The applied measuring yoke contains a robust "U" shaped laminated Fe-Si iron core with a magnetizing coil. The excitation current was produced by a digital function generator and a power amplifier used in voltage regulated current generator mode. The pick-up coil was around the middle of the sample. The permeameter was completely controlled by a personal computer in which a 16 bit input-output data acquisition card accomplished the measurements. The applied maximal excitation field strength was 2150 A/m. In each case 200 minor hysteresis loops were measured. Each minor hysteresis loops were recorded by measuring 1000 points of them. The excitation magnetic field was increased in steps and there was 5 seconds delay between the increase of the excitation and the data acquisition for hysteresis loops to ensure the sample's perfect magnetic accommodation.

All the magnetic measurements were completed by using 5 Hz sinusoidal excitation frequency. Because of the relatively low excitation frequency and small thickness of the samples (0.35 - 0.5 mm) the completed magnetic measurements can be considered as pseudo-static and the effect of eddy-currents to the magnetization curves is supposed to be not significant.

The permeameter allowed us to derive all the traditional magnetic properties from the hysteresis loop like maximal induction, remanent induction, coercivity, relative permeability and hysteresis loss values.

The measured samples were soft magnetic alloys with strongly different magnetic properties. Samples E1, E2, E3, E4, E5A, E5B, E6, E7 and E8 are iron-silicon, S11 is an iron-nickel alloy (permalloy). The chemical composition of the samples was measured by energy dispersive X-ray analyzer (EDS). The obtained compositions can be found in table 1. Samples E5A and E5B were grain-oriented with cube texture the others were non-oriented.

From the metal sheets 150x30 mm specimens were cut. The thickness of each FeSi and FeNi samples was 0.35 mm and 0.5 mm respectively. The samples were longitudinally magnetized according to their rolling direction.

As an illustration the measured hysteresis loops of three of the investigated samples are shown in figure 1.

The relative initial (μ_0) and maximal permeability (μ_m) and the maximal induction (B_m) values were determined from the measured magnetization curves. These results are summarized in table 2.

This set of specimens was used as model materials for testing the usability of the proposed form of the hyperbolic model of magnetization.

Table 1. Chemical composition of the tested samples (figures in volume %). The rest is Fe in all cases.

Sample	Al	Si	Mn	P	Ni
E1	-	2.41	1.12	-	-
E2	1.61	2,14	-	-	-
E3	-	2.10	0.93	-	-
E4	-	4.82	-	0.86	-
E5A	1.11	3.46	-	-	-
E5B	1.61	3.41	-	-	-
E6	-	1.45	-	-	-
E7	-	0.43	0.91	-	-
E8	1.53	0.51	0.96	-	-
S11	-	1.29	0.87	-	76.24

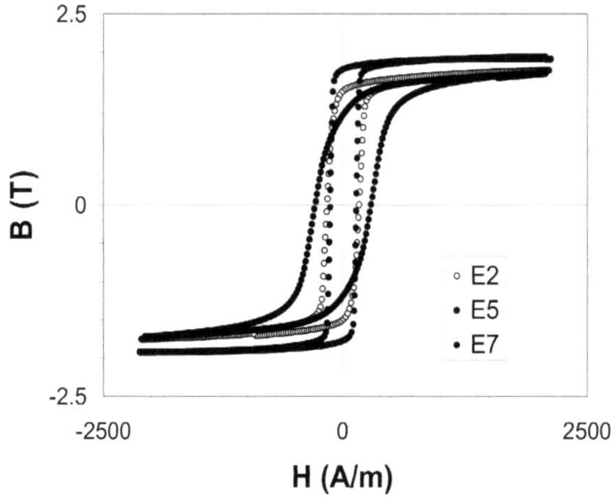

Figure 1. Illustrative hysteresis loops of three tested samples.

3. Results and discussion

The experimental normal curves were generated as the loci of the crossover points of the up-going and down-going parts of the measured minor hysteresis loops.

The theoretical normal magnetization curve derived from the hyperbolic model of magnetization (2) was fitted to the measured normal curves. The five fitting parameters (N_1, α_1, A_2, N_2, α_2) of the hyperbolic model were determined by applying the Levenberg-Marquardt iteration method. Some of the measured and the fitted normal magnetization curves are presented in Figure 2 and 3.

As it can be seen the fitted normal magnetization curves are practically perfectly fit to the experimental points from zero fields to the near saturation range. The calculated deterministic coefficients (R^2) were always better than 0.999 in all tested cases (table 3).

It can be noted that the fitting is always significantly better than in the previous form of the model in which only one irreversible magnetization part was supposed [5].

Table 2. Saturation induction and permeability values of the tested samples.

Sample	B_m (T)	μ_0	μ_m
E1	1.71	2730	9303
E2	1.88	3942	10117
E3	1.46	2297	9568
E4	1.89	9264	16330
E5A	1.91	12351	22024
E5B	2.03	13239	23231
E6	1.86	583	4546
E7	1.89	450	2834
E8	1.76	400	2596
S11	1.10	17657	28809

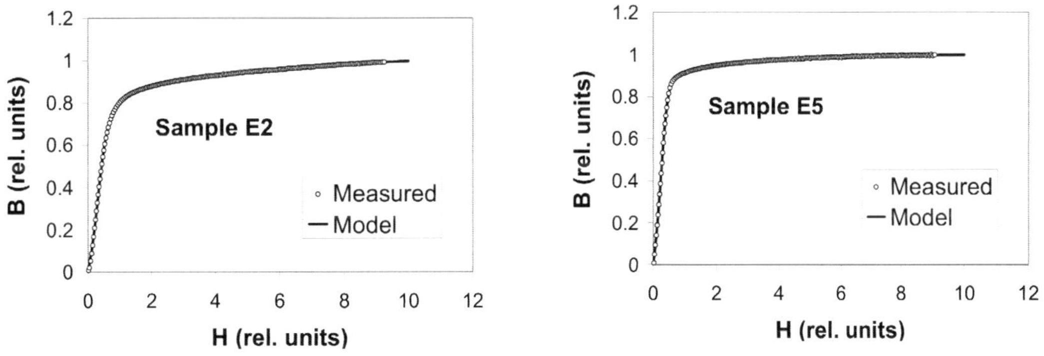

Figure 2. The experimental and fitted normal magnetization curves of the E2, E5 samples.

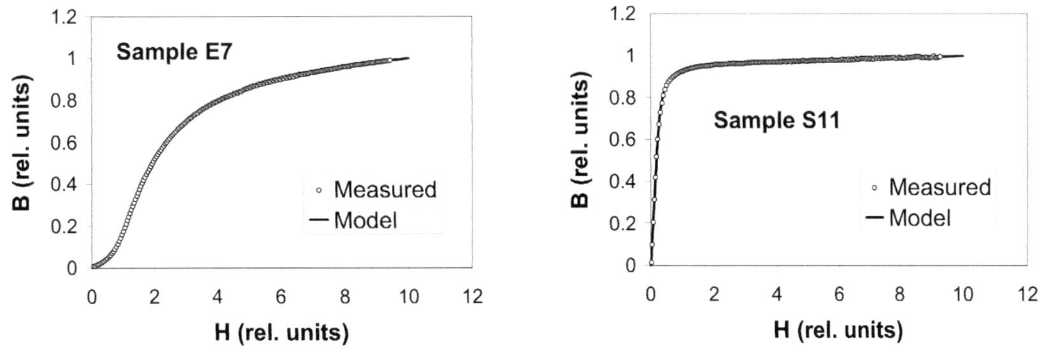

Figure 3. The experimental and fitted normal magnetization curves of the E7, S11 samples.

The applied hyperbolic model can perfectly describe the near saturation range. Therefore it can be used for determining the real saturation state of magnetic materials. Earlier several another phenomenological approaches were attempted to use for determination the saturation induction value

(B_S). One of the most commonly used equations for describing the high-field region of the normal magnetization curves is the so called 'law of approach to saturation' (3) [6].

$$B = B_S\left(1 - \frac{a}{H} - \frac{b}{H^2}\right), \tag{3}$$

The saturation induction values of the tested samples were determined by the hyperbolic model (B_S-Hip) and by the law of approach (B_S–LA) (3) as well. The obtained results are summarized in table 3. As it can be seen the two applied extrapolation methods resulted practically the same saturation induction values.

However the law of approach can describe only the near saturation region while the hyperbolic model can model the whole normal magnetization curve from zero to the saturation field.

Table 3. Deterministic coefficients (R^2) of the fitted normal magnetization curves and the saturation induction values determined by the hyperbolic model and by the "law of approach".

Sample	R^2	B_S-Hip (T)	B_S–LA (T)
E1	0.99971	1.80	1.82
E2	0.99971	1.99	1.97
E3	0.99964	1.57	1.57
E4	0.99985	1.89	1.89
E5A	0.99986	1.91	1.91
E5B	0.99977	2.03	2.04
E6	0.99995	1.97	1.99
E7	0.99986	2.08	2.10
E8	0.99983	2.04	2.07
S11	0.99931	1.11	1.12

4. Summary

The hyperbolic model of magnetization was applied for eight different iron-silicon alloys and for an iron–nickel alloy. The tested samples were soft magnetic materials with intentionally strongly different magnetic properties.

The hyperbolic model of magnetization presents a closed form mathematical description of magnetic hysteresis loops and normal magnetization curves partially based on physical principles, rather than simply on the mathematical curve-fitting of observed data. This model can be applied for a lot of magnetic materials using only some fitting parameters. In the applied form of the model two hyperbolic components were supposed.

The normal magnetization curves of the samples were measured and the theoretical normal magnetization curves calculated from the hyperbolic model were fitted to them. It can be concluded that the calculated curves nearly perfectly fit onto the measured normal magnetization curves from the low field (Rayleigh region) to the near saturation range.

The calculated deterministic coefficients (R^2) were always better than 0.999 in cases of the tested nine samples. It demonstrates that the application of two hyperbolic components is sufficient and necessary for describing the normal magnetization curves of soft magnetic alloys.

The application of the model can have several additional advantages. Among others it can be used for determination of the real magnetic saturation state of the material and it can help to understand the physical sub processes of the magnetization.

Acknowledgements

The author gratefully acknowledges to the assistance of Professor Takacs from the University of Oxford. Our discussions were greatly helped the theoretical considerations of the hyperbolic model. This work was supported by Hungarian Scientific Research Found (project 80173 CK).

References

[1] Flatley T W and Henretty D A 1995 *NASA Technical Reports Server* (N95-27801) 405
[2] Takacs J 2003 *Mathematics of Hysteresis Phenomena* (Weinheim: Wiley-VCH Verlag)
[3] Bozorth R M 1951 *Ferromagnetism* (Toronto, New York, London: D.Van Nostrand Inc.)
[4] Fiorillo F 2004 *Measurements and Characterization of Magnetic Materials*, (Amsterdam: Elsevier Pub.Co.)
[5] Mészáros I 2006 *J. of Electrical Engineering* **57** 151
[6] B.D. Cullity B D 1972 *Introduction to Magnetic Materials*, (Philippines: Addison-Wesley Pub. Co. Inc.)

Journal of Physics: Conference Series **268** (2011) 012021 doi:10.1088/1742-6596/268/1/012021

Detailed structure of electromagnetic pulses passing through one-dimensional photonic crystal

G Nyitray
Pollack Mihaly Faculty of Engineering, Department of Energetics, Pecs, Hungary
E-mail: nyitray@fizika.ttk.pte.hu

Abstract. The electromagnetic wave propagation in one-dimensional photonic cyrstals is studied by a finite-difference time domain method. We consider a conventional perfect photonic crystal structures consists of identical alternating layers of high and low refractive indices. When the central frequency is close to the edge of the photonic band gap, the electromagnetic pulse undergoes a large group-velocity dispersion. By using transfer matrix method we examined the transmittance spectrum of terahertz wave filter based on one-dimensional photonic crystals with line defect. We show that is identical structure can be sandwiched between different substrate. Simulation results demonstrate that the terahertz wave filter can achieve a narrow transmission band and high transmission centered at λ_0, having a flat passband at the central frequency, and sharp edges.

1. Introduction

Photonic crystals (PCs) or photonic bandgap (PBG) materials are a novel class of artificially fabricated structures which have the ability to control and manipulate the propagation of electromagnetic (EM) waves. Properly designed photonic crystals can prohibit the propagation of light, or allow it along only certain directions, or localize light in specified areas. They can be constructed in one, two, and three dimensions with either dielectric or metallic materials. The ability of PCs to control the propagation of light has its origin in their photonic band structure. Just as electron waves traveling in the periodic potential of a crystal are arranged into energy bands separated by bandgapes. We expect the analogous phenomenon to occur when EM waves propagate in a medium in which the dielectric constant varies periodically in space. Photonic bandgap materials are the structures which show such a phenomenon, that is, produce a forbidden frequency gap in which all propagation states are prohibited. It was Yablonovitch who first discovered a dielectric structure containing an absolute gap in the frequency spectrum [1]. Similar gaps can be seen in figure 1, which shows the transmission spectrum of a conventional one-dimensional photonic crystal [2]. A conventional perfect one-dimensional photonic crystal structures consists of identical alternating layers high and low refractive indices. The investigation of these materials is a topic of intensive studies by many groups, both theoretically and experimentally.

The most widely used computer simulation tehniques, which can be applied to PCs are the plane-wave (PW) method, the transfer matrix method (TMM), and the finite-difference time-domain (FDTD) method.

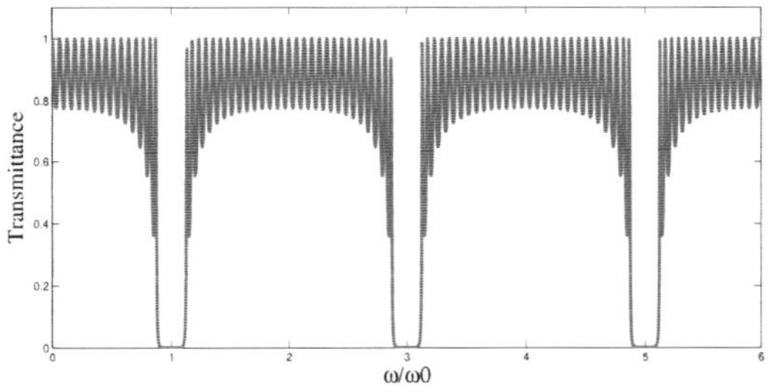

Figure 1. The transmission spectrum of a conventional 1D photonic crystal. The transmittance was calculated by TMM. ($n_1 = 1,4$; $n_2 = 2$; N=31; $n_1 d_1 = n_2 d_2 = \frac{\lambda_0}{4}$).

1.1. The PW method

The PW method is used to calculate the $\omega(\boldsymbol{k})$ dispersion relation of infinite systems by transforming the problem into an eigenvalue problem [3]. The system must be infinite and frequency independent. Since we study finite systems the PW method is not directly relevant to our work.

1.2. The TMM in one-dimension

The TMM was first used to calculate the band structure of a PC by Pendry and MacKinnon [4]. The simplified TMM for 1D structure is called by TLTMM method [5]. In the transmission line transfer matrix method, the reflection coefficient at the surface of the first layer is obtained by starting the calculations from the last layer using the impedance matching concept. The tangential electric and magnetic field components on the boundary surfaces are continuous. We consider an isotropic and homogeneous multilayered medium as shown in figure 2.

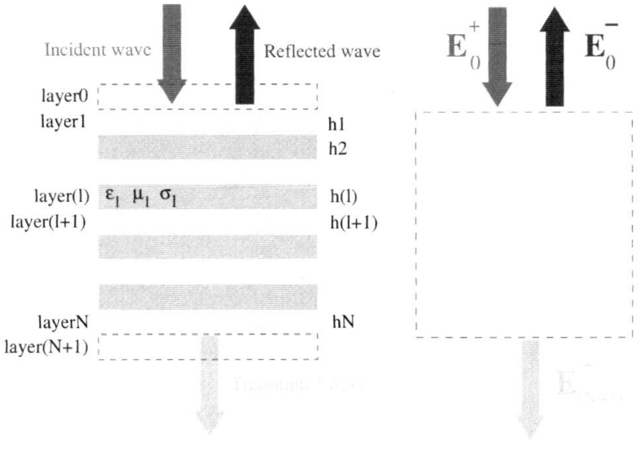

Figure 2. A multilayer dielectric structure (one-dimensional photonic crystal) as a frequency selected surface.

When the light wave normally incident the spectra of TE and TM are the same. Now, we may define r and t as the total reflected and transmitted coefficients of structure, respectively, according to

$$r = \frac{E_0^-}{E_0^+} \qquad in \qquad l = 0, \tag{1}$$

$$t = \frac{E_{N+1}^+}{E_0^+} \qquad in \qquad l = N + 1. \tag{2}$$

Also, we may define reflectance (R), transmittance (T) and absorption (A) as follows:

$$R = rr^*, \tag{3}$$

$$T = \frac{n_{N+1}}{n_0} tt^*, \tag{4}$$

$$A = 1 - (R + T), \tag{5}$$

where n_{N+1} and n_0 are the refractive indices of the $(N + 1)$'th and 0'th media. The $[I]$ discontinuity transfer matrix of a layer and $[L]$ amplitude transmission matrix may be defined with reference to figure 3.

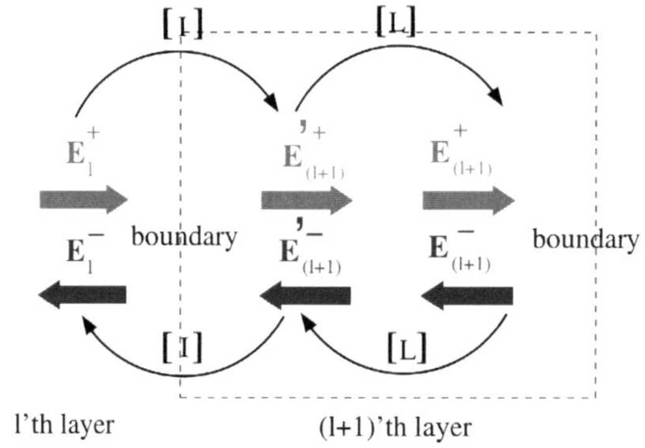

Figure 3. Forward and backward traveling waves.

The $E^{\pm}$ and $E'^{\pm}$ are the forward and backward traveling waves at the end and beginning of each layer, respectively, $[I]_{(l+1)l}$ is the transmission matrix of the discontinuity between the l'th and $(l + 1)$'th layer. $[L]_{(l+1)}$ is the wave amplitude transmission matrix of the $(l + 1)$'th layer

$$\begin{bmatrix} E'^+_{(l+1)} \\ E'^-_{(l+1)} \end{bmatrix} = [I]_{(l+1)l} \begin{bmatrix} E_l^+ \\ E_l^- \end{bmatrix}, \tag{6}$$

$$\begin{bmatrix} E^+_{(l+1)} \\ E^-_{(l+1)} \end{bmatrix} = [L]_{(l+1)} \begin{bmatrix} E'^+_{(l+1)} \\ E'^-_{(l+1)} \end{bmatrix}. \tag{7}$$

Therefore

$$\begin{bmatrix} E^+_{(l+1)} \\ E^-_{(l+1)} \end{bmatrix} = [L]_{(l+1)}[I]_{(l+1)l} \begin{bmatrix} E_l^+ \\ E_l^- \end{bmatrix}, \tag{8}$$

and the transfer matrix of a layer may be defined as:

$$[T]_{(l+1)l} = [L]_{l+1}[I]_{(l+1)l}, \quad l = 0, 1, \ldots, N - 1. \tag{9}$$

Consequently, the waves on the two outer sides of the multilayered medium between the 0'th and $(N+1)$'th layers are related as:

$$\begin{bmatrix} E^{+}_{(\mathrm{N}+1)} \\ E^{-}_{(\mathrm{N}+1)} \end{bmatrix} = [T]_{(N+1)0} \begin{bmatrix} E^{+}_{0} \\ E^{-}_{0} \end{bmatrix}, \tag{10}$$

where the total transfer matrix can be obtained by multiplaying all individual transfer matrixes in sequence

$$[T]_{(N+1)0} = \prod_{N+1}^{0} [L][I] = \begin{bmatrix} T_{11} & T_{12} \\ T_{21} & T_{22} \end{bmatrix}. \tag{11}$$

Now, equation (10) is divided by E^{+}_{0}

$$\begin{bmatrix} t \\ 0 \end{bmatrix} = [T]_{(N+1)0} \begin{bmatrix} 1 \\ r \end{bmatrix}, \tag{12}$$

which is expressed in terms of r and t according to the definitions of equations (1) and (2). The outer half space is assumed matched ($E^{-}_{N+1} = 0$).

$$\begin{bmatrix} t \\ 0 \end{bmatrix} = \begin{bmatrix} T_{11} & T_{12} \\ T_{21} & T_{22} \end{bmatrix} \begin{bmatrix} 1 \\ r \end{bmatrix} \tag{13}$$

The reflection responses and the transmission coefficient will satisfy the recursions:

$$r = -\frac{T_{21}}{T_{22}}, \tag{14}$$

$$t = T_{11} - T_{12}\frac{T_{21}}{T_{22}}. \tag{15}$$

If the media is lossless and nonmagnetic the $[L]$ matrix can be written in the following forms:

$$[L]_{(l+1)} = \begin{bmatrix} \exp\{-j\frac{2\pi}{\lambda}\sqrt{\mu_r\varepsilon_r}\, h_r\} & 0 \\ 0 & \exp\{j\frac{2\pi}{\lambda}\sqrt{\mu_r\varepsilon_r}\, h_r\} \end{bmatrix} = \begin{bmatrix} \exp\{-j\frac{2\pi}{\lambda}\, n_r h_r\} & 0 \\ 0 & \exp\{j\frac{2\pi}{\lambda}\, n_r h_r\} \end{bmatrix},$$

where $n_r = \sqrt{\varepsilon_r}$ is the refractive index of the layer. It should be mentioned that the optical thicknesses are typically chosen to be quarter-wavelength long, that is, $n_r h_r = \frac{\lambda_0}{4}$ at operating wavelength λ_0. The $[I]$ matrix is given by

$$[I]_{(l+1)l} = \frac{1}{1 - r_{(l+1)l}} \begin{bmatrix} 1 & -r_{(l+1)l} \\ -r_{(l+1)l} & 1 \end{bmatrix},$$

where

$$r_{(l+1)l} = \frac{\eta_{(l+1)} - \eta_l}{\eta_l + \eta_{(l+1)}},$$

and η_l, $\eta_{(l+1)}$ the wave impedance of the l'th and $(l+1)$'th layer.

1.3. The FDTD method in one-dimension

The FDTD method employes finite differences as approximations to both the spatial and temporal derivatives that appear in Maxwell's equations [6]. FDTD and related techniques are marching-in-time procedures that simulate the continuous actual electromagnetic waves in a finite spatial region by sampled-data numerical analogs propagating in a computer data space. Time-stepping continues as the numerical wave analogs propagate in the space lattice to causally connect the physics of the modeled region [7]. Although we only have one spatial dimensions, time can be thought of as another dimensions. Thus this is really a sort of two-dimensional problem as it can be seen in figure 4.

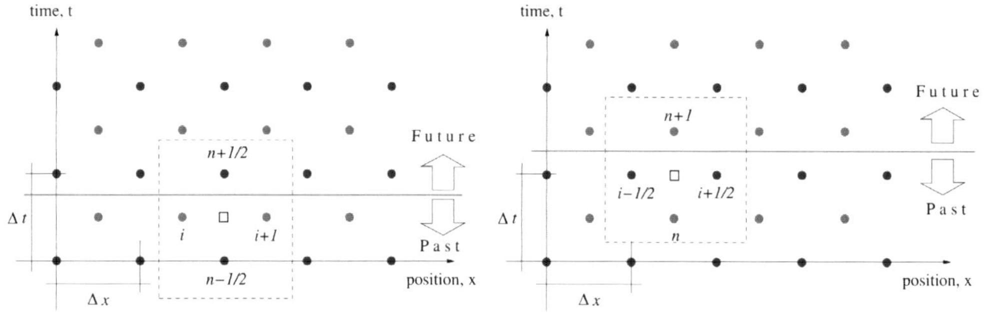

Figure 4. Time-space chart of the one-dimensional Yee-algorithm.

Δx is the spatial offset between sample points and Δt is the temporal offset. The index i corresponds to the spatial step, while the index n corresponds to the temporal step. Figure 4 shows the arrangement of the electric and magnetic field sample points, also known as nodes. The electric-field nodes are shown as red circles and the magnetic-field nodes as blue circles. Assume that all the fields below the red line are known—they are considered to be in the past—while the fields above the red line are future fields and hence unknown. The FDTD algorithm provides a way to obtain the future fields from the past field. From the discretized form of Faraday's law

$$\mu_0\,\mu_r\,\frac{\partial\,H_y}{\partial\,t}\bigg|_{(i+1/2)\Delta x}^{n\,\Delta t} = \frac{\partial\,E_z}{\partial\,x}\bigg|_{(i+1/2)\Delta x}^{n\,\Delta t} \tag{16}$$

we obtain an equation for the unknown magnetic fields:

$$H_y\bigg|_{i+1/2}^{n+1/2} = H_y\bigg|_{i+1/2}^{n-1/2} + \frac{1}{\mu_0\,\mu_r}\frac{\Delta\,t}{\Delta\,x}\left\{E_z\bigg|_{i+1}^{n} - E_z\bigg|_{i}^{n}\right\}. \tag{17}$$

Equation (17) shows that the future value of H_y depends on only its previous value and the neighboring electric fields. After applying to all the magnetic-field nodes, the dividing line between future and past values has advanced a half-time step. From the discretized form of Ampere's law

$$\varepsilon_0\,\varepsilon_r\,\frac{\partial\,E_z}{\partial\,t}\bigg|_{i\,\Delta x}^{(n+1/2)\,\Delta t} = \frac{\partial\,H_y}{\partial\,x}\bigg|_{i\,\Delta x}^{(n+1/2)\,\Delta t} \tag{18}$$

we obtain an equation for unknown electric field:

$$E_z\Big|_i^{n+1} = E_z\Big|_i^{n} + \frac{1}{\varepsilon_0\,\varepsilon_r}\frac{\Delta t}{\Delta x}\left\{H_y\Big|_{i+1/2}^{n+1/2} - H_y\Big|_{i-1/2}^{n+1/2}\right\}. \tag{19}$$

After applying to all the electric-field nodes in the grid, the dividing line between what is known and what is unknown moves forward another one-half temporal step [8].

In most instances one is interested in modelling a problem which exists in an open domain. If we realize absorbing boundary conditions the grid which contain only a finite number of nodes can behave as if it were infinite.

1.3.1. Accuracy and numerical stability It is worth noting that the FDTD scheme fulfill of four Maxwell's equation, although only two equations are directly employed. The ratio $c\,\Delta t/\Delta x$ is often called the Courant number which plays an important role in determining the stability of a simulation (c is the light velocity in free space). In FDTD simulations there are restrictions on how large temporal step can be. If the temporal step is too large, the algorithm produces unstable results. The restriction can be expressed $c\,\Delta t \leq \Delta x$ or $S_c \leq 1$, where $S_c = c\,\Delta t/\Delta x$. It turns out that the optimum ratio for the Courant number is also the maximum ratio $S_c = 1$ [8], which is called the "magic time step".

2. Comparison of different numerical techniques

We will discuss the capabilities and disadvantages of the tree most successful and widely used numerical techniques [9]. The PW method is usually used to treat infinite periodic systems, giving their dispersion properties. Although it can be applied only in systems with nondispersive components, the PW method is the fastest and easiest to apply. The TMM is able to calculate the band structure of systems with dispersive components but it is less easy to apply than the PW method. Among the most important advantages of the TMM is its ability to treat samples with almost arbitrary internal structure and arbitrary material combinations, giving the complex transmission and reflection amplitudes, that is, magnitude, phase, and polarization information.

Among the drawbacks of the TMM is the necessity of discretization of the unit cell, which introduces some numerical artifacts and some constraints into the shape size of the components inside the unit cell.

The FDTD method, like the TMM, can also model finite slabs with almost arbitrary internal structure and material combinations. Its main advantages compared to the TMM is that it can give the transmission properties over a wide spectral range with just a single calculation. It also can give time-domain pictures of the fields and the currents over the entire computational domain. Moreover, it can treat defects with no additional computational complications.

Concerning the disadvantages of the FDTD method, part of them stem from the inherent discretizations required. In the case of dispersive materials, though, one encounters additional problems, coming from the time scale that the dispersion model introduces, as the time step of the method (Δt) cannot be much larger than the characteristic time scale of the dispersion model.

Apart from the three methods that we have described, additional methods have been applied to the study of PCs, although less extensively. Some of those are the variations of the PW, TMM, and FDTD methods. Among the existing methods, one worth mentioning is the multiple-scattering (MS) or photonic-KKR method, which is a vectorial extension of the well-known electronic band structure calculation method KKR, and its modification known as the layered-MS method. Their main disadvantages are the heavy formalism, the difficulties in the computational procedure, and the large calculation times.

3. Wave propagation in a one-dimensional photonics crystals

Let us consider a one-dimensional photonic crystal consist of 150 layers (see figure 5). Each layer is isotropic and homogeneous. Furthermore we assume that the material is nonmagnetic.

The structure of one-dimensional PCs with linear media and finite length is denoted as $(\mathrm{HL})^{75}$, where H stands for layers with the same high dielectric constant $\varepsilon_\mathrm{h} = 9$ and the same thickness $d = 1.5 \times 10^{-5}$ m (15 μm), while L stands for layers with the same low dielectric constant $\varepsilon_\mathrm{l} = 1$ and the same thickness of d, respectively, and 75 stands the number of periods. In the one-dimensional FDTD computation used, the computation domain is set as 1000 cells and each cell length is set as $\Delta h = 3$ μm. Thus one period of the PC occupy 10 cells. The whole PC is set from cell No. 100 to cell No. 900. The length of the computational domain and the photonic crystal are 3 mm and 2.25 mm. The other domain is free space. The time interval is $\Delta t = \Delta h/c = 10^{-14}$ s (10 fs) (c is the light velocity in free space).

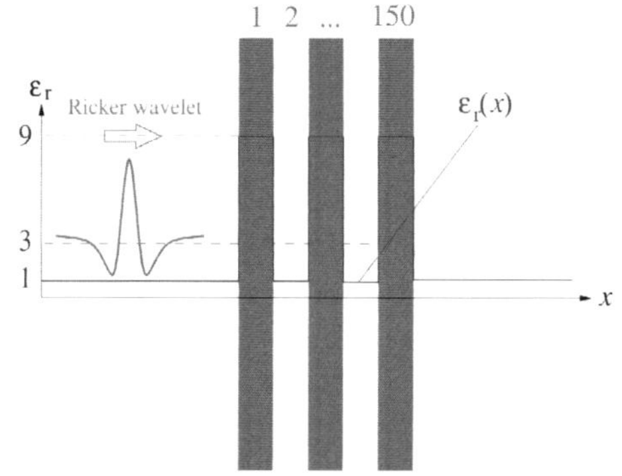

Figure 5. Wavelet-shaped DC electric pulse approaches to the entrance of the one-dimensional photonic crystal. The pulse is normally incident from the left.

3.1. Pulse excitation

One of the features of the FDTD technique is that it allows the modelling of broad range of frequencies using a single simulation. Therefore it is generally advantageous to use pulsed sources rather than harmonic source. The Gaussian pulse is potentially an acceptable source expect that it contains a dc component. Source with dc components also have the possibility of introducing artifacts which are not physical. Therefore we consider a different pulsed source which has no dc component. The Ricker wavelet is equivalent to the second derivative of a Gaussian. The operating wavelength and frequency of the wavelet are $\lambda_0 = 9 \times 10^{-5}$ m and $\nu_0 = 3.\dot{3} \times 10^{12}$ Hz.

Figure 6 shows temporal-spatial evolution of the electric field. It is clear that, in this case the group-velocity dispersion becomes so large that the pulse entirely disperses. Figure 7, 8, 9, 10 and 11 are the snapshots of the pulse taken at time step 25, 125, 325, 625 and 849.

It should be mentioned that by introducing the Kerr nonlinearity into a PC material, the balance of the large dispersion and the self phase modulation (due to the Kerr nonlinearity) can produce a soliton-like pulse propagating inside the PC structure [10].

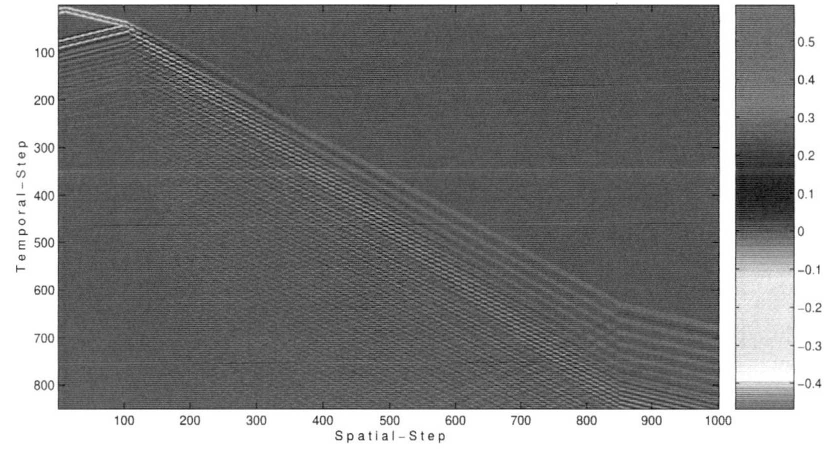

Figure 6. Computed electric-field distribution (color map) of a Ricker wavelet as a function of space and time. The dashed lines show the boundaries of the photonic crystal.

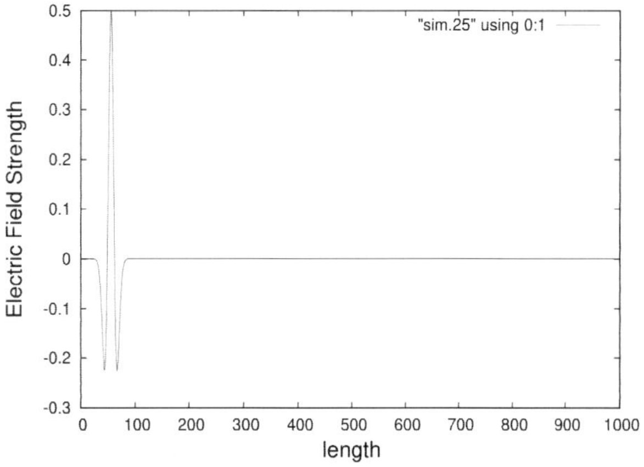

Figure 7. Ricker wavelet pulse approaches to the entrance of the photonic crystal.

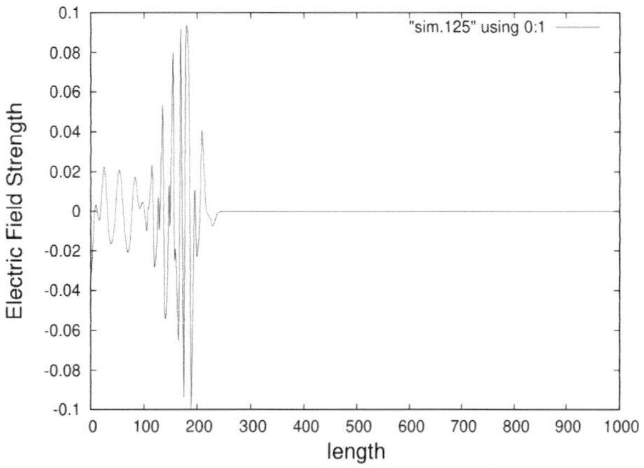

Figure 8. Ricker wavelet pulse propagating in the photonic crystal.

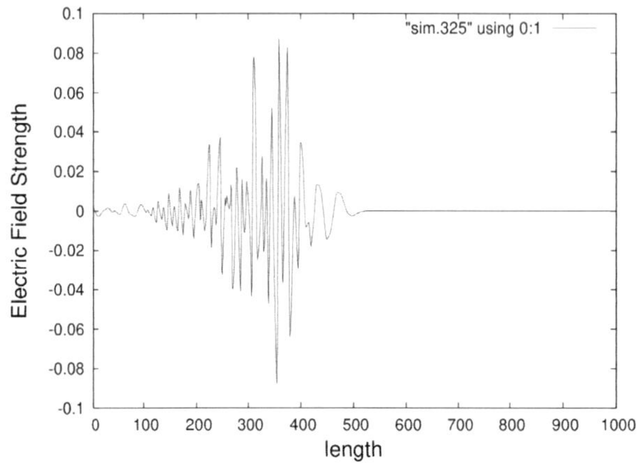

Figure 9. Ricker wavelet pulse propagating in the photonic crystal.

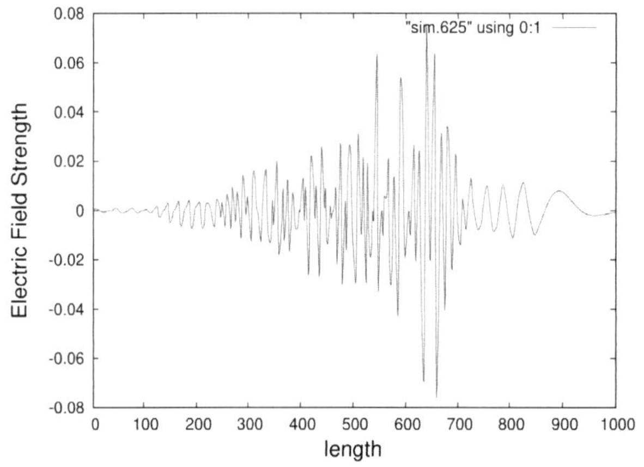

Figure 10. Ricker wavelet pulse propagating in the photonic crystal.

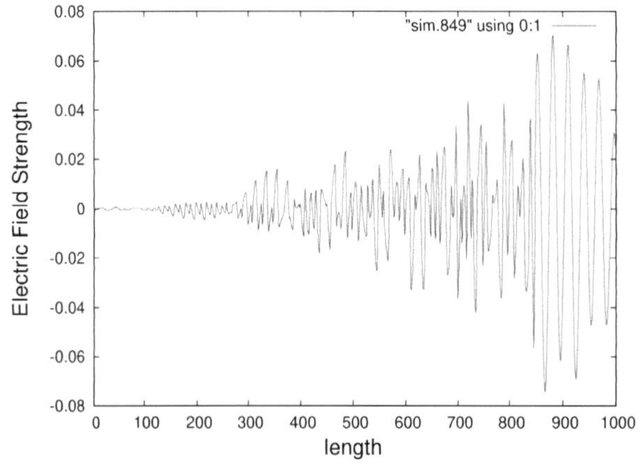

Figure 11. Ricker wavelet pulse propagating in the photonic crystal.

4. The transmittance spectrum of the slightly modified Li filter

Terahertz wave, which refers to the frequencies from 0.1 THz to 10 THz, lie in the frequency gap between the infrared and microwave, have received considerable attention during the past decades. Spectroscopy in terahertz radiation could provide novel information in chemistry and biochemistry. Recently developed methods of THz time-domain spectroscopy (THz TDS) and THz tomography have been shown to be able to perform measurements on, and obtain images of, samples which are opaque in the visible and near-infrared regions of the spectrum. The utility of THz-TDS is limited when the sample is very thin, or has a low absorbance, since it is very difficult to distinguish changes in the THz pulse caused by the sample from those caused by long term fluctuations in the driving laser source or experiment. However, THz-TDS produces radiation that is both coherent and spectrally broad, so such images can contain far more information than a conventional image formed with a single-frequency source. A primary use of submillimeter waves in physics is the study of condensed matter in high magnetic fields, since at high fields (over about 15 teslas), the Larmor frequencies are in the submillimeter band. Terahertz radiation could let art historians see murals hidden beneath coats of plaster or paint in centuries-old building, without harming the artwork.

Up to now, some kinds of THz wave filter based on photonic crystal, metamaterial and surface plasmon, have been reported [11],[12]. Recently, a novel terahertz filter has been designed by Li [13]. The device can be called Li filter. A conventional perfect photonic crystal structures consists of identical alternating layers of high (n_1) and low (n_2) refractive indices. The configuration of the Li filter using one-dimensional photonic crystal is shown in figure 12.

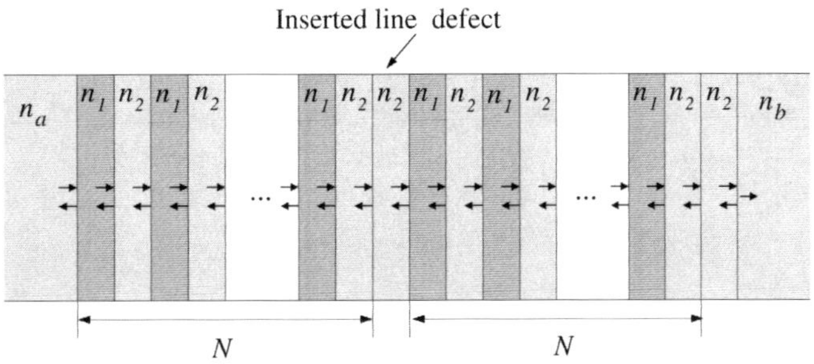

Figure 12. Schematic diagram of the terahertz wave filter (Li filter).

It consists of a line defect layer inserted into a periodic structure of alternating layers of high/low index materials. To design a linear defect with specified defect states located at the prescribed frequencies, a quarter-wave layer n_2 is implanted in the perfect photonic crystal (namely a quarter-wave layer n_2 is inserted between the two groups $(n_1n_2)^N(n_1n_2)^N$, resulting in $(n_1n_2)^N n_2(n_1n_2)^N$, where N represents the periodicity of the cross, n_1 and n_2 are the refractive index of the dielectric materials, respectively.) Adding another n_2 layer on the right, the structure $(n_1n_2)^N n_2(n_1n_2)^N n_2$ will act as $2n_2$, that is, a half-wave absentance layer at λ_0. Li demonstrated that if identical structure is sandwiched between the same substrate, then it will act as an absentance layer, opening up a narrow transmission window at λ_0, in the middle of its reflecting band (see figure 13) [13]. The optical thicknesses of the layers are quarter-wavelength long at the operating wavelength λ_0. The THz frequencies were selected for THz wave wireless communication applications. The frequency is 0.3 THz. The operating wavelength is taken to be $\lambda_0 = 1$ mm. The refractive indices of the layers are selected $n_a = n_b = 1.52$ (glass), $n_1 = 2.1$ (zirconium oxide), $n_2 = 1.4$ (magnesium fluoride). The periodicity of the cross N is equal to 10.

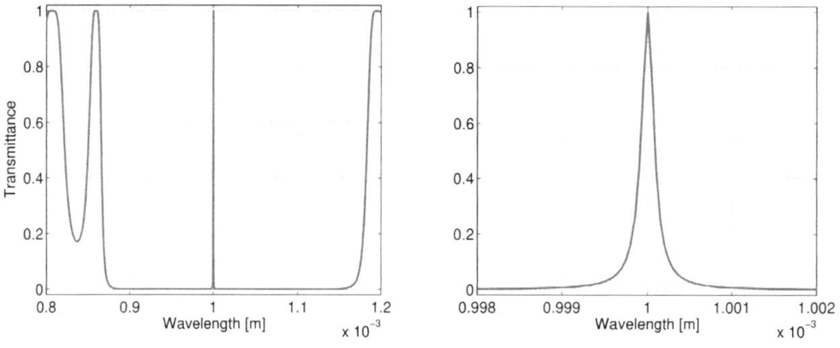

Figure 13. The transmission spectrum of THz wave narrow bandpass filter ($N = 10$). As Li suggested the filter is sandwiched between the same substrate ($n_a = n_b = 1.52$).

Simulation results demonstrate that the Li filter can achieve a narrow transmission band and high transmission centered at λ_0, having a flat passband at the central frequency, and sharp edges. We will show that is identical structure can be sandwiched between different substrate. The refractive indices of the layers are selected $n_a = 1$, (air) $n_b = 2.315$ (titanium dioxide), $n_1 = 2.1$ (zirconium oxide), $n_2 = 1.4$ (magnesium fluoride). As can be seen in figure 14, our simulation results show that the slightly modified filter also exhibit good transmission performance at the central frequency.

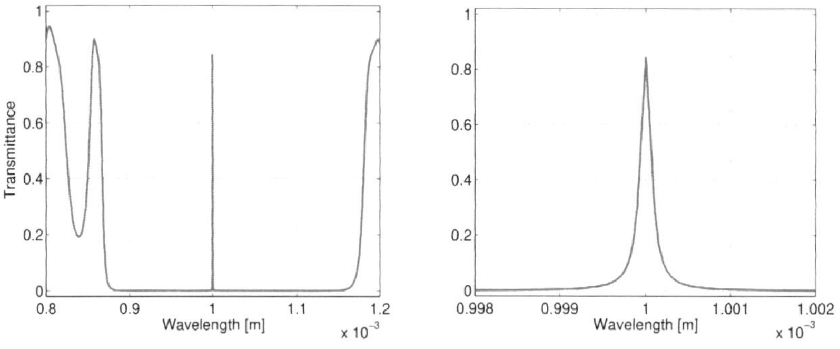

Figure 14. The transmission spectrum of THz wave narrow bandpass filter ($N = 10$). The filter is sandwiched between different substrate ($n_a = 1$, $n_b = 2.315$).

By the changing of the refractive indices outside the filter the transmission bandwith reamines the same. However, the magnitude of the transmission peak becomes lower, and it is about 85%.

5. Conclusion

We presented a brief review of the three most successful and widely used numerical techniques. We presented the key ideas and equations of the TMM and FDTD method and discussed their capabilities and disadvantages. We have been given several numerical calculations starting from the pulse propagation through one-dimensional photonic crystal and arriving the transmission characteristic of the slightly modified THz wave filter. We have shown the temporal-spatial evolution of the Ricker wavelet pulse within the PC. The effect of the group-velocity dispersion is connected with the center frequency of the electromagnetic pulse: when it is close to the edge of the photonic band gap, the pulse undergoes a large group velocity dispersion. The simulation result hase shown that the slightly modified terahertz wave filter preserves the required transmission performance at the central frequency, adjustable bandpass, sharp edges, and good rejection of the sideband frequencies.

Acknowledgments

The author thanks A. Amman for his valuable discussion.

References

[1] Yablonovitch E, Gmitter T J and Leung K M 1991 *Phys. Rev. Lett.* **67** 2295
[2] Shen X, Chen X, Jiang M and Shie D 2005 *Optoelectronic Letters* **1** 0201
[3] Villeneuve P R and Piche M 1994 *Progr. Quantum Electron.* **18** 153
[4] Pendry J B 1994 *Journal of Modern Optics*, **41** 209
[5] Oriazi H and Afsahi M 2007 *Progress in Electromagnetic Research* **74** 217
[6] Yee K S 1996 *IEEE.Trans. Antennas and Propagation*, **14** 302
[7] Taflove A and Hagness S C 2000 *Computational Electrodynamics: The Finite-Difference Time-Domain Method* 2nd ed. (Boston, London: Artech House)
[8] Schneider J B, Lecture notes by John Schneider, *www.eecs.wsu.edu/~schneidj/*
[9] Engheta N and Ziolkowski R W 2006 *Metamaterials Physics and Engineering Explorations* (Wiley Interscience, A John Wiley & Sons, Inc. , Publication)
[10] Fang Y and and Zhou J 2007 *Journal of Russion Laser Research* **28** 372
[11] Libon I H et al. 2000 *Appl. Phys. Lett.* **76** 2821
[12] Wu D et al. 2003 *Appl. Phys. Lett.* **83** 201
[13] Li J 2010 *Opt. Commun.* **283** 2647

Journal of Physics: Conference Series **268** (2011) 012022 doi:10.1088/1742-6596/268/1/012022

Damping of materials and members in structures

F Orban

Department of Mechanical Design. Pollack Mihaly Faculty of Engineering
University of Pécs, Pécs, Hungary

E-mail: orb@witch.pmmf.hu

Abstract. The state of a structure subject to oscillatory deformation can be described by the combination of kinetic and potential energy. In the case of real structures there is also an energy dissipative element as some of the energy is lost per deformation cycle. The energy dissipation is caused by material damping which basically depends on three factors: amplitude of stress, number of cycles and geometry. In the case of non-homogeneous stress distribution the geometry of the structure influences the vibration damping. In this paper the influence of the geometry will be investigated with special regard to the cross-section. The examinations can be executed experimentally, theoretically and by the help of computer programs using FEM. In most cases the main goal is to increase the damping of the structure.

1. Measures of damping

Damping of structures is a very complex phenomenon, which refers to two basic reasons a) material damping b) friction damping at the connections. When a structure is subject to oscillatory deformations the state of the structure can be described by the combination of kinetic and potential energy. In the case of real structures some of this energy is lost per deformation cycles and this is called material damping.

Damping is the conversion of mechanical energy of a vibrating structure into thermal energy. If we want to quantify the level of damping in a structure the absorbed energy per cycle must be determined. By plotting the force versus displacement for a given cycle of motion a hysteresis curve is generated (figure 1).

At linear damping the hysteresis loop is an ellipses (figure 1a). In general metals have linear damping in the case when the stress amplitude less than the fatigue limit. At nonlinear damping the hysteresis loop is peaked when damping is a result of friction (figure 1b) One possibility to quantify the level of damping is to determine the area captured within the hysteresis loop

$$D = \oint F \cdot dy \, , \tag{1}$$

then the specific damping coefficient is:

$$\psi = \frac{D}{U} \, , \tag{2}$$

where U is the stored energy during loading.

For an unforced damped single degree of freedom (SDOF) system the general equation of motion becomes:

$$m \cdot \ddot{y} + k \cdot \dot{y} + \frac{y}{c} = 0, \tag{3}$$

where m is the mass, k is viscous damping constant and c is the spring constant.

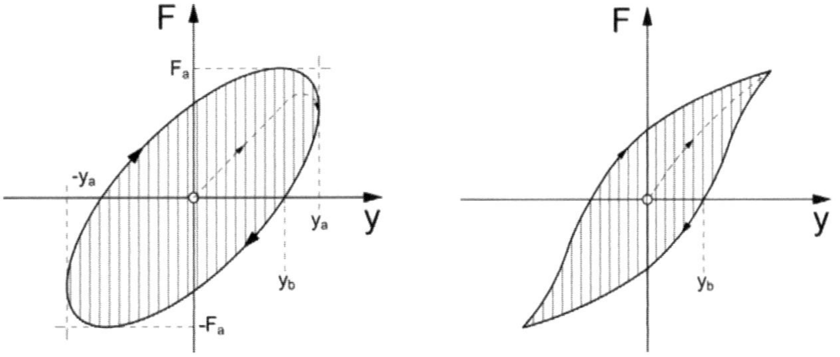

Figure 1. Hysteresis loop a) linear damping b) non linear damping.

In the under-damped case when $\beta < \alpha$

$$\alpha^2 = \frac{1}{m \cdot c}; \quad \beta = \frac{k}{2m}, \tag{4}$$

$$\ddot{y} + 2\beta \cdot \dot{y} + \alpha^2 \cdot y = 0. \tag{5}$$

The solution is given by.

$$y(t) = A \cdot e^{-\beta t} \cdot \sin(\gamma t + \varepsilon_0), \tag{6}$$

where

$$\gamma = \sqrt{\alpha^2 - \beta^2}, \tag{7}$$

$$\frac{A_1}{A_2} = \frac{e^{-\beta T}}{e^{-\beta(t+T)}} = e^{\beta T}, \tag{8}$$

$$\delta = \ln \frac{A_1}{A_2} = \frac{1}{k} \ln \frac{A_i}{A_{i+k}}. \tag{9}$$

The log decrement of transient response δ at linear damping is irrespective of time. There is a simple interrelationship between the specific damping coefficient ψ and the log decrement of transient response δ. It is possible to express the variable ψ as

$$\Psi = -\int_{U_i}^{U_{i+T}} \frac{dU}{U} = \ln \frac{U_i}{U_{i+T}}. \tag{10}$$

The potential energy for a spring is

$$U = \frac{y^2}{2c}, \tag{11}$$

$$\psi = \ln\frac{U_i}{U_{i+T}} = \ln\frac{A_i^2}{A_{i+T}^2} = 2\ln\frac{A_i}{A_{i+T}} = 2\delta\,. \tag{12}$$

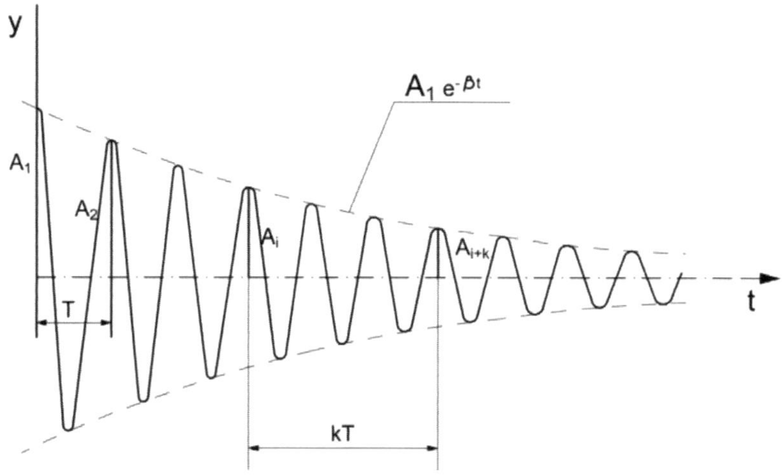

Figure 2. Transient response of a classically under-damped system.

The level of damping can be subjectively determined by noting the sharpness of resonant peak at f_1 (figure 3). For a quantitative measure of damping the Half-Power Bandwidth Method can be employed.

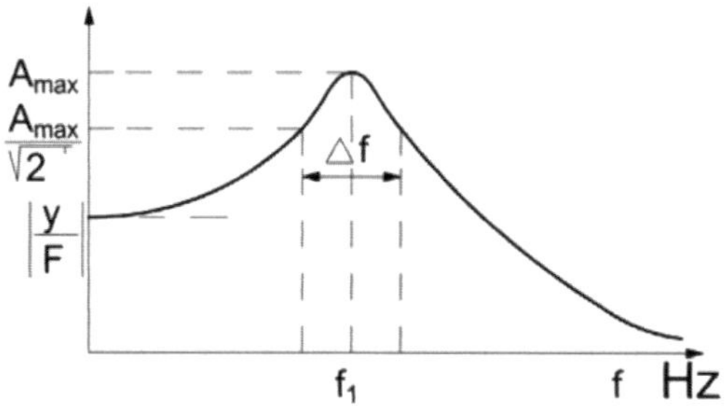

Figure 3. Compliance transfer function for a SDOF system.

The damping of the structure η can be determined from the ratio Δf to f_1. The resonant peak value is A_{max} and in this case the loss factor can be expressed as

$$\eta = \frac{\Delta f}{f_1}\,. \tag{13}$$

2. Material damping

Material damping depends on many factors. The most important of these factors are: type of materials, stress amplitude, internal forces, the number of cycles, sizes of geometry, the quality of surfaces and temperature. The factors were examined in Lazan book [1]. Damping depends mainly on the stress amplitude as

$$D = J \cdot \sigma_a^n\,, \tag{14}$$

where J and n are constants.

The value of n can be between 2 and 4 but generally n = 2.3 is used. Damping increases with the number of cycles and finally a fatigue collapse happens. The values of J are very different for the same material according different authors.

According to [2] $J = 2.326 \cdot 10^{-8}$ for low carbon steel and therefore

$$D = 2.321 \cdot 10^{-8} \cdot \sigma^{2.3} . \tag{15}$$

At the uniaxial stressed state the stored energy can be calculated:

$$U = \frac{\sigma^2}{2 \cdot E} , \tag{16}$$

$$\psi = \frac{D}{U} = 9.769 \cdot 10^{-3} \cdot \sigma^{0.3} . \tag{17}$$

According to Equation (17) the values computed for ψ are summarised in table 1.

Table 1. Specific damping coefficient values refers to σ_{MAX}.

σ N/mm^2	20	40	60	80	100	120	140
ψ %	2.4	2.9	3.3	3.7	3.9	4.1	4.3

The relationship between measures of damping can be expressed:

$$\eta = \frac{\psi}{2\pi} = \frac{\delta}{\pi} = 2\,\xi . \tag{18}$$

The values of the viscous damping ratio ξ is listed in table 2 according to [3].

Table 2. Dynamic properties of materials under standard conditions.

Representative Damping Ratios	
System	Viscous Damping Ratio ξ
Metals (in elastic range)	< 0.01
Continuous Metal Structures	0.02 to 0.04
Metal Structure with Joints	0.03 to 0.07
Aluminium / Steel Transmission Lines	≈ 0.0004
Small Diameter Piping Systems	0.01 to 0.02
Large Diameter Piping Systems	0.02 to 0.03
Auto Shock Absorbers	≈ 0.30
Rubber	≈ 0.05
Large Buildings during Earthquakes	0.01 to 0.05

3. Material damping in the structure

If the stress distribution is not homogeneous, the geometry of the structure influences the damping properties. The damping, which belongs to the maximum stress amplitude is

$$D_{am} = J \cdot \sigma_{am} \cdot \tag{19}$$

The structural damping can be calculated as

$$D_s = D_{am} \cdot V \cdot \alpha_k \cdot \alpha_h , \tag{20}$$

where V is the volume of the structure, α_k is the cross section factor, α_h is the length factor.

The cross section factor for a rectangular cross section (figure 4a) is

$$\alpha_k = \frac{1}{A_0 \cdot y^n{}_{\max}} \cdot 2 \int_{y=0}^{b/2} a \cdot y^n \cdot dy = \frac{1}{n+1} , \tag{21}$$

where $A_0 = a \cdot b$.

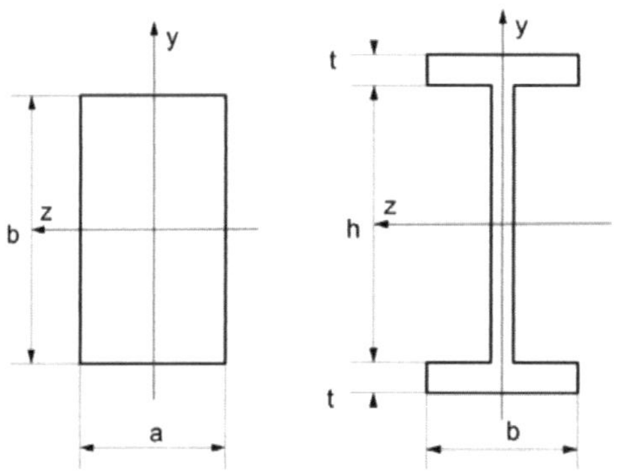

Figure 4. a) rectangular, b) I-section.

In the case of an I-section the cross section factor is:

$$\alpha_k = \frac{2}{A_0 \left(\frac{h}{2}\right)^n} \left[t_w \int_0^{h/2} y^n \cdot dy + b \cdot t \cdot \left(\frac{n}{2}\right)^n \right] = 1 - \frac{n}{n+1} \cdot \frac{t_w \cdot h}{A_0} . \tag{22}$$

It can be seen that the cross-section factor for an I-section is twice as big as for the rectangular section. The length factor is calculated as

$$\alpha_h = \frac{1}{l} \int_0^l \left(\frac{\sigma_{ax}}{\sigma_{am}} \right)^n dx . \tag{23}$$

A cantilever beam with concentrated load is plotted in figure 5. If the load is concentrated on the cantilever beam then the length factor is

$$\alpha_h = \frac{1}{l} \int_0^l \left(\frac{x}{l} \right)^n dx = \frac{1}{n+1} . \tag{24}$$

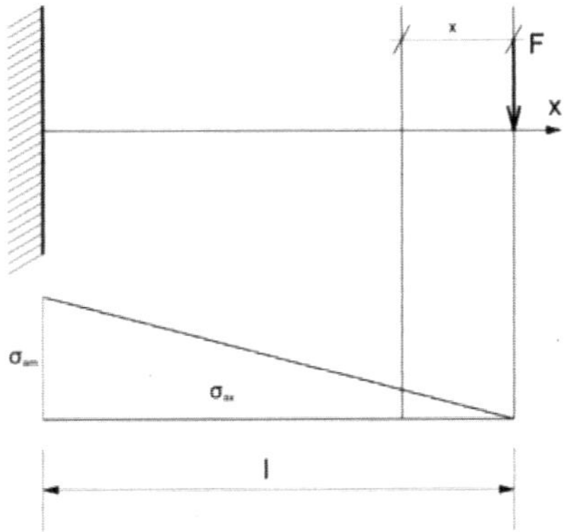

Figure 5. Maximum stress distribution at cantilever beam.

A cantilever beam with distributed load can be seen in figure 6.

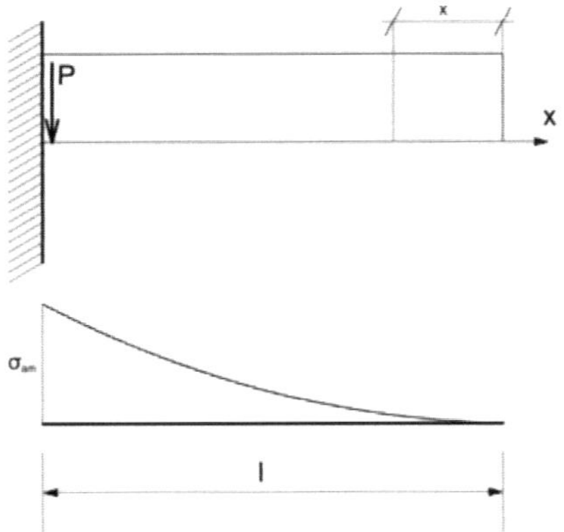

Figure 6. Maximum stress distribution at cantilever beam.

If the load is distributed load the length factor is

$$\alpha_h = \frac{1}{l}\int^{l}\left(\frac{x^2}{l^2}\right)^n dx = \frac{1}{2n+1}. \tag{25}$$

A simply supported beam with concentrated load is plotted in figure 7.

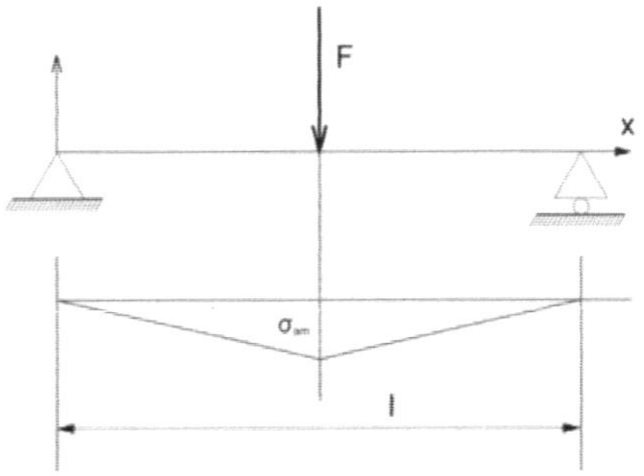

Figure 7. Maximum stress distribution at simply supported beam.

If the load is concentrated the length factor for a simple beam can be determined as

$$\alpha_h = \frac{1}{l} 2 \int_0^{l/2} \left(\frac{2x}{l} \right)^n dx = \frac{1}{n+1} . \tag{26}$$

Simply supported beam with concentrated load is plotted in figure 8.

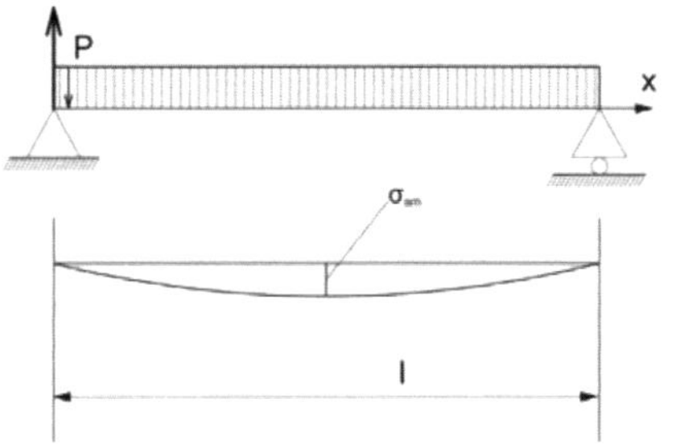

Figure 8. Maximum stress distribution at simply supported beam.

In this case the length factor can be determined as

$$\alpha_h = \frac{1}{l} \int_0^l \left[\frac{4}{l^2} \left(l_x - x^2 \right) \right]^n dx = \frac{2}{3} \left(\frac{3}{4} \right)^n + \frac{1}{6} . \tag{27}$$

4. Damping analysis with FEM

The effect of the geometry on the damping will be analyzed by finite element method (FEM) in this section. The GEOSTAR program controls the complete operations of the COSMOSM package and all modules are accessed from there.

The FREQUENCY/BUCKLING option of the program calculates the frequencies without damping The POST DYNAMIC analysis (ASTAR) is used to examine the damping and for this it is necessary to define the damping ratio ξ. Load excitations may be applied as concentrated loads at specified nodes or as distributed pressure applied to specified element faces. The response of the structure at certain nodes may be calculated for displacement.

The log decrement of a transient response is calculated with the help of the displacement time curve.

$$\delta = \frac{1}{k} \ln \frac{A_i}{A_{i+k}} , \tag{28}$$

$$\delta = \beta \cdot T , \tag{29}$$

$$\beta = \frac{\delta}{T} = 2\pi\xi \cdot f_1 . \tag{30}$$

The calculated values β and T are suitable for measuring damping. The force-time diagram shown in figure 9 will applied as an excitation. The examined cross-sections can be seen in figure 10.

Six models were examined according to table 3.

In the examined cases the cross-section area were the same and the maximum stresses were the same for a given beam, in the case of the cantilever beam 75 MPa and for the simply supported beam 20 MPa.

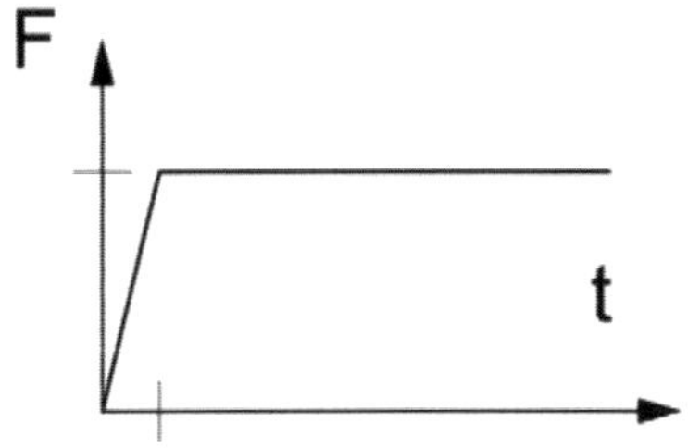

Figure 9. Force-time diagram.

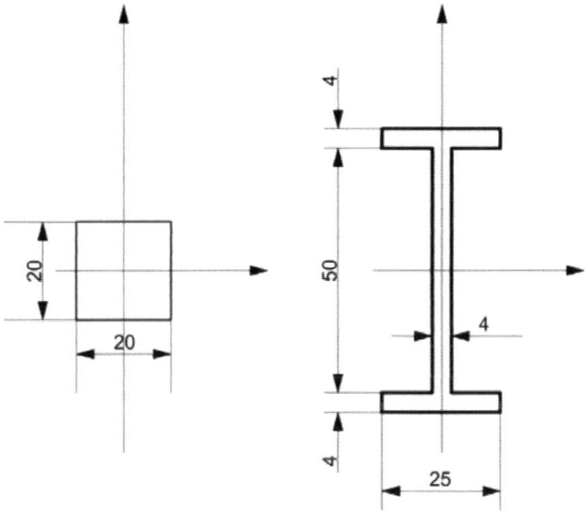

Figure 10. Dimensions of the cross-section.

Table 3. The investigated models

Model	Type of beam	Loading	Cross-section	Max stress
1.	cantilever	Force end of the beam	□	75 MPa
2.	cantilever	Force end of the beam	I	75 MPa
3.	cantilever	distributed load	□	75 MPa
4.	simply supported	distributed load	□	20 MPa
5.	simply supported	distributed load	I	20 MPa
6.	simply supported	Force end of the beam	I	20 MPa

Computation of the damping is illustrated in figure 11, where the beam is fixed and the cross-section is rectangular.

The damping results for a fixed beam with an I-section can be seen in figure 12.

The results are listed in table 4.

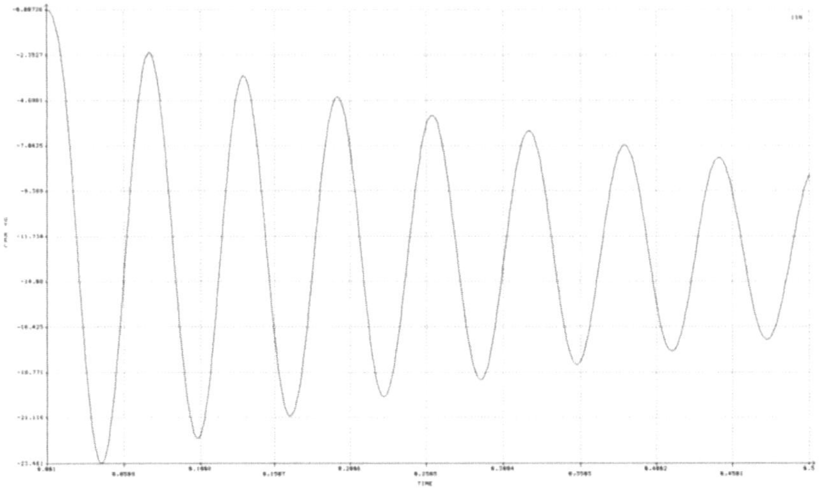

Figure 11. Damped vibration for a fixed beam with rectangular cross-section.

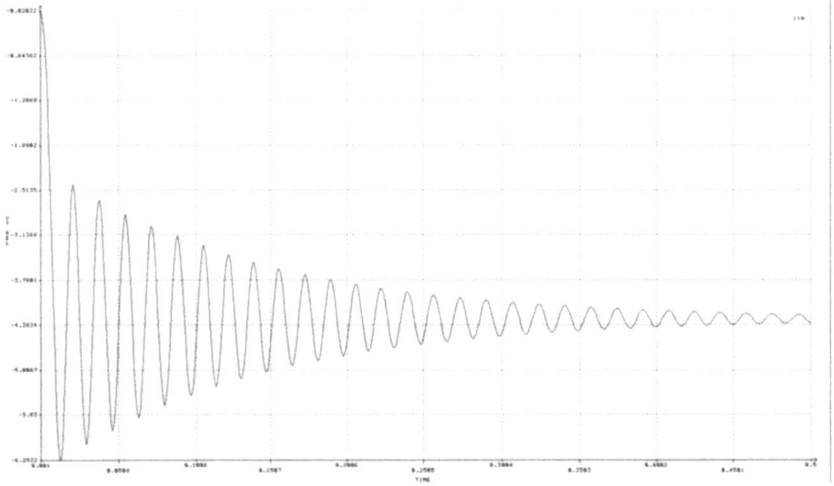

Figure 12. Damped vibration for a fixed beam with an I-section.

Table 4. Calculated results for β eigenvalue frequency.

Model	β	f_1 [H_z]
1.	2.08	16.06
2.	4.05	60.2
3.	1.77	16.06
4.	5.78	45.3
5.	20.67	170
6.	17.36	170

The cross section factor α_k for I section (figure 4b) is twice as high as for the rectangular section (figure 4a). According to the FEM calculations the results are similar. The effects of the load distribution are smaller.

5. Measuring the damping in a laboratory

Different methods exist to measure the characteristics of damping. In one method the starting displacement is applied and then the system is left free to move (figure 13). In this case a damped vibration can be observed. To determine the changing of amplitudes the log decrement δ can be calculated, however the changing of amplitudes and strain changing near the fixed point is the same. Strain-gauge was used to measure the strains. Figure 14 illustrates the acceleration of beam end and strain changing. From the measured values the value of β can be calculated. The first 10 amplitudes according to measurements are listed in table 5.

Figure 13. Arrangement of measurement.

Table 5. Measurement results.

Strain (μ/m)	Time (s)	Acceleration m s^{-2}
355.68	0.44	-.0.72
307.44	0.54	-0.1
269.04	0.64	-0.089
220.08	0.74	-0.074
268.8	0.85	-0.086
232.08	0.95	-0.075
240.96	1.07	-0.075
250.56	1.17	-0.078
247.92	127	-0.078
233.76	1.37	-0.074

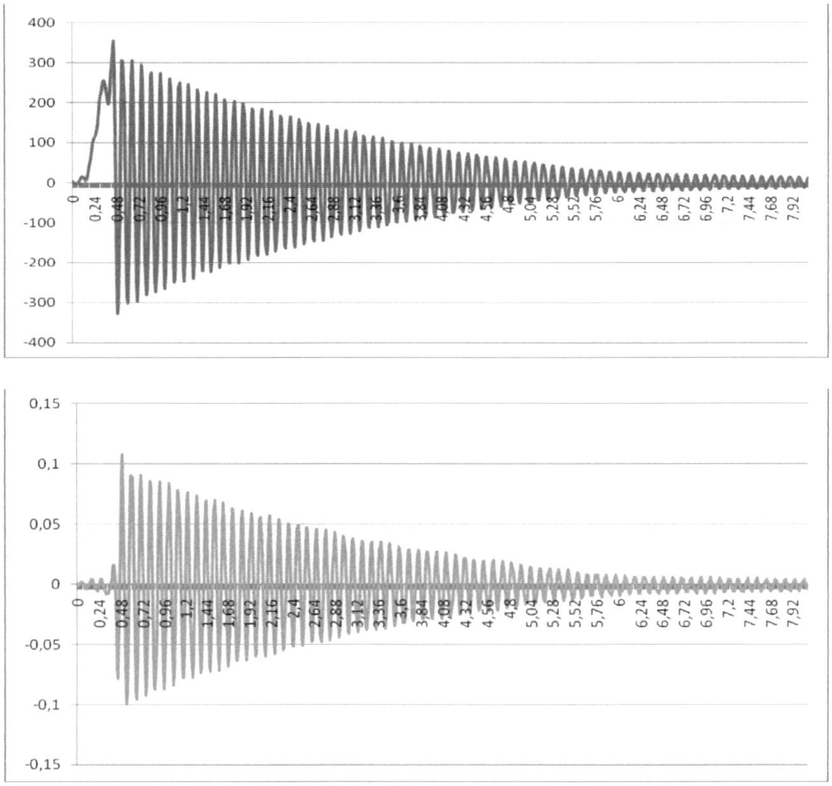

Figure 14. Acceleration of the beam end and strains changing near the fixed point

The model of the experiment can be seen in figure 15. The parameters for the model are l =550 mm, a= 10 mm, m_{beam} = 0.434 kg, m = 0.5 kg.

$$E = 2.1 \cdot 10^5 \text{ MP}, \quad c = \frac{l^3}{3IE}, \tag{31}$$

$$\alpha = \sqrt{\frac{1}{m_{red} \cdot c}}. \tag{32}$$

The eigenvalue frequency is

$$f = \frac{\alpha}{2\pi} = 10.55 \text{ Hz}, \qquad T = 0.094\, s, \tag{33}$$

$$\delta = \frac{1}{k} \ln \frac{A_1}{A_{10}} = \frac{1}{9} \ln \frac{355.86}{233.76} = 0.04669 \tag{34}$$

and the results of the measurements:

$$T = \frac{1.37 - 0.44}{9} = 0.103 \sec, \tag{35}$$

$$f = \frac{1}{T} = 9.708 \text{ Hz}. \tag{36}$$

The calculated and measured values are very close to each other. In this case the damping ratio becomes:

$$\xi = \frac{\delta \cdot f}{\alpha} = 0.007. \tag{37}$$

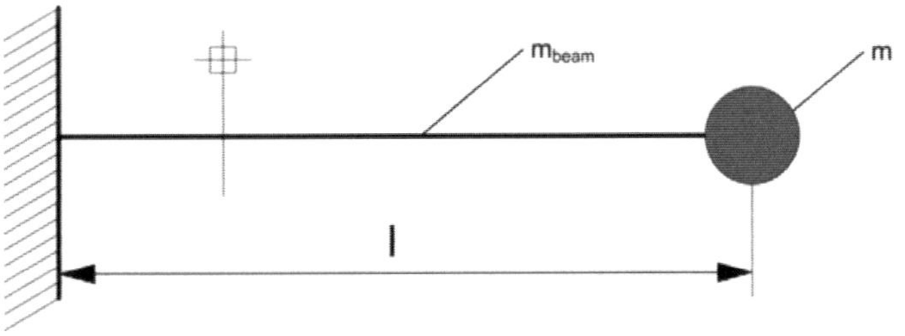

Figure 15. The model of the experiment.

6. Calculating structural damping

Let us determine the structural damping for the model shown in figure16.

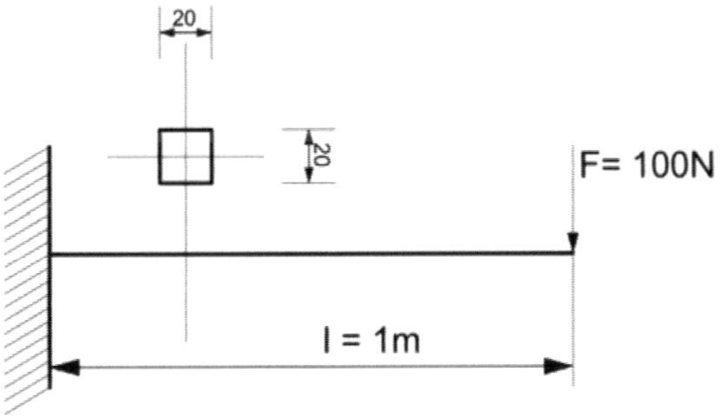

Figure 16. Dimensions of the cantilever beam.

$$\sigma_{MAX} = \frac{M_{MAX}}{W} = \frac{100}{1.333} = 75 \text{ N/mm}^2. \tag{38}$$

The maximum displacement at the end of the beam is:

$$y_{MAX} = \frac{F \cdot l^3}{3IE} = \frac{100 \cdot 1000^3}{3 \cdot 1.333 \cdot 10^4 \cdot 2.1 \cdot 10^5} = 12.5 \text{ mm}. \tag{39}$$

The damping ratio according to the measurements is ξ =0.007. The loss factor is η = 0.014, the log decrement is δ = 0.044 and the absorbed energy is

$$U = W = \frac{1}{2} F \cdot y_{MAX} = 625 \text{ N mm}. \tag{40}$$

The structural damping result is

$$D_s = \psi \cdot U = 2\delta \cdot U = 55 \text{ N mm}. \tag{41}$$

The damping of the structure can be increased by using sandwich beams.

Journal of Physics: Conference Series **268** (2011) 012022 doi:10.1088/1742-6596/268/1/012022

7. Sandwich beams

Comparing the damping properties of this model the dimensions of this beam are selected according to figure 17.

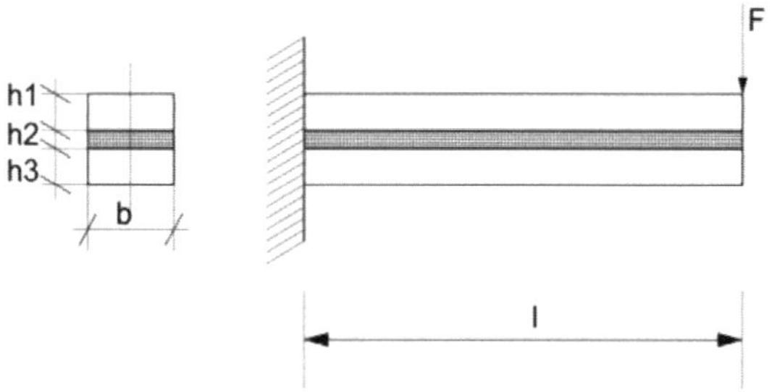

Figure 17. Sandwich beam.

Calculation of the maximum stress data is shown in figure 18.

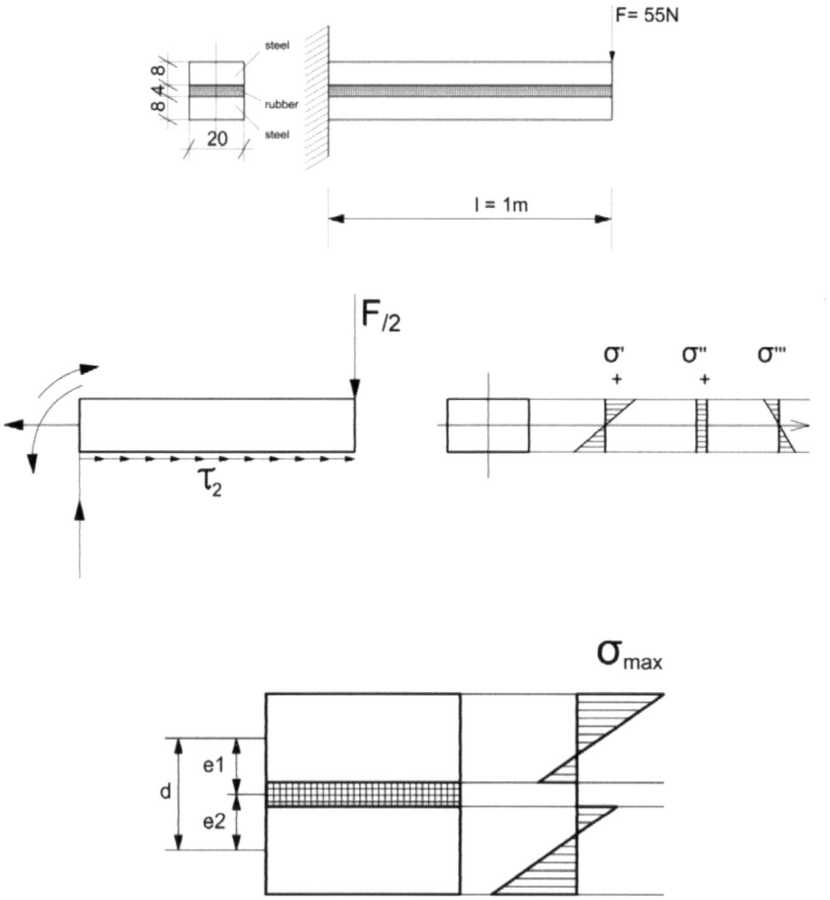

Figure 18. Stress distribution of the sandwich beam.

To calculate the normal stresses the following expressions are applied:

$$\sigma' = \frac{F \cdot l}{2W_1}, \quad \sigma'' = \frac{\tau_2 \cdot b \cdot l}{2A_1}, \quad \sigma''' = \frac{\tau_2 \cdot b \cdot l \cdot h_1}{4W_1}, \tag{42}$$

$$\sigma_{MAX} = 128.9 + 24.75 - 74.25 = 79.4 \text{ N/mm}. \tag{43}$$

The maximum displacement at the end of beam is

$$y_{MAX} = 19.52 \text{ mm}.$$

The loss factor can be calculated using the formula for sandwich beams developed by Ungar [4] (also in Farkas [5])

$$\eta = \frac{\eta_2 \cdot XY}{1 + (2+Y) + (1+Y)\left(1+\eta_2^2\right) \cdot X^2}, \tag{44}$$

where η_2 is the loss factor of the rubber, X is the shear parameter and Y is the stiffness parameter of the laminates.

$$X = g_0 \left(\frac{C_d \cdot L}{2\pi} \right), \tag{45}$$

$$g_0 = \frac{2G_d \cdot b}{h_2 \sum A_i E_i}, \tag{46}$$

$$Y = \frac{d^2 \cdot b}{B_f} \cdot \frac{E_1 \cdot h_1 \cdot E_3 \cdot h_3}{E_1 \cdot h_1 + E_3 \cdot h_3}. \tag{47}$$

For steel $E_1 = E_2 = 2.1 \cdot 10^5$ MPa , for rubber G_d = 4.5 MPa and η_2 = 1.8, thus the loss factor of a sandwich beam is η= 0.0738. The calculated absorbed energy is

$$U = W = \frac{1}{2} F \cdot y_{MAX} = 536.8 \text{ N mm}, \tag{48}$$

$$\psi = 2\pi \cdot \eta = 0.4637. \tag{49}$$

The structural damping of a sandwich beam is:

$$D_s = \psi \cdot U = 248.9 \text{ N mm}. \tag{50}$$

The dissipative energy is more than four times higher then it is in the steel beam case, which has similar dimensions.

8. Other structures

Damping of different structures can be investigated similarly to the beams, however it is impossible to determine the damping of a structure theoretically. A possible way to compare different structures and determine which structure has better damping ability is to determine the value of β or T.

Figure 19 shows the first vibration mode shape in the case of the square plate stiffened by flat ribs. The cover plate is simply supported around the contour.

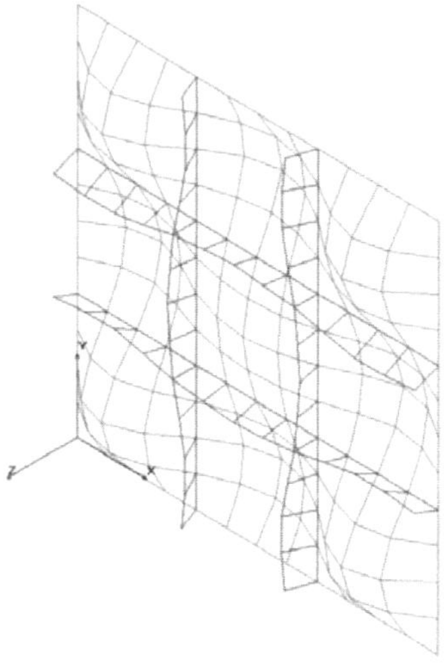

Figure 19. The first vibration mode shape at stiffened plates.

The calculation has been evaluated by FEM. The first natural frequency is f_1 = 49.27 Hz and the time of period is T_1 = 0.02 sec. Dimensions of the stiffened plates are: cover plate is 1410x1410x2 mm; flat ribs are 1410x90x3 mm. To examine a flat plate the dimensions are 1410x1410x2.5 mm, the first natural frequency is f_1 = 6.17 Hz and the time of period is T_1 = 0.16 sec. The examined two structures have the same volume. The stiffened plate has better damping ability according expression (30).

References

[1] Lazan B J 1968 *Damping of materials and members in structural mechanics* (Oxford Pergamon Press Oxford)
[2] Ponomarjov S D 1966 *Static design of mechanical structures* (in Hungarian) (Műszaki Könyvkiadó Budapest)
[3] Cremer L and Heckl M 1988 *Structure borne sound* (Springer Verlag New York)
[4] Ungar E E 1962 Loss factors of visco-elastically damped beam structures *J Acoust. Soc. Amer* **18** 1082
[5] Farkas J 1984 *Optimum design of metal structures* (Budapest. Akadémiai Kiadó Chichester: Ellis Horwood)

5th International Workshop on Multi-Rate Processes and Hysteresis (MURPHYS 2010) IOP Publishing
Journal of Physics: Conference Series **268** (2011) 012023 doi:10.1088/1742-6596/268/1/012023

A geometric classification of traveling front propagation in the Nagumo equation with cut-off

N Popović

University of Edinburgh, School of Mathematics and Maxwell Institute for Mathematical Sciences, James Clerk Maxwell Building, King's Buildings, Mayfield Road, Edinburgh, EH9 3JZ, United Kingdom

E-mail: `Nikola.Popovic@ed.ac.uk`

Abstract. An important category of solutions to reaction-diffusion systems of partial differential equations is given by traveling fronts, which provide a monotonic connection between rest states and maintain a fixed profile when considered in a co-moving frame. Reaction-diffusion equations are frequently employed in the mean-field (continuum) approximation of discrete (many-particle) models; however, the quality of this approximation deteriorates when the number of particles is not sufficiently large. The (stochastic) effects of this discreteness have been modeled via the introduction of (deterministic) 'cut-offs' that effectively deactivate the reaction terms at points where the particle concentration is below a certain threshold. In this article, we present an overview of the effects of such a cut-off on the front propagation dynamics in a prototypical reaction-diffusion system, the classical Nagumo equation. Our analysis is based on the method of geometric desingularization ('blow-up'), in combination with dynamical systems techniques such as invariant manifolds and normal forms. Using these techniques, we categorize front propagation in the cut-off Nagumo equation in dependence of a control parameter, and we classify the corresponding propagation regimes ('pulled,' 'pushed,' and 'bistable') in terms of the bifurcation structure of a projectivized system of equations that is obtained from the original traveling front problem, after blow-up. In particular, our approach allows us to determine rigorously the asymptotics (in the cut-off parameter) of the correction to the front propagation speed in the Nagumo equation that is due to a cut-off. Moreover, it explains the structure of that asymptotics (logarithmic, superlinear, or sublinear) in dependence of the front propagation regime. Finally, it enables us to calculate the corresponding leading-order coefficients in the resulting expansions in closed form.

1. Introduction

Front propagation in reaction-diffusion systems represents a fundamental topic in non-equilibrium physics. Traveling fronts are monotonic connections between rest states that retain a fixed profile in an appropriately defined ('co-moving') frame; they are frequently relevant in the asymptotic (long-time) limit of such systems. For the sake of definiteness, we will only consider front propagation in the family of scalar reaction-diffusion equations that is given by

$$\frac{\partial \phi}{\partial t} = \frac{\partial^2 \phi}{\partial x^2} + f(\phi), \tag{1}$$

where $(t, x) \in \mathbb{R}^+ \times \mathbb{R}$, $\phi(t, x) \in \mathbb{R}$, and f denotes a smooth function with $f(\phi^-) = 0 = f(\phi^+)$; without loss of generality, we assume that the two rest states at ϕ^- and ϕ^+ have been rescaled to 1 and 0, respectively. Equation (1) has found widespread use throughout the sciences, and has *e.g.* been applied in population dynamics, genetics, thermal combustion, mathematical neuroscience, chemical kinetics, and fluid dynamics; see *e.g.* [1, 2] for details and references.

As is well-known [3, 4], front solutions propagating between the rest states at 1 and 0 in (1) can be classified in terms of the reaction function f, which is said to be of type I if it is strictly positive on $(0, 1)$, while it is of type II if it changes sign on that interval exactly once. Generically, one assumes that f is negative on some subinterval $(0, \alpha)$ of $(0, 1)$, with $\alpha \in (0, 1)$, whereas it is positive on $(\alpha, 1)$; moreover, one requires $\int_0^1 f(\phi)\,d\phi > 0$ for the front propagation speed c to be positive [2], as specified in detail below.

For type I-functions with $f'(0) > 0$, zero is an unstable rest state of (1); the corresponding front solution can be either 'pulled' or 'pushed.' Pulled fronts propagate at the linear spreading speed $c^* = 2\sqrt{f'(0)}$ of perturbations about the zero rest state in (1). (Intuitively speaking, the front is 'pulled' along by the linearization of f about zero.) A prototypical example is provided by the Fisher-Kolmogorov-Petrowskii-Piscounov (FKPP) equation [5], which is obtained for $f(\phi) = \phi(1-\phi)$ in (1). By contrast, the propagation speed of pushed fronts, termed $c^\dagger$, is higher than c^*, as is for example the case in the classical Nagumo equation [1, 2], which is defined by

$$\frac{\partial \phi}{\partial t} = \frac{\partial^2 \phi}{\partial x^2} + \phi(1-\phi)(\phi-\gamma), \tag{2}$$

with γ a real parameter: for $\gamma \in (-\frac{1}{2}, 0)$, equation (2) supports front solutions with $c^\dagger = \frac{1}{\sqrt{2}} - \sqrt{2}\gamma$, while the propagation speed obtained from linearization about zero is given by $c^* = 2\sqrt{-\gamma}$. (The nonlinear dynamics behind the front drives, or 'pushes,' front propagation in this regime.)

***Remark* 1.** In both regimes, traveling fronts exist for a continuum of speeds, with $c \geq c^*$; fronts propagating at different speeds are distinguished by their decay properties at 0 [6, 7]. However, for sufficiently localized initial conditions, one particular front is 'selected' by the system: in the pulled regime, that front propagates at the linear spreading speed c^*, while the propagation speed $c^\dagger$ in the pushed regime is higher, as stated above. □

For type-II reaction functions with $f'(0) < 0$, zero is a (meta)stable rest state of (1). If, additionally, $f(\alpha) = 0$ for some $\alpha \in (0, 1)$, the front is called 'bistable' and propagates at a speed c that is not obviously related to the linearization of f at 0; a prototypical example is again given by the Nagumo equation, with $\gamma \in (0, \frac{1}{2})$ in (2). Moreover, the front propagation speed is unique, *i.e.*, there exists exactly one traveling front solution in that regime.

The reader is referred to the reviews in [8] and [9] for an exhaustive discussion of this classification, and of the physical implications thereof.

***Remark* 2.** If $f'(0) = 0$, the rest state at 0 is neutrally stable, and a transition occurs between the generic types I and II, as is the case in the family of scalar equations with degenerate polynomial potential $f(\phi) = \phi^m(1-\phi)$ in (1), for $m \geq 2$; that family, which includes the Zeldovich equation ($m = 2$), was analyzed in [4] and [10], as well as rigorously in [7] and [11]. □

Reaction-diffusion equations of the form in (1) are often employed in the large-scale (continuum) approximation of discrete, N-particle systems [12, 13]. The quality of this approximation, while oftentimes surprisingly good, deteriorates as the number of particles N decreases, since stochastic fluctuations need to be taken into account at a microscopic scale. (In particular, the front propagation speed is frequently misestimated, even when the qualitative front profile is accurately reproduced.)

Recently [5, 14], it has been suggested to model the stochastic effects of this discreteness by introducing a deterministic 'cut-off' at the zero rest state in (1):

$$\frac{\partial \phi}{\partial t} = \frac{\partial^2 \phi}{\partial x^2} + f(\phi)\theta(\phi, \varepsilon). \tag{3}$$

Simply speaking, the cut-off function θ dampens the reaction amplitude whenever the particle concentration ϕ in (3) lies below a certain threshold. In N-particle systems, no particles are available to react when ϕ is below the threshold N^{-1}; correspondingly, the reaction terms in (1) are 'cut-off' in an N^{-1}-neighborhood of 0. In the simplest case, where θ is the Heaviside cut-off H, with

$$H(\phi - \varepsilon) \equiv 0 \quad \text{for} \quad \phi < \varepsilon \quad \text{and} \quad H(\phi - \varepsilon) \equiv 1 \quad \text{for} \quad \phi \geq \varepsilon, \tag{4}$$

the function f is simply set to zero for $\phi < \varepsilon \equiv N^{-1}$.

A cut-off substantially changes the characteristics of the front propagation dynamics of the family of reaction-diffusion equations in (1): in [5], Brunet and Derrida studied traveling fronts in the cut-off FKPP equation, with $f(\phi) = \phi(1 - \phi)$ in (3). They found that the correction Δc to the linear spreading speed $c^* = 2$ that is induced by a Heaviside cut-off is given by $\Delta c = -\pi^2(\ln \varepsilon)^{-2}[1 + o(1)]$, for ε sufficiently small; moreover, they conjectured that this leading-order asymptotics is universal over a wide range of cut-off functions θ. Finally, they argued that a Heaviside cut-off always selects a unique front solution for (3), even when the corresponding equation without cut-off supports an entire family of solutions. (A comprehensive review of the physical aspects of Brunet and Derrida's conjecture can *e.g.* be found in [8].)

Starting with [5], a growing body of work has emerged on the front propagation dynamics of cut-off reaction-diffusion systems and, in particular, on the change in front propagation speed that is caused by a cut-off. Thus, it has been shown that the effects of a cut-off depend strongly on the type of reaction function f in (3): specifically, cut-offs reduce the propagation speed of pulled fronts, whereas they speed up pushed and bistable fronts [10]. Similarly, the correction Δc to the propagation speed that is due to a cut-off is logarithmic in ε in the pulled propagation regime, which implies that the front speed converges slowly to its continuum value c^* as $\varepsilon \to 0$ [3, 6]. By contrast, Δc is sublinear and superlinear in ε in the pushed and bistable regimes, respectively; correspondingly, the convergence to the continuum speed is fast in those regimes [10, 15, 16, 17].

Recently, these results have been made rigorous by us, using the method of geometric desingularization ('blow-up') [18, 19]. Front propagation in the cut-off FKPP equation, as well as in certain generalizations thereof, was studied geometrically in [6]; in particular, we confirmed the leading-order expansion for Δc, as stipulated by Brunet and Derrida [5], and we gave a proof of their conjecture for a very general family of cut-off functions θ. In the follow-up article [15], we investigated the effects of a cut-off on bistable traveling fronts, and we obtained corresponding results for the asymptotics of Δc: we proved that Δc is positive, and we explained the occurrence of fractional powers of ε in the resulting expansion. Analogous results on the effects of a cut-off on the front propagation dynamics in the pushed regime will be given in the upcoming article [16].

In the present article, we provide a systematic overview of front propagation in the Nagumo equation, as given in (2), in the presence of a cut-off. In particular, we classify the resulting front solutions geometrically, in terms of the dynamics of a rescaled ('projectivized') version of (2) in the various propagation regimes, *i.e.*, in dependence of the parameter γ. Without loss of generality, we will only consider the case where θ is the Heaviside cut-off H defined in (4) here:

$$\frac{\partial \phi}{\partial t} = \frac{\partial^2 \phi}{\partial x^2} + \phi(1 - \phi)(\phi - \gamma)H(\phi - \varepsilon). \tag{5}$$

(Other choices of cut-off functions can be studied in a similar fashion; see [6, 15, 20, 21] for details.) As is well-known, equation (2) (without cut-off) supports traveling front solutions that propagate between (ϕ^- =)1 and (ϕ^+ =)0 for an entire (real) interval of γ-values. (The third rest state of (2), which is located at $\phi^\circ = \gamma$, is of no interest to us here.) The stability of these rest states is determined by the value of the parameter γ [22]: specifically, the rest states at 1 and γ are stable, while the rest state at 0 is unstable for $\gamma \in [-1, 0)$ in (2); conversely, for $\gamma \in [0, 1)$, the rest states at 1 and 0 are stable, while the rest state at γ is unstable.

Traveling front solutions to (2) are conveniently analyzed in the co-moving frame that is defined by the traveling wave variable $\xi = x - ct$. Writing $u(\xi) = \phi(t, x)$ for the corresponding front, we obtain the traveling front equation

$$u'' + cu' + u(1-u)(u-\gamma)H(u-\varepsilon) = 0, \tag{6}$$

where c denotes the front propagation speed. Since we are interested in front solutions that propagate with non-zero speed from 1 into 0, we will restrict ourselves to $\gamma \in [-1, \frac{1}{2})$ in (6): multiplying the traveling front equation

$$u'' + cu' + u(1-u)(u-\gamma) = 0 \tag{7}$$

that corresponds to equation (2) with u', integrating the result over ξ, and taking into account that $u' \to 0$ as $\xi \to \pm\infty$, one finds

$$c\int_{-\infty}^{\infty} [u'(\xi)]^2\, d\xi = \int_0^1 u(1-u)(u-\gamma)\, du = \frac{1}{12}(1-2\gamma).$$

Hence, the front propagation speed c that is realized in the absence of a cut-off must change sign (and, indeed, become negative) when $\gamma > \frac{1}{2}$. In particular, c vanishes for $\gamma = \frac{1}{2}$; thus, $\xi = x$, and the corresponding solution to (7) becomes stationary. (A detailed analysis of the effects of a linear cut-off in (2) in that case can be found in [21].) Beyond this so-called 'Maxwell point,' *i.e.*, for $\gamma \in (\frac{1}{2}, 1)$, the traveling front reverses direction. Correspondingly, the rest state at 1 becomes 'dominant' [2] and loses stability, as observed also independently in [23].

This article is organized as follows. In Section 2, we revisit the geometric framework that was first introduced in [6], and we adapt that framework for our study of front propagation in the cut-off Nagumo equation; see also [15, 16, 20, 21]. Then, in Section 3, we discuss the propagation regimes that can be realized in equation (5). For the sake of completeness, we first review the bistable and pushed regimes that were analyzed in detail in [15] and [16], respectively, and we summarize the main results obtained there. Then, we show how the pulled propagation regime can be studied in the context of theory previously developed in [6]. Specifically, we introduce a coordinate transformation that reduces equation (5) to an FKPP-like equation with cut-off. We deduce the existence of traveling fronts in the presence of a cut-off in the resulting, transformed equation, and we infer the leading-order ε-asymptotics of the propagation speed c of these fronts. Reverting to the original coordinates, we obtain corresponding results for equation (5). Then, we discuss three boundary cases that limit the pulled, pushed, and bistable regimes. Finally, we combine these findings and interpret them, completing our classification of front propagation in the cut-off Nagumo equation. In Section 4, we summarize our results, and we discuss possible topics for future research.

2. Geometric framework

In this section, we review the geometric framework on which our classification of the front propagation dynamics of equation (5) is based. In particular, we outline the construction of a singular heteroclinic connection in the corresponding traveling front equation (6); details can be found in [6], where that same framework was first introduced for the study of the FKPP equation with cut-off, as well as in [15, 16, 20, 21].

2.1. Background

As indicated in the Introduction, our classification of front propagation in the cut-off Nagumo equation draws on geometric methods from dynamical systems theory and, in particular, on the so-called blow-up technique (geometric desingularization) [18, 19]. Blow-up complements the geometric singular perturbation theory initiated by Fenichel [24]; it is essentially a sophisticated coordinate transformation that regularizes the dynamics in neighborhoods of singularities, allowing one to apply standard dynamical systems techniques. Moreover, it identifies the relevant asymptotics near these singularities and, in particular, the nature of the dependence of that asymptotics on the corresponding perturbation parameter.

Beginning with [7], we have shown that the blow-up technique is ideally suited to the study of front propagation in degenerate reaction-diffusion systems, *i.e.*, in equations in which the reaction terms exhibit non-hyperbolic behavior (at 0), as is the case for $f(\phi) = \phi^m(1-\phi)$ $(m \geq 2)$ in (1) [20], or, alternatively, in cut-off equations of the type in (3), where the degeneracy is caused by a cut-off at the zero rest state: in addition to introducing a non-hyperbolicity (in the form of zero eigenvalues) in the traveling front equation (6), the cut-off also renders the corresponding vector field discontinuous at $u = \varepsilon$.

As in [6, 15, 16, 20, 21], our study of front propagation in (5) is based on a geometric (phase space) analysis of the extended, equivalent first-order system that is obtained from (6) by introducing the new variable $v = u'$ and by appending the trivial ε-dynamics:

$$u' = v, \tag{8a}$$
$$v' = -cv - u(1-u)(u-\gamma)H(u-\varepsilon), \tag{8b}$$
$$\varepsilon' = 0. \tag{8c}$$

Fronts that propagate between the rest states at 1 and 0 in (5) correspond to heteroclinic connections between the equilibrium points $Q_\varepsilon^+ := (0,0,\varepsilon)$ and $Q_\varepsilon^- := (1,0,\varepsilon)$ of (8), for ε sufficiently small. (The third equilibrium point, which is located at $Q_\varepsilon^\circ := (\gamma,0,\varepsilon)$, is the equivalent of the rest state at γ in (5), and is of no interest to us here.) For future reference, we note that, in the singular limit as $\varepsilon \to 0^+$, these equilibria reduce to $Q_0^- = (1,0,0)$ and $Q_0^+ = (0,0,0)$, which are equilibrium points for the first-order system that is obtained from (7), in the absence of a cut-off:

$$u' = v, \tag{9a}$$
$$v' = -cv - u(1-u)(u-\gamma). \tag{9b}$$

Correspondingly, the value of the front propagation speed c for which a heteroclinic connection is realized in (9) is given by c_0. (Here, we note that c_0 can equal either $c^*(= 2\sqrt{-\gamma})$ or $c^\dagger(= \frac{1}{\sqrt{2}} - \sqrt{2}\gamma)$, depending on the front propagation regime under consideration; see Sections 3.1 through 3.4 below for details.) The properties of Q_0^- and Q_0^+ in dependence of $\gamma \in [-1, \frac{1}{2})$ and c_0, in the context of the first-order system (9), as well as the eigenvalues $\lambda_\mp^- = -\frac{c_0}{2} \mp \sqrt{\frac{c_0^2}{4} + 1 - \gamma}$ and $\lambda_\mp^+ = -\frac{c_0}{2} \mp \sqrt{\frac{c_0^2}{4} + \gamma}$ of the associated linearization about these points, are summarized in tables 1 and 2, respectively. (In particular, we remark that the stability of $Q_0^\mp$ is the converse of that of the corresponding rest states at $\phi^\mp$ of the partial differential equation (2).)

The degeneracy that is introduced by the cut-off into (8) can be removed by 'blowing up' the origin Q_0^+ to an invariant two-sphere in three-space; the dynamics on that sphere is conveniently studied by introducing (local) coordinates, which are essentially rescalings of the state variables u and v and the parameter ε. Two such coordinate systems ('charts') are required here: a 'family chart,' which corresponds to an ε-rescaling that is uniformly valid in any bounded neighborhood of the north pole of the sphere, and a 'phase-directional chart,' which projectivizes the dynamics

Table 1. The equilibrium point Q_0^- in dependence of γ and c_0.

γ	$\lambda_\mp^-$	type	stability
-1	$-1 \mp \sqrt{3}$	saddle	unstable
$(-1, -\frac{1}{2})$	$-\sqrt{-\gamma} \mp \sqrt{1-2\gamma}$	saddle	unstable
$-\frac{1}{2}$	$-\frac{3}{\sqrt{2}}, \frac{1}{\sqrt{2}}$	saddle	unstable
$(-\frac{1}{2}, 0)$	$(\gamma-1)\sqrt{2}, \frac{1}{\sqrt{2}}$	saddle	unstable
0	$-\sqrt{2}, \frac{1}{\sqrt{2}}$	saddle	unstable
$(0, \frac{1}{2})$	$(\gamma-1)\sqrt{2}, \frac{1}{\sqrt{2}}$	saddle	unstable

Table 2. The equilibrium point Q_0^+ in dependence of γ and c_0.

γ	$\lambda_\mp^+$	type	stability
-1	-1 (double)	degenerate node	stable
$(-1, -\frac{1}{2})$	$-\sqrt{-\gamma}$ (double)	degenerate node	stable
$-\frac{1}{2}$	$-\frac{1}{\sqrt{2}}$ (double)	degenerate node	stable
$(-\frac{1}{2}, 0)$	$-\frac{1}{\sqrt{2}}, \sqrt{2}\gamma$	node	stable
0	$-\frac{1}{\sqrt{2}}, 0$	saddle-node	stable
$(0, \frac{1}{2})$	$-\frac{1}{\sqrt{2}}, \sqrt{2}\gamma$	saddle	unstable

near the equator. The corresponding rescaled coordinates will be denoted by (U, V, ε) and (u, W, E), respectively.

The phase-space of the cut-off first order system (8) naturally decomposes into three regions, an outer region (where $u > \varepsilon$ and the flow is unaffected by the cut-off), an inner region (where $u < \varepsilon$ and the vector field is cut-off), and an intermediate region, with $u = O(\varepsilon)$, that represents the boundary between the two. The flow of (8) in the inner and intermediate regions (after blow-up) will be analyzed separately in the family and phase-directional charts, respectively, while the dynamics in the outer region is described in the original (u, v, ε)-variables. In particular, blow-up resolves the singular nature of the transition between the inner and outer regions; specifically, it regularizes the singular limit as $\varepsilon \to 0^+$ in the intermediate region, allowing for the construction of a singular heteroclinic connection Γ between the relevant equilibria Q_0^- and Q_0^+ of (8). Then, invariant manifold theory can be applied to show that this connection will persist, for ε sufficiently small, in the intersection of the unstable manifold $\mathcal{W}^u(Q_\varepsilon^-)$ of Q_ε^- with the stable manifold $\mathcal{W}^s(Q_\varepsilon^+)$ of Q_ε^+. The resulting persistent heteroclinic corresponds to a traveling front solution for the cut-off Nagumo equation (5); moreover, the corresponding persistence proof determines the ε-asymptotics of the propagation speed c of that front, as discussed in detail in Section 3 below.

***Remark* 3.** The blow-up transformation introduced above, in an informal fashion, can more formally be written as a mapping of the form

$$\varphi : \begin{cases} \mathbb{S}^2 \times [0, r_0] \to \mathbb{R}^3, \\ (\bar{u}, \bar{v}, \bar{\varepsilon}, \bar{r}) \mapsto (\bar{r}\bar{u}, \bar{r}\bar{v}, \bar{r}\bar{\varepsilon}), \end{cases}$$

where $\mathbb{S}^2$ denotes the 2-sphere in $\mathbb{R}^3$, while r_0 is a positive constant. In particular, the inverse mapping φ^{-1} maps the origin Q_0^+ to $\mathbb{S}^2$, as required; cf. [6], where this transformation was employed in the analysis of the FKPP equation with cut-off. It was subsequently applied in [25] as well as in [15, 16, 20, 21]; for reference, we remark that the phase-directional and family charts defined above were labeled K_1 and K_2 there, respectively. To keep our presentation accessible, we will not pursue that more formal approach here; a detailed, rigorous introduction to the blow-up technique can *e.g.* be found in [18]. □

2.2. Outer region

The flow in the outer region, where $u > \varepsilon$, is governed by the first-order system (9) that is obtained from equation (7) (without cut-off), as $H \equiv 1$ there; recall the definition of H in (4). The traveling front solution that propagates between the rest states at 1 and 0 in (2), with propagation speed c_0, corresponds to a heteroclinic connection Γ between the two equilibrium points $Q_0^- = (1, 0, 0)$ and $Q_0^+ = (0, 0, 0)$ which is realized for $c = c_0$ in (9), as discussed above. Moreover, that connection coincides with the unstable manifold $\mathcal{W}^u(Q_0^-)$ of Q_0^-. In general, for $\varepsilon \in (0, \varepsilon_0]$, with ε_0 positive and sufficiently small, the unstable manifold $\mathcal{W}^u(Q_\varepsilon^-)$ of Q_ε^- is analytic in the state variables u and v and in the parameter c. Hence, writing $c = c_0 + (c - c_0) = c_0 + \Delta c$, with $\Delta c = o(1)$, we may assume an expansion of the form

$$v(u, c) = \sum_{j=0}^{\infty} \frac{1}{j!} \frac{\partial^j v}{\partial c^j}(u, c_0)(\Delta c)^j \tag{10}$$

for that manifold.

***Remark* 4.** The expansion in (10) is only implicitly ε-dependent, as it follows from (9) that any such dependence must enter through c. Correspondingly, in the extended system (8), the unstable manifold $\mathcal{W}^u(\ell^-)$ of the line of equilibria $\ell^- = \{(1, 0, \varepsilon) \,|\, \varepsilon \in [0, \varepsilon_0]\}$ is a trivial foliation in ε, with fibers $\mathcal{W}^u(Q_\varepsilon^-)$. □

As shown in [15, 20], only the first two terms in (10) play a role to the order considered here: the leading-order term $v(u, c_0)$ again corresponds to $\mathcal{W}^u(Q_0^-)$ or, equivalently, to the orbit Γ, while the next-order (linear) term in Δc is found as the solution $\frac{\partial v}{\partial c}(u, c_0)$ of the variational equation that is associated with (9), taken along $v(u, c_0)$. That equation is obtained as follows: we first rewrite (9) with u as the independent variable, *i.e.*, we divide (9b) (formally) by (9a):

$$v \frac{dv}{du} = -cv - u(1 - u)(u - \gamma). \tag{11}$$

Then, we differentiate (11) with respect to c and substitute c_0 and $v(u, c_0)$, as well as (11), into the resulting equation to find

$$\frac{\partial}{\partial u}\Big(\frac{\partial v}{\partial c}(u, c_0)\Big) = -1 + \frac{u(1 - u)(u - \gamma)}{[v(u, c_0)]^2} \frac{\partial v}{\partial c}(u, c_0). \tag{12}$$

For the variational equation (12) to have a closed-form solution, it is necessary that c_0 and $v(u, c_0)$ are known explicitly, as discussed in detail in [15] and [20]. In particular, equation (12) can be solved exactly in the bistable and pushed propagation regimes; the relevant results from [15] and [16] will be recalled in Sections 3.1 and 3.2 below; cf. also [20, 21]. (Since no closed-form solution to the traveling front equation (7) seems to be known in the pulled propagation regime [6], it follows that (12) can have no exact solution in that case, either.)

Next, we introduce a section for the flow of (8), as follows: for $\rho \geq \varepsilon_0$ and $v_0 > c_0$ positive and fixed, we write Σ^- for the hyperplane

$$\Sigma^- = \{(\rho, v, \varepsilon) \,|\, (v, \varepsilon) \in [-v_0, 0] \times [0, \varepsilon_0]\} \tag{13}$$

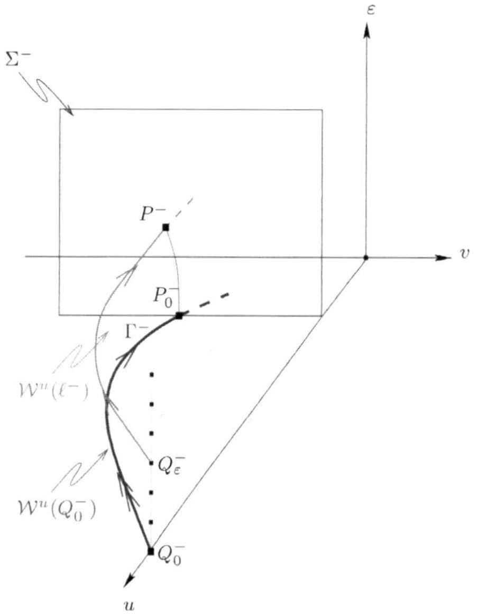

Figure 1. The geometry in the outer region.

in (u, v, ε)-space, and we denote the point of intersection of Γ with Σ^- by $P_0^- = (\rho, v_0^-, 0)$, where $v_0^- = v(\rho, c_0)$. Similarly, we write P^- for the point of intersection of $\mathcal{W}^u(Q_\varepsilon^-)$ with Σ^-; in particular, we suppress the dependence of that point on the parameters c and ε, for brevity. Finally, the segment of the singular heteroclinic connection Γ that is located in this outer region, *i.e.*, in $\{u \geq \rho\}$, will be labeled Γ^-; the resulting geometry is summarized in figure 1.

2.3. Inner region

In the inner region, with $u < \varepsilon$, the dynamics of equation (5) is cut-off, as the Heaviside cut-off H vanishes there; see again (4). The flow in this region is desingularized via the family rescaling

$$u = \varepsilon U, \quad v = \varepsilon V, \quad \text{and} \quad \varepsilon = \varepsilon; \tag{14}$$

the governing equations that are obtained from (8), in the rescaled (U, V, ε)-coordinates, read

$$U' = V, \tag{15a}$$
$$V' = -cV, \tag{15b}$$
$$\varepsilon' = 0. \tag{15c}$$

Although for ε fixed, all points on the U-axis are equilibria of (15), we only consider points on the line $\ell^+ = \{(0, 0, \varepsilon) \,|\, \varepsilon \in [0, \varepsilon_0]\}$ here, as only the latter can correspond to Q_ε^+ in the original (u, v, ε)-variables.

In the singular limit as $\varepsilon \to 0^+$ in (8), the front propagation speed c reduces to c_0; cf. Section 2.1. The unique solution of the resulting singular equation in this inner region, *i.e.*, of

$$\frac{dV}{dU} = -c_0, \quad \text{with} \quad V(0) = 0, \tag{16}$$

is given by $V(U) = -c_0 U$; the orbit corresponding to that solution, which we denote by Γ^+, yields precisely the stable manifold $\mathcal{W}^s(Q_0^+)$ of the origin Q_0^+ in the family chart.

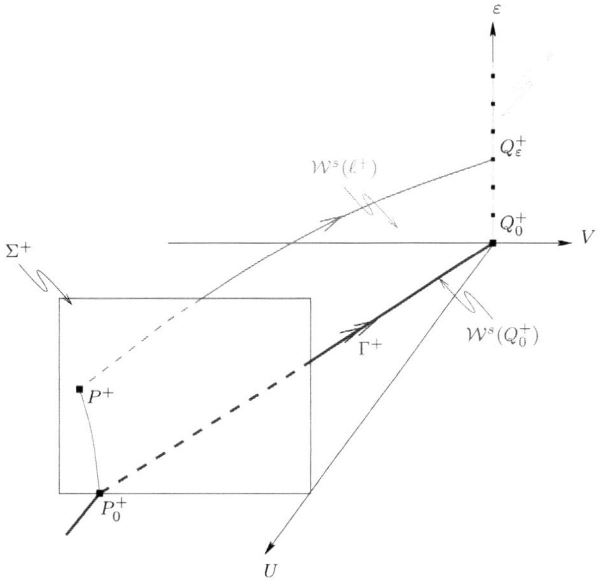

Figure 2. The geometry in the inner region.

Next, we define the section Σ^+ for the flow of (15) via

$$\Sigma^+ = \{(1, V, \varepsilon) \mid (V, \varepsilon) \in [-v_0, 0] \times [0, \varepsilon_0]\}, \tag{17}$$

with $v_0 > 0$ as in (13). Since $u < \varepsilon$ is equivalent to $U < 1$, by (14), Σ^+ naturally separates the inner region, where the dynamics of (8) is governed by the simplified (cut-off) equations in (15), from the outer region, where that dynamics is unaffected by the cut-off, and is described by (9). Correspondingly, Γ^+ gives the portion of the singular orbit Γ that is located in $\{U < 1\}$. The point of intersection of Γ^+ with Σ^+ will be labeled $P_0^+ = (1, v_0^+, 0)$; in particular, we note that $v_0^+ = -c_0$. Finally, we write P^+ for the point of intersection of $\mathcal{W}^s(Q_\varepsilon^+)$ with Σ^+; see figure 2 for a summary of the geometry in this inner region.

2.4. Intermediate region

The dynamics in the intermediate region, where $u = O(\varepsilon)$, is again governed by the first-order system (9) (without cut-off), since $H \equiv 1$ there; recall Section 2.2. Hence, in the corresponding phase-directional coordinates

$$u = u, \quad v = uW, \quad \text{and} \quad \varepsilon = uE, \tag{18}$$

the governing equations are given by

$$u' = uW, \tag{19a}$$
$$W' = -cW - W^2 - (1-u)(u-\gamma), \tag{19b}$$
$$E' = -EW. \tag{19c}$$

Since c evaluates to c_0 in the singular limit as $\varepsilon \to 0^+$ in (9), we find that the two principal equilibria of (19) in this intermediate region are located at $P_- = (0, \lambda_-^+, 0)$ and $P_+ = (0, \lambda_+^+, 0)$, where $\lambda_\mp^+ = -\frac{c_0}{2} \mp \sqrt{\frac{c_0^2}{4} + \gamma}$, as defined in Section 2.2. Since, moreover, the rescaling in (18)

Table 3. The equilibrium point P_- in dependence of γ and c_0.

γ	P_-	$\lambda_-^+,\ -c_0-2\lambda_-^+,\ -\lambda_-^+$	type
-1	$(0,-1,0)$	$-1,\ 0,\ 1$	saddle-node
$(-1,-\frac{1}{2})$	$(0,-\sqrt{-\gamma},0)$	$-\sqrt{-\gamma},\ 0,\ \sqrt{-\gamma}$	saddle-node
$-\frac{1}{2}$	$(0,-\frac{1}{\sqrt{2}},0)$	$-\frac{1}{\sqrt{2}},\ 0,\ \frac{1}{\sqrt{2}}$	saddle-node
$(-\frac{1}{2},0)$	$(0,-\frac{1}{\sqrt{2}},0)$	$-\frac{1}{\sqrt{2}},\ \frac{1}{\sqrt{2}}(1+2\gamma),\ \frac{1}{\sqrt{2}}$	saddle
0	$(0,-\frac{1}{\sqrt{2}},0)$	$-\frac{1}{\sqrt{2}},\ \frac{1}{\sqrt{2}},\ \frac{1}{\sqrt{2}}$	saddle
$(0,\frac{1}{2})$	$(0,-\frac{1}{\sqrt{2}},0)$	$-\frac{1}{\sqrt{2}},\ \frac{1}{\sqrt{2}}(1+2\gamma),\ \frac{1}{\sqrt{2}}$	saddle

Table 4. The equilibrium point P_+ in dependence of γ and c_0.

γ	P_+	$\lambda_+^+,\ -c_0-2\lambda_+^+,\ -\lambda_+^+$	type
-1	$(0,-1,0)$	$-1,\ 0,\ 1$	saddle-node
$(-1,-\frac{1}{2})$	$(0,-\sqrt{-\gamma},0)$	$-\sqrt{-\gamma},\ 0,\ \sqrt{-\gamma}$	saddle-node
$-\frac{1}{2}$	$(0,-\frac{1}{\sqrt{2}},0)$	$-\frac{1}{\sqrt{2}},\ 0,\ \frac{1}{\sqrt{2}}$	saddle-node
$(-\frac{1}{2},0)$	$(0,\sqrt{2}\gamma,0)$	$\sqrt{2}\gamma,\ -\frac{1}{\sqrt{2}}(1+2\gamma),\ -\sqrt{2}\gamma$	saddle
0	$(0,0,0)$	$0,\ -\frac{1}{\sqrt{2}},\ 0$	saddle-node
$(0,\frac{1}{2})$	$(0,\sqrt{2}\gamma,0)$	$\sqrt{2}\gamma,\ -\frac{1}{\sqrt{2}}(1+2\gamma),\ -\sqrt{2}\gamma$	saddle

projectivizes the flow of (9) in a small-u neighborhood of Q_ε^+, as $W=\frac{v}{u}=\frac{u'}{u}$, the points $P_\mp$ correspond to the two eigendirections of the linearization about Q_0^+; recall table 2, where the stability of Q_0^+ is characterized in dependence of $\gamma\in[-1,\frac{1}{2})$.

***Remark* 5.** Other equilibrium points for (19) are found at $(\gamma,0,E)$ and $(1,0,E)$; however, these points, which correspond to the original equilibria Q_ε° and Q_ε^- of (9), respectively, after transformation to phase-directional coordinates, are of no interest to us in this intermediate region. □

Linearizing (19) about the two points $P_\mp$, we find the eigenvalues $\lambda_\mp^+$, $-c_0-2\lambda_\mp^+$, and $-\lambda_\mp^+$, respectively. Consequently, the character of these equilibria again depends substantially on the front propagation regime under consideration, in accordance with the change in stability of Q_0^+ as γ is varied in $[-1,\frac{1}{2})$; see tables 3 and 4 for a summary of the stability properties of P_- and P_+, respectively, in dependence of γ and c_0. The corresponding dynamics in the intermediate region is illustrated in figure 3; in particular, the geometry of the resulting local flow about P_- and P_+ is determined by the ratios of the eigenvalues $\lambda_\mp^-$ and $\lambda_\mp^+$, respectively. As will become clear in the following, the relevant equilbrium point here will always be P_-. We note that, in Sections 3.1 through 3.4 below, we will discuss in detail the various front propagation regimes that are realized in the cut-off Nagumo equation (2), as indicated already in figure 3; the corresponding bifurcation structure of the vector field in (19) will be classified in Section 3.5.

Next, we determine the portion of the singular heteroclinic connection Γ that lies in this intermediate region. To that end, we note that the two hyperplanes $\{u=0\}$ and $\{E=0\}$ are invariant under the flow of (19). Since, in $\{E=0\}$, these equations are equivalent to the first-order system (9), the existence of a singular solution follows from the fact that the

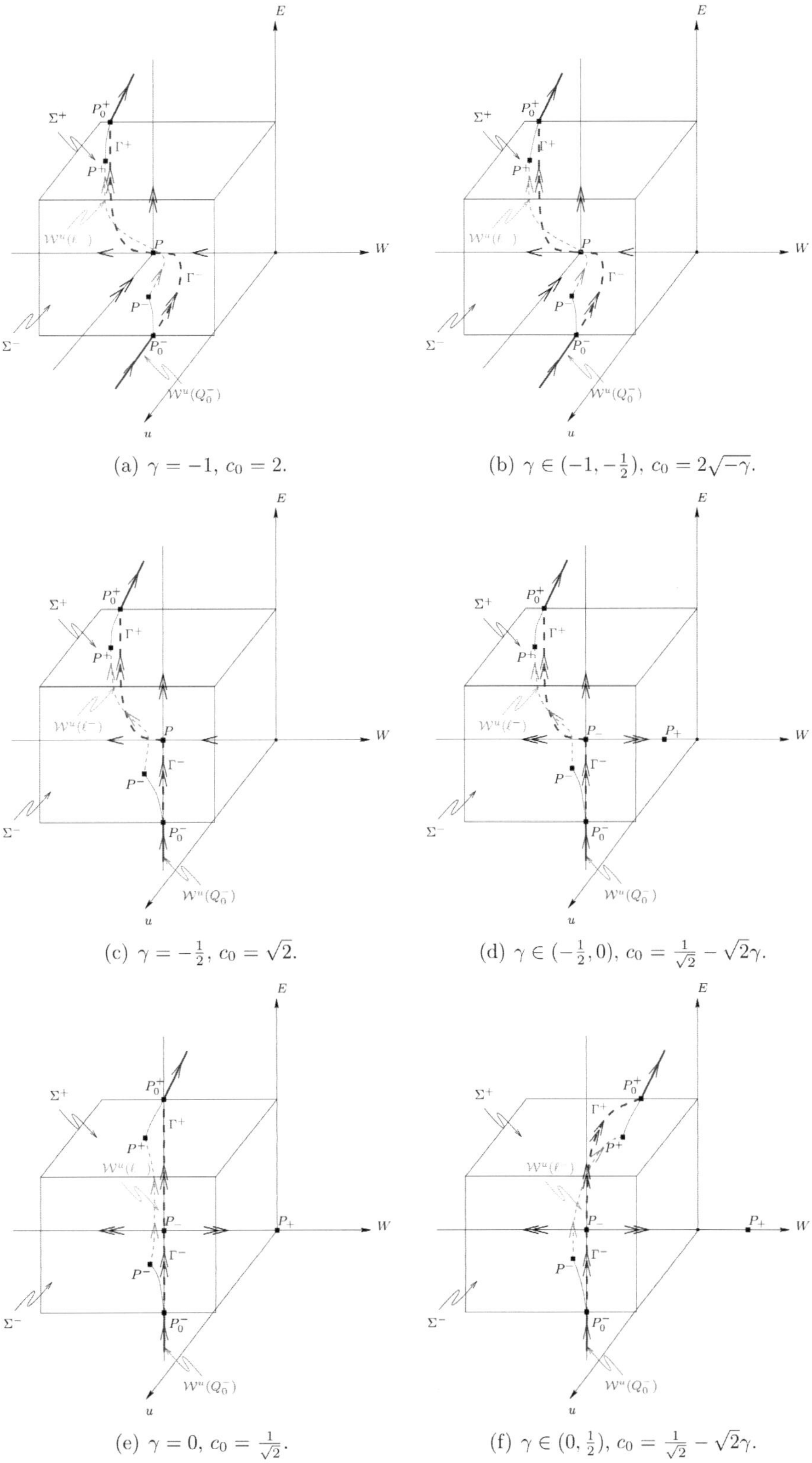

(a) $\gamma = -1$, $c_0 = 2$.

(b) $\gamma \in (-1, -\frac{1}{2})$, $c_0 = 2\sqrt{-\gamma}$.

(c) $\gamma = -\frac{1}{2}$, $c_0 = \sqrt{2}$.

(d) $\gamma \in (-\frac{1}{2}, 0)$, $c_0 = \frac{1}{\sqrt{2}} - \sqrt{2}\gamma$.

(e) $\gamma = 0$, $c_0 = \frac{1}{\sqrt{2}}$.

(f) $\gamma \in (0, \frac{1}{2})$, $c_0 = \frac{1}{\sqrt{2}} - \sqrt{2}\gamma$.

Figure 3. The geometry in the intermediate region in dependence of γ and c_0.

Nagumo equation supports traveling fronts in the absence of a cut-off for any $\gamma \in [-1, \frac{1}{2})$. Correspondingly, the portion of Γ that is located in $\{E = 0\}$ is given by the equivalent of the segment Γ^-, as defined in Section 2.2, in (u, W, E)-coordinates. By contrast, for $u = 0$, we find the singular equation

$$\frac{dW}{dE} = \frac{c_0 W + W^2 - \gamma}{EW}, \tag{20}$$

which is obtained by rewriting (19) with E as the independent variable. In particular, equation (20) is separable and can hence be solved (in implicit form):

$$\ln E - \frac{1}{2}\ln\left|c_0 W + W^2 - \gamma\right| - \frac{c_0}{\sqrt{4\gamma + c_0^2}}\operatorname{arctanh}\left(\frac{2W + c_0}{\sqrt{4\gamma + c_0^2}}\right) \equiv \text{constant.}$$

The corresponding orbit, which we denote by Γ^+, as in Section 2.3, determines the portion of the singular heteroclinic connection Γ that is located in $\{u = 0\}$.

For future reference, we note that the hyperplanes Σ^- and Σ^+, which were introduced in Sections 2.2 and 2.3, respectively, are given by

$$\begin{aligned} \Sigma^- &= \{(\rho, W, E) \,|\, (W, E) \in [-v_0, 0] \times [0, 1]\} \quad \text{and} \\ \Sigma^+ &= \{(u, W, 1) \,|\, (u, W) \in [0, \rho] \times [-v_0, 0]\} \end{aligned} \tag{21}$$

after transformation to phase-directional coordinates, as can be seen from (18) and the change of coordinates between the family chart and the phase-directional chart:

$$u = \varepsilon U, \quad W = \frac{V}{U}, \quad \text{and} \quad E = \frac{1}{U}. \tag{22}$$

(Here, ρ and v_0 are defined as before.) Moreover, we remark that these two sections for the flow of (19) naturally separate the outer region from the intermediate region and the intermediate region from the inner region, respectively. Correspondingly, the respective points of intersection of Γ^- and Γ^+ with Σ^- and Σ^+ are given by the equivalents, in (u, W, E)-coordinates, of P_0^- and P_0^+, respectively, as defined in Sections 2.2 and 2.3.

In sum, the singular heteroclinic connection Γ is therefore given as the union of the orbits Γ^- and Γ^+ and the singularities at Q_0^-, P_-, and Q_0^+, which completes our discussion of the singular limit as $\varepsilon \to 0^+$ in (8).

Finally, we recast the governing equations in this intermediate region in a form that is more convenient for the following analysis of the persistence of Γ: for $\varepsilon \in (0, \varepsilon_0]$, with ε_0 positive and small, we first rewrite (19) in terms of the new variables $Z = W + \frac{1}{2}c_0$ and $c = c_0 + \Delta c$:

$$u' = -(\tfrac{1}{2}c_0 - Z)u, \tag{23a}$$
$$Z' = (\tfrac{1}{2}c_0 - Z)\Delta c - Z^2 + \tfrac{1}{4}c_0^2 + \gamma - (1 + \gamma)u + u^2, \tag{23b}$$
$$E' = (\tfrac{1}{2}c_0 - Z)E. \tag{23c}$$

Then, we transform the resulting vector field by dividing out a factor of $\frac{1}{2}c_0 - Z$ (which is positive in the Z-regime considered here, cf. Section 2.4) from the right-hand sides in (23):

$$u' = -u, \tag{24a}$$
$$Z' = \Delta c - \frac{Z^2 - (\frac{1}{4}c_0^2 + \gamma)}{\frac{1}{2}c_0 - Z} - \frac{(1 + \gamma)u - u^2}{\frac{1}{2}c_0 - Z}, \tag{24b}$$
$$E' = E. \tag{24c}$$

Since this last transformation corresponds to a rescaling of ξ that leaves the phase portrait of (23) unchanged, the prime now denotes differentiation with respect to some new independent variable ζ. The transformed equations in (24) are the starting point for the geometric classification of front propagation in equation (5) that is given in the following section.

3. Propagation regimes

In this section, we classify the propagation regimes (bistable, pushed, and pulled) that are realized in the cut-off Nagumo equation in dependence of the value of the control parameter $\gamma \in [-1, \frac{1}{2})$; then, we discuss the boundary cases that limit these three 'canonical' regimes.

3.1. Bistable regime

For $\gamma \in (0, \frac{1}{2})$, the Nagumo equation (2) (without cut-off) displays classical bistable reaction kinetics. It is well-known [1, 2] that a solution to the corresponding traveling front equation (6) is given by

$$u(\xi) = \frac{1}{1 + \mathrm{e}^{\frac{1}{\sqrt{2}}(\xi - \xi^-)}}, \tag{25}$$

where $\xi^- \geq 0$ denotes an arbitrary phase. That solution propagates between the stable rest states at 1 and 0, with propagation speed $c^\dagger = \frac{1}{\sqrt{2}} - \sqrt{2}\gamma$; the corresponding orbit, in the context of the equivalent first-order system (8), is given by

$$v(u, c^\dagger) = \frac{1}{\sqrt{2}} u(u-1), \tag{26}$$

as can be seen directly from (25). The reader is referred to [15] for details; in particular, we note that the relevant equilibrium point of the 'projectivized' dynamics in the intermediate region, which is governed by (19), is located at $P_- = (0, -\frac{1}{\sqrt{2}}, 0)$, as $\frac{v}{u} = W \to -\frac{1}{\sqrt{2}}$ for $u \to 0^+$, by (18) and (26). (Correspondingly, the singular orbit Γ^-, which is defined by the equivalent, in (u, W, E)-coordinates, of (26), coincides with the stable manifold $\mathcal{W}^s(P_-)$ of P_-.)

The effects of a cut-off in the bistable propagation regime were studied geometrically in [15], where it was proven that the cut-off Nagumo equation supports a traveling front solution for a unique value of c, with $\varepsilon \in (0, \varepsilon_0]$ sufficiently small; cf. [15, Theorem 2.1]. (That value reduces to $c^\dagger$ in the singular limit as $\varepsilon \to 0^+$.) In particular, it was shown there that the correction $\Delta c(\varepsilon) = c - c^\dagger$ to the front propagation speed that is induced by the Heaviside cut-off H satisfies

$$\Delta c = \frac{\Gamma(4)}{\Gamma(1+2\gamma)\Gamma(3-2\gamma)} \frac{\sqrt{2}\gamma}{(1+2\gamma)^{2\gamma}} \varepsilon^{1+2\gamma} + o(\varepsilon^{1+2\gamma}) \equiv K_\gamma \varepsilon^{1+2\gamma}[1 + o(1)], \tag{27}$$

where $\Gamma(\cdot)$ denotes the standard Gamma function [26, Section 6.1].

***Remark* 6.** As observed in [15], the exponent of ε in the leading-order asymptotics of Δc is given by the ratio of two of the eigenvalues of the linearization of (19) at P_-. Specifically, for $c_0 = c^\dagger$, these eigenvalues evaluate to $\frac{1}{\sqrt{2}}(1 + 2\gamma)$ and $\frac{1}{\sqrt{2}}$; recall table 3. □

The proof of these results, as given in [15], relies on a detailed analysis of the transition through the intermediate region under the flow of (24), and is performed entirely in the phase-directional chart. The crucial step in the argument, which is outlined below, consists in approximating the corresponding transition map between the two sections Σ^- and Σ^+; cf. (21).

First, the equations in (24) are simplified via a (smooth, near-identity) normal form transformation [27] of the form $\widehat{Z} = \widehat{Z}(u, Z) = Z + O(u)$, which eliminates any u-dependence from (24b), whereas the dynamics of u and E remains unaffected:

$$\widehat{Z}' = \Delta c - \frac{\widehat{Z}^2 - [\frac{1}{4}(c^\dagger)^2 + \gamma]}{\frac{1}{2}c^\dagger - \widehat{Z}} = \Delta c - \frac{\widehat{Z}^2 - \frac{1}{8}(1+2\gamma)^2}{\frac{1}{2\sqrt{2}}(1-2\gamma) - \widehat{Z}}; \tag{28}$$

here, we have additionally substituted in $c^\dagger = \frac{1}{\sqrt{2}} - \sqrt{2}\gamma$. Next, we recall the definition of P^- and P^+, *i.e.*, of the points of intersection of $\mathcal{W}^u(Q_\varepsilon^-)$ and $\mathcal{W}^s(Q_\varepsilon^+)$ with Σ^- and Σ^+, respectively, from Sections 2.2 and 2.3. The points corresponding to P^- and P^+, after transformation to $(u, \widehat{Z}, E)$-variables, are labeled $\widehat{P}^-$ and $\widehat{P}^+$, respectively.

***Remark* 7.** We emphasize that the normal form transformation defined above removes the u-dependence in (24) to arbitrary order, as there is no resonance in the γ-regime considered here. In particular, the singular orbit Γ^- is rectified by that transformation, and is given by $\widehat{Z}^- = -\frac{1}{2\sqrt{2}}(1+2\gamma)$. □

Equation (28) is separable and can be integrated (in implicit form); the limits of integration in the resulting solution are fixed by the requirement that the point $\widehat{P}^-$ is mapped to $\widehat{P}^+$ in the transition through the intermediate region, *i.e.*, between Σ^- and Σ^+, under the flow of (28). The corresponding ζ-values can be found as follows: since (28) is autonomous, we can, without loss of generality, set $\zeta^- = 0$. To determine $\zeta^+ = -\ln\frac{\varepsilon}{\rho}$, we solve (24a) exactly, noting that u evaluates to ρ and ε in Σ^- and Σ^+, respectively; recall (13) and (17). In sum, we find [15, Proposition 2.2]

$$\begin{aligned} -\ln\frac{\varepsilon}{\rho} - \frac{1}{2}\ln\left|2\widehat{Z}^2 + 2\Delta c\widehat{Z} - \tfrac{1}{\sqrt{2}}(1-2\gamma)\Delta c - \tfrac{1}{4}(1+2\gamma)^2\right|\Big|_{\widehat{Z}^-}^{\widehat{Z}^+} \\ - \frac{\frac{1}{\sqrt{2}}(1-2\gamma)+\Delta c}{\sqrt{\frac{1}{2}(1+2\gamma)^2 + \sqrt{2}(1-2\gamma)\Delta c + (\Delta c)^2}} \\ \times \operatorname{arctanh}\left(\frac{2\widehat{Z}+\Delta c}{\sqrt{\frac{1}{2}(1+2\gamma)^2 + \sqrt{2}(1-2\gamma)\Delta c + (\Delta c)^2}}\right)\Bigg|_{\widehat{Z}^-}^{\widehat{Z}^+} = 0; \quad (29) \end{aligned}$$

by construction, the relation in (29) yields a necessary condition on the parameter $\Delta c (= c - c^\dagger)$ for a connection between the two points $\widehat{P}^-$ and $\widehat{P}^+$ to persist, for $\varepsilon \in (0, \varepsilon_0]$ positive and small.

It remains to obtain estimates for the respective $\widehat{Z}$-coordinates $\widehat{Z}^-$ and $\widehat{Z}^+$ of these points that are sufficiently accurate to the order considered here. To that end, one has to account for the global dynamics in the outer and inner regions, respectively, *i.e.*, one needs to expand the manifolds $\mathcal{W}^u(Q_\varepsilon^-)$ and $\mathcal{W}^s(Q_\varepsilon^+)$ introduced in Sections 2.2 and 2.3 to the corresponding order.

Thus, given $v(u, c^\dagger)$, as defined in (26), the v-coordinate of P^- (in the outer region) satisfies $v^- = -\frac{1}{\sqrt{2}}\rho(\rho-1) + O(\Delta c)$; cf. the expansion in (10). Here, the $O(\Delta c)$-terms can be evaluated exactly due to the fact that a traveling front solution for (5) is known in the absence of a cut-off, as well as that the solution of the variational equation (12) along the corresponding orbit of (9) can also be determined in closed form in this case: with $v(u, c^\dagger)$ as before, that equation reduces to

$$\frac{\partial}{\partial u}\left(\frac{\partial v}{\partial c}(u, c^\dagger)\right) = -1 + 2\frac{u-\gamma}{u(1-u)}\frac{\partial v}{\partial c}(u, c^\dagger). \quad (30)$$

Equation (30) can be solved by variation of constants, which gives

$$\frac{\partial v}{\partial c}(u, c^\dagger) = \frac{1}{3-2\gamma}u^{-2\gamma}(1-u)F(3-2\gamma, -2\gamma; 4-2\gamma; 1-u) \quad (31)$$

for the unique solution satisfying the boundary condition $\frac{\partial v}{\partial c}(1, c^\dagger) = 0$; cf. [15, Lemma 2.2]. (Here, $F(\cdot,\cdot;\cdot;\cdot)$ denotes the hypergeometric function; see *e.g.* [26, Section 15].)

Journal of Physics: Conference Series **268** (2011) 012023 doi:10.1088/1742-6596/268/1/012023

Similarly, in the inner region, the v-coordinate v^+ of P^+ can be determined from the corresponding solution of (15) with $V(0) = 0$: since, clearly, $V(U) = -cU$, it follows that $v(u) = -cu$ and, hence, that $v^+ = -c = -c^\dagger - \Delta c$ after 'blow-down,' *i.e.* in the original (u, v, ε)-variables; cf. (14).

Consequently, after application of the normal form transformation defined above, it follows that $\widehat{Z}^-$ and $\widehat{Z}^+$ can be estimated as

$$\begin{aligned}\widehat{Z}^-(\rho, \Delta c) &= -\frac{1}{2\sqrt{2}}(1+2\gamma) + \frac{1}{\rho^{1+2\gamma}}\frac{\Gamma(1+2\gamma)\Gamma(3-2\gamma)}{\Gamma(4)}[1+o(1)] \quad \text{and} \\ \widehat{Z}^+(\Delta c, \varepsilon) &= -\frac{1}{2\sqrt{2}}(1-2\gamma) - \Delta c[1+o(1)];\end{aligned} \tag{32}$$

see [15, Lemmas 2.2 and 2.4]. (Here, $o(1)$ denotes higher-order terms in ρ, Δc, and ε.)

Substituting (32) into (29), expanding the result in terms of Δc and ε and applying the Implicit Function Theorem at $(\Delta c, \varepsilon) = (0, 0)$, one can show that the (unique) value of $\Delta c = \Delta c(\varepsilon, \gamma)$ for which Γ persists is indeed well-defined; moreover, the persistence proof sketched above also yields the leading-order ε-asymptotics of Δc, cf. (27), as can be seen by solving the corresponding relation that is obtained from (29) to leading order. Finally, since the persistent heteroclinic connection between Q_ε^- and Q_ε^+ that is realized for $c = c^\dagger + \Delta c(\varepsilon)$ in (8) corresponds to a traveling front solution of the cut-off Nagumo equation (5), the existence and uniqueness of such fronts in this bistable regime follows.

3.2. Pushed regime

When $\gamma \in (-\frac{1}{2}, 0)$, equation (2) supports pushed fronts in the absence of a cut-off, as outlined in the Introduction. The corresponding propagation speed is again given by $c^\dagger = \frac{1}{\sqrt{2}} - \sqrt{2}\gamma$ [2], as in the bistable regime considered in Section 3.1. (We note that, for $\gamma \in (-\frac{1}{2}, 0)$, that speed is higher than the speed $c^* = 2\sqrt{-\gamma}$ that is predicted by linearization at the zero rest state, in accordance with the definition of pushed traveling fronts [3].)

Results obtained previously in this pushed regime [4], which relied on a modification of the integral variational principle developed by Benguria and Depassier [3], indicate that the correction to $c^\dagger$ that is due to the Heaviside cut-off H in (6) is again given by (27), as was the case in the bistable regime.

This expectation is confirmed by our geometric analysis of the first-order system (8) for $\gamma \in (-\frac{1}{2}, 0)$; a rigorous, in-depth discussion of pushed front propagation in the cut-off Nagumo equation will be given in the upcoming article [16]. In particular, it is shown there that the existence and uniqueness analysis developed in [15], for the bistable regime, remains mostly valid in the pushed regime, *i.e.*, that only minor modifications are required. Thus, the singular orbit Γ can be constructed exactly as before; moreover, the proof of persistence of that orbit for $\varepsilon \in (0, \varepsilon_0]$ positive and small remains valid when $\gamma \in (-\frac{1}{2}, 0)$. Specifically, the Z-dynamics in the intermediate region is still governed by the normal form equation (28); see Section 2.4 above. (In particular, we remark that the eigenvalues of the linearization about the relevant equilibrium point P_- are again not in resonance.) Correspondingly, the argument presented in [15, Section 3], and outlined in Section 3.1, carries over almost verbatim to the pushed regime.

Finally, we remark that the correction Δc to $c^\dagger$ that is due to the cut-off H in (5) is negative in this regime, as is evident from (27) with $\gamma \in (-\frac{1}{2}, 0)$, which contrasts with the bistable regime and is in agreement with the general observation made in the Introduction that cut-offs reduce the propagation speed of pushed fronts; see also [15, 16, 20, 21].

3.3. Pulled regime

In the pulled regime, with $\gamma \in (-1, -\frac{1}{2})$, the propagation speed of traveling fronts in the absence of a cut-off is determined by linearization of equation (2) at the zero rest state: $c^* = 2\sqrt{-\gamma}$.

To find the correction $\Delta c = c - c^*$ that is induced by a Heaviside cut-off in (5) in this regime, one could proceed as in Sections 3.1 and 3.2. In particular, the dynamics in the outer region is again governed by the corresponding traveling front equation without cut-off, or, equivalently, by the first-order system (9); in contrast to the bistable and pushed regimes, however, no exact expression seems to be available for the traveling front solution that propagates between the rest states at 1 and 0 in this case. The governing equations in the inner region, where the flow of (8) is cut-off, are given by (15), as before, since the corresponding reaction terms are canceled by the Heaviside cut-off H. In the intermediate region, equation (19b) reduces to

$$Z' = \Delta c - \frac{Z^2}{\sqrt{-\gamma} - Z} - \frac{(1+\gamma)u - u^2}{\sqrt{-\gamma} - Z}, \tag{33}$$

after transformation to the new variables $Z = W + \sqrt{-\gamma}(= W + \frac{1}{2}c^*)$ and $c = 2\sqrt{-\gamma} + \Delta c(= c^* + \Delta c)$. Here, we note that there exists only one relevant equilibrium point in this pulled regime, since $P_- = P_+$ for $c = c^*$ in (19). We denote that point by P, and we remark that P corresponds to the one eigendirection of the linearization of (9) at Q_0^+, which is a degenerate node now; recall table 2. Finally, the relevant normal form equation is given by $\widehat{Z}' = \Delta c - \frac{\widehat{Z}^2}{\sqrt{-\gamma} - \widehat{Z}}$; see also [6, Equation (34b)]. (In particular, comparing this equation to the general normal form in (28), one finds that the constant term cancels.) The associated transition map Π from Σ^- to Σ^+ can then be approximated to leading order, as was done in the proof of [6, Proposition 3.2].

***Remark* 8.** As the eigenvalues of the linearization of (19) about P are given by $-\sqrt{-\gamma}$, 0, and $\sqrt{-\gamma}$ in this regime, recall tables 3 and 4, they are clearly in resonance. However, since none of the u-dependent terms in (33) are resonant, those terms can still be eliminated by a standard normal form transformation, as in [6, Proposition 3.2]. □

Instead of studying the pulled propagation regime in (2) directly, as outlined above, we take an alternative approach here: we show that, for $\gamma \in (-1, -\frac{1}{2})$, the cut-off Nagumo equation (5) is equivalent to an FKPP-type equation, which was analyzed geometrically in [6]. In particular, for the FKPP equation proper, with $f(\phi) = \phi(1-\phi)$ in (1), it was proven there that the correction Δc to the front propagation speed c^* that is due to a cut-off is given by

$$\Delta c(\varepsilon) = -\frac{\pi^2}{(\ln \varepsilon)^2} + o[(\ln \varepsilon)^{-2}] \equiv \frac{K_{-1}}{(\ln \varepsilon)^2}[1 + o(1)], \tag{34}$$

in accordance with a conjecture made previously by Brunet and Derrida [5]; see the Introduction and [6] for details. Moreover, in Section 4 of that article, it was shown that the leading-order asymptotics in (34) remains valid for more general cut-off equations of the type

$$\frac{\partial \phi}{\partial t} = \frac{\partial^2 \phi}{\partial x^2} + [\phi - g(\phi)]H(\phi - \varepsilon), \tag{35}$$

where H is as defined in (4) and the function g satisfies the following assumptions:

(i) $g(\phi) = O(\phi^2)$ as $\phi \to 0^+$,
(ii) there exists a $\phi^- > 0$ such that $g(\phi^-) = \phi^-$,
(iii) $g'(\phi^-) > 1$, and
(iv) $0 < g(\phi) < \phi$ for all $\phi \in (0, \phi^-)$.

(The above conditions ensure that zero is again a rest state of the traveling front equation $u'' + cu' + [u - g(u)]H(u - \varepsilon) = 0$ that is associated with (35), as well as that the rest state at ϕ^- corresponds to a hyperbolic saddle equilibrium of that same equation. In particular, the condition in (iv) implies that there exist no other equilibria, as well as that a trapping region can be constructed for the flow of the corresponding vector field in the intermediate region.)

The following is the main result of this section (and, indeed, of this article):

Proposition 1. *For $\gamma \in (-1, -\frac{1}{2})$ and $\varepsilon \in (0, \varepsilon_0]$, equation (5) supports a unique traveling front solution propagating between the rest states at 1 and 0. Moreover, the corresponding propagation speed c can be written as $c = c^* + \Delta c(\varepsilon)$, where $c^* = 2\sqrt{-\gamma}$ and*

$$\Delta c(\varepsilon) = -\frac{\pi^2\sqrt{-\gamma}}{(\ln\varepsilon)^2} + o[(\ln\varepsilon)^{-2}] \equiv \frac{K_\gamma}{(\ln\varepsilon)^2}[1 + o(1)]. \tag{36}$$

To transform the cut-off Nagumo equation (5) into the framework of [6, Section 4], we first rewrite the corresponding traveling front equation (6) as

$$u'' + cu' - \gamma\Big(u - \frac{\gamma+1}{\gamma}u^2 + \frac{1}{\gamma}u^3\Big)H(u-\varepsilon) = 0; \tag{37}$$

then, we introduce the new variable $\Xi = \sqrt{-\gamma}\xi$ in (37). Finally, rescaling the front propagation speed c via $c = \sqrt{-\gamma}C$, we obtain

$$\ddot{u} + C\dot{u} + \Big[u - \Big(\frac{\gamma+1}{\gamma}u^2 - \frac{1}{\gamma}u^3\Big)\Big]H(u-\varepsilon) = 0, \tag{38}$$

where overdots denote differentiation with respect to Ξ.

***Remark* 9.** Alternatively, instead of rescaling ξ, we could have rescaled time and space in the original reaction-diffusion equation (5) via $T = -\gamma t$ and $X = \sqrt{-\gamma}x$, which would have given $\Xi = X - CT = \sqrt{-\gamma}(x - \sqrt{-\gamma}Ct) = \sqrt{-\gamma}\xi$, as before. □

Defining $g(u) = \frac{\gamma+1}{\gamma}u^2 - \frac{1}{\gamma}u^3$, one sees that the transformed equation (37) satisfies the conditions in (i), (ii), and (iii) above, with $\phi^- = 1$. Hence, it only remains to verify the condition in (iv): however, while $g(u) < u$ is certainly satisfied, *i.e.*, while (37) has no rest states in the (open) interval $(0, 1)$, as the roots of $g(u) = u$ are located at γ, 0, and 1, the condition that g is non-negative on that interval does not hold for $\gamma > -1$. The latter condition is imposed in [6, Section 4] to ensure the existence of a trapping region for the flow in the intermediate region which, in turn, guarantees the existence of a singular heteroclinic connection Γ in that region. Nevertheless, we can construct such a trapping region explicitly here; the corresponding argument is analogous to that given in the proof of [6, Lemma 2.5], and is based on a phase plane analysis of the first-order system

$$\dot{u} = v, \tag{39a}$$

$$\dot{v} = -2v - \Big[u - \Big(\frac{\gamma+1}{\gamma}u^2 - \frac{1}{\gamma}u^3\Big)\Big] \tag{39b}$$

that is obtained from (38) for $\varepsilon = 0$, *i.e.*, in the absence of a cut-off. (Here, we note that the singular front propagation speed C_0 is given by $C^* = 2$ now.)

Lemma 1. *For $\gamma \in (-1, -\frac{1}{2})$ and $u \in (0, \rho)$, with $\rho \geq \varepsilon_0$ as in (21), the flow of (39) is trapped in the wedge that is bounded by $\{v = 0\}$ and $\{v = -u - \frac{\gamma+1}{\gamma}u^2\}$.*

Proof. Evaluating the right-hand sides in (39) for $v = 0$, we find

$$(0,1)\cdot\Big(0, -u + \frac{\gamma+1}{\gamma}u^2 - \frac{1}{\gamma}u^3\Big)^T = -u\Big(1 - \frac{\gamma+1}{\gamma}u + \frac{1}{\gamma}u^2\Big),$$

which is negative when $\gamma \in (-1, -\frac{1}{2})$ and $u \in (0, 1)$. Similarly, for $v = -u - \frac{\gamma+1}{\gamma}u^2$, the corresponding expression is given by

$$\Big(1 + 2\frac{\gamma+1}{\gamma}u, 1\Big)\cdot\Big(-u - \frac{\gamma+1}{\gamma}u^2, u + 3\frac{\gamma+1}{\gamma}u^2 - \frac{1}{\gamma}u^3\Big)^T = -u^3\Big[2\Big(\frac{\gamma+1}{\gamma}\Big)^2 + \frac{1}{\gamma}\Big], \tag{40}$$

which is positive in that same (γ, u)-regime. Hence, a trapping region for (39) is defined by $\{v = 0\}$ and $\{v = -u - \frac{\gamma+1}{\gamma}u^2\}$.

It remains to show that the flow of (39) actually enters the region that is bounded by these two curves and, hence, that it has to remain trapped there. Now, the origin Q_0^+ is a degenerate node for (39), with a strong stable manifold $\mathcal{W}^{ss}(Q_0^+)$ that satisfies $v(u) = -u - \frac{\gamma+1}{\gamma}u^2 + \left[\frac{1}{2\gamma} + \left(\frac{\gamma+1}{\gamma}\right)^2\right]u^3 + O(u^4)$. Denoting the point of intersection of that manifold with the section Σ^- defined in (21) by P_0^{ss} and taking into account that $\left[\frac{1}{2\gamma} + \left(\frac{\gamma+1}{\gamma}\right)^2\right]$ is negative for $\gamma \in (-1, -\frac{1}{2})$, we find $v_0^{ss} < -[1 + \frac{\gamma+1}{\gamma}\rho]\rho$ for the v-coordinate of P_0^{ss}. Since $\mathcal{W}^{ss}(Q_0^+)$ coincides with the stable manifold of the equilibrium point $P = (0, -1, 0)$ for

$$\dot{u} = uW, \tag{41a}$$

$$\dot{W} = -2W - W^2 - \left[1 - \left(\frac{\gamma+1}{\gamma}u - \frac{1}{\gamma}u^2\right)\right], \tag{41b}$$

$$\dot{E} = -EW, \tag{41c}$$

after transformation of (39) to (u, W, E)-coordinates, we conclude as in [6, Lemma 2.5] that the segment Γ^- of the singular heteroclinic orbit Γ, which is located in the outer region and which corresponds precisely to the unstable manifold $\mathcal{W}^u(Q_0^-)$ of Q_0^-, has to enter the trapping region constructed above, which completes the proof. □

***Remark* 10.** The definition of the trapping region in Lemma 1 is motivated by the fact that the linearization of (41) in the (u, W)-plane about P has eigenvectors $(0, 1, 0)^T$ and $(1, -\frac{\gamma+1}{\gamma}, 0)^T$, from which the curves bounding that region can easily be obtained. □

***Remark* 11.** The proof of Lemma 1 implies, in particular, that the singular orbit Γ^- must be tangent to the W-axis as $\Gamma^- \to P$ in the intermediate region; see also [6, Lemma 2.5]. Specifically, for $E = 0$, the flow of (41) is trapped in the wedge that is bounded by the u-axis $\{W = 0\}$ and the curve defined by $\{W = -1 - \frac{\gamma+1}{\gamma}u\}$. Moreover, when $\gamma \in (-1, -\frac{1}{2})$, the equivalents W_0^{ss} and W_0^- of v_0^{ss} and v_0^- (the W-coordinates of the points of intersection of $\mathcal{W}^{ss}(Q_0^+)$ and Γ^-, respectively, with Σ^-) must satisfy $W_0^- > W_0^{ss}$. Since the relevant eigenvalues of the linearization of (41) at P are given by -1 and 0, the statement follows. □

In sum, we conclude that the results of [6, Section 4] can be applied to the rescaled Nagumo equation (5) (with cut-off) in this pulled propagation regime. In particular, it follows that, for $\gamma \in (-1, -\frac{1}{2})$ and $\varepsilon \in (0, \varepsilon_0]$ sufficiently small, equation (38) supports a unique traveling front solution that propagates between the rest states at 1 and 0, with speed

$$C = C^* + \Delta C(\varepsilon) = 2 - \frac{\pi^2}{(\ln \varepsilon)^2} + o[(\ln \varepsilon)^{-2}].$$

Recalling $c = \sqrt{-\gamma}C$, we obtain (36), as claimed, which completes the proof of Proposition 1.

3.4. Boundary cases

In this section, we study the cut-off Nagumo equation (5) for the three values -1, $-\frac{1}{2}$, and 0 of γ that bound the open intervals $(-1, -\frac{1}{2})$, $(-\frac{1}{2}, 0)$, and $(0, \frac{1}{2})$ corresponding to the pulled, pushed, and bistable front propagation regimes, respectively, discussed in Sections 3.1 through 3.3 above. In particular, we show that the dynamics of (8) in one of these boundary cases is complicated by the occurrence of resonances in the rescaled equations that describe the dynamics in the intermediate region, *i.e.*, in phase-directional coordinates.

3.4.1. Case 1: $\gamma = -1$ For $\gamma = -1$, equation (5) corresponds precisely to the classical Kolmogorov-Petrowskii-Piscounov (KPP) equation discussed in [6], with reaction terms $f(\phi) = \phi(1-\phi^2)$. (Alternatively, that equation goes by the name Newell-Whitehead equation [22].) In particular, the trapping region constructed in Lemma 1 coincides with the corresponding region that was found in [6, Lemma 2.5] in this case. Hence, the result of Proposition 1 remains valid, and is an immediate consequence of [6, Theorem 1.1] now; correspondingly, the correction Δc to the linear spreading speed c^* that is due to the cut-off is again given by (34).

3.4.2. Case 2: $\gamma = -\frac{1}{2}$ The case where $\gamma = -\frac{1}{2}$ separates the pushed and pulled propagation regimes, and can be analyzed via an adaptation of the analysis developed in [6] and [15]; recall our discussion of the corresponding regimes in Sections 3.2 and 3.3 above.

As before, the conditions in (i), (ii), and (iii) imposed in that section will hold in this case, as well, whereas (iv) will again fail. However, the proof of Lemma 1 will no longer yield a trapping region for the flow of the transformed first-order system (39), as the expression in (40) will evaluate to zero when $\gamma = -\frac{1}{2}$. Hence, the results of [6, Section 4] are not applicable here. Nevertheless, we can still prove an analogue of Proposition 1, as follows: first, we note that the unique connection between Q_0^- and Q_0^+ in (9), in the absence of a cut-off, can be found in closed form, with $v(u, \sqrt{2})$ as defined in (26). (The front propagation speed is given by $c_0 = \sqrt{2}$ now; in particular, the linear spreading speed $c^* = 2\sqrt{-\gamma}$ equals $c^\dagger = \frac{1}{\sqrt{2}} - \sqrt{2}\gamma$ when $\gamma = -\frac{1}{2}$.) Hence, the singular orbit Γ^- is again known explicitly, and is given by $W(u) = \frac{1}{\sqrt{2}}(u-1)$, after transformation to (u, W, E)-coordinates; recall (18). Since the singular equations that are obtained from (19) for $u = 0$, *i.e.*, in the (W, E)-plane, can be solved exactly, as in [6, Section 2.1], one can show that the segment of Γ^+ that is located in this intermediate region can be found in terms of the so-called Lambert W-function [28]. In sum, it follows that a singular heteroclinic Γ can be constructed as in [6, Section 2.1].

Therefore, it only remains to prove that Γ will persist as a heteroclinic connection between Q_ε^- and Q_ε^+, for ε positive and sufficiently small and a unique value of c in (8). The required persistence argument follows the discussion in Sections 3.1 and 3.2, and will only be outlined here; cf. also the proof of [6, Proposition 3.1].

First, we note that the normal form equation for $\widehat{Z}$ and the resulting (implicit-form) solution are again given by (28) and (29), respectively, with γ there replaced by $-\frac{1}{2}$ here. (Thus, the equations governing the dynamics in the intermediate region, as found in the bistable and pushed regimes, are still valid in this boundary case.) Similarly, the solution of the variational equation (30) can be determined in closed form, as before:

$$\frac{\partial v}{\partial c}(u, \sqrt{2}) = \frac{u}{(u-1)^3}\Big(-\frac{1}{3}u^3 + \frac{3}{2}u^2 - 3u + \ln u + \frac{11}{6}\Big) = u\Big(-\frac{11}{6} - \ln u\Big) + O(u^2 \ln u).$$

Correspondingly, estimates for the respective $\widehat{Z}$-coordinates of $\widehat{P}^-$ and $\widehat{P}^+$ are given by

$$\widehat{Z}^-(\rho, \Delta c) = -\frac{11}{6}\Delta c[1 + o(1)] \quad \text{and} \quad \widehat{Z}^+(\Delta c, \varepsilon) = -\frac{1}{\sqrt{2}} - \Delta c[1 + o(1)] \tag{42}$$

now; recall the discussion in Section 3.1 and, in particular, (32). (Here, $o(1)$ denotes higher-order terms in ρ, Δc, and ε, as above.) Substituting (42) into (29), expanding the result in terms of Δc and ε, recalling that $\zeta^+ = -\ln\frac{\varepsilon}{\rho}$, and taking into account that Δc is once again negative, we have

$$-\ln\frac{\varepsilon}{\rho} - \frac{1}{2}\ln\big|-1 + o(1)\big| + \frac{1}{2}\ln\big|\sqrt{2}\Delta c[1 + o(1)]\big| \\ - \frac{1}{\sqrt[4]{2}\sqrt{-\Delta c}}\Big\{\frac{\pi}{2} - \frac{7}{3}\sqrt[4]{2}\sqrt{-\Delta c}[1 + o(1)]\Big\} = 0. \tag{43}$$

Retaining only the leading-order terms in (43), we find that Δc must necessarily satisfy

$$\sqrt{-\Delta c}\big(-\ln\varepsilon+\ln\sqrt{-\Delta c}\big)=-\frac{\pi}{2\sqrt[4]{2}}[1+o(1)].$$

Solving this relation to lowest order in Δc and ε and noting that the solution can again be written in terms of the Lambert W-function, we obtain $\sqrt{-\Delta c}=\frac{\pi}{2\sqrt[4]{2}}(\ln\varepsilon)^{-1}[1+o(1)]$. In sum, it follows that the result of Proposition 1 carries over almost verbatim to this case; however, while the leading-order asymptotics of $\Delta c(\varepsilon)=c-c^*$ is again logarithmic in ε, as in (36), the corresponding coefficient differs by a factor of $\frac{1}{4}$ now:

$$\Delta c(\varepsilon)=-\frac{\pi^2}{4\sqrt{2}(\ln\varepsilon)^2}+o[(\ln\varepsilon)^{-2}]\equiv\frac{K_{-\frac{1}{2}}}{(\ln\varepsilon)^2}[1+o(1)]. \tag{44}$$

***Remark* 12.** The discrepancy between the leading-order asymptotics of Δc, as stated in (44), and the expression obtained by evaluating (36) for $\gamma\to-\frac{1}{2}^-$ is due to $\widehat{Z}^-=O(\Delta c)$; recall (42). Since the argument in the proof of [6, Proposition 3.1] assumes $\widehat{Z}^-=O(1)$ (in our notation), the results of [6, Section 4] no longer apply when $\gamma=-\frac{1}{2}$. □

3.4.3. Case 3: $\gamma=0$ For $\gamma=0$, equation (2) reduces to the so-called Zeldovich equation, which is also known as the Huxley equation, see *e.g.* [1] or [22]. The corresponding cut-off equation was analyzed in detail in [20], where it was shown that the correction Δc to the front propagation speed $c_0=\frac{1}{\sqrt{2}}$ that is induced by a Heaviside cut-off is given by

$$\Delta c(\varepsilon)=-\frac{3}{\sqrt{2}}\varepsilon^2+o(\varepsilon^2)\equiv K_0\varepsilon^2[1+o(1)], \tag{45}$$

to leading order in ε. The proof of (45) is based on the same geometric framework that was introduced in Section 2 above; however, the analysis of the flow in the intermediate region is complicated by the occurrence of resonances between the eigenvalues of the linearization of the vector field in (19) at P_-. Specifically, it follows from standard theory [27] that the corresponding u-dependent terms are resonant; hence, they cannot be removed via a normal form transformation, as was the case in the canonical propagation regimes (pulled, pushed, and bistable) discussed above; rather, these terms give rise to logarithmic ('switchback') terms (in ε) in the transition through the intermediate region. Correspondingly, the expansion for Δc in (45) will contain an $O(\varepsilon^4\ln\varepsilon)$-term, as well as higher-order terms of the form $\varepsilon^j(\ln\varepsilon)^k$, with $j\geq k$ and $j\geq 4$ [20]. In particular, while the restriction of the unstable manifold $\mathcal{W}^u(Q_\varepsilon^-)$ of Q_ε^- to the outer region is analytic in c and ε, the logarithmic dependence of Δc on ε gives rise to switchback terms in the expansion for $v(u,c)$ in (10), resulting in a loss of smoothness when that expansion is considered in terms of ε alone; see [20, Section 5] for details. For a more general discussion of logarithmic switchback from a geometric point of view, the reader is referred to [29].

3.5. Classification

Given our discussion of the various propagation regimes that can be realized in the cut-off Nagumo equation (5) in dependence of γ, see Sections 3.1 through 3.4, we now combine the asymptotics obtained separately in these regimes to arrive at a comprehensive geometric classification of the global front propagation dynamics of (5). In fact, the equivalence of the traveling wave equation (6) and the first-order system (8) allows for such a classification in terms of the local flow about the origin Q_0^+: in the absence of a cut-off, that point is a hyperbolic saddle equilibrium for (8) in the bistable propagation regime, when $\gamma \in (0, \frac{1}{2})$, while it becomes a proper (stable) node if $\gamma \in (-\frac{1}{2}, 0)$ and a degenerate stable node for $\gamma \in [-1, -\frac{1}{2}]$, *i.e.*, in the pushed and pulled regimes, respectively; cf. table 2. (In particular, we find that Q_0^+ undergoes a saddle-node bifurcation at $\gamma = 0$.)

This transition from a hyperbolic saddle via a regular node to a degenerate node is reflected in the dynamics of the governing equations that are obtained from (8) in phase-directional coordinates (after blow-up). Thus, a crucial step in our analysis consists in the accurate description of the flow in the corresponding intermediate region. (Here, we recall that the cut-off equations in (8) reduce to (9) in that region, as the dynamics is unaffected by the cut-off there.) In particular, it follows that front propagation in the cut-off Nagumo equation is classifiable in terms of the normal form system that corresponds to the equations in (19), after removal of any non-resonant (higher-order) terms. The resulting dynamics is determined by the parameter γ and the front propagation speed c_0 that is realized in the absence of a cut-off, as illustrated in figure 3. Recalling that the equilibrium points $P_- = (0, \lambda_-^+, 0)$ and $P_+ = (0, \lambda_+^+, 0)$ for (19) correspond to the two eigendirections of the linearization of the equivalent first-order system (9) about Q_0^+, see Section 2.4, we can classify the flow in the intermediate region as follows.

When $\gamma \in (0, \frac{1}{2})$, *i.e.*, when Q_0^+ is a hyperbolic saddle, the unique decaying mode of that linearization is given by P_-, since $\lambda_-^+ < 0 < \lambda_+^+$; similarly, for $\gamma \in (-\frac{1}{2}, 0)$, the stronger decay rate at the node Q_0^+ corresponds to P_-, as $\lambda_-^+ < \lambda_+^+ < 0$ then. The relevant front propagation speed $c^\dagger$ exceeds the linear spreading speed c^* in these two regimes. Finally, when $\gamma = 0$, P_+ lies at the origin, as λ_+^+ evaluates to zero; hence, P_- corresponds to the unique strongly decaying mode at the saddle-node Q_0^+, while the front propagation speed is still given by $c^\dagger$.

Now, if $c^\dagger$ remained the selected speed for $\gamma \in [-1, -\frac{1}{2})$, the points P_- and P_+ would pass through each other and exchange position, *i.e.*, the stronger rate of decay would be given by P_+ in that scenario. Since, however, the Z-direction would then become attracting for P_-, cf. table 3, while the Z-coordinates of P^- and P^+ (the points of intersection of $\mathcal{W}^u(Q_\varepsilon^-)$ and $\mathcal{W}^s(Q_\varepsilon^+)$ with Σ^- and Σ^+, respectively) would still satisfy $Z^- > Z^+$, recall Section 3.1 and, in particular, (32), it would follow by continuity that no connection between P^- and P^+ would be possible. (A similar argument can be found in [25], where the correction Δc in the bistable Schlögl equation [1] was shown to be negative, via phase space considerations.)

Consequently, the cut-off traveling front equation (6) would have no solution for any $\varepsilon \in (0, \varepsilon_0]$ if $c^\dagger$ was the relevant front propagation speed when $\gamma \in [-1, -\frac{1}{2})$; rather, the traveling front must select the lower speed c^* then. (To state it differently, solvability of (6) is maintained by adjusting the propagation speed so that the dynamics in this intermediate region is 'frozen' at a degenerate node, with $P_- = P_+$.) For completeness, we note that $c^\dagger$ equals c^* when $\gamma = -\frac{1}{2}$, *i.e.*, that no distinction has to be made between the two in that case; correspondingly, the two equilibria P_- and P_+ collapse onto one point P, as the eigenvalues λ_-^+ and λ_+^+ coalesce.

A distinguishing characteristic of the dynamics in the intermediate region is the potential occurrence of resonances among the eigenvalues of the linearization of (19) about P_-. In particular, these resonances are realized in the pulled propagation regime, as zero is always an eigenvalue, while the remaining two eigenvalues are negatives of each other. Correspondingly, the ε-asymptotics of Δc contains logarithmic switchback terms [29, 30] of the form $(\ln \varepsilon)^{-k}$, with $k \geq 2$, in that regime; see also [31, 32], where such terms were reported in the context of the so-called Lagerstrom model problem from fluid dynamics [30]. However, the corresponding normal form equations are simpler than those found in the boundary cases where $\gamma = 0$ [20] or $\gamma = \frac{1}{2}$ [21], see Section 3.4.3, and are in fact explicitly integrable; recall Section 3.3. Finally, we remark that no resonances occur in the bistable and pushed propagation regimes. Hence, any higher-order terms can be removed from the relevant normal form equations, as discussed in Sections 3.1 and 3.2; moreover, Δc is regular in ε, *i.e.*, no logarithmic switchback can occur

Table 5. Correction Δc to c_0 due to a cut-off in (5) in dependence of γ.

γ	regime	c_0	Δc	K_γ	sign
-1	pulled	2	$O[(\ln \varepsilon)^{-2}]$	$-\pi^2$	negative
$(-1, -\frac{1}{2})$	pulled	$2\sqrt{-\gamma}$	$O[(\ln \varepsilon)^{-2}]$	$-\pi^2\sqrt{-\gamma}$	negative
$-\frac{1}{2}$	pulled	$\sqrt{2}$	$O[(\ln \varepsilon)^{-2}]$	$-\frac{1}{4\sqrt{2}}\pi^2$	negative
$(-\frac{1}{2}, 0)$	pushed	$\frac{1}{\sqrt{2}} - \sqrt{2}\gamma$	$O(\varepsilon^{1+2\gamma})$	$\frac{\Gamma(4)}{\Gamma(1+2\gamma)\Gamma(3-2\gamma)}\frac{\sqrt{2}\gamma}{(1+2\gamma)^{2\gamma}}$	negative
0	pushed	$\frac{1}{\sqrt{2}}$	$O(\varepsilon^2)$	$-\frac{3}{\sqrt{2}}$	negative
$(0, \frac{1}{2})$	bistable	$\frac{1}{\sqrt{2}} - \sqrt{2}\gamma$	$O(\varepsilon^{1+2\gamma})$	$\frac{\Gamma(4)}{\Gamma(1+2\gamma)\Gamma(3-2\gamma)}\frac{\sqrt{2}\gamma}{(1+2\gamma)^{2\gamma}}$	positive

in those regimes.

The transition between the various propagation regimes in dependence of γ is reflected in the leading-order asymptotics of Δc, as observed already in [15, 16] as well as in [4, 10, 17]. Thus, for instance, the power of ε in the expansion in (27) tends to zero as $\gamma \to -\frac{1}{2}^+$. Correspondingly, the leading-order coefficient K_γ is no longer given as in (27) in that limit, since one can show that the estimate for $\widehat{Z}^-$ in (32) breaks down; the correct expression for $K_{-\frac{1}{2}}$ can be found in (45). Similarly, K_γ vanishes as $\gamma \to 0$, where the limit can be taken either from the left or from the right. Hence, the leading term in the ε-asymptotics of Δc in (27) is of the order $O(\varepsilon^2)$ in that case; see table 5 for a summary of the asymptotic structure of Δc for $\gamma \in [-1, \frac{1}{2})$. In general, we remark that Δc becomes markedly more singular in the transition between the pushed and pulled regimes, as an expansion in (fractional) powers of ε is replaced by a logarithmic ε-dependence. Correspondingly, the correction Δc to the front propagation speed that is due to a cut-off is weaker in the bistable [15] and pushed propagation regimes [16] (sublinear or superlinear in the cut-off parameter ε, respectively) than it is in the pulled regime [6]. (This distinction is again obvious from the dynamics in the intermediate region: since the Z-direction is strongly repelling for P_- in the former two regimes, while it is merely weakly repelling in the latter, Δc has to be asymptotically weaker if a connection between P^- and P^+ is to be achieved for $\varepsilon \in (0, \varepsilon_0]$.) Moreover, our analysis confirms that the correction Δc is negative for $\gamma \in [-1, 0]$, *i.e.*, when the reaction function f in (5) is of type I and the rest state at 0 is unstable, whereas it is positive when f is of type II or, equivalently, when $\gamma \in (0, \frac{1}{2})$ and the zero state is (meta)stable, as observed previously [3]. In fact, the sign of Δc can be inferred directly from the dynamics in the intermediate region, as found already in [25]: since the W-coordinates of P_- and P_0^+ are given by λ_-^+ and $-c_0$, respectively, it follows from table 5 in combination with figure 3 that Δc is positive (negative) for $\lambda_-^+ < -c_0$ ($\lambda_-^+ \geq -c_0$). Finally, we remark that our predictions are supported by numerical simulations of the cut-off traveling front equation (6) (data not shown), which indicate, in particular, that satisfactory agreement between c and its first-order truncation in (36) can only be expected for very small values of ε; see also [33], where bounds derived for c, in the context of the cut-off KPP equation, were shown to be valid merely up to $\varepsilon = o(10^{-7})$.

Next, we emphasize that the determination of the leading-order ε-asymptotics of Δc in the bistable and pushed propagation regimes requires explicit knowledge of the linear term (in Δc) in the expansion for $\mathcal{W}^u(Q_\varepsilon^-)$ defined in (10) [15, 16]. That term can be obtained from the solution of the variational equation (11), which is known explicitly for any $\gamma \in [-\frac{1}{2}, \frac{1}{2})$ and which depends significantly on the value of γ: specifically, $\frac{\partial v}{\partial c}(u, c_0)$ has a branch point at zero for $\gamma \in (0, \frac{1}{2})$, *i.e.*, $\frac{\partial v}{\partial c} \to \infty$ as $u \to 0^+$, while the corresponding limit is well-defined for

$\gamma \in [-\frac{1}{2}, 0]$; cf. equation (31). In particular, when $\gamma = 0$, we have $\frac{\partial v}{\partial c} = -\frac{1}{3}(u-1)$ and, hence, $\lim_{u \to 0^+} \frac{\partial v}{\partial c} = \frac{1}{3}$ [20]. In the pulled regime, *i.e.*, for $\gamma \in [-1, -\frac{1}{2}]$, on the other hand, knowledge of the constant term (in Δc) in (10) – or, rather, of the small-u asymptotics of that term – is sufficient. Correspondingly, the leading-order coefficient in the asymptotics of Δc is universal, *i.e.*, cut-off independent, in the pulled propagation regime [6], whereas it depends substantially on the choice of θ in the bistable [15] and pushed regimes [16], as suggested previously in [5] and in [10, 17], respectively.

Finally, our analysis shows that the cut-off equation (5) supports a unique front solution, irrespective of the propagation regime under consideration. In particular, as the Heaviside cut-off H defined in (4) cancels the reaction terms in (5) in the inner region, the origin Q_ε^+ is a saddle-node in the limit as $\varepsilon \to 0^+$, with a unique stable manifold that corresponds to the singular front in (7).

***Remark* 13.** As an aside, we remark that the above classification allows for a geometric interpretation of the 'critical' front speed in equation (7) (without cut-off) [7, 34], in the context of the equivalent first-order system (9): in the bistable regime, only orbits that lie on the (unique) stable manifold of the saddle equilibrium at Q_0^+ can asymptote into Q_0^+ as $\xi \to \infty$, *i.e.*, the critical speed is simply the propagation speed of the unique traveling front solution to (7). By contrast, any orbit of (9) is asymptotic to Q_0^+ when that point is a node: in the pushed regime, orbits on the strong stable manifold of Q_0^+ exhibit the strongest (exponential) decay in ξ; the critical speed is defined as the speed of the corresponding front solution in that case. In the pulled regime, all (relevant) orbits decay at the same (algebro-exponential) rate, as Q_0^+ is a degenerate node; correspondingly, the critical speed is given by the minimum speed for which (7) supports propagating fronts. An in-depth discussion of critical front speed phenomena can *e.g.* be found in [6, 7, 34]. □

4. Discussion

In this article, we have classified the effects of a Heaviside cut-off on the front propagation dynamics of the classical Nagumo equation [1, 2] in dependence of a control parameter. We have outlined a constructive, geometric proof for the existence and uniqueness of traveling fronts in the presence of a cut-off; moreover, we have deduced the leading-order asymptotics (in the cut-off parameter ε) of the correction Δc to the front propagation speed c_0 that is induced by the cut-off. Finally, we have determined explicitly the corresponding coefficients in the resulting expansions; our findings are summarized in table 5. Our analysis is based on the geometric framework that was first proposed in [6], in the context of the cut-off Fisher-Kolmogorov-Petrowskii-Piscounov (FKPP) equation.

We remark that equation (2), as well as its cut-off counterpart in (5), constitute ideal prototypes for a systematic geometric classification of front propagation in scalar reaction-diffusion equations of the type in (1) (with and without cut-off). First, they realize the three 'canonical' front propagation regimes (pulled, pushed, and bistable) observed in these equations, as well as the respective boundary cases, as the real parameter γ is varied in $[-1, \frac{1}{2})$; cf. Sections 3.1 through 3.4 above. Second, they are 'minimal' representatives of the family of equations in (1), in the sense that other members of that family, such as the Schlögl equation [10, 25] or the (real) Ginzburg-Landau equation [10, 17], are equivalent to equation (2) under linear coordinate transformations; see [15, 16] for details.

To the best of our knowledge, our work [6, 15, 16, 20, 21] represents the first instance of a rigorous proof for the existence and uniqueness of front solutions in the cut-off Nagumo equation (5) or, more generally, in the family of equations in (3). The leading-order ε-asymptotics of Δc, on the other hand, was derived previously by Benguria, Depassier, and co-workers [3, 4] as well as by Méndez, Campos, and others [10], via an integral variational principle [3]. Analogous results in the framework of matched asymptotic expansions [30] were *e.g.* obtained by Brunet and Derrida [5] and by Kessler *et al.*; see [17] and the references therein. Our analysis confirms their findings, and contextualizes them in terms of the dynamics of the cut-off first-order system (8) in the intermediate region, after blow-up; in particular, it explains the asymptotic structure of the correction $\Delta c = c - c_0$, identifying the source of the fractional or logarithmic powers of ε that can occur in the corresponding expansions; moreover, it gives a geometric interpretation for the sign of that correction, as obtained previously in [4, 5, 10]. Finally, we remark that our results concur with those of [17], where an analogous classification was given, based on classical 'mode counting' arguments: as emphasized there, the pushed propagation regime, which exhibits 'nonlinear (type II) marginal stability' (in their terminology), represents an intermediate case between the 'metastable' (bistable) and the 'unstable' (pulled) regimes, in that it is more singular than the former, but more regular than the latter.

Given the existence and uniqueness of front solutions in the cut-off equation (3), a natural next step is the investigation of the stability of these fronts. The underlying theory centers on the Evans function, which has been widely applied in the study of front propagation in reaction-diffusion systems without cut-off [35]. More recently [36, 37], a generalized Evans function has been defined in situations where the front solution decays only algebraically at the zero rest state and the essential spectrum of the associated stability operator has a branch point at the origin. The corresponding analysis is based on geometric desingularization and can be expected to be valuable in the context of equation (3), as well.

An alternative approach to our geometric methodology, which has been widely applied in the analysis of front propagation in the presence of a cut-off [4, 33], relies on the integral variational principle developed by Benguria and Depassier [3]; roughly speaking, the traveling front equation that is associated with (3) corresponds to the Euler-Lagrange equation in that approach, while the Lagrange multiplier is given by the front propagation speed c. A preliminary discussion of the relationship between the two approaches, in the context of results obtained by Benguria *et al.* [4] in the bistable and pushed regimes, can be found in [20]; nevertheless, a more complete understanding of that relationship is still lacking.

An important aspect of the front propagation dynamics of equation (1) (with and without cut-off) concerns the process of front speed selection: the selection mechanism is termed either linear or nonlinear, depending on whether the front is pulled or pushed. While no general selection criteria seem to be known, preliminary results in this direction have *e.g.* been obtained in [38, 39]; however, no precise characterization of the underlying mechanisms has as yet been developed. The classification outlined in Section 3.5, which is based on geometric desingularization (blow-up), might provide a starting point for such a characterization.

A natural extension of the geometric framework developed here, in the context of the scalar equation (1), is the analysis of multi-component systems in more than one spatial dimension. As is well-known, front interactions in such systems can lead to the emergence of novel patterns, in particular when stochastic fluctuations are accounted for. Thus, Kessler and Levine [40] reported diffusive instabilities in a coupled two-species model with cut-off, while Zemskov and Méndez [41] observed bifurcation patterns in piecewise linear activator-inhibitor systems. The resulting two-parameter singular perturbation problems are bound to yield interesting asymptotics and warrant further investigation.

Another topic of substantial recent interest is the relationship between the deterministic (cut-off) equation (3) and the discrete (many-particle) systems that it approximates. Previous work [13] has shown that the approximation provided by (3) is in good (qualitative) agreement with the underlying discrete dynamics as long as the number of particles in the system is sufficiently large, which makes an efficient numerical implementation crucially important. Although several *ad hoc* simplifications have been proposed to reduce the required computational effort, no systematic analysis seems to be available to date.

Deterministic cut-offs have been successfully employed in a number of applications and, in particular, in the life sciences; however, the front propagation dynamics of the resulting cut-off equations has not been considered systematically thus far. Thus, the mean-field description of the evolution of RNA virus populations proposed in [14] predicted a divergent mean fitness; after introduction of a cut-off, the front propagation speed was found to depend logarithmically on a threshold concentration below which autocatalytic viral growth ceases. Similarly, in certain models of Darwinian evolution [42], a standard continuum approximation gave rise to finite-time singularities in the propagation speed, while a cut-off reproduced that speed correctly. Finally, the review of pattern-generating mechanisms in [43] has shown that cut-offs in the growth term can generate complex branching patterns. Additionally, there exist numerous biological models in which the inclusion of a cut-off will lead to a more realistic description of the modeled phenomena; see *e.g.* [44, 45].

Undoubtedly, there exist a variety of asymptotic techniques that can be brought to bear on the front propagation dynamics of equation (3); however, as has become evident through recent work in this area, geometric desingularization (blow-up) is particularly well-suited to a systematic analysis of the (more general) family of equations in (1) [6, 7, 15, 16, 20, 21, 34]. One advantage of the geometric approach lies in its highly visual and intuitive appeal, which contrasts with many of the classical techniques; moreover, it supplies the rigor that the latter sometimes lack. The method of matched asymptotics [30], for instance, generally only produces formal expansions for the correction Δc to the front propagation speed that is due to a cut-off, while our approach systematically uncovers the structure of these expansions, relating them to well-defined geometric objects. The integral variational principle [3], on the other hand, allows for considerable freedom in the choice of suitable trial functions, especially in the pulled propagation regime [33]; correspondingly, it often only yields bounds for Δc. Similarly, the expansions for Δc obtained in [4], in the pushed and bistable regimes, are formal. By contrast, the geometric counterparts found in [6] and in [15, 16], respectively, are rigorous and exact. Finally, these alternative approaches typically only give the leading-order asymptotics of Δc, whereas geometric desingularization can in principle be applied to evaluate that asymptotics to arbitrary order [20].

While the existence and uniqueness of front solutions in the presence of a cut-off and the asymptotics of the corresponding front propagation speed are naturally studied in the context of the traveling front equation (6), other questions, such as the convergence of general solutions to traveling fronts, the time asymptotics of the associated transients, the rate of convergence of the front propagation speed to its asymptotic value, as well as, to some extent, the stability properties of front solutions, require an understanding of the dynamics of the partial differential equation (3). Starting with [46], there have been numerous attempts at developing a geometric (invariant manifold) theory for the latter. Thus, preliminary results in the context of certain hyperbolic-parabolic systems of partial differential equations were obtained in [47], based on the concepts of similarity variables and weighted spaces. It remains to be seen if geometric desingularization can be combined with these or similar techniques to investigate the front propagation dynamics of cut-off reaction-diffusion equations of the type in (3).

References

[1] Gilding B H and Kersner R 2004 *Travelling waves in nonlinear diffusion-convestion reaction* (Progress in Nonlinear Differential Equations and their Applications vol 60) (Basel: Birkhäuser Verlag)
[2] Keener J and Sneyd J 1998 *Mathematical physiology* (Interdisciplinary Applied Mathematics vol 8) (New York: Springer-Verlag)
[3] Benguria R D and Depassier M C 1996 *Phys. Rev. Lett.* **77** 1171
[4] Benguria R D, Depassier M C and Haikala V 2007 *Phys. Rev. E* **76** 051101
[5] Brunet E and Derrida B 1997 *Phys. Rev. E* **56** 2597
[6] Dumortier F, Popović N and Kaper T J 2007 *Nonlinearity* **20** 855
[7] Popović N and Kaper T J 2006 *J. Dynam. Differential Equations* **18** 103

[8] Panja D 2004 *Phys. Rep.* **393** 87
[9] van Saarlos W 2003 *Phys. Rep.* **386** 29
[10] Méndez V, Campos D and Zemskov E P 2005 *Phys. Rev E* **72** 056113
[11] Sherratt J A and Marchant B P 1996 *IMA J. Appl. Math.* **56** 289
[12] Brunet E and Derrida B 2001 *J. Stat. Phys.* **103** 269
[13] Brunet E, Derrida B, Mueller A H and Munier S 2006 *Phys. Rev. E* **73** 056126
[14] Tsimring T L, Levine H and Kessler D A 1996 *Phys. Rev. Lett.* **76** 4440
[15] Dumortier F, Popović N and Kaper T J 2010 *Phys. D* **239** 1984
[16] Dumortier F, Popović N and Kaper T J 2010 Wave speeds for pushed fronts in reaction-diffusion equations with cut-offs *Preprint*
[17] Kessler D A, Ner Z and Sander L M 1998 *Phys. Rev. E* **58** 107
[18] Dumortier F 1993 *Bifurcations and periodic orbits of vector fields* (NATO ASI Series C vol 408) ed D Schlomiuk (Dordrecht: Kluwer Academic Publishing) 19
[19] Krupa M and Szmolyan S 2001 *SIAM J. Math. Anal.* **33** 286
[20] Popović N 2010 A geometric analysis of front propagation in a family of degenerate reaction-diffusion equations with cut-off *Preprint*
[21] Popović N 2010 A geometric analysis of front propagation in an integrable Nagumo equation with a linear cut-off *Preprint*
[22] Chen Z X and Guo B Y 1992 *IMA J. Appl. Math.* **48** 107
[23] Armero J, Sancho J M, Casademunt J, Lacasta A M, Ramírez-Piscina L and Sagués F 1996 *Phys. Rev. Lett.* **76** 3045
[24] Fenichel N 1979 *J. Differential Equations* **31** 53
[25] Popović N 2007 *Fluids and waves* (Contemp. Math. vol 440) (Providence, RI: Amer. Math. Soc.) 187
[26] *Handbook of mathematical functions with formulas, graphs, and mathematical tables* 1974 ed M A Abramowitz and L A Stegun (New York: Dover Publications, Inc.) 9th printing
[27] Chow S N, Li C and Wang D 1994 *Normal forms and bifurcation of planar vector fields* (Cambridge: Cambridge University Press)
[28] Corless R M, Gonnet G H, Hare D E G, Jeffrey D J and Knuth D E 1996 *Adv. Comput. Math.* **5** 329
[29] Popović N 2005 *J. Phys.: Conf. Series* **22** 164
[30] Lagerstrom P A 1988 *Matched asymptotic expansions. Ideas and techniques* (Applied Mathematical Sciences vol 76) (New York: Springer-Verlag)
[31] Popović N and Szmolyan P 2004 *J. Differential Equations* **199** 290
[32] Popović N and Szmolyan P 2004 *Nonlinear Anal.* **59** 531
[33] Benguria R D, Depassier M C and Loss M 2008 *Eur. Phys. J. B* **61** 331
[34] Dumortier F, Popović N and Kaper T J 2007 *J. Math. Anal. Appl.* **326** 1007
[35] Sandstede B 2002 *Handbook of dynamical systems vol 2* (Amsterdam: North-Holland) 983
[36] Popović N and Szmolyan P 2010 Evans functions and blow-up for degenerate shock waves *Preprint*
[37] Sandstede B and Scheel A 2004 *Discrete Contin. Dyn. Syst.* **10** 941
[38] Benguria R D and Depassier M C 1994 *Phys. Rev. Lett.* **73** 2272
[39] Lucia M, Muratov C B, and Novaga M 2004 *Comm. Pure Appl. Math.* **57** 616
[40] Kessler D A and Levine H 1998 *Nature* **394** 556
[41] Zemskov E P and Méndez V 2005 *Eur. Phys. J. B* **48** 81
[42] Cohen E and Kessler D A 2006 *J. Stat. Phys.* **122** 925
[43] Golding I, Kozlovsky Y, Cohen I and Ben-Jacob E 1998 *Phys. A* **260** 510
[44] Hallatschek O and Nelson D R 2008 *Theor. Pop. Biol.* **73** 158
[45] Yin J and McCaskill J S 1992 *Biophys. J.* **61** 1540
[46] Henry D 1981 *Geometric theory of semilinear parabolic equations* (Lecture Notes in Mathematics vol 840) (Berlin: Springer-Verlag)
[47] van Baalen G, Popović N and Wayne C E 2008 *SIAM J. Math. Anal.* **39** 1951

Journal of Physics: Conference Series **268** (2011) 012024 doi:10.1088/1742-6596/268/1/012024

Extending vector hysteresis operators

V Recupero

Politecnico di Torino, Dipartimento di Matematica
Corso Duca degli Abruzzi, 24, 10129 Torino - Italy

E-mail: vincenzo.recupero@polito.it

Abstract. In some recent papers we studied how to extend to BV a hysteresis operator defined on Lipschitzian inputs, preserving suitable continuity properties. More precisely we considered the so called strict metric defined by means of the essential variation. This approach may have some drawbacks from the physical point of view, therefore in the present paper we show how to extend a general hysteresis operator with respect to a notion of convergence which takes into account of the pointwise variation rather than the essential variation.

1. Introduction

In two recent papers ([1, 2]) we studied how to extend to BV a hysteresis operator defined on Lipschitzian inputs on the interval $[0,T]$, preserving suitable continuity properties. More precisely we considered the so called strict metric defined by

$$d_s(u,v) := \|u - v\|_{L^1} + |\,\mathrm{V}_e(u,[0,T]) - \mathrm{V}_e(v,[0,T])|. \tag{1.1}$$

Here $\mathrm{V}_e(u,[0,T])$ denotes the essential variation, which is defined by

$$\mathrm{V}_e(u,[0,T]) := \inf\{\mathrm{V}_p(w,[0,T]) \ : \ w = u \text{ a.e. in } [0,T]\},$$

where $\mathrm{V}_p(w,[0,T])$ is the pointwise variation:

$$\mathrm{V}_p(w,[0,T]) := \sup \sum_{j=1}^{m} |w(t_j) - w(t_{j-1})|,$$

the supremum being taken over all subdivisions $0 = t_0 < t_1 < \cdots < t_m = T$ of the interval $[0,T]$. The strict metric defined in (1.1) is natural if we consider the problem of approximating a BV function by convolution: indeed if $u \in BV$ and u_ε is its convolution, then u_ε converges to u with respect to the strict metric. Moreover two functions differing on a set of Lebesgue measure zero have the same essential variation, thus this is the natural notion of variation in the framework of Lebesgue equivalence classes of integrable functions. The essential variation is considered also in other works on hysteresis (see, e.g., [3, 4]).

However this approach may have some drawback if we consider a situation like the following one. Let us consider the hysteresis play operator which can be described by means of the following simple mechanical model: a cylinder of length $2r$ which can move along the y-axis when driven by a piston. The position of the piston at the time t is denoted by the coordinate

Journal of Physics: Conference Series **268** (2011) 012024 doi:10.1088/1742-6596/268/1/012024

$u(t)$, whereas $y(t)$ denotes the position of a point of the cyilinder, for instance its center. The operator $u \longmapsto y$ is the relation defining the play operator $y := \mathsf{P}(u)$. Assume that the initial position of the cylinder is $y(0) = 0$ and the movement $u(t)$ of the piston is given by

$$u(t) = \begin{cases} 0 & \text{if } t < T/2 \\ 2r & \text{if } t = T/2 \\ 0 & \text{if } t > T/2 \end{cases}$$

i.e. at the time $T/2$ the input u moves instantaneously to the position $2r$ and comes back to 0. From the point of view of Lebesgue measure the input u is equivalent to the constant function $v(t) \equiv 0$, indeed the singleton $T/2$ has measure zero. But this description is non physical if we interpret the term "instantaneously" as a movement that is so fast that it cannot be measured by the available measure instruments: we expect that after the time $T/2$ the output of the play operator is $y(t) = r$, whereas $\mathsf{P}(v) \equiv 0$ is the output associated to $v(t) \equiv 0$.

This example suggests that we should consider the pointwise variation rather than the essential variation. The aim of this note is to prove an extension theorem for a general hysteresis operator which is continuous with respect the convergence of the pointwise variations, more precisely we consider the following notion of convergence:

$$u_n \to u \quad \text{if and only if} \quad \begin{cases} u_n(t) \to u(t) & \forall t \in [0,T] \\ \mathrm{V}_p(u_n, [0,T]) \to \mathrm{V}_p(u, [0,T]) \end{cases} \tag{1.2}$$

We will prove our results in the framework of vector hysteresis operators acting between curves in a Hilbert spaces. We use a procedure which is analogous to the one used in [2], but we have to modify the proofs in order to treat with the pointwise variation rather than with the essential variation. In particular we also need an approximation property of BV with respect to the convergence defined in (1.2).

Let us remark that the problem of extending a hysteresis operator to spaces of non regular functions was dealt by several authors: see, e.g., the monographs [5, 3, 6, 7], the papers [4, 8, 9, 10] and the references therein. In most of these references the extension was obtained for particular cases of hysteresis operators or for scalar hysteresis operators. Our result holds for a general vector hysteresis operators having suitable continuity properties and applies to a large variety of particular cases occurring in the literature.

2. Preliminaries

In the sequel $\mathbb{N}$ denotes the set of strictly positive integers and $T > 0$. We say that a real function $f : [0,T] \longrightarrow \mathbb{R}$ is *increasing* if $(f(s) - f(t))(s - t) \geq 0$ for every $s, t \in [0,T]$. We recall that χ_S denotes the characteristic function of a set S: $\chi_S(t) = 1$ if $t \in S$ and $\chi_S(t) = 0$ if $t \notin S$. The interior of a subset $S \subseteq \mathbb{R}$ is denoted by $\mathring{S}$. Throughout the paper we assume that $\mathcal{H}$ is a real Hilbert space endowed with the scalar product $\langle \cdot, \cdot \rangle$ and we set $\|u\|_{\mathcal{H}} := \langle u, u \rangle^{1/2}$, $u \in \mathcal{H}$. The Lebesgue measure on $\mathbb{R}$ is denoted by $\mathcal{L}^1$, and by $L^1(0,T;\mathcal{H})$ we indicate the space of Lebesgue $\mathcal{H}$-valued integrable function defined on $[0,T]$ (cf., e.g., [11, Part Three]). If $f : [0,T] \longrightarrow \mathcal{H}$ the $\|f\|_\infty := \sup_{t \in [0,T]} \|f(t)\|_{\mathcal{H}}$, by $\mathrm{Cont}(f)$ we denote the set of continuity points of f, whereas $\mathrm{Discont}(f) := [0,T] \smallsetminus \mathrm{Cont}(f)$. As usual we set $\mathrm{Lip}(f) := \sup_{t \neq s} \|f(t) - f(s)\|_{\mathcal{H}}/|t - s|$ and $Lip([0,T];\mathcal{H}) := \{f : [0,T] \longrightarrow \mathcal{H} \ : \ \mathrm{Lip}(f) < \infty\}$. By the symbol $W^{1,p}(0,T;\mathcal{H})$ we denote the Sobolev spaces of $\mathcal{H}$-valued functions, $p \in [1, \infty]$. It is well known that $W^{1,\infty}(0,T;\mathcal{H}) \cap C([0,T];\mathcal{H}) = Lip([0,T];\mathcal{H})$ (see, e.g., the appendix of [12]).

Definition 2.1. If J is a subinterval of $[0,T]$, the symbol $St(J;\mathcal{H})$ denotes the set of *$\mathcal{H}$-valued step maps on J*, i.e. maps $f : J \longrightarrow \mathcal{H}$ such that J can be partitioned into a finite number of (possibly degenerate) intervals $J_1, \ldots, J_m$ and f is constant on each J_j for $j = 1, \ldots, m$. A function $f : J \longrightarrow \mathcal{H}$ is called *regulated on* J if at each point $t \in]\inf J, \sup J[$ there exist the one sided limits $f(t-)$ and $f(t+)$ in $\mathcal{H}$ and if there exist $f(\inf J+)$ and $f(\sup J-)$ in $\mathcal{H}$. We denote by $Reg(J;\mathcal{H})$ the set of regulated maps on J.

It is well known that $St([0,T];\mathcal{H})$ is dense in $Reg([0,T];\mathcal{H})$ with respect to the supremum norm $\|\cdot\|_\infty$. In particular every step function is Lebesgue measurable and $\mathrm{Discont}(f)$ is a countable set (see, e.g., next Proposition 3.1). We also remark that if $f \in St(J,\mathcal{H})$ and J is compact, then f is bounded.

Definition 2.2. We recall that a *subdivision* of a nondegenerate subinterval $J \subseteq [0,T]$ is a family $(s_j)_{j=0}^m$, $m \in \mathbb{N}$, with the property that $s_0 < \cdots < s_m$ and $s_j \in J$ for $j = 0, \ldots, m$. The set of all subdivisions of J is indicated by $\mathfrak{S}(J)$. If $f : J \longrightarrow \mathcal{H}$ and $\mathfrak{s} = (s_j)_{j=0}^m \in \mathfrak{S}(J)$, the *variation of* f *with respect to* $\mathfrak{s}$ is defined by

$$\mathrm{V}(f,\mathfrak{s}) := \sum_{j=1}^m \|f(s_j) - f(s_{j-1})\|_{\mathcal{H}}.$$

If J is nondegenerate the *pointwise variation of* u *on* J is defined by

$$\mathrm{V}_p(f,J) := \sup\{\mathrm{V}(f,\mathfrak{s}) \ : \ \mathfrak{s} \in \mathfrak{S}(J)\},$$

otherwise we set $\mathrm{V}_p(f,J) = 0$. We define $BV_p([0,T];\mathcal{H}) := \{f : [0,T] \longrightarrow \mathcal{H} \ : \ \mathrm{V}_p(f,[0,T]) < \infty\}$, the set of functions with pointwise bounded variation on $[0,T]$.

It is well known and clear from the definition that the pointwise variation is lower semicontinuous with respect to the pointwise convergence, i.e. $\mathrm{V}_p(f,J) \le \liminf_{n\to\infty} \mathrm{V}_p(f_n,J)$ whenever $f_n(t) \to f(t)$ for every $t \in J$. If $f : [0,T] \longrightarrow \mathcal{H}$, $\mathrm{V}_p(f,[0,T]) < \infty$, and $t, t_0 \in [0,T]$, the inequality $\|f(t)\|_{\mathcal{H}} \le \|f(t_0)\|_{\mathcal{H}} + \mathrm{V}_p(f,[0,T])$ yields the boundedness of f. Moreover it is well known that $f \in Reg([0,T];\mathcal{H})$, therefore we can define the maps $f_-, f_+ : [0,T] \longrightarrow \mathcal{H}$ by setting

$$f_-(t) := f(t-), \quad f_+(t) := f(t+), \qquad t \in [0,T], \tag{2.1}$$

with the convention that $f(0-) := f(0)$ and $f(T+) := f(T)$. It is easy to check that $\mathrm{V}_p(f_+,[0,T]) = \mathrm{V}_p(f,[0,T]) - \|f(0+) - f(0)\|_{\mathcal{H}}$ and $\mathrm{V}_p(f_-,[0,T]) = \mathrm{V}_p(f,[0,T]) - \|f(T) - f(T-)\|_{\mathcal{H}}$. Let us observe that if $g_1, g_2 : [0,T] \longrightarrow \mathcal{H}$ are two functions in the same $\mathcal{L}^1$-equivalence class and $\mathrm{V}_p(g_j,[0,T]) < \infty$, $j = 1,2$, then every $t \in]0,T]$ is a left Lebesgue point of g_j, hence

$$(g_1)_-(t) = \lim_{h\nearrow 0} \frac{1}{h}\int_{t-h}^t g_1(s)\,\mathrm{d}s = \lim_{h\nearrow 0} \frac{1}{h}\int_{t-h}^t g_2(s)\,\mathrm{d}s = (g_2)_-(t) \qquad \forall t \in]0,T].$$

In the same manner we see that $(g_1)_+ = (g_2)_+$ on $[0,T[$. This remark allows us to formulate the following

Definition 2.3. Let $f : [0,T] \longrightarrow \mathcal{H}$ be given. If there is no $\mathcal{L}^1$-representative g of f such that $\mathrm{V}_p(g,[0,T]) < \infty$, we set $\mathrm{V}_e(f,[0,T]) := \infty$. Otherwise if $g : [0,T] \longrightarrow \mathcal{H}$ is such that $f = g$ $\mathcal{L}^1$-a.e. in $[0,T]$ and $\mathrm{V}_p(g,[0,T]) < \infty$, we set

$$\mathrm{V}_e(f,[0,T]) := \mathrm{V}_p(g_-,[0,T]) + \|f(T) - f(T-)\|_{\mathcal{H}} \ (= \mathrm{V}_p(g_+,[0,T]) + \|f(0+) - f(0)\|_{\mathcal{H}}), \tag{2.2}$$

where g_- is defined in (2.1). The real extended number $\mathrm{V}_e(f,[0,T])$ is called *essential variation of* f. The space of maps of bounded essential variation on $[0,T]$ is denoted by $BV_e([0,T];\mathcal{H})$.

Let us recall that $W^{1,p}(0,T;\mathcal{H}) \subseteq BV_e([0,T];\mathcal{H})$ and $\mathrm{V}_e(f,[0,T]) = \int_0^T \|f'(t)\|_{\mathcal{H}}\,\mathrm{d}t$ for every $f \in W^{1,p}(0,T;\mathcal{H})$, where f' denotes the distributional derivative of f (see the appendix of [12]).

Remark 2.1. Observe that in the previous definition the essential variation of f takes into account of the values of f in 0 and T, therefore the points 0 and T have a privileged role in comparison with the singletons of $]0,T[$. In this way our definition coincide with the one in [4]. One could also define the essential variation neglecting the values $f(0)$ and $f(T)$ and setting $\mathrm{V}_e(f,[0,T]) := \mathrm{V}_p(g_-,]0,T[)$ where $g = f$ $\mathcal{L}^1$-a.e. and has pointwise bounded variation. In this way we obtain that $\mathrm{V}_e(f,[0,T]) = \inf\{\mathrm{V}_p(g,[0,T]) \ : \ f = g \ \mathcal{L}^1\text{-a.e. in } [0,T]\}$, as we wrote in the Introduction in order to make the presentation clearer. In [2, Section 4.4] it is shown the the first approach can be reduced to the second one, and vice versa.

Let us now recall the notion of arc length (see [13, Section 2.5.16]). If $u \in BV_p([0,T];\mathcal{H})$ we define $\ell_u : [0,T] \longrightarrow [0,T]$ by

$$\ell_u(t) := \begin{cases} \frac{T}{\mathrm{V}_p(u,[0,T])}\,\mathrm{V}_p(u,[0,t]) & \text{if } \mathrm{V}_p(u,[0,T]) \neq 0 \\ 0 & \text{if } \mathrm{V}_p(u,[0,T]) = 0 \end{cases}, \qquad t \in [0,T]. \tag{2.3}$$

The function ℓ_u is increasing, $\ell_u(0) = 0$ and $\ell_u(T) = T$. Moreover $\mathrm{Discont}(\ell_u) = \mathrm{Discont}(u)$ and

$$[0,T] \smallsetminus \ell_u([0,T]) = \bigcup_{t \in \mathrm{Discont}(u)} [\ell_u(t-),\ell_u(t+)] \smallsetminus \{\ell_u(t)\}.$$

Moreover ℓ_u is Lipschitz continuous if $u \in \mathit{Lip}([0,T];\mathcal{H})$. If $t_1 < t_2$ we have $\|u(t_1) - u(t_2)\|_{\mathcal{H}} \leq \mathrm{V}_p(u,[t_1,t_2]) = \mathrm{V}_p(u,[0,t_2]) - \mathrm{V}_p(u,[0,t_1])$ therefore

$$\|u(t_1) - u(t_2)\|_{\mathcal{H}} \leq \frac{\mathrm{V}_p(u,[0,T])}{T}|\ell_u(t_1) - \ell_u(t_2)| \qquad \forall t_1, t_2 \in [0,T]. \tag{2.4}$$

This inequality yields that $u(\ell_u^{-1}(\sigma))$ is a singleton for every $\sigma \in \ell_u([0,T])$, therefore there is a unique function $U : \ell_u([0,T]) \longrightarrow \mathcal{H}$ such that $U \circ \ell = u$. From (2.4) it also follows that $\mathrm{Lip}(U) \leq \mathrm{V}_p(u,[0,T])/T$. In order to extend U to all of $[0,T]$ we define $\widetilde{u} : [0,T] \longrightarrow \mathcal{H}$ by setting

$$\begin{aligned} \widetilde{u}(\sigma) &:= (1-\lambda)u(t-) + \lambda u(t) \quad \text{if } \sigma = (1-\lambda)\ell_u(t-) + \lambda\ell_u(t),\ t \in [0,T],\ \lambda \in [0,1], \\ \widetilde{u}(\sigma) &:= (1-\lambda)u(t) + \lambda u(t+) \quad \text{if } \sigma = (1-\lambda)\ell_u(t) + \lambda\ell_u(t+),\ t \in [0,T],\ \lambda \in [0,1]. \end{aligned}$$

It is clear that $\widetilde{u}$ extends U and that $\mathrm{Lip}(\widetilde{u}) = \mathrm{Lip}(U)$. The function $\widetilde{u}$ may be regarded as a kind of *reparametrization of u by the normalized arc length.* We summarize the previous discussions in the following

Proposition 2.1. *If $u \in BV_p([0,T];\mathcal{H})$ and $\ell_u : [0,T] \longrightarrow [0,T]$ is its "normalized" arc length defined by* (2.3)*, then there exists a unique function $\widetilde{u} \in \mathit{Lip}([0,T];\mathcal{H})$ such that*

$$u = \widetilde{u} \circ \ell_u \qquad \forall t \in [0,T], \tag{2.5}$$

$$\widetilde{u} \textit{ is affine on } [\ell_u(t-),\ell_u(t)] \qquad \forall t \in \mathrm{Discont}(u), \tag{2.6}$$

$$\widetilde{u} \textit{ is affine on } [\ell_u(t),\ell_u(t+)] \qquad \forall t \in \mathrm{Discont}(u). \tag{2.7}$$

Moreover $\mathit{Lip}(\widetilde{u}) \leq \mathrm{V}_p(u,[0,T])/T$.

3. Various kind of convergences and approximations

In this section we discuss some kind of convergences and approximations in the spaces $BV_p([0,T];\mathcal{H})$ and $BV_e([0,T];\mathcal{H})$.

Definition 3.1. Let us consider $u_n, u : [0,T] \longrightarrow \mathcal{H}$. We say that:

(i) $u_n \to u$ *strictly* if $\|u_n - u\|_{L^1(0,T;\mathcal{H})} + |\operatorname{V}_e(u_n,[0,T]) - \operatorname{V}_e(u,[0,T])| \to 0$ as $n \to \infty$,

(ii) $u_n \to u$ *uniformly strictly* if $\|u_n - u\|_\infty + |\operatorname{V}_p(u_n,[0,T]) - \operatorname{V}_p(u,[0,T])| \to 0$ as $n \to \infty$,

whenever these convergences makes sense.

Let us start recalling a classical argument showing how to approximate a function $u \in Reg([0,T];\mathcal{H})$ by a sequence of step functions (cf. [14, Chapter 2, Section 1] and [4, Proposition 1.3]). We define iteratively a sequence of subdivisions of $[0,T]$. If $n \in \mathbb{N}$, since u is regulated, for every $t \in [0,T]$ there exists an interval $J_n(t) = [a_n(t), b_n(t)]$ such that $t \in \overset{\circ}{J_n}(t)$, $\mathcal{L}^1(J_n(t)) < 1/n$ and

$$s \in [a_n(t), t[\,\cap\, [0,T] \implies \|f(t-) - f(s)\|_{\mathcal{H}} < 1/4n, \tag{3.1}$$
$$s \in \,]t, b_n(t)] \,\cap\, [0,T] \implies \|f(s) - f(t+)\|_{\mathcal{H}} < 1/4n. \tag{3.2}$$

Since $[0,T]$ is compact, we find a subdivision $\mathfrak{s}_n = (t_0^n, t_1^n, \ldots, t_{m_n}^n)$ of $[0,T]$ such that

$$[0,T] \subseteq \bigcup_{j=1}^{m_n} [a_n(t_j^n), b_n(t_j^n)[. \tag{3.3}$$

It is easy to see that we can assume that $a_n(t_{j-1}^n) \le a_n(t_j^n)$ and $b_n(t_{j-1}^n) \le b_n(t_j^n)$ and we can restrict the intervals $[a_n(t_j^n), b_n(t_j^n)[$ in such a way they are mutually disjoint (hence $b_n(t_{j-1}^n) = a_n(t_j^n)$). In the next step for $n+1$ we can assume that the subdivision $\mathfrak{s}_{n+1}$ contains all the point of the previous one: $\{t_0^n, \ldots, t_{m_n}^n\} \subseteq \{t_0^{n+1}, \ldots, t_{m_{n+1}}^{n+1}\}$. It also follows that

$$\operatorname{osc}_f(t) := \lim_{r \searrow 0} \sup_{|s-t| \le r} \|f(s) - f(t)\|_{\mathcal{H}} < 1/n \qquad \forall t \in [0,T] \smallsetminus \{t_0^n, \ldots, t_{m_n}^n\},$$

therefore

$$\operatorname{Discont}(u) \subseteq \bigcup_{n \in \mathbb{N}} \{t_0^n, t_1^n, \ldots, t_{m_n}^n\}.$$

Now we can define for every $n \in \mathbb{N}$ the step function v_n by setting for every $t \in [0,T]$

$$v_n(t) := \begin{cases} u(t_j-) & \text{if } t \in [a_n(t_j^n), t_j^n[\\ u(t_j) & \text{if } t = t_j^n \\ u(t_j+) & \text{if } t \in \,]t_j^n, b_n(t_j^n)[\end{cases}. \tag{3.4}$$

It is clear that $\operatorname{V}_p(v_n,[0,T]) \le \operatorname{V}_p(u,[0,T])$ and $\|v_n - u\|_\infty < 1/n$. Therefore we infer the following

Proposition 3.1. *For every $u \in Reg([0,T];\mathcal{H})$ there exists a sequence $v_n \in St([0,T];\mathcal{H})$ such that $v_n \to u$ uniformly and $\operatorname{V}_p(v_n,[0,T]) \to \operatorname{V}_p(u,[0,T])$ as $n \to \infty$.*

Indeed by semicontinuity with respect ot the pointwise convergence we have that $\operatorname{V}_p(u,[0,T]) \le \liminf_{n\to\infty} \operatorname{V}_p(v_n,[0,T]) \le \limsup_{n\to\infty} \operatorname{V}_p(v_n,[0,T]) \le \operatorname{V}_p(u,[0,T])$.

Thus we have seen that $St([0,T];\mathcal{H})$ is dense in $Reg([0,T];\mathcal{H})$ with respect to the uniform strict convergence. In order to obtain a more regular approximation we need a modification of the previous argument. We partition every $J_n(t)$ in the following way: we consider $\rho_n(t)$ such that $0 < \rho_n(t) < \min\{(t - a_n(t))/3, (b_n(t) - t)/3\}$ and we set

$$A_n^1(t) := [t - \rho_n(t), t[, \quad A_n^2(t) := [a_n(t), t - \rho_n(t)[,$$
$$B_n^1(t) :=]t, t + \rho_n(t)[, \quad B_n^2(t) := [t + \rho_n(t), b_n(t) - \rho_n(t)[, \quad B_n^3(t) := [b_n(t) - \rho_n(t), b_n(t)],$$

hence

$$J_n(t) = A_n^2(t) \cup A_n^1(t) \cup \{t\} \cup B_n^1(t) \cup B_n^2(t) \cup B_n^3(t).$$

It is enough to consider the unique continuous function $u_n : [0, T] \longrightarrow \mathcal{H}$ such that

$$\begin{cases} u_n(t) = v_n(t) \text{ if } t \in A_n^2(t_j^n) \cup B_n^2(t_j^n) \\ u_n(t) = v(t_j^n) \text{ if } t = t_j^n \\ u_n \text{ is affine on } A_n^1(t_j^n) \\ u_n \text{ is affine on } B_n^1(t_j^n) \\ u_n \text{ is affine on } B_n^3(t_j^n) \quad (\text{only for } j \neq m_n) \end{cases},$$

instead if $t \in B_n^3(t_{m_n}^n)$ we define $u_n(t) := v_n(t)$. It is clear that $u_n \in Lip([0, T]; \mathcal{H})$ and we still have that $\mathrm{V}_p(u_n, [0, T]) \leq \mathrm{V}_p(u, [0, T])$. Of course we lose uniform convergence, but the pointwise convergence still holds, indeed we have the following

Proposition 3.2. *If $u \in Reg([0, T]; \mathcal{H})$ then there exists a sequence $u_n \in Lip([0, T]; \mathcal{H})$ such that $u_n(t) \to u(t)$ for every $t \in [0, T]$ and $\mathrm{V}_p(u_n, [0, T]) \to \mathrm{V}_p(u, [0, T])$ as $n \to \infty$. In particular $u_n \to u$ in $L^1(0, T; \mathcal{H})$ as $n \to \infty$.*

Proof. Fix $t \in [0, T]$. We first prove that $u_n(t) \to u(t)$. If $t = t_j^n$ for some n and some j, then by construction $u_n(t)$ is definitively equal to $u(t)$ and we are done. Thus let us consider the case when $t \neq t_j^n$ for every n and every j, in particular $t \in \mathrm{Cont}(u)$. If for every n there exists j_n such that $t \in A_n^2(t_{j_n}^n) \cup B_n^2(t_{j_n}^n)$ for every n then $\|u_n(t) - u(t)\|_{\mathcal{H}} < 1/n$ for every n and we are done. If instead for every n there exists $j_n \in \{1, \ldots, m_{n-1}\}$ such that $t \in B_n^3(t_{j_n}^n)$, then, as $u_n(t)$ belongs to the line segment $[u(t_j^n+), u(t_{j+1}^n-)]$ and $b_n(t_j^n) = a_n(t_{j+1}^n)$,

$$\begin{aligned} \|u_n(t) - v_n(t)\|_{\mathcal{H}} &= \|u_n(t) - u(t_j^n+)\|_{\mathcal{H}} \leq \|u(t_j^n+) - u(t_{j+1}^n-)\|_{\mathcal{H}} \\ &\leq \|u(t_j^n+) - u(b(t_j^n))\|_{\mathcal{H}} + \|u(a(t_{j+1}^n)) - u(t_{j+1}^n-)\|_{\mathcal{H}} < 1/2n, \end{aligned}$$

thus also in this case we obtain the desired convergence.

Now let us consider the case when there is a sequence $j_n \in \{1, \ldots, m_n\}$ such that $t \in A_n^1(t_{j_n}^n) \cup B_n^1(t_{j_n}^n)$. We can assume that $t_{j_n}^n \to \bar{t} \in [0, T]$, at least for a subsequence which we do not relabel. Therefore, as $\rho_n(s) \to 0$ for every s, we infer that $t = \bar{t}$, so that $\bar{t} \neq t_j^n$ for every n and every j and $t \in \mathrm{Cont}(u)$. It follows that $u(t_{j_n}^n) \to u(t)$ and $u(t_{j_n}^n-) \to u(t)$ as $n \to \infty$. Since $u_n(t)$ belongs to the line segment $[u(t_{j_n}^n), u(t_{j_n}^n-)]$ for every n, we deduce that $u_n(t) \to u(t)$. All the other cases can be reduced to the previous ones by considering subsequences.

The convergence of the variation is proved as above. Concerning the L^1-convergence, we observe that $\|u_n\|_\infty \leq \|u_n(0)\|_{\mathcal{H}} + \mathrm{V}_p(u_n, [0, T])$, therefore the sequence $\|u_n\|_\infty$ is bounded and the dominated convergence theorem applies. □

In the papers [1, 2] we considered the problem of extending a hysteresis operators in a continuous way with respect to the strict convergence. In order to consider the pointwise variation one could modify the notion of strict convergence by declaring that $u_n \to u$ if and only if $\|u_n - u\|_{L^1(0,T;\mathcal{H})} \to 0$ and $\mathrm{V}_p(u_n, [0,T]) \to \mathrm{V}_p(u, [0,T])$ as $n \to \infty$. This definition has a drawback: the limits are not unique, for instance if u_n is defined by $u_n := \chi_{\{T/2\}}$ if n is odd, $u_n := -\chi_{\{T/2\}}$ if n is even, then u_n is not pointwise convergent, but has several limits in the previous sense, for instance any function of the kind χ_{t_0} with $t_0 \in [0,T]$.

Since we consider the pointwise variation, we do distinguish functions which differ on a set of Lebesgue measure zero. Therefore a suitable convergence we may consider is the one we introduce in the following

Definition 3.2. We say that $u_n : [0,T] \longrightarrow \mathcal{H}$ is *pointwise strictly convergent* to u if $\mathrm{V}_p(u_n, [0,T]) \to \mathrm{V}_p(u, [0,T])$ and $u_n(t) \to u(t)$ for every $t \in [0,T]$ as $n \to \infty$.

Observe that this convergence is not induced by a metric, but the limits are unique. The previous discussion shows that $Lip([0,T];\mathcal{H})$ is dense in $BV_p([0,T];\mathcal{H})$ with respect to this convergence. Let us conclude this section with the following

Lemma 3.1. *Assume that $u, u_n \in BV_p([0,T];\mathcal{H}) \cap C([0,T];\mathcal{H})$ for every n and $u_n \to u$ pointwise strictly on $[0,T]$. Then $u_n \to u$ uniformly on $[0,T]$.*

Proof. First of all let us observe that $\|u_n\|_\infty \le \|u_n(0)\|_{\mathcal{H}} + \mathrm{V}_p(u_n, [0,T])$, therefore the sequence $\|u_n\|_\infty$ is bounded and by the dominated convergence theorem we find that $u_n \to u$ in $L^1(0,T;\mathcal{H})$. Since u and u_n are continuous, the pointwise and essential variations coincide, therefore we have that $u_n \to u$ strictly on $[0,T]$. Thus we can apply [2, Corollary 4.2] and infer the uniform convergence. □

4. Extension of hysteresis operators

In this section we show how to extend a hysteresis operator $\mathsf{R} : Lip([0,T];\mathcal{H}) \longrightarrow BV_p([0,T];\mathcal{H}) \cap C([0,T];\mathcal{H})$ which is continuous with respect to the pointwise strict convergence.

Lemma 4.1. *Assume that $u, u_n \in BV_p([0,T];\mathcal{H})$ and let ℓ_u and ℓ_{u_n} be their renormalized arc lengths defined by Proposition 2.1. If $u_n \to u$ pointwise strictly on $[0,T]$ then $\ell_{u_n}(t) \to \ell_u(t)$ for every $t \in [0,T]$.*

Proof. Fix $t \in [0,T]$. By the lower semicontinuity we have that $\mathrm{V}_p(u, [c,d]) \le \liminf_{n\to\infty} \mathrm{V}_p(u_n, [c,d])$ for every $c, d \in [0,T]$, $c \le d$. On the other hand, exploiting again the lower semicontinuity and the convergence $\mathrm{V}_p(u_n, [0,T]) \to \mathrm{V}_p(u, [0,T])$, we infer that

$$\begin{aligned}\limsup_{n\to\infty} \mathrm{V}_p(u_n, [0,t]) &= \limsup_{n\to\infty} \big(\mathrm{V}_p(u_n, [0,T]) - \mathrm{V}_p(u_n, [t,T])\big) \\ &\le \mathrm{V}_p(u, [0,T]) - \liminf_{n\to\infty} \mathrm{V}_p(u_n, [t,T]) \\ &\le \mathrm{V}_p(u, [0,T]) - \mathrm{V}_p(u_n, [t,T]) = \mathrm{V}_p(u, [0,T]),\end{aligned}$$

thus we have proved that $\mathrm{V}_p(u_n, [0,t]) \to \mathrm{V}_p(u, [0,t])$ as $n \to \infty$. Therefore $\lim_{n\to\infty} \ell_{u_n}(t) = \ell_u(t)$. □

Now we need to prove a generalization of [2, Lemma 4.1].

Lemma 4.2. *Let $v \in C([0,T];\mathcal{H})$ be such that $\mathrm{V}_p(v, [0,T]) < \infty$ and let $\beta : [0,T] \longrightarrow [0,T]$ be an increasing function satisfying $\beta(0) = 0$ and $\beta(T) = T$. Moreover assume that*

$$\mathrm{V}_p(v, [\beta(t-), \beta(t)]) = \|v(\beta(t)) - v(\beta(t-))\|_{\mathcal{H}} \qquad \forall t \in \mathrm{Discont}(\beta), \tag{4.1}$$

$$\mathrm{V}_p(v, [\beta(t), \beta(t+)]) = \|v(\beta(t+)) - v(\beta(t))\|_{\mathcal{H}} \qquad \forall t \in \mathrm{Discont}(\beta). \tag{4.2}$$

Then $\mathrm{V}_p(v \circ \beta, [0,T]) = \mathrm{V}_p(v, [0,T])$.

Proof. We prove the lemma when $\beta(0) = \beta(0+)$ and $\beta(T-) = \beta(T)$, the other cases needing very slight modifications. The inequality $\mathrm{V}_p(v \circ \beta, [0,T]) \le \mathrm{V}_p(v, [0,T])$ is obvious, hence $\mathrm{V}_p(v, [0,T])$ is an upper bound for $\{\sum_{j=1}^{n} \|v(\beta(t_j)) - v(\beta(t_{j-1}))\|_{\mathcal{H}} \ : \ n \in \mathbb{N},\ 0 = t_0 \le \cdots \le t_n = T\}$. Let $\varepsilon > 0$ be arbitrarily fixed. There exists a subdivision $(t_j)_{j=0}^{n}$ of $[0,T]$ such that

$$\mathrm{V}_p(v, [0,T]) < \sum_{j=1}^{n} \|v(t_j) - v(t_{j-1})\|_{\mathcal{H}} + \varepsilon/2. \tag{4.3}$$

For every $\sigma \in \mathrm{Discont}(\beta)$ there is a possibly empty subset $E_\sigma \subseteq \{t_j\}$ contained in $[\beta(\sigma-), \beta(\sigma+)]$. Adding the points $\beta(\sigma-), \beta(\sigma), \beta(\sigma+)$ to E_σ, the sum in (4.3) can only increase (of course it can happen that $\beta(\sigma)$ is one of the points $\beta(\sigma-), \beta(\sigma+)$). Moreover, thanks to the assumptions (4.1)-(4.2) we can also replace E_σ by $\{\beta(\sigma-), \beta(\sigma), \beta(\sigma+)\}$ without affecting such a sum. Therefore we can assume that (4.3) holds for a subdivision (t_j) such that

$$\begin{aligned}\{t_j\}_{j=0}^{n} = {} & \{s_0^1, \dots, s_{k_1}^1\} \cup \{\beta(\sigma_1-), \beta(\sigma_1), \beta(\sigma_1+)\} \\ & \cup \{s_0^2, \dots, s_{k_2}^2\} \cup \{\beta(\sigma_2-), \beta(\sigma_2), \beta(\sigma_2+)\} \cup \cdots\cdots \\ & \cdots\cdots \cup \{s_0^m, \dots, s_{k_m}^m\} \cup \{\beta(\sigma_m-), \beta(\sigma_m), \beta(\sigma_m+)\} \cup \{s_0^{m+1}, \dots, s_{k_{m+1}}^{m+1}\}\end{aligned}$$

where $T = s_{k_{m+1}}^{m+1}$ and

$$\begin{aligned}& \sigma_i \in \mathrm{Discont}(\beta) \qquad \forall i = 1, \dots, m; \\ & \{s_0^i, \dots, s_{k_i}^i\} \subseteq \beta([0,T]) \qquad \forall i = 1, \dots, m+1, \\ & s_{h-1}^i < s_h^i < \beta(\sigma_i) < s_{h-1}^{i+1} < s_h^{i+1} \qquad \forall i = 1, \dots, m,\ \forall h = 0, \dots, k_i\end{aligned}$$

Hence, selecting

$$\tau_h^i \in \beta^{-1}(s_h^i) \qquad i = 1, \dots, m+1, \quad h = 0, \dots, k_i,$$

we can write

$$\begin{aligned}\sum_{j=1}^{n} \|v(t_j) - v(t_{j-1})\|_{\mathcal{H}} = {} & \sum_{i=1}^{m} \Bigg(\sum_{h=1}^{k_i} \|v(\beta(\tau_h^i)) - v(\beta(\tau_{h-1}^i))\|_{\mathcal{H}} \\ & + \|v(\beta(\sigma_i-)) - v(\beta(\tau_{k_i}^i))\|_{\mathcal{H}} + \|v(\beta(\sigma_i)) - v(\beta(\sigma_i-))\|_{\mathcal{H}} \\ & + \|v(\beta(\sigma_i+)) - v(\beta(\sigma_i))\|_{\mathcal{H}} + \|v(\beta(\tau_0^{i+1})) - v(\beta(\sigma_i+))\|_{\mathcal{H}} \Bigg) \\ & + \sum_{h=1}^{k_{m+1}} \|v(\beta(\tau_h^{m+1})) - v(\beta(\tau_{h-1}^{m+1}))\|_{\mathcal{H}}.\end{aligned}$$

The continuity of v yields that for every $i = 1, \dots, m$ there exist $\hat{\sigma}_i, \tilde{\sigma}_{i+1}$ very near σ_i, such that $\hat{\sigma}_i < \sigma_i < \tilde{\sigma}_{i+1}$, $\|v(\beta(\tilde{\sigma}_i)) - v(\beta(\sigma_i-))\|_{\mathcal{H}} < \varepsilon/(4m)$, and $\|v(\beta(\sigma_i+)) - v(\beta(\hat{\sigma}_{i+1}))\|_{\mathcal{H}} < \varepsilon/(4m)$, so that

$$\begin{aligned}\sum_{j=1}^{n} \|v(t_j) - v(t_{j-1})\|_{\mathcal{H}} \le {} & \sum_{i=1}^{m} \Bigg(\sum_{h=1}^{k_i} \|v(\beta(\tau_h^i)) - v(\beta(\tau_{h-1}^i))\|_{\mathcal{H}} \\ & + \|v(\beta(\tilde{\sigma}_i)) - v(\beta(\tau_{k_i}^i))\|_{\mathcal{H}} + \|v(\beta(\sigma_i)) - v(\beta(\tilde{\sigma}_i))\|_{\mathcal{H}} \\ & + \|v(\beta(\sigma_i+)) - v(\beta(\sigma_i))\|_{\mathcal{H}} + \|v(\beta(\tau_0^{i+1})) - v(\beta(\hat{\sigma}_{i+1}))\|_{\mathcal{H}} + \varepsilon/m \Bigg) \\ & + \sum_{h=1}^{k_{m+1}} \|v(\beta(\tau_h^{m+1})) - v(\beta(\tau_{h-1}^{m+1}))\|_{\mathcal{H}}.\end{aligned}$$

That is, we have found a subdivision $(\theta_j)_{j=0}^r$ such that $\mathrm{V}_p(v,[0,T]) < \sum_{j=1}^n \|v(\beta(\theta_j)) - v(\beta(\theta_{j-1}))\|_{\mathcal{H}} + \varepsilon$, and the lemma is proved. □

Corollary 4.1. *If $u \in BV_p([0,T];\mathcal{H})$ and $\widetilde{u}$ is the reparametrization of u defined in Proposition 2.1, then $\mathrm{V}_p(u,[0,T]) = \mathrm{V}_p(\widetilde{u},[0,T])$.*

Let us recall the following lemma (see [2, Lemma 4.7]).

Lemma 4.3. *Let $c, d \in \mathbb{R}$ be such that $c < d$ and assume $x, y \in \mathcal{H}$. Then the affine map $w : [c,d] \longrightarrow \mathcal{H}$ defined by $w(t) := x + t(y-x)/(d-c)$ is the only minimizer of the functional $v \longmapsto \|v'\|^2_{L^2(]c,d[;\mathcal{H})}$ in the set $\{v \in Lip([c,d];\mathcal{H}) \ : \ v(c) = x, \ v(d) = y\}$.*

Now we can prove the following proposition connecting the convergences of u_n and $\widetilde{u}_n$. The previous preparatory lemmas enable us to exploit the argument of [2, Proposition 4.10], but in the present situation the convergence of ℓ_{u_n} holds everywhere in $[0,T]$, rather than almost everywhere. In the proof we will use the notions of weak and weak star convergence, denoted respectively by $\rightharpoonup$ and $\overset{*}{\rightharpoonup}$ (see, e.g., [15, Chapter 3]).

Proposition 4.1. *Assume that $u, u_n \in BV_p([0,T];\mathcal{H})$ for every $n \in \mathbb{N}$ and $u_n \to u$ pointwise strictly on $[0,T]$. Let $\widetilde{u}$ and $\widetilde{u}_n$ be their reparametrizations defined defined by Proposition 2.1. Then*

$$\widetilde{u}_n \to \widetilde{u} \qquad \text{in } W^{1,p}(0,T;\mathcal{H}) \quad \forall p \in [1,+\infty[. \tag{4.4}$$

In particular $\widetilde{u}_n \to \widetilde{u}$ poinwise strictly on $[0,T]$.

Proof. Now observe that $\widetilde{u}(0) = u(0)$ and $\widetilde{u}_n(0) = u_n(0)$, therefore

$$\widetilde{u}_n(0) \to \widetilde{u}(0) \qquad \text{in } \mathcal{H} \tag{4.5}$$

as $n \to \infty$, since u_n is pointwise convergent to u. We also have

$$\begin{aligned} \|\widetilde{u}_n(\sigma)\|_{\mathcal{H}} &\le \|\widetilde{u}_n(0)\|_{\mathcal{H}} + \int_0^T \|\widetilde{u}_n'(\tau)\|_{\mathcal{H}} \,\mathrm{d}\tau \\ &= \|\widetilde{u}_n(0)\|_{\mathcal{H}} + \mathrm{V}_p(\widetilde{u}_n,[0,T]) \\ &= \|\widetilde{u}_n(0)\|_{\mathcal{H}} + \mathrm{V}_p(u_n,[0,T]) \qquad \forall \sigma \in [0,T] \end{aligned} \tag{4.6}$$

and, by Proposition 2.1 and Corollary 4.1,

$$\|\widetilde{u}_n'\|_{L^\infty(0,T;\mathcal{H})} \le \frac{\mathrm{V}_p(\widetilde{u}_n,[0,T])}{T} = \frac{\mathrm{V}_p(u_n,[0,T])}{T}. \tag{4.7}$$

Hence (4.5)-(4.7) imply that $(\widetilde{u}_n)$ is bounded in $W^{1,p}(0,T;\mathcal{H})$ for every $p \in [1,\infty]$. Hence there exists $\widehat{u} \in Lip([0,T];\mathcal{H})$ such that, at least for a subsequence which we do not relabel,

$$\widetilde{u}_n \overset{*}{\rightharpoonup} \widehat{u} \qquad \text{in } W^{1,p}(0,T;\mathcal{H}) \quad \forall p \in [1,\infty]. \tag{4.8}$$

This convergence, together with (4.5) implies that $\widetilde{u}_n(0) \to \widehat{u}(0) = \widetilde{u}(0)$ in $\mathcal{H}$, from which we infer that

$$\widetilde{u}_n(\sigma) \rightharpoonup \widehat{u}(\sigma) \qquad \forall \sigma \in [0,T], \tag{4.9}$$

indeed for every $x \in \mathcal{H}$

$$\langle u_n(\sigma) - u(\sigma), x\rangle = \langle u_n(0) - u(0), x\rangle + \int_0^\sigma \langle u_n'(\tau) - u'(\tau), x\rangle \,\mathrm{d}\tau \to 0.$$

Now for every $x \in \mathcal{H}$ and for every $n \in \mathbb{N}$ define $f_n^x : [0,T] \longrightarrow \mathbb{R}$ by $f_n^x(\sigma) := \langle \widetilde{u}_n(\sigma), x\rangle$ and $f^x : [0,T] \longrightarrow \mathbb{R}$ by $f^x(\sigma) := \langle \widehat{u}(\sigma), x\rangle$. We have seen that $f_n^x \to f^x$ pointwise in $[0,T]$. Estimate (4.6) implies that $\|f_n^x\|_\infty$ is bounded, and for every pair $\sigma, \tau \in [0,T]$ we have, thanks to (4.7), that

$$|f_n^x(\sigma) - f_n^x(\tau)| \le \|x\|_{\mathcal{H}}\|\widetilde{u}_n(\sigma) - \widetilde{u}_n(\tau)\|_{\mathcal{H}} \le \|x\|_{\mathcal{H}} \operatorname{Lip}(\widetilde{u}_n)|\tau - \sigma| \le \|x\|_{\mathcal{H}} \frac{\mathrm{V}_p(\widetilde{u}_n, [0,T])}{T}|\sigma - \tau|,$$

thus $(f_n^x)_n$ is equicontinuous and $f_n^x \to f^x$ uniformly on $[0,T]$ for every $x \in \mathcal{H}$. But $\ell_n(t) \to \ell_u(t)$ for every $t \in [0,T]$, hence for every $x \in \mathcal{H}$ we have that $f_n^x(\ell_n(t)) \to f^x(\ell_u(t))$ for every $t \in [0,T]$, i.e.

$$\widetilde{u}_n(\ell_n(t)) \rightharpoonup \widehat{u}(\ell_u(t)) \quad \text{in } \mathcal{H}, \quad \forall t \in [0,T]. \tag{4.10}$$

On the other hand by the assumptions we know that $\widetilde{u}_n(\ell_{u_n}(t)) = u_n(t) \to u(t)$ for every $t \in [0,T]$, hence, by construction of $\widetilde{u}$ and by the continuity of $\widetilde{u}$ and $\widehat{u}$, we get that $\widehat{u} = \widetilde{u}$ on $\ell_u([0,T])$. Observe now that $\widetilde{u}_n' \rightharpoonup \widehat{u}'$ in $L^2(0,T;\mathcal{H})$, therefore by the lower semicontinuity of the norm with respect to the weak convergence

$$\begin{aligned}
\|\widehat{u}'\|^2_{L^2(0,T;\mathcal{H})} &\le \liminf_{n\to\infty} \|\widetilde{u}_n'\|^2_{L^2(0,T;\mathcal{H})} \le \liminf_{n\to\infty} \int_0^T \left(\frac{\mathrm{V}_p(\widetilde{u}_n,[0,T])}{T}\right)^2 \mathrm{d}\sigma \\
&= \liminf_{n\to\infty} \frac{\mathrm{V}_p(\widetilde{u}_n,[0,T])^2}{T} = \frac{\mathrm{V}_p(\widetilde{u},[0,T])^2}{T} \\
&= \frac{1}{T}\left(\int_0^T \|\widetilde{u}'(\sigma)\|_{\mathcal{H}}\,\mathrm{d}\sigma\right)^2 \le \|\widetilde{u}'\|^2_{L^2(0,T;\mathcal{H})}.
\end{aligned}$$

Thus by Lemma 4.3 and by (2.6)–(2.7), we infer that $\widetilde{u} = \widehat{u}$ on $[0,T]$, so that

$$\widetilde{u}_n \rightharpoonup \widetilde{u} \qquad \text{in } W^{1,p}(0,T;\mathcal{H}). \tag{4.11}$$

Now we prove that $\widetilde{u}_n' \to \widetilde{u}'$ in $W^{1,p}(0,T;\mathcal{H})$ for every $p \in [1,+\infty[$. For every $n \in \mathbb{N}$ we have that, using also the Hölder inequality,

$$\|\widetilde{u}_n'\|^p_{L^p(0,T;\mathcal{H})} = \int_0^T \|\widetilde{u}_n'(\sigma)\|^p_{\mathcal{H}}\,\mathrm{d}\sigma \le \int_0^T \left(\frac{\mathrm{V}_p(\widetilde{u}_n,[0,T])}{T}\right)^p \mathrm{d}\sigma = \frac{\mathrm{V}_p(\widetilde{u}_n,[0,T])^p}{T^{p-1}}.$$

Hence, since $\mathrm{V}_p(\widetilde{u}_n,[0,T]) \to \mathrm{V}_p(\widetilde{u},[0,T])$ as $n \to \infty$, we get that

$$\limsup_{n\to\infty} \|\widetilde{u}_n'\|^p_{L^p(0,T;\mathcal{H})} \le \frac{\mathrm{V}_p(\widetilde{u}_n,[0,T])^p}{T^{p-1}} = \frac{1}{T^{p-1}}\left(\int_0^T \mathrm{V}_p(\widetilde{u},[0,T]\,\mathrm{d}\sigma\right)^p \le \int_0^T \|\widetilde{u}'(\sigma)\|^p_{\mathcal{H}}\,\mathrm{d}\sigma.$$

Therefore we have shown that

$$\limsup_{n\to\infty} \|\widetilde{u}_n'\|_{L^p(0,T;\mathcal{H})} \le \|\widetilde{u}'\|_{L^p(0,T;\mathcal{H})}, \tag{4.12}$$

But we also have

$$\widetilde{u}_n' \rightharpoonup \widetilde{u}' \qquad \text{in } L^p(0,T;\mathcal{H}) \tag{4.13}$$

as $n \to \infty$. Hence, as $L^p(0,T;\mathcal{H})$ is uniformly convex for $p \in]1,+\infty[$, thanks to [15, Proposition III.30] we have that (4.12)–(4.13) imply that

$$\widetilde{u}_n' \to \widetilde{u}' \qquad \text{in } L^p(0,T;\mathcal{H}) \tag{4.14}$$

for every $p \in]1,+\infty[$ as $n \to \infty$. Since $[0,T]$ is bounded we get that (4.14) holds also for $p = 1$ and we are done. From (4.14) follows that $\widetilde{u}_n \to \widetilde{u}$ in $L^p(0,T;\mathcal{H})$ for every $p \in [1,\infty[$, and we are done. □

Now we are in position to extend a general rate independent operator $\mathsf{R} : Lip([0,T];\mathcal{H}) \longrightarrow BV_p([0,T];\mathcal{H}) \cap C([0,T];\mathcal{H})$, i.e. an operator such that

$$\mathsf{R}(u \circ \phi) = \mathsf{R}(u) \circ \phi \qquad \forall u \in Lip([0,T];\mathcal{H}) \tag{4.15}$$

whenever $\phi : [0,T] \longrightarrow [0,T]$ is a Lipschitz continuous increasing map such that $\phi([0,T]) = [0,T]$. If $u \in BV_p([0,T];\mathcal{H})$ then by Proposition 3.2 there exists a sequence $u_n \in Lip([0,T];\mathcal{H})$ such that $\lim_{n\to\infty} \mathrm{V}_p(u_n,[0,T]) = \mathrm{V}_p(u,[0,T])$ and u_n is pointwise convergent to u. Thanks to the rate independence we find that

$$\mathsf{R}(u_n) = \mathsf{R}(\widetilde{u}_n \circ \ell_{u_n}) = \mathsf{R}(\widetilde{u}_n) \circ \ell_{u_n} \qquad \forall n \in \mathbb{N}. \tag{4.16}$$

If we assume that R is continuous with respect to the pointwise strict convergence, then $\mathsf{R}(\widetilde{u}_n) \to \mathsf{R}(\widetilde{u})$ pointwise strictly because by Proposition 4.1 we know that $\widetilde{u}_n \to \widetilde{u}$ pointwise strictly on $[0,T]$. Moreover $\mathsf{R}(\widetilde{u})$ and $\mathsf{R}(\widetilde{u}_n)$ are continuous functions, therefore by Lemma 3.1 it follows that the convergence of $\mathsf{R}(\widetilde{u}_n)$ is uniform. On the other hand, thanks to Lemma 4.1 ℓ_{u_n} is pointwise convergent to ℓ_u, therefore

$$\lim_{n\to\infty} \mathsf{R}(u_n)(t) = (\mathsf{R}(\widetilde{u}) \circ \ell_u)(t) \qquad \forall t \in [0,T]. \tag{4.17}$$

Let us compute the variations. Since $\mathsf{R}(u_n) = \mathsf{R}(\widetilde{u}_n) \circ \ell_{u_n}$ and ℓ_{u_n} is continuous, we find that

$$\mathrm{V}_p(\mathsf{R}(u_n),[0,T]) = \mathrm{V}_p(\mathsf{R}(\widetilde{u}_n),[0,T]) \qquad \forall n, \tag{4.18}$$

therefore by the continuity of R we infer that

$$\lim_{n\to\infty} \mathrm{V}_p(\mathsf{R}(u_n),[0,T]) = \mathrm{V}_p(\mathsf{R}(\widetilde{u}),[0,T]). \tag{4.19}$$

However the pointwise variation of $\mathsf{R}(\widetilde{u})$ is not the same of $\mathsf{R}(\widetilde{u})\circ\ell_u$, because ℓ_u is not continuous. A sufficient condition for the variations to be the same is given by Lemma 4.2: if

$$\begin{aligned} \mathrm{V}_p(\mathsf{R}(\widetilde{u}),[\ell_u(t-),\ell_u(t)]) &= \|\mathsf{R}(\widetilde{u})(\ell_u(t-)) - \mathsf{R}(\widetilde{u})(\ell_u(t))\|_{\mathcal{H}} \qquad \forall t \in \mathrm{Discont}(\ell_u), \\ \mathrm{V}_p(\mathsf{R}(\widetilde{u}),[\ell_u(t),\ell_u(t+)]) &= \|\mathsf{R}(\widetilde{u})(\ell_u(t+)) - \mathsf{R}(\widetilde{u})(\ell_u(t))\|_{\mathcal{H}} \qquad \forall t \in \mathrm{Discont}(\ell_u), \end{aligned}$$

then $\mathrm{V}_p(\mathsf{R}(\widetilde{u}),[0,T]) = \mathrm{V}_p(\mathsf{R}(\widetilde{u}) \circ \ell_u,[0,T])$. Indeed these conditions holds true by virtue of (2.6)–(2.7). Summarizing we have proved the following

Theorem 4.1. *Assume that* $\mathsf{R} : Lip([0,T];\mathcal{H}) \longrightarrow BV_p([0,T];\mathcal{H}) \cap C([0,T];\mathcal{H})$ *is a rate independent operators which is continuous with respect to the pointwise strict convergence. Define* $\overline{\mathsf{R}} : BV_p([0,T];\mathcal{H}) \longrightarrow BV_p([0,T];\mathcal{H})$ *by*

$$\overline{\mathsf{R}}(u) := \mathsf{R}(\widetilde{u}) \circ \ell_u, \qquad u \in BV_p([0,T];\mathcal{H}), \tag{4.20}$$

where $\widetilde{u}$ *is the reparametrization defined defined by Proposition 2.1. Then* $\overline{\mathsf{R}}$ *is the unique continuous extension of* R *in the following sense: if* $u_n \to u$ *pointwise strictly on* $[0,T]$ *then* $\overline{\mathsf{R}}(u_n)(t) \to \overline{\mathsf{R}}(u)(t)$ *for every* $t \in [0,T]$. *Moreover if we assume that*

$$\mathrm{V}_p(v,[c,d]) = \|v(d) - v(c)\|_{\mathcal{H}} \implies \mathrm{V}_p(\mathsf{R}(v),[c,d]) = \|\mathsf{R}(v)(d) - \mathsf{R}(v(c))\|_{\mathcal{H}} \tag{4.21}$$

whenever $0 \le c < d \le T$, *then* $\overline{\mathsf{R}}$ *is continuous with respect to pointwise strict convergence.*

In [2] we called *locally isotone* an operator satisfying (4.21). The argument of [2, Proposition 4.2] shows that condition (4.21) is also necessary in order to obtain that $\overline{\mathsf{R}}$ is continuous with respect to pointwise strict convergence. If $\mathcal{H} = \mathbb{R}$ this condition is a slight generalization of the well known local monotonicity.

In the scalar case $\mathcal{H} = \mathbb{R}$ the hypotheses of Theorem 4.1 are satisfied by most of the concrete hysteresis operators occurring in applications (see, e.g., [1, Section 5] and [16, Section 5]). In the multidimensional case the condition (4.21) is very restrictive: for instance the vector play operator satisfies it if and only if its characteristic is a vector subspace or is the intersection of two parallel half spaces (cf. [2]). However the vector play operator is continuous with respect to the pointwise strict convergence on $Lip([0,T];\mathcal{H})$ ([2, Theorem 3.3]), therefore it can be extended in a weaker sense to BV_p by means of formula (4.20).

Acknowledgments
I would like to thank the organizers of the International Workshop MURPHYS 2010 for giving me the chance to take part of this stimulating scientific meeting.

References

[1] Recupero V 2009 BV-extension of rate independent operators *Math. Nachr.* **282** 86
[2] Recupero V BV solutions of rate independent variational inequalities *Ann. Sc. Norm. Super. Pisa Cl. Sc. (5)* in press.
[3] Visintin A 1994 *Differential Models of Hysteresis* (Berlin: Springer-Verlag)
[4] Krejčí P and Laurençot P 2002 Generalized variational inequalities *J. Convex Anal.* **9** 159
[5] Krasnosel'skii M A and Pokrovskii A V 1989 *Systems with Hysteresis* (Heidelberg: Springer-Verlag)
[6] Krejčí P 1996 *Hysteresis, Convexity and Dissipation in Hyperbolic Equations* (Tokyo: Gakkatosho)
[7] Brokate M and Sprekels J 1996 *Hysteresis and Phase Transitions* (New York: Springer-Verlag)
[8] Logemann H and Mawby A D 2003 Extending hysteresis operators to spaces of piecewise continuous functions *J. Math. Anal. Appl.* **282** 107
[9] Iyer R and Ekanayake D 2008 Extension of hysteresis operators of Preisach-type to real Lebesgue measurable functions *Physica B* **403** 437
[10] Krejčí P and Liero M 2009 Rate independent Kurzweil processes, *Appl. Math.* **54** 117
[11] Lang S 1993 *Real and Functional Analysis - Third Edition* (New York: Springer Verlag).
[12] Brezis H 1973 *Operateurs maximaux monotones et semi-groupes de contractions dans les espaces de Hilbert* (Amsterdam: North-Holland Publishing Company)
[13] Federer H 1969 *Geometric Measure Theory* (Berlin: Springer-Verlag)
[14] Bourbaki N 1958 *Fonctions d'une variable rèelle"* (Paris: Hermann)
[15] Brezis H 1983 *Analyse Fonctionelle - Théorie et applications* (Paris: Masson)
[16] Recupero V 2009 Sobolev and strict continuity of general hysteresis operators *Math. Methods Appl. Sci.* **32** 2003

Journal of Physics: Conference Series **268** (2011) 012025 doi:10.1088/1742-6596/268/1/012025

Measuring of rotating magnetic flux in an integrated environment

Z Sari and A Ivanyi

Department of Information Technology, Faculty of Engineering, University of Pécs, Rokus 2. Pécs, Hungary

E-mail: ski@morpheus.pte.hu

Abstract. The paper describes a measuring method and the establishment of a measuring environment for two-dimensional rotating magnetic fluxes and fields. The measuring method relies on measuring currents and voltages in coils attached to the magnetic specimen. For the processing and generating of measuring signals, a PC-based system running LabView is applied extended with a NI ELVIS II board, and two high performance KIKUSUI bipolar power supplies. The measurement results are satisfactory and show that the measuring environment established is capable of non-destructively capture the two-dimensional magnetic characteristics of the specimen investigated.

1. Introduction

Since there are many application areas where rotating magnetic fields and fluxes appear, and the intrinsic vectorial nature of hysteresis can not be neglected, the goal of the paper is to present a possible approach of two-dimensional measurement and investigation of hysteresis characteristics under rotating field conditions. The measuring environment (hardware/software), the two-dimensional set-up, and the theory behind the measurement is discussed along with the brief description of the applicability of hardware and software used. Measurement results are presented in order to prove the applicability of the method.

2. The method of two-dimensional measurement

The schematic of the measurement setup can be seen in figure 1. In the figure it can be seen, that the cross-shaped specimen is equipped by four coils; two primary coils (x and y directions) providing the excitation current [1], and two diagonally attached perpendicular secondary coils for the measuring of induced voltage. The theory behind the measurement can be derived from the quasi-stationary Maxwell equations as follows [2].

The set of equations needed is

$$\nabla \times \boldsymbol{H} = \boldsymbol{J}_s + \sigma \boldsymbol{E}, \tag{1}$$

$$\nabla \times \boldsymbol{E} = -\frac{\partial \boldsymbol{B}}{\partial t}, \tag{2}$$

where $\boldsymbol{H}$ is the magnetic field strength, $\boldsymbol{E}$ is the electric field, $\boldsymbol{B}$ is the magnetic flux density and $\boldsymbol{J}_s$ is the external current density. The eddy current term $\sigma\boldsymbol{E}$ in (1) can be omitted due to the thin (0.27 mm) sheet specimen and low (max. 50Hz) frequency applied during this measurement.

2.1. Theory of measurement

As figure 1 shows, the whole measurement procedure is conducted by a PC equipped with National Instruments LabView 8.6 software and a NI ELVIS II board. The NI ELVIS II board is responsible for interfacing the signal flow between the computer and the actual measuring device. The specimen measured consists of two perpendicularly attached Fe-Si 3.1%(wt) transformer sheets, which have thickness of 0.27mm. For the purpose of excitation there are two primary excitation coils (x and y directions) driven by two KIKUSUI PBX 20-20 bipolar power supplies as current sources[1]. The secondary coils (u' and u'') attached are responsible for measuring the induced voltage by the aid of the NI ELVIS II board.

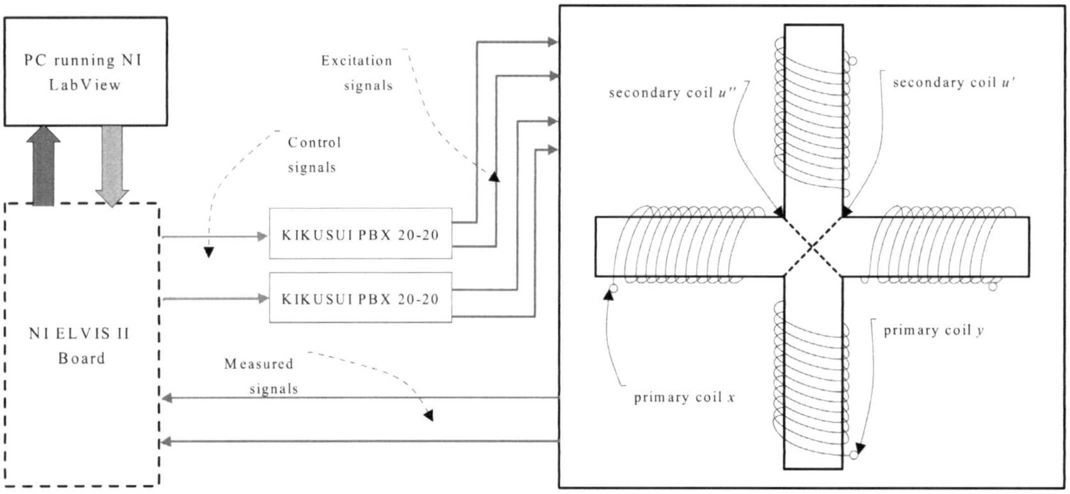

Figure 1. The schematics of the measurement set-up.

This set-up enables to produce various two-dimensional excitations by creating nearly homogenous magnetic field strength aligned to different directions at the center of the specimen. The actual excitation is generated in the form of two current signals (x and y directions respectively) with the aid of the two KIKUSUI PBX 20-20 programmable bipolar power supplies.

The relation of the $\boldsymbol{H}$ field to the excitation current can be derived from (1) like

$$\nabla \times \boldsymbol{H} = \boldsymbol{J}_s, \tag{3}$$

which can be written in integral form on a properly selected surface Ω in the investigated domain as follows

$$\int_\Omega \nabla \times \boldsymbol{H} \, d\Omega = \int_\Omega \boldsymbol{J}_s \, d\Omega, \tag{4}$$

and by the application of Stokes' theorem, (4) has the form

[1] The frequency of the excitation signals was 10 Hz.

$$\oint_l \boldsymbol{H}\ dl = \int_\Omega \boldsymbol{J}_s\ d\Omega, \tag{5}$$

where the right hand side is basically the sum the excitation currents in the coil-domain, and assuming that the field strength is constant along the path of the integral (due to the symmetrical set-up of Epstein-frame) the left hand side integral can be substituted by $\boldsymbol{H}l$, thus

$$\boldsymbol{H}l = N_p I, \tag{6}$$

where N_p is the number of turns of the primary excitation coil, l is the length of the frame around the specimen, and I is the primary excitation current.

In a similar way the flux density $\boldsymbol{B}$ can be derived from the integral form of (2)

$$\int_\Omega \nabla \times \boldsymbol{E}\ d\Omega = -\frac{\partial}{\partial t}\int_\Omega \boldsymbol{B}\ d\Omega, \tag{7}$$

applying Stokes' theorem we have

$$\int_l \boldsymbol{E}\ dl = -\frac{\partial}{\partial t}\int_\Omega \boldsymbol{B}\ d\Omega, \tag{8}$$

where the left hand side of the equation is the voltage u_t of one turn of the measuring (secondary) coil, and the integral on the right hand side can be substituted by $\boldsymbol{B}A_s$ under the assumption that the flux density $\boldsymbol{B}$ is homogenous in the central region of the specimen, and spread flux can be neglected. Now (8) can be written as

$$u_t = -\frac{\partial}{\partial t}\boldsymbol{B}A_s. \tag{9}$$

Since the measuring coil has N_s turns, the actual measured voltage u of the secondary coil is N_s times higher than u_t, thus finally we have

$$\frac{1}{N_s}u = -\frac{\partial}{\partial t}\boldsymbol{B}A_s, \tag{10}$$

where A_s is the cross-sectional area of the the sheet that the specimen is made of, N_s is the number of turns of the secondary coil. Writing (10) in integral form leads to

$$\boldsymbol{B} = \frac{1}{N_s A_s}\int u\ dt. \tag{11}$$

At this point (6) and (11) is enough to calculate the field strength and flux density inside the specimen under homogenous assumptions. It is important to note here, that the components of the flux density $\boldsymbol{B}$ obtained by (11) are not aligned with the x and y directions of the specimen[2], due to the diagonally attached measuring coils. As figure 2 shows, the diagonal component B' of the flux density has a contribution from both B_x and B_y, thus we actually measure the superposition of them.

The orientation of the components of the flux density $\boldsymbol{B}$ calculated by (11) depends on the angle α between the measuring (secondary) coils and the axes of the coordinate system attached to the specimen investigated. According to the set-up in figure 1, the axes of the coordinate system (aligned with primary excitation coils) are parallel to the horizontal and vertical part of the cross shaped sheet, and the two measuring coils are situated along the diagonals in the center.

[2] It could be useful to introduce a different symbol for this transformed $\boldsymbol{B}$, but for the sake of easy readability it will be omitted.

Since the measuring (secondary) coils are attached diagonally to the specimen (see figure 1), the calculation of actual flux density $\boldsymbol{B} = [B_x, B_y]^T$ (corresponding to the field strength $\boldsymbol{H} = [H_x, H_y]^T$) requires the solution of the following linear system

$$\begin{bmatrix} \cos\alpha & \sin\alpha \\ -\sin\alpha & \cos\alpha \end{bmatrix} \begin{bmatrix} B_x \\ B_y \end{bmatrix} = \begin{bmatrix} B' \\ B'' \end{bmatrix} \tag{12}$$

where B' and B'' are the components of the measured flux density calculated by (11).

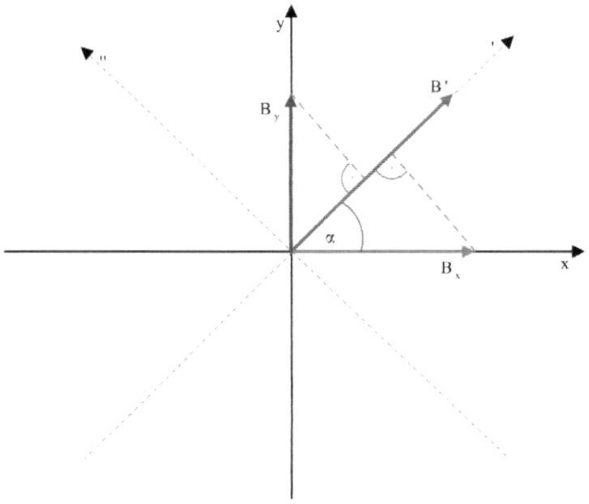

Figure 2. Transformation of the geometry.

2.2. *The compensated measurement procedure*

The goal of the vectorial measurement is to obtain circular (uniform magnitude) flux density, and since the excitation is directly related to the field strength $\boldsymbol{H}$ instead of flux density $\boldsymbol{B}$, an iterative compensation procedure is required in order to achieve the above goal.

During the iterative measuring procedure we get a more and more uniform magnitude flux density by adjusting the input signal in each iteration step according to the error function $\varepsilon(\phi)$ of the preceding iteration. The error function can be defined as

$$\varepsilon^{(i)}(\phi) = \|B_m^{(i)}(\phi) - B_{ref}\|, \tag{13}$$

where $B_m^{(i)}(\phi) = \|\boldsymbol{B}^{(i)}(\phi)\|$ is the magnitude of the rotating flux density at angle ϕ at the i-th iteration step, and B_{ref} is the desired magnitude, which is about to be achieved by the iterative measuring procedure. The iteration stops if the total absolute error falls below a given threshold ϵ, like

$$\int_0^{2\pi} \varepsilon(\phi)\, d\phi < \epsilon. \tag{14}$$

The schematics of the measuring algorithm can be seen in figure 3.

As it can be seen in the figure, after the i-th measurement step is accomplished and the processing of the measured data had been carried out, the error function of the actual iteration step is calculated according to (13). If (14) holds, the iteration terminates, if not, the algorithm adjusts the excitation signal as follows

$$H_m^{(i+1)}(\phi) = H_m^{(i)}(\phi) - \eta\varepsilon^{(i)}(\phi) \tag{15}$$

Journal of Physics: Conference Series **268** (2011) 012025 doi:10.1088/1742-6596/268/1/012025

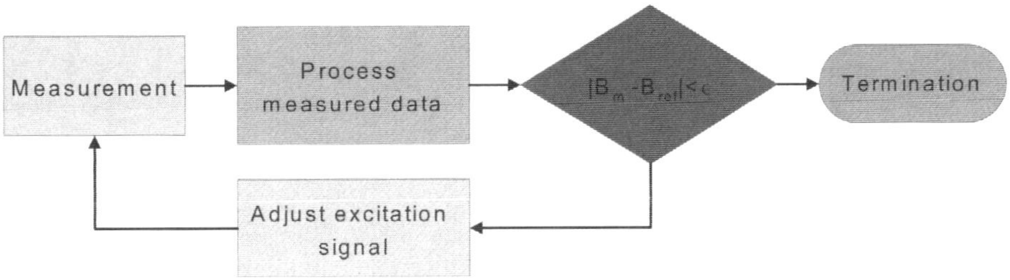

Figure 3. The schematics of one iteration cycle of the compensated measurement.

where $H_m^{(i+1)}(\phi) = \|\boldsymbol{H}^{(i+1)}(\phi)\|$ and $H_m^{(i)}(\phi) = \|\boldsymbol{H}^{(i)}(\phi)\|$ are the magnitude of the rotating field strength at angle ϕ at the $(i+1)$-th and i-th iteration steps respectively, and η is a properly selected feedback coefficient.

3. Results and discussion

In this section various measurement results obtained by the method outlined above is presented.

3.1. Scalar measurements along x and y axes

The specimen investigated is constructed by attaching two Fe-Si sheets with slightly different characteristics. As a consequence of this the set-up shows anisotropic behaviour. In figure 4 the separate hysteresis curves corresponding to the x and y directions can be seen.

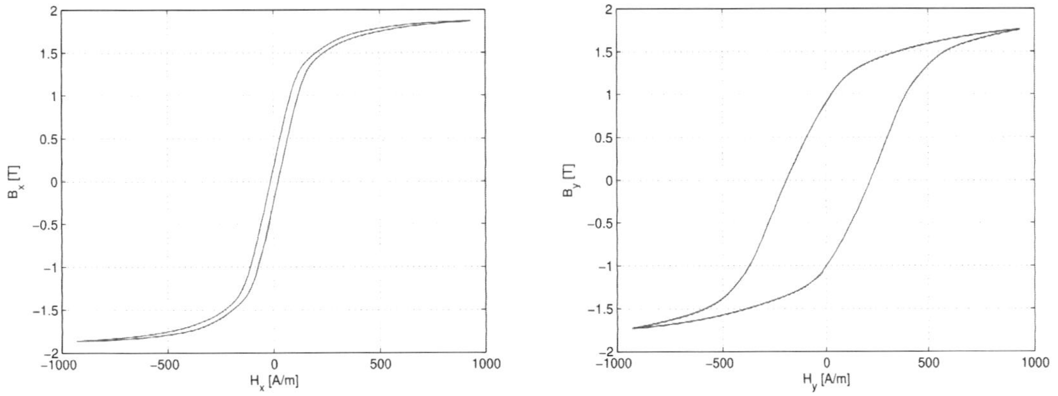

Figure 4. Hysteresis characteristics along the x and y axes respectively.

As it is apparent from the figure the sheet aligned to the x axis has a sharper characteristic with smaller coercivity and higher differential permeability. These differences certainly affect the vectorial measurement results as well, as it is shown in the next section.

3.2. Vector measurements

The results correspond to different reference flux density B_{ref} values showing the circular (almost uniform magnitude) flux pattern and the field strength excitation. The allowed deviation of flux patterns from the perfectly uniform circle is controlled by the error threshold ϵ defined in (14).

Figures 5,6, and 7 show the circular flux patterns and the corresponding two-dimensional excitation. It is apparent from the figures, that the investigated specimen shows anisotropic behavior, the field strength patterns are not symmetrical. In order to obtain a circular, uniform magnitude flux density, significantly higher magnitude field strength is required to be applied to the y direction.

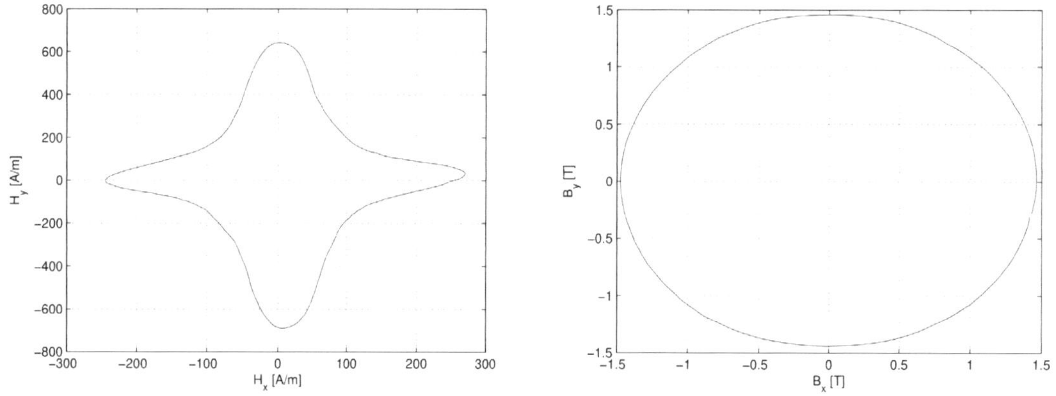

Figure 5. Rotating field and flux density pattern at reference flux B_{ref}=1.45 T.

This magnitude difference between the directions is higher in the case of lower reference flux density values (figure 5), and the difference is lower in the higher magnitude flux density region (figure 7) since as the specimen is getting close to saturation, the differences between the x and y hysteresis characteristics (see figure 4) begin to vanish.

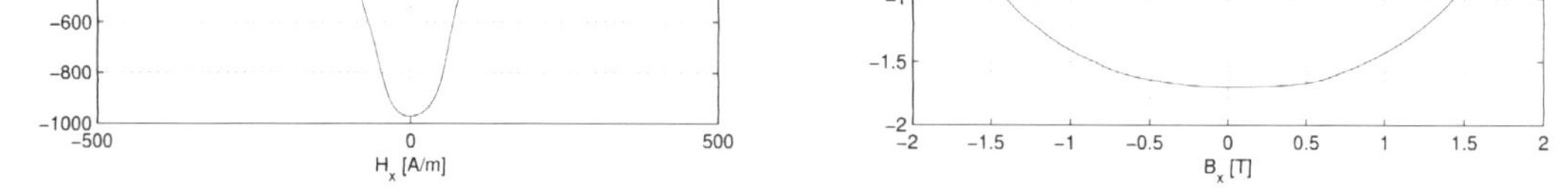

Figure 6. Rotating field and flux density pattern at reference flux B_{ref}=1.725 T.

It can also be seen in the figures, that the axis of easy magnetization has a slight declination from the x axis of the coordinate system attached to the specimen. This phenomenon is getting stronger with the increasing of the reference flux density B_{ref}, as it is clear from figure 7.

4. Conclusion and future works

A possible measuring environment is established, and a non-destructive way of hysteresis measurement method is applied for two-dimensional vector hysteresis measurements in order to investigate the magnetic field strength and flux density patterns. Both the hardware-software environment - consisting of NI LabView, NI ELVIS II board and the two programmable power supplies - and the theory applied proved to be acceptable for the purpose of successful vectorial measurements, investigations. The PC-based measuring environment provides great flexibility and programmability along with high accuracy.

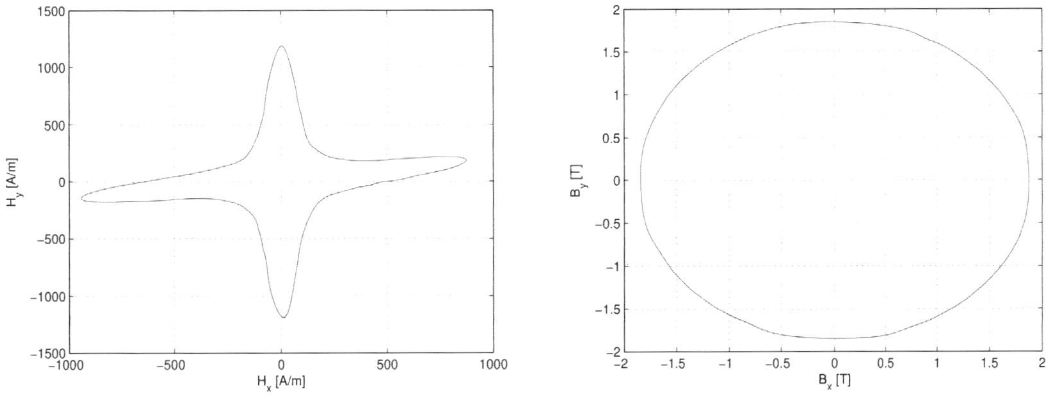

Figure 7. Rotating field and flux density pattern at reference flux B_{ref}=1.845 T.

As a future goal it is planned that the environment is extended by a programmable Finite Element package in order to carry out more complicated modelling/simulation tasks, investigate more complex flux patterns and to validate various vector hysteresis model implementations.

5. References

[1] Enokizono M 1992 *Two-dimensional magnetic measurement and its properties: JSAEM studies in applied electromagnetics* vol.1 (Oita)

[2] Ivanyi A 1997 *Hysteresis models in electromagnetic computation* (Budapest, Akadémiai Kiadó)

Journal of Physics: Conference Series **268** (2011) 012026 doi:10.1088/1742-6596/268/1/012026

Modelling of droplet heating and evaporation: recent results and unsolved problems

S S Sazhin[1*], I N Shishkova[2], A Elwardany[1], I G Gusev[1] and M R Heikal[1]

[1]Sir Harry Ricardo Laboratories, School of Computing, Engineering and Mathematics, University of Brighton, Brighton, BN2 4GJ United Kingdom

[2]Low Temperature Department, Moscow Power Engineering Institute, Moscow 111250, Russia

*E-mail: S.Sazhin@brighton.ac.uk

Abstract. The most recent results referring to the hydrodynamic and kinetic modelling of droplet heating and evaporation are briefly summarised. Two new solutions to the heat conduction equation, taking into account the effect of the moving boundary during transient heating of an evaporating droplet, are discussed. The first solution is the explicit analytical solution to this equation, while the second one reduces the solution of the differential transient heat conduction equation to the solution of the Volterra integral equation of the second kind. It has been pointed out that the new approach predicts lower droplet surface temperatures and slower evaporation rates compared with the traditional approach. A simplified model for multi-component droplet heating and evaporation, based on the analytical solution of the species diffusion equation inside droplets, is reviewed. A new algorithm, based on simple approximations of the kinetic results for droplet radii and temperatures, suitable for engineering applications, is discussed.

1. Introduction

The importance of accurate and CPU efficient modelling of droplet heating and evaporation in various engineering and environmental applications is well recognised. These models are typically based on a number of rather restrictive assumptions, including the assumption that the thermal conductivity of droplets is infinitely large and, in the case of multi-component droplets, the diffusivity of species within them is infinitely large or small. Kinetic effects have been ignored. In a series of our previous papers, a number of simplified models of droplet heating and evaporation have been developed, where the effects of the finite thermal conductivity of droplets and recirculation within them, and kinetic effects in the evaporation process have been taken into account [1]-[6]. The format of most of these models was such that they could be incorporated into CFD codes and used for modelling real-life engineering processes. The results of these papers are summarised in a number of reviews, including [7]-[9].

The aim of this paper is to summarise some results of the most recent developments of these models. These new or further developed models take into account the impact of moving boundary and kinetic effects on the evaporation process and the finite rate of species diffusion in multi-component droplets. The models are presented in the format in which they can be potentially implemented into engineering CFD codes. This review is essentially an updated version of the previous review [9]. Particular emphasis is put on unsolved problems on which we are currently working.

2. Effects of the moving boundary

The models for droplet heating used in [1, 2] were implicitly based on the assumptions that the evaporation rate of droplets is small and the value of droplet radius R_d does not change during any time step (although this radius changes from one step to another). This means that the effect of a moving boundary on droplet heating was ignored. This is a well known approach used in all available CFD codes. We attempted to take into account the changes in droplet radius R_d during the time steps. Two approaches were developed. Firstly it was assumed that R_d is a linear function of time. Secondly, we calculated the evolution of droplet temperature for an *a priori* fixed function $R_d(t)$. In both cases the droplet temperature ($T \equiv T(t, R)$) was found from the solution of the heat conduction equation in the form [10, 11]:

$$\frac{\partial T}{\partial t} = \kappa \left(\frac{\partial^2 T}{\partial R^2} + \frac{2}{R}\frac{\partial T}{\partial R} \right) \tag{1}$$

for $0 \leq t < t_e$, $0 \leq R < R_d(t)$, where κ is the liquid thermal diffusivity ($\kappa = k_l/(c_l \rho_l)$), k_l is the thermal conductivity, c_l is the specific heat capacity, ρ_l is the density, R is the distance from the centre of the droplet. The effect of thermal radiation is ignored.

Remembering the physical background to the problem, we look for the solution of this equation in the form of a twice continuously differentiable function $T \equiv T(t, R)$ for $0 \leq t < t_e$, $0 \leq R < R_d(t)$. This solution should satisfy the boundary condition:

$$\left(k_l \frac{\partial T}{\partial R} + hT \right)\Big|_{R=R_d(t)} = hT_g + \rho_l L \dot{R}_d(t), \tag{2}$$

T is finite and continuous at $R \to 0$, $T_s = T(R_d(t), t)$ is the droplet's surface temperature, L is the latent heat of evaporation. We took into account that $\dot{R}_d(t) \equiv \mathrm{d}R_d/\mathrm{d}t \leq 0$. Effects of swelling are ignored. Equation (2) is the energy balance condition at $R = R_d(t)$. The initial condition is taken in the form:

$$T(t = 0) = T_0(R), \tag{3}$$

where $0 \leq R \leq R_{d0} = R_d(t = 0)$. The value of $R_d(t)$ is controlled by fuel vapour diffusion from the droplet surface, and can be found from the equation [7]:

$$\dot{R}_d = -\frac{k_g \ln(1 + B_M)}{\rho_l c_{pg} R_d}, \tag{4}$$

where $B_M = Y_{vs}/(1 - Y_{vs})$ is the Spalding mass transfer number, Y_{vs} is the mass fraction of fuel vapour near the droplet surface, which is ultimately controlled by gas temperature [7].

In the model described in [1, 2], equation (1) with the boundary condition (2) and fixed R_d and h during the time step was solved analytically. This solution was used for the determination of the distribution of temperature inside droplets at the end of the time step. The value of R_d at the end of this time step was obtained from equation (4). The distribution of temperature at the beginning of the next time step was recalculated for the new droplet radius. This model was generalised by assuming that $R_d(t)$ is the linear function of t during the time step [12]:

$$R_d(t) = R_{d0}(1 + \alpha t). \tag{5}$$

Remembering (4), the expression for α can be presented as:

$$\alpha = -\frac{k_g \ln(1 + B_M)}{\rho_l c_{pg} R_{d0}^2}. \tag{6}$$

Assuming that α (rather than R_d) is fixed during the time step, the analytical solution to equation (1) subject to the above boundary and initial conditions was obtained in the form [12]

$$T(R) = \frac{1}{R\sqrt{R_d(t)}} \exp\left[-\frac{\alpha R_{d0} R^2}{4\kappa R_d(t)}\right] = \left[\sum_{n=1}^{\infty} \Theta_n(t) \sin\left(\lambda_n \frac{R}{R_d(t)}\right) + \frac{\mu_0(t)}{1+h_0} \frac{R}{R_d(t)}\right], \quad (7)$$

where

$$\Theta_n(t) = \Theta_n(0) \exp\left[\frac{\kappa\lambda_n^2}{\alpha R_{d0}^2}\left(\frac{1}{1+\alpha t} - 1\right)\right] + f_n \int_0^t \frac{\mathrm{d}\mu_0(\tau)}{\mathrm{d}\tau} \exp\left[\frac{\kappa\lambda_n^2}{\alpha R_{d0}^2}\left(\frac{1}{1+\alpha t} - \frac{1}{1+\alpha\tau}\right)\right] \mathrm{d}\tau, \quad (8)$$

$$\mu_0(t) \equiv \tilde{\mu}(t)\sqrt{R_d(t)} \exp\left[\frac{R_d'(t) R_d(t)}{4\kappa}\right], \quad (9)$$

$\tilde{\mu}(t) = M(t) R_d^2(t)$, $M(t) = \frac{h}{k_l} T_g + \frac{\rho_l}{k_l} L \dot{R}_d(t)$, $f_n = -\frac{\sin\lambda_n}{||v_n||^2 \lambda_n^2}$, $v_n(\xi) = \sin\lambda_n \xi$ $(n = 1, 2, ...)$.

$$|| v_n ||^2 = \frac{1}{2}\left(1 - \frac{\sin 2\lambda_n}{2\lambda_n}\right) = \frac{1}{2}\left(1 + \frac{h_0}{h_0^2 + \lambda_n^2}\right). \quad (10)$$

λ_n are positive solutions to the equation

$$\lambda \cos\lambda + h_0 \sin\lambda = 0 \quad (11)$$

presented in ascending order,

$$h_0 = \frac{h(t)}{k_l} R_d(t) - 1 - \frac{R_d'(t) R_d(t)}{2\kappa} \quad (12)$$

is assumed to be constant during the time step,

$$\Theta_n(0) = q_n + \mu_0(0) f_n, \quad q_n = \frac{1}{|| v_n ||^2} \int_0^1 W_0(\xi) v_n(\xi) \mathrm{d}\xi.$$

$$W_0(\xi) = R_{d0}^{3/2} \xi T_0(\xi R_{d0}) \exp\left[\frac{R_d'(0) R_{d0}}{4\kappa} \xi^2\right]. \quad (13)$$

If Assumption (5) is replaced by the assumption that $T_{d0}(R) = \mathrm{const}$ then the solution of equation (1) subject to the above-mentioned boundary and initial conditions can be presented as:

$$T(t, R) = T_{d0} + \frac{\sqrt{\kappa}}{2R\sqrt{\pi}} \int_0^t \frac{\nu(\tau)}{\sqrt{t-\tau}} \left\{\exp\left[-\frac{(R - R_d(\tau))^2}{4\kappa(t-\tau)}\right] - \exp\left[-\frac{(R + R_d(\tau))^2}{4\kappa(t-\tau)}\right]\right\} \mathrm{d}\tau, \quad (14)$$

where $\nu(\tau)$ is the solution to the following integral equation:

$$\frac{\nu(t)}{2} + \int_0^t \nu(\tau) \left\{\left[\frac{\partial G(t, \tau, R)}{\partial R}\right]\Big|_{R=R_d(t)} + H(t) G(t, \tau, R_d(t))\right\} \mathrm{d}\tau = \mu_0(t), \quad (15)$$

where

$$G(t, \tau, R) = \frac{\sqrt{\kappa}}{2\sqrt{\pi(t-\tau)}} \left\{\exp\left[-\frac{(R - R_d(\tau))^2}{4\kappa(t-\tau)}\right] - \exp\left[-\frac{(R + R_d(\tau))^2}{4\kappa(t-\tau)}\right]\right\} \quad (16)$$

(note that $G(t, \tau, R = 0) = 0$), $H(t) = \frac{h(t)}{k_l} - \frac{1}{R_d(t)}$, $\mu_0(t) = -\frac{h}{k_l} R_d(t) T_{d0} + \mu(t)$, $\mu(t) = M(t) R_d(t)$, $M(t) = \frac{h}{k_l} T_g + \frac{\rho_l}{k_l} L \dot{R}_d(t)$.

Solution (14) was applied iteratively. At first $R_d(t)$ was found using the conventional approach when $R_d =$ const during the time step. Then this $R_d(t)$ was substituted into Solution (14). At the next step this solution for $T(R, t)$ was used to obtain the updated value of $R_d(t)$ etc. The calculations were performed for $R_{d0} = 10$ μm, and gas temperature was taken equal to 1000 K. We assumed that fuel can be approximated as n-dodecane ($M_f = 170$ kg/kmole) and $p = 3000$ kPa (typical values for Diesel engine-like conditions). The time step 0.008 ms was used for the conventional approach while the time step 0.01 ms was used for the new approach based on Solution (7). The predictions of Solutions (7) and (14) practically coincided which gives additional confidence in our new model. The new approach predicts slightly smaller droplet surface temperatures and slightly longer evaporation times compared with the conventional approach $((t_{e(\text{new})} - t_{e(\text{conventional})})/t_{e(\text{new})}$ = (2.49 ms −2.38 ms)/2.49 ms = 0.044=4.4%). This increase in the evaporation time is certainly large enough to warrant taking into account the effect considered in this paper in the practical modelling of droplet heating and evaporation for engineering applications.

Species diffusion inside droplets

The analysis will be restricted to the models of multi-component droplet heating and evaporation based on the analysis of individual components (e.g. [13, 14, 15]), applicable in the case when a small number of components needs to be taken into account. Moreover, at this stage only the simplest case of bi-component droplets is considered, as in [15], which allows us to develop a better understanding of the underlying physics of the processes involved. The new model, originally described in [16, 17], is much simpler than the vortex models (e.g. [15]) which makes it potentially attractive for implementation into computational fluid dynamics (CFD) codes. The predictions of the model agreed with the measured time evolution of droplet temperatures in monodisperse bi-component (ethanol/acetone) droplet streams.

The new model is based on the analytical solution (7) for $R_d =$ const combined with equation (4) as in the case of the mono-component droplets considered in the previous section. These equations are complemented by the equation for species mass fraction ($Y_{li} \equiv Y_{li}(t, R)$) inside droplets [7]:

$$\frac{\partial Y_{li}}{\partial t} = D_l \left(\frac{\partial^2 Y_{li}}{\partial R^2} + \frac{2}{R} \frac{\partial Y_{li}}{\partial R} \right), \tag{17}$$

where $i = 1, 2$, D_l is the liquid mass diffusivity. Equation (17) is solved with the following boundary condition [7]:

$$\alpha_Y (\epsilon_i - Y_{lis}) = -D_l \left. \frac{\partial Y_{li}}{\partial R} \right|_{R=R_d-0}, \tag{18}$$

and the initial conditions $Y_{li}(t = 0) = Y_{li0}(R)$, where $Y_{lis} = Y_{lis}(t)$ are liquid components' mass fractions at the droplet's surface,

$$\alpha_Y = |\dot{m}_d| / \left(4\pi \rho_l R_d^2\right), \tag{19}$$

$\dot{m}_d$ is the droplet evaporation rate determined by equation (4), ϵ_i are species evaporation rates determined from the following equation:

$$\epsilon_i = Y_{vis} / \sum_i Y_{vis}, \tag{20}$$

where the subscript $_v$ indicates the vapour phase.

Equations (1) and (17), originally derived for spherically symmetric stationary droplets, were generalised to the case of moving droplets by replacing k_l and D_l with the so called effective thermal conductivity and diffusivity [7, 18]:

$$k_{\text{eff}} = \chi k_l, \qquad D_{\text{eff}} = \chi_Y D_l,$$

where the coefficients χ and χ_Y vary from 1 to 2.72 depending on Reynolds, Prandtle and Schmidt numbers (see [7] for details); in the case of stationary droplets $\chi = \chi_Y = 1$.

The molar fractions of species i in the vapour phase at the surface of the droplet were determined from the following equation:

$$X_{vi} = X_{li}\gamma_i \exp\left[\frac{L_i M_i}{R_u}\left(\frac{1}{T_{bi}} - \frac{1}{T_s}\right)\right], \tag{21}$$

where X_{li} is the liquid molar fraction at the surface of the droplet, γ_i is the activity coefficient, L_i M_i and T_{bi} are the latent heat of evaporation, molar mass and boiling temperature of species i respectively, R_u is the universal gas constant, T_s is the droplet surface temperature. The activity coefficient for the ethanol-acetone mixture was estimated following [19].

Although equation (17) looks identical to (1), Solution (5) to equation (1) cannot be used for equation (17), due to different boundary conditions (equations (2) and (19)). The analytical solution to equation (17), subject to the above boundary and initial conditions was obtained in the form [16, 17]:

$$Y_{li} = \epsilon_i + \frac{1}{R}\left\{\left[\exp\left[D_l\left(\frac{\lambda_0}{R_d}\right)^2 t\right]\left[q_{i0} - \epsilon_i(0)Q_0\right]\right]\sinh\left(\lambda_0\frac{R}{R_d}\right)\right.$$
$$\left. + \sum_{n=1}^{\infty}\left[\exp\left[-D_l\left(\frac{\lambda_n}{R_d}\right)^2 t\right]\left[q_{in} - \epsilon_i(0)Q_n\right]\right]\sin\left(\lambda_n\frac{R}{R_d}\right)\right\}, \tag{22}$$

where

$$Q_n = \begin{cases} -\frac{1}{||v_0||^2}\left(\frac{R_d}{\lambda_0}\right)^2(1+h_0)\sinh\lambda_0 & \text{when} \quad n=0 \\ \frac{1}{||v_n||^2}\left(\frac{R_d}{\lambda_n}\right)^2(1+h_0)\sin\lambda_n & \text{when} \quad n\geq 1 \end{cases} \tag{23}$$

$$q_{in} = \frac{1}{||v_n||^2}\int_0^{R_d} RY_{li0}(R)v_n(R)\mathrm{d}R, \tag{24}$$

$n \geq 0$, λ_0 is the solution to the equation

$$\tanh\lambda = -\frac{\lambda}{h_0}, \tag{25}$$

while λ_n ($n \geq 1$) are solutions to equation (11),

$$h_0 = -\left(1 + \frac{\alpha R_d}{D_l}\right), \tag{26}$$

(note that h_0 used in this section is different from the one used in the previous section; the same symbol is used in both cases to highlight similarities between two solutions),

$$v_n = \begin{cases} \sinh\left(\lambda_0 R/R_0\right) & \text{when} \quad n=0 \\ \sin\left(\lambda_n R/R_0\right) & \text{when} \quad n\geq 1. \end{cases} \tag{27}$$

Solution (22) was incorporated into the numerical scheme for simulating monodisperse ethanol-acetone droplet heating and evaporation in the experiment described in [15]. The results referring to average droplet temperatures were shown to be reasonably close to both experimental data and the predictions of the more rigorous and complex vortex model. This could justify the application of this model instead of the vortex model for modelling the processes of multi-component droplet heating and evaporation.

3. Kinetic effects

In a series of our previous papers (see [5] and the references therein) it was demonstrated that, even in the case of droplet heating and evaporation into a high-pressure gas (Diesel engine-like conditions), the kinetic effects may not be negligible. In these papers, two regions of gas above the surface of the evaporating droplet were considered: the kinetic and hydrodynamic regions. In the most comprehensive model, described in [5], it was assumed that gas consisted of two components, fuel vapour and background gas (air), and both mass and heat transfer processes in the kinetic region were taken into account. The analysis of the processes in this region was based on the numerical solution of the Boltzmann equations for fuel vapour and background gas with the relevant boundary conditions. The CPU requirements of this solution, however, make it impossible to incorporate it into engineering computational fluid dynamics (CFD) codes designed to model fuel droplet evaporation in a realistic engineering environment, taking into account all complexities of the processes in Diesel engines (see [20, 21]).

It seems the only feasible way to apply the results of [5] to the analysis of Diesel fuel droplet heating and evaporation within these CFD codes would be to approximate the results of kinetic calculations using simple approximate analytical formulae. The selection of the approximations should be based on finding a compromise between their accuracy and simplicity. The preliminary results of our analysis in this direction were reported in [22, 23]. Some more recent results are presented in this section, based on [6, 9]. The most important ranges of the values of fuel droplet radii and gas temperatures were considered, while the initial value of droplet temperature was assumed equal to 300 K. Although the applicability of the approximations found was restricted to the very specific problem of modelling the processes in Diesel engines, it is anticipated that this approach can be generalised to a much wider range of engineering applications.

Based on the direct comparison of the results of kinetic and hydrodynamic computations for gas temperatures T_g=750 K, 1000 K and 1500 K and initial droplet radii R_0=5 μm and R_0=20 μm, it was shown that the best approximation for droplet radii, predicted by the kinetic model, can be achieved using the following formula:

$$\overline{\Delta R} = a_{1R} \exp\left(-\frac{\overline{R}}{t_{1R}}\right) + a_{2R} \exp\left(-\frac{\overline{R}}{t_{2R}}\right), \tag{28}$$

where

$$\overline{\Delta R} = \left(R_{dk} - R_{dh}\right)/R_{d0}, \quad \overline{R} = R_{dh}/R_{d0},$$

R_{dk} and R_{dh} are droplet radii predicted by the kinetic and hydrodynamic models respectively, R_{d0} is the initial droplet radius, the coefficients a_{1R}, a_{2R}, t_{1R} and t_{2R} are fitting constants. The effects of droplets on gas temperature were ignored, following [5].

A slightly less accurate, but simpler approximation was found in the form:

$$\overline{\Delta R} = a_{3R} \exp\left(-\frac{\overline{R}}{t_{3R}}\right), \tag{29}$$

where a_{3R} and t_{3R} are new fitting constants.

All these fitting constants are presented in [6, 9] for several combinations of initial droplet radii and gas temperatures. As follows from our analysis, the first approximation is almost ideal, while there are some noticeable deviations between the predictions of Formula (29) and the results of calculations. These deviations can be ignored in many practical applications. Similar closeness between the results of calculations and approximations was demonstrated for other combinations of R_{d0} and T_g considered above. Approximations (28) and (29) can be interpolated for the whole range of gas temperatures from 750 K to 1500 K, using the following interpolation formula for the coefficients a_{1R}, a_{2R}, t_{1R} and t_{2R} (we restricted this interpolation to Approximation (28) only):

$$s_R = b_{0R} + b_{1R}\tilde{T} + b_{2R}\tilde{T}^2, \tag{30}$$

where $\tilde{T} = T_g/T_{\rm cr}$, $T_{\rm cr} = 659$ K is the critical temperature of n-dodecane $C_{12}H_{26}$ (approximation of Diesel fuel), s_R stands for a_{1R}, a_{2R}, t_{1R} or t_{2R}. The values of coefficients b_{iR}, where $i = 0, 1, 2$, for $R_{d0} = 20$ μm and $R_{d0} = 5$ μm are given in [6, 9]. The values of coefficients a_{1R}, a_{2R}, t_{1R} and t_{2R}, predicted by Approximation (30) for $R_{d0} = 20$ μm and $R_{d0} = 5$ μm and $T_g = 750$ K, 1000 K and 1500 K are also given in [6, 9]. The maximal difference between the values of fitting constants did not exceed about 1%. This justifies the application of Approximation (30).

Based on the direct comparison of the results of kinetic and hydrodynamic calculations for the same gas temperatures and initial droplet radii as in the previous section, we were able to show that a good approximation for the droplet temperatures predicted by the kinetic model can be achieved using the following formula:

$$\overline{\Delta T} = a_{1T} \exp\left(-\frac{\overline{T}}{t_{1T}}\right) + a_{2T}, \tag{31}$$

where

$$\overline{\Delta T} = \frac{T_k - T_h}{T_{\rm cr} - T_0}, \quad \overline{T} = \frac{T_h - T_0}{T_{\rm cr} - T_0},$$

T_k and T_h are droplet temperatures predicted by the kinetic and hydrodynamic models respectively, T_0 is the initial droplet temperature, the coefficients a_{1T}, t_{1T} and a_{2T} are fitting constants. These constants are presented in [6, 9] for the same combinations of initial droplet radii and gas temperatures as in the previous section.

The closeness of the plots of $\overline{\Delta T} = \frac{T_k - T_h}{T_{\rm cr} - T_0}$ versus $\overline{T} = \frac{T_h - T_0}{T_{\rm cr} - T_0}$ obtained using the kinetic model (calculation) and approximate equation (31) was demonstrated for various combinations of R_{d0} and T_g considered above.

As in the case of droplet radii, Approximation (31) with the values of coefficients given in [6, 9], can be interpolated for the whole range of gas temperatures from 750 K to 1500 K, using Approximation (30) for the coefficients a_{1T}, a_{2T} and t_{1T}, replacing subscript R with T everywhere. As a result, equation (30) is replaced by:

$$s_T = b_{0T} + b_{1T}\tilde{T} + b_{2T}\tilde{T}^2, \tag{32}$$

where s_T stands for a_{1T}, a_{2T} or t_{1T}. The values of these coefficients are given in [6, 9].

The values of coefficients a_{1T}, t_{1T} and a_{2T}, predicted by Approximation (32) for $R_{d0} = 20$ μm and $R_{d0} = 5$ μm, and for $T_g = 750$ K, 1000 K and 1500 K are given in [6, 9]. The maximal difference between the values of fitting constants did not exceed about 1%. This justifies the application of Approximation (32).

4. Conclusions

Recently developed approaches to the hydrodynamic and kinetic modelling of fuel droplet heating and evaporation are reviewed. Two new solutions to the heat conduction equation, describing transient heating of an evaporating droplet, are discussed. The first solution is the explicit analytical solution to this equation, while the second one reduces the solution of the differential transient heat conduction equation to the solution of the Volterra integral equation of the second kind. Both solutions take into account the effect of the reduction of the droplet radius due to evaporation. It is anticipated that the analytical solution can be used in computational fluid dynamics (CFD) codes to describe droplet heating and evaporation over a small time step when all parameters, except droplet radius and temperature, can be assumed constant. This can be considered as the generalisation of the approach currently used in all research and commercial CFD codes, in which it is assumed that droplet radius is constant during the time step. It has been pointed out that the new approach predicts lower droplet surface temperatures and slower evaporation rates compared with the traditional one.

A simplified model for bi-component droplet heating and evaporation is developed and applied for the analysis of the observed average droplet temperatures in a monodisperse spray. The model takes into account all key processes, which take place during this heating and evaporation, including the distribution of temperature and diffusion of liquid species inside the droplet and the effects of the non-unity activity coefficient (ideal and non-ideal models). The effects of recirculation in the moving droplets on heat and mass diffusion within them are taken into account using the effective thermal conductivity and the effective diffusivity models. The previously obtained analytical solution of the transient heat conduction equation inside droplets is incorporated in the numerical code alongside the original analytical solution of the species diffusion equation inside droplets.

A new algorithm, based on simple approximations of the kinetic results for droplet radii and temperatures, suitable for engineering applications, is discussed. These approximations are based on the presentation of the kinetic effects as corrections to the results of the hydrodynamic calculations. These corrections are presented in the form of simple algebraic expressions. The analysis described is applicable to a wide range of problems related to droplet heating and evaporation, although the main emphasis is on the application to Diesel engines.

5. Unsolved problems

The complete list of unsolved problems, referring to droplet heating and evaporation, is open-ended. In what follows we will restrict ourselves to the problems on which we are currently working.

5.1. Combined model, taking into account the effects of moving boundary, multi-component droplets and thermal radiation

The models described in Sections 2 and 3 took into account either the effects of a moving boundary, or the effects of multi-component droplets, but not a combination of the two. Also, at this stage, the contribution of thermal radiation was not taken into account (cf. the original model described in [1]). We are currently working on the model in which all these effects are taken into account simultaneously.

5.2. Effects of non-elastic collisions in the kinetic model

One of the most important limitations of the kinetic model described in Section 3, is that it is based on the assumption that collisions between molecules are elastic. This could be acceptable in the case of mono-atomic molecules (e.g. Argon). However, in the case of complex hydrocarbons, such as n-dodecane (approximation of Diesel fuel) it is not at first obvious that the contribution of internal degrees of freedom can be ignored. We are currently working on the development of a new model for inelastic collisions which would be simple enough to take into account the effect of such collisions in the kinetic model.

5.3. Evaporation coefficient of complex hydrocarbons

One of the parameters of the kinetic model is the evaporation coefficient of molecules at the surface of the droplets. Possible approaches to experimental measurements of this coefficient and its crude estimates are discussed in our previous papers (e.g. [7, 22]). The most accurate estimate of this coefficient should be based on molecular dynamics (MD) simulations. MD technique is reasonably well developed for relatively simple molecules, but nobody, to the best of our knowledge, has attempted to apply this technique to complex hydrocarbon molecules, such as n-dodecane. At the moment we are in the process of developing such a technique and are working on its application to calculation of the evaporation coefficient of n-dodecane droplets.

5.4. Approximation of the results of kinetic calculations using simple formulae

Once the developments of the kinetic model, described above, have been completed, the results of kinetic calculations would need to be approximated by simple formulae, similar to those described in Section 4. This would enable us to incorporate these approximations into computational fluid dynamic (CFD) codes and use them in various engineering and environmental applications.

Acknowledgements

The authors are grateful to the European Regional Development Fund Franco-British INTERREG IVA (Project C5, Reference 4005), EPSRC (Project EP/H001603/1) and the Russian fund for fundamental research (Grant 10-08-00614) for their financial support.

References

[1] Sazhin S S, Krutitskii P A, Abdelghaffar W A, Mikhalovsky S V, Meikle S T and Heikal M R 2004, *Int J Heat Mass Transfer* **47** 3327
[2] Sazhin S S, Abdelghaffar W A, Krutitskii P A, Sazhina E M and Heikal M R 2005, *Int J Heat Mass Transfer* **48** 4215
[3] Sazhin S S, Kristyadi T, Abdelghaffar W A and Heikal M R 2006, *Fuel* **85** 1613
[4] Sazhin S S, Krutitskii P A, Martynov S B, Mason D, Heikal M R and Sazhina E M 2007, *Int. J. Thermal Science* **46**(5) 444
[5] Sazhin S S and Shishkova I N 2009, *Atomization and Sprays* **19**(5) 473
[6] Sazhin S S, Shishkova I N and Heikal M 2010, *International J of Engineering Systems Modelling and Simulation* **2**(3) 169
[7] Sazhin S S 2006, *Progress in Energy and Combustion Science* **32** 162
[8] Sazhin S S 2009, *Pollack Periodica* **4**(1) 5
[9] Sazhin S S, Shishkova I N, Gusev I G, Elwardany A, Krutitskii P A and Heikal M R 2010, "Fuel droplet heating and evaporation: new hydrodynamic and kinetic models". *Proceedings of the 14th International Heat Transfer Conference*, Washington 8-13 August 2010, paper IHTC14-22320
[10] Kartashov E M 2001 *Analytical Methods in the Thermal Conductivity Theory of Solids*, (Vyshaya Shkola) (in Russian)
[11] Carslaw H S and Jaeger J C 1986 *Conduction of heat in solids*, Oxford: (Clarendon Press), Chap. 9
[12] Sazhin S S, Krutitskii P A, Gusev I G and Heikal M R 2010, *Int. J. of Heat and Mass Transfer* **53** 2826
[13] Lage P L C, Hackenberg M and Rangel R H 1995, *Combustion and Flame* **101** 36
[14] Torres D J, O'Rourke P J and Amsden A A 2003, *Combustion Theory and Modelling* **7** 67
[15] Maqua C, Castanet G and Lemoine F 2008, *Fuel* **87** 2932
[16] Sazhin S S, Elwardany A, Krutitskii P A, Castanet G, Lemoine F, Sazhina E M and Heikal M R 2010, *International J of Heat and Mass Transfer* **53** 4495
[17] Sazhin S S, Krutitskii P A, Elwardany A, Castanet G, Lemoine F and Heikal M R 2010 "An analytical solution to the spherically symmetric species diffusion equation: application to modelling of heating and evaporation of bi-component droplets". *Proceedings of the ILASS-Europe 2010*, Brno, Czech Republic, 6-8 September 2010
[18] Abramzon B and Sirignano W A 1989, *Int. J. Heat and Mass Transfer* **32** 1605
[19] Campbell S W, Willsak R A and Thodos G 1987, *J. Chem. Eng. Data* **32** 357
[20] Reitz R D and Rutland C J 1995, *Prog. Energy Combustion Science* **21** 173
[21] Sazhina E M, Sazhin S S, Heikal M R, Babushok V I, and Johns R 2000, *Combustion Science and Technology* **160** 317
[22] Sazhin S S, Shishkova I N, Kryukov A P, Levashov V Yu, and Heikal M R 2008 "Evaporation of droplets into a background gas in the presence of heat flux: kinetic and hydrodynamic modelling", *Proceedings of the 19th International Symposium on Transport Phenomena*, 17-20 August, 2008, Reykjavik, Iceland, paper 48
[23] Sazhin S S, Shishkova I N, Kryukov A P, Levashov V Yu, and Heikal M R 2009 "A Simple algorithm for kinetic modelling of Diesel fuel droplet evaporation", *Progress in Computational Heat and Mass Transfers; Proceedings of 6th ICCHMT* May 18-21, 2009, Guangzhou, China, paper 151, pages 386-391

Journal of Physics: Conference Series **268** (2011) 012027 doi:10.1088/1742-6596/268/1/012027

Vibration of transformer sheets using non-linear characteristics

A Schiffer and A Ivanyi

University of Pecs, Rokus 2, Pecs, Hungary

E-mail: adamsc.pte@gmail.com

Abstract. Intensive research has been carried out over the years to reduce acoustic noise resulting from the vibration of transformer plates. The noise of transformer sheets excited by electromagnetic field can be mostly attributed to induced forces between the plates. The vibration generated from transformer structures causes abnormal vibration, breakage of the machine, and noise. Vibration measurement and analysis of the structure is important to prevent this phenomenon. This chapter presents the research resulted by measurement and Finite Element Method of the resonance of the transformer sheets.

The goal of this research is to introduce the coupling of the developed nonlinear hysteresis model and the Finite Element Method. The COMSOL and MATLAB based finite element simulation method combines the electromagnetic, structural mechanics and acoustic models. The moving mesh (ALE) geometry of the transformer sheets leads on a valid model and assures the calculations of the electromagnetic properties in different deformed shapes. The measurement of the natural frequencies of the plates yields the relation between the excitation frequency and the resonance of the system.

The goals of the developed model are:

- to derive the magnetic field intensity H and with hysteresis nonlinearity the magnetic flux density B properties inside the transformer sheets with $d = 0.35$ mm and $l = 150$ mm;
- to compute the induced current density J_i inside the plates;
- to calculate the Lorentz force F distribution along the sheets;
- to derive the plane stresses the horizontal and vertical displacements along the sheets are calculated;
- applying the ALE moving mesh the deformation of plates can be simulated.

For modelling the nonlinear characteristic of the magnetic material $M = \mathcal{H}\{H\}$ a novel developed hysteresis model is used. The advantages of the hysteresis model are its easy identification, past memory representation and numerical simplicity [1]. So, the hysteresis characteristic can be determined as

$$B = \mathcal{H} \quad H, \frac{\partial B}{\partial H}, \psi_i(H_i, B_i) \quad . \tag{1}$$

1. The developed scalar hysteresis model

In this section a novel scalar hysteresis model is constructed on the basis of a special modification field, determined from the permeability vectors. The model properties are determined from measured data points. The aim of this research is to develop a new hysteresis model, based on finite number of measured data points and on the introduced special vector field that is identified as modification vector field.

The hysteresis model presented below is introduced by the aid of a modification vector field $\boldsymbol{\Psi}$, that can be derived from the permeability vectors belonging to the ascending and the descending measured loops. The vector field $\boldsymbol{\Psi}$ contains spatial vector components. Introducing this modification vector field the next point, $y(t_i + \Delta t)$ on the hysteresis curve can easily be determined from the actual value of the input signal $s(t_i)$ and from the differential permeability $\hat{\mu}_{diff}(s_i, y_i)$ as the past local memory representation of the model. So, the hysteresis characteristic can be determined as

$$y = \mathcal{H}\left\{s, \partial y/\partial s, \boldsymbol{\Psi_i}\left(s_i, y_i\right)\right\}, \tag{2}$$

where $\partial y/\partial s = \mu_{diff}$ is the local differential permeability and a vector component $\boldsymbol{\Psi}_i(s_i, y_i)$ of the vector field $\boldsymbol{\Psi}$ can be determined from the measured ascending and descending first order reversal curves, where i represents the actual measured point, s is the input and y is the output signal. Without accommodation any measured data points $\boldsymbol{\rho}(s_i, y_i)$ on the hysteresis loop can be achieved from the major loop only from the direction, defined by the local differential permeability vector

$$\boldsymbol{\mu}_{diff,i} = \left.\frac{\partial y}{\partial s}\right|_{s=s_i, y=y_i}, \tag{3}$$

which one represents the memory of the hysteresis in this model, because applying different local differential permeability vectors as the local memory of the hysteresis model, different output can be achieved.

1.1. Identification of the scalar model

Denote $\hat{\boldsymbol{\mu}}_{diff}\left(s_i, y_i\right) : G \to \mathrm{R}^2\ G \subset \mathrm{R}^2,\ \ G = \left\{\left(s_i, y_i\right), i = 1, \cdots, Z\right\}$ where Z is the number of measured points and $\boldsymbol{\rho}\left(s_i, y_i\right)$ represents the measured data. From the measured points of the FORC the local differential permeability vectors $\hat{\boldsymbol{\mu}}_{diff}\left(s_{xx}, \cdot\right)$ can be calculated along $s_{xx} = const$ lines as

$$\hat{\boldsymbol{\mu}}_{diff}\left(s_{xx}, y_i\right) = \left(\Delta s_{xx}, \Delta y_i\right), \tag{4}$$

where $\Delta f_i = f_i - f_{i-1}$ specifies the segment of validity. The s_{xx} means an equidistant inner point along the s axis, so $s_{xx} = k\left(s_{\max} - s_{\min}\right)/K | k, K \in \mathrm{N}$. If $\hat{\boldsymbol{\mu}}_{diff}\left(s_i, y_i\right)$ and $\hat{\boldsymbol{\mu}}_{diff}\left(d_{i+1}, y_{i+1}\right)$ values are known the relation between the calculated modification vectors $\boldsymbol{\Psi}_i$ and the local differential permeability vectors are $\boldsymbol{\mu}_{i+1} = \left(\boldsymbol{\Psi}_i + \boldsymbol{\mu}_i\right)/2$, where from the modification vector $\boldsymbol{\Psi}_i$ can be determined as

$$\boldsymbol{\Psi}_i\left(H_i, B_i\right) = \left(2\Delta H_{i+1} - \Delta H_i, 2\Delta B_{i+1} - \Delta B_i\right). \tag{5}$$

In case if the measured points $\boldsymbol{\rho}\left(s_i, y_i\right)$ are equidistant in $\boldsymbol{s}$ then the vector

$$\boldsymbol{s} \in \mathrm{R},\ \ s_k = s_{\min} + k\left(s_{\max} - s_{\min}\right)/K \tag{6}$$

is defined, where K the number of the equidistant points in $\boldsymbol{s}$. The limiting points are specified as $s_0 = -s_{sat},\ \ s_K = s_{sat}$. Let introduce matrix form to denote the measured set of the output signal as

$$\begin{gathered}\mathbf{y} \in \mathrm{R}^2, y_{k,n} = y_i : \exists\ \ \boldsymbol{\rho}\left(s_i, y_i\right), \\ s_i \in \mathbf{s},\ \ k \in [0, K] \subset \mathrm{Z}^+,\ \ n \in [1, N] \subset \mathrm{N},\end{gathered} \tag{7}$$

where N is the number of the FORC. Applying this nomination the k -th rows in $\mathbf{y}$ means the k -th assignment in the matrix form $\mathbf{s}$ and the n -th column means the n -th ascending or descending curve. All modification vectors $\mathbf{\Psi}_k^*$ must also be stored in the same sequence as above. For the reversal curves from descending ones $\mathbf{s}^-, \mathbf{y}^-, \mathbf{\Psi}_k^-$, to the positive loops $\mathbf{s}^+, \mathbf{y}^+, \mathbf{\Psi}_k^+$ can be defined. To the simulation of the hysteresis loop at any s_j, y_j point any $s_k^* \in \mathbf{s},\ y_{k,n}^* \in \mathbf{y}$ measured points can be found as $\left|s_k^* - s_j^*\right| = \min_k$, $\left|y_{k,n}^* - y_j^*\right| = \min_n$, and then $\mathbf{\Psi}_k^*(s_k, y_{k,n})$ can be determined for (k,n) pairs with $* \in [+,-]$. So the sign $'*'$ has been defined for positive and negative reversal curves

$$\mathbf{\Psi}^* : \mathrm{R}^2 \to \mathrm{R}^2,\ \ \mathbf{\Psi}^*(s_j, y_j) = \mathbf{\Psi}_k^*(s_k, y_k)\ , \qquad \left|s_k^* - s_j^*\right| = \min_k, \ \ \left|y_k^* - y_j^*\right| = \min_n\ . \tag{8}$$

The direction of the change in $\boldsymbol{s}$ can be determined as $* = sign(\Delta s_j)$, while $\mathbf{\Psi}^*(s_j, y_j)$ can be interpolated if $s_j \notin \mathbf{s}^*$.

If $\mathbf{s}^*$ contains equidistant elements, $\Delta s_{i+1} - \Delta s_i = 0$, then from the measured values the modification vector at point of (s_i, y_i) is

$$\mathbf{\Psi}_k^*(s_i, y_i) = (\Delta s_{i+1}, 2\Delta y_{i+1} - \Delta y_i)\,. \tag{9}$$

Calculating the required y_i from the modification vector field $\mathbf{\Psi}$

$$\Delta y_{i+1} = (\mathbf{\Psi}^*(s_i, y_i)[2] + \Delta y_i)/2, \tag{10}$$

so

$$y_{i+1} = (\mathbf{\Psi}^*(s_i, y_i)[2] + \Delta y_i)/2 + y_i. \tag{11}$$

By definition $\mathbf{\Psi}^*(s_i, y_i)[2]$ means the second, $\mathbf{\Psi}^*(s_i, y_i)[1]$ the first scalar component of the modification field vector $\mathbf{\Psi}^*(s_i, y_i)$. The prehistory of the material is represented by the local $\hat{\boldsymbol{\mu}}_{diff}(s_i, y_i)$ permeability vector through the tangent of the curve. Figure 1 shows the vector representation of the scalar hysteresis model. The modification field vector $\mathbf{\Psi}^*(H_i, B_i)$ is calculated from $\hat{\boldsymbol{\mu}}_{diff}(s_i, y_i)$ and $\hat{\boldsymbol{\mu}}_{diff}(s_{i+1}, y_{i+1})$ vectors according to (11).

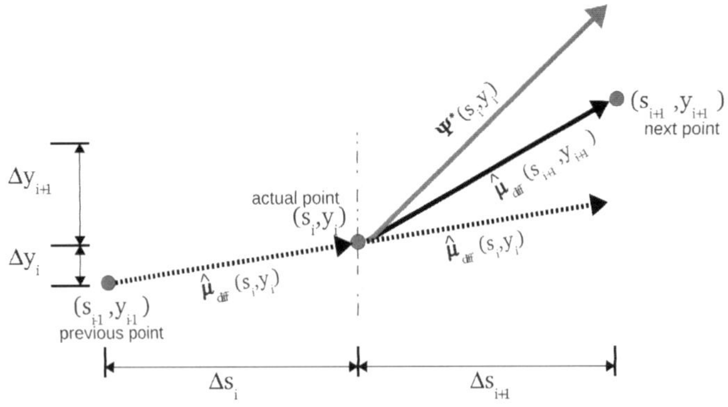

Figure 1. Representation of the model.

1.2. Application of the scalar model

In this section the introduced hysteresis model is applied in two cases. First, if the simulation data set, $\mathbf{s}$ has equidistant difference, coincident with the measured data, and second, if the steps in $\mathbf{s}$ are not equidistant and different from the measured values.

Generally, if the measured points in $\mathbf{s}^*$ have equidistant difference, $s_j \in \mathbf{s}^*$, any y_{j+1} nonlinear system output value can be calculated from y_j , and from the local measured $\hat{\boldsymbol{\mu}}^*_{diff}(s_j, y_j)$ and from the determined $\boldsymbol{\Psi}^*(s_j, y_j)$

$$y_{j+1} = (\boldsymbol{\Psi}^*(s_j, y_j)[2] + \Delta y_j)/2 + y_j \tag{12}$$

Simulating a hysteresis loop if $s_j \notin \mathbf{s}^*$ and Δs_j is different from the measured Δs_i data then $\boldsymbol{\Psi}^*(s_j, y_j)[2]$ must be scaled with a scalar as

$$\begin{aligned} &\eta\, \boldsymbol{\Psi}^*(s_j, y_j)[2] \ , \\ &y_{j+1} = (\eta\, \boldsymbol{\Psi}^*(s_j, y_j)[2] + \Delta y_j)/2 + y_j \ . \end{aligned} \tag{13}$$

The scalar η can be determined from the simulated $\Delta s_{j+1}, s_j, y_j$ and from the measured $\boldsymbol{\Psi}^*(s_j, yj)[1]$ values as

$$\eta = \frac{\Delta s_{j+1}}{\boldsymbol{\Psi}^*(s_j, y_j)[1]} \ , \tag{14}$$

where $\boldsymbol{\Psi}^*(s_j, y_j)[1]$ can be determined from (8) according to the measured data.

1.3. Turning points

In the last section the model has been specified when the local permeability vectors $\hat{\boldsymbol{\mu}}^*_{diff}(s_j, y_j)$ are known. This should be different when the direction of the excitation signal s is changing, it means that increasing excitation values turn to decreasing excitation values and opposite. This is the point on the major loop where the reversal curves start and called as *turning point*. In these points the modification value $\boldsymbol{\Psi}^*(s_j, y_j)$ is different from the previous vectors, the modification vectors has to be specified at the turning point. Otherwise the modification vectors $\boldsymbol{\Psi}^*(s_{sat}, y_{sat})$ must be calculated according to (13) and when the excitation is over saturation $s_{j+1} > s_{sat}$, the vector field must keep the saturation condition of the hysteresis curve, so the direction of $\boldsymbol{\Psi}^*(s_{sat}, y_{sat})$ must be the same as the local differential permeability vector. For the exact simulation the turning point has to be specified. If $s_{j+1} \notin s^*$ and $s_{j+1} < -s_{\max}$, $s_{j+1} > s_{\max}$ the hysteresis loop is in saturation condition,

$$\left.\frac{\partial^2 y}{\partial s^2}\right|_{s=\pm s_{\max}} = 0, \tag{15}$$

and

$$\left.\frac{\partial y}{\partial s}\right|_{s=\pm s_{\max}} = \mu_0. \tag{16}$$

In the case of saturation condition and the input point of the nonlinear system is outside of the measured range in the identification process, then the modification vector is not changing. Mathematically if $\boldsymbol{\Psi}^*(s_j, \cdot) : |s_{j+1}| > s_{\max}$ then from (12) $\Delta y_{j+1} = \Delta y_j$. When the excitation input signal s has local maximum or minimum then $* = sign(\Delta s_{j+1})$ is changing. The reversal curves are measured from reversal points. Because of at the first point on the reversal curve $\Delta y_{j+1} = \Delta s_{j+1} = 0$ it means that $\boldsymbol{\Psi}^*(s_j, y_j)$ defines the reversible permeability in that points. This cannot be expressed as a function with respect to the excitation sigal because of it depends on the domain configuration that comes from the material history. In saturation case the direction of the reversible permeability in the ascending, and the differential permeability in the descending loops are just opposite [2]. Figure 2 shows the components $\boldsymbol{\Psi}^*$ of the vector field B on the major loop. The turning points specified above and can be seen in figure 3. Using the turning points the model ensures the behaviour on the major loop, so the simulation data of the hysteresis loop s_j, y_j must be inside the ascending and descending major loops. Otherwise when $s_j \notin \mathbf{s}^*$ and $s_j < -s_{\max}$ or $s_j > s_{\max}$ these turning point conditions ensure that the model remain in between saturation condition [1].

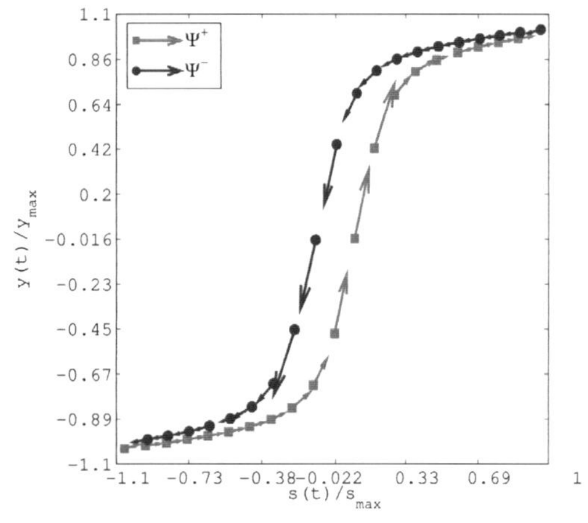

Figure 2. Modification vectors $\mathbf{\Psi}^\bullet(\mathbf{s}^\bullet, \mathbf{y}^\bullet)$ on the major loop.

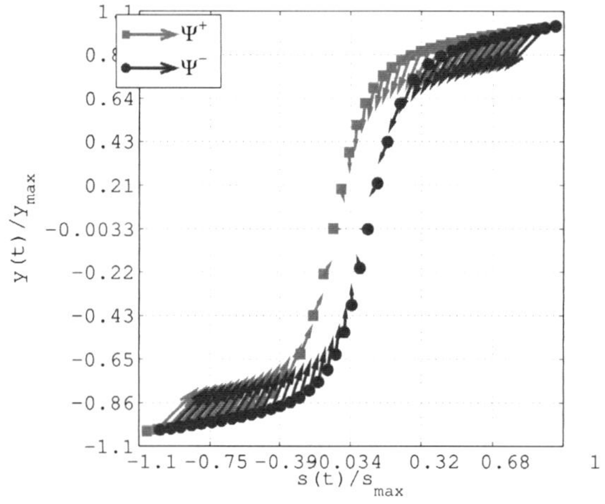

Figure 3. Turning points of $\mathbf{\Psi}^\bullet(\mathbf{s}^\bullet, \mathbf{y}^\bullet)$ vector fields.

After the $\mathbf{\Psi}^*_k$ modification vector fields are derived from the ascending and descending curves for the resulted simulation the accuracy of the model is investigated. Figure 4 shows the simulated and measured data points under decreasing sinusoidal excitation process. It can be seen that the simulated results and the measurement points are very close. The measured values are for checking the simulation validity. The accommodation property of the hysteresis curve can be ensued from the property of the real material and the construction philosophy of the simulation based on measured data, as it can be seen in figure 5.

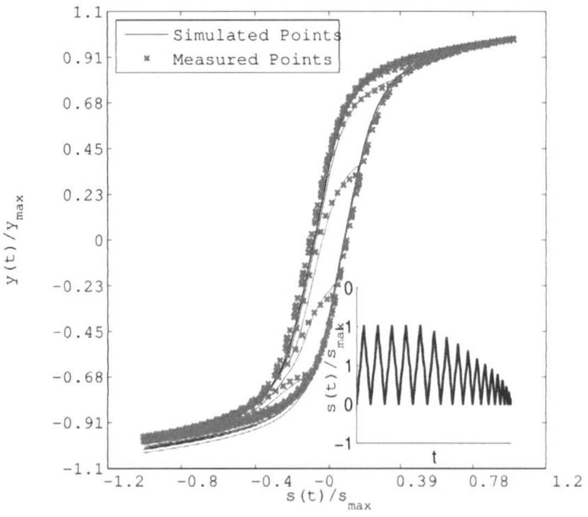

Figure 4. Simulation result and measurement data.

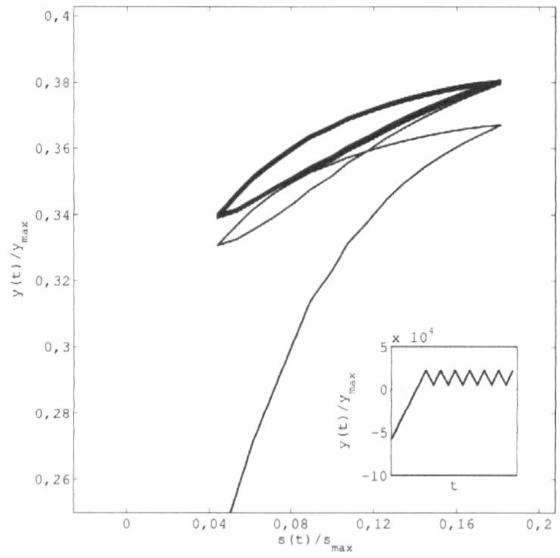

Figure 5. Representation of the accommodation property.

2. Simulation of the transformer sheets

Electrical machines and transformers have a laminated core from ferromagnetic materials. A part of the vibration and noise arising from the magnetic core that caused by the magnetic forces and magnetostriction. The magnetostriction can be defined as the deformation of the ferromagnetic material in the presence of magnetic field. The magnetostrictive behaviour is so complex and varies for different materials. It also depends on the the applied magnetic field.

The unwanted noise and the vibrations in electrical machines and transformers in mainly classified into three groups: mechanical noise, aerodynamic noise and magnetic noise. The laminated core of the transformers is a stack of laminations of ferromagnetic material. Due to their high permeability these materials can produce high magnetic induction with weak magnetic fields.

The magnetic noise of the transformers is caused by magnetic forces and magnetostriction. The magnetic forces are tending to deform the geometry of the magnetic core. Combined with periodic magnetic field excitation it leads to vibration.

This is widely accepted that vibration and the sound from transformer come from the following items [3]:

- Magnetostriction of the core. It means that the specimen extends along the magnetic flux density $\boldsymbol{B}$ direction but shorten toward that direction. The magnetostriction has direct and close connection with the magnetic flux density $\boldsymbol{B}$ and the winding voltage;
- Electrodinamics of the specimen and the windings due to leak magnetic flux. This is an important property but negligible for little dry-type transformers that for the measurement were used;
- Oil pumps and transformer cooling fan vibration. The frequency range of those cooling equipments are usually below than 100 Hz, that is why those frequencies easily can be isolated from other vibration signals. In the measurement it was not needed to use cooling equipment that is why we can ignore those vibration sources.

Figure 6 shows the representation of the computational problem. Applying the above field equations to the two infinite large plates in z-direction with sinusoidal excitation of the magnetic field is investigated. The deformation of H_y, M_y, B_y waves can be calculated along direction z. The magnetic forces $\boldsymbol{F}$' and $\boldsymbol{F}$" acting on the plates separately are induced along in direction z. In the cross-section of the plates the current density vectors $\boldsymbol{J}$' and $\boldsymbol{J}$" are inside of the material.

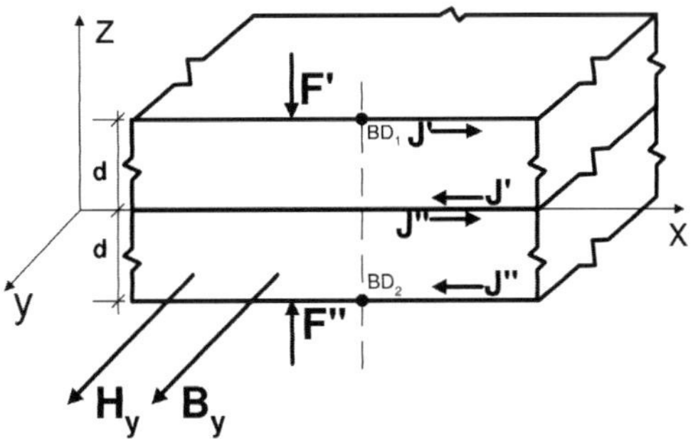

Figure 6. The schematic of the infinite plates.

2.1. Equations of the electromagnetic field

The well-known Maxwell's equation system defines relations between electromagnetic field quantities, such as electric field intensity $\boldsymbol{E}$, magnetic field intensity $\boldsymbol{H}$, electric flux density $\boldsymbol{D}$ and the magnetic flux density $\boldsymbol{B}$. The electric current density $\boldsymbol{J}$ and the electric charge density ρ are the source of the electromagnetic field.

In the studied eddy current field problem the electric and magnetic field are coupled, so the field quantities are depending on the time variation. So it is supposed that $\partial/\partial t \neq 0$. In the eddy-current field the region Ω_c of the ferromagnetic specimen with nonlinear hysteresis characteristic is surrounded by the nonmagnetic and non-conducting region (e.g. air) Ω_n. The equations are discussed in eddy-current and eddy-current free regions [4],[5],[6],[7],[8].

The classical differential form of the Maxwell's equations are the following in the eddy-current free region Ω_n:

$$\nabla \times \boldsymbol{H} = \boldsymbol{J}_0 \text{ in } \Omega_n, \tag{17}$$

$$\nabla \cdot \boldsymbol{B} = 0 \text{ in } \Omega_n, \tag{18}$$

where $\boldsymbol{J}_0$ is the source current density of the excitation coils.

In the eddy current region Ω_c the Maxwell's equations can be written as:

$$\nabla \times \boldsymbol{H} = \boldsymbol{J} \text{ in } \Omega_c, \tag{19}$$

$$\nabla \times \boldsymbol{E} = -\frac{\partial \boldsymbol{B}}{\partial t} \text{ in } \Omega_c, \tag{20}$$

$$\nabla \cdot \boldsymbol{B} = 0 \text{ in } \Omega_c, \tag{21}$$

$$\nabla \cdot \boldsymbol{J} = 0 \text{ in } \Omega_c, \tag{22}$$

where $\boldsymbol{J}$ is the eddy current density.

2.2. The constitutive relations

The constitutive relations are the follows:

$$\boldsymbol{B} = \begin{cases} \mu_0 \boldsymbol{H} & \text{in air } \Omega_n, \\ \mu_0 \mu_r \boldsymbol{H} & \text{in magnetically linear material } \Omega_c, \\ \mathcal{H}\{\boldsymbol{H}\} & \text{in magnetically non} - \text{linear material } \Omega_c, \end{cases} \tag{23}$$

and

$$\boldsymbol{J} = \sigma \boldsymbol{E} \text{ in air } \Omega_c. \tag{24}$$

$$\boldsymbol{H} = \begin{cases} \nu_0 \boldsymbol{B} & \text{in air } \Omega_n, \\ \nu_0 \nu_r \boldsymbol{B} & \text{in magnetically linear material } \Omega_c, \\ \mathcal{H}^{-1}\{\boldsymbol{B}\} & \text{in magnetically non} - \text{linear material } \Omega_c. \end{cases} \tag{25}$$

The inverse form of the relation (24) is the following:

$$\boldsymbol{E} = \rho \boldsymbol{J} \text{ in air } \Omega_c. \tag{26}$$

Here μ is the permeability, ν is the reluctivity, $\sigma = 7.47 \cdot 10^5$ Sm^{-1} is the conductivity and ρ is the resistivity of the material, $\mu_0 = 4\pi 10^{-7}$ H/m is the permeability of vacuum [2], [8].

2.3. The boundaries of the regions

The two regions Ω_c and Ω_n are coupled at the interface Γ_{nc}. The Γ_B signs the closing boundary of the eddy current free region, where the normal component of the magnetic flux density b is disappearing.

These conditions can be formulated as

$$\boldsymbol{B} \cdot \boldsymbol{n} = -b \text{ or } \boldsymbol{B} \cdot \boldsymbol{n} = 0 \text{ on } \Gamma_B, \tag{27}$$

where $\boldsymbol{n}$ is the outer normal unit vector of the region. The interface between the conducting and non-conducting regions Γ_{nc} can be described by the following terms

$$\boldsymbol{H}_c \times \boldsymbol{n}_c + \boldsymbol{H}_n \times \boldsymbol{n}_n = \boldsymbol{0} \text{ on } \Gamma_{nc}, \tag{28}$$

and

$$\boldsymbol{B}_c \times \boldsymbol{n}_c + \boldsymbol{B}_n \times \boldsymbol{n}_n = \boldsymbol{0} \text{ on } \Gamma_{nc}, \tag{29}$$

and

$$\boldsymbol{J} \times \boldsymbol{n}_c = \boldsymbol{0} \text{ on } \Gamma_{nc}, \tag{30}$$

where $\boldsymbol{n}_c,\boldsymbol{n}_n$, $\boldsymbol{H}_c$, $\boldsymbol{H}_n$, $\boldsymbol{B}_c$, $\boldsymbol{B}_N$ are the outer normal unit vector of the conducting and non-conducting regions. The magnetic field intensity and the magnetic flux density vectors in the corresponding region on the boundary, where $\boldsymbol{n}_n = -\boldsymbol{n}_c$ along Γ_{nc}.

Furthermore the initial conditions of the magnetic field must be initialized as

$$\boldsymbol{B}_n(t=0) = \boldsymbol{B}_{n,0} \text{ on } \Omega_N \textit{ and } \boldsymbol{B}_c(t=0) = \boldsymbol{B}_{c,0} \text{ on } \Omega_c. \tag{31}$$

To solve an electromagnetic field problem it is usual that the partial differential equation system of the studied phenomenon is simplified to a potential formulation. This section shows the determination of the potential formulations. In the investigated electromagnetic field problem the specimens carrying the eddy currents and surrounded by a non-conducting medium which is free of eddy currents. The solutions of the Maxwell's equations are handled by scalar and vector potentials [4],[5],[6],[7],[8],[9].

2.4. Eddy current magnetic fields

As it was written the eddy current field is defined by Maxwell's equations (19),(20),(21), constitutive relations (23),(24) and inverse relations (25),(26). In this work the eddy current field is described by magnetic vector potential $\boldsymbol{A}$ that can be coupled with an electric scalar potential denoted by V [4],[5],[6],[7],[8],[9].

In the eddy current region the field quantities are depending on the time variation, so $\partial/\partial t \neq 0$.

The magnetic flux density vector $\boldsymbol{B}$ can be described by the curl of the magnetic vector potential $\boldsymbol{A}$ as:

$$\boldsymbol{B} = \nabla \times \boldsymbol{A}. \tag{32}$$

Basically the vibration of the transformer sheets is a two dimensional problem the electric scalar potential can be selected as $V = 0$ in the $\boldsymbol{A}, V-$ potential formulation. In this case the electric field intensity is the following in two dimensional case:

$$\boldsymbol{E} = -\frac{\partial \boldsymbol{A}}{\partial t}. \tag{33}$$

Journal of Physics: Conference Series **268** (2011) 012027 doi:10.1088/1742-6596/268/1/012027

Substituting (32) and (33) into (19) and using the constitutive relations (26) and (25) the partial differential equation of the $\boldsymbol{A}, V-$ potential formulation in two dimensional case is the following:

$$\nabla \times (\nu \nabla \times \boldsymbol{A}) + \sigma \frac{\partial \boldsymbol{A}}{\partial t} = \boldsymbol{0} \text{ in } \Omega_c. \tag{34}$$

2.5. Coupling static magnetic field and eddy current field formulations

In the presented problem the specimens carrying the eddy currents and surrounded by non-conducting medium where a static magnetic field is present (figure 7). The potential formulations of the static and eddy current fields must be coupled because the static field is induced by the eddy currents and source currents of coils.

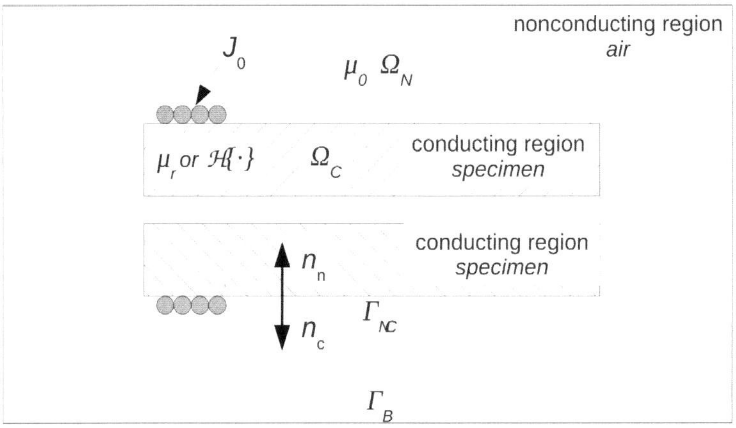

Figure 7. Representation of the eddy current field problem.

In this work the static magnetic field Γ_n is described by magnetic vector potential $\boldsymbol{A}$, this is a more usual way than applying reduced magnetic scalar potential Φ.

The eddy current field in Γ_c the magnetic vector potential $\boldsymbol{A}$ is coupled with electric scalar potential V, so it called $\boldsymbol{A}, V - \boldsymbol{A}$ formulation.

The magnetic vector potential $\boldsymbol{A}$ is used in this formulation in all of the the regions Ω_c and Ω_n. The electric scalar potential V is used only in the region Ω_c. The interface conditions are defined as well.

The summarized equations are the following in two dimensional assignment:

$$\nabla \times (\nu_0 \nabla \times \boldsymbol{A}) + \sigma \frac{\partial \boldsymbol{A}}{\partial t} = -\nabla \times \boldsymbol{I} \text{ in } \Omega_c, \tag{35}$$

$$\nabla \times (\nu \nabla \times \boldsymbol{A}) = \boldsymbol{J}_0 \text{ in } \Omega_n, \tag{36}$$

$$\boldsymbol{n}_c \times \boldsymbol{A} + \boldsymbol{n}_n \times \boldsymbol{A} = \boldsymbol{0} \text{ on } \Gamma_{nc}, \tag{37}$$

$$(\nu_0 \nabla \times \boldsymbol{A} + \boldsymbol{I}) \times \boldsymbol{n}_c + (\nu \nabla \times \boldsymbol{A}) \times \boldsymbol{n}_n = \boldsymbol{0} \text{ on } \Gamma_{nc}. \tag{38}$$

2.6. The weak formulation

The finite element simulation is calculated by using the weak form of the presented potential formulations. The weak form of a partial differential equation is obtained by using the weighted residual method. The Galerkins method is a particular form of residual method and it is widely used in electromagnetism. The weighted residual method is applied to minimize the residual of a partial differential equation and based on the inner product of the partial differential equation and a weighting function $\boldsymbol{W}$ [4],[5],[6],[7],[8],[9]. The weak formulations of $\boldsymbol{A}, V-\boldsymbol{A}$ formulation are based on the partial differential equations (35) and (36) and on the interface condition (37):

$$\begin{aligned}
&\int_{\Omega_c} \boldsymbol{W}_k \cdot [\nabla \times (\nu_0 \nabla \times \boldsymbol{A})] \mathrm{d}\Omega \\
&+ \int_{\Omega_c} \boldsymbol{W}_k \cdot \left(\sigma \frac{\partial \boldsymbol{A}}{\partial t}\right) \mathrm{d}\Omega \\
&+ \int_{\Omega_n} \boldsymbol{W}_k \cdot [\nabla \times (\nu \nabla \times \boldsymbol{A})] \mathrm{d}\Omega \\
&+ \int_{\Gamma_{nc}} \boldsymbol{W}_k \cdot [(\nu_0 \nabla \times \boldsymbol{A} + \boldsymbol{I}) \times \boldsymbol{n}_c + (\nu \nabla \times \boldsymbol{A}) \times \boldsymbol{n}_n] \mathrm{d}\Gamma \\
&= \int_{\Omega_n} \boldsymbol{W}_k \cdot \boldsymbol{J}_0 d\Omega - \int_{\Omega_c} \boldsymbol{W}_k \cdot (\nabla \times \boldsymbol{I}) \mathrm{d}\Omega.
\end{aligned} \tag{39}$$

2.7. Numerical modelling of electromagnetic field computation

The analysis of electromagnetic field problems requires a solid technique for the treatment of the nonlinearity. The system of nonlinear equations is solved by *fixed point* (FP) iteration technique that shows great advantages with respect to other methods [4],[5],[7],[8],[10].

In principle the *polarization method* that is used to solve the nonlinear equations. It based on the decomposition of the nonlinear system into a linear and a nonlinear part. The decomposition of the hysteretic relationship $\boldsymbol{B} = \mathcal{H}\{\boldsymbol{H}\}$ is linking the magnetic field intensity $\boldsymbol{H}$ and the magnetic flux density $\boldsymbol{B}$ into a liner term and a residual nonlinearity $\boldsymbol{I}$ as

$$\boldsymbol{H} = \nu_{FP} \boldsymbol{B} + \boldsymbol{I}, \tag{40}$$

where the residual $\boldsymbol{I}$ is computed by iteratively and starting from any trial value. The constant $\nu_{FP} = 1/\mu_{FP}$ controls the convergence of the process. The μ_{FP} can be estimated from the minimum and maximum slopes of the hysteresis curve as:

$$\mu_{FP} = \frac{\mu_{max} + \mu_{min}}{2}. \tag{41}$$

This scheme requires that the field computation gives as the magnetic flux density $\boldsymbol{B}$ as result. In this computation the magnetic vector potential $\boldsymbol{B} = \nabla \times \boldsymbol{A}$ was used , where from the magnetic flux density $\boldsymbol{B}$ can be calculated and involves the magnetic field intensity as unknown.

The new residual $\boldsymbol{I}$ is computed making use the direct relationship between $\boldsymbol{H}$ and $\boldsymbol{B}$ as

$$\boldsymbol{I} = \boldsymbol{H} - \nu_{FP} \boldsymbol{B} = \boldsymbol{H} - \nu_{FP} \mathcal{H}\{\boldsymbol{H}\}. \tag{42}$$

The nonlinear iteration steps can be summarized as follows:

(i) the iteration is started by an arbitrary value of $\boldsymbol{I}^{(0)}$;

(ii) in the nth iteration step (n>0) the magnetic flux density $\boldsymbol{B}^{(n)}$ is calculated from the Maxwell's equations using $\boldsymbol{I}^{(n)}$;

Journal of Physics: Conference Series **268** (2011) 012027 doi:10.1088/1742-6596/268/1/012027

(iii) the magnetic field intensity $\boldsymbol{H}$ can be calculated as

$$\boldsymbol{H}^{(n+1)} = \nu_{FP}\boldsymbol{B}^{(n)} + \boldsymbol{I}^{(n)}; \tag{43}$$

(iv) the nonlinear term must be updated ny using the magnetic field intensity and the direct hysteresis model $\boldsymbol{B} = \mathcal{H}\{\boldsymbol{H}\}$ as

$$\boldsymbol{I}^{(n+1)} = \boldsymbol{H}^{(n+1)} - \nu_{FP}\mathcal{H}\{\boldsymbol{H}^{(n+1)}\}; \tag{44}$$

(v) steps (1)-(4) must be repeated until the convergence that criterion can be defined as

$$\|\boldsymbol{I}^{(n+1)} - \boldsymbol{I}^{(n)}\| < \varepsilon, \tag{45}$$

where ε is a small positive number. In this work $\varepsilon = 10^{-4}$ has been chosen.

2.8. Solution of the problem

The electromagnetic problem has been solved by finite element method. The $\boldsymbol{A}$, $\boldsymbol{V} - \boldsymbol{A}$ potential formulations are used for the simulation in the time domain. The commercial finite element code, the COMSOL Multiphysics was used to building up the finite element model. The COMSOL Multiphysics package is a finite element analysis and solver package for various physics and engineering applications. It has an extensive and well-managed interface to MATLAB where the programming and the postprocessing calculations were done. The COMSOL Multiphysics also allows to couple different systems of partial differential equations using the weak form. It has many modules which from the AC/DC Module, Structural Mechanics Module, Moving Mesh module. The AC/DC module is used to simulate the model in static analysis type, when $\partial/\partial t = 0$. The transient model, when $\partial/\partial t \neq 0$ has used the static model and computed in MATLAB using the fixed point technique. The finite element mesh contains 75088 elements and 37655 mesh points, number of boundary elements is 3232, the number of degrees of freedom is 482867.

The simulation result using the polarization method can be seen in figure 8. The results corresponds to different time points applying $f_e = 50$ Hz excitation frequency. The changes of the Lorenz force in time leads to the vibration of the plates. The calculated displacement of the upper transformer sheet can be seen in figure 9. It can be seen that the maximum distance of the displacement is $6 \cdot 10^{-9}$ m.

3. Conclusion and future work

In this paper the calculation method of the transformer sheet vibration was discussed. Applying the developed hysteresis model and the moving mesh property in the finite element calculation it leads reliable results in the electromagnetic field calculation.

The postprocessing of the nonlinear field problem has been yielded to the calculation of the magnetic force between the plates. It is planned as a future goal that the simulated vibration results are compared with and without the moving mesh property. The difference between the two calculation methods verifies the necessity to taking into account the deformation of the sheets in the electromagnetic FEM calculations.

References

[1] Adam Schiffer and Amalia Ivanyi 2006 *New dynamical hysteresis model* Electromagnetic Fields in Mechatronics, Electrical and Electronic Engineering vol 27 pp 90-95 (IOS Press)

[2] Szabo Zs 2002 Hysteresis models from elementary operators and integral equations *PhD Thesis* (Budapest: University of Technology and Economics).

[3] Weiser B and Pfutzner H 2000 *IEEE Trans. Magn.* **36** 3759-77

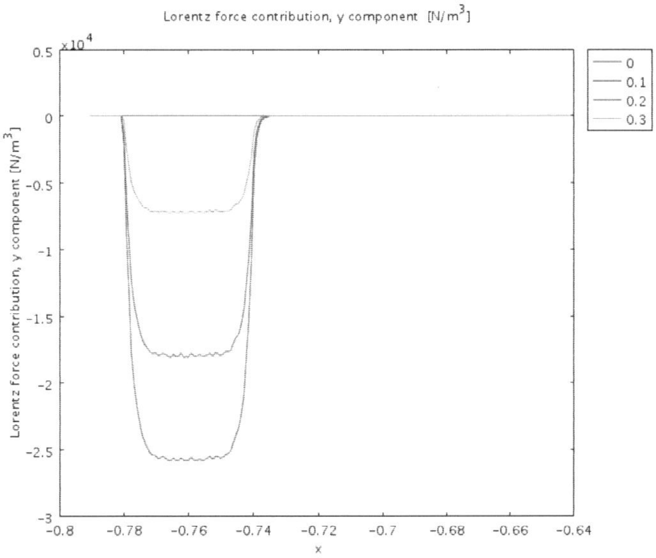

Figure 8. Lorenz force distribution of the upper transformer sheet applying moving mesh (ALE) geometry at different times.

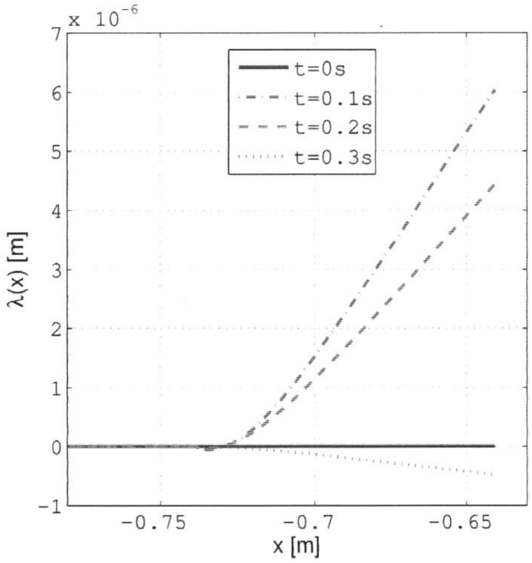

Figure 9. Displacement of the upper transformer sheet.

[4] Ivanyi A and Kuczmann M 2008 *The finite element method in magnetics* (Budapest: Akademiai Kiado)
[5] Jin J 2002 *The finite element method in electromagnetics* (New York: John Wiley and Sons)
[6] Bastos J P A and Sadowski N 2003 *Electromagnetic modeling by finite element methods* (New York: Marcel Dekker Inc.)
[7] Ivanyi A 1998 *Magnetic field computation with R-functions* (Budapest: Akademiai Kiado)
[8] Ivanyi A 2003 *Continous and discrete simulations in electrodinamics* (in Hungarian) (Budapest: Akademiai Kiado)
[9] Bíró O, Preis K and Richter K R 1996 *IEEE Trans. Magn.* **32** 651-54
[10] Bottauscio O, Chiampi M, Chiarabaglio D and Repetto M 2000 *Physica B* **275** 34-39

A generalized model for rate-independent ferromagnetic hysteresis phenomena

A Stadler
STS Spezial-Transformatoren Stockach GmbH & Co. KG,
Am Krottenbühl 1, 78333 Stockach, Germany

E-mail: stadler@sts-trafo.de

Abstract. In this paper a new approach is presented, how to calculate the inductance and the core losses of chokes considering DC premagnetization. The formulation is based on the well known Hodgdon model. Generalized material functions are applied that can be easily adapted to measured data. To obtain these data, an experimental set-up was developed according to the standards. It is shown that the measurement of symmetrical *B-H* loops is sufficient to calibrate the model. By means of the adapted dynamical model also unsymmetrical hysteresis loops can be simulated with high accuracy in steady state. The influence of DC premagnetization on the inductance and the core losses is investigated under practical conditions using commercial ferrite ring core specimens. The results are verified by measurements and illustrated in various figures and tables.

1. Introduction

Considering climate protection requirements and rapidly rising energy prices, energy saving technology has become essential to realize the ecological and economical objectives. Improving the power efficiency is now a permanent challenge of research and development in power electronics. Here, the focus is put especially on small to medium power electronic systems (up to some kW), due to their widespread applications and high cost pressure. A typical commercial 300W power supply is exemplified in figure 1.

Figure 1. Commercial 300W PC power supply (total losses ~60W, losses caused by inductive components ~20W).

The total efficiency of such a component is usually ~ 80% , thus leading to a total power loss of ~60W. About one third (~20W) of the total power loss is generated by the inductive components. The example depicts that chokes and transformers are one of the main contributors to losses in switch mode power supplies (SMPS). This problem will become much more critical with increasing switching frequencies due to new generation power switches (e.g. silicon carbide technology [1]). Consequently, further optimization has to be based on the passives and particularly on the inductive components.

The aim of the development process is to find an optimum component design related to the desired inductance value(s). Therefore simulations of the assembly of core and coil have to be as accurate as possible. Due to the strong nonlinear behavior of the ferrite material usually preferred in that power range, the calculation of the inductance(s) and the core losses can lead to considerable problems, particularly under DC premagnetization. Conventional approximate formulas based on previously measured data or information supplied by the manufacturer are often inadequate for practical applications. To close this gap, the new approach of this paper is to calculate the inductance(s) and the core losses by means of a time domain simulation of the surrounded loop(s) in the *B-H* plane. Therefore the material behavior has to be known in detail, including its dependencies on the influencing quantities. Of course, the formulation of the hysteresis phenomena has to be in agreement with physics.

A measurement test set-up was assembled to obtain the power loss and to derive the material properties. Figure 2 shows the equivalent circuit according to the standards IEC 60404-6 (2003).

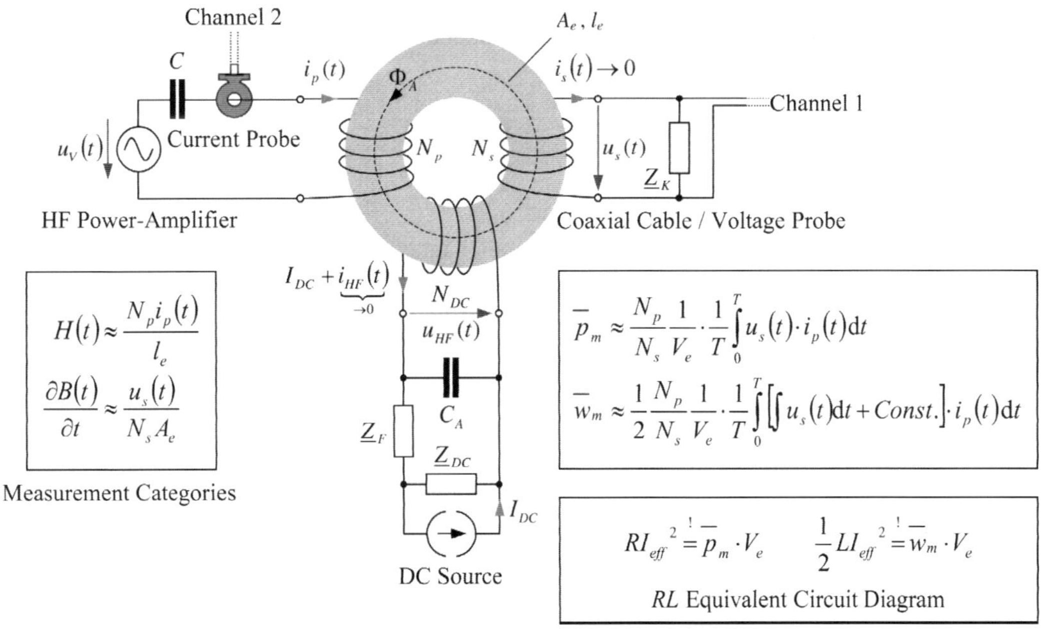

Figure 2. Measurement test set-up according to the standards IEC 60404-6 (2003).

Three independent windings are realized on the ring core specimens. The primary winding conducts the magnetizing high frequency (HF) current $i_p(t)$ that is related to the *H*-field by Oersted's law. The time derivation of the magnetic flux density can be determined via the secondary voltage $u_s(t)$ evaluating Faraday's law of electromagnetic induction. A third winding conducts the DC current I_{DC} that is required for a certain DC premagnetization. Of course, I_{DC} could be also applied to the primary winding, but in the configuration presented here, a current probe is used to measure the primary current. The measurement accuracy of these components is very sensitive to DC bias and the sensor has to be prevented from running into saturation. If the current probe is replaced by a (low inductive) shunt resistor, a third winding is optional. During the measurement process, the data of at least one period of $i_p(t)$ and $u_s(t)$ is being sampled at the two channels of a digital oscilloscope, transmitted to a PC and post processed using Matlab™ language. Finally, the desired values of the mean power loss per volume $\overline{p_m}$ and the mean stored magnetic energy per volume $\overline{w_m}$ are calculated via numerical integration following the formulas denoted in figure 2. In a previous work [2], the measurement accuracy of this experimental set-up has been determined by a detailed error analysis. It fulfils the requirements up to the MHz-range.

2. Transient hysteresis modeling based on Hodgdon's constitutive relations

2.1. *Model formulation*

Sophisticated dynamical models are required for a realistic simulation of the nonlinear behavior of inductive components. Arbitrary transitions from one state (B_i, H_i) in the *B-H* plane to another state (B_{i+1}, H_{i+1}) have to be described. Therefore a replication of surrounded symmetrical hysteresis loops as given e.g. in [3], [4] and [5] is not sufficient. Adequate models have been published by the authors L. O. Chua, M. L. Hodgdon, D. C. Jiles & D. L. Atherton and F. Preisach, their advantages and disadvantages have been investigated in [6], [7], [8] and [9]. The approach of this paper is based on the previous work of B. D. Coleman and M. L. Hodgdon published in [10], [11], [12], [13] and [14]. According to the authors, the time derivation of the magnetic flux density can be expressed by the following differential equation:

$$\frac{\partial B}{\partial t} = \alpha \left| \frac{\partial H}{\partial t} \right| [\mathrm{f}(H) - B] + \frac{\partial H}{\partial t} \mathrm{g}(H). \tag{1}$$

In (1), the material functions $\mathrm{f}(H)$ and $\mathrm{g}(H)$ as well as the positive constant α remain as the degrees of freedom and are used for model calibration. If static (rate independent) hysteresis is assumed, the time derivations can be reduced, thus leading to an expression for the differential permeability:

$$\frac{\partial B}{\partial H} = \alpha\, \mathrm{s}\left(\frac{\partial H}{\partial t} \right) [\mathrm{f}(H) - B] + \mathrm{g}(H). \tag{2}$$

Equation (2) contains the signum-function defined by

$$\mathrm{s}(\eta) = \begin{cases} 1 & \eta > 0 \\ 0 \quad \text{if} & \eta = 0 \\ -1 & \eta < 0. \end{cases} \tag{3}$$

An alternate formulation of (2) is obtained by the interchange of *B* and *H* in (2):

$$\frac{\partial H}{\partial B} = \tilde{\alpha}\, \mathrm{s}\left(\frac{\partial B}{\partial t} \right) [\tilde{\mathrm{f}}(B) - H] + \tilde{\mathrm{g}}(B). \tag{4}$$

The following symmetrical conditions have to be kept by the material functions in (2) and (4):

$$\begin{aligned} &\mathrm{f}(H) = -\mathrm{f}(-H), \quad \mathrm{f}(0) = 0 \quad \text{and} \quad \mathrm{g}(H) = \mathrm{g}(-H) \quad \text{resp.} \\ &\tilde{\mathrm{f}}(B) = -\tilde{\mathrm{f}}(-B), \quad \tilde{\mathrm{f}}(0) = 0 \quad \text{and} \quad \tilde{\mathrm{g}}(B) = \tilde{\mathrm{g}}(-B). \end{aligned} \tag{5}$$

To give a better understanding of the physical meaning of the equations described above, a small signal analysis of (2) is shown below in detail.

Journal of Physics: Conference Series **268** (2011) 012028 doi:10.1088/1742-6596/268/1/012028

2.2. Small signal analysis

To begin with, the material functions $\mathrm{f}(H)$ and $\mathrm{g}(H)$ are replaced by its first order Taylor approximations close to the origin $H=0$:

$$\mathrm{f}(H) \approx f_1 H \quad \text{and} \quad \mathrm{g}(H) \approx g_0 \,. \tag{6, 7}$$

Due to formality reasons, the following abbreviations are introduced to denote the different branches of the hysteresis loop:

$$B(H) = \begin{cases} B_1(H) & \partial H/\partial t > 0 & & H > 0, \\ B_2(H) & \partial H/\partial t < 0 & & H > 0, \\ B_3(H) & \partial H/\partial t < 0 & & H < 0, \\ B_4(H) & \partial H/\partial t > 0 & & H < 0. \end{cases} \quad \text{if} \quad \text{and} \tag{8}$$

Following the notation (8), two different partial differential equations are obtained from (2):

$$\frac{\partial B_{1,4}}{\partial H} + \alpha B_{1,4} = \alpha f_1 H + g_0 \quad \text{and} \quad \frac{\partial B_{2,3}}{\partial H} - \alpha B_{2,3} = -\alpha f_1 H + g_0 \,. \tag{9, 10}$$

The equations (9) and (10) can be solved analytically. Superpositions of the homogeneous and particular solutions lead to

$$\begin{aligned} B_1(H) &= c_1 \mathrm{e}^{-\alpha H} + f_1 H - \frac{f_1}{\alpha} + \frac{g_0}{\alpha}, \\ B_2(H) &= c_2 \mathrm{e}^{\alpha H} + f_1 H + \frac{f_1}{\alpha} - \frac{g_0}{\alpha}, \\ B_3(H) &= c_3 \mathrm{e}^{\alpha H} + f_1 H + \frac{f_1}{\alpha} - \frac{g_0}{\alpha} \quad \text{and} \\ B_4(H) &= c_4 \mathrm{e}^{-\alpha H} + f_1 H - \frac{f_1}{\alpha} + \frac{g_0}{\alpha} \end{aligned} \tag{11}$$

with the unknown coefficients c_1, c_2, c_3 and c_4. If the magnetic field strength H oscillates between $-\hat{H}$ and $+\hat{H}$ for a long time, the solutions have to fulfill the steady state conditions

$$-B_1(0) = B_2(0) = B_3(0) = -B_4(0) := B_r \tag{12}$$

depending on the remanent flux density B_r. With the coefficients specified by

$$c_1 = -c_2 = -c_3 = c_4 = -B_r + \frac{f_1}{\alpha} - \frac{g_0}{\alpha} \tag{13}$$

and the remanent flux density

$$B_r = \left(1 - \frac{2\mathrm{e}^{-\alpha\hat{H}} - 2\mathrm{e}^{-3\alpha\hat{H}}}{1 - \mathrm{e}^{-4\alpha\hat{H}}}\right)\left(\frac{f_1}{\alpha} - \frac{g_0}{\alpha}\right), \tag{14}$$

(11) now can be described by an entire formula:

$$B(H) = \mathrm{s}\left(\frac{\partial H}{\partial t}\right)\left[\left(-B_r + \frac{f_1}{\alpha} - \frac{g_0}{\alpha}\right)\mathrm{e}^{-\alpha \mathrm{s}\left(\frac{\partial H}{\partial t}\right) H} - \frac{f_1}{\alpha} + \frac{g_0}{\alpha}\right] + f_1 H \,. \tag{15}$$

With aid of the exponential approximation for small arguments η

$$e^{\eta} \approx 1 + \eta + \frac{1}{2}\eta^2 , \tag{16}$$

(15) can be further reduced:

$$B(H) \approx s\left(\frac{\partial H}{\partial t}\right)\left[\frac{\alpha^2}{2}\left(-B_r + \frac{f_1}{\alpha} - \frac{g_0}{\alpha}\right)H^2 - B_r\right] + (\alpha B_r + g_0)H . \tag{17}$$

The hysteresis loop for small signal excitations observed by Lord Rayleigh [15] has the form

$$B_{Ray}(H) = \mu_0\left[s\left(\frac{\partial H}{\partial t}\right)\frac{\nu}{2}\left(H^2 - \hat{H}^2\right) + \left(\mu_i + \nu\hat{H}\right)H\right] \tag{18}$$

with the hysteresis coefficient ν, the initial permeability μ_i and the amplitude permeability μ_a according to

$$\mu_i = \frac{1}{\mu_0}\frac{\partial B}{\partial H}\bigg|_{H\to 0} \quad \text{and} \quad \mu_a = \mu_i + \nu\hat{H} . \tag{19, 20}$$

By means of a comparison of (17) and (18), these material parameters can be calculated in turn:

$$\nu = \frac{2\alpha}{\mu_0}\frac{f_1 - g_0}{2 + \alpha^2\hat{H}^2}, \quad \mu_a = \frac{\alpha\nu\hat{H}^2}{2} + \frac{g_0}{\mu_0} \quad \text{and} \quad \mu_i = \mu_a - \nu\hat{H} . \tag{21, 22, 23}$$

By (21), (22) and (23), it is demonstrated that the empirical relation (2) shows the expected physical behavior (18) for symmetrical small signal excitations. The surrounded loops can be directly associated with common material parameters.

2.3. Large signal analysis

Now (2) is solved also for large signal excitations up to the saturation level. Considering (5), the material functions can be expressed by the Fourier series

$$f(H) = \sum_{n=1}^{+\infty} F_n \sin(p_n H) \quad \text{with} \quad p_n = \frac{n\pi}{H_{max}} \quad \text{and} \tag{24}$$

$$g(H) = G_0 + \sum_{n=1}^{+\infty} G_n \cos(p_n H) . \tag{25}$$

Therein H_{max} denotes an arbitrary positive value that should never be reached during the simulation. Analogously (9) and (10), two different partial differential equations are obtained from (2):

$$\frac{\partial B_{1,4}}{\partial H} + \alpha B_{1,4} = G_0 + \sum_{n=1}^{+\infty}\left[\alpha F_n \sin(p_n H) + G_n \cos(p_n H)\right] \quad \text{and} \tag{26}$$

$$\frac{\partial B_{2,3}}{\partial H} - \alpha B_{2,3} = G_0 + \sum_{n=1}^{+\infty}\left[-\alpha F_n \sin(p_n H) + G_n \cos(p_n H)\right] . \tag{27}$$

As an advantage of the harmonic approaches (24) and (25), an analytical solution can be found for (26) and (27) as well.

If the homogeneous and particular parts are superimposed, the solutions for all of the different branches of the hysteresis loop are given by

$$\begin{aligned} B_1(H) &= c_1\,\mathrm{e}^{-\alpha H} + B_0 + \sum_{n=1}^{+\infty}\left[A_n \sin(p_n H) + B_n \cos(p_n H)\right], \\ B_2(H) &= c_2\,\mathrm{e}^{\alpha H} - B_0 + \sum_{n=1}^{+\infty}\left[A_n \sin(p_n H) - B_n \cos(p_n H)\right], \\ B_3(H) &= c_3\,\mathrm{e}^{\alpha H} - B_0 + \sum_{n=1}^{+\infty}\left[A_n \sin(p_n H) - B_n \cos(p_n H)\right] \quad \text{and} \\ B_4(H) &= c_4\,\mathrm{e}^{-\alpha H} + B_0 + \sum_{n=1}^{+\infty}\left[A_n \sin(p_n H) + B_n \cos(p_n H)\right]. \end{aligned} \tag{28}$$

The set of equations (28) encloses the unknown coefficients c_1, c_2, c_3 and c_4 and the abbreviations

$$A_n = \frac{\alpha^2 F_n + p_n G_n}{p_n^{\;2} + \alpha^2}, \quad B_0 = \frac{G_0}{\alpha} \quad \text{and} \quad B_n = \frac{-\alpha p_n F_n + \alpha G_n}{p_n^{\;2} + \alpha^2}. \tag{29, 30, 31}$$

If a symmetrical excitation with the amplitude $\hat{H}$ of the magnetic field strength is assumed, the conditions

$$-B_1(0) = B_2(0) = B_3(0) = -B_4(0) := B_r \tag{32}$$

are fulfilled in steady state by the relations

$$c_1 = -c_2 = -c_3 = c_4 = -B_r - B_0 - \sum_{n=1}^{+\infty} B_n\,. \tag{33}$$

With $\hat{H}$ as a parameter, the remanent flux density is given by

$$B_r = -B_0 - \sum_{n=1}^{+\infty} B_n + \frac{2\,\mathrm{e}^{-\alpha\hat{H}} - 2\,\mathrm{e}^{-3\alpha\hat{H}}}{1 - \mathrm{e}^{-4\alpha\hat{H}}}\left[B_0 + \sum_{n=1}^{+\infty} B_n \cos\left(p_n \hat{H}\right)\right]. \tag{34}$$

Thus the different solutions (28) for each of the branches (8) can be condensed into one entire expression:

$$B_{Hodg.}(H) = \mathrm{s}\left(\frac{\partial H}{\partial t}\right)\left[\left(-B_r - B_0 - \sum_{n=1}^{+\infty} B_n\right)\mathrm{e}^{-\alpha\,\mathrm{s}\left(\frac{\partial H}{\partial t}\right) H} + B_0 + \sum_{n=1}^{+\infty} B_n \cos(p_n H)\right] + \sum_{n=1}^{+\infty} A_n \sin(p_n H). \tag{35}$$

The final equation (35) offers the opportunity to determine the material functions (24) and (25) directly by means of numerical curve fitting (e.g. by least squares method) to a vector of measured symmetrical subloops of different excitation levels. These data can be easily obtained from the measurement test set-up depicted in figure 2. In contrast to [12], [13] and [14], where distinctive analytical functions are suggested for $\mathrm{f}(H)$ and $\mathrm{g}(H)$, the presented model is not restricted. Of course, the stability and closure criteria given in [13] also have to be fulfilled by the generalized material functions (24) and (25). This requires a numerical check after model calibration up to the saturation value of the magnetic field strength H_S that defines the saturation loop. Finally, all simulation has to be limited by $|H| \le H_S$ to guarantee stability and closed loops in steady state.

2.4. Adaptation of the model to measured data

Figure 3 shows a set of symmetrical subloops of different excitation levels. The data is obtained from the measurement test set-up shown in figure 2 using commercial R16 ring core specimens. For $f = 10\text{kHz}$, rate-independent hysteresis can still be assumed. To be able to investigate the influence of the temperature, it is possible to measure the samples inside a heating cabinet. According to the manufacturer's data sheet, the investigated MMG™ F49 material is adequate for power applications. These materials usually show lower core losses at high temperatures. As expected, thinner loops are obtained at $T_c = 100°\text{C}$ (lower picture) in comparison to room temperature $T_c = 25°\text{C}$ (upper picture).

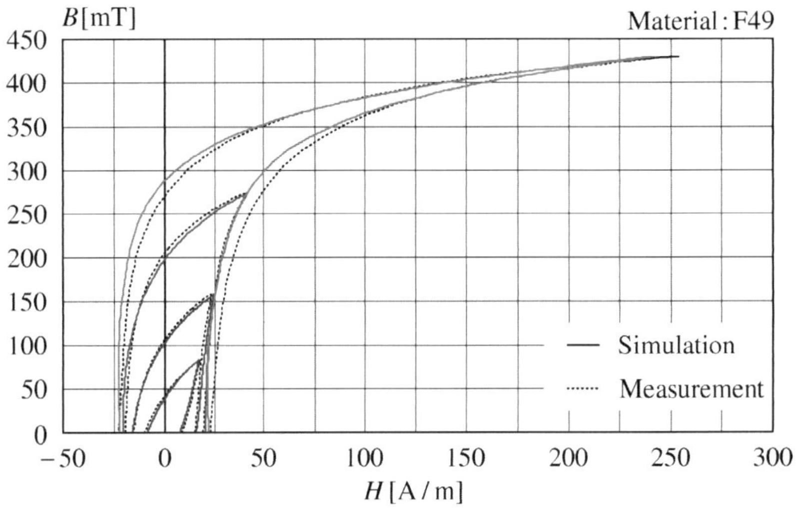

Parameters: $f = 10\text{kHz}$, $T_c = 25°\text{C}$

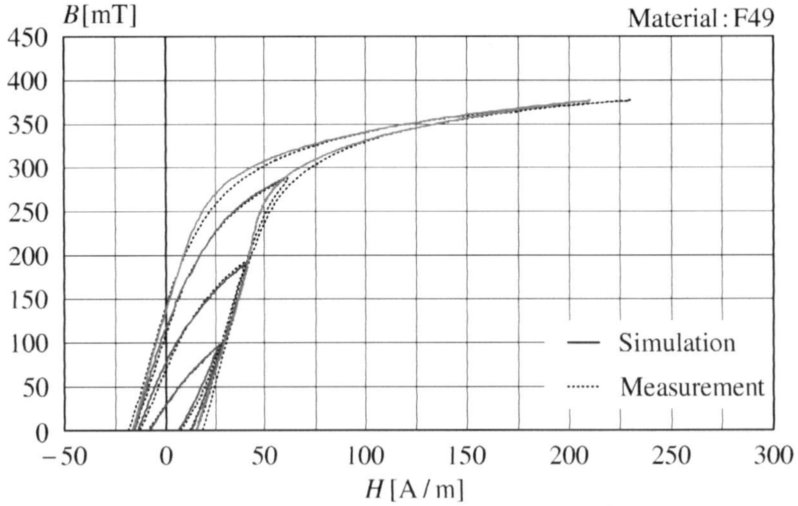

Parameters: $f = 10\text{kHz}$, $T_c = 100°\text{C}$

Figure 3. Measured symmetrical *B-H* loops used for model calibration (MMG™ F49 ferrite [16]).

To adapt the model to the measured subloops depicted in figure 3, an optimum value for the positive constant α in (2) has to be found. Therefore the differential permeability is determined in the remanent point $(B = B_r, H = 0)$ of the saturation loop. We obtain

$$\left.\frac{\partial B}{\partial H}\right|_{(B_r,H=0)} = \alpha B_r + \mathrm{g}(0) = \alpha B_r + \mu_0\mu_i \quad \rightarrow \quad \alpha = B_r^{-1}\left(\left.\frac{\partial B}{\partial H}\right|_{(B_r,H=0)} - \mu_0\mu_i\right). \tag{36}$$

Now the unknown coefficients F_n, G_0 and G_n of the generalized material functions (24) and (25) can be determined by means of numerical curve fitting of (35). To reduce the number of necessary iterations, it is recommended to use adequate starting parameters. These are directly given by the small signal analysis described above. A Fourier expansion of (6) and (7) leads to

$$\mathrm{f}(H) = f_1 H = \sum_{n=1}^{+\infty} F_n \sin(p_n H) \quad \text{with} \quad p_n = \frac{n\pi}{H_{max}} \quad \text{and} \tag{37}$$

$$\mathrm{g}(H) = g_0 = G_0 + \sum_{n=1}^{+\infty} G_n \cos(p_n H) \quad \text{with} \tag{38}$$

$$F_n = \frac{2 f_1}{p_n}(-1)^{(n-1)}, \quad G_0 = g_0 \quad \text{and} \quad G_n = 0. \tag{39, 40, 41}$$

Consequently the starting parameters (39), (40) and (41) only depend on f_1 and g_0 that can be derived from the data sheet values of ν, μ_a and μ_i by solving the nonlinear system of equations (21), (22) and (23). Finally, the starting parameters have to be varied until the least squares condition

$$\left|B^2_{Meas.}(H_{Meas.}) - B^2_{Hodg.}(H_{Meas.})\right| = \min! \tag{42}$$

is fulfilled. In praxis, the calculation can easily be done with aid of the Matlab™ function 'lsqcurvefit.m'. Adequate values for H_{max} in (38) are

$$H_S \le H_{max} \le 10 H_S\,. \tag{43}$$

A comparison of the measured and the simulated results (dotted lines) in figure 3 shows the good agreement between the measured data and the calibrated model. Due to the simultaneous fitting of all subloops, the mean errors are equally distributed over the single loops, no runaways can be detected. For the example evaluation, the Fourier series (24) and (25) is truncated after $N = 10$ addends. A time domain resolution of 512 sampling points per period is sufficient for an adequate representation of the subloops. Consequently, the amount of calculation time for the fitting process is only some seconds on a standard office PC.

3. Results

3.1. Simulations based on the adapted model

During the investigation it was found that the alternate representation (4) is more adequate for simulations up to the saturation level. Therefore the alternate positive constant $\tilde{\alpha}$ has to be determined instead of α. It is given by the inverse differential permeability in the coercive point $(B = 0, H = H_{CB})$ of the saturation loop:

$$\left.\frac{\partial H}{\partial B}\right|_{(B=0,H=H_{CB})} = -\hat{\alpha} H_{CB} + \tilde{\mathrm{g}}(0) = -\tilde{\alpha} H_{CB} + (\mu_0\mu_i)^{-1} \quad \rightarrow \quad \tilde{\alpha} = H_{CB}^{-1}\left[(\mu_0\mu_i)^{-1} - \left.\frac{\partial H}{\partial B}\right|_{(B=0,H=H_{CB})}\right]. \tag{44}$$

By the interchange of B and H in (24) and (25), the Eigenvalues of the Fourier expansions have to be replaced by

$$\tilde{p}_n = \frac{n\pi}{B_{max}}. \tag{45}$$

Finally, the derivations of the alternate model (4) are substituted by finite differences

$$\left.\frac{\partial H}{\partial B}\right|_{(B=B_i,H=H_i)} \approx \frac{H_{i+1}-H_i}{B_{i+1}-B_i} \approx \tilde{\alpha}\,\mathrm{s}\left(\frac{B_{i+1}-B_i}{t_{i+1}-t_i}\right)\left[\tilde{\mathrm{f}}(B_i)-H_i\right]+\tilde{\mathrm{g}}(B_i) \tag{46}$$

and expression (46) is solved for the desired value H_{i+1} of the magnetic field strength in the future point:

$$H_{i+1} = H_i + \underbrace{\left\{\tilde{\alpha}\,\mathrm{s}(B_{i+1}-B_i)\left[\tilde{\mathrm{f}}(B_i)-H_i\right]+\tilde{\mathrm{g}}(B_i)\right\}}_{\tilde{\mathrm{h}}(B_i,B_{i+1},H_i)} \cdot \underbrace{(B_{i+1}-B_i)}_{\Delta B}. \tag{47}$$

Annotations: In practical simulations, the magnetic devices are usually excited by voltage- and not by current sources. In case of voltage sources, the value B_{i+1} of the magnetic flux density in the future point is identified, in case of current sources the value H_{i+1} of the magnetic field strength is fixed. Therefore the alternate notation (47) is more adequate in praxis than those based on the original representation (2). Moreover no inversion of the "modification" $\tilde{\mathrm{h}}(B_i, B_{i+1}, H_i)$ is necessary during evaluation. Another inherent problem is given by the determination of the DC premagnetization B_{DC}. Due to the transformer measurement principle depicted by figure 2, the measured average value of $B(t)$ over each period has to be zero. Of course, a Hall sensor could be used to measure B_{DC}, but as a disadvantage, the ring core specimens have to equipped with an air gap, thus leading to considerable problems due to the parasitic coupling of the stray field. To avoid these problems, the 'commutation curve' is utilized here as a representation of the nonlinear relation between B_{DC} and H_{DC}. The curve shown in figure 4 is given by the closure points of the ascendant and descendant branches of a set of symmetrical subloops of different excitation levels. It is usually very close to the initial magnetization curve. Figure 5 shows the run of the normalized alternate material functions $\tilde{\mathrm{f}}(B)$ and $\tilde{\mathrm{g}}(B)$.

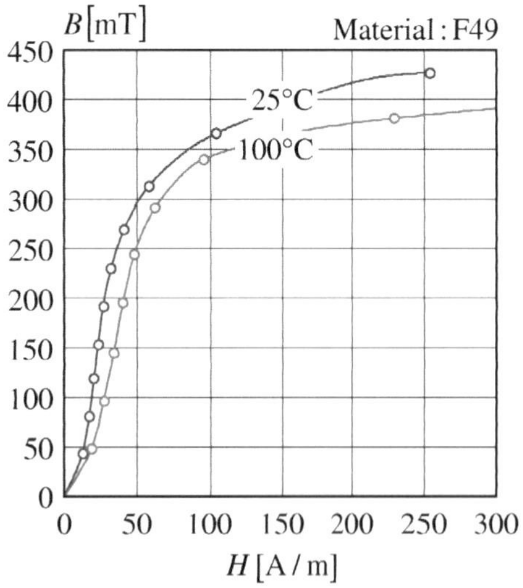

Figure 4. Measured commutation curves.

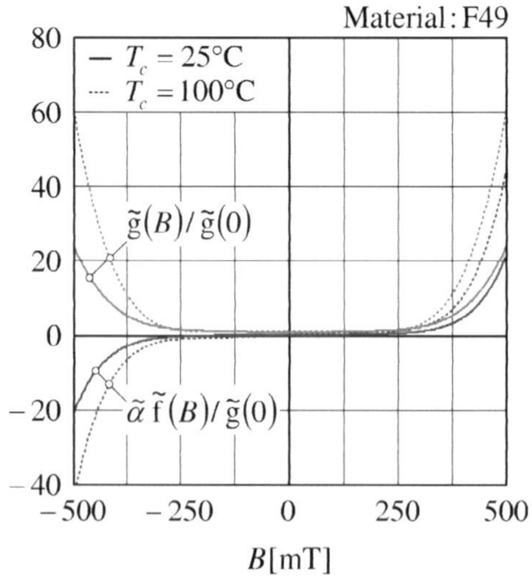

Figure 5. Material functions $\tilde{\mathrm{f}}$ and $\tilde{\mathrm{g}}$.

The numerical values of the parameters $\tilde{\alpha}$ and $\tilde{g}(0)$ are given in table 1. In figure 6 a comparison between time domain simulations and measured results is presented in steady state.

Table 1. Numerical values of the material parameters $\tilde{\alpha}$ and $\tilde{g}(0)$.

$f = 10\text{kHz}$	Material: F49 $T_c = 25°C$	$T_c = 100°C$
$\tilde{\alpha}\left[(\text{Ws/m}^2)^{-1}\right]$	12.511	13.112
$\tilde{g}(0)\left[(\text{Vs/Am})^{-1}\right]$	260.185	329.612

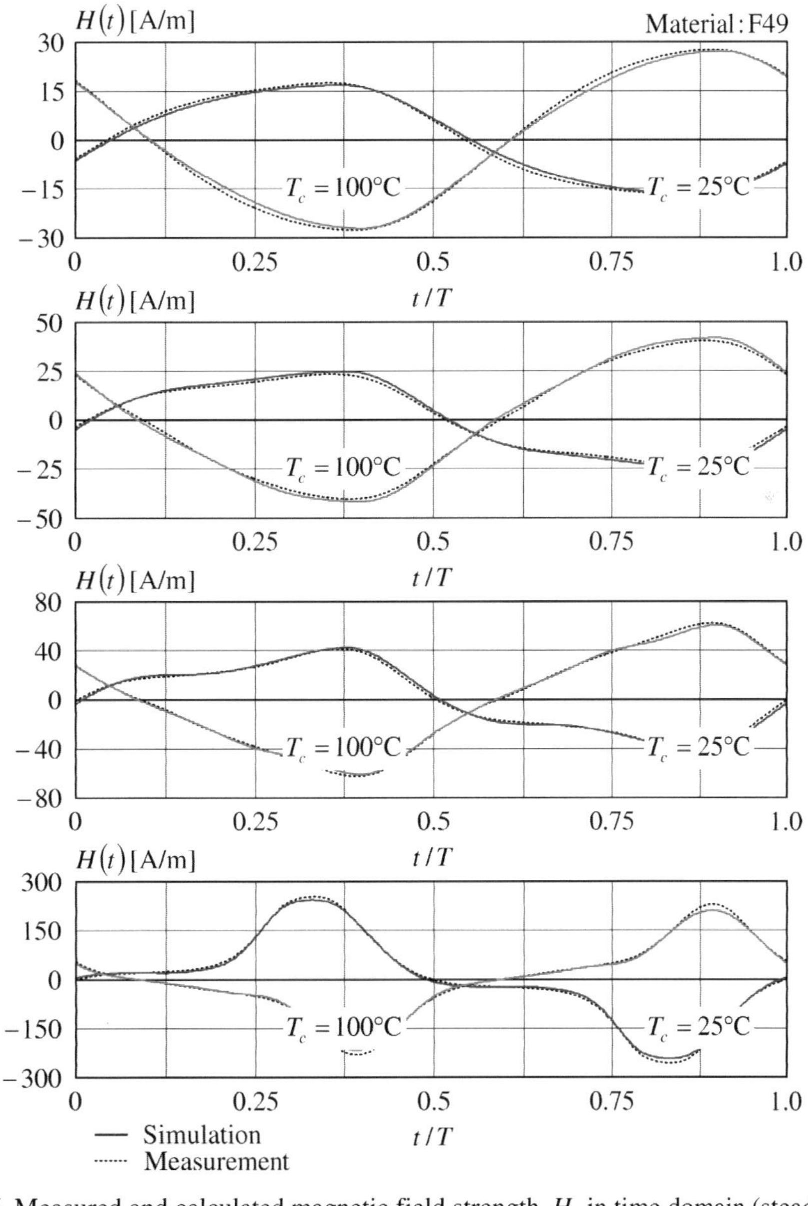

Figure 6. Measured and calculated magnetic field strength H in time domain (steady state).

Journal of Physics: Conference Series **268** (2011) 012028 doi:10.1088/1742-6596/268/1/012028

To give an impression of the achieved accuracy, the mean power loss per volume and per cycle

$$\overline{p}_m = \frac{1}{T}\int_0^T p_m(t)\mathrm{d}t \quad \text{with} \quad p_m(t) = H(t)\cdot\frac{\partial B(t)}{\partial t} \tag{48}$$

is evaluated numerically as a function of the maximum flux density $\hat{B}$ and compared to measurement points in figure 7. The results show that a very good replication of the hysteresis phenomena is obtained from the calibrated model up to the saturation level. This is more or less a direct consequence of the generalized material functions (24) and (25). Furthermore simulations of $\overline{p}_m$ and the average stored magnetic energy per volume and per cycle

$$\overline{w}_m = \frac{1}{T}\int_0^T w_m(t)\mathrm{d}t \quad \text{with} \quad w_m(t) = \frac{1}{2}B(t)\cdot H(t) \tag{49}$$

that is associated with the inductance of the component, are conducted based on the calibrated model. In table 2 a comparison between calculated and measured points of $\overline{p}_m$ and $\overline{w}_m$ is listed.

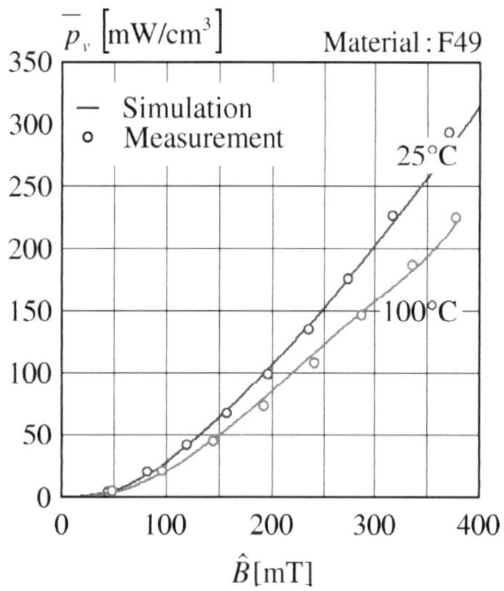

Figure 7. Measured and calculated mean power loss per volume $\overline{p}_v$ $(f = 10\text{kHz})$ as a function of the maximum flux density $\hat{B}$.

Table 2. Simulation of $\overline{p}_v$ and $\overline{w}_m$ in time domain considering different values of the applied DC premagnetization B_{DC} (parameters : $\hat{B} = 100\text{mT}$, $f = 20\text{kHz}$).

$\overline{p}_v$ [mW/cm³]	$T_c = 25°C$		$T_c = 100°C$	
	Measurement	**Simulation**	**Measurement**	**Simulation**
$B_{DC} = 0$mT	66.1899	56.0969	52.7062	40.8264
200mT	67.8200	69.1616	55.5901	46.2143
300mT	75.2274	72.1820	56.0040	48.4272
350mT	80.1502	73.5691	55.3041	61.3847
$\overline{w}_m$ [Ws/m³]	$T_c = 25°C$		$T_c = 100°C$	
	Measurement	**Simulation**	**Measurement**	**Simulation**
$B_{DC} = 0$mT	0.4363	0.4998	0.6775	0.7187
200mT	1.2452	1.1623	1.1528	1.1853
300mT	1.9330	1.7065	1.9705	2.1506
350mT	2.8466	2.4852	3.4908	3.7068

Therein the AC excitation level defined by $\hat{B} = 0.5(B_{max} - B_{min})$ is fixed at $\hat{B} = 100\text{mT}$. All results are obtained for a frequency of $f = 20\text{kHz}$ (the model was adapted to symmetrical loops measured at $f = 10\text{kHz}$). It is found that the description of the physical phenomena by the model is precise enough to calculate results under DC premagnetization with high accuracy, even if none of these information is used for model adaptation. Figure 8 shows the surrounded loops with the DC premagnetization B_{DC} as a parameter. The DC offset of the magnetic field strength H_{DC} is automatically given by (47), if the desired DC premagnetization is applied to $B(t)$ in time domain. Hence no use of the commutation curves shown in figure 4 is required.

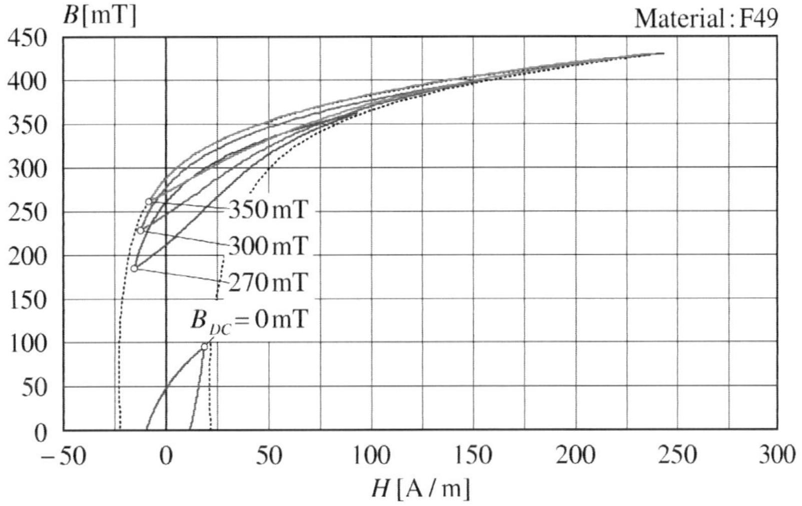

Parameters: $\hat{B} = 100\text{mT}$, $f = 20\text{kHz}$, $T_c = 25°\text{C}$

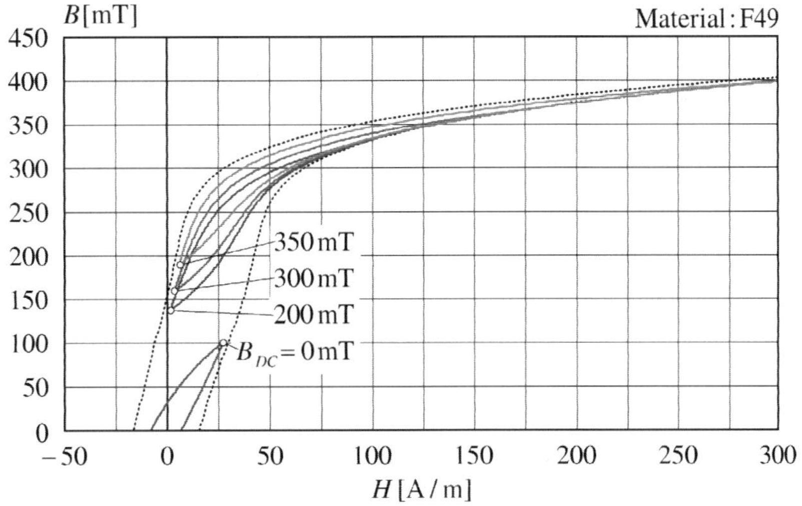

Parameters: $\hat{B} = 100\text{mT}$, $f = 20\text{kHz}$, $T_c = 100°\text{C}$

Figure 8. Simulated *B-H* loops for different values of the applied DC premagnetization B_{DC}.

4. Conclusion

A generalized approach is presented based on the well known Hodgdon dynamical model. It is shown how the model is calibrated using a set of measured symmetrical subloops – data that can be easily acquired by means of the shown experimental set-up. It is found that the physical behavior of the model is precise enough to simulate hysteresis loops also under DC premagnetization, even if none of these information is used for the adaptation of the model. The conditions are discussed how to obtain stability and closed loops in steady state. Of course, all of the verified results presented by the paper refer to the steady state in case of periodical excitation. No fast transient processes are investigated. This is due to the fact that fast transients usually are not of interest in the calculation of the choke inductance and the core losses. But regarding to the surveys of dynamical models to be found in literature, it can be supposed that Hodgdon's constitutive relations are not adequate for a precise modeling of these effects. The well known approaches of D. C. Jiles & D. L. Atherton as well as F. Preisach seem to be more promising here.

Acknowledgments
The author would like to thank the German Academic Exchange Service (DAAD) for the financial support.

References

[1] Reimann T, Petzoldt J, Zverev I, Mitlehner H and Friedrichs P 2002 *Conf. Rec. of the 10th Int. Power Elec. and Mo. Con. Conf. EPE-PEMC* **1** 1-12
[2] Stadler A 2010 *Simulation und Messung der Kernverluste weichmagnetischer Materialien: Ein praktisches Verfahren zur Berechnung der Kernverluste bei beliebigen Kernformen mit Hilfe der Finite Elemente Methode (FEM)* (Saarbrücken: Südwestdeutscher Verlag für Hochschulschriften SVH)
[3] Mohammed I A, Al-Hashemy B A R and Tawfik M A 1997 *IEEE Trans. on Mag.* **33** 686-91
[4] Rivas J, Zamarro J M, Martin E and Pereira C 1981 *IEEE Trans. on Mag.* **17** 1498-1502
[5] Udpa S S and Lord W 1985 *IEEE Trans. on Mag.* **21** 2370-3
[6] Iványi A 1997 *Hysteresis models in electromagnetic computation* (Budapest: Akadémiai Kiadó)
[7] Andrei P, Caltun O and Stancu A 1998 *IEEE Trans. on Mag.* **34** 231-41
[8] Ossart F and Meunier G 1990 *IEEE Trans. on Mag.* **26** 2837-9
[9] Takach M D and Lauritzen P O 1995 *Proc. of the 10th Appl. Power Elec. Conf. and Exp. APEC* **1** 560-6
[10] Boley C D and Hodgdon M L 1989 *IEEE Trans. on Mag.* **25** 3922-4
[11] Caltun O, Papusoi C, Kappel W, Andrei P, Stancu A and Hatmanu D 1995 *Rom. Univ. Sci. Ann.* **41** 37-46
[12] Coleman B D and Hodgdon M L 1987 *Arch. for Rat. Mech. and Anal.* **1** 375-96
[13] Hodgdon M L 1988 *IEEE Trans. on Mag.* **24** 218-21
[14] Hodgdon M L 1988 *IEEE Trans. on Mag.* **24** 3120-2
[15] Lord Rayleigh 1887 *Phil. Mag.* **23** 225
[16] MMG Magdev Ltd. 2008 Data sheets *http://www.magdev.co.uk*

Journal of Physics: Conference Series **268** (2011) 012029 doi:10.1088/1742-6596/268/1/012029

Explosion safety in industrial electrostatics

S V Szabó, I Kiss and I Berta

Department of Electric Power Engineering, Group of High Voltage Engineering and Equipment, Budapest University of Technology and Economics, Hungary

E-mail: szabo.sandor@vet.bme.hu

Abstract. Complicated industrial systems are often endangered by electrostatic hazards, both from atmospheric (lightning phenomenon, primary and secondary lightning protection) and industrial (technological problems caused by static charging and fire and explosion hazards.) According to the classical approach protective methods have to be used in order to remove electrostatic charging and to avoid damages, however no attempt to compute the risk before and after applying the protective method is made, relying instead on well-educated and practiced expertise. The Budapest School of Electrostatics - in close cooperation with industrial partners - develops new suitable solutions for probability based decision support (Static Control Up-to-date Technology, SCOUT) using soft computing methods. This new approach can be used to assess and audit existing systems and - using the predictive power of the models - to design and plan activities in industrial electrostatics.

1. Introduction

Many industrial processes are extremely sensitive to electrostatic hazards, and the consequences of the events can be very severe, ranging from huge financial loss to possible human injury. Therefore there is a clear need on part of the industry to provide the necessary toolset to effectively protect against the effects of this sort of risk. In order to design effective protection, a method for gauging the risk is needed, so that the risk before and after applying the protective measure can be accurately assessed, thereby enabling the expert to choose the best set of protective devices for a given process. The complex nature of electrostatic risk however provides a challenge for those trying to model it.

Electrostatic risk can arise from atmospheric sources, including both primary and secondary effects of lightning, or from industrial electrostatics, where as a by-product of certain processes a charge accumulation and eventual discharge may occur, leading to the inflammation or explosion of the surrounding atmosphere. In both cases, the factors influencing the probability of these events are various, their interdependence complex and highly nonlinear. Also, the severity of the outcome in the different scenarios can vary over a large range, from mild inconvenience to debilitating consequences.

In case of both types of hazard it is clear that some form of protection against these damages are needed, however electrostatic protection usually does not constitute an integral part of the design effort for the process, but is rather added as a mere afterthought. It is difficult to appreciate the severity of risk in a complex industrial system, because only the effects or symptoms of the electrostatic charging can be detected.

From the above properties it can be seen that specialized and carefully controlled methodologies are needed to accurately assess and thus control electrostatic risk [1].

2. The optimisation of the protection and/or prevention strategy

When faced with the task of designing an optimal protection or prevention method for a production line or industrial process, ideally the designer has at their disposal a wide variety of sources to gather information about the system. Generally the underlying physical processes can be identified and therefore for many hazards exact models can be obtained and these can be validated against experimental data. Complex interactions of several hazards are described poorly using these exact models.

It is evident that model computations provide an efficient and valuable tool in the hands of the expert, because they aid the understanding of the phenomenon as well as having their use in simulations for the more integrated effects, thus leading to the development of new protection schemes and new process technologies.

The traditional approach to designing protective measures relies on well-educated and experienced experts applying the accepted knowledge in the field in order to reduce electrostatic risk [2]. Because there is no quantitative risk assessment or target value, over- (and in the unlucky case) under engineering of the protective methods is possible.

3. Novel approach to electrostatic hazards

The authors propose that a novel approach to assessing the electrostatic risk in a given situation that relies on a more formal protocol and focuses on quantitative risk calculations can create a situation in which overall risk can be more accurately controlled than using the traditional approach, leading to both efficient use of resources and increased safety (by choosing the most appropriate protective method) and savings (by preventing spurious measures).

Figure 1 and figure 2 illustrate the differing approaches taken by the traditional and the novel method to electrostatic risk assessment. It can be readily seen, that the specific strategy of the risk taker is explicitly included in determining the acceptable risk level, along with the particular characteristics of the possible damage.

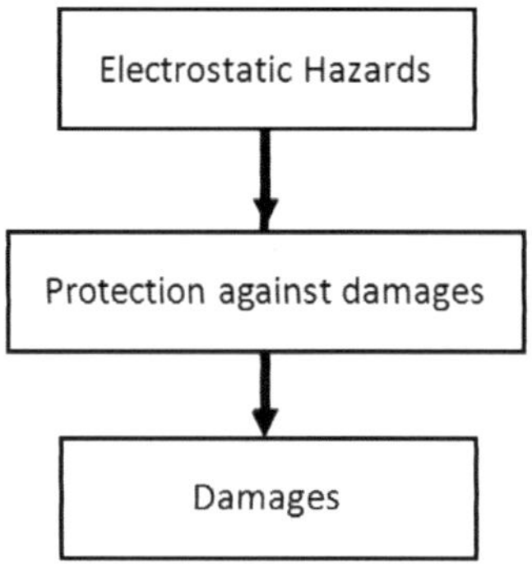

Figure 1. Traditional approach to electrostatic risk management.

It is also very important, that the parameter that describes the acceptable risk is in the dimension EUR/year, which is readily understood by all parties involved in decisions regarding process planning. Should the calculated risk be higher than the acceptable, it is natural that protective or preventive measures have to be taken. There is usually a range of preventive measures that can influence the calculated risk in a favourable way, including those that prevent the damage by making the event less frequent, and those that decrease the cost associated with each of these occurrences (often including measures that do not increase process safety in the eye of the engineers, such as purchasing additional insurance.)

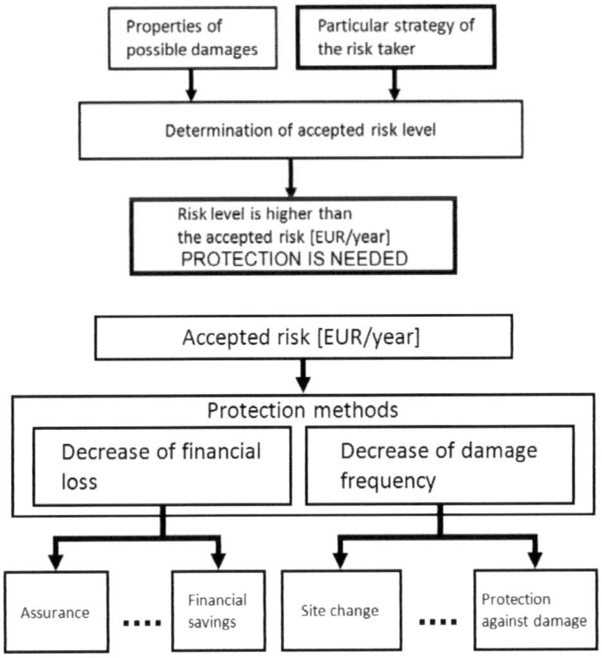

Figure 2. Novel approach to electrostatic risk management.

The proposed novel approach, as its central element, involves the use of an auditing process. To start this process, the client gives a declaration of its objectives regarding process safety, based on which an independent auditor examines all relevant aspects of the system and gives assurance that the given objectives are met. During the audit a fixed protocol and high standards for documentation are used in order to ensure the integrity of the process. Also very important as a basis for estimation is the use of a database of electrostatic hazards, their models and associated experimental data.

4. Fault diagnostics

While all methods outlined above to manage the risk associated with a given process in important, the part relevant in our current paper is the usage of technological methods to decrease the frequency and severity of damage events. Through the process of fault diagnostics a given electrostatic hazard is identified, evaluated and managed. This means that based on quantitative examination probability based calculations have to be performed in order to make a decision regarding the protective or preventive measures; finally an audit of the modified process has to be carried out in order to verify and document the gains of the diagnostic process and the decreased risk estimate.

During the diagnostic process the current risk level and the risk reduction in case of the different protective measures should be estimated. The focus of fault diagnostics is to determine the optimal set of measures that ensure the appropriate risk level. It is clear that applying all candidate measures is very inefficient due to primarily to financial reasons – however quantitative analysis of the appropriate risks lets us determine which measures are necessary, and which are the ones that are worth or not worth using.

All of the above support that the key factor in successfully managing electrostatic risk is the ability to accurately estimate the exact risk level, and thus be able to quantify in a robust manner the gains (in the form of risk reductions) from applying different measures. This information is crucial, because it allows the export to make informed and well supported decisions even though no direct measurement of the involved probabilities can be carried out.

Journal of Physics: Conference Series **268** (2011) 012029 doi:10.1088/1742-6596/268/1/012029

5. Event and fault trees

Event and fault trees are widely used in the industry to estimate the probabilities of events in cases where we have no prior information of these probabilities, however through knowledge of the physical model describing the events leading to the events is available. An event tree describes - through the use of logical operators - the relationship of the top event to basic events. In figure 3 an example event tree can be seen.

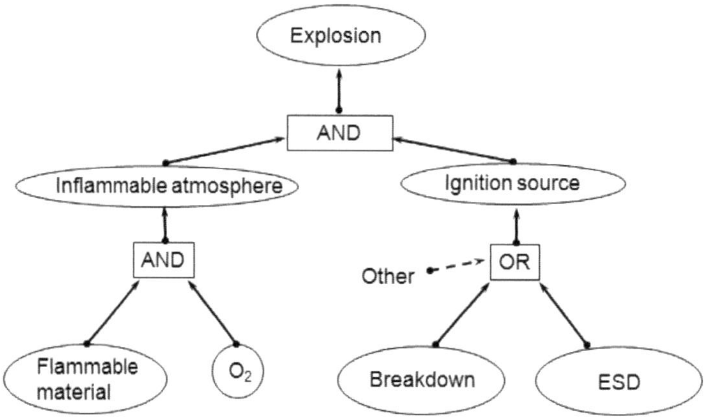

Figure 3. Event tree for an electrostatic explosion.

It is clear that by assuming the independence of these events and applying the 'and' (conjunction) and 'or' operations from probability calculus, the probability to the top event can be connected to the probabilities of base events - this is done is case of fault trees. Naturally, we hope that the probabilities of the base events are better known or easier to estimate than the probabilities of the top event.

A newer proposal is the usage of fuzzy numbers to represent the probabilities that are involved in the calculation. The rational for this is that the probabilities of the base events are generally only estimates, and fuzzy numbers provide a means for representing the inherent inaccuracy. When the calculation is performed using fuzzy numbers, the resulting probability for the top event is also a fuzzy number. This has several advantages. By examining the variance of the fuzzy result, the inaccuracy of the estimate is revealed, which quantifies the trust that should be placed in that value. A more interesting use is that the sensitivity of the inaccuracy of the top event probability to the base event inaccuracies can be studied, which can guide the expert in choosing to measure and/or control the relevant parameters better.

Possibly the greatest difficulty in using fault trees is the estimation of the probabilities of basic events. This estimate may be obtained by using parameters derived from similar processes, or by statistical evaluation of a significant number of measurements. Very often, a physical model for the system can be constructed, giving an estimate of the parameter. Very often however the only means to estimate a parameter remains evaluation by an expert based on experience.

6. On-line fault diagnostics

In the previous sections we have discussed the process of fault diagnostics as performed by field experts. This sort of off-line diagnostics treats all conditions of the industrial process as being static - in case of changing circumstances naturally the worst condition is assumed. It is evident that the diagnosis of faults can be done in an on-line manner where subsequent preventive actions are based on real time data about the process. This, of course, requires that an adequate expert system is constructed that is made aware of the observations about the system, and according to a set if inference rules is able to determine when and what intervention to the process is necessary.

The schematic diagram of such a system is presented in figure 4. It is clear that because observations are not exact, and they do not reveal the complete state of the system that is being observed, and additionally our model of the system is almost certainly imperfect, the evaluation of the observations can only be done by soft computing methods, such as fuzzy logic.

The system of figure 4 uses knowledge of the system to - via inverse reasoning - substantiate probable faults from the observations. Such a system - because of its on-line property - promises timely evaluation of faults, and therefore increased safety. Also, the soft computing element makes it able to adapt to situations in which imperfect observations and knowledge is present.

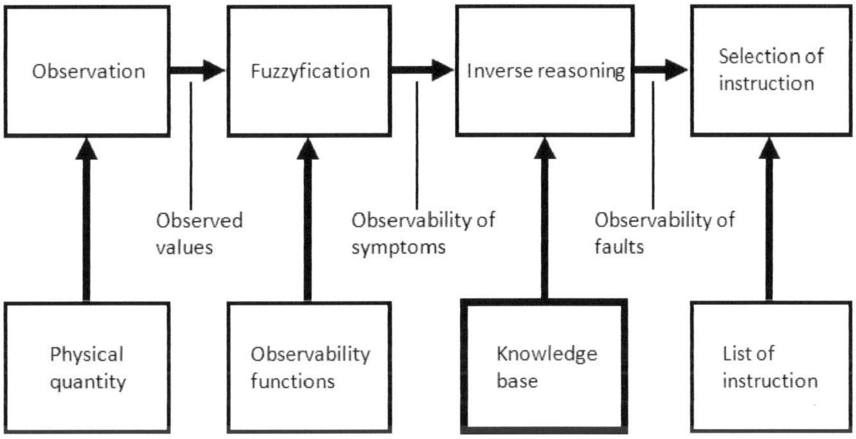

Figure 4. Schematic of an on-line diagnostic system.

7. The SCOUT system

In the previous sections we have discussed how new quantitative methods, many of them based on soft computing methods, allow us to more precisely assess electrostatic risk. The SCOUT (Static Control Up-to-date Technologies) is a procedure of risk estimation based on audits that not only help for planning, but also guidelines for decision makers and auditors.

The SCOUT system incorporates a preparatory pre-audit step, based on which sets of complex of protective measures may be proposed, and risk estimation for these alternatives can be carried out.

Based on the data collected in the preparatory phase a decision can be made that is both technically and economically optimal. According to the SCOUT procedure, the chosen technological measures are then applied, and their effectiveness is verified and documented via a post-audit (third party audit).

8. Preventive lightning protection

When designing lightning protection solutions, we must take into account not only the primary, but also the secondary effects of lightning that through conductive, inductive and capacitive means are coupled with our electrical systems.

The traditional means of primary lightning protection are the lightning arrestors. There is less agreement on how and how many of these should be placed on a structure to adequately elongate the mean strokeless period. The electrogeometric theory can be used to estimate this period (interval), and is based on the attractive volume concept - the volume, which once lightning enters is attracted to the structure rather than the ground. From the difference of the attractive volumes of the structure and its arrestors, an equivalent land area can be calculated, which receives the lightning strikes with the same frequency as the structure under examination [3].

Modern electronic equipment is usually very sensitive to the secondary effects of lightning. These are usually dealt with by reducing the coupling between a lightning and the sensitive equipment, and installing protective circuit elements that dissipate the energy of the lightning strike. Some equipment, such as radio transmitters and receivers, or radar equipment pose a difficult problem, when designing lightning protection.

Often the level of protection that is desired for a structure is very different for different time intervals. As an example, a transmission line should generally only have very basic protection, however when maintenance work is performed on it, then a lightning strike may endanger the maintenance workers, thus much better protection would be preferable.

A possible solution to these problems is preventive lightning protection. The preventive lightning protection method means avoiding damage of a lightning strike with special preventive actions. The preventive actions can be of various types, and the primary goal of preventive lightning protection is to decrease the risk of damage due to lightning for the duration of the thunderstorm. The preventive action shall be initiated before the beginning of the lightning activity, and shall be discontinued after the end of the thunderstorm [4].

This system uses on-line information from lightning detectors or meteorological radars to give warning to operators about the increased probability of a lightning strike, thereby the system can be transferred to a state in which potential damages are reduced (third party audit).

9. Conclusion

It was established that for both sources of electrostatic risk (both atmospheric and industrial) new methods for estimating and controlling the risk level are available. In the case of atmospheric effects, preventive lightning protection, and in the industrial case, new methods of computation, fault diagnostics are available. The latter serve as a basis for the new SCOUT procedure of industrial process auditing and a new method for helping the work of decision makers and risk takers.

In both the atmospheric and industrial cases, the usage of soft computing, such as fuzzy numbers, fuzzy inference systems and expert systems seems very useful in creating algorithms that can cope with the imperfect conditions of complex applications.

It is crucial to recognize the importance of the human element in applying protective measures – without the precise work of the experts and continuing human support no protection can operate properly, even though many technical equipment and devices are used. Also critical is that well-educated partners are available, who understand the characteristics of risk, and that meaningful information about the risk level is provided to support their decisions. This can be achieved by the ability to provide quantitative estimates in dimensions that are universally understood: EUR/year.

References

[1] Berta I 2005 *Journal of Electrostatics* **63** 6-10 679
[2] Glor M 2005 *Journal of Electrostatics* **63** 2-3 447
[3] Horváth T 2006 *Understanding lightning and lightning protection* Research Studies Press (Wiley & Sons) http://www.wiley.com/go/horvath
[4] Berta I 2009 *Journal of Electrostatics* **67** 2-3 235

Journal of Physics: Conference Series **268** (2011) 012030 doi:10.1088/1742-6596/268/1/012030

Parallel algorithms for finding cliques in a graph

S Szabó

Institute of Mathematics and Informatics, University of Pécs, Ifjúság u. 6. 7624 Pécs, HUNGARY

E-mail: sszabo7@hotmail.com

Abstract. A clique is a subgraph in a graph that is complete in the sense that each two of its nodes are connected by an edge. Finding cliques in a given graph is an important procedure in discrete mathematical modeling. The paper will show how concepts such as splitting partitions, quasi coloring, node and edge dominance are related to clique search problems. In particular we will discuss the connection with parallel clique search algorithms. These concepts also suggest practical guide lines to inspect a given graph before starting a large scale search.

1. Introduction

In a naive but intuitive level a graph is simply a collection of points on the plane some of which are connected by an arc while others are not. In this paper we do not need a more sophisticated graph concept. The points are called nodes or vertices invariably. The arcs are termed edges. If two nodes are connected by an edge, then we may express this fact by saying that the nodes are adjacent or neighbors of each other. The graphs we use will have finitely many points. A node is never connected to itself, that is, the graphs do not admit loops. Two distinct nodes are connected with one edge or not connected at all. In other words, the graphs do not have double edges.

Let Γ be a graph. The set of vertices of Γ is denoted by V. Suppose U is a subset of V. If each two distinct points of U are always connected by an edge of Γ, then we have a subgraph Δ of Γ that is called a clique. The set of vertices of Δ is U. If U has k elements we say that Δ is a clique of size k or simply that Δ is a k-clique in Γ. When U is the empty set, then although the graph Δ does not have any nodes or edges, by definition Δ is considered to be a clique of size zero. When U contains one single node of Γ, then by definition Δ considered to be a clique of size one. A clique Δ in Γ is called maximal if it cannot be extended to a larger clique by adding a further point of Γ. Maximal cliques do not make any appearance in this paper. We mention them just to contrast them to the concept of maximum cliques we define next. A clique Δ in Γ is a maximum clique in Γ if the size of Δ is the largest possible among the cliques in Γ.

It is an empirical fact that finding cliques in graphs can be used in mathematical modeling very much in the same fashion as say solving differential equations. In the course of modeling the graphs are man made objects designed cleverly to express pertinent aspects of the problem in the same spirit as one sets up differential equations. We resist the temptation to detail these problems for it would lead off from our main track. The interested reader may find applications in [1], [2], [3], [4], [5], [6], [7], [8]. The clique search problems of course must be labeled and categorized. Otherwise they could not be used systematically.

Problem 1 *Given a graph Γ and a positive integer k. Decide if Γ has a clique of size k.*

Problem 2 *Given a graph Γ and a positive integer k. List all k-cliques in Γ.*

Problem 1 is called the decision version of the k-clique search problem. Problem 2 is called the listing version or generation version of the k-clique search problem. There are analogous problems to list all maximum cliques in a given graph or exhibit one maximum clique or simply just find the size of the maximum cliques.

By the theory of the complexity of computations, Problem 1 is an NP complete problem. (The acronym "NP" stands for "nondeterministic polynomial".) Loosely speaking the theory predicts that in this family of clique search problems some hard instances are lurking. The theory does not predict where these instances are located.

There are algorithms and well tested programs for solving these problems. (See for example [9], [10], [11], [12], [13], [14], [15], [16], [17], [18].) Parallel programs are also available. (see for instance [19].) In this paper we propose further parallel algorithms. We divide the original problem to smaller problems that can be processed independently. To achieve a reasonable load balancing we try to choose many times more subproblems than processors are available and try to choose subproblems of similar sizes hoping that their running times do not vary too widely.

On the practical side if someone has to carry out a clique search in a hurry, then please use well tested existing programs. (The author does the same.) However, before embarking on a large scale clique search it is advisable to carry out a thorough inspection of the original graph to detect deletable nodes and edges. In the paper we present results relevant to such an inspection and we do recommend using these ideas.

Section 2 describes a parallel algorithm based on splitting partitions. Section 3 provides us with a way to construct splitting partitions using biclique search in a bipartite graph. Section 4 contains two extensions of the coloring idea the triangle free partitions and a new type of edge coloring to estimate the clique size in a graph. The second parallel algorithm is presented in Section 5 and it relies on quasi-coloring of the graph. Section 6 and Section 7 are devoted to the same theme dominance relation defined on nodes and edges to spot nodes and edges that are redundant during clique searches. Finally, Section 8 shows how normal partitions can be used for a parallel clique search.

2. Splitting partitions

Let Γ be a graph and let C_1, C_2, C_3 be a partition of the vertex set V of Γ.

Definition 1 *We say that the partition C_1, C_2, C_3 is a splitting partition of Γ if $C_1 \neq \emptyset$, $C_3 \neq \emptyset$ and if there is no edge of Γ whose one end point is in C_1 and the other end point is in C_3.*

For a splitting partition C_1, C_2, C_3 of Γ we define two subgraphs Γ_1, Γ_3 of Γ. Let Γ_1 be the subgraph of Γ spanned by the vertex set $C_2 \cup C_3$. Let Γ_3 be the subgraph of Γ spanned by the vertex set $C_1 \cup C_2$.

Proposition 1 *If Δ is a clique of Γ, then either Δ is a clique in Γ_1 or Δ is a clique in Γ_3.*

Proof. Assume on the contrary that Δ is neither a clique in Γ_1 nor is a clique in Γ_3. As Δ is not a clique in Γ_1, there is a node a of Δ such that $a \in C_1$. As Δ is not a clique in Γ_3, there is a node b of Δ such that $b \in C_3$. The facts that Δ is a clique in Γ and a, b are nodes of Δ give that $\{a, b\}$ is an edge of Δ and so it is an edge of Γ. This violates that the partition C_1, C_2, C_3 is a splitting partition of Γ. □

A word of warning. Proposition 1 does not imply that if Δ is a clique in Γ, then Δ is a clique only in one of Γ_1 or Γ_3. In fact Δ can be a clique in both Γ_1 and Γ_3. Therefore when one lists all maximum cliques (or k-cliques) in Γ_1 and Γ_3 one may meet a given clique in Γ twice.

Let a be a node of Γ. We say that a has full degree in Γ if a is adjacent to each point of Γ distinct from a.

Proposition 2 *Let C_1, C_2, C_3 be a splitting partition of Γ. If there is node in C_1 which is full degree in Γ_1 or there is a node in C_3 which is full degree in Γ_3, then a maximum clique in Γ cannot be a clique in both of Γ_1 and Γ_3.*

Proof. For the sake of definiteness assume that there is a node $a \in C_1$ which is full degree in Γ_1. Let Δ be a maximum clique in Γ and assume on the contrary that Δ is a clique in Γ_1 and Γ_3. Then Δ is a clique in the subgraph Γ' of Γ spanned by C_2. Note that a is not a node of Δ and $\Delta \cup \{a\}$ is a clique in Γ. Thus Δ cannot be a maximum clique in Γ. □

If C_1, C_2, C_3 is a partition of the set of nodes of Γ, then after rearranging rows and columns, the adjacency matrix $\mathbf{M}$ of Γ can be written in the form

$$\begin{bmatrix} \mathbf{M}_{1,1} & \mathbf{M}_{1,2} & \mathbf{M}_{1,3} \\ \mathbf{M}_{2,1} & \mathbf{M}_{2,2} & \mathbf{M}_{2,3} \\ \mathbf{M}_{3,1} & \mathbf{M}_{3,2} & \mathbf{M}_{3,3} \end{bmatrix}.$$

The block $\mathbf{M}_{i,j}$ is a $|C_i|$ by $|C_j|$ matrix. Here $|C_i|$ stands for the number of the elements of the subset C_i. If the partition C_1, C_2, C_3 is a splitting partition, then each entry of $\mathbf{M}_{1,3}$ and $\mathbf{M}_{3,1}$ is zero. The adjacency matrices of Γ_1 and Γ_3 are

$$\begin{bmatrix} \mathbf{M}_{2,2} & \mathbf{M}_{2,3} \\ \mathbf{M}_{3,2} & \mathbf{M}_{3,3} \end{bmatrix}, \quad \begin{bmatrix} \mathbf{M}_{1,1} & \mathbf{M}_{1,2} \\ \mathbf{M}_{2,1} & \mathbf{M}_{2,2} \end{bmatrix}$$

respectively. In Proposition 1 we may choose the clique Δ to be a maximum clique and we get that finding a maximum clique in Γ can be reduced to finding a maximum clique in two smaller graphs Γ_1 and Γ_3.

Let $\rho = \min\{|C_1|, |C_2|\}$. The larger is ρ the larger is the reductive power of the splitting partition. The reason for this is that the number of vertices of the larger of the graphs Γ_1 and Γ_3 is equal to

$$\max\{|C_2| + |C_3|, |C_1| + |C_2|\} = \max\{|V| - |C_1|, |V| - |C_3|\} = |V| - \rho.$$

As a consequence we try to choose splitting partitions for which ρ is as large as possible.

Problem 3 *Given a graph Γ construct a C_1, C_2, C_3 splitting partition with possibly large value of ρ.*

There is a rather trivial way to construct splitting partitions for which $\rho = 1$. Suppose that a is a node of Γ such that a has at least one non-neighbor in Γ. The set of neighbors of a in Γ we will denote by $N(a)$. Set

$$C_1 = \{a\}, \quad C_2 = N(a), \quad C_3 = V \setminus N(a).$$

Then C_1, C_2, C_3 is a splitting partition of Γ with $\rho = 1$.

Next we describe a greedy algorithm to try to construct splitting partition with $\rho \geq 2$. We introduce three lists L_1, L_2, L_3.

(1) Initially set $L_1 = \emptyset$, $L_2 = \emptyset$, $L_3 = V$.

(2) Pick an $a \in L_2 \cup L_3$. Let L_a be the set of all the neighbors of a in L_3. ($L_a = L_3 \cap N(a)$.) We choose such an a that minimizes $|L_a|$.

(3) If $|L_3| - |L_a| - 1 \geq |L_1| + 1$, then set $L_1 = L_1 \cup \{a\}$, $L_2 = (L_2 \cup L_a) \setminus \{a\}$, $L_3 = L_3 \setminus [L_a \cup \{a\}]$ and continue at step (2).

(4) If $|L_3| - |L_a| - 1 < |L_1| + 1$, then stop.

After the procedure stops set

$$C_1 = L_1, \quad C_2 = L_2, \quad C_3 = L_3.$$

If $C_1 \neq \emptyset$ and $C_3 \neq \emptyset$, then the partition C_1, C_2, C_3 is a splitting partition of Γ. It may or may not happen that $\rho \geq 2$. If for each vertex a in Γ there is exactly one node b such that $\{a, b\}$ is not an edge of Γ, then the algorithm ends up with a splitting partition with $\rho = 1$. If each node of Γ has full degree in Γ, that is if Γ itself is a clique, then the output of this algorithm is identical to the initial values and the algorithm does not produce a splitting partition in this case.

A possible parallel maximum clique (or k-clique) algorithm is now the following. Let us start with the graph Γ. The greedy algorithm provides us with a splitting partition. If the reducing factor $\rho \geq 2$, then we replace Γ by the smaller graphs Γ_1 and Γ_3. We repeat this procedure with Γ_1 and Γ_3. If the reducing factor ρ is no longer larger than one, then we do not try to split the graph into smaller ones. We end up with a long list of graphs $\Gamma'_1, \ldots, \Gamma'_s$ instead of Γ. The new graphs are smaller than the original graph Γ. These graphs then can be processed independently of each other. The independent processes can use any serial clique search algorithm.

3. Bipartite graphs and bicliques

We now turn to a more systematic non-greedy way to construct splitting partitions. For this we need bipartite graphs and bicliques in bipartite graphs. A graph Λ is called a bipartite graph if the vertex set of Λ can be partitioned into two subsets V_1, V_2 such that Λ has no edge whose end points are in V_1 and Λ has no edge whose end points are in V_2. We will talk about bicliques in bipartite graphs. Let Λ be a bipartite graph whose vertex set is partitioned into V_1, V_2. Let $U_1 \subseteq V_1$, $U_2 \subseteq V_2$. If each vertex of U_1 is adjacent to each vertex of U_2, then the subgraph Ω spanned by $U_1 \cup U_2$ in Λ is called a biclique of Λ. The biclique Ω is called a (k_1, k_2)-biclique if $|U_1| = k_1$, $|U_2| = k_2$. The next biclique search problem is relevant for us.

Problem 4 *Given a bipartite graph Λ and given two integers k_1, k_2. Decide if Λ has a (k_1, k_2)-biclique. If Λ does contain a (k_1, k_2)-biclique, then exhibit one.*

There are algorithms to solve this problem in the literature. See for instance [20].

Let Γ be a graph. We define a bipartite graph Λ whose bicliques can be used to construct splitting partitions of Γ. Let V be the vertex set of Γ and let $\mathbf{M}$ be the adjacency matrix of Γ. Each entry in the main diagonal of $\mathbf{M}$ is zero. Replace each of these zeros by one to get a new matrix $\mathbf{M}'$. The rows and columns of $\mathbf{M}$ and $\mathbf{M}'$ are labeled by the elements of V. The vertex set of Λ is partitioned into V_1, V_2. We define V_1, V_2 by

$$V_1 = \{(v, 1) : \ v \in V\}, \quad V_2 = \{(v, 2) : \ v \in V\}.$$

We connect the nodes $(v, 1)$, $(v', 2)$ of Λ by an edge if the entry of $\mathbf{M}'$ in the (v)-th row and (v')-th column is equal to zero.

Let ρ be a positive integer and suppose that the biclique search algorithm has found a (ρ, ρ)-biclique Ω in Λ. The vertex set of Ω is partitioned into U_1, U_2. There are subsets C_1, C_3 of V such that

$$U_1 = \{(u, 1) : \ u \in C_1\}, \quad U_2 = \{(u, 2) : \ u \in C_3\}.$$

Proposition 3 *If $\rho \geq 1$, then the sets C_1, $C_2 = V \setminus (C_1 \cup C_2)$, C_3 form a splitting partition of the graph Γ.*

Proof. Plainly $\rho \geq 1$ implies that $C_1 \neq \emptyset$, $C_3 \neq \emptyset$. We need to prove that $C_1 \cap C_3 = \emptyset$. This will imply that C_1, C_2, C_3 is a partition of V. Then we should prove that Γ has no edge $\{a, b\}$ with $a \in C_1$, $b \in C_3$.

In order to prove $C_1 \cap C_3 = \emptyset$ assume on the contrary that there is an $a \in C_1$, $a \in C_3$. This means that $(a, 1) \in U_1$, $(a, 2) \in U_2$. Since Ω is a biclique of Λ, the nodes $(a, 1)$, $(a, 2)$ of Λ are adjacent. This gives that the entry in the (a)-th row and (a)-th column of $\mathbf{M}'$ is zero. But the main diagonal of $\mathbf{M}'$ is filled with one. This contradiction verifies the claim that $C_1 \cap C_3 = \emptyset$.

Next suppose that Γ has an edge $\{a, b\}$ for which $a \in C_1$, $b \in C_3$. This means that $(a, 1) \in U_1$, $(b, 1) \in U_2$. Since Ω is a biclique of Λ, the nodes $(a, 1)$, $(b, 1)$ of Λ are adjacent. So the entry of $\mathbf{M}'$ in the (a)-th row and (b)-th column is zero. On the other hand note that $a \neq b$ as Γ has no loops. Since $\{a, b\}$ is an edge of Γ, the entry in the (a)-th row and (b)-th column in $\mathbf{M}$ is one. Using that $a \neq b$, it follows that this entry is the same in $\mathbf{M}$ and $\mathbf{M}'$. ($\mathbf{M}$ and $\mathbf{M}'$ differ only at the main diagonal.) This contradiction completes the proof. □

To illustrate the argument above we work out a small example in details. Let Γ be the graph depicted in figure 1. The adjacency matrix of this Γ graph is given in table 1. Using the adjacency matrix of Γ we construct the bipartite graph Λ. This Λ graph is depicted in figure 2. An inspection shows that the subgraph Ω spanned by the subsets

$$U_1 = \{(4, 1), (5, 1)\}, \quad U_2 = \{(2, 2), (6, 2)\}$$

form a $(2, 2)$-biclique of Λ. From the subsets U_1, U_2 one can read off that the sets

$$C_1 = \{4, 5\}, \quad C_2 = \{1, 3\}, \quad C_3 = \{2, 6\}$$

form a splitting partition of Γ. One can see that the 2 by 2 block at the upper right corner in the rearranged adjacency matrix of Γ is blank. As it supposed to be. The graph Γ can be replaced by the smaller Γ_1 and Γ_3 graphs. We included figure 3 to exhibit these graphs.

4. Coloring

Coloring is a well studied subject in graph theory. (See for example [21].) First we quickly glance over the basic concepts pertinent to clique search. Let Γ be a graph. We color the nodes of Γ with t distinct colors. A coloring of the nodes is called a legal or well coloring if each node is colored with one of the t colors and if adjacent nodes do not receive the same color.

Let C_i be the set of the nodes of Γ that are colored with the (i)-th color. The set C_i is termed the (i)-th color class. The sets $C_1, \ldots, C_t$ form a partition of the vertex set V of Γ. This means that $C_1 \cup \cdots \cup C_t = V$ and $C_i \cap C_j = \emptyset$ for each i, j, $1 \leq i < j \leq t$. Clearly, two distinct nodes in C_i cannot be adjacent in Γ. We call a set C_i edge free if it does not contain edges from Γ. A partition $C_1, \ldots, C_t$ is called an edge free partition if each C_i is edge free.

In a legal coloring of Γ with t colors the color classes $C_1, \ldots, C_t$ form an edge free partition of V. Conversely, if $C_1, \ldots, C_t$ is an edge free partition of V, then coloring the elements of C_i with the (i)-th color we end up with a legal coloring of Γ.

The essential link between the coloring and the clique search is the following observation. Let Δ be a clique of size k in Γ and let $C_1, \ldots, C_t$ be the color classes of a well coloring of Γ. Each color class can contain at most one node of Δ. From this it follows that $k \leq t$. In other words, if Γ admits a legal coloring with $k - 1$ colors, then Γ cannot contain a clique of size k.

Determining the least number of colors a graph can be well colored with is computationally hard. (It is an NP complete problem.) However, there are various greedy algorithms to exhibit if not optimal but still useful well colorings to provide upper bounds for the clique size can appear in a given graph. We propose now two variants of the coloring idea to obtain further upper bounds for the possible clique size.

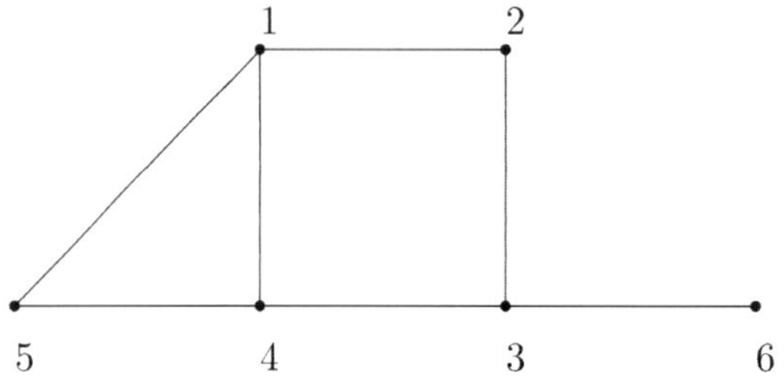

Figure 1. The Γ graph in the first example.

Table 1. The adjacency matrix of Γ in the first example and the rearranged adjacency matrix of Γ.

	1	2	3	4	5	6
1	×	•		•	•	
2	•	×	•			
3		•	×	•		•
4	•		•	×	•	
5	•			•	×	
6			•			×

	4	5	1	3	2	6
4	×	•	•	•		
5	•	×	•			
1	•	•	×		•	
3	•			×	•	•
2			•	•	×	
6				•		×

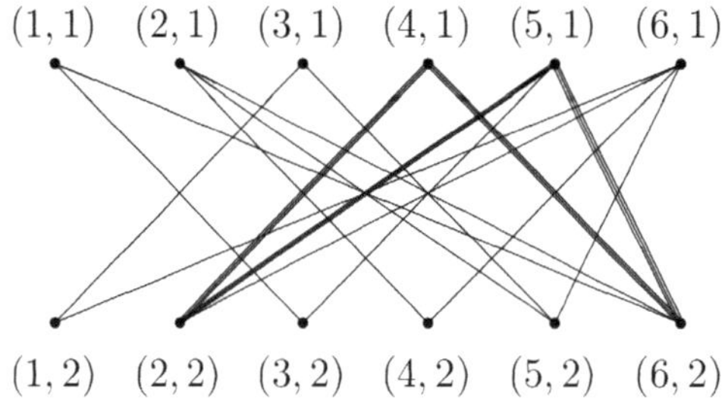

Figure 2. The bipartite graph Λ in the first example. A $(2, 2)$-biclique is highlighted.

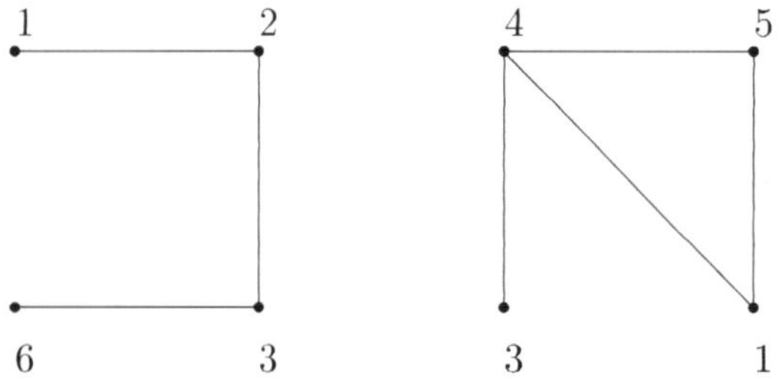

Figure 3. The graph Γ can be replaced by the smaller Γ_1 and Γ_3 graphs.

Definition 2 *A subset C_i of the vertex set V of the graph Γ is called triangle free if C_i does not contain a triangle of Γ. A partition $C_1, \ldots, C_t$ of V is called a triangle free partition if each C_i is triangle free.*

Proposition 4 *Suppose the graph Γ admits a triangle free partition $C_1, \ldots, C_t$. Let Δ be a clique of size k in Γ. Then $k \leq 2t$.*

Proof. As $C_1 \cup \cdots \cup C_t = V$, each node of Δ falls into one of the classes $C_1, \ldots, C_t$. Each C_i can contain at most two nodes of Δ, since C_i does not contain any triangle of Γ. This implies $k \leq 2t$, as required. □

One can describe the triangle free partitions in terms of coloring the nodes of a hyper graph. Recall that a 3-uniform hyper graph Ξ on the set of vertices V is a collection E of hyper edges, where each hyper edge contains three distinct vertices from V. We color the nodes of Ξ using t colors. We say that a coloring is a well coloring or a legal coloring if each node receives exactly one color and if the three nodes of a hyper edge never receive the same color. Let C_i be the set of nodes of Ξ colored with the (i)-th color. We call C_i the (i)-th color class of the coloring.

For a given graph Γ we define a 3-uniform hyper graph Ξ. The nodes of Ξ are the nodes of Γ. Three distinct nodes a, b, c of V form a 3-hyper edge of Ξ if a, b, c are mutually adjacent in Γ. Now the color classes $C_1, \ldots, C_t$ of Ξ form a triangle free partition of Γ.

Problem 5 *Given a graph Γ. Construct a triangle free partition of Γ with possibly few members of the partition.*

Here is a greedy algorithm for solving this problem. Suppose the vertex set V of Γ has n nodes. We start with $n+1$ lists $L, L_1, \ldots, L_n$. Set $L = V$, $L_1 = \cdots = L_n = \emptyset$. While $L \neq \emptyset$ repeat the following. Choose and $a \in L$. If for each edge $\{b, c\}$ in L_i, either $\{a, b\}$ or $\{a, c\}$ is not an edge of Γ, then place a to L_i. If a cannot be placed to L_i, then try to place a to L_{i+1}.

Numerical experiments indicate that greedily constructed triangle free partitions usually provide better upper bounds than greedily constructed edge free partitions. However, this improvement comes for a higher price we should pay in computational resources.

Instead of the nodes of Γ one may try to color the edges of Γ using t distinct colors. For such coloring we single out certain edge colored subgraph of Γ and call these prohibited configurations. Suppose that the edges $\{x, y\}$, $\{u, v\}$ of Γ receive the same color. Since Γ does not have loops, $x \neq y$ and $u \neq v$. If the nodes x, y, u, v are all distinct and they are pair-wise adjacent, then this configuration is prohibited. If two of the nodes x, y, u, v are identical, say $y = u$, and the other two nodes x, v are adjacent, then this configuration is prohibited. The prohibited configurations are depicted in figure 4.

Definition 3 *We call a coloring of the edges of a graph Γ a legal edge coloring of Γ if each edge is colored with one of the t colors and no prohibited configuration occurs.*

Proposition 5 *Suppose that the edges of Γ are legally colored using t colors. Let Δ be a clique of size k in Γ. Then $k(k-1) \leq 2t$.*

Proof. Let C_i be the set of all the edges of Γ that are colored with the (i)-th color. We call C_i the (i)-th color class of Γ. The sets $C_1, \ldots, C_t$ form a partition of the edge set E of Γ since each edge receives exactly one color. As $C_1 \cup \cdots \cup C_t = E$, each edge of Δ falls into one of the color classes. If two distinct edges of Δ falls into the same color class, then this color class will contain a prohibited configuration. By the definition of the legal coloring, Γ does not contain any prohibited configuration. Consequently each color class contains at most one edge of Δ. Since Δ has $(1/2)k(k-1)$ edges, it follows that $(1/2)k(k-1) \leq t$, as required. □

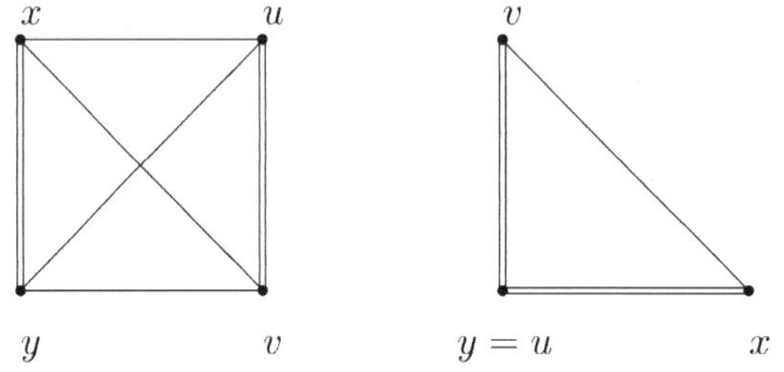

Figure 4. Prohibited configurations. The double line edges are colored with the same color.

The edge coloring of Γ can be expressed in terms of coloring the nodes of a derived graph Θ. The nodes of Θ are the edges of Γ. Temporarily we extend Γ to Γ' by connecting each node of Γ to itself, that is, introducing loops into Γ. Two distinct edges $\{x, y\}$, $\{u, v\}$ of Γ will be adjacent nodes in Θ if the nodes x, y, u, v are mutually adjacent in Γ'. Now a legal coloring of the nodes of Θ results a legal coloring of the edges of Γ.

Problem 6 *Given a graph Γ. Construct a legal coloring of edges of Γ with possibly few colors.*

A straight forward greedy algorithm starts with two lists $L = E$ and $L_1 = \emptyset$, where E is the edge set of Γ. In a typical situation there are $s + 1$ lists $L, L_1, \dots, L_s$. While $L \neq \emptyset$ repeat the following. Choose an edge e of L_1 and test if e can be placed to L_i by checking if for each edge $f \in L_i$ the edges e, f colored with the same color form a prohibited configuration. If e cannot be placed to any of the lists $L_1, \dots, L_s$, then open a new list L_{s+1} for e. In either cases set $L = L \setminus \{e\}$.

Alternatively one may start constructing a legal edge coloring of Γ with a legal node coloring of Γ. Suppose $C_1, \dots, C_s$ are the color classes of a legal vertex coloring of Γ. We arrange the situation such that $|C_1| \leq \dots \leq |C_s|$. Choose an $a \in C_s$. If $\{a, b\}$ is an edge of Γ such that $b \in C_i$, then color the edge $\{a, b\}$ with color i, that is, place $\{a, b\}$ to the list L_i. After repeating this for each $a \in C_s$ we get $s - 1$ lists $L_1, \dots, L_{s-1}$. Set $L = E \setminus (L_1 \cup \dots \cup L_{s-1})$ and start the earlier algorithm with the initial lists $L, L_1, \dots, L_{s-1}$.

Numerical experiments show that greedy edge colorings can provide tighter bounds than greedy node colorings. But this improvement comes for higher computational price.

5. Quasi coloring

Let Γ be a graph with vertex set V and let $C_1, \dots, C_t$ be a partition of V. After rearranging rows and columns the adjacency matrix $\mathbf{M}$ of Γ can be written in the form

$$\begin{bmatrix} \mathbf{M}_{1,1} & \mathbf{M}_{1,2} & \dots & \mathbf{M}_{1,t} \\ \mathbf{M}_{2,1} & \mathbf{M}_{2,2} & \dots & \mathbf{M}_{2,t} \\ \vdots & \vdots & \ddots & \vdots \\ \mathbf{M}_{t,1} & \mathbf{M}_{t,2} & \dots & \mathbf{M}_{t,t} \end{bmatrix},$$

where the block $\mathbf{M}_{i,j}$ is a $|C_i|$ times $|C_j|$ matrix. The rows of $\mathbf{M}_{i,j}$ are labeled with the elements of C_i and the columns of $\mathbf{M}_{i,j}$ are labeled by the elements C_j. The set C_i is called independent in Γ if any two distinct vertices in C_i are not adjacent. If C_i is an independent set in Γ, then each entry of the matrix $\mathbf{M}_{i,i}$ is zero, that is, $\mathbf{M}_{i,i}$ is a zero matrix. If each $\mathbf{M}_{i,i}$ is a zero matrix, then each C_i is an independent set in Γ and so the nodes of Γ can be well colored with t colors. The sets $C_1, \dots, C_t$ can be chosen to be color classes of the coloring. In a typical situation the

Journal of Physics: Conference Series **268** (2011) 012030 doi:10.1088/1742-6596/268/1/012030

$\mathbf{M}_{i,i}$ matrices are not necessarily zero matrices. Let 2σ be the overall number of the non-zero entries in the $\mathbf{M}_{i,i}$ matrices. The closer is σ to zero, the closer is the partition $C_1, \ldots, C_t$ to be the color classes of a well coloring of Γ.

Definition 4 *We will refer to a partition $C_1, \ldots, C_t$ of the vertex set of Γ as a (σ, t)-quasi coloring of Γ.*

We propose a parallel k-clique search algorithm based on quasi colorings. Given a graph Γ and a positive integer k. We would like to decide if Γ has a clique of size k. Suppose we have a $(\sigma, k-1)$-quasi coloring of Γ at our disposal. Let $\{a_1, b_1\}, \ldots, \{a_\sigma, b_\sigma\}$ be the edges of Γ that correspond to the 2σ non-zero entries in the matrices $\mathbf{M}_{1,1}, \ldots, \mathbf{M}_{k-1,k-1}$. Let Γ_i be the subgraph of Γ spanned by the set $N(a_i) \cap N(b_i)$.

Proposition 6 *If Γ has a clique of size k, then at least one of the graphs $\Gamma_1, \ldots, \Gamma_\sigma$ has a clique of size $k-2$.*

Proof. Assume on the contrary that Γ_i does not contain a clique of size $k-2$ for each i, $1 \leq i \leq \sigma$.

There is a clique of size k in Γ. Let Δ be such a clique. If $\{a_1, b_1\}$ is an edge of Δ, then there is a clique Δ_1 of size $k-2$ in $N(a_1) \cap N(b_1)$. Hence Δ_1 is a clique of size $k-2$ in Γ_1. This contradicts the indirect assumption. Thus $\{a_1, b_1\}$ is not an edge of any clique of size k in Γ. This means that we may delete the edge $\{a_1, b_1\}$ from Γ without disturbing the number of cliques of size k in Γ. Let $\Gamma^{(1)}$ be the graph we get from Γ after deleting the node $\{a_1, b_1\}$. (We do not delete any of the nodes a_1 or b_1.)

There must be a clique of size k in $\Gamma^{(1)}$. Let Δ be such a clique. If $\{a_2, b_2\}$ is an edge of Δ, then there is a clique Δ_2 of size $k-2$ in $N(a_2) \cap N(b_2)$. Hence Δ_2 is a clique of size $k-2$ in Γ_2. This contradicts the indirect assumption. Thus $\{a_2, b_2\}$ is not an edge of any clique of size k in $\Gamma^{(1)}$. This means that we may delete the edge $\{a_2, b_2\}$ from $\Gamma^{(1)}$ to get the graph $\Gamma^{(2)}$.

Continuing in this way after σ steps we get that there must be a clique of size k in $\Gamma^{(\sigma)}$. On the other hand note that in the adjacency matrix $\mathbf{M}^{(\sigma)}$ of $\Gamma^{(\sigma)}$ the $\mathbf{M}^{(\sigma)}_{i,i}$ blocks are zero matrices as we have deleted the edges $\{a_1, b_1\}, \ldots, \{a_\sigma, b_\sigma\}$. In other words the nodes of $\Gamma^{(\sigma)}$ can be well colored using $k-1$ colors. Thus $\Gamma^{(\sigma)}$ cannot have any clique of size k. This contradiction completes the proof. □

Problem 7 *Let Γ be a graph and let t be a positive integer. Construct a (σ, t)-quasi coloring of Γ keeping the value of σ possibly small.*

Here is a straight forward greedy algorithm in which we maintain $t+1$ lists $L, L_1, \ldots, L_t$. At the beginning of the computation $L = V$, $L_1 = \cdots = L_t = \emptyset$. While $L \neq \emptyset$ let us iterate the following procedure. Pick an $a \in L$. Let d_i be the number of neighbors of a in L_i. Let d_j be the least among $d_1, \ldots, d_t$. Then set $L = L \setminus \{a\}$, $L_j = L_j \cup \{a\}$. When $L = \emptyset$ set $C_1 = L_1, \ldots, C_t = L_t$. Clearly, the sets $C_1, \ldots, C_t$ form a partition of V. A moment of contemplation will convince the reader that the number of times this procedure reads an entry of the adjacency matrix is at most $(1/2)n(n-1)$, where n is the number of the nodes of Γ.

Alternatively one can construct a (σ, t)-quasi coloring starting with a well coloring of Γ. Let $C_1, \ldots, C_s$ be the color classes of a well coloring of Γ. Suppose that the situation is arranged such that $|C_1| \geq \cdots \geq |C_s|$. Initialize the lists $L, L_1, \ldots, L_t$ by setting

$$L = C_{t+1} \cup \cdots \cup C_s, \qquad L_1 = C_1, \ldots, L_t = C_t.$$

Then distribute the elements of L among the lists $L_1, \ldots, L_t$ using the algorithm above.

A parallel k-clique search algorithm can be summarized in the following way. Let us start with the graph Γ. Construct a $(\sigma, k-1)$-quasi coloring for Γ. Locate the edges $\{a_1, b_1\}, \ldots, \{a_\sigma, b_\sigma\}$. Based on these edges construct the graphs $\Gamma_1, \ldots, \Gamma_\sigma$. By Proposition 6, one of $\Gamma_1, \ldots, \Gamma_\sigma$ has a clique of size $k-2$ if and only if Γ has a clique of size k. In short we reduced the problem into a large number of smaller problems that can be processed independently of each other. The independent processes are free to use any serial $(k-2)$-clique search algorithm.

The algorithm above can also be used to list all k-cliques in Γ. Namely, list all $(k-2)$-cliques in Γ_i then augment them with the nodes a_i, b_i to get the k-cliques in Γ. We should be aware of that in this way we may meet a given k-clique in Γ more than once. If this is undesirable, then one should use the graphs $\Gamma_1^*, \ldots, \Gamma_\sigma^*$ in place of the graphs $\Gamma_1, \ldots, \Gamma_\sigma$. Here Γ_i^* is constructed from Γ_i by deleting the edges $\{a_1, b_1\}, \ldots, \{a_{i-1}, b_{i-1}\}$. (But not deleting the end points of the edges.) The next proposition justifies this claim.

Suppose that Δ_i is a clique of size $k-2$ in Γ_i^*. Obviously Δ_i is a clique in Γ_i too. Set $\Delta^{(i)} = \Delta_i \cup \{a_i\} \cup \{b_i\}$. By the construction of Γ_i, $\Delta^{(i)}$ is a clique of size k in Γ.

Proposition 7 $\Delta^{(i)} = \Delta^{(j)}$ *implies* $i = j$.

Proof. Assume on the contrary $i \neq j$ and $\Delta^{(i)} = \Delta^{(j)}$. We may assume that $i < j$ since this is only a matter of swapping the roles of i and j. From $\Delta^{(i)} = \Delta^{(j)}$ it follows that $\{a_i, b_i\}$ is an edge of $\Delta^{(j)}$. Hence $\{a_i, b_i\}$ is an edge of Γ_j^*. But by the construction of Γ_j^*, $\{a_i, b_i\}$ is not an edge of Γ_j^*. □

6. Dominating nodes

We introduce a binary relation between the nodes of a graph.

Definition 5 *Let Γ be a graph and let a, b be distinct nodes of Γ. We say that b dominates a if a and b are not connected by an edge of Γ and $N(a) \subseteq N(b)$.*

An example of dominated nodes can be seen in figure 5. One can observe that a dominated node can be dropped from the graph during the search for a maximum clique. This is the very reason why we introduced the concept of dominance We spell out this observation formally as a proposition.

Proposition 8 *If a is dominated by b, then a can be canceled from Γ when we are deciding if Γ contains a clique of size k for any given fixed k.*

Proof. Let Δ be a clique of size k of Γ. Let Γ' be the graph we get from Γ after deleting the node a. Only one of the following four possibilities may hold in connection with a, b, Δ

(i) $a \in \Delta$, $b \in \Delta$,
(ii) $a \in \Delta$, $b \notin \Delta$,
(iii) $a \notin \Delta$, $b \in \Delta$,
(iv) $a \notin \Delta$, $b \notin \Delta$.

If $a \notin \Delta$, then $\Delta \subseteq \Gamma'$ and so Γ' contains a clique of size k. For the remaining part of the proof we may assume that $a \in \Delta$. Note that $b \in \Delta$ cannot hold since a and b are not connected by edge in Γ. Thus $b \notin \Delta$. Since $a \in \Delta$, there is a clique Δ_1 of size $k-1$ in $N(a)$. Now $\Delta_1 \subseteq N(b)$ as $N(a) \subseteq N(b)$. Clearly, $\Delta_1 \cup \{b\}$ is a clique of size k. Thus Γ' contains a clique of size k. □

We would like to emphasize that listing all maximum cliques (or k-cliques) in Γ' does not necessarily provide a complete list of the maximal cliques (or k-cliques) of Γ. In other words the reduction offered by Proposition 8 is not applicable when one needs complete lists of maximum

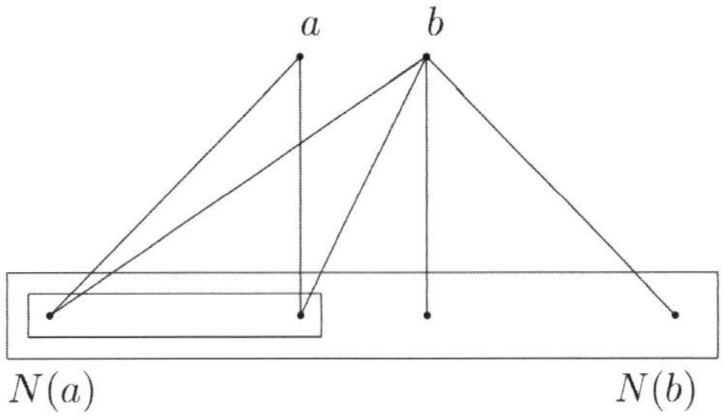

Figure 5. Node a is dominated by node b.

or k-cliques. On the other hand getting rid of certain maximal cliques (or k-cliques) promises that an algorithm visits a smaller portion of the search space.

The dominance relation is obviously not symmetric. Further the dominance relation is not anti symmetric either as it may happen that a and b are distinct nodes of Γ such that $N(a) = N(b)$, a and b are not connected by an edge in Γ. In this case a and b are mutually dominate each other in the same time a and b are not equal.

Proposition 9 *The dominance relation is transitive.*

Proof. Let a, b, c be vertices of the graph Γ. Suppose that a and b are not connected by edge in Γ and $N(a) \subseteq N(b)$. Further suppose that b and c are not connected and $N(b) \subseteq N(c)$. If a and c are not connected by edge in Γ, then c dominates a. For the remaining part of the proof we may assume that there is an edge between a and c. This means that $c \in N(a)$. Using $N(a) \subseteq N(b)$ it follows that $c \in N(b)$. Therefore b and c are connected by edge in Γ. This is an outright contradiction. □

Let a, b, c distinct nodes of Γ such that b dominates a. Let Γ' be the graph we get from Γ after deleting c.

Proposition 10 *Node b dominates node a in Γ' too.*

Proof. Assume that b dominates a in Γ. This means that a and b are not connected distinct nodes such that $N(a) \subseteq N(b)$. For a node u of Γ' the set of neighbors of u in Γ' is denoted by $N'(u)$. It is plain that a and b are not connected in Γ'. We should verify that $N'(a) \subseteq N'(b)$. We distinguish the following four cases

(i) $c \notin N(a)$, $c \notin N(b)$,
(ii) $c \notin N(a)$, $c \in N(b)$,
(iii) $c \in N(a)$, $c \notin N(b)$,
(iv) $c \in N(a)$, $c \in N(b)$.

If $c \notin N(a)$ and $c \notin N(b)$, then $N'(a) = N(a)$ and $N'(b) = N(b)$. From $N(a) \subseteq N(b)$ it follows that $N'(a) \subseteq N'(b)$.

If $c \notin N(a)$ and $c \in N(b)$, then $N'(a) = N(a)$ and $N'(b) = N(b) \setminus \{c\}$. Subtracting $\{c\}$ from both sides of $N(a) \subseteq N(b)$ implies that $N'(a) \subseteq N'(b)$.

If $c \in N(a)$ and $c \notin N(b)$, then $N(a) \subseteq N(b)$ gives the contradiction that $c \in N(b)$. Therefore this case is not possible.

If $c \in N(a)$ and $c \in N(b)$, then $N'(a) = N(a) \setminus \{c\}$ and $N'(b) = N(b) \setminus \{c\}$. Subtracting $\{c\}$ from both sides of $N(a) \subseteq N(b)$ we get that $N'(a) \subseteq N'(b)$. □

In order to illustrate the results in this section let us consider an example. Let Γ be the graph depicted in figure 6. The adjacency matrix of Γ is given by table 2. An inspection reveals that

node 2 dominates nodes 1, 5, 9,

node 3 dominates node 4,

node 5 dominates nodes 1, 2, 9,

node 6 dominates node 4,

node 7 dominates nodes 3, 4, 6.

By Proposition 8, we may delete nodes 1, 2, 3, 4, 6, 9 from the graph when we are looking for a maximum clique. The remaining nodes are 5, 7, 8. They form a clique of size 3. Therefore the size of the maximum clique in the original graph is 3.

The graph in the example above is exceedingly small. However it illustrates the point well. Namely, before embarking on determining the size of a maximum clique in a given graph we should inspect the graph to spot dominating nodes. When found the dominated node can be deleted and we should start to look for dominating nodes in the reduced graph.

Problem 8 *Given a graph Γ and a node a of Γ. Decide if there is node b in Γ that dominates node a.*

A proposed algorithm for solving Problem 8 works with two lists L_1 and L_2. We fill in L_1 with the neighbors of a. The order of the elements in L_1 is irrelevant. We fill in L_2 with the non-neighbors of a. Again, the ordering of the elements in L_2 is irrelevant. (We do not place a to L_2.) Pick a $b \in L_2$. Note that if for each $c \in L_1$ the edge $\{c, b\}$ is edge of Γ, then b dominates a. In order to verify the claim note that the nodes a, b are not adjacent as $b \notin N(a)$. Further $N(a) \subseteq N(b)$ holds since $c \in N(a)$ implies $c \in N(b)$.

The algorithm above reads an entry of the adjacency matrix of Γ to check if two nodes are adjacent or not. Let s be the number of times the algorithm reads a single entry of the adjacency matrix. Let n be the number of nodes of Γ.

Proposition 11 $s \leq n + (1/4)n^2$.

Proof. Let n_1 be the length of L_1 and let n_2 be the length of L_2. Clearly $n_1 + n_2 = n - 1$. Filling in L_1 and L_2 needs n readings of the adjacency matrix. For an element $b \in L_2$ there are n_2 choices. For an element $c \in L_1$ there are n_1 choices. After at most $n_1 n_2$ readings of the adjacency matrix we will reach a definite conclusion if there is a $b \in L_2$ such that $\{b, c\}$ is an edge of Γ for each $c \in L_1$.

The solution of the optimization problem, maximize $n_1 n_2$ subject to $n_1 + n_2 = n - 1$, is $n_1 = n_2 = (1/2)(n - 1)$. Therefore $s \leq n + n_1 n_2 \leq n + (1/4)(n - 1)^2 \leq n + (1/4)n^2$. □

Problem 8 is formulated solely for the purpose to incorporate the algorithm that solves it into the Carraghan-Pardalos clique search algorithm. (The description of the algorithm with a complete program can be found in [11].) The author has experimented with this modified clique search algorithm. The result is that there are graphs for which the extra work of testing dominance pays off and there are graph for which the modified algorithm slows down very slightly.

Another way to utilize this algorithm is simply to inspect the initial adjacency matrix of Γ to test if there are nodes to delete before one starts the proper clique search. By Proposition 11, the full inspection can be accomplished in at most $n^2 + (1/4)n^3$ looks up of a single entry of the adjacency matrix. When a dominated node is found we can delete the node at once. Since

Table 2. The adjacency matrix of Γ in the second example

	1	2	3	4	5	6	7	8	9
1	×		•	•		•	•		
2		×	•	•		•	•	•	
3	•	•	×		•			•	
4	•	•		×	•				
5			•	•	×	•	•	•	
6	•	•			•	×			•
7	•	•			•		×	•	•
8		•	•		•		•	×	•
9						•	•	•	×

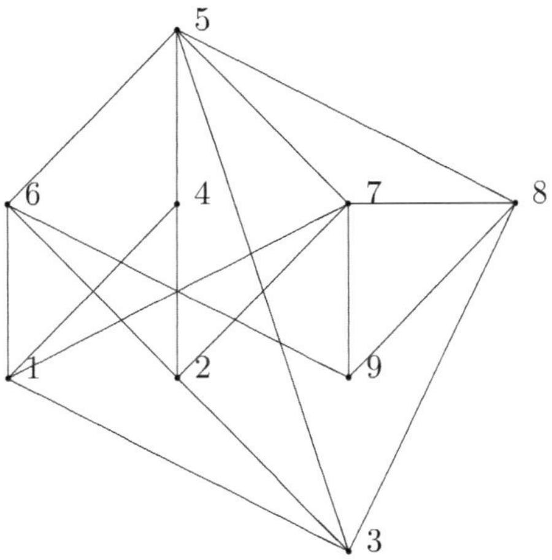

Figure 6. The graph Γ in the second example.

by Proposition 10, the dominated nodes in the original graph remain dominated in the reduced graph.

The edge dominance problem can be put in a different form. Given a node b of Γ. Find all nodes a of Γ that is dominated by b. This problem can be solved in a similar way. Initially set the list L to be the non-neighbors of b. ($L = V \setminus [N(b) \cup \{b\}]$.) Note that if $a \in L$ and for each $c \in L$, $\{c, a\}$ is not edge of Γ, then b dominates a.

Let $m = |L|$, $n = |V|$. The list L can be initialized with n looks up of the adjacency matrix of Γ. The node $c \in L$ can be chosen in m ways. Similarly, the node $a \in L$ can be chosen in m ways. Therefore after at most $n + m^2$ looks up we have a list of the nodes dominated by b.

In challenging clique search instances the nodes usually have more neighbors than non-neighbors. In the $m < n/2$ particular case the worst case estimate for this algorithm is better than the worst case estimate for the previous algorithm. The author has been experimenting with the dominance spotting in the original graph for a considerable time. There are cases when no dominated nodes are spotted at all and the initial graph cannot be reduced in this manner However, there are cases when the clique search problem with the original graph is out of the feasibility range while it is feasible for the reduced graph.

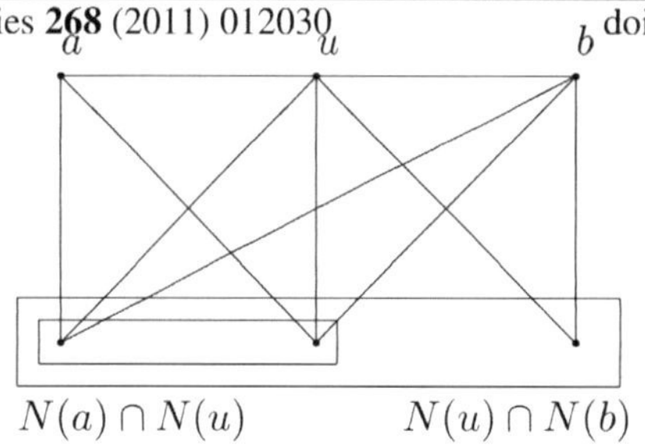

Figure 7. Edge $\{a, u\}$ is dominated by edge $\{u, b\}$. The dominance is short.

7. Dominating edges

We introduce dominance relations between edges of a graph.

Definition 6 *Let Γ be a graph and let a, u, b be distinct nodes of Γ such that $\{a, u\}$, $\{u, b\}$ are edges of Γ. We say that edge $\{u, b\}$ dominates edge $\{a, u\}$ if $b \notin N(a) \cap N(u)$ and $N(a) \cap N(u) \subseteq N(u) \cap N(b)$.*

The dominance in Definition 6 is illustrated by figure 7. We will introduce another edge dominance later so let us call the above defined edge dominance short edge dominance. Let Γ' be the graph we get from Γ after deleting the edge $\{a, u\}$. (But not deleting the nodes a or u.)

Proposition 12 *If Γ contains a clique of size k, then so does Γ'.*

Proof. Let Δ be a clique of size k in Γ. If $\{a, u\}$ is not an edge of Δ, then Δ is a clique of size k in Γ' too. For the remaining part of the proof we may assume that $\{a, u\}$ is an edge of Δ. Now there is a clique Δ_1 of size $k - 2$ in $N(a) \cap N(u)$. Simply we can choose Δ_1 to be $\Delta \setminus (\{a\} \cup \{u\})$. Using $N(a) \cap N(u) \subseteq N(u) \cap N(b)$ we get that Δ_1 is a clique of size $k - 2$ in $N(u) \cap N(b)$. It follows that $\Delta_2 = \Delta_1 \cup \{u\} \cup \{b\}$ is a clique in Γ. Note that $a \notin \Delta_1$, $u \notin \Delta_1$. Consequently Δ_2 is a clique of size k in Γ'. □

Proposition 13 *If edge $\{a, u\}$ is dominated by edge $\{u, b\}$ and edge $\{b, u\}$ is dominated by edge $\{u, c\}$, then edge $\{a, u\}$ is dominated by edge $\{u, c\}$.*

Proof. We have to verify that $c \notin N(a) \cap N(u)$ and $N(a) \cap N(u) \subseteq N(u) \cap N(c)$. In order to show that $c \notin N(a) \cap N(u)$ assume on the contrary that $c \in N(a) \cap N(u)$. The fact that edge $\{a, u\}$ is dominated by edge $\{u, b\}$ means that $N(a) \cap N(u) \subseteq N(u) \cap N(b)$. Thus $c \notin N(u) \cap N(b)$. But as edge $\{b, u\}$ is dominated by edge $\{u, c\}$, $c \notin N(b) \cap N(u)$. This contradiction shows that $c \notin N(a) \cap N(u)$, as required.

In order to show that $N(a) \cap N(u) \subseteq N(u) \cap N(c)$ note that $N(a) \cap N(u) \subseteq N(u) \cap N(b)$ since node $\{a, u\}$ is dominated by edge $\{u, b\}$. Further $N(b) \cap N(u) \subseteq N(u) \cap N(c)$ since edge $\{b, u\}$ is dominated by edge $\{u, c\}$. Thus $N(a) \cap N(u) \subseteq N(u) \cap N(c)$, as required. □

Let a, b, c, u, v be nodes of Γ such that $\{a, u\}$ is dominated by $\{u, b\}$ and $\{a, v\}$ is dominated by $\{v, c\}$. Let Γ' be the graph we get from Γ after deleting the edge $\{a, u\}$. (But not deleting the nodes a or u.)

Proposition 14 *Node $\{a, v\}$ is dominated by $\{v, b\}$ in Γ'.*

Proof. From the assumptions of the proposition we know that

$$\{a, u\}, \{u, b\} \text{ are edges of } \Gamma, \tag{1}$$

$$a, b \text{ are not adjacent in } \Gamma, \tag{2}$$

$$N(a) \cap N(u) \subseteq N(u) \cap N(b). \tag{3}$$

Further

$$\{a, v\}, \{v, c\} \text{ are edges of } \Gamma, \tag{4}$$

$$a, c \text{ are not adjacent in } \Gamma, \tag{5}$$

$$N(a) \cap N(v) \subseteq N(v) \cap N(c). \tag{6}$$

Using these we would like to verify that

$$\{a, v\}, \{v, c\} \text{ are edges of } \Gamma', \tag{7}$$

$$a, c \text{ are not adjacent in } \Gamma', \tag{8}$$

$$N'(a) \cap N'(v) \subseteq N'(v) \cap N'(c). \tag{9}$$

Here $N'(x)$ stands for the set of neighbors of x in Γ'.

By (5), a and c are not adjacent in Γ. Deleting the edge $\{a, u\}$ is not going to make a and u adjacent. This verifies (8).

By (4), $\{a, v\}$, $\{v, c\}$ are edges of Γ. Deleting the edge $\{a, u\}$ cannot alter this fact. This verifies (7).

In order to verify (9) note that only the following cases may occur in connection with u, $N(a) \cap N(v)$, $N(v) \cap N(c)$.

(i) $u \notin N(a) \cap N(v)$, $u \notin N(v) \cap N(c)$,
(ii) $u \notin N(a) \cap N(v)$, $u \in N(v) \cap N(c)$,
(iii) $u \in N(a) \cap N(v)$, $u \notin N(v) \cap N(c)$,
(iv) $u \in N(a) \cap N(v)$, $u \in N(v) \cap N(c)$.

By (6), the third case is not possible. In the first case

$$N'(a) \cap N'(v) = N(a) \cap N(v), \quad N'(v) \cap N'(c) = N(v) \cap N(c)$$

and so by (6), (9) holds. In the fourth case

$$N'(a) \cap N'(v) = [N(a) \cap N(v)] \setminus \{u\}, \quad N'(v) \cap N'(c) = [N(v) \cap N(c)] \setminus \{u\}$$

and so by (6), (9) follows. In the second case

$$N'(a) \cap N'(v) = [N(a) \cap N(v)] \setminus \{u\}, \quad N'(v) \cap N'(c) = N(v) \cap N(c)$$

and (9) follows. □

Definition 7 *Let Γ be a graph and let x, y, u, v be distinct point of Γ such that $\{x, y\}$, $\{u, v\}$ are edges of Γ. We say that edge $\{u, v\}$ dominates edge $\{x, y\}$ if each of the following holds*

(i) *$\{u, x\}$ or $\{u, y\}$ is not edge of Γ,*
(ii) *$\{v, x\}$ or $\{v, y\}$ is not edge of Γ,*
(iii) *$N(x) \cap N(y) \subseteq N(u) \cap N(v)$.*

The edge dominance in Definition 7 can be seen in figure 8. The edge dominance defined here will be referred as long edge dominance. Let Θ be the subgraph of Γ spanned by the set of vertices $\{x, y, u, v\}$. In the situation described in the definition $\{x, y\}$, $\{u, v\}$ are edges of Θ. Then any of the remaining pairs $\{x, u\}$, $\{x, v\}$, $\{y, u\}$, $\{y, v\}$ is either an edge of Θ or not. Out of the overall 16 cases conditions (i) and (ii) hold in 7 cases and do not hold in 9 cases.

Let us use the notation Γ' for the graph we get from Γ after deleting the edge $\{x, y\}$. (We do not delete the nodes x or y.)

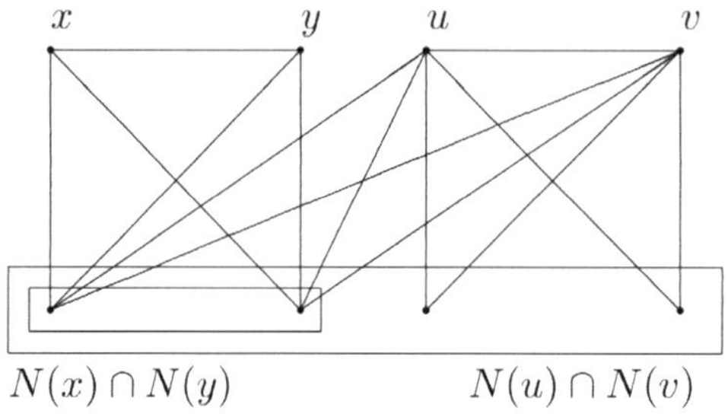

Figure 8. Edge $\{x, y\}$ is dominated by edge $\{u, v\}$. The dominance is long.

Proposition 15 *If Γ contains a clique of size k, then so does Γ'.*

Proof. Let Δ be a clique of size k in Γ. If x is not a node of Δ, then edge $\{x, y\}$ is not an edge of Δ. Consequently, Δ is a clique of size k in Γ' too. Similarly, if y is not a node of Δ, then Δ is a clique of size k in Γ' again. In both cases Γ' contains a clique of size k, as required. For the rest of the proof we may assume that x, y are nodes of Δ. The nodes of Δ that are distinct from x and y are coming from $N(x) \cap N(y)$ and so $\Delta_1 = \Delta \setminus (\{x\} \cup \{y\})$ is a clique of size $k - 2$ in $N(x) \cap N(y)$. Using $N(x) \cap N(y) \subseteq N(u) \cap N(v)$ we can see that Δ_1 is a clique of size $k - 2$ in $N(u) \cap N(v)$. Each node of Δ_1 is connected to u and v. This means that $\Delta_2 = \Delta_1 \cup \{u\} \cup \{v\}$ is a clique in Γ'. We will show that $u \notin \Delta_1$, $v \notin \Delta_1$ This will provide that Δ_2 is a clique of size k in Γ'.

In order to prove that $u \notin \Delta_1$ assume on the contrary that $u \in \Delta_1$. Plainly $u \in \Delta$. Using the fact that $\{x, y\}$ is an edge of Δ we get that $\{u, x\}$, $\{u, y\}$ are edges of Δ. Then $\{u, x\}$, $\{u, y\}$ are edges of Γ. But this is not the case. Thus $u \notin \Delta_1$. A similar argument gives that $v \notin \Delta_1$. □

Proposition 16 *If node $\{r, s\}$ is dominated by edge $\{x, y\}$ and node $\{x, y\}$ is dominated by edge $\{u, v\}$, then node $\{r, s\}$ is dominated by edge $\{u, v\}$.*

Proof. It is enough to verify that

$$\{u, r\} \text{ or } \{u, s\} \text{ is not an edge of } \Gamma \tag{10}$$

$$\{v, r\} \text{ or } \{v, s\} \text{ is not an edge of } \Gamma \tag{11}$$

$$N(r) \cap N(s) \subseteq N(u) \cap N(v) \tag{12}$$

In order to prove (10) assume on the contrary that $\{u, r\}$, $\{u, s\}$ are edges of Γ. This means that $u \in N(r) \cap N(s)$. As edge $\{r, s\}$ is dominated by edge $\{x, y\}$, it holds $N(r) \cap N(s) \subseteq N(x) \cap N(y)$ and so $u \in N(x) \cap N(y)$. As edge $\{x, y\}$ is dominated by edge $\{u, v\}$, it holds that $\{u, x\}$ or $\{u, y\}$ is not an edge of Γ and so $u \notin N(x) \cap N(y)$. This contradiction implies (10), as required. The statement (11) can be verified in a similar way.

To prove (12) note that $N(r) \cap N(s) \subseteq N(x) \cap N(y)$ since edge $\{r, s\}$ is dominated by edge $\{x, y\}$. Further, $N(x) \cap N(y) \subseteq N(u) \cap N(v)$ since edge $\{x, y\}$ is dominated by edge $\{u, v\}$. □

We would like to point out that the dominance concepts we introduced are not independent of each other. The dominance of nodes and the (short) dominance of edges are related in the following way. Let us suppose that edge $\{a, u\}$ is (short) dominated by edge $\{u, b\}$ in Γ. Working in $N(u)$ one can see that node a is (node) dominated by b.

The short and long dominance of edges are related in the next way. Let us suppose that edge $\{x, y\}$ is (long) dominated by edge $\{u, v\}$ in Γ. Assume that the nodes y and u are identical but the nodes x, y, v are distinct. By the definition of long dominance

$$\{x, y\}, \ \{u, v\} \ \text{are edges of} \ \Gamma, \tag{13}$$

$$\{x, u\} \ \text{or} \ \{y, u\} \ \text{is not an edge of} \ \Gamma, \tag{14}$$

$$\{x, v\} \ \text{or} \ \{y, v\} \ \text{is not an edge of} \ \Gamma, \tag{15}$$

$$N(x) \cap N(y) \subseteq N(u) \cap N(v). \tag{16}$$

For the sake of clarity let us denote the identical y and u by z and set $x = a$, $v = b$. Now conditions (13), (14), (15), (16) reduce to

$$\{a, z\}, \ \{z, b\} \ \text{are edges of} \ \Gamma, \tag{17}$$

$$\{a, z\} \ \text{or} \ \{z, z\} \ \text{is not an edge of} \ \Gamma, \tag{18}$$

$$\{a, b\} \ \text{or} \ \{z, b\} \ \text{is not an edge of} \ \Gamma, \tag{19}$$

$$N(a) \cap N(z) \subseteq N(z) \cap N(b). \tag{20}$$

As Γ does not have any loop, $\{z, z\}$ cannot be an edge of Γ and so condition (18) drops out. By (17), $\{z, b\}$ is an edge of Γ and so by (19), $\{a, b\}$ is not an edge of Γ. We are left with the following conditions

$$\{a, z\}, \ \{z, b\} \ \text{are edges of} \ \Gamma, \tag{21}$$

$$\{a, b\} \ \text{is not an edge of} \ \Gamma, \tag{22}$$

$$N(a) \cap N(z) \subseteq N(z) \cap N(b). \tag{23}$$

Therefore edge $\{a, z\}$ is (short) dominated by edge $\{z, b\}$.

We consider the following problem.

Problem 9 *Given a node a of Γ. Find all nodes u of Γ for which there is a node b of Γ such that the edge $\{a, u\}$ is (short) dominated by the edge $\{u, b\}$.*

We describe an algorithm that solves this problem. The algorithm maintains four lists L_1, L_2, L_3, L_4. Initially, L_1 contains the neighbors of a and L_2 contains the non-neighbors of a. (We do not place a to L_2.) Pick a $u \in L_1$. Then fill in L_3 with the elements of $L_1 \cap N(u)$ and fill in L_4 with the elements of $N(u) \cap L_2$.

Proposition 17 *If there is a $b \in L_4$ such that $\{b, c\}$ is an edge of Γ for each $c \in L_3$, then edge $\{a, u\}$ is (short) dominated by edge $\{u, b\}$.*

Proof. We will verify that

$$\{a, u\}, \ \{u, b\} \ \text{are edges of} \ \Gamma, \tag{24}$$

$$\{a, b\} \ \text{is not an edge of} \ \Gamma, \tag{25}$$

$$N(a) \cap N(u) \subseteq N(u) \cap N(b). \tag{26}$$

Using $u \in L_1$ and $L_1 = N(a)$ we get that $\{a, u\}$ is an edge of Γ. Using $b \in L_4$ and the definition of L_4 we get that $\{u, b\}$ is an edge of Γ. These prove (24).

From $b \in L_4$ it follows that $b \in L_2$. The elements of L_2 are not adjacent to a and so $\{a, b\}$ is not an edge of Γ. This settles (25).

To prove (26) choose a $c \in N(a) \cap N(u)$. In particular it holds that $c \in N(a)$. Using $N(a) \cap N(u) = L_1 \cap N(u) = L_3$ we get that $c \in L_3$. By the definition of b, $\{c, b\}$ is an edge of Γ and so $c \in N(b)$. Combining $c \in N(a)$ and $c \in N(b)$ implies $c \in N(u) \cap N(b)$, as required. □

Let s be the number of times the above algorithm reads single entries of the adjacency matrix of Γ during execution and let n be the number of nodes of Γ.

Proposition 18 $s \leq n + n^2 + (2/9)n^3$

Proof. Let n_1 be the length of L_1 and let n_2 be the length of L_2. Note that $n_1 + n_2 = n - 1$. The lists L_1 and L_2 can be filled in using n looks up of the adjacency matrix. For a fixed u the lists L_3 and L_4 can be filled in using at most n look ups. Clearly the length of L_3 is at most n_1 and the length of L_4 is at most n_2 in any time during the execution. For $u \in L_1$ there are at most n_1 choices. For $b \in L_4$ there are at most n_2 choices. For $c \in L_3$ there are at most n_1 choices. (We included figure 9 to assist in following the argument.)

An upper estimate for s is

$$n + n_1(n + n_1 n_2) = n + n_1 n + n_1^2 n_2 \leq n + n^2 + n_1^2 n_2.$$

The solution of the optimization problem, maximize $n_1^2 n_2$ subject to $n_1 + n_2 = n - 1$, is $n_1 = (2/3)(n-1)$, $n_2 = (1/3)(n-1)$. Therefore $s \leq n + n^2 + (2/9)(n-1)^3 \leq n + n^2 + (2/9)n^3$. □

We formulated Problem 9 with the definite purpose to incorporate the algorithm that solves it into the Carraghan-Pardalos clique search algorithm. But in the same time the algorithm above can be used to inspect the initial adjacency matrix to spot edges to delete. When we spot a deletable edge we delete it immediately. This might look as a rather simple minded approach. The deleted edge may dominate a number of further edges and so these edges will not be deleted later. We included Proposition 14 to reassure the reader that in fact this is not going to happen. By Proposition 18, the complete inspection looks up entries of the adjacency matrix at most $n^2 + n^3 + (2/9)n^4$ times.

8. Normal partitions

Let Γ be a graph with vertex set V.

Definition 8 *A partition $C_1, \ldots, C_t$ of V is called a normal partition of Γ if for each i, $1 \leq i \leq t$ there is an element $c_i \in C_i$ such that the elements of $C_i \setminus \{c_i\}$ are not adjacent to c_i and the elements of $C_{i+1} \cup \cdots \cup C_t$ are adjacent to c_i.*

The element c_i is called the leading element of the class C_i.

Consider the graph Γ in the first example. We remind the reader that figure 1 depicts a geometrical representation of Γ and table 1 contains two forms of the adjacency matrix of Γ. Let

$$\begin{array}{lll} C_1 = \{1, 3, 6\}, & C_2 = \{4, 2\}, & C_3 = \{5\}, \\ c_1 = 1, & c_2 = 4, & c_3 = 5. \end{array}$$

It is a routine exercise to check that C_1, C_2, C_3 is a normal partition of Γ with leading elements c_1, c_2, c_3. The list

(1) 3 6 || (4) 2 || (5)

could represent this partition in a more concise way.

Consider the graph Γ in the second example. For the reader convenience figure 6 presents Γ in a geometric form with points and arcs while table 2 presents the adjacency matrix of Γ. Set

$$\begin{array}{lll} C_1 = \{2, 1, 5, 9\}, & C_2 = \{8, 4, 6\}, & C_3 = \{3, 7\}, \\ c_1 = 2, & c_2 = 8, & c_3 = 3. \end{array}$$

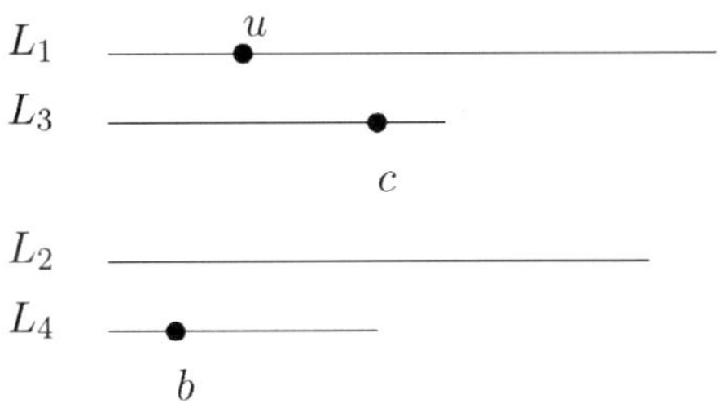

Figure 9. The lists L_1, L_2, L_3, L_4 in the proof of Proposition 18.

One can verify that C_1, C_2, C_3 is a normal partition of Γ with leading elements c_1, c_2, c_3. The list

$$(2)\ 1\ 5\ 9\ ||\ (8)\ 4\ 6\ ||\ (3)\ 7$$

could also be used to record this partition.

The normal partition $C_1, \ldots, C_t$ of Γ can be viewed as a (σ, t)-quasi coloring of Γ. The adjacency matrix $\mathbf{M}$ of Γ is cut into blocks $\mathbf{M}_{i,j}$ as we have seen in Section 5. Here 2σ is the overall number of ones in the $\mathbf{M}_{i,i}$ blocks and $\{a_1, b_1\}, \ldots, \{a_\sigma, b_\sigma\}$ are the edges of Γ corresponding to these ones. (Two 1's in symmetrical positions with respect to the main diagonal represent the same edge.) The subgraph of Γ spanned by $N(a_i) \cap N(b_i)$ is denoted by Γ_i. Let Γ' be the graph we get from Γ after deleting each of the edges $\{a_1, b_1\}, \ldots, \{a_\sigma, b_\sigma\}$. (But not deleting any of the end points of the edges.)

Proposition 19 *If Γ has a clique of size k, then at least one of $\Gamma_1, \ldots, \Gamma_\sigma$ has a clique of size $k-2$ or Γ' has a clique of size k.*

Proof. Assume that none of $\Gamma_1, \ldots, \Gamma_\sigma$ has a clique of size $k-2$. Let Δ be a clique of size k of Γ. If for some i, $1 \leq i \leq \sigma$, $\{a_i, b_i\}$ is an edge of Δ, then Γ_i contains a clique of size $k-2$. This is not the case so $\{a_i, b_i\}$ is not an edge of Δ for each i, $1 \leq i \leq \sigma$. After deleting these edges from Γ the subgraph Δ is still a clique of size k in the remaining graph, that is, Δ is a clique in Γ'. □

Proposition 20 *The size of the maximum clique in Γ' is t.*

Proof. As the elements of $C_2 \cup \cdots \cup C_t$ are adjacent to c_1 and $c_2, \ldots, c_t \in C_2 \cup \cdots \cup C_t$, it follows that c_1 is adjacent to each of $c_2, \ldots, c_t$. Similarly, as the elements of $C_3 \cup \cdots \cup C_t$ are adjacent to c_2 and $c_3, \ldots, c_t \in C_3 \cup \cdots \cup C_t$, it follows that c_2 is adjacent to each of $c_3, \ldots, c_t$. Continuing in this way finally we have that any two distinct nodes among $c_1, \ldots, c_t$ are adjacent. Thus there is a clique of size t in Γ'.

Let $\mathbf{M}'$ be the adjacency matrix of Γ'. Note that the blocks $\mathbf{M}'_{i,i}$ of $\mathbf{M}'$ are zero matrices and so Γ' can be well colored using t colors. Consequently Γ' cannot have a clique of size $t+1$. □

Problem 10 *Given a graph Γ. Construct a normal partition $C_1, \ldots, C_t$ for Γ.*

One can construct a normal partition of a given graph using the following greedy procedure. Let V be the vertex set of the given graph Γ. Set $i = 0$, $L = V$. While $L \neq \emptyset$ choose an $a \in L$ for which $|N(a)|$ is the largest possible. Increment i by one and set $c_i = a$, $C_i = L \setminus N(a)$, $L = N(a)$.

By Proposition 19, when we are looking for a k-clique in Γ we should check if the graphs $\Gamma_1, \dots, \Gamma_\sigma$ has a clique of size $k-2$ or the graph Γ' has a clique of size k. If $k = t$, then by Proposition 20, Γ' has a clique of size k. If $k > t$, then again by Proposition 20, we do not need to check Γ' at all. Thus we need to work only with the graphs $\Gamma_1, \dots, \Gamma_\sigma$.

The normal partition $C_1, \dots, C_t$ helps us to detect dominated nodes. To see how let $C_1 = \{a_1, \dots, a_s\}$, where $a_1 = c_1$. Consider the block $\mathbf{M}_{1,1}$ of the adjacency matrix $\mathbf{M}$ of Γ. The first row and the first column of $\mathbf{M}_{1,1}$ is filled with zeros as a_1 is not adjacent to $a_2, \dots, a_s$. If the second row and second column is filled with zeros, then node a_1 dominates a_2. Indeed, a_1 and a_2 are not adjacent. Further $N(a_2) \subseteq N(a_1)$ holds as $c_1 = a_1$ is adjacent to the elements of $C_2 \cup \cdots \cup C_t$. In short if the second row of $\mathbf{M}_{1,1}$ contains only zeros, then the node a_2 can be canceled from the graph Γ when we are looking for a k-clique in Γ.

Suppose there are $2r$ non-zero elements in the second row and the second column of the matrix $\mathbf{M}_{1,1}$ and $\Gamma_1, \dots, \Gamma_r$ are the graphs among $\Gamma_1, \dots, \Gamma_\sigma$ that correspond to ones in the second row and second column of $\mathbf{M}_{1,1}$. After completing the clique search for the graphs $\Gamma_1, \dots, \Gamma_r$ we can delete the ones from the second row and second column of $\mathbf{M}_{1,1}$ for the remaining part of the work. Therefore the node a_2 can be deleted from the graphs $\Gamma_{r+1}, \dots, \Gamma_\sigma$ to get smaller graphs to work with. We can repeat this reduction at any time when the non-zero elements in a given row and in the corresponding column of the matrix are completely processed. In fact these reductions can be carried out at the beginning of the computation and we end up with a large collection of graphs that can be processed independently of each other.

References

[1] I M Bomze M Budinich P M Pardalos M Pelillo *The Maximum Clique Problem* Handbook of Combinatorial Optimization Vol 4 Kluwer Academic Publishers 1999

[2] S Butenko W E Wilhelm Clique-detection models in computational biochemistry and genomics *European Journal of Operational Research* **173** (2006) 1–17

[3] D S Johnson M A Trick *Cliques Colorings and Satisfiability* 2nd DIMACS Implementation Challenge American Mathematical Society 1996

[4] J Hasselberg P M Pardalos V Vairaktarakis Test case generators and computational results for the maximum clique problem *J Global Optim* **3** (1993) 463–482

[5] P Kaski P R J Östergård *Classification Algorithms for Codes and Designs* Springer 2006

[6] D E Knuth Dancing links in *Millennial Perspectives in Computer Science* J Davies B Roscoe J Woodcock Eds Palgrave Macmillan Basingstoke 2000 pp 187–214

[7] P M Pardalos J Xue The maximum clique problem *J. Global Optim* **4** (1994) 301–328

[8] `ftp://dimacs.rutgers.edu/pub/challenge/graph/benchmarks/cliques/` (DIMACS clique benchmark instances)

[9] L Babel Finding maximum cliques in arbitrary and in special graphs *Computing* **46** (1991) 321–341

[10] E Balas J Xue Weighted and unweighted maximum clique algorithms with upper bounds from fractional coloring *Algorithmica* **15** (1996) 397–412

[11] R Carragham P M Pardalos An exact algorithm for the maximum clique problem *Oper Res Lett* **9** (1990) 375–382

[12] J Konc D Janežič An improved branch and bound algorithm for the maximum clique problem *MATCH Communications in Mathematical and Computer Chemistry* **58** (2007) 569–590

[13] D Kumlander A new exact algorithm for the maximum-weight clique problem based on a heuristic vertex-coloring and back-track search *Proceeding of The Fourth International Conference on Engineering Computational Technology* Civil-Comp Press 2004

[14] P R J Östergård A fast algorithm for the maximum clique problem *Discrete Applied Mathematics* **120** (2002) 197–207

[15] E C Sewell A branch and bound algorithm for the stability number of a sparse graph *INFORMS J Comput* **10** (1998) 438–447

[16] E Tomita T Seki An efficient branch-and-bound algorithm for finding a maximum clique *Lecture Notes in Computer Science* 2631 (2003) 278–289

[17] D R Wood An algorithm for finding a maximum clique in a graph *Operations Research Letters* **21** (1997) 211–217

[18] `ftp://dimacs.rutgers.edu/pub/challenge/graph/solvers/` (The `dfmax` clique solver)

[19] P M Pardalos J Rappe M G C Resende An exact parallel algorithm for the maximum clique problem in *High Performance Algorithms and Software in Nonlinear Optimization* R De Leone A Murli P M Pardalos G Toraldo (eds) Kluwer 279–300 Dordrecht 1998

[20] Y Zhang E J Chester M A Langston On finding bicliques in bipartite graphs: A novel approach with applications to the integration of diverse biological data types *Proceedings of the 41st Hawaii International Conference on System Sciences* (2008)

[21] M Molloy B Reed *Graph Coloring and The Probabilistic Method* Springer 2002

Journal of Physics: Conference Series **268** (2011) 012031 doi:10.1088/1742-6596/268/1/012031

Mode switching and hysteresis in the edge tone

I Vaik and G Paál

Budapest University of Technology and Economics, Department of Hydrodynamic Systems, Műegyetem rkp. 1, Budapest, H-1122

E-mail: vaik@hds.bme.hu

Abstract. The edge tone flow is one of the most simple aero-acoustic flow formulation. Despite it has been researched for several decades it is not yet fully understood. The flow consists of a planar free jet and a wedge shaped object placed oppositely to the jet. Under certain circumstances the jet starts to oscillate as it impinges on the wedge. The jet can take different shapes, these are called the stages of the edge tone. As the main parameters of the configuration varies the frequency of oscillation changes continuously within each stage. At some points the jet jumps between stages making a sudden jump in the frequency of oscillation as well. Also it was found that at certain parameters the jet jumps between stages without any change in the conditions. This study throws light on some very interesting non-linear phenomena according to the stage jumps of the edge tone. Both experimental and numerical techniques were used during the research.

1. Introduction

The edge tone is a simple flow configuration producing a remarkably complex behaviour. It consists of a planar free jet and a wedge-shaped object, placed oppositely to the jet exit (figure 1). The main parameters of the configuration is the nozzle wedge distance (h) and the mean exit velocity of the jet (u), but also the exit velocity profile of the jet has a mentionable effect on the evolved flowfield.

It was observed that under certain conditions the jet oscillates around the wedge (also called edge) in a more or less periodic manner. The oscillating flow creates an oscillating force on the wedge which creates a dipole sound source, thus this flow provides the excitation mechanism for the resonator in many wind instruments. [1, 2, 3, 4]

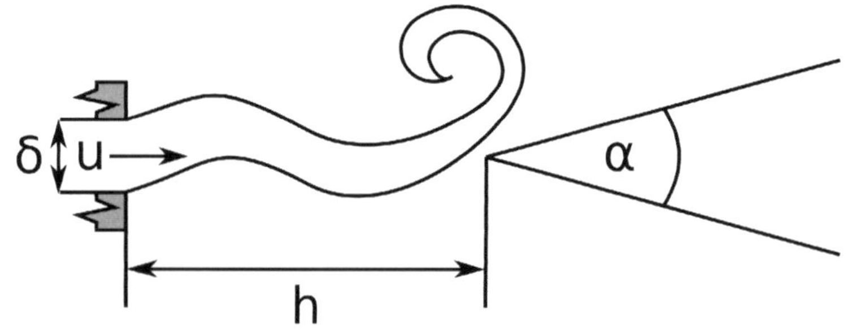

Figure 1. Sketch of the edge tone.

The frequency of oscillation varies continuously as the main parameters change. The frequency of oscillation is linearly proportional with the mean exit velocity of the jet, and inversly proportional with the nozzle-wedge distance. However, at certain parameter values the evolved flow changes qualitatively, resulting in a sudden change in the frequencies as well. These qualitatively different looks of the evolved flow are called the stages of the edge tone. Higher stages can be superposed onto lower ones, thus creating a nonperiodic oscillation or can be present purely. The ordinal numbers of the stages correspond roughly to the number of half waves in the shape of the jet between the nozzle and the wedge. Figures 2, 3 and 4 present three different appearances of the edge tone. The velocity fields calculated by a commercial computational fluid dynamics (CFD) solver are presented as vector plots.

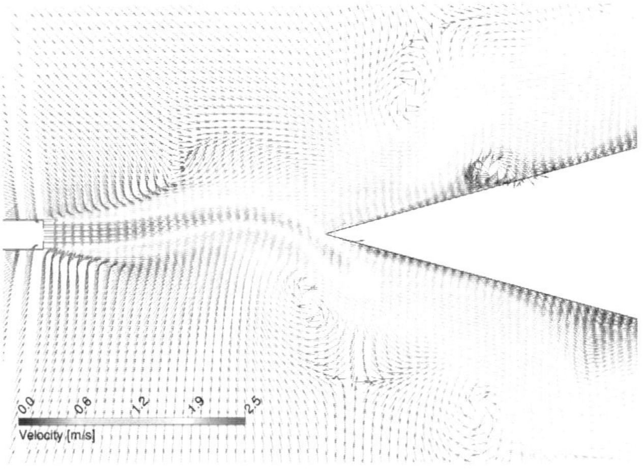

Figure 2. Stage I. at Re = 150.

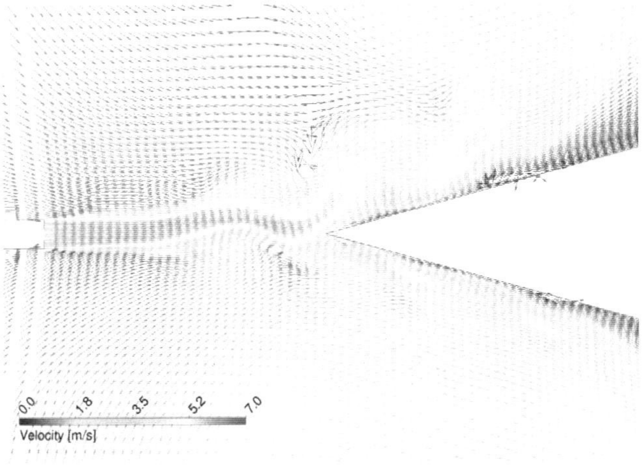

Figure 3. Stage II. at Re = 400.

The authors made an extensive experimental research on the edge tone, and also investigated this flow phenomenon with the help of CFD simulations. Several attributes of the phenomenon (such as the linear frequency dependence on the mean exit velocity and the inverse proportonality between the frequency and the nozzle-wedge distance) were reproduced in both ways. Four stages of the edge tone have been reproduced experimentally and three with numerical flow simulations. [5, 6]

In the numerical simulations geometrical sizes were chosen to be identical with those found in Powell's work [1]. During the measurements geometrical sizes were about three times larger,

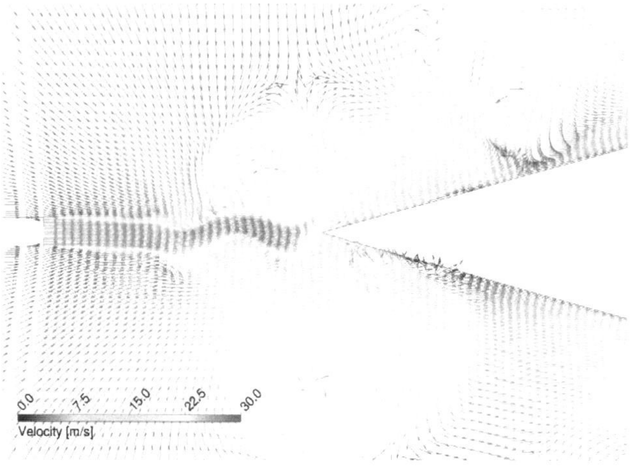

Figure 4. Stage III. at Re = 1800.

because this way - according to the similarity rules - the frequency of oscillation was about 10 times lower, and so the observation of the visualised flow was possible with a high speed digital camera taking still pictures at a maximum frame rate of 90 fps. To be able to compare the results from the literature, from the CFD simulations and from the experiments nondimensional numbers were introduced as the following:

- Reynolds number, $Re = \frac{u \cdot \delta}{\nu}$: for the mean exit velocity of the jet, where δ is the width of the jet and ν is the viscosity of the air;
- $\frac{h}{\delta}$: for the nozzle-wedge distance, where h is the nozzle-wedge distance and δ is the width of the jet;
- Strouhal number, $St = \frac{f \cdot \delta}{u}$: for the frequency of oscillation (f).

2. CFD simulations

A commercially available CFD code based on a finite volume technique (ANSYS-CFX) was used to calculate the flow field. Without going into the details of the solver ([7]) only the setup is discussed here.

As a first step of all numerical solution of partial differential equation systems a spatial discretisation of the domain is needed. The virtual geometry and the numerical mesh was created in ANSYS ICEM CFD. The domain was much larger than the area of interest to avoid the unwanted effects of the boundary conditions on the flow phenomenon. ANSYS CFX is a 3D solver, calculations of 2D flows is possible with a one cell layer mesh by defining symmetry boundary conditions on the bottom and top surfaces. Thus a mesh with one layer of hexahedra was created. After a careful mesh study the resulting mesh contained only 36300 volume elements.

The second step is to define the physics of the flow and to set the numerical discretisation of the PDE-s. Second order temporal and spatial discretisation of the Navier-Stokes equation was used in the solver. Because of the low Reynolds number (the maximum Reynolds number used during the simulations or the experiments was not greater than 3000) neither compressibility nor turbulence modelling was required. With the addition of a passive scalar variable for the "smoke" it was possible to model the flow visualisation of the experiments. A simple transport equation was used to calculate the density of smoke in the domain.

As a transient calculation one has to define the initial state of the flow. It was also investigated to what extent the initial conditions effects the evolved flow. It was found that the least numerical error and instability is introduced into the computation if the initial condition is a medium is at rest, so with 0 m/s velocity values and constant pressure distribution.

Details of the CFD simulations (more about the mesh, the setup and results discussed from several aspects) can be found in [5].

3. Experimental work

The sketch of the measurement rig can be seen in figure 5. Shop air was introduced into a 57 l cylindrical reservoir through a mass flow rate sensor (Sensortehcnics, Honeywell AWM700, working on a heated element principle producing voltage output). This mass flow rate sensor was used to measure the mean exit velocity of the jet. 3/4" reinforced flexible plastic tubes were used in the rig to connect the shop air system of the laboratory trough the mass flow rate sensor to the reservoir. A pressure reducing valve was used to reduce the pressure to 0.5 bar, thus eliminating the effect of the exhausting main shop air reservoir of the laboratory. The mass flow rate sensor was placed between two throttle valves, as the first calibrations showed that the sensor is sensitive to the inner pressure. With the careful parallel settings of these two valves it was possible to regulate the mass flow rate (and thus the mean exit velocity of the jet) without changing the inner pressure in the sensor. A long copper pipe was inserted just before the mass flow rate sensor to ensure undisturbed inflow into the sensor. Two nozzles were used to create a top hat or a parabolic exit velocity profile for the jet (figure 6). Former was formed by two quarter cylinders ensuring quick contraction and so creating a top hat velocity profile. For the creation of parabolic velocity profile the nozzle was built up of two 150 mm long parallel plates with a 15 mm rounded entry. The dimensions of the nozzle for the top hat case were: $\delta = 3.31$ mm with a height of 76.1 mm; and for the parabolic case: $\delta = 3.09$ mm with a height of 63.4 mm. So the aspect ratio was high enough to get a two-dimensional flow in the middle. The steel wedge was 150 mm high, so twice as high as the nozzle to avoid the end effects. The nozzle wedge distance was adjustable between 5 mm and 54 mm, so the dimensionless nozzle-wedge distance h/δ was adjustable between 1.67 and 17.7. A pressure transducer (Sensortechnics, 113LP01D-PCB) was built into the steel wedge at a distance of 26.2 mm from the tip of the wedge. The output signal was recorded with a 16 bit National Instruments A/D converter. The measuring rage of this sensor is [-100 Pa; 100 Pa] and the amplitude of the pressure fluctuation to be measured was about ± 0.05 Pa for the lower range of Reynolds number. The output of the sensor for the full range is [1 V;6 V], so for the 0.05 Pa amplitude pressure oscillation the fluctuation of the output voltage was in the range of 0.00125 V resulting an only 4 step resolution using the 16 bit A/D converter. The highest pressure during the measurements was about 4 Pa, what results about 0.1 V difference in the output signal. In order to be able to digitalize such small analogue voltage signals with the 16 bit A/D converter a special amplifier was used, which extended the [3.3 V;3.7 V] voltage region to the [-10 V;10 V] region. In the top hat jet case it was possible to visualize the flow with an incense stick. The smoke filament was illuminated by floodlight and the image was recorded with a high speed digital camera (LaVision ImagerCompact) taking pictures with a maximum frequency of 90 Hz with a spatial resolution of 320x240 pixels. No visualization was made in the parabolic case. Details of the measurements can be found in [6].

4. Detection of a mode switch

As discussed in the abstract and in the introduction, this paper aims to describe the mode switching events of the edge tone. There are several kinds of mode switching in the edge tone if the main parameters are changed continuously. Some of the borders of these modes are 'permanent' (the mode stays stable if the parameter is changed), some of them can show hysteretic behaviour, some of them exhibit a random mode switching within a certain parameter range.

Journal of Physics: Conference Series **268** (2011) 012031 doi:10.1088/1742-6596/268/1/012031

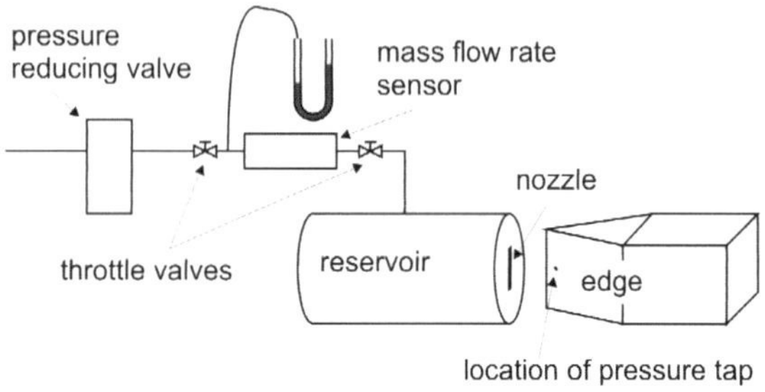

Figure 5. Sketch of the measurement rig.

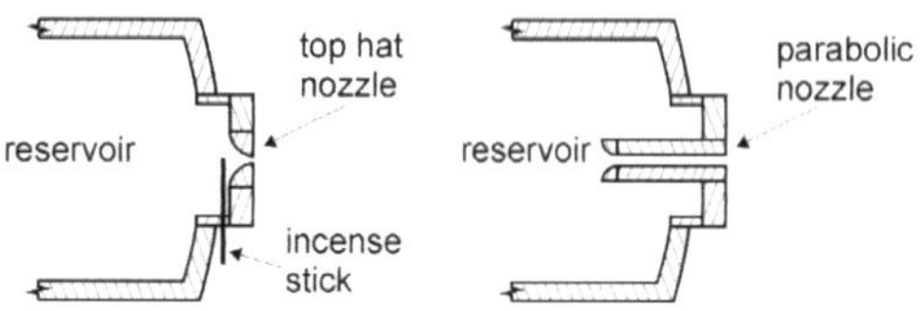

Figure 6. Sketch of the nozzles.

It was possible to take a picture series of the visualised flow field (only in the top hat case), but with our experimental rig it was not possible to properly trigger the camera to record the stage jump, so that it was rather fortuitous to record the jump even if the parameters were set to the proper values.

The analogue output signal of the pressure transducer was continuous and it was possible to digitalize and record this signal even long enough to record the pressure history during a mode switch. The qualitative change of this signal shows evidently the mode switch. To find the qualitative change in the recorded pressure histories a sliding window Fourier transformation technique can be used.

The process was the following: FFT was carried out from a Δt s long piece of the full signal starting at each $n \cdot \tau$ s time (usuall τ was much less than Δt, about 10-50% of Δt), thus creating a spectrum with a $1/\Delta t$ Hz frequency resolution for that period. Then all of the spectra can be visualised in one contour plot: the amplitude is plotted as colours against the frequency on the x axis, and the time (the start time of the 'window') on the y axis. So the colours on a horizontal cut of the figure at t s on the y axis represent the spectrum of the pressure signal for the $[t, t + \Delta t]$ period.

5. Observed phenomena

5.1. Stage jump in the CFD simulation

During the numerical simulations the jet was much more stable than in the experiments. This is because the CFD simulations are lack of any flow disturbances that throw the jet off it's 'balance'. With the high computational cost required for a long transient simulation this made the CFD simulation of a stage jump very hard. Still a jump from a pure first stage into a first and second stage coexistence was simulated. Figure 7 shows a pressure history at a point on the surface of the wedge at a distance of 2 mm from the tip of the wedge. The qualitative change of the signal can clearly be observed. The described moving window FFT method can be used for this signal too (see figure 8), the stage jump is nicely observable.

5.2. Mode switching and hysteresis during the change of the Reynolds number in experiments

There is a qualitative difference between the top hat and the parabolic case: In the top hat case when a higher stage of the edge tone evolves the lower stage(s) never switch(es) off, so a configuration when two or more stages coexist can be found. Pure 2nd or 3rd stage oscillation cannot be produced. In the parabolic jet case the same coexistence can be found, but there are also pure higher stage oscillations. Since it has more modes, the parabolic case is discussed here.

With a constant geometry ($h/\delta = 10$) the following can be observed when the Reynolds number is increased from zero:

- At first there is no oscillation at all. The wedge cuts the jet in half, and the flow is steady;

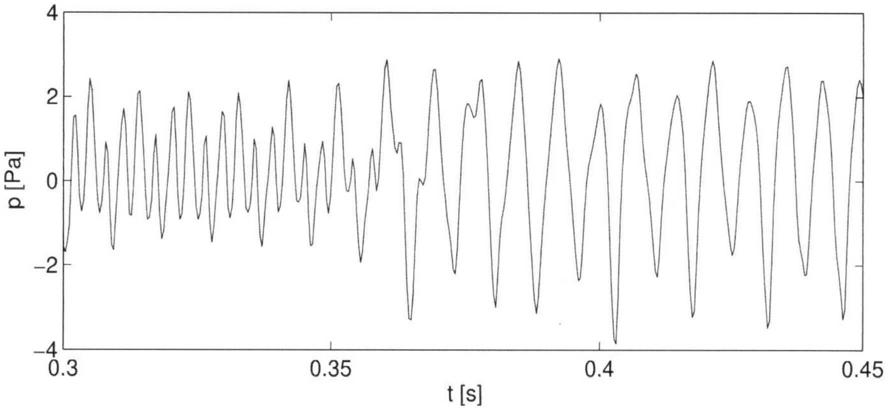

Figure 7. Pressure history of a stage jump in a CFD simulation; Re = 250.

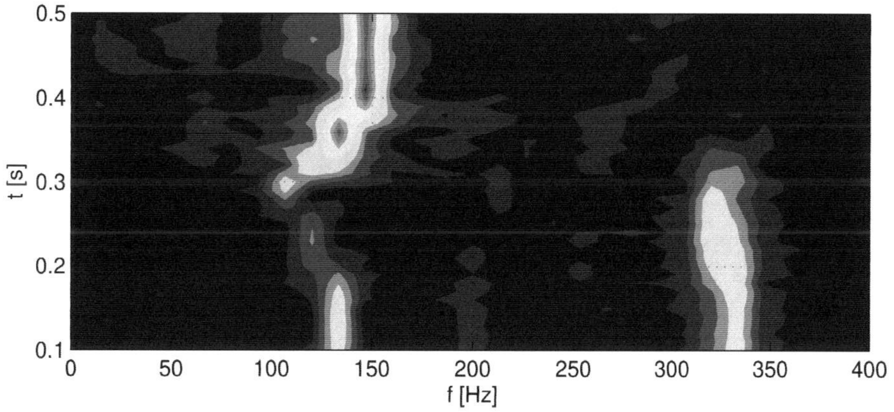

Figure 8. Moving window FFT of the pressure history of figure 7; Re = 250.

- At $Re \approx 85$ the first stage of the edge tone evolves;
- Then increasing the Reynolds number, at around $Re \approx 180$ the second stage appears next to the first stage, the two of them coexisting;
- Then further increasing the Reynolds number qualitatively different behaviour can be observed: from about $Re \approx 360 - 420$ the first stage disappears and only the second stage exists;
- Further increasing the Reynolds number ($Re \approx 550 - 750$) the second stage disappears and the third stage of the edge tone evolves, but at the same time the first stage appears again both coexisting;
- At the highest Reynolds numbers a pure third stage oscillation was observed (from about $Re \approx 650 - 1000$).

5.3. Hysteresis in experiments
When decreasing the Reynolds number the same process can be observed in the opposite direction, but with one big difference: there is no pure second stage oscillation when decreasing the Reynolds number from a thrid and first stage coexistent mode. So the edge tone behaves

differently in the [350; 600] Reynolds number region if the it approached from lower Reynolds number of higher Reynolds number. Figures 9 and 10 show this hysteretic behaviour.

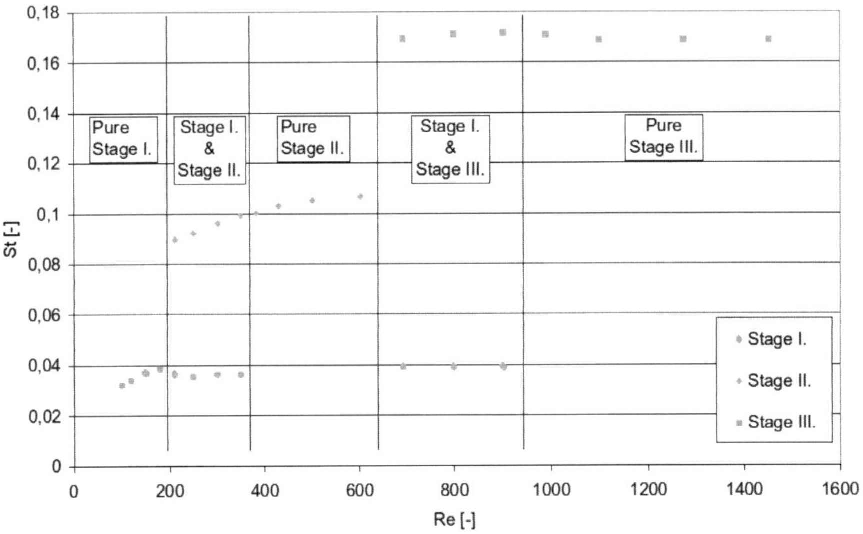

Figure 9. Hysteretic behaviour, increasing Reynolds number.

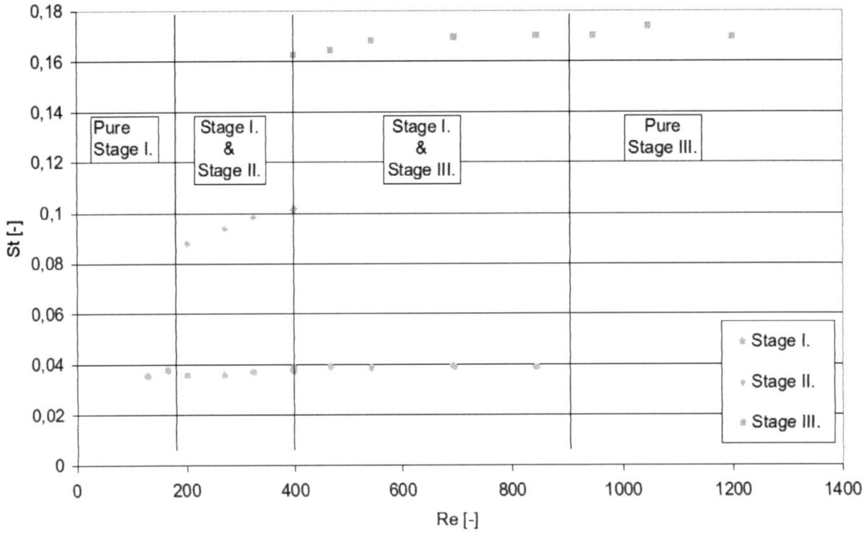

Figure 10. Hysteretic behaviour, decreasing Reynolds number.

5.4. Stage jumps and mode switching in experiments

The following pictures show the results of the moving window FFT for some mode switches:

- Figure 11: A jump from a first and second stage coexistence into a pure first stage mode can be seen here. One can clearly observe the qualitative change in the spectrum at about $t = 6$

s. It was found that the frequency of the first stage is about 10% lower if the second stage is also present than in pure first stage oscillation. This phenomenon can also be observed in this picture. Also another typcal nonlinear phenomenon can be found: in the two stage coexistence mode there is a peak at the difference of the frequencies of the two stages;

- Figure 12: The first stage switches on and off next to the second stage, without any change in the conditions;
- Figure 13: The first stage switches on next to the second stage;
- Figure 14: The first stage switches off;
- Figure 15: The jet jumps from a pure second stage oscillation into a first and third stage coexistence;
- Figure 16: The jet jumps back from a first and third stage coexistence mode into a pure second stage coexistence;
- Figure 17: The first stage switches on next to the third stage;
- Figure 18: The first stage switches off and on next to the third stage.

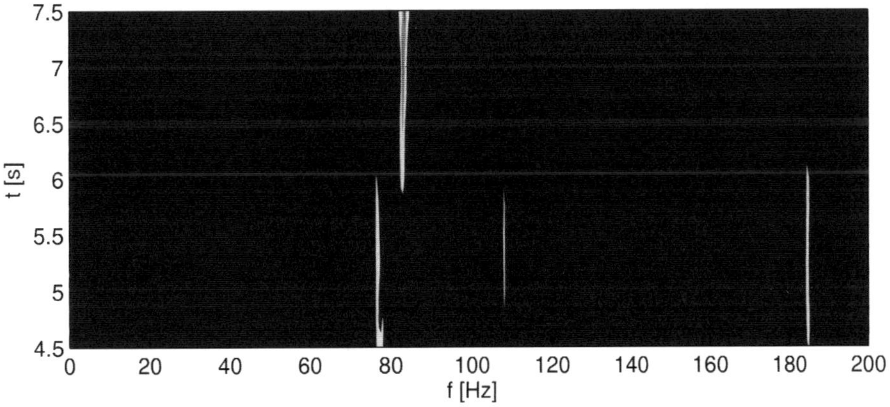

Figure 11. Mode switching: jump from first and second stage coexistence into pure first stage.

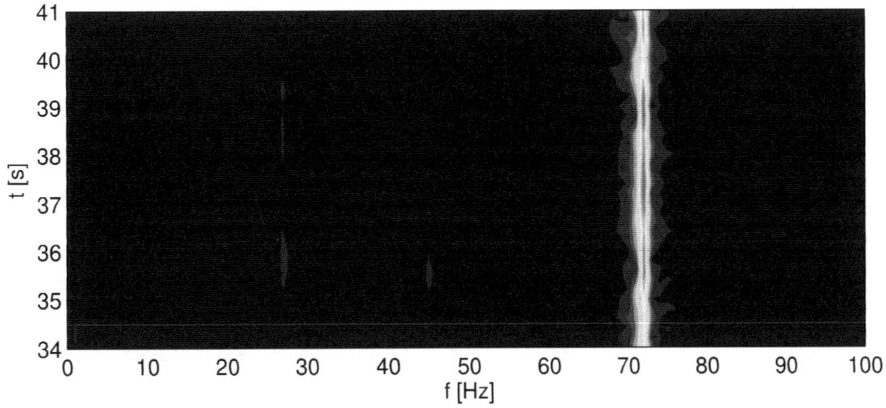

Figure 12. Mode switching: the first stage switches on and off next to the second stage.

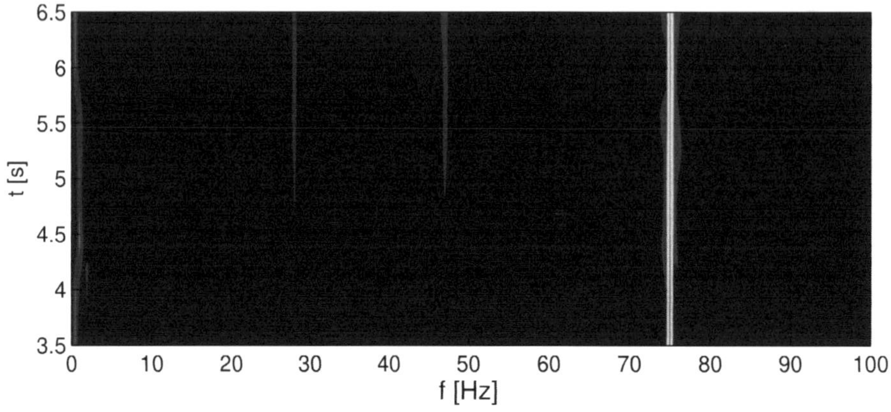

Figure 13. Mode switching: the first stage switches on next to the second stage.

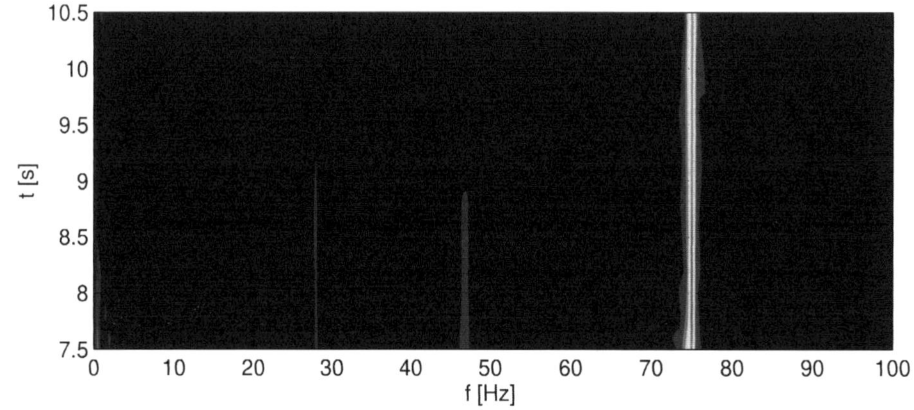

Figure 14. Mode switching: the first stage switches off resulting a pure second stage oscillation.

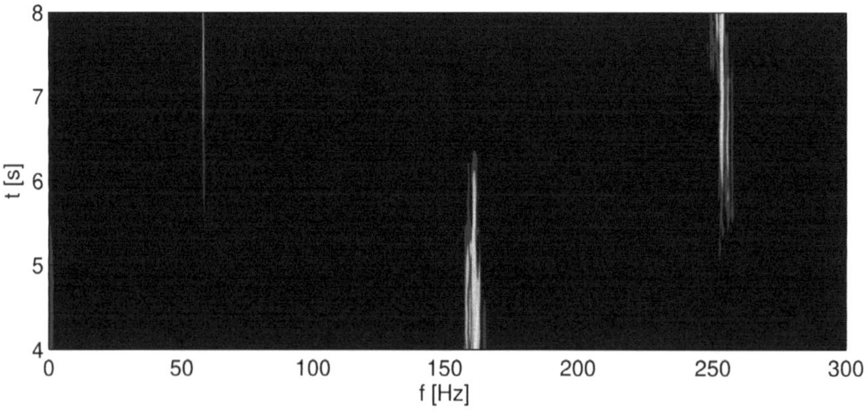

Figure 15. Mode switching: jump from the second stage into the first and third stage coexistence.

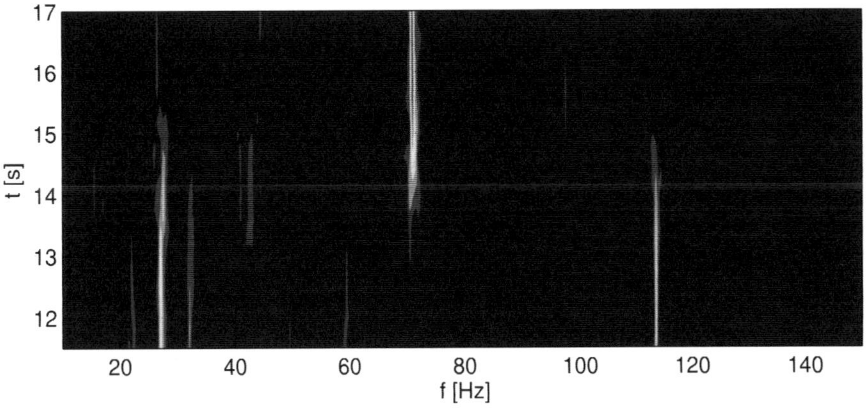

Figure 16. Mode switching: jump from the first and third stage coexistence into a pure second stage oscillation.

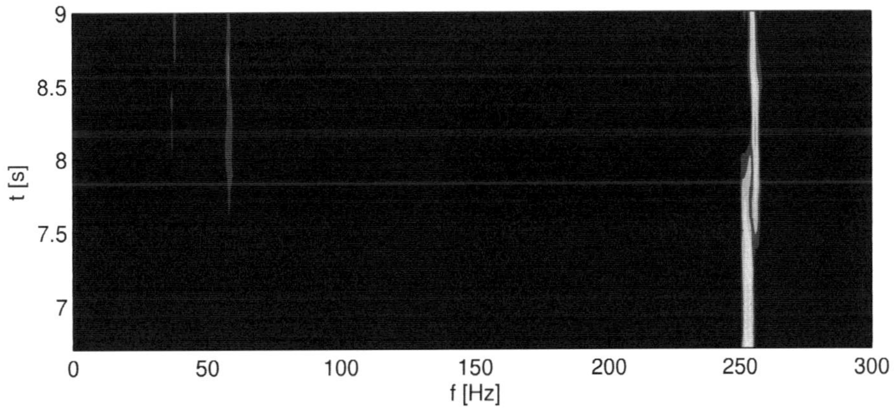

Figure 17. Mode switching: jump from a pure third stage oscillation into a first and third stage coexistence mode.

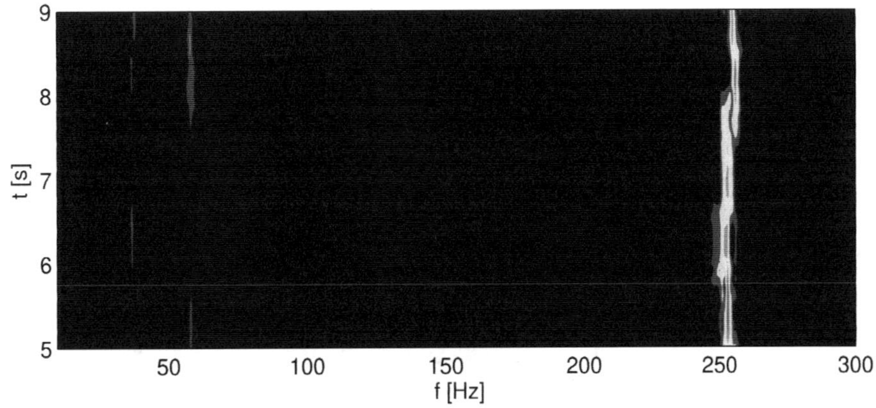

Figure 18. Mode switching: the first stage switches on and off next to the third stage.

6. Conclusions

The edge tone has several qualitatively different appearances, and the jut jumps between stages if the main parameters are varied. Pure stages or the coexistence of the stages are both possible. There are certain conditions where the edge tone is not stable in any of its appearances, and switches between modes without having any change in the conditions of the flow. The edge tone also displays hysteresis. Stage jumps and mode switching were reproduced both numerically and experimentally and the hysteretic behaviour of the edge tone was also observed in the measurements.

7. Acknowledgement

This work is connected to the scientific program of the 'Development of quality-oriented and harmonized R+D+I strategy and functional model at BME' project. This project is supported by the New Hungary Development Plan (Project ID: TMOP-4.2.1/B-09/1/KMR-2010-0002).

References

[1] Powell A 1961 On the edge tone *Journal of the Acoustical Society of America* **33**, 395-409
[2] Brown G B 1937 The vortex motion causing edge tones *Proc. Phys. Soc. Lond.* **49** 493-507
[3] Jones A T 1942 Edge tones *Journal of the Acoustical Society of America* **14** 131-9
[4] Curle N 1953 The mechanics of edge tones *Proc. Roy. Soc. A.* **216** 412-24
[5] Paál G and Vaik I 2007 Unsteady flow phenomena in the edge tone *Int. J. Heat Fluid Flow* **28** 575-86
[6] Vaik I and Paál G 2009 Experiments on the edge tone *Proceedings of the Conference on Modelling Fluid Flow (CMFF'09), 9-12 September 2009, Budapest, Hungary*
[7] www.ansys.com

Journal of Physics: Conference Series **268** (2011) 012032 doi:10.1088/1742-6596/268/1/012032

Micromagnetic analysis and development of high sensitivity spin-valve magnetic sensors

M Volmer[1] and J Neamtu[2]

[1]Transilvania University of Brasov, Physics Department, Brasov, Romania
[2]Advanced Research Institute for Electrical Engineering, Bucharest, Romania

E-mail: volmerm@unitbv.ro

Abstract. In this paper we present a micromagnetic analysis and development of low-field magnetic sensors based on the planar Hall effect (PHE) in Permalloy ($Ni_{80}Fe_{20}$) thin films and spin-valve structures. We used these sensors for field measurements, biomolecules detection in lab-on-a-chip applications and angle positioning devices. Both field and angular PHE measurements were performed. Some distortions and hysteretic effects were observed at low magnetic fields for the angular dependence of the PHE voltage. Micromagnetic simulations were performed to explain the field and the angular dependence of the PHE voltage. We used dc biasing fields to improve the response of the PHE sensor and we compare these results with our micromagnetic simulations.

1. Introduction

Magnetoresistance effect (MR) which can be found in ferromagnetic (Fe, Co, Ni or alloys like Permalloy) based thin films is an attractive solution for the fabrication of magnetic sensors. The resistance behaviour of such thin films (3*d* ferromagnetic alloys) is anisotropic with respect to the applied field direction, the MR being positive when the magnetic field is parallel to the current (longitudinal) and negative when the magnetic field is perpendicular to the current direction (transversal). This is the anisotropic magnetoresistance effect (AMR) and it was discovered in 1857 by William Thomson (Lord Kelvin). The AMR effect arises from anisotropic scattering due to spin orbit-orbit interaction. It is worth to mention that the MR effect in ferromagnetic thin films is determined by the sample magnetization rather than the external magnetic field, H. The resistance produced by scattering reaches a maximum value when the magnetization direction is parallel or antiparallel (i.e. 0° or 180°) to the current direction and minimum when the magnetization is perpendicular to the current.

The magnetoresistance ratio $\Delta R_{AMR}/R_0$ is relatively large for ternary Fe-Co-Ni alloys containing 70 to 90 atomic percent of Ni, amounting to at most 2.5 to 3% in 30 nm Permalloy ($Ni_{81}Fe_{19}$) films [1]. This is because when the film thickness becomes comparable or smaller than the mean free path of the carriers we should take in account the scattering processes at the surfaces and interfaces which lower the amplitude of the MR effects. Here, $\Delta R_{AMR}=R_l-R_t$ is the amplitude of the AMR effect calculated in the saturated state when the applied magnetic field is parallel and perpendicular to the current respectively. Usually, R_0 is the resistance in the saturated state when H is applied perpendicular to the current direction but in the film plane. For many applications in read heads and other sensors, Permalloy (Py) is the preferred choice due to its favourable soft magnetic properties. In the ternary Fe-Co-Ni diagram the Permalloy composition lies close to both to zero magnetostriction and the zero crystalline anisotropy line. Therefore, the magnetic behaviour of the prepared Permalloy films is dominated by the uniaxial in-plane anisotropy induced by a field applied during deposition. Thus, for a field along the hard axis, i.e. perpendicular to the anisotropy direction, the change of the magnetisation will result from a coherent magnetization rotation process. The maximum sensitivity and linearity is achieved when the magnetization is at 45° with respect to the current direction. The 45° alignment is commonly achieved using the Barber pole biasing technique, i.e. by patterning diagonal stripes of highly conductive metal (gold) onto the more resistive AMR material, as shown in figure 1(a). The current will then run perpendicular to these "barber pole" stripes while the magnetization vector remains preferentially along the long direction of the MR device. The application of an external magnetic field will rotate the magnetization with a resulting change in resistance as shown in figure 1(b).

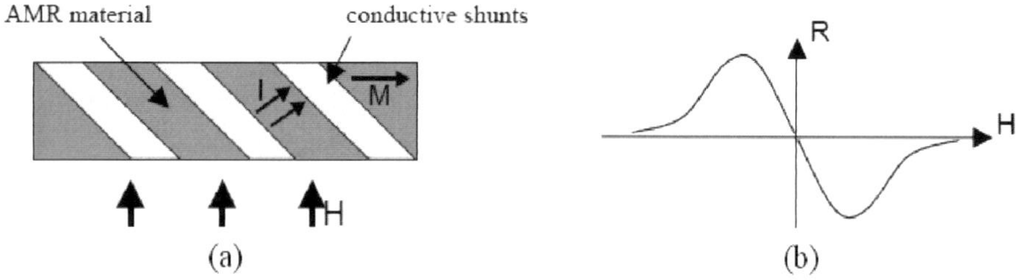

Figure 1. (a) Barber-pole structure of conductive shunts that constrain the current to run at 45° to the rest position for the magnetization. (b) Resistance versus field for a properly biased AMR device.

To develop practical applications, such as contactless potentiometer or magnetic field sensors, it must to minimize the thermal drift and to optimize the MR response. For this purpose it is convenient to operate the device with the two or four active arms of a Wheatstone bridge. Each arm corresponds to one MR element. We have to mention that many other materials can be used for this application. Granular manganese perovskites, having ferromagnetic transitions above room temperature, are good candidates for use as magnetoresistive sensors [2]. Using thick films of $La_{2/3}Sr_{1/3}MnO_3$ prepared by screen printing on polycrystalline Al_2O_3 it can be developed a room temperature AMR sensor to build a contactless potentiometer.

Because of the AMR effect will appear an electric field perpendicular to the applied current, in a Hall effect geometry, even when the magnetic field is in the film plane [3, 4]. This is the so-called planar Hall effect (PHE). Using this setup we get direct access to the anisotropic part of the resistance with the advantage of a reduced thermal drift of the output signal.

In a single domain approximation, the PHE voltage is determined by $U_{PHE} = CM^2 j \sin 2\theta$, where C is a constant determined by the structure properties, j is the current density, M is the saturation magnetization and θ is the angle between the current and the magnetization vector that, in turn, is determined by the value and direction of the external magnetic field [4, 5, 6].

Although the signal derived from the PHE is small, there is a higher signal-to-noise ratio (S/N) and a better thermal stability when compared to GMR spin valve sensors, hence it has the potential of detecting very small fields produced by various sources, single micro- or nanoparticles [7, 8]. The planar Hall voltage depends on the magnetization quadratically at the small applied fields and parabolically at the applied fields above the saturated magnetization. If the magnetization is initially oriented along the driving current inside the sensor, a rotation with angle θ produces a variation of the PHE voltage proportional with $\sin 2\theta$. This property can be used to build magnetic sensors. In order to obtain a coherent rotation of the magnetization inside the PHE sensor under the action of an applied magnetic field, a magnetic biasing can be used, as can be seen in the layout design presented in figure 2(a). This layout is also used to perform micromagnetic simulations regarding the field dependence of the PHE voltage. The micromagnetic simulations are using a method that minimises the free energy of the system in magnetic field based on the Stoner-Wohlfarth model [6, 9, 10].

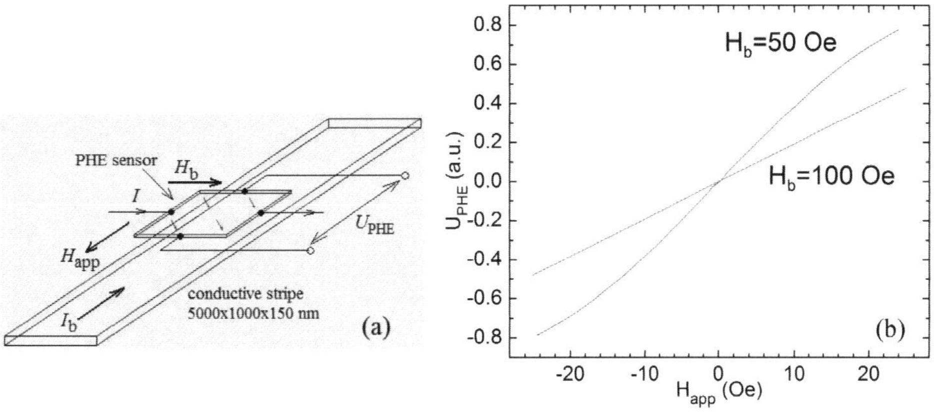

Figure 2. (a) The setup used for biasing and micromagnetic simulations; (b) Results of micromagnetic simulations for field dependence of the PHE voltage.

The current that is flowing through the conductive stripe generates a biasing magnetic field, H_b, which induces a single domain structure in sensor. The distance between the sensor's surface and the conductive stripe is 200 nm. Applying a magnetic field, H_{app}, along a direction perpendicular on H_b, will produce a rotation of the magnetization and a voltage appears in a Hall effect setup, figure 2(a). Using micromagnetic simulations to find the rotation angle of the magnetisation, we can plot the field dependence of the PHE for different values of the biasing field, figure 2(b). From these very simple simulations it comes that the sensitivity and linearity of the field dependence of the PHE voltage can be tuned. In a single domain approximation, a lower biasing field gives a higher sensitivity but nonlinearity effects appear. On the other hand, the real magnetic films do not behave like single domain structures for low values of the biasing field. So, we expect to have a more complex field dependence of the measured PHE voltage accompanied, also, by some hysteretic effects.

2. Results and discussion

In what follows we will present some results obtained on Permalloy thin films and Permalloy based multilayered (ML) structures. The measurement system consists in a programmable current source Keithley 6221 and a nanovoltmeter 2182A. To generate and control the magnetic field we used a bipolar 4 quadrants BOP 10-100 MG power source that drives a 1.5 T electromagnet. The field is measured using a Lake Shore 475 DSP gaussmeter.

2.1. Field dependence of the PHE

Disk shape structures of $Ni_{80}Fe_{20}$(15, 5 nm) and Co(30nm)/Cu(7 nm)/$Ni_{80}Fe_{20}$(70 nm), 5 mm diameter, were deposited by magnetron sputtering onto oxidised Si substrate. No uniaxial anisotropy axes were defined during the samples deposition. Four Au contacts were deposited, symmetrically, by thermal evaporation to form the electrodes for the driving current and PHE signal, like in figure 2(a).

2.1.1. Si/SiO$_2$/Ni$_{80}$Fe$_{20}$(15.5 nm)

Because the sensor is a disk shape structure there is no shape anisotropy in the film plane and because there was no field applied during the deposition there is no uniaxial anisotropy.

In figure 3 we present the field dependence of the PHE effect measured when θ=45° and no biasing, H_b=0. Because there are no anisotropy effects the magnetization process is mainly due to domain wall movement, i.e., the measured PHE voltage is a measure of the sample magnetization at square power. In this case, θ=45°, we get direct access to the anisotropic part of the resistance with the advantage of a reduced thermal drift of the output signal.

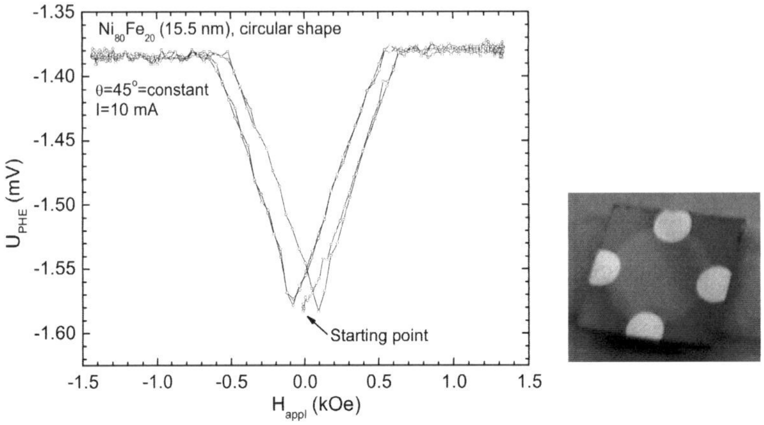

Figure 3. The field dependence of the PHE for a disk shape Permalloy film; the film is also shown.

This sensor can be used for field measurement by applying a biasing field, B_0=0.02 T, over this direction, θ=45°, using a small magnet, and an instrumentation circuit to compensate the offset and to amplify the signal, like in figure 4.

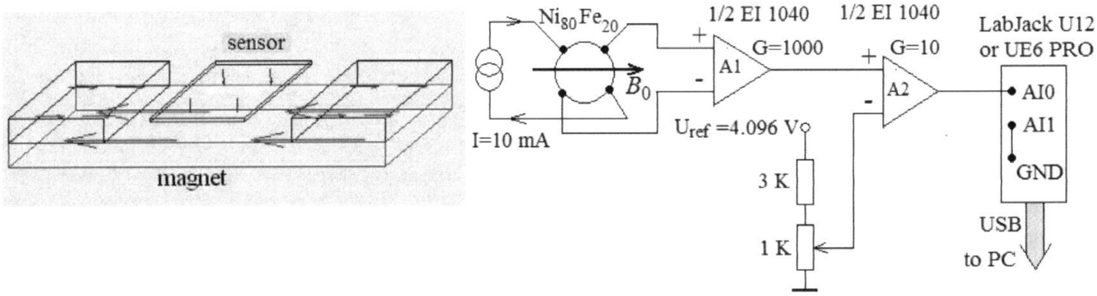

Figure 4. The biasing concept and the basic instrumentation schematic used to build a digital teslameter.

The field dependence, presented in figure 5, obtained after four scans between -0.02 to 0.02 T, reveals a very good linearity and a small hysteresis effect.

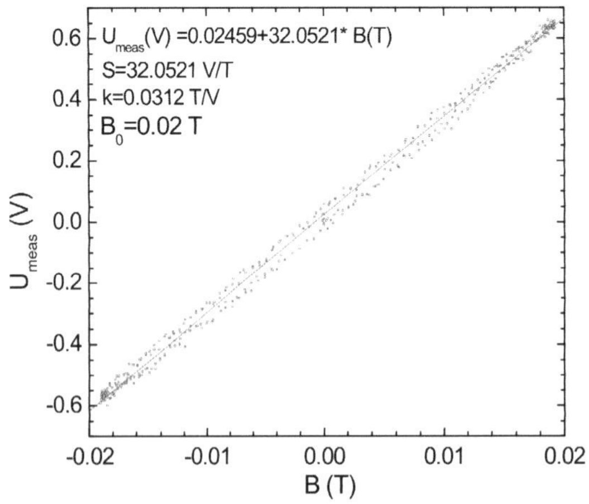

Figure 5. The field dependence of the measured voltage, U_{meas}, using biasing and signal conditioning.

The setup presented in figure 4 represents the basic schematics of a digital gaussmeter which is using a thin film of Permalloy.

When is used the biasing setup presented in figure 2(a), the magnetization keeps, in principle, the same value, but rotates under the action of the applied field. If the rotation angle is small, the variation of the PHE is proportional with θ. figure 6(a) presents, the field dependence of the PHE voltage for different values of the biasing field, H_b. It is to be noted the hysteresis effects that appear even for a higher biasing field. For lower biasing fields, the magnetization do not saturates and the film will not behave like a single domain structure. So, when H_b increases, the magnetization increases until H_b reaches the saturation value. If during the film deposition a small field, 100 Oe, is applied in the film plane along two opposite contacts, a uniaxial anisotropy axis will be defined with about 5 Oe the amplitude of the anisotropy field. In this case, the hysteresis effects disappear, figure 6(b).

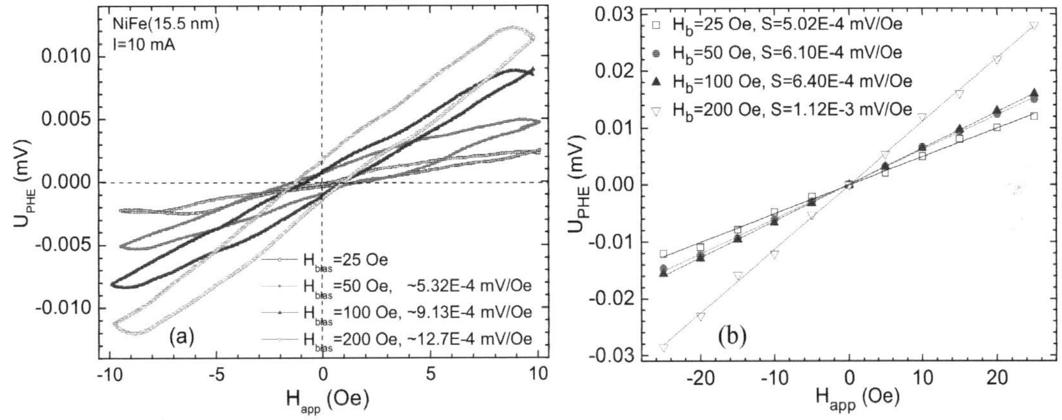

Figure 6. (a) the field dependence of the PHE voltage for a disk without uniaxial anisotropy and (b) the same dependence when there is a uniaxial anisotropy axis in the film plane over θ=0.

To simulate the response of this structure, we considered a biasing setup like in figure 2(a) but instead of a square shape single domain structure, we used a disk shape structure, figure 7, divided in a mesh with magnetic single domains [6, 9, 10].

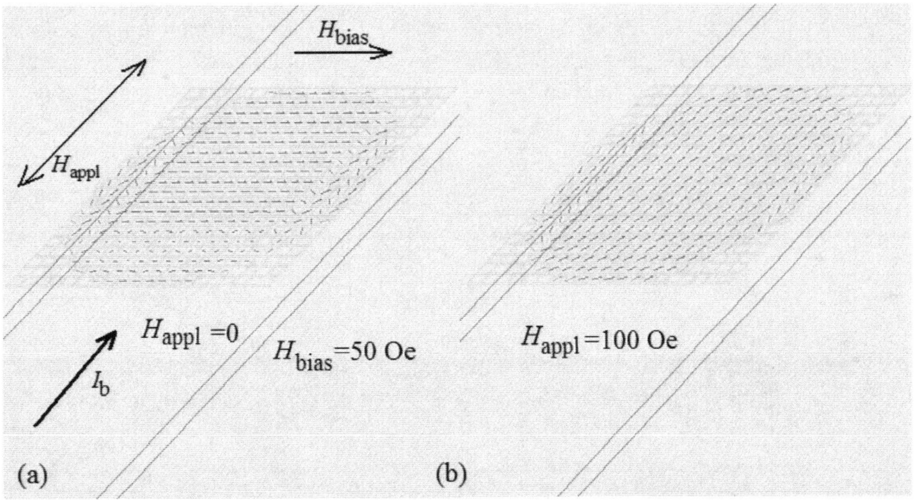

Figure 7. The structure (disc of Permalloy) used for micromagnetic simulations placed above the conductive stripe and (a) the magnetic moments orientations for H_{appl}=0 and (b) for H_{appl}=100 Oe.

Figure 7 illustrates this setup and presents the magnetic moments orientations for a biasing field of 50 Oe (I_b=5.25 mA) without an applied field, figure 7(a), and for an applied field of 100 Oe, figure 7(b). We see that the structure polarized at 50 Oe is not behaving like a single domain and explains the behaviour observed in figure 6. We performed this kind of micromagnetic simulations for three values of the biasing field and the results are presented in figure 8.

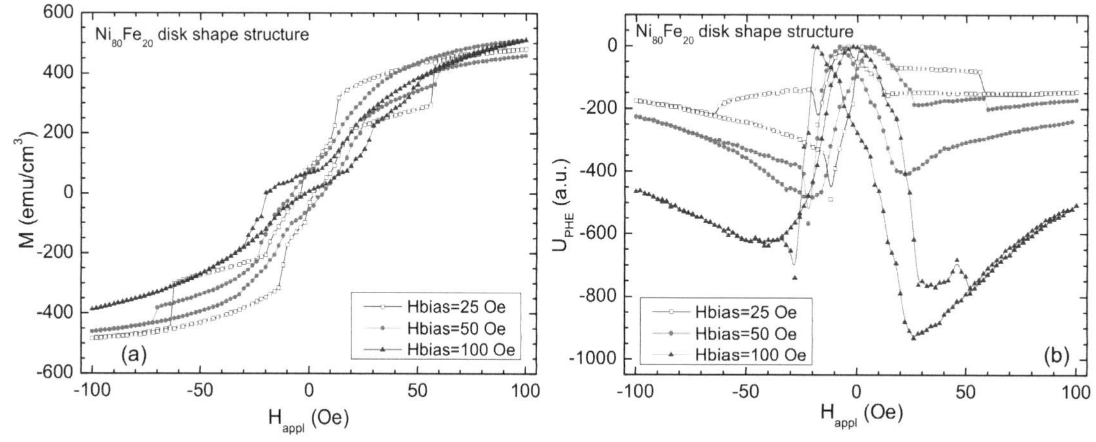

Figure 8. Micromagnetic simulations regarding the field dependences of magnetization (a) and planar Hall effect voltage (b) for different values of the biasing field.

From these simulations we see how the hysteresis width decreases when the biasing field increases and the magnetization reversal processes will be mainly due to rotation of the magnetization rather than domain wall displacements, figure 8(a). The amplitude of the calculated PHE voltage increases when H_b increases and the saturation effect is obtained for high fields, figure 8(b). The calculated data for PHE signal presents a 180° phase shift and is not normalized to zero.

2.1.2. Si/SiO$_2$/ Co(30nm)/Cu(7 nm)/Ni$_{80}$Fe$_{20}$(70 nm)

In this subsection we present the same type of measurements and simulations made on a ML structure of the type Co/Cu/NiFe. In this case, the coupling effects between the magnetic layers trough the nonmagnetic layer (Cu) and the higher anisotropy of the Co layer will play important roles on the magnetic field and angular behaviour of the PHE. Figure 9 presents the field dependences of the PHE voltage for θ=45° without a biasing field and with a biasing field applied like in figure 2(a).

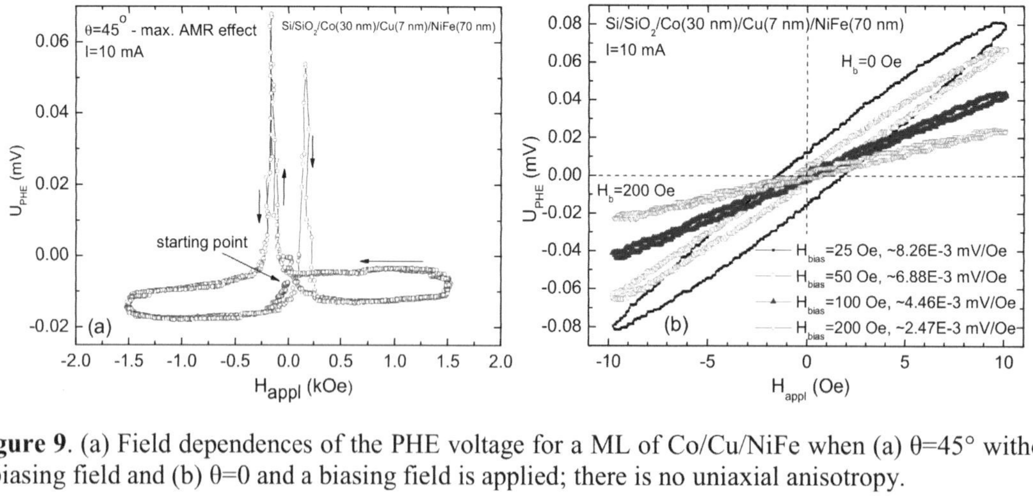

Figure 9. (a) Field dependences of the PHE voltage for a ML of Co/Cu/NiFe when (a) θ=45° without a biasing field and (b) θ=0 and a biasing field is applied; there is no uniaxial anisotropy.

Compared with the dependence presented for $Si/SiO_2/Ni_{80}Fe_{20}$(15.5 nm) in subsection 2.1.1, we see in this case a higher coercive field of about 200 Oe (100 Oe for $Ni_{80}Fe_{20}$ 15.5 nm) and the influence of the Co layer which lowers the sensitivity at higher biasing fields. When the biasing field is higher, the Co layer tends to keep the orientation of the NiFe layer because of his anisotropy and the positive magnetostatic coupling between them. Again, when the biasing field is low there are hysteretic effects because of the reversal mechanism of magnetisation which is mainly due to domain wall movement.

To simulate the behaviour of this ML, we considered a design like in figure 7 which now consists in two magnetic layers separated by the nonmagnetic layer. The mesh structure and the interactions between the magnetic layers are described in [6, 9]. The orientations of the magnetic moments when the applied field is 100 Oe are presented in figure 10(a) for H_b=50 Oe and figure 10(b) for H_b=200 Oe.

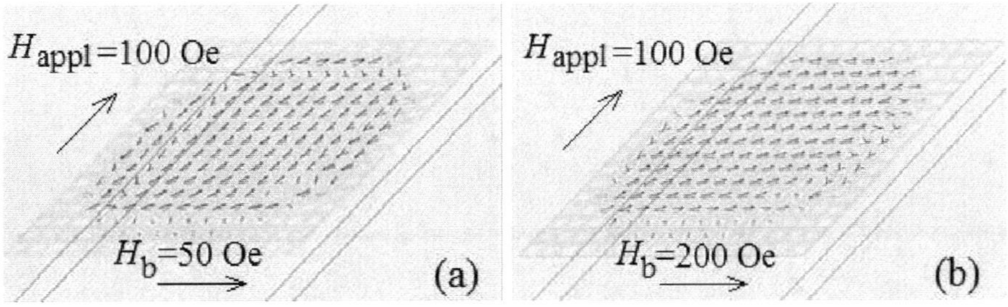

Figure 10. The magnetic moments orientations for H_{appl}=100 Oe when (a) H_b=50 Oe and (b) H_b=200 Oe; the red arrows (upper layer) denote the NiFe layer and blue arrows denote the Co layer.

Figure 11 presents the field dependences of the magnetization and PHE voltage obtained by micromagnetic simulations for three values of the biasing field applied over θ=0.

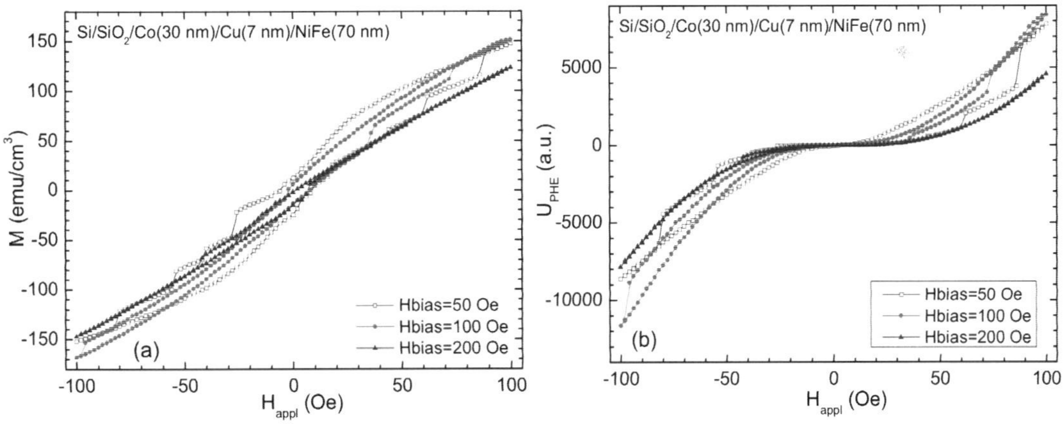

Figure 11. Micromagnetic simulations regarding the field dependences of magnetization (a) and planar Hall effect voltage (b) for different values of the biasing field.

From these simulations is revealed the influence of the Co layer; the ML do not saturates at 100 Oe and consequently the PHE do not saturates. The agreement between measured and simulated PHE characteristics is mainly a qualitative one and shows a decreasing of the PHE sensitivity when H_b increases. Also, we can see from these simulations, a decreasing of the hysteresis width both for magnetization and PHE curves.

2.2. Angular dependence of the PHE

The equation that describes the PHE predicts a periodic dependence on the angle, θ, between the current direction and the film magnetization. This dependence, in sin2 θ, suggests another applications of the PHE: contactless potentiometer, magnetic compass, etc.

In figure 12 we present the angular dependences of the PHE voltage measured for the ML Si/SiO_2/ Co(30nm)/Cu(7 nm)/$Ni_{80}Fe_{20}$(70 nm) for different values of the applied field; there is no biasing field in this case. The applied magnetic field rotates with respect to the current direction [6].

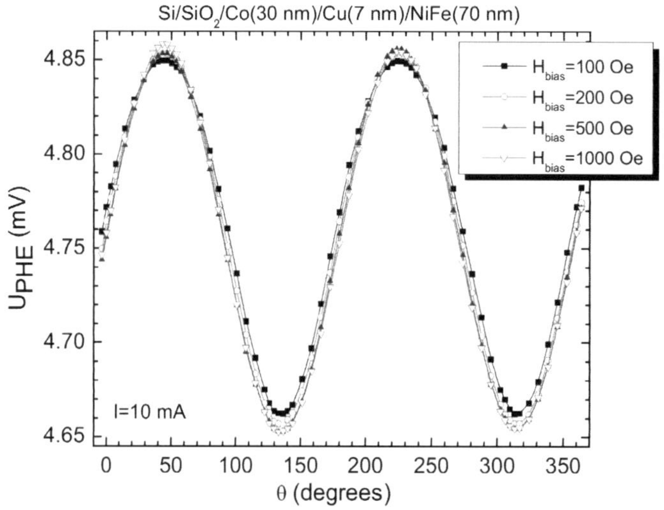

Figure 12. Angular dependence of the PHE, measured for different values of the applied field.

These measurements show a very good agreement with the equation that predicts this behaviour. We see that the amplitude of the angular dependences depends on value of the applied field for lower values (100 and 200 Oe). This is because the magnetization needs a field higher than 200 Oe to saturate. If H_{appl}>500 Oe the curves that describe the angular dependence of the PHE will have the same amplitude because the film magnetization saturates. These aspects are important for practical applications because the signal becomes field independent when H_{appl} is higher than a critical value.

The micromagnetic simulations [6] performed on this structure in order to describe the angular dependence of the PHE are presented in figure 13.

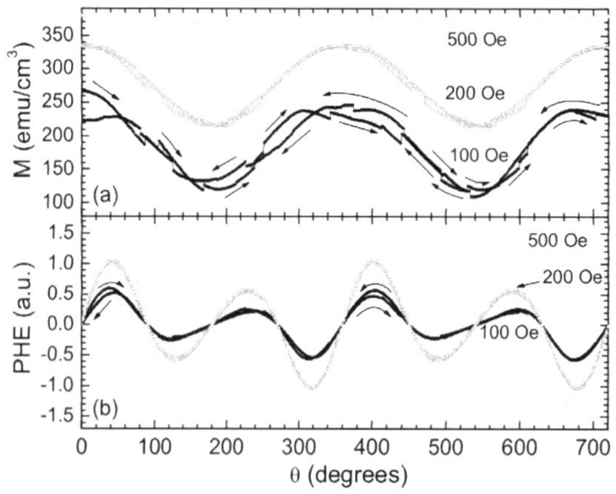

Figure 13. Micromagnetic simulations of the angular dependence of the magnetization (a0 and (b) PHE voltage for different values of the applied rotating field; the arrows are guides for the eyes.

These simulations are in very good agreement with the experimental measurements and show how, at low fields, the magnetization cannot follow accurately the applied field. When H>500 Oe, the angular dependence of the film magnetization shows a flat line which means that the magnetization has a constant value and follow the orientation of the rotating field. For this reason we observed distortions of these angular dependences at fields lower than 200 Oe [6].

3. Conclusions

In this paper we presented some measurements regarding the field and angular behaviour of the PHE in thin films and ML structures. We used disk shape samples in order to avoid the shape anisotropy in the film plane.

Micromagnetic simulations were used to explain some aspects observed in these measurements. Further works are necessary to improve the quality of the micromagnetic simulations.

Using samples with a small uniaxial anisotropy field the hysteresis effect can be lowered and by carefully adjusting the biasing field we can tune the sensor's sensitivity. The same sensor can be used both for field and angular measurements.

Acknowledgments

This work was funded by the Romanian National Authority for Scientific Research through Contract 71-127 and the CNCSIS Contract PCCE-ID-76

References

[1] Kools J C S, Coehoorn R, Folkerts W, De Nooijer M C and Somers G H K 1998 *Philips J. Research* **51** 125
[2] Balcells L, Calvao E, Fontcuberta J 2002 *J. Magn. Magn. Mater.* **242-245** 1166
[3] Prados C, Garcia D, Lesmes F, Freijo J J, and Hernando A 1995 *Appl. Phys. Lett.* **67** 718
[4] Montaigne F, Schuhl A, Nguyen Van Dau F, Encinas A 2000 *Sensors and Actuators* **81** 324
[5] Epshtein E M, Krikunov A I, Ogrin Yu F 2003 *J. Magn. Magn. Mater.* **258-259** 80
[6] Volmer M and Neamtu J 2008 *Physica B: Condensed Matter* **403** 350
[7] Chuia K M, Adeyeye A O, Li Mo-Huang 2007 *J. Magn. Magn. Mater.* **310** e992
[8] Schuhl A, Nguyen Van Dau F, and Childress J R 1995 *Appl. Phys. Lett.* **66** 2751
[9] Volmer M and Neamtu J 2006 *Physica B: Condensed Matter* **372** 198
[10] Oti John O *SimulMag Version 2.0j, Micromagnetic Simulation Software, User's Manual*, Electromagnetic Technology Division, National Institute of Standards and Technology Boulder, Colorado 80303.

AUTHOR INDEX